# Innovative Technology in Architecture and Design (ITAD 2020)

IOP Conference Series: Materials Science and Engineering
Volume 907

Kharkiv, Ukraine
21 - 22 May 2020

ISBN: 978-1-7138-4754-0
ISSN: 1757-8981

**Printed from e-media with permission by:**

Curran Associates, Inc.
57 Morehouse Lane
Red Hook, NY 12571

CURRAN ASSOCIATES INC.
proceedings.com

**Some format issues inherent in the e-media version may also appear in this print version.**



Institute of Physics
Dirac House, Temple Back
Bristol BS1 6BE UK

Phone: 44 1 17 929 7481
Fax: 44 1 17 920 0979

techtracking@iop.org

**Additional copies of this publication are available from:**

Curran Associates, Inc.
57 Morehouse Lane
Red Hook, NY 12571 USA
Phone: 845-758-0400
Fax: 845-758-2633
Email: curran@proceedings.com
Web: www.proceedings.com

# TABLE OF CONTENTS

Preface

Peer review declaration

Modern specific of Japanese urbanism as a result of the country's cultural mentality distinctiveness ............... 1
*G V Shevtsova, O O Gorbyk, A Y Kubko*

The actualization of the garden city model in modern urban planning with participation practices (on the example of Grabnyk park renewal in the city of Rivne)........ 10
*M Y Blinova, S A Ilchenko*

Sustainable Mobility in the Context of Humanization of the Urban Environment: a regional experience........ 20
*N N Korshunova, E B Morozova, O E Dolinina*

The influence of traditional architecture on the design of social housing by masters of modernism........ 28
*I Kudryashova, A Martynenko, O Remizova*

The architectural heritage of the second half of the 1940s - 1950s in the context of humanization processes and urban aesthetic comfort (on the example of Rivne city)........ 38
*M I Isupova*

Information Field of the Modern city silhouette........ 45
*L S Martyshova*

Mobile housing on the islands in post-industrial Dnieper........ 55
*V P Myronenko, T A Tsymbalova, O V Myronenko*

The Basic Principles of Ecovillage Design........ 62
*O O Blahovestova, O O Pechertsev, S O Dansheva*

Thesaurus of Parametric Formation Paradigm Architectural Space........ 70
*D Sopov, E Protsenko, V Mironenko*

Types of ecological relationships in the system "nature, artificial environment, humans". Peculiarities of modern eco interior development........ 78
*Svitlana Zymina*

Schools of architecture as an instrument of architectural policy on the examples of European countries........ 86
*A O Yesipov, V P Myronenko, N V Ivanova*

Multi-Level Viewing Points in the Urban Structure of a City........ 95
*V Kuzmych, Yu Petrovska*

Innovative Approaches in the Period of Mass Industrial Development (on the Example of Residential Areas of Kharkiv)........ 102
*O Buriak, O Vihdorovich, Y Gayevyi, A Golovchenko*

Architectural and City-Planning Aspects of Innovation Hubs Formation in the Context of Creative Urban Regeneration (On the Case of Ukrainian Cities)........ 113
*K T Holubchak, O S Sleptsov, R Tomlins*

Innovative Component in the Creation of the Most Outstanding Greenhouses in the World of XIX - XXI Centuries. At the Junction of Architecture and Engineering ........ 122
*L. Kachemtseva, N. Khoroian, O. Gella, S. Nikiforova*

Visual-communication environment of a modern city ........ 128
*V P Dubinskiy, A A Nesen*

Problems of Architecture and Professional Development of an Architect beyond “Limits to Growth” ........ 135
*O O Fomenko, S M Danylov, A M Izbash*

The problem of formation of innovative directions in the profession “Architect” ........ 141
*Oksana Fomenko, Serhii Danylov, Oleg Sleptcov, Andrii Izbash*

Innovations in the Architectural Environment: the Impact of Society on the Positive Perception and Conflict of the Incomprehensible ........ 151
*Nataliia Mezhenna, Daria Filippova*

Symbolism and functional organization of architecture of the Orthodox religious and educational complex ........ 159
*V Khaidukov, O Sleptsov*

Conceptual modeling to control the quality parameters of the BIM model ........ 167
*Olga Kysil, Sergii Mikhalchenko, Valery Tovbych*

The conception and development tendencies of the helicourts network design in the world ........ 173
*N H Semyroz, S S Kysil, O S Sleptsov, O O Safronova, T V Bulhakova*

Innovative concepts in the Kharkiv projects of J. Steinberg of the capital period ........ 181
*O V Maimeskul, A S Mamon, M V Rusanova*

Reactive powder concrete incorporating metakaolin and fly ash for monumental architectural objects ........ 191
*L Dvorkin, V Zhitkovsky, N Lushnikova, M Fursovych*

Design features of low-rise buildings walls erected from permanent formwork ComBlock units ........ 200
*K V Spirande, Yu V Bondarenko, M V Iakymenko, R M Shemet, V V Bielka*

Peculiarities of erecting walls for houses and buildings by technology of monolithic construction using permanent formwork unit ........ 210
*S V Butnik, V A Viatkin, M N Dzhalalov, I V Hovorukha, V V Bielka*

Role of potential-determining ions and water molecules in the structure formation of cement compositions ........ 218
*A Plugin, T Kostyuk, D Bondarenko, O Pluhin, V Arutyunov*

Architectural construction using technologies based on knowledge about the flow of Bingham plastic fluids in various pipelines ........ 226
*A A Zadorozhnyi*

Influence of genesis of mineral components of asphalt concrete for its durability in aggressive environments ........ 231
*Sergii Yefremov*

Nanomodified cement composites for thin walled architectural structures ........ 239
*Andrii Plugin, Kateryna Pushkarova, Maryna Sukhanevych*

Radiation protective properties of fine-grained concretes and their radiation resistance ................................ 247
*D V Anopko, O A Honchar, M O Kochevykh, L O Kushnierova*

Modification of the structure of the stabilized basalt fiber ................................................................ 256
*V I Gots, P P Palchyk, S A Tymoshenko, S P Palchyk*

Influence of whitening additives on the properties of decorative slag-alkaline cements................................. 261
*V I Gots, A G Gelevera, O N Petropavlovsky, N V Rogozina, V V Smeshko*

Resource saving and energy saving at the simultaneous production of two types of cement.......................... 268
*A A Riazanov, V I Vinnichenko, A N Riazanov, R Z Rakhimov, I V Nedoseko*

Innovative building materials in creation an architectural environment........................................................ 275
*V A Abyzov, K K Pushkarova, M O Kochevykh, O A Honchar, N L Bazeliuk*

Processes of structure formation and neoformation of basalt fiber in an alkaline environment....................... 285
*O Yu Berdnyk, O V Lastivka, A A Maystrenko, N O Amelina*

Temperature-time monitoring of concrete hardening of a wind farm foundation ........................................... 292
*O V Kabus, O OKulaienko, A V Ryabushko, V P Sopov, O V Usherov-Marshak*

Application of the easy concentration effect in concrete technology ............................................................ 298
*Alexandra Shishkina, Alexander Shishkin*

Evaluation of the frost resistance of concrete in real operating conditions ..................................................... 304
*S D Lapovska, G Iu Krasnianskyi, V I Klapchenko, I O Aznaurian*

Optimization of the composition of hydrophobized cellular concrete according to its moisture-transporting and water-holding characteristics ........................................................................................ 313
*S D Lapovska, V I Klapchenko, G Iu Krasnianskyi, Yu G Gasan, I O Kuznetsova*

A source of thermal energy for roasting phosphogypsum – incinerators flue gas........................................... 321
*V I Vinnichenko, O P Krot, O Y Krot*

Special Features of Using Universal Technological Sets of Small-sized Equipment for Renovation of Buildings ...................................................................................................................................... 328
*I A Emeljanova, D O Chayka, E S Lebedeva, D Yu Subota*

The composite facing material for electromagnetic felds shielding ............................................................... 334
*V A Glyva, L O Levchenko, O V Panova, O M Tykhenko, M M Radomska*

Features of hardening and structure formation of concrete with high-dispersion mineral additives ............... 341
*L N Butskaya, V A Hurkalenko, K V Latorets, O V Makarenko, L O Pershina*

Contact formation mechanism between squeezed crystalline solids ............................................................. 347
*M A Volosyuk, E M Protsenko*

Utilization of chemical waste and by-products as one of technological progress directions........................... 354
*A I Gabitov, V A Ryazanova, L Z Rolnik, A S Salov, V A Timofeev*

Innovative Thermal Insulation Materials with Dual Properties....................................................................... 359
*K Plakhotnikov, O Starkova, O Dedenova, D Bondarenko, O Demina*

Comparative evaluation of quality of hard and special paving grade bitumen's ............................................ 366
*Y Pyrig, A Galkin, S Oksak*

Manufacture of construction materials by energy-saving technology through the example of the Bashkir region ........ 374
*A I Gabitov, V A Ryazanova, A S Salov, A M Gaisin, A A Timofeev*

The research of tubular mixer with improved design ........ 379
*S Epoyan, G Sukhorukov, V Volkov, O Haiduchok*

Fillers for modification of polyester powder coating ........ 386
*V I Gots, O V Lastivka, O O Tomin, S A Tymoshenko*

Ecological efficiency photocatalytic concrete ........ 393
*V P Sopov, N S Shyshko, V I Kondrashchenko, Zhang Yihe*

Tensile strength of the pipe material determination according to the ultimate tensile strength ........ 403
*B Younis, L Saienko, I Kazimahomedov, S Salem*

Compatibility analysis of superplastisizers with different types of cements ........ 411
*S M Tolmachov, O A Belichenko, D S Tolmachov*

Complex multifunctional additive for anchoring grout based on alkali-activated portland cement ........ 418
*P V Krivenko, O M Petropavlovskyi, I I Rudenko, O P Konstantynovskyi, A V Kovalchuk*

Research of technologies for restoration of the concrete protective layer of reinforced concrete constructions during the reconstruction of the buildings and structures ........ 427
*O M Galinsky, O S Molodid, N V Sharikina, R O Plokhuta*

Modeling of thermal conductivity of reed products ........ 434
*Yu V Tsapko, A Yu Tsapko, O P Bondarenko*

Reduction of formaldehyde in ambient air through application of special acrylic co-polymer in paint-and-lacquer coatings ........ 443
*R F Runova, V V Pipa, Yu L Nosovsky, M A Laboda, I B Monfort*

Express-method of determination hidroactiviti of Portland cement ........ 449
*A P Nikolaev, E V Kondrachenko, V I Kondrachenko*

Investigation of adhesive-strength characteristics of fire-retardant epoxy polymers modified with metal-containing additives ........ 454
*O Hryhorenko, Ye Zolkina, N Saienko, Y Popov, R Bikov*

Theoretical substantiation and calculations of water flow to ranney water intakes and drainages under protection from submergence of the urban territories and buildings by ground water ........ 460
*S V Telyma, E O Oliynyk, O S Voloshkina, I M Vildman*

A study of the alkali-silica reaction in recycled glass concrete ........ 467
*V P Sopov, O I Korkh, M Yu Izbash*

Development of the tram network in the big city (on the example of Lviv) ........ 473
*B S Posatskyi, T M Mazur, E I Korol*

Improvement of the safety of multi-floor housing ........ 480
*T Tkachenko, V Mileikovskyi, V Dziubenko, O. Tkachenko*

Typological and Functional Planning Features of Sports and Tourist Hotels ........ 487
*D S Borodai, A S Borodai, S P Borodai, Ya O Borodai*

Principles of reconstruction of multifunctional stadiums on the example of the Kiev stadium of CSC Armed Forces of Ukraine ........ 494
*A N Chub, I I Kreizer*

Utilization of sorted secondary PET waste - raw materials in the context of sustainable development of the modern city ........ 500
*M S Malovanyy, S D Synelnikov, O A Nagurskiy, K M Soloviy, I S Tymchuk*

“Green” principles of sustainable development of road and transport infrastructure of the cities of Ukraine ........ 505
*N V Vnukova, G M Zhelnovach, O V Kozlovskyi*

Introduction of innovative water treatment technologies - the key to sustainable development of the modern city ........ 514
*K.I Petrushka, M S Malovanyy, I M Petrushka, V D Pohrebennyk, V I Mokryy*

Aspects of preserving historic landscapes of palace and park ensembles of the Romantic period ........ 520
*E T Cherkasova, A A Zavorina*

The Modern Shopping Centers of Kharkiv in Visitors’ Perception ........ 527
*S A Smolenska, A S Borysenko*

Problems of Formation of Architectural and Spatial Composition Organization of the Environment of Donbas Mono City ........ 536
*I V Honcharova, V P Myronenko, O V Hubanov*

Principles of architectural and planning arrangements for school complexes on difficult terrain ........ 543
*O Sleptsov, A Karban*

Architectural and Urban Planning Features of Ecotopia ........ 550
*A Fedak*

The Modern Trends of an Architectural Parterre Formation ........ 559
*V P Dubinskiy, O V Skorobohatko*

The formation of residential complexes in the structure of Ukrainian largest cities ........ 569
*I V Ladigina, N G Dubina, E V Bizhko*

Peculiarities of Formation of a Systems Approach in Domestic Urban Planning ........ 576
*I V Ladigina, A O Rudenko*

Problems of environmental safety of water disposal as a factor of sustainable urban development ........ 583
*V.O. Iurchenko, E.S. Lebedeva, Yu.S. Levashova, A. V. Kovalenko*

Creation of modern Orthodox churches in Ukraine - “as languages of Ukrainian autochthony” in the context of urban development” ........ 590
*O S Sleptsov, Y U Dunaevskiy*

The conceptual framework of sustainable development and the role of environmental and technogenic safety in achieving it ........ 600
*O Hunchenko, O Voloshkina, I Korduba, M Kravchenko, P Stefanovych*

Integrated management systems as a basis for sustainable development ........ 609
*Okcana Hunchenko*

Formation of atrium spaces in the urban environment ........ 616
*P A Solobay, O V Vakhnichenko*

Modern trends at natural and wastewater treatment plants reconstruction ........ 623
*S Martynov, O Kvartenko, V Kovalchuk, A Orlova*

Effect of technological parameters on clarification low-concentration suspension by dissolved air flotation ........ 631
*S Epoyan, O Syrovatsky, O Haiduchok, A Titov*

Acoustic loading in modern city as negative factor of sustainable development ........ 637
*A I Samokhvalova, V O Iurchenko, N G Onyshchenko, N O Kosenko*

**Author Index**

IOP Conf. Series: Materials Science and Engineering **907** (2020) 011001 doi:10.1088/1757-899X/907/1/011001

**IV International Scientific and Practical Conference "Innovative Technologies in Architecture and Design" May 21-22, 2020 Kharkiv National University of Civil Engineering and Architecture, Ukraine**

The purpose of the conference was to discuss the results of fundamental and applied research in architecture, design and construction, aimed at humanizing the architectural environment, creating innovative technologies for the construction of buildings and structures, modern building materials and products, and applying the latest design technologies.

The conference was scheduled for April 2020, but due to the COVID-19 pandemic, it was postponed to May 21-22. Since by this time the situation with COVID-19 had not improved (interstate borders were closed, movements were limited and mass events were prohibited), it was decided to hold the conference online. This decision was also based on the fact that a number of the presented papers could lose their relevance when the conference was postponed. In addition, the online conference was also of scientific and practical interest in terms of disclosing new communication opportunities between scientists from different countries and educational institutions.

The Zoom communication platform was used as a means of videoconferencing, which allows you to enter the conference from any device (PC, laptop, mobile phone, tablet).

The conference was attended by 315 specialists from universities, leading architecture schools and technical universities of Ukraine, Belarus, China, England and Russia, as well as representatives of industrial enterprises and organizations in the field of construction. 122 papers were presented.

At the plenary session on May 21, 5 reports from Ukraine, Belarus, Russia and China were presented. The reports discussed the issues of sustainable urban development in the context of the humanization of the architectural environment, the impact of pandemics on the development of the architectural environment, the development of new types of environmentally friendly building materials to create a comfortable environment for human existence, the use of "green technologies", etc. For each plenary presentation, a separate video conference (40 minutes) was organized to enable a detailed discussion of the problem.

On May 22, the conference was carried out in five sections: "Humanization of the architectural environment in the context of European integration", "Innovative approaches in the formation of modern architecture and design", "Architectural structures and concrete", "Latest building materials and modern technologies in construction", "Architectural design and sustainable development of a modern city ". In each section, several Zoom meetings were organized for 40 minutes to present 4 reports. At the end of each section, one conference was organized to discuss the reports.

The main topics for consideration at the conference were innovative technologies, previously known technologies, but in a new quality or technology of today, which contribute to rethinking the ways of interaction, communication, solution and implementation of project tasks in the practice of architects and designers. The main task was to show the indissoluble unity of architecture and materials science, to reveal the ways of their interaction, which is an incentive for the creation of new technologies of building materials and the formation of a new architectural worldview.

The conference outlines thematic areas covering the humanization of the architectural environment in the context of European integration, innovative approaches in the formation of modern architecture and design, the use of architectural structures and the latest building materials in the professional activities of architects and designers, as well as aspects of ergonomics, urban and technical systems that allow to reveal in the most complete and meaningful way actual problems in application to innovative technologies in architecture and design.

The work of students, postgraduates, teachers and researchers of universities, as well as representatives of industrial organizations and enterprises that operate with various materials and technologies, and can clearly demonstrate ready-made elements of products or processes in the form of animated, interactive content.

The conference was attended by representatives of 40 universities, academies and research institutes from Ukraine (Kharkov, Kiev, Lvov, Odessa, Dnipro, Sumy, Lutsk, Rovno, Ivano-Frankovsk, Kropyvnytskyi, Krivoy Rog, Cherkassy), Russia (Moscow, Belgorod, Kazan, Ufa), Belarus (Minsk), China (Beijing, Chengdu), England (Coventry), as well as representatives of 10 manufacturing enterprises.

The greatest interest was aroused by reports on new building materials that cleanse the atmosphere of cities that use secondary resources, on technologies for the construction of buildings and structures, on modern design technologies.

IOP Conf. Series: Materials Science and Engineering **907** (2020) 011002 doi:10.1088/1757-899X/907/1/011002

# Peer review declaration

All papers published in this volume of IOP Conference Series: Materials Science and Engineering have been peer reviewed through processes administered by the Editors. Reviews were conducted by expert referees to the professional and scientific standards expected of a proceedings journal published by IOP Publishing.

- **Type of peer review: Other**

An open way of reviewing was used: two reviewers - min requirements - PhD specialists on the topic of the report.

- **Conference submission management system:**

The conference was managed by an international organizing committee, the composition of which is given in the appendix. Conference materials (program, abstracts) were presented in hard copy: program brochure, book of abstracts, and in electronic form (* .pdf format) on the conference website https://itad.com.ua/. Additional information about the conference was presented on the website of the Kharkiv National University of Civil Engineering and Architecture http://kstuca.kharkov.ua/nauka/konferencii-ta-seminari.

- **Number of submissions received:** 129
- **Number of submissions sent for review:** 129
- **Number of submissions accepted:** 123
- **Acceptance Rate (Number of Submissions Accepted / Number of Submissions Received X 100):** 95.3
- **Average number of reviews per paper:** 2
- **Total number of reviewers involved:**67
- **Any additional info on review process:**

The review process was carried out in accordance with the instructions for the reviewer:

**REMINDER TO REVIEWER**
**BASIC REQUIREMENTS FOR REVIEW**
Technical criteria
• Scientific novelty: especially scientific rigor, precision and correctness.
• Clarity of expression; exchange of ideas; readability and discussion of concepts.
• Sufficient discussion of the context of the work and appropriate references.
Quality criteria
• Originality: is the work relevant and new?
• Motivation: Does the problem under consideration have valid motivation? All documents must clearly demonstrate scientific interest in the results.
• Repetition: Has a significant portion of the manuscript already been published?
• Duration: is the content of the work of sufficient scientific interest?

Presentation criteria
• Title: Is it adequate and appropriate for the content of the article?
• Abstract: Does it contain basic information about the article? Is it complete? Is it suitable for inclusion in an abstraction service?
• Diagrams, figures, tables and captions: are they important and understandable?
• Text and math: are they concise and easy to understand? If you recommend shortening, please suggest what to skip.
• Conclusion: Does the article contain a carefully written conclusion summarizing what was learned and why it is interesting and useful?

- **Contact person for queries:** prof. Viktor Sopov E-mail: vpsopov@gmail.com

**APPENDIX**

**ORGANIZING COMMITTEE OF CONFERENCE**

*Chairman of the organizing committee:*

Goncharenko Dmytro - Doctor of Technical Sciences, Professor, Vice-Rector for Scientific and Pedagogical Work of KhNUCEA;

*Co-chairs of the organizing committee:*

Myronenko Viktor - Dr. Architect, Prof. KhNUCEA, Ukraine;
Sopov Viktor - Doctor of Technical Sciences, Prof. KhNUCEA, Ukraine.

*Members of the organizing committee:*

Buryak Alexander - Dr. Architect, Prof. KhNUCEA, Ukraine;
Gonchar Olena - Dr. Pedagogical Sciences, Prof. KhDADM, Ukraine;
Huk Valeriy - Doctor of Technical Sciences, Prof. KhNUCEA, Ukraine;
Dubinsky Volodymyr - Dr. Architect, Prof. KhNUCEA, Ukraine;
Izbash Mykhailo - Doctor of Technical Sciences, Professor KhNUCEA, Ukraine;
Kondrashchenko Valeriy - Doctor of Technical Sciences, Professor Russian University of Transport, Russian Federation;
Kostyuk Tetyana - Doctor of Technical Sciences, Prof. KhNUCEA, Ukraine;
Kravets Volodymyr - Dr. Architect, Prof. KhNUCEA, Ukraine;
Miletska Malgozhata - Dr. Architect, Prof. University of Natural Sciences, Lublin, Poland;
Morozova Olena - Dr. Architect, Prof. Belarusian National Technical University, Minsk, Belarus;
Solobay Petro - Dr. Architect, Prof. KhNUCEA, Ukraine;
Usherov-Marshak Alexander - Doctor of Technical Sciences, Prof. KhNUCEA, Ukraine;
Fischer Hans-Bertram - Dr. Eng. Bau House University, Weimar, Germany;
Fomenko Oksana - Dr. Architect, Prof. KhNUCEA, Ukraine;
Chuck Marek - Doctor of Technical Sciences, Prof. University of Warmia and Mazury, Olsztyn, Poland;
Cherkasova Kateryna - Dr. Architect, Prof. KhNUCEA, Ukraine;
Cherkes Bogdan - Dr. Architect, Prof. Lviv Polytechnic National University, Ukraine;
Zhang Ihe - Ph.D., Professor, Director of the Institute of Materials Science and Materials Technology of the Chinese Geological University, Beijing, China;
Shilo Alexander - Doctor of Arts, Prof. KhNUCEA, Ukraine.

IOP Conf. Series: Materials Science and Engineering **907** (2020) 012001 doi:10.1088/1757-899X/907/1/012001

# Modern specific of Japanese urbanism as a result of the country's cultural mentality distinctiveness

**G V Shevtsova[1] O O Gorbyk[1] and A Y Kubko[2]**

[1]Kyiv National University of Construction and Architecture, Department of Architectural Basis and Design, 31 Povitroflotskyi Prospect str., Kyiv 03037, Ukraine
[2] Vernadsky National Library of Ukraine, Department of Bibliometric and Scientometric, 3 Golosyivskyi Prospect str., Kyiv 03039, Ukraine

nekosanka@hotmail.com

**Abstract.** The article analyzes the urbanistic experience of modern Japan in terms of the country's architectural history. The following features of Japanese modern urbanistic are highlighted: dynamism and mobility – the ability of permanent development and modification (so called architectural metabolism); communicativeness of city space; the cohesion of the concepts of interior and exterior; freedom and novelty of architectural forms; humane and psychological comfort of urban environment. Based on the analysis of the unique climatic and seismological conditions of Japan, its geographical, ethnic and historical characteristics, it is revealed that all indicated above modern urbanistic features originate in the prerequisites of Japanese mentality formation directly related to architectural traditions. Since ancient times, Japanese culture has easily adapted to new circumstances modifying both obsolete traditional phenomena and foreign influences to a fundamentally different level. During the industrial revolution, thanks to the active borrowing of European and American engineering experience, such an approach leads in Japan to a qualitative leap of urbanism and design, forming an imaginary gap between the historical and modern architecture. In reality, urban planning and architecture of modern Japan demonstrate a deep genetic connection with historical tradition, which is present here as a permanent global evolutionary vector of ideas.

## 1. Introduction

The ultra-modern Japanese urbanism more than half a century ago has crossed the line when buildings exist separately: the megalopolis functions as a precisely calibrated mechanism, where all buildings are merged together with engineering and transport networks. At the same time, the architecture of the highest class is still present in such cities showing original forms of buildings and their ingenious structural embodiment. The most important thing is that a person feels comfortable in this urban environment. At first glance, this is the quintessence of modern technology and novelty of architectural design that seemed to appear from scratch. However, analysing the specifics of modern Japanese urbanism in the light of the centuries-old architectural and construction traditions of this country, one can come to completely different conclusions. We set the task to highlight in the history of Japanese architecture the origins of the principles and trends that led to the formation of this modern specificity.

## 2. Japanese modern city in the point of view of local traditional architecture

For completing this task let us analyse and compare some contemporain and traditional features of Japanese architecture and urbanism.

### *2.1. Urbanistic features of Japanese megacities*

The urban environment of Japan is constantly evolving and changing, adapting to new conditions and needs, replacing obsolete parts, and modernizing itself discretely. The Japanese metropolis is a structure of accented dynamic type where changes occur so quickly that could be tracked over a very short period, sometimes within only a few months. In the middle of the 20th century young Japanese architects of the Kenzo Tange entourage correlated this feature of the Japanese urbanism with the principles of growth and development of living organisms, designating it as a phenomenon of architectural metabolism. Nakagin Capsule Tower (1972) built on the outskirts of the central Ginza district (Figure 1a) in Tokyo by architect Kisho Kurokawa, so-called «hotel of the future», is traditionally considered the quintessence of the idea. The composition of the building is based on a group of cubic blocks of apartments fixed on a vertical axial trunk. According to the author's intention, apartments could be replaced or remodelled when necessary. In fact, the foundation for this idea was laid several years earlier by Kenzo Tange himself, having built not far from the location of the future Nakagin Tower the building of Shizuoka Press newspaper office (1967) that is similar in compositional idea (Figure 1b).

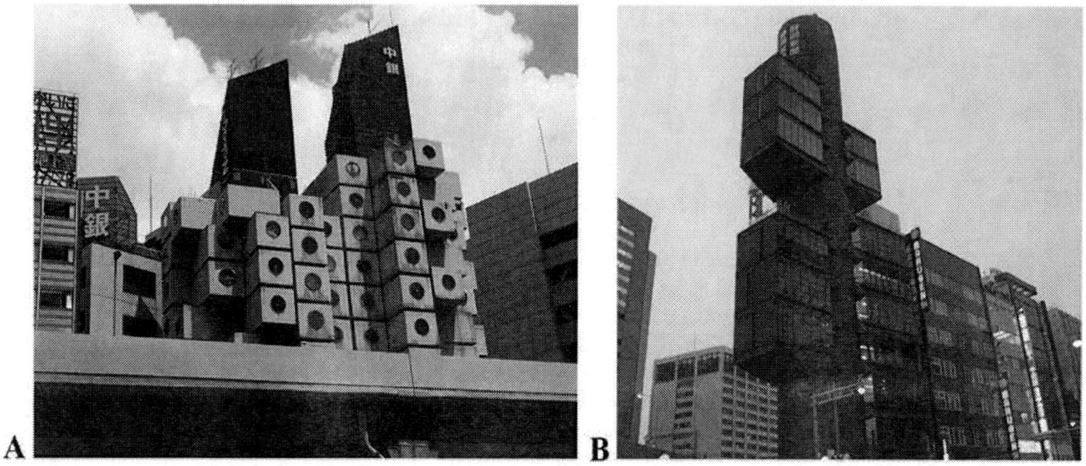


**Figure 1**. Urban metabolism ideas in Tokyo: a – Nakagin Capsule tower by Kisho Kurokawa; b – Shizuoka Press building by Kenzo Tange.

The dynamism (metabolism) of the urban environment is directly intertwined in Japan with the communicative nature of urban space: the entire urban system exists here as a harmonious mechanism of architecture and transport networks permeating buildings at all levels: ground, underground and aboveground. Modern Tokyo has reached the point where freeways and metro lines, aerial monorails and other communications already are not just ordinary spatial connections, but rather tending to the function of arteries. Therefore, here appeared a coherent, meaningful structure that is directly providing the well-being of citizens with correctness and accuracy of functioning. The next feature of Japanese cities that we have highlighted is the tendency to merge the external and internal space of the metropolis, the constant diffusion of buildings' interior and exterior. What is urbanism in the Japanese sense? This is the existence of such cities, of which we can say: "in the depth of Tokyo (Osaka, Yokohama, Nagoya, etc.)." Not in the building, not in the street between the buildings, not in the subway galleries, but inside the city, as inside single spatial system, open and closed at the same time, where it is difficult to distinguish between the internal and the external. The halls of the buildings have natural light and are so open that there is doubt that this is an interior. The streets and squares are so closed that there is doubt that this is the exterior space (Figure 2).

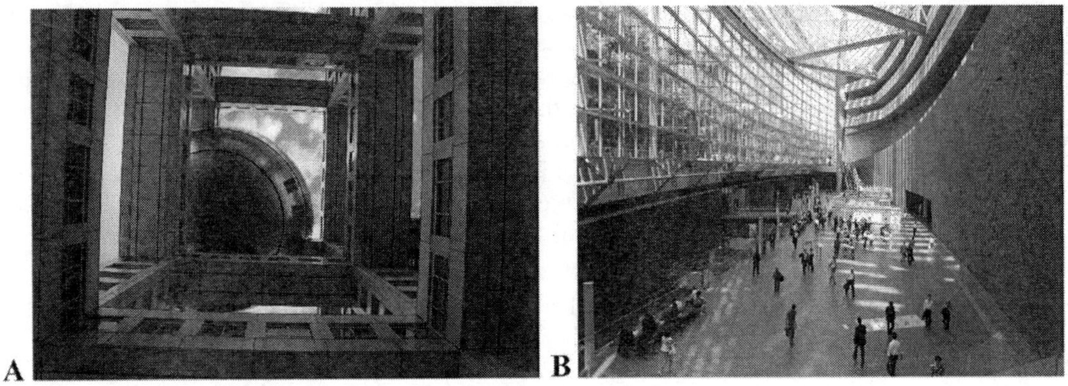


**Figure 2**. Communicativeness and interior to exterior diffusion of Japanese urban space: a – Fuji Television building in Odaiba (Tokyo, 1996) by Kenzo Tange; b – International Forum building in Ginza (Tokyo, 1996) by Rafael Viñoly.

Among the important features of the Japanese metropolis, one should also note the unusual freedom and novelty of architectural forms, as well as the psychological comfort of the urban environment, which at the first glance seems to be unsuitable for a human because of its disproportion with him. The latter characteristic is achieved by fitting into a large urban scale of small and agreeable for human architectural forms and design details: sculptures, benches, lamps, signboards, wooden pedestrian flooring, etc. (Figure 3).

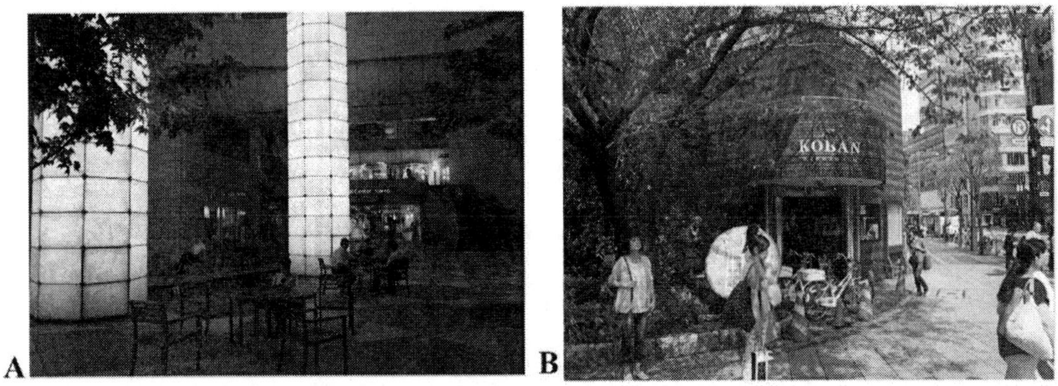


**Figure 3**. Human scale details in Tokyo megacity at Shinbashi (a) and Ginza (b) central districts.

Kenzo Tange, the architectural genius who had carefully studied the history of Japanese architecture all his life, was one of the first scientists systematically developing the theory of Japanese urban construction. As the main ideologist of the city's constant metabolic development possibility, and, in general, the adherent of determining the advanced science and technology achievements in the service of architecture, he nevertheless noted that tradition should be present in modern projects as a catalyst for ideas [1].

### *2.2. Natural and historical background of Japanese architectural tradition*

The territory of Japan is originally distinguished by a very special geographical and climatic specificity, which, in turn, entailed the formation of the specific cultural mentality of the ethnos. This volcanic archipelago is located on the tectonic split of the Earth's crust, which determines the constancy of volcanic eruptions, earthquakes and tsunamis. The force of chthonic nature is supplemented in Japan by climatic problems: seasons of summer rains and autumn sea typhoons, permanently causing mudflows and floods, as well as unusually stuffy and humid summer heat is combined here with rather severe winters and, in many areas, with heavy snowfalls. Despite a rather large area, Japan has a very specific relief: only small valleys and narrow coastal zones are suitable for living here, making up only about 30 percent of the total land. Remaining area of the archipelago is occupied by wooded mountains unsuitable for living and practically not amenable to land reclamation.

The insular location of Japan, in some meaning, cuts it off from the mainland cultural space, bringing to existence separate, peculiar forms of civilization. The differences begin already at the very point of the formation of the Japanese ethnos since entire population of the islands migrated here in one way or another. According to studies of archaeologists [2, p15] and ethnographers (analysis of mythological systems and cosmogonic [3, pp35-7]), the resettlement in Japanese land of different ethnic groups took place at different times and from completely different territories, and the development of civilization occurred with a significant delay compared to mainland. The relief features at the initial stages of development contribute to the formation of separate settlements (clans) in Japan. However, by the 3rd century BC we can talk about the formation of a more or less uniform system of beliefs and rituals, which subsequently, after significant transformation, gets the general name Shinto. This is an animistic religion of an explicit community-agricultural type. Its ceremonies primarily aimed to ensure the well-being of the whole clan, and not of an individual. Shinto rituals carry two main goals of appeasing the natural forces: the prevention of natural disasters and increasing fertility of the ground. The most remarkable feature of Shinto is the complete absence of bloody victims that is uncharacteristic to primary religions. Moreover, among its main postulates, Shinto is aiming to avoid the "pollution" associated with blood and death. Violation of the ban directly entails the wrath of the gods, or of the nature forces that for Shinto religion is almost the same. A real opportunity for everyone to die unexpectedly from a rage of nature leads to the formation in Japanese culture of the idea of temporality, inconstancy of life on the one hand, and hard material functionalism on the other. At the same time, the beauty of Japanese nature with clear difference between the seasons forms a habit of contemplation, attention to detail, a clear understanding of the charm and inevitable mortality of every moment of life. It leads to the development of the aesthetics of nature's wilting, as well as giving the aesthetic component to things simple and functional. Introduced in Japan since the 5th-6th centuries Buddhist traditions, with their constant idea of not being attached to the things, quite successfully harmonized with this authentic foundation.

Japanese ability of cultural influences' absorption is a special topic for reflection. There were long periods of close contact with China and other countries of the East, and later of the West, while Japanese culture actively borrowed the worldview, state, religious and technical (including architectural) experience from other countries. In the same time, Japanese so quickly and globally modified and adapted all foreign experience to the local conditions that finally we should talk about birth on their basis of completely innovative ideas. It should be noted that the reasons of this Japanese mentality phenomenon are still practically unexplored. It is possible that the initial ethnic heterogeneity of the population that already was present at very early stages of development, contributed to the formation of Japanese ability to mutual adaptation, cooperation, synthesis of philosophical ideas and cultural habits. The historical periods of active continental influences alternated in Japan with the periods of relative isolation of the country while Japanese ethnos had time to fully comprehend and creatively interpret the accumulated foreign experience. In any case, it is clear that the natural conditions and cultural-historical features of Japan are the main shaping factors of its mentality, which, in turn, forms the specificity of the architectural tradition of this unique country.

### *2.3. The specifics of Japanese architectural tradition*

We would like to highlight some key features of traditional Japanese architecture and urban planning, which from the very early stages of their existence developed quite apart and with serious differences from continental counterparts. At the same time, it is impossible to deny the significant influence of the Chinese (and Korean) architectural tradition, which, however, was transmitted discretely to Japan. The historical periods of extensive foreign liaisons were several times replaced by the epochs of Japan's closure from alien influences, during which the accumulated arsenal of foreign architectural techniques changed significantly, architecture has evolved constructively and compositionally, adapting to local conditions and generating new original styles [4, pp14-16].

Mobility and metabolism of architecture and urban space in general, is a very ancient Japanese feature. In particular, in Japan from at least 5th c. were traditions of moving the capital from place to place, associated with the above-described features of Shinto beliefs: the concept of "pollution" of a long used territory with emanations of disease and death, leading to misfortunes [3, pp 156-8]. In other words, in the case of the death of the ruler (especially premature or violent death), as well as in the case of any other calamities (epidemics, crop failures, fires, repeated natural disasters, etc.), the place of the old capital was considered unsuitable for life, "unclean", "unhappy." In this case, it was necessary to find a new place for the capital. Political reasons could also cause the transfer of the capital. In particular – the need to create a new state centre "according to the Chinese administrative type", as it was when the capital was transferred from Asuka to Heijio-kyo (now Nara), or the emperor's escape from the power of monasteries, as it was when the capital was transferred from Heijo-kyo to Heian-kyo (now Kyoto). The early history of Japan very convincingly illustrates this process [5, p28]. Despite the lightness and mobility of traditional wooden building structures in Japan, the transfer of settlements from place to place required serious preparatory work, primarily earthwork. It was difficult to find suitable for a new city vast and comparatively flat place in mountainous Japan. In addition to the landscape, many other factors, both practical and empirical, played a role, for example, it was necessary to comply with the Chinese laws of geomancy (feng shui) when searching the territory for the new settlement. Because of this, when transferring capitals, quite often it was necessary to practice the digging of the surrounding mountains, filling up of sea bays and others serious artificial transformations of the landscape. Especially large-scale works were carried out during the construction of Heijo-kyo [4, p60] (now Nara, the border of the 8th – 9th c.) and Edo [4, p152] (now Tokyo, the beginning of the 17th c.). Shinto traditions entailed not only multiple transfers of capitals from place to place, but also the constant need for updating and rebuilding of religious buildings. In ancient times, main Shinto shrines were usually rebuilt every 20-30 years. These traditions are still preserved in the imperial sanctuary of Ise and other temples of its subordination [5, p21].

The idea of a human dwelling was also perceived in Japan as something volatile, temporary. Usually, such an attitude to housing is characteristic of nomadic cultures, however, due to the climatic-seismological and cultural characteristics described above, it is also relevant for Japan [6]. The seismic activity of the wooded Japanese islands, combined with the lack of soft natural stone, led to the historical absence of stone construction. As a defence against the nature force, the Japanese chose the path of adaptation rather than resistance: they developed lightweight architecture — a wooden frame with an almost weightless filling based on clay with straw, wooden planks, etc. The consequences of the collapse of such buildings were less catastrophic, and the restoration work did not require excessive efforts and funds [4, p115]. Since the early Middle Ages anti-seismic framework system, so called "nuki", have also been developing in Japan. "Nuki" construction is based on a number of movably horizontal ties (beams and railings) that pierce the wooden poles of buildings at different levels. A modular construction system spread at about the same time, based on a unified step of load-bearing poles with tatami between them [4, p110], further accelerated and emphasized the development of the idea of interchangeability of building elements allowing to modify easily the parts of the building, as well as to dismantle old components and add new ones. Thus, Japanese architecture in some meaning began to resemble a giant spatial construction set [7].

The communicative nature of the building inner space, the possibility of its free flow into the external environment, is also a national feature of Japanese architecture. Already during the Heian period (8th-12th c.), the style of residential architecture of aristocrats, so-called Shinden-zukuri, was fully formed in Japan. The estate of this type was a series of buildings interconnected by covered galleries on piles [4, p77]. Thus, spatially communicative pedestrian links were established between all the buildings of the estate, and the entire complex could already be considered not as a set of separate buildings, but as a single architectural organism. Dwellings in Japan are traditionally surrounded by one or more belts of verandas and had inner courtyards and gardens. Thus, the transition from the interior of the building to the external environment and back was carried out gradually. Direct sunlight was also unable to penetrate inside the building. These architectural features undoubtedly appeared due to the hot and excessively humid climate of Japan (Figure 4).

**Figure 4**. Aristocratic estate of Heian period: drawing from the scroll of Nenchugyouji-emaki (Annual rituals) dating from the late Heian period.

These features are authentic for Japanese architecture, that is, they appeared before the Chinese building traditions actively spread in Japan since the 7th c. They are present, in particular, in the spatial structure of the ancient Shinto shrines in Ise and Izumo (dated from 5th-6th c.). The sanctuary in Ise (Ise-jingu) has retained its authentic form until our days (Figure 5). The sanctuary in Izumo (Izumo-taisha) is believed to be the replica of that time aristocratic residences. Here we see the transfer of the principles of residential architecture to the sanctuary through the idea of Deity's dwelling. Izumo-taisha's structure has somewhat changed, but the changes primarily affected the scale of the building, and not its architectural and spatial characteristics [8, pp84-6] (see Figure 5).

A  B 

**Figure 5**. Interior to exterior diffusion in the structure of early Shinto shrines: a – Ise-jingu shrine (main pavilion transferred after rebuilding to Nishina-Shinmeigu sanctuary); b – Izumo-taisha shrine.

Starting from the 8th-9th c. these specific traditions continue to develop already under the influence of Chinese architecture in the Buddhist temples and monasteries. Between the buildings of the Buddhist sacred complex, spatial pedestrian connections were organized in the form of bridges, galleries, covered walkways (Figure 6a). Monasteries and temples were constantly expanding: more and more new buildings were added to the layout of the complex, some of them were the individual dwellings of high rank monks or the so-called "subordinate temples" that, in turn, were also surrounded with subordinate structures. Thus, the temple complex could grow to the size of a small town with its own infrastructure, street grid, etc. Such is, for example, the spatial organization of the mountain monasteries of the Tendai and Shingon Buddhist schools (9th-11th c.), the headquarters of them are located at Hie and Koya Mountains, respectively. These traditions were further developed in the early Middle Ages (12th-13th c.), with the spread of Zen Buddhism in Japan, that introduced the Chinese traditions of the strict regular layout of sacral complexes. Zen monasteries were usually planned as a system of symmetrical rectangular courtyards closed with galleries connecting buildings (Figure 6b).

IOP Conf. Series: Materials Science and Engineering **907** (2020) 012001 doi:10.1088/1757-899X/907/1/012001

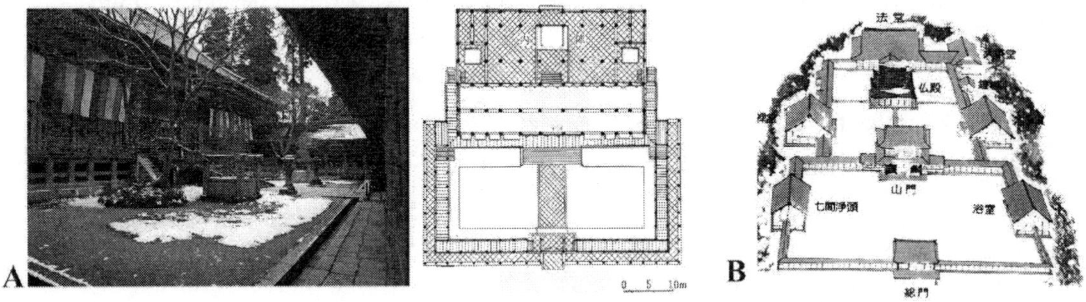


**Figure 6**. Layout of Buddhist monasteries: a – Konponchudo pavilion in Hiezan Enryakuji monastery (Shiga prefecture, Heian period); b - typical layout of Zen monastery (Zuiryuji monastery, Toyama prefecture).

As a result, the structure of the monastery complex became so complicated with various spatial connections that it is sometimes difficult to determine where the interior of the building ends and the exterior begins. Thus, the entire complex, which in principle consists of separate buildings, may well be considered a single architectural structure. Good examples of such a phenomenon can be well-preserved complexes of Japanese Zen monasteries Zuiryuji (Toyama Prefecture), Eihoji (Gifu Prefecture), Tofukuji (Kyoto) and others [4, pp119-26].

## 3. Results and discussions

The unusual freedom and innovative design of Japanese megalopolises' architecture also have historical reasons. By the beginning of the era of modern construction, in Japan was no usual for Europe structure of the historical city. Almost no stone construction was developed in Japan until the late-19th c. The centres of large cities were built up with two-three-story wooden houses along the front of the streets, in some cases – supplemented by the distant dominant of the castle. Temples and gardens were separate territories inside or on the periphery of the town and, as a rule, did not play a special role in city planning. This situation does not at all resemble the long-established relief-dominant planning of monumental European cities of the 11-19th c. [9, pp91-102]. Thus, from the era of wooden structures, Japan almost immediately stepped into époque of glass and reinforced concrete [10]. Therefore, thanks to the industrial revolution, Japanese architects gained Western engineering and construction experience, and with new building capabilities, materials and structures, at the same time they gained and unattainable for Europe freedom of construction without the need of new buildings' harmonisation with the ensembles of old towns [11]. Some European architectural trends also were adopted to local conditions. The brick buildings of European historicism were quite widespread in Japan during the Meiji period (1868-1912) but quickly found their seismic unreliability and thus received serious structural modifications [5, pp108-11]. It is not surprising because by this moment in Japan a huge historical experience of anti-seismic construction had already been accumulated. Obviously, the Japanese were able to create such original works of modern architecture, because they were not initially constrained by the inertia of the consciousness of European stone-brick architecture, gravitating to the traditional cubic shape of buildings. Small in size old sections of the city naturally integrated in the new urban system. Old quarters and oases of temples are still alive and functioning, inscribed in a new urban context as grains in a pea pod.

New, sprouting from the old, this is perhaps the main slogan of the Japanese urbanism. Indeed, even such seemingly ultra-modern Japanese urban planning methods as the expansion of the territory by means of artificial islands creation (for example, the construction of Odaiba coastal district in Tokyo), as noted above, have long local historical tradition of landscape transformation. The psychological comfort of the environment of Japanese modern cities mentioned above fits in perfectly with this picture. According to the latest research by psychologists, a finely detailed and diverse urban environment is the most comfortable for the human psyche, because it activates the most ancient brain mechanisms associated with a sense of security and abundance of food [12]. In Japanese metropolitan areas, this effect is achieved through the inclusion in a large-scale urban environment of a variety of architectural design details commensurate with human scale. Close attention to detail is an original Japanese trait, one of the fundamental features of this ancient distinctive culture. It would be strange if this point did not find its creative reflection in a modern urban environment.

## 4. Conclusions

Thus, it is obvious that the current trends in the development of architecture and urban planning in Japan, and, in particular, the idea of the Japanese "urban metabolism" actually have deep historical roots and are an authentic feature of Japanese architecture. The principles and basic ideas of Japanese urban specificity are directly related to the mentality of the Japanese ethnos that is in turn formed due to a complex of atypical geographical, climatic-seismological and historical backgrounds of the region. In particular, the ideas of simplicity and functionality of life and non-attachment to the things, caused by the harsh climatic and seismological situation and the characteristics of the Shinto cult, brought to life the communicativeness and mobility of the architectural space. These phenomena originated at the turn of our era, and are clearly visible in the historical customs of Shinto shrines' rebuilding and capitals' moving, as well as in the compositional paradigm of estates' and monasteries' complexes with galleries and connecting passages. From the traditional climate-related presence of verandas and inner courtyards in Japanese buildings, there also follows a tendency towards the cohesion of the external and internal spaces of the modern city. The psychological comfort of Japanese urban environment is achieved through the introduction of small-scale design details. This feature appeared due to original for Japanese ethnos attention to detail.

At the same time, the initial heterogeneity of the Japanese ethnos led to the unique ability of culture to easy and creative adaptation of foreign influences and old techniques to new local needs. The historical architectural trends described above on examples of ancient temples, monasteries and residential complexes, during the industrial revolution developed and enlarged to the citywide level. Based on the new engineering knowledge obtained from Europe and America, the Japanese, free of historical stereotypes of monumental stone building, made a qualitative evolutionary leap in architecture. This led to an imaginary gap in the methods of architectural shaping of antiquity and modernity. In fact, there was a phenomenon that could be compared with the evolution of plant growth, when a leaf appears from a bud, a flower from a pod, a tree from a sprout: all them are not resembling to their origin, but genetically representing the essence of it. In this ability, and not just in the possibility of constant changes, in our opinion, lies the true meaning of the phenomenon of Japanese urban metabolism. In this point the historical tradition is something like vector of evolution determining general principles of Japanese contemporain architecture development.

## References

[1] Tange K and Kawazoe N 1965 *Ise – Origin of Japanese Architecture* (Cambridge, Massachusetts: MIT Press); Tange K and Gropius W 1960 *Tradition and Creation in Japanese Architecture* (New Haven: Yale University Press, Tokyo: Zokeisha Publication Ltd); Tange K 1959 An Approach to Tradition *The Japan Architect* **1-2** 55-8; Tange K 1956 Contemporary Architectural Creation and the Japanese Architectural Tradition *Shinkenchiku* **6** 25-33

[2] Kidder J E 1959 *Japan before Buddhism* (London: Thames and Hadson)

[3] Nakorchevskiy A A 2003 *Shinto* (Sankt-Petersburg: Azbuka)

[4] Shevtsova G 2011 *The History of Japanese Architecture and Art* (Kyiv: Grani-T)

[5] Young D and Young M 2004 *Introduction to Japanese architecture* (Singapore: Periplus)

[6] Kenzo Tange 1965 Urbanisation and Japanese Culture *Change and continuity in Japanese Culture* **10** 1-16

[7] Shevtsova G V 2011 Architectural Tradition as a Vector of Japanese Megacities' Development *Contemporary Architecture of the World: Collection of Scientific Articles of the Russian Academy of Architecture and Construction Sciences, Research Institute of Theory and History of Architecture and Urban Planning* **1** 291-300

[8] Watanabe Y 1964 *Shinto art: Ise and Izumo shrines* (New-York: Weatherhill / Tokyo: Heibonsha)

[9] Gorbyk O O 2018 *World History of Architecture in abstracts and pictures (monuments' handbook). Part 1. Architecture of primitive age and traditional architecture. The architecture of the ancient world. Architecture of Antiquity and Early Christianity* (Kyiv: Phoenix)

[10] Xue Ch and Xiao J 2014 Japanese modernity deviated: Its importation and legacy in the Southeast Asian architecture since the 1970s *Habitat International* **44** pp 227-36 http://www.sciencedirect.com/science/article/pii/S0197397514000873

[11] Mezhenna N 2012 Integration of object of historical urban environment in composition of contemporary building *Scientific enquiry in the contemporary world: Theoretical basics and innovative approach* **4** 82-3; Serra J, Llinares C, Iñarra S, Torres A and Llopis J 2020 Improvement of the integration of visually impacting architectures in historical urban scene, an application of semantic differencial method *Environmental Impact Assessment Review* **81** 106353 http://www.sciencedirect.com/science/article/pii/S0195925519302926

[12] Natali R 2018 The Psychological Impact of Architectural Design *CMC Senior Theses* (2018)1767 https://scholarship.claremont.edu/cmc_theses/1767

IOP Conf. Series: Materials Science and Engineering **907** (2020) 012002 doi:10.1088/1757-899X/907/1/012002

# The actualization of the garden city model in modern urban planning with participation practices (on the example of Grabnyk park renewal in the city of Rivne)

**M Y Blinova and S A Ilchenko**

Department of Design of Architectural Environment, Kharkiv National University of Civil Engineering and Architecture, Kharkiv, Ukraine

blinova.marja@gmail.com

**Abstract.** The participation practices in Ukrainian urban planning are a fairly new phenomenon. The growing interest in such approaches primarily reflects positive changes in the development of cities. On the other hand, this interest draws attention to problems of implementation of these approaches in actual urban planning practice. It is important to note here that although participation practices are well regulated in countries with developed democracy, under conditions of developing democracy in Ukraine there are still no well-functioning mechanisms of interaction between government and society at local levels. Regarding participation of local communities in city development, the discussion about the place of a professional architect in urban planning processes that include participation practices becomes particularly important. In this article we uncover effectiveness of participation practices and limitations of their application in the context of Ukraine. We also determine the differences in understanding effectiveness and limitations of these practices in Ukraine and developed democracies.

## 1. Introduction

Issues of applying participation practices in urban planning have become topical since 2014 due to certain changes in Ukrainian legislation, primarily with the increasing role of local authorities [1]. It should be noted that if 10 years ago the participation practices were considered a unique experience of foreign countries, now it is possible to analyze their application in Ukraine.

In this context, it seems relevant to consider existing examples, determine effectiveness of participation of local communities in city development, establish the framework for application and impact of this participation on the final result when developing urban projects. Moreover, participation practices are interesting precisely for their influence on the morphology of the architectural environment of a city, i.e. how the initial requests of local urban communities are realized in specific design decisions. The answer to this question implies searching for algorithms (ways?) to include such practices in the process of urban planning and project implementation, as well as determining the degree of influence of local community requests on the relevance of project proposals and finding balance in the interaction between society and experts.

In the context of this research, particularly interesting are the fields which developed an understanding of the need for participation practices. Primarily, these are sociology and urban sociology, as well as anthropology and social geography. United by the topic of local community involvement in planning and city development, all these fields demonstrate some vulnerability in terms of practical implementation of local residents' requests. In our opinion, the problem lies not so much in the opportunity for public participation in urban planning, but in its effectiveness - and in the very possibility of implementing such plans.

Numerous works on this subject in the framework of this study can be divided into two conceptual blocks.

The first block includes the works studying inclusion of participation practices in urban planning processes and their criticism. It covers the period from 1960 to 2011, preceding the processes of legislative consolidation of the role of local communities in urban planning processes.

For example, Turner, J. (1969) [2], (1972) [3] outlines the prerequisites for returning to local communities their right to self-determination, using participation practices as an instrument. In his works, the author focuses on the economic feasibility of the citizen participation in development projects: "*When dwellers control the major decisions and are free to make their own contribution to the design, construction or management of their housing, both the process and the environment produced stimulate individual and social well-being. When people have no control over, nor responsibility for key decisions in the housing process, on the other hand, dwelling environments may instead become a barrier to personal fulfillment and a burden on the economy.*" Turner, J. (1972) [3]. Similar suggestions for small settlements where participation practices are seen as a development resource have been formulated by Chambers, R. (1983) [4].

The criticism of such approaches, which arose at about the same time, was focused on the methods of their application. The main conflicts between participatory democracy and professional experience, as well as problems of involving citizens in decision-making, are described by Burke, E. (1968) [5].

The threat of tokenism, i.e. the policy of taking only superficial or symbolic actions that give the appearance of equality when applying practices was identified by Arnstein, S. (1969) [6].

The subject of ability of a territorial neighborhood community to be self-sufficient, introduced into the theory of urban planning in the post-Soviet space by Glazychev, V. Egorov, M. (1995) [7], is similar to the communicative theory of urban planning built on the involvement of the parties interested and involved in this process, stated by Innes, J.E (1995) [8]. Healey, P. (1996) [9].

A comprehensive and detailed retrospective of participation practices is presented in research by Gella O I (1999) [10], where author examines the prerequisites for emergence of the "participation practices" phenomenon and traces its development in the context of the environmental approach from 1960 to 1990, preceding the legislative consolidation of the local communities role in urban planning. The first block of the works under consideration is concisely summed up in the statement: "*... the tenets of participation practice have undergone little change even though criticism regularly spells out prescriptions, ...*" p.197 mentioned in a detailed review of the critique of participation practices by Arboleda, G. (2015) [11].

The second block of sources is interesting in the context of the legislative shift to local self-government. It is determined by The Localism Act (2011) in the UK and the Laws On Cooperation of Territorial Communities (2014) in Ukraine [1]. The British authors Holman N. and Rydin Y. (2013) [12], focusing on creating networks and specific forms of social capital necessary to achieve participation, as well as on attracting professional skills in neighborhood planning, subsequently turn to the concept of social capital in the context of The Localism Act, and the impact of the local agenda on the implementation of urban planning projects - Holman N and Rydin Y (2013) [12]. The effects of local communities influencing urban planning practices in the context of changes to the system of government are described by Clifford B and Tewdwr-Jones M (2013) [13]. An article by Ukrainian researcher Levytska O and Zapototska V (2018) [14] covering the participation practices in urban planning, presents relevant examples of urban projects involving the community in Kyiv, Lviv, and Ivano-Frankivsk. This material, which examines the processes of strategic planning, reveals the main features and problems of the application of similar practices in Ukraine. The article gives a rather detailed picture of the participation of local residents in urban space development and examines the relevance of these practices to the world trends in urban development. The materials of the second block are united by the topic of the change in the situation of spatial planning, due to the increasing role of local communities in urban development processes.

However, despite such a strong interest in influence of participation practices on the urban environment and a wealth of foreign experience in their application, the question of the frontiers of their use and the impact on the space morphology in Ukrainian realities remains relevant.

*We strive to identify not only the effectiveness of the participation practices and the limits of their application in Ukraine, but also the mechanism for including local community requests in real projects by the experts in the field of architecture and planning.* We use the Grabnyk park renewal in the city of Rivne as an example and analyze the various historical stages of its development in the context of place identity and how place attachment can affect the planning process.

## 2. Methods

The method we use is qualitative research with the interpretation of the urban planning processes under consideration in the context of their influence on the structure and morphology of the place. Primary data were obtained as a result of a direct examination and assessment of the situation during the field study. These data include sociological polls conducted by an initiative group. The secondary data were gathered by studying the literature and reviewing documents, maps, master plans, as well as analyzing the project proposals of the initiative group. A small historical retrospection has become an integral part of this work. It allowed to determine the impact of the original concept of the place development on the city residents and the impact of this concept on the project proposals that followed up. The data obtained are interpreted in the context of the analytical concept of place identity frames that considers them as: *"... a project of manufactured unity in which a coherent vision of place is distilled from discordant views, potentially to mobilise a population despite tensions and divisions."* Bradley Q (2017) [15]. The research on this topic is intended to lay the foundation for further studies on the role of place attachment in urban planning processes.

## 3. Discussion

The theoretical environment, in which a specific example of the Grabnyk park renewal in the city of Rivne will be considered, includes a number of works. The effectiveness criteria for community participation in urban planning are discussed in the research carried out by Brown G and Sean Yeong Wei Chin (2013) [16]. Pointing to the absence of the direct relationship between the application of inclusive practices and the end result, the authors highlight the following main effectiveness criteria:

- reaching a consensus;
- the impact of the cooperation process compared to the alternative processes;
- stakeholder satisfaction with the process and outcome;
- achieving other social capital benefits, such as improving relations between stakeholders, improving their skills and knowledge.

Thus, the authors mainstream the issue of confidence between the parties, which is built in the process of interaction between the local community, planners and local governments.

Residents set their sociocultural context and provide knowledge about the local lifestyle, planners show how this information was used to develop design decisions, and local governments demonstrate transparency in financing and organizing processes.

Further, mention should be made of the work of Bradley Q (2017), devoted to the study of the impact of collective identity and place attachment on urban planning and design processes [15]. The author, having studied various cases in the UK, identifies a certain approach that allows achieving maximum efficiency of the design decisions, i.e. local community involvement into the design processes and studying the place attachment formed in it, the specificity of the local lifestyle and sociocultural context. Accordingly, the stronger features of this community are, the stronger its sociocultural identity is manifested, the higher will be the support of the proposed design solutions in referendum. In developed democracy, the main efficiency rule for participation practices is the compliance of project proposals with the identified place identity.

We suggest considering a project of the Grabnyk park renewal in the city of Rivne as an example demonstrating the effectiveness and scope of the application of participation practices in urban planning in Ukraine.

Several stages can be distinguished in the development of this site. The first stage includes the planning scheme of 1926 [24], which served as a basis for developing the street network and public space of this area. The radial pattern of the streets oriented towards the spacious zones of parks correlates quite accurately with the model of the garden city by Howard E (1898) [17], which was popular at that time. Although we only have a geodetic map of the land allocation, it provides a sufficient insight into the spatial frame of the area (Figure 1).

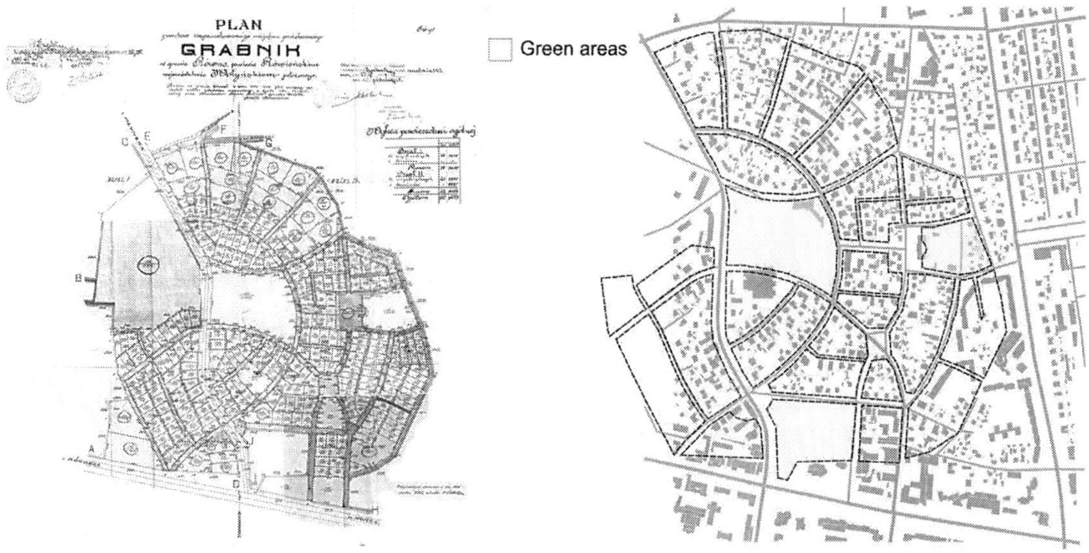


**Figure 1.** The geodetic plan of land allocation (1926) [18] and its boundaries in the current development.

The second stage involves the development of this territory in the 1970-80s. At that time, some public buildings were built in the area, such as the Palace of Pioneers and Schoolchildren (architect M. Chabak), the Children's Music School for 460 students (architect M. Chabak, M. Sholudko), and the Children's Art School (architect M. Chabak). This period is also characterized by dividing the city into distinct functional areas - the plot was allocated for the extracurricular children activity from across the city. The conservation of the green zone is a distinctive feature of the design solution: the volumes of buildings are located around the perimeter of the existing park space, without invading its territory. Objects with an unusual typology, such as a semicircular amphitheater with an open stage and a lighting tower with a viewing platform, included in the complex, determined the individuality of the place.

IOP Conf. Series: Materials Science and Engineering **907** (2020) 012002 doi:10.1088/1757-899X/907/1/012002

This period of development is of particular interest not only because of the constructions discussed, but also because of the project for the development of the region, still unrealized due to unforeseen events. It was designed to enlarge the scale of development and change its character. If we apply the project proposals of the 1970-1980s to the existing situation, we can see that the park could have lost its key role in the area, as the document suggested the shift of emphasis (Figure 2).

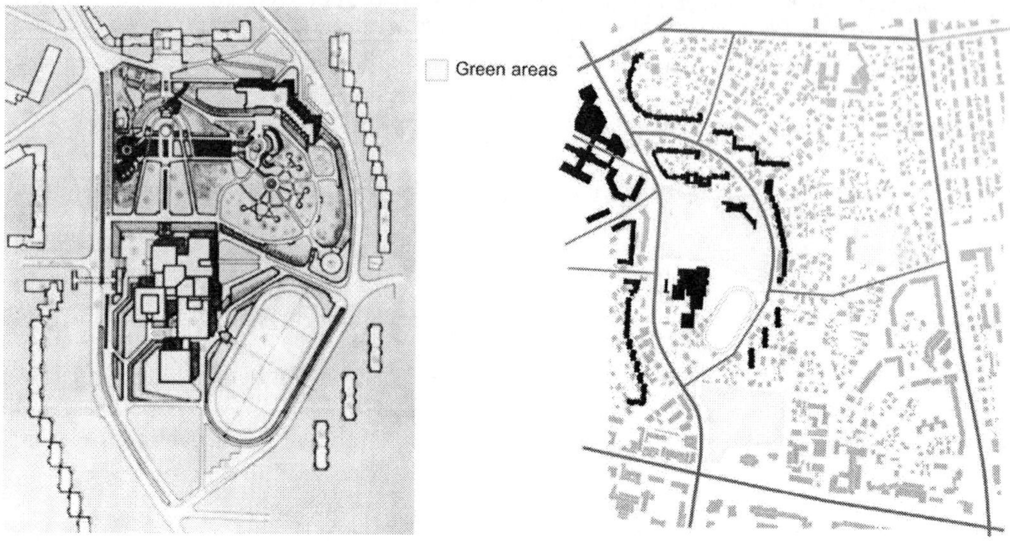


**Figure 2.** The project proposals for the development of the 1980s district [19] and the plan illustrating enlargement of the scale when superimposed on the actual buildings.

The third, modern ("post-industrial") stage of the Grabnyk district development can be described as a positive example of balancing social and professional components of urban planning.

*3.1 Timeline*

The information that appeared in the summer of 2016 about the planned construction of an aquapark in the parkland, caused a protest movement of citizens. A number of public organizations and active residents of Rivne opposed the construction and put forward the initiative of the park renewal as an alternative to the park development. A group was created on the Facebook social network, where people discussed the possibilities of the park renewal and highlighted the activities undertaken by the activists in support of maintaining the recreation area.

Within a year, the initiative residents submitted three projects of park restoration (on a participative budget) that won and received funding (Figure 3). To develop a concept for the park renewal, the initiative group organized a workshop "Park near the Palace of Children and Youth - the course for restoration" in the summer of 2017. Rivne architects and city planners took part in the event. As a result, several conceptual ideas for the development of the park were proposed, which were presented to the public at a city picnic (October 2017) [20, 21].

In January-February 2018, the Rivne Reform Council and the Civic network OPORA Rivne organized the public hearings to further discuss and detail the concept of the park renewal. They were attended by city authorities, the administration of the Palace of Children and Youth, the representatives of the Music school №2 and the public organizations, as well as the architects, the authors of public projects, the area residents and the media.

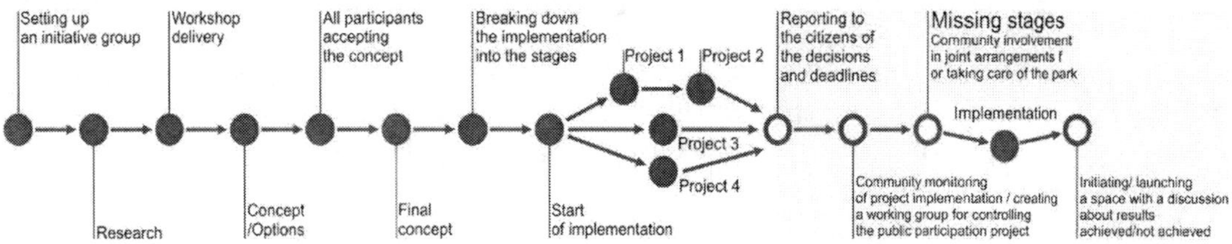


**Figure 3.** The algorithm for implementing the concept of the Grabnik park renewal (2017) and the stages not addressed by the strategy.

All residents were able to explore the working concept of the future park - online or at the information stands in the Palace of Children and Youth, INFOHUB and the Prostor Interaction Platform. In the same locations, public discussions of the concept took place. As a result of public consultations, a working group was created to develop the final concept of the park. It included the area residents, government officials, representatives of the Palace of Children and Youth, the Music school №2, public organizations, urban architects and the authors of public projects.

In February-March, the working group held five meetings to discuss the concept of each of the park zones [23] (Figure 4), and subsequently, the main stages of the park renewal were approved:

- modernization of the park lighting system;
- making the surface of the parkways, installing benches and rubbish bins, opening a second entrance to the park, arranging a picnic area, a training area, creating a climbing wall and ramps, areas for winter activities;
- moving a garbage dump, arranging a playground, a dog walking area, a universal sports ground, creating a network of alleys and a wide walking alley, a trampoline court and a skatepark;
- arranging the seats on the slope, a concert and educational zone and an amphitheater.

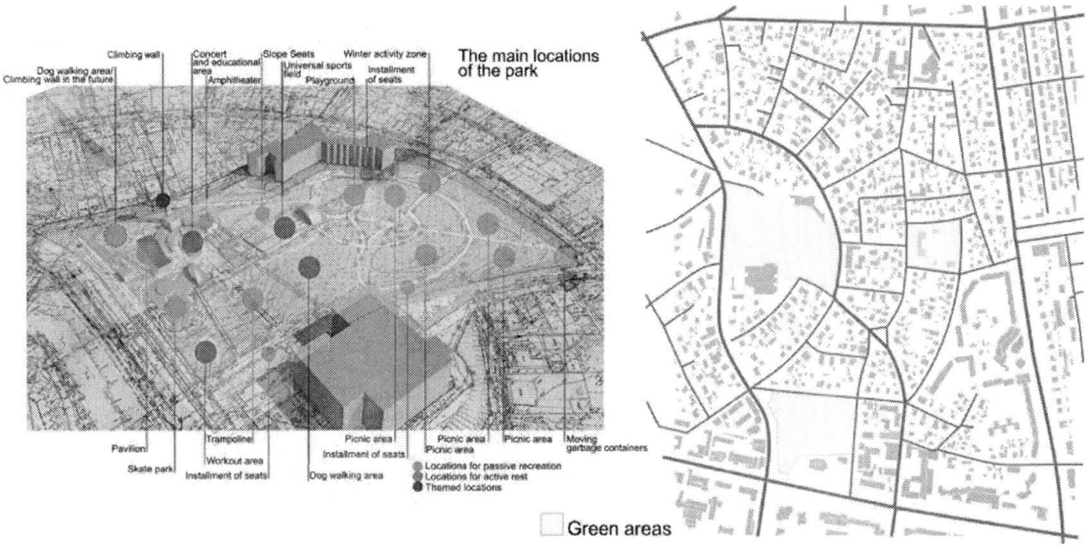


**Figure 4.** The project proposal for the Grabnyk Park renewal (2017) [22] and its place in the general structure of the district

Currently, only the first 2 projects of the public budget have been implemented due to the long-term process of land allocation (2-3 years), finding an agreement on the candidature of the manager, selection of contractors and obtaining technical conditions. Part of the park obtained a hard surface with barrier-free access, lighting and drainage have been restored in transit directions. A partial replacement of worn paving is planned for 2020. In the context of a continuous decrease in funding, the final completion of the project is raising serious doubts.

Conducting the population survey after the implementation of the first projects, the initiative group participants drew attention to the change in the discourse regarding the park. From a dangerous, deserted, useless area, it is gradually becoming a park and the key area of the district.

## 4. Interpretation of Results

First, we will describe the character of participation and its extent. The participation of the local community that arose from the spontaneous protest movement can be characterized as organic, as opposed to the induced participation organized by the governing bodies. Considering the forms of participation in the context of the "ladder of citizen participation" Arnstein S (1969) [6], the extent of citizens' influence within a specific example of urban planning has been very high. In this case, there is no duality inherent in a "symbolic event" (Tokenism). This shows that the influence on the process has been distributed between the city administration and the local community.

At the same time, the application of participation practices, in this case, confirmed the problems identified by Levytska O and Zapototska V (2018) [14]. Based on the results of a sociological survey conducted by an initiative group in the research phase [24], the responses of the majority of respondents turned out to be distinctive: they covered the local issues incompletely, focused on details and lacked a comprehensive vision of the problem, having unreasonably high expectations. The involvement of residents of the remote areas seems to have distorted the real needs of the local community. This was especially manifested in the varied picture of the entertainment proposed (functional content). It is important to note that the requests of representatives of the local community were pragmatic in comparison with the desires of "outsiders" (who talked about rides, fountains, etc.), and reduced to the natural needs of security, accessibility and order. In total, 71% of respondents described the place as being desolated and cluttered which was supposed to be the main drawback of the existing state of the park, but this was hardly worth conducting a survey. Problems of lack of infrastructure, which seem logical at first glance, were important only for 7%. The idea of placing a toilet in the park gained less than 3%. Such a reaction can be interpreted rather as a spontaneous answer to an incomprehensible question. It can also be concluded that both the members of the initiative group and the respondents lacked understanding of the purpose of such a survey. It turned out that the participants in the process: the city residents, activists and the city administration lacked understanding the purpose of interaction and the role of each side. The very concept of participatory budgeting is only becoming a part of real life, and its application is hindered by the lack of joint work skills and the problem of the transition from an authoritarian style of decision-making to democratic practices.

Application of participation practices in the UK focuses on identifying local specifics, distinctive characteristics and values of the place, which are subsequently captured in local planning strategies Bradley Q (2017) [15]; that is, the process of local community participation is not aimed at discussing individual projects, but at forming a planning strategy in the context of discourse about place attachment. With this approach, the opinions of citizens and their understanding of the place are constructive and unique information that defines the scope of local planning. Here we return to the problem of "efficiency frontiers" with the involvement of residents. In the case of the Grabnyk park, this is very clearly visible: the discourse about the uniqueness of the place and its sociocultural identity arises only at the stage of attracting professional architects for the implementation of the project.

The workshop with the architects and city planners participating in the initiative group turned out to be effective and allowed to form two initial project proposals for the park renewal [20] [21] that served as the basis for the follow-up dialogue with local residents. As has been repeatedly noted by many researchers, in particular Bradley, Q. (2017) [15], *"The complex emotional interconnections between people and place mean that plans to change a particular environment can be perceived as a threat to personal autonomy or identity..."*, therefore, the ability to create an acceptable solution by integrating a wide and diverse knowledge base is one of the criteria for the effectiveness of participatory practices Brown G, and Sean Yeong Wei Chin (2013) [16]. Removing contradictions between participants in the planning process and incorporating needs and requests relevant to the development of the place in the Grabnyk park development project meet this criterion.

In this context, particularly important is the opinion of the initiative group that the change in focus of the development of this area and decision against water park construction is beneficial , as due to this the park has regained its recreational function . In this case, the trend towards the renovation of an existing object can be considered an indicator of the effectiveness of participation practices.

By its structure, the object we are considering is in full compliance with the principles of New urbanism: the place includes a noticeable green center of the district, most of the dwellings are located within a five-minute walk from it, there is an elementary school nearby, some relatively narrow streets, public buildings framing view axes. All this is a consequence of the original layout of 1926, which is based on the urban concept of the "garden city" by E. Howard. The preservation of the park as the green core of the district, as well as unlocking the value of existing modernist buildings by the local community, formed by professional architects during the discussion, led to the perception of these buildings as a specific marker of the place and a symbol of the local identity of the place. Place attachment, in its turn, serves as the basis for public support for local development projects. In this context, issues of the efficient use of the park complex - its available volumes and the spaces surrounding it - have come to the foreground. Architects deliberately narrowed their arsenal to the landscape and architectural design tools, their proposal focused on creating green spaces, small forms and open spaces. In general, the primary tools used to solve the problem can be structured as follows:

- architectural and planning: returning the central role of the district to the park space, creating a network of alleys in transit zones;
- landscape: the renewal of green spaces and lawns;
- architectural and design: arranging ramps and seats on the slope; creating a concert and educational zone, an amphitheater, a picnic area, trampoline and training area;
- marketing: the installation of a climbing wall, creating a skate park and winter activity zones as markers of the area;
- technological: modernization of the lighting and wastewater systems, elimination of landfills, getting rid of the garbage dump, installing benches and garbage bins;
- management: preservation of the architectural heritage - a unique complex of modernist buildings, refusal from building up the park, organization of the financing and implementation processes.

## 5. Conclusion

We can characterize an important part of the social capital available to the local community as unique knowledge about the life of the locus. Lefebvre, H. (1991) [25]. It should be noted that practices of participation as tools are not factories of ideas and do not generate any new meanings.

*Thus, a detailed analysis of our case example allows us to conclude that the condition for effective application of participation practices is a clear statement of the purpose of their application. Community participation in identifying features and resources that are characteristic and important for the place is effective in phases of discourse and defining planning strategies, with maintaining control at planning and implementation stages.*

In Ukrainian realities, with a predominance of the administrative-command approach in planning, such practices are more tokenistic, and do not involve communities in the assessment of the needs, planning, design and implementation. Designing in several stages is the most effective means of communicating the meaning of project proposals to the community and receiving feedback. Step-by-step iterations determine the verification, consolidation of the meanings and goals of specific projects in the local community and, definitely, build trust in project proposals. The role of city administration in such events should not be limited to the position of an outside observer. It must provide legal basis for the process without interfering with decision-making. Turner J (1969) [2].

The design process and the execution of work should have information support for monitoring by the local community, - the results of the monitoring phase significantly affect the criterion of satisfaction of the parties with the process and the result. Greg Brown & Sean Yeong Wei Chin (2013) [16].

In the context of structure and morphology of urban spaces, the case we are considering demonstrates the intuitive need of the participants (citizens) to return to the urban concept of the "garden city" by E. Howard, formed by the original planning grid of 1926. Based on the concept of the "identity frame" of a place Martin's (2003) [26], we can trace its influence on the formation of group identity of the community. A similar influence is manifested in the adoption of a certain morphology and structure of the urban environment as a norm, and the subsequent correlation of urban development programs with this norm. Place attachment largely determines the community's attitude to urban planning projects. Changing a familiar environment is guaranteed to cause a protest reaction. In this context, the Grabnyk park renewal can be defined, rather, as the creation of a public space (de facto) that completed the time-consuming process of forming this city area. Development resources, in this case, were spatial structure of the place, public buildings of modernist period, green spaces, and local community. This process made it possible to create a discourse of the place and identity of the community, which, unfortunately, were not framed in the local development strategy as *"... a text through which the social construction of place and community can be read and analysed, and where the influence of place attachment and place identity on planning policy for house-building may be discerned and scrutinised."* Bradley, Q (2017) [15] Nevertheless, the Grabnyk park renewal in the city of Rivne can be considered a unique example of development of an area in the context of its identity. This is an example that demonstrates the effect of place attachment on the planning process and the positive role of professional architects in decision making.

The process of updating the Grabnyk park in Rivne is an inspiring example of participation practices in the Ukrainian context.

**References**

[1]. Zakon Ukrayiny «Pro spivrobitnytstvo terytorial'nykh hromad» vid 17.06.2014 r. no. 1508-VII [The Law of Ukraine «On cooperation of local communities»], Available at: http://zakon4.rada.gov.ua/laws/show/1508-18 [in Ukrainian]
[2]. Turner J 1969 Architecture that works Ekistics **27:158** pp 40–44.
[3]. Turner J 1972 *Freedom to build: Dweller control of the housing process* (New York: Macmillan)
[4]. Chambers R 1983 *Rural development: Putting the last first* (New York: John Wiley and Sons)
[5]. Burke E 1968 Citizen participation strategies *J. of the American Institute of Planners* **34:5** pp 287– 294.
[6]. Arnstein S 1969 A ladder of citizen participation *J. of the American Institute of Planners* **35:4** pp 216–224.
[7]. Glazychev V, Egorov M et al. 1995 *"Urban environment. Technology of development: a board book"* (Moscow: Ladya) [in Russian].
[8]. Innes J 1995 Planning Theory's Emerging Paradigm: Communicative Action and Interactive Practice *J. of Planning Education and Research* **14:3** pp 183–189.
[9]. Healey P 1996 The Communicative Turn in Planning Theory and its Implications for Spatial Strategy Formation. *Environment and Planning* B **23:2** pp 217–234.
[10]. Gella O I 1999 *Environmental approach and programs of participation in architecture 1960-1990*: Dis. Cand. Arch. Science: 72.01. + 72.02. / Hella Olena Ivanivna - Kharkiv
[11]. Arboleda G 2015 Participation Practice and its Criticism: Can They Be Bridged? *Housing and Society* **41:2** pp 195–227.

[12]. Holman N and Rydin Y 2013 What can social capital tell us about planning under localism? *Local Government Studies* **39:1** pp 71–88
[13]. Clifford B and Tewdwr-Jones M 2013 *The Collaborating Planner?* (Bristol, UK: Policy Press)
[14]. Levytska O and Zapototska V 2017 Public participation in urban planning: German and Ukrainian experience *J. of Geography and Environmental Management* **45:2** pp 18–27
[15]. Bradley Q 2017 Neighbourhood planning and the impact of place identity on housing development in England *Planning Theory & Practice* **18:2** pp 233-248
[16]. Brown G, and Sean Yeong Wei Chin 2013 Assessing the Effectiveness of Public Participation in Neighbourhood Planning *Planning Practice & Research* **28:5** pp 563-588
[17]. Howard E 1898 *To-morrow: A Peaceful Path to Reform* (London: Swan Sonnenschein & Co)
[18]. Mykhaylyshyn O 2013 *Architecture and Urban Planning of Western Volyn 1921-1939* Rivne : Dyatlyk M p 93 [in Ukrainian]
[19]. Sholudko M 2019 Prerequisites for the developmeni of modern tendency in architectural objects of the equal 1970-1990 *Architectural education and science in Ukraine and in the world: experience and prospects of development* pp 299–305
[20]. Starting concept 1 of updating the park Grabnik Rivne 2017 Available at: https://drive.google.com/file/d/0B3PQqgxy2_NTbDBTV2hjcUFnU00/view [in Ukrainian]
[21]. Starting concept 2 of updating the park Grabnik, Rivne 2017 Available at: https://drive.google.com/file/d/0B3PQqgxy2_NTbi15cldDV2F6MjQ/view [in Ukrainian]
[22]. Final concept of updating the park Grabnik Rivne 2018 Available at: https://drive.google.com/drive/folders/1_YAgjTAAMBVAfDnUbbWrKNYThEffmuki [in Ukrainian]
[23]. Working group meeting report 2017, Available at: https://drive.google.com/file/d/11OACtT1v9-AIlizn6PPP4B1K4sjNYvHi/view?fbclid=IwAR33DBEDc6n7cQSN2V5RPHoy8mLMW41am1EZ-vm_rY4atT5cz5o2OhXah8 [in Ukrainian]
[24]. Opinion polls concept of updating the park Grabnik 2017 Available at: https://drive.google.com/file/d/0BxtvnZ3OqFQIT2hjeUtPaGc0OWM/view?fbclid=IwAR3MpYiUkFp iWx7zWZcwhhC3QceYoZmcqE-lA7THeqfK-fLoVi-c01Fy0jo [in Ukrainian]
[25]. Lefebvre H 1991 *The Production of Space* (Oxford: Blackwell)
[26]. Martin D 2003 Place-Framing as Place Making: constituting a neighbourhood for organising and activism. *Annals of the Association of American Geographers* **93:3** pp 730-750

IOP Conf. Series: Materials Science and Engineering **907** (2020) 012003 doi:10.1088/1757-899X/907/1/012003

# Sustainable Mobility in the Context of Humanization of the Urban Environment: a regional experience

**N N Korshunova and E B Morozova O E Dolinina**
Department of Architecture, Engineering Academy, RUDN University, Moscow, 115419, Russia

dolinina_oe@pfur.ru

**Abstract.** The purpose of the research is to identify modern approaches to the spatial development of Belarusian cities. The article focuses on the role of transit-oriented design in spatial development of cities. The principles of 'green urban construction' and their implementation in pilot projects for urban areas reconstruction in Polotsk-Novopolotsk agglomeration are considered. The introduction of developed territory transformation models into urban planning practice was suggested. The participation of Belarus in international initiatives and projects relating to urban and transport planning are assessed. Organizational and institutional prerequisites for using the European experience of mobility management have been established. The scientific novelty consists in the study of the concept of sustainable mobility as a tool for territorial development and humanization of the environment with a focus on the conditions in Belarus.

## 1. Introduction

The city is not just a place on the map but also a reality that is formed by the feelings of each person; then it is transformed into a general idea that determines the image of a settlement. The quality characteristic of a city determines its *attractiveness or competitiveness as a location for various types of activities* [1]. Moreover, the city should be attractive not only for enterprises but also for the population, which is one of the most valuable resources for production facilities. It happens due to the globalization and an increasing role of qualified staff and innovations. Another target force cities are fighting for is visitors and tourists. It brings profit to local businesses and is an important carrier and distributor of information about a city, thus 'putting it on the map'.

The most acceptable criteria among the many criteria of competitiveness for describing cities in the CIS countries are the following indicators: *productivity (efficiency), employment, standard (quality) of life* **Ошибка! Источник ссылки не найден.** The latter criterion directly depends on the problems of the spatial organization of the city such as lagging in the improvement of infrastructure facilities, excessive load on them, an imbalance of proportions in the general city plan structure and so on. The quality of life means not only economically integral space but also comfortable, aesthetically attractive and continuous urban environment. That is why the main message of modern spatial development is aimed at *'humanization'* of the city due to which it becomes not only convenient for people's lives but also contributes to their professional implementation, social and cultural development .

The quality of urban life is formed by many components, the most important of which is of course the accessibility of all its districts and main objects of urban infrastructure. The transport system determines the future of the city since it is intended not only for the current movement of people but also is a way to introduce urban development in the right direction. *Transit-oriented design* is one of the most effective directions of urban planning, which justifies the intensification of the use of urban territories and the need to implement interdepartmental strategic documents on spatial development [3]. According to it the transport framework does not only affect the architecture of the city but also, in fact, controls the usage and limits the growth of the territory, ensuring its connectivity and permeability. Thus following the principles of transit-oriented development it contributes to achieving sustainable mobility, which increases the competitiveness of cities by humanizing the environment and improving the quality of life.

The current planning system, which is typical for Belarusian cities, is not effective enough to implement the principles of transit-oriented design since the spatial development of the city is determined by master plans, the most obvious drawback of which is their failure to make the city convenient for life and attractive to businesses and investors as they are not integrated sufficiently into the urban management system and act separately from city services and the budget process [3]. The urban planning based on people's standard work schedules does not correspond to their more complex and diverse needs, including the free mode of doing business, the active use of information technologies.

The problem of sustainable mobility cannot be solved successfully within the traditional transport planning, which is aimed at solving the problems of different means of transport, building infrastructure facilities and increasing the capacity of the road network. When planning sustainable mobility, efforts are focused on the development of integrated transport systems, intersectoral cooperation and the promotion of environmentally friendly movements. That's why new tools such as *sustainable urban mobility plans* (SUMPs), which are based on the principles of sustainable development and system cross-sectoral integration, have proved successful in many European countries [4].

In the EU, where this innovation originates, the development of SUMP has become mandatory for cities with the population of more than 100 thousand people. It is held under the motto "Planning for people". The following official definition has been adopted: "A sustainable urban mobility plan is a strategic plan designed to meet the mobility needs of people and businesses in and around cities to improve the quality of life. It is based on existing planning practices and takes due account of the principles of integration, participation and evaluation" [4].

The SUMP concept is a pan-European success story and it is based *on several important principles*:
- coordination of development strategies in various sectors, including transport, land use, environment, economic development, social policy, health and security;
- coordination of actions by State bodies at various levels, as well as the interests of neighboring territories;
- involvement of residents and representatives of other parties interested in the planning process.

## 2. Implementation of SUMP in Belarus

The first attempt to develop and implement SUMP in Belarus was made in 2015-2017 as part of the project "From energy efficiency to urban mobility: introducing an approach for local participation in the development of sustainable urban mobility plan in Polotsk" [5]. The project working group, which controlled the development of the mobility plan, included representatives of the Polotsk District Executive Committee, city municipal services, several public organizations and businesses. The developers of the project were BelNIIPgradostroitelstva (the Institute of Urban and Regional Planning) and the Belarusian Association of Transport Experts and Surveyors (BAES). As a result of the implementation of recommendations developed by experts together with residents of the city the first segment of bicycle infrastructure was implemented. It was the first time in Belarus when the bike path was created through the traffic lanes and was separated with delineators, with bicycle parking appearing on its route. "From energy efficiency to urban mobility" project was the first in the field of transport in Belarus implemented with the involvement of the local population. Strengthening the capacities of citizens and social organizations, support to local authorities and other parties concerned contributed to the further participation of Polotsk in urban mobility planning projects.

Since 2016 the "Green cities" Project which was funded by the Global Environmental Fund and implemented by UNDP in cooperation with the Ministry of Natural Resources has been developed in three pilot cities – Novogrudok, Polotsk and Novopolotsk [6]. It is the implementation of initiatives related to sustainable mobility and energy efficiency that the promotion of the principles of green urban development in Belarus is based on. The most important tasks of the project are to establish intersectoral interaction, to develop partnerships and cooperation between cities, even at the international level. In particular, the project has developed unique documents for Belarus – a Integrated Sustainable Urban Mobility Plan for Polotsk and Novopolotsk and a Green Urban Development Plan for each city [7].

The goal of the green urban development plan *"Novogrudok. Save the future"* is a sustainable development in order to improve the quality of life. The activities will be implemented in three priority areas: comfortable and unique urban environment, territory transformation based on historical and cultural potential, the use of alternative resources in circular economy and the activities of urban facilities.

*"Polotsk. Let's build bridges",* a developed green urban development plan for the district center of Vitebsk region, solves the problem of bridge reconstruction and improving the links between the center and the suburbs. The plan provides for three themes: a diverse city, cozy environment and harmony with the nature. The action program involves strengthening community centers for the development of self-sufficient areas, encouraging the use of public transport and creating energy-efficient engineering infrastructure.

*"Novopolotsk is new again. Version 2.0"* plan is designed to increase the attractiveness of the city as a place to live and invest. The project involves the formation of a unified water-green system, the development of a closed-loop economy infrastructure, improving the comfort of living with the help of smart technologies and transit-oriented development.

It should be noted that the proposed models have been developed according to Polotsk and Novopolotsk agglomeration development. The formation of agglomerations does not require administrative integration of communities but involves coordination of their plans for social, economic, territorial and infrastructural development. The potential for agglomeration development wasn't revealed in the plan for sustainable urban mobility for Polotsk which had been adopted previously, but its implementation was the basis of the first mobility plan in Belarus for the two cities [9].

The main guidelines for the development of new quality mobility systems in accordance with the Integrated Sustainable Mobility Plan developed by experts of the Belarusian Transport Workers Union are to be the *following principles of green urban development [10]*:
- Compactness: a) building without gaps; b) connectivity of urban areas between each other and the public transport system; c) service and job availability;
- High density: at least 15,000 people per km2, that is 150 people/ha;
- Mixed land use: multi-purpose territory use which is characterized by a combination of several types of functional land use;

- Social balance: the availability of housing in different price ranges in all areas of the city, which provides social and economic stability due to the fact that people with different income levels live in the same part of the city;
- Sufficient street space and an efficient street network: the density of streets, pedestrian and bicycle paths should be at least 18 km of streets per km2;
- Transit-oriented development: organization of blocks or neighborhoods around public transport stops due to dense and multifunctional development with a gradual decrease in the intensity of land use from the center to the district;
- Accessibility of green areas: placing of at least 90% of planned and existing housing within the walking distance (400 m) of green areas.

Polotsk and Novopolotsk, which are developing as conurbations, illustrate a characteristic global trend: the transition to agglomeration development [11]. Territorial proximity, established economic, social and cultural links make these "paired" cities partners in solving many problems. That's why it is necessary to develop a common position on the territorial development of the agglomeration. Communities are recommended to integrate the main aspects of urban planning and transport policy in order to determine the vector that will be used for building and placing attraction points. The Integrated Sustainable Urban Mobility Plan for Polotsk and Novopolotsk allows cities to solve regional problems of demand for public transport and congestion caused by excessive use of personal cars. Joint efforts contribute to the creation of comfortable urban environment and reduction of forced trips of citizens.

The development of new territories, the revitalization of industrial areas and the conversion of dying enterprises require mutual agreement [8]. For example, the village of Ekiman which is situated at the junction of the cities has territorial reserves and can turn into 'an ecological corridor', a well-equipped zone with a set of objects and public spaces attractive to businesses and the population, and it can become a kind of 'bridge of activities' between Polotsk and Novopolotsk (figure 1).

An equally important component of the joint activities of the cities concerns the outline of the structure and milestones of the formation of an integrated transport system and mechanisms for mobility management. The development of a tram network connecting the cities as a coherent whole will be the best solution for the reasons of productivity, social and environmental parameters, but it involves significant external resources, namely bank loans guaranteed by the government.

## 3. Case study: pilot projects for Polotsk-Novopolotsk agglomeration

Taking into account the recommendations of the Integrated Sustainable Mobility Plan, MLA+ experts developed pilot projects for the transformation of two residential areas in the Polotsk-Novopolotsk agglomeration, illustrating the application of the principles of green urban development in Belarusian cities [12],[13][13]. The provisions were supplemented by the corporate principles of the consulting bureau: *inclusion of the object in the city structure, its identity, structure and flexibility*. On the results of workshops with the participation of residents, representatives of the administration and local businesses the overall goals of the projects have been determined. Further, the versions of the transformation proposed by the architects were combined into an integrated version that takes into account the advantages of all the concepts and suggestions made by the residents.

The transformation of urban areas which are experimental sites for green development will be an example of consistent implementation of Green Urban Development Plans [7]. The conceptual models of transformation are assumed to be used in the future by city administrations as initial documentation for making decisions about their further spatial development, namely architectural and engineering preparation for the implementation of investment projects, design and survey work, preparation of acts of land selection

IOP Conf. Series: Materials Science and Engineering **907** (2020) 012003 doi:10.1088/1757-899X/907/1/012003

for construction objects. Following the transformation models helps to increase social and economic value of territories, ensure environmental diversity, sustainable mobility and equal access to basic urban services. Orientation to the proposed models also contributes to the formation of a continuous, clear and diverse urban water and green space network, well-planned and filled public areas, visual identification of the block boundaries and the status of land within districts.

The first transformation model is based on the *concept of a compact city* [12]. The territory under consideration, which is a residential area of mainly estate development “Gromy”, is adjacent to the Central district of Polotsk but it is separated from the district by railway tracks and a production zone along them (Figure 1). The transformation project requires the preservation of the district's identity in the urban village format. An alternative vision of land use provides for the streamlining of the road and street network, the formation of a well-maintained public space system and functional differentiation of development due to the redevelopment of stagnant enterprises which increases the attractiveness of the district within the agglomeration.

Only the industrial zone along the railway and the garage cooperative are undergoing intensive transformation to create new places of residence and work. Their reconstruction will increase the permeability and connectivity of the urban fabric. A new level of mobility in the private sector will be provided by a developed network of intra-district public transport which is not linked to railway crossings. The reconstruction of the main streets will improve the quality of walking and cycling. The formation of a continuous green network will not only reveal the recreational potential of the territory but it will also solve existing environmental problems with water disposal. It is emphasized that the proposed scenario of spatial development is associated with support for small businesses and an increase in people's self-organization concerning territory management.

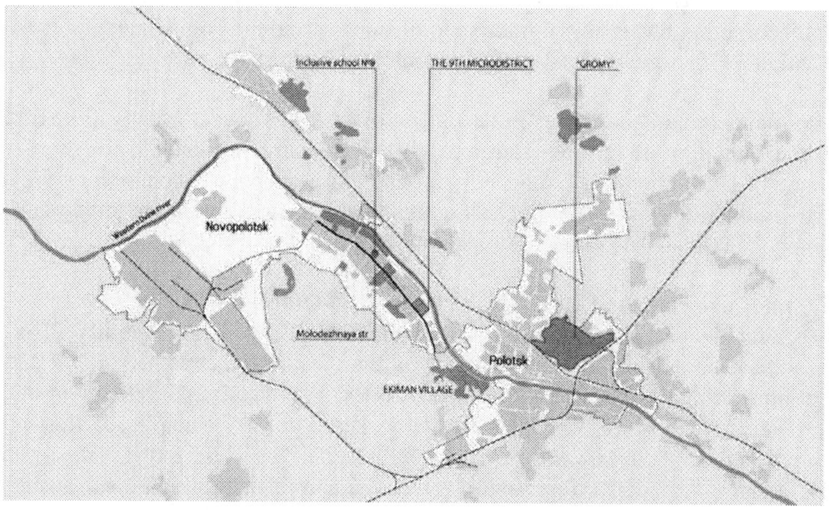


**Figure 1.** Polotsk-Novopolotsk agglomeration. Source: own work.

This transformation model will be relevant for historically formed large Belarusian cities, whose structure is divided by railway tracks and whose significant share in the land balance is occupied by estate buildings with a homogeneous plan structure. The developed model is applicable to such areas as Selhoztehnika, Pustynki, Zalinejnyj (Orša), Leninskij, Chapaeva, Festivalnyj (Mogilev), Peskovatik, Železnodorožnyj (Vitebsk), Staroborisov (Borisov) etc.

The second pilot site for testing the principle *of transit-oriented development* was the 9th microdistrict of Novopolotsk which is adjacent to the village of Ekiman [13]. The 9th microdistrict is located at an equal distance from the centers of both cities and is adjacent to the main transport corridor called Molodežnaâ street. This creates all the prerequisites for the formation of a new center of polycentric agglomeration here, where objects and places of employment for residents of both Novopolotsk and Polotsk can be located. In contrast to the project discussed above this site is not fully developed, but it already has negative features typical of the microdistrict development, namely monotony, abandoned territories, chaotic landscaping, etc. According to the authors' intention the transformed area should have a unique character due to the inclusion of modern typological units of development in the planning framework which will emphasize the peculiarities of the area and reveal its natural potential.

The second model is relevant for most cities in Belarus, whose characteristic is a dynamic growth by the "greenfield development" type due to the development of free territories for residential development [14]. Unfortunately, the prevailing type of development remains the microdistrict which, in practice, is characterized by secondary implementation of social infrastructure. It leads to gaps in the urban fabric and prevents the formation of quality urban environment (districts of Mihalova, Kamennaâ Gorka (Minsk), Microdistrict-7 (Bobrujsk), Vulka (Brest), etc.).

Both transformation projects are aimed at strengthening the connectivity of territories and the planned areas themselves are considered to be attractions in the context of the Polotsk-Novopolotsk agglomeration. According to the Integrated Sustainable Mobility Plan the sustainable development of the agglomeration is associated with the organization of a high-speed tram route between cities, the implementation of which was made possible owing to the "Green cities" project. The project also involved the implementation of some local technical activities that improve the general conditions of mobility in the agglomeration, particularly the construction of a bicycle path using the optimal route through the village of Ekimany. In addition to formation of bicycle road network the development of bicycle infrastructure in Polotsk and Novopolotsk involves the installation of five bicycle garages for 100 seats, bicycle parking, the opening of bicycle cafes, repair shops and rental outlets. According to the project public transport stops will be replaced, information boards will be installed, adaptive control system to synchronize traffic lights will be implemented.

## 4. Belarus and the international initiatives for sustainable mobility

A number of completed studies, the Integrated Sustainable Mobility Plan for the Cities of Polotsk and Novopolotsk in particular, became the basis for a constructive dialogue between the Novopolotsk City Executive Committee and the European Bank for Reconstruction and Development, during which a memorandum of cooperation was signed and work on an investment project to finance a new tram line connecting two important centers of the northern region of Belarus was started.

Issues of sustainable mobility were also a priority for *the "Novopolotsk - from an inclusive school to an inclusive city"* project funded by the European Union. One of the goals of the project was the organization of gradual formation of accessible urban environment based on the principles of universal design on the site of and its surrounding area. They plan to use the creation of continuous inclusive routes in the reconstruction of other parts of the city.

The development of public transport as well as the development of an innovative tool such as plans of sustainable urban mobility are identified as priorities in the *National Action Plan for the Development of the Green Economy* [15]. This document focuses on a more active participation in the European Union's Covenant of Mayors for Climate and Energy initiative, which brings together cities and districts that have made voluntary commitments to reduce greenhouse gas emissions and adapt to climate change [16]. Polotsk was the first city in Belarus to join this initiative in 2011, 45 other cities followed.

In 2015 Belarus along with 193 countries approved *the 2030 Agenda for Sustainable Development* and expressed support for its implementation by achieving 17 sustainable development goals (SDGs) [17]. 78.4% of the Belarusian population lives in cities, while the localities occupy only 11% of its territory. Being the most urbanized post-Soviet country, the Republic must pay special attention to SDG 11: "Sustainable cities and communities" [18]. A number of Belarusian cities have joined the international initiative *CIVITAS*, which is a network of local authorities, public organizations and transport experts from hundreds of cities across Europe. The main goal of CIVITAS is to help cities develop environmentally friendly and sustainable transport [19].

The sustainable mobility theme is recognized by Belarusian cities as particularly acute which is proved by the activation of their participation in *the international European Mobility Week campaign* [15]. Only in the last five years the number of participants from Belarus has grown from 3 to 76, which made it possible to enter the top ten leaders of the campaign. The most interesting initiative is the *World Bank's Eco2 Cities: Ecological Cities as Economic Cities"* (Eco2 Cities: ecological cities as economic cities). Its main aim is to help cities in developing countries achieve greater environmental and economic sustainability. Eco2 Cities is an approach to sustainable urban development based on universal principles, as well as the use of a set of tools that simplify self-assessment process and ensure integrated decision-making.

In February 2018 the World Bank has approved the partnership framework with Belarus for 2018-2022. The projects aimed at meeting SDGs will get 570 million dollars also the willingness was expressed to provide an additional loan of 100 million dollars for energy-efficient projects. The collaboration between Belarus and the World Bank on urbanization and sustainable urban development has not yet been established but it is of great promise. Belarusian cities have great opportunities to use international technical and humanitarian assistance in other programs aimed directly at the transport sector: *CIVITAS, Smart Energy, and Marco Polo* [15].

## 5. Conclusions

It is impossible to imagine an urbanized society without constant communication and movement. The relatively new concept of "sustainable mobility" symbolizes the human freedom in his/her spatial movement which does not harm the environment and the health of other people, the interests of future generations are taken into consideration. Globalization and large-scale urbanization along with technological innovations and the digital revolution are rapidly changing the world around us, the structure of the economy and the way people live. In the foreseeable future the transition of Belarus to a new technological economy will become more active. As a result, comfortable urban environment, the quality of which is one-third dependent on mobility conditions, becomes a more valuable equity and a source of economic growth. Unfortunately, strategic documents aimed at integrated development (integration of territorial, social and economic planning) of cities and communities in Belarus have not yet been developed. This situation hinders the development of many activities at the city level. The activities are associated with improving the quality of people's life which is an integral indicator of sustainable development.

Sustainable Urban Mobility Plans (SUMPs) are recognized as the most promising tool for urban development, their development methodology having proved to be effective. The methodology is quite applicable in Belarus but to activate SUMP development this new tool should be supported by appropriate institutions as well as by the adoption of appropriate legislation.

IOP Conf. Series: Materials Science and Engineering **907** (2020) 012003 doi:10.1088/1757-899X/907/1/012003

## 6. References


[1] Nikolaeva N 2001 The competitiveness of the city: the view of foreign scientists *Marketing in Russia and abroad* 6 pp 69-75 (in Russian)

[2] Bevilacqua C, Cappellano F and Zingali L 2013 TOD -Transit Oriented Development: a sustainable tool towards smart living *7th Conf. of International Forum on Urbanism* Available at: https://www.researchgate.net/publication/273895589_TOD_-Transit_Oriented_Development_a_sustainable_tool_towards_smart_living (accessed 20 May 2020)

[3] Vardevanyan P 2016 Urban streets planning based on new principles of sustainable mobility *Architecture* (Minsk: BNTU) 9 pp 122-127 (in Russian)

[4] Rupprecht S, Brand L, BöhlerBaedeker S, Brunner L M 2019 *Second edition of the Guidelines for developing and implementing a Sustainable Urban Mobility PlanPolotsk* (Germany: ELTIS) p 166

[5] Astapenia P, Bibikov D, Vishnevetskaya T, Gorbunov P, Rodnenkov A, Khoruzhiy E 2017 *Sustainable urban mobility planning. Analysis of current practices and recommendations for improving legislation* p 56 (in Russian)

[6] UNDP 2015 *Belarus: Supporting Green Urban Development in Small and Medium-Sized Cities in Belarus* p 132

[7] Ministry of Foreign Affairs of the Republic of Belarus 2020 *Global Agenda-2030: Cities in Belarus on a sustainable development path* Available at: http://sdgs.by/en/news_events/news/af9ae709a01f687.html (accessed 20 May 2020)

[8] Belarusian Transport Union (BST) 2019 Integrated plan for sustainable mobility for the cities of Polotsk and Novopolotsk By joint efforts to a new quality of life *UNDP project "Belarus: Support for Green Urban Planning in Small and Medium Cities of Belarus"* p 103 (in Russian)

[9] Katsigris E, Kalosha V 2019 *Belarus Green Cities Mid-Term Review* p 119

[10] Sysoeva V 2019 Approaches to the planning, design and regulation of green cities *Architecture* (Minsk: BNTU) 12 pp 93-100 (in Russian)

[11] UNDP 2015 *National Human Development Report 2015 - Regional Competitiveness in the Republic of Belarus* p 21 Available at: http://hdr.undp.org/sites/default/files/resume_eng.pdf (accessed 20 May 2020)

[12] MLA+ 2019 Public discussion of the transformation model of Novopolotsk within the boundaries of Molodezhnaya, Yakub Kolos and Vasilevtsy streets *UNDP project "Belarus: Support for Green Urban Planning in Small and Medium Cities of Belarus"* p 38 (in Russian)

[13] MLA+ 2019 Public discussion of the development prospects of the territory of the city of Polotsk within the boundaries of the Gromy district *UNDP project "Belarus: Support for Green Urban Planning in Small and Medium Cities of Belarus"* p 37 (in Russian)

[14] Rachkevich T and Protasova J 2016 City block or microdistrict? *Architecture* (Minsk: BNTU) 9 pp 151-156 (in Russian)

[15] Dolinina O and Leonchik V 2019 *Sustainable mobility in liveable cities. Global challenges and new approaches: transport planning guide* (Minsk: Colorgrad) p 36 (in Russian)

[16] CIUDAD 2012 *Tips for ENPI cities on how to develop a sustainable energy action plan* Available at: http://sdgs.by/en/news_events/news/af9ae709a01f687.html (accessed 20 May 2020)

[17] UNDP 2018 *Belarus: Agenda - 2030. The future of our planet is in our hands* Available at: https://www.by.undp.org/content/belarus/en/home/presscenter/pressreleases/2018/02/06/belarus-agenda-2030-the-future-of-our-planet-is-in-our-hands/ (accessed 20 May 2020)

[18] Azemsha, S, Yasinskaya V and Hryshchanka T 2020 Sustainable development in Belarus: Goals for transport and safety indicator movements *Journal of Sustainable Development of Transport and Logistics* 5(1) pp 19-27

[19] Bezruchonak A 2019 Geographic Features of Zero-Emissions Urban Mobility: The Case of Electric Buses in Europe and Belarus European Spatial Research and Policy 26 issue 1 pp 81-99

IOP Conf. Series: Materials Science and Engineering **907** (2020) 012004 doi:10.1088/1757-899X/907/1/012004

# The influence of traditional architecture on the design of social housing by masters of modernism

**I Kudryashova, A Martynenko, O Remizova**
Department of Architecture Fundamentals, Kharkiv National University of Civil Engineering and Architecture, 61002, Ukraine

kudryashovaiv21@gmail.com

**Abstract**. The work is devoted to the study of the influence of folk architecture on the professional practice of housing design by modern architects. The study focuses on the use of traditional heritage techniques in the design of social housing in modernism in the period 1910-1970. The procedure of reproducing the experience of traditional (vernacular) architecture is described as a process of translation, which involves the decoding of folk heritage by professional architects in order to identify the characteristics and techniques relevant to the "modern movement". The task of designing housing for the working class is observed throughout the period of modernism in various regional areas, from European countries and America to Egypt, Algeria and India. Historical analysis of the architecture of modernism helped to identify such valuable for its representatives' qualities and properties of traditional architecture as: functionality, utilitarianism and archetypal. It has been proven that modernists were inspired by the architecture of traditional Mediterranean cities. It is shown that in the period of modernism, along with engineering innovations, architects introduced traditional design systems, materials, technologies and rethought their compositional potential, which significantly diversified the morphology of social construction and enriched its symbolic language. Thanks to the borrowing of traditional features, modern architecture has become clear to the consumer, and such morphological elements as a flat roof, white walls, simple rectangular columns and so on have become a new and quite understandable language of modern architecture.

**Relevance of research.** From the beginning of their work, the architects of the modern era declared a break with the academic culture of the past and refusal to rethink the historical heritage. Denying the decorative approach of eclecticism and secession, a new generation of architects fought for the assertion of a new aesthetic in architecture. They were interested in the process of mastering new building materials, structural systems, construction technologies, which appeared due to colossal technological and industrial progress, updating outdated notions about form and tectonics in architecture. Modernists used a rational approach in design and construction, highly elevated the ethical function of art. Among the ethical principles of modernism should be noted the idea of social housing, the impact of architecture on human health in particular and society in general [1; 2]. All this has to be reflected in the new avant-garde style or architectural language and its symbolism.

**The state of the issue.** Rethinking the legacy of modernist architecture, modern scholars find evidence that it is more closely linked to traditional architecture than previously thought. The relationship between the professional architecture of modernism and non-professional folk architecture is evidenced by the research of a number of scholars. Among them: Emma Dummett [3], Francesco Passanti [4], Sigfried Giedion [5], Sheila Crane [6], Jean-Francois Lejeune [7; 8], Francis E. Lyn [9], Tom Avermaete [10],

Gio Ponti [11]. But the question of the influence of traditional architecture on the formation of methods of housing design was almost not raised by representatives of modernism.

Therefore, the aim of our article is to study the influence of folk architecture on the professional practice of designing social housing by modernist architects in the period 1910-1970.

**From a methodological point of view**, the procedure of reproducing the experience of traditional (vernacular) architecture should be described as a process of translation that involves decoding folk heritage by professional architects through identifying symbolic and morphological features and syntactic techniques relevant to the construction of "modern movement" language.

**Results of the research**. At the beginning of the XX century architects first faced design tasks for the working class. Architects faced the difficult task of typifying and standardizing social housing to reduce its cost and mass construction. Hermann Muthesius noted that the appeal of architects to tradition served as a catalyst for design during this period. The object of interest and study was primarily the spatial and functional organization of buildings of folk architecture, which served as a model of rationality and efficiency in construction. Traditional architecture has also been used as a starting point in the central concept of housing standardization [12]. For example, Le Corbusier first characterized "folk art" as a norm, a measure, the standard of which - man [13]. Le Corbusier's work for many years was accompanied by the idea of finding a system of proportionality in architecture. The result of his developments was the measuring system Modulor, in which a person with a raised hand determines the height of living spaces 2.10-2.20 m, adopted "... in all harmonious works, both folk architects and professional architects" [14]. Modular measuring system is laid down in the original design of the cottage, called "Barn Le Corbusier" ("Cabanon Le Corbusier") built in 1952. The building, like a hut, has modest dimensions (in terms of 3.66 m by 3.66 m, height 2.26 m) and is equipped with everything you need (bed, closet, desk, sink, toilet) (Figure 1), which is very similar to the traditional hut of different European countries.

**Figure 1.** Le Corbusier. "Barn Le Corbusier", Roquebrune-Cap-Martin, France, 1951; Source: [15]

In the early twentieth century in many European countries, solving the problem of mass housing for industrial workers became a state task. The implementation of social housing construction programs included the development of building statutes and sanitary norms. Historian Carlos Sambricio noted that the search for a model of a typical budget building was to bring the standard of public housing, solving problems of building a reasonable price and its hygiene [16]. The architecture of housing construction in the north-western countries of Europe (England, Denmark, Finland, Sweden, etc.) acquired a special character.

English architects of the beginning of the century continued to consistently develop the typology of the cottage. The basis for the development of social housing was a traditional village house - cottage - a building on two or three floors with a separate entrance from the street and a small garden. In the period between the First and Second World Wars, an individual or paired cottage became widespread. In the postwar period, the cottage replaced the more economical terraces - block houses on two or three floors (they were built by English architects Richard Shephard, Hugh Wilson, J. Lewis Womersley, Sir Frederick Ernest Gibberd), which were widely used in the XVII - XIX centuries in urban development (Figure 2, 3). The typology of the terrace house had an impact on housing construction in Belgium, the Netherlands, and Germany. Until the middle of the XX century. England remained mostly two-story. It is important to emphasize that the multi-storey gallery-type building housed the basic residential unit of an English country house - the apartments are located on the traditional two levels (Philip Powell, Hidalgo Moya, Jean Conder) [17]. In 1905 the German architect Hermann Mutezius (1861-1927) published a three-volume book called The English House (Das Haus Englische). In this book he described the arrangement of a traditional English house, its functional structure and encouraged fellow architects to follow the example of the British, who in the development of mass housing willingly relied on the traditions of their ancestors [18].

**Figure 2.** Block house "terrace house", consisting of a number of apartments of the traditional type, which retain a separate entrance from the street, UK. Source: [17 p 127]

**Figure 3.** Terraces building in London, UK. Source: [17 p 129]

In Germany in the 1930s and 1940s, residential architecture for the poor and middle classes was conservative, even when attempts were made to apply new constructive methods. For example, in the half-timbered houses of the architect P. Schmitthenner elements of factory production were sometimes used. But the layout, interior design and exterior of the buildings reproduced the features of a traditional German burgher house. The houses built by the Atelier Breuhaus architects (house in Dortmund), Hans Gerson and Oskar Gerson (Nordquist house in Great Flottbek, Hamburg), Thiloo Schoder (Wilhelm Ernst Meyer House – single-family house in Gera), Martin Elsaesser (Heinrich Landfried house in Heidelberg) and others also had a traditional look.

The use of the traditions of folk architecture in Sweden in the 1930's and 1940's was conditioned by the search for national identity and was called the "new empiricism". This direction arose at the junction of the spread of "international style" and the wave of national romanticism of the early twentieth century, and was implemented in mass housing construction [19]. The features of Swedish national architecture are most noticeable in low-rise construction. Architects turned to tradition to search the most appropriate architectural solutions in cold climates, methods of interaction between the building and the landscape, ways to use local materials (especially wood). Architects Gunnar Asplund, Sven Markelius, Ralph Erskine, Sven Backström, Leif Reinius and others took an active part in this direction.

For example, in his own wooden house in Kevinge (1945), S. Markelius used the following spatial planning features of a traditional Swedish house: L-shaped composition of the volume of a one-story building, gabled roof. Also, the architect, according to tradition, organically inscribed the house in the landscape. S. Markelius placed the building as close as possible to the street and facing north, thus giving the lion's share of the site to a large garden, which opened to the rays of the southern sun (figure 4, 5).

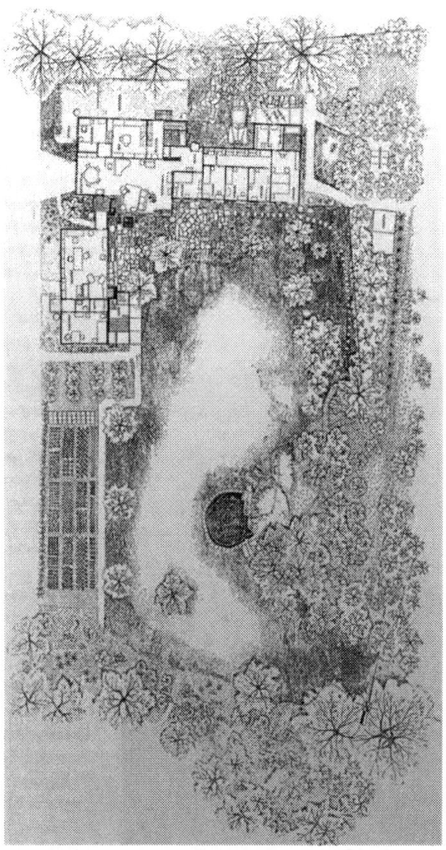

**Figure 4.** S. Markelius. Villa Markelius in Kevinge, a suburb of Stockholm, Sweden, 1945. Source: [20 p 17]

**Figure 5.** S. Markelius. Villa Markelius in Kevinge, a suburb of Stockholm, Sweden, 1945. Source: [20 p 16]

The famous Finnish modernist architect Alvar Aalto (1898-1976) executed several projects of typical residential buildings using the traditions of building a Finnish house. Among them: wooden standard housing in Varkaus (1937), in Tampere (1940-1941), VOK Standard Houses, A. Ahlstrom Company, Pihlava, Pori (1946) (Figure 6), Enso -Gutzeit Company, Summa, Hamina (1958–1964).

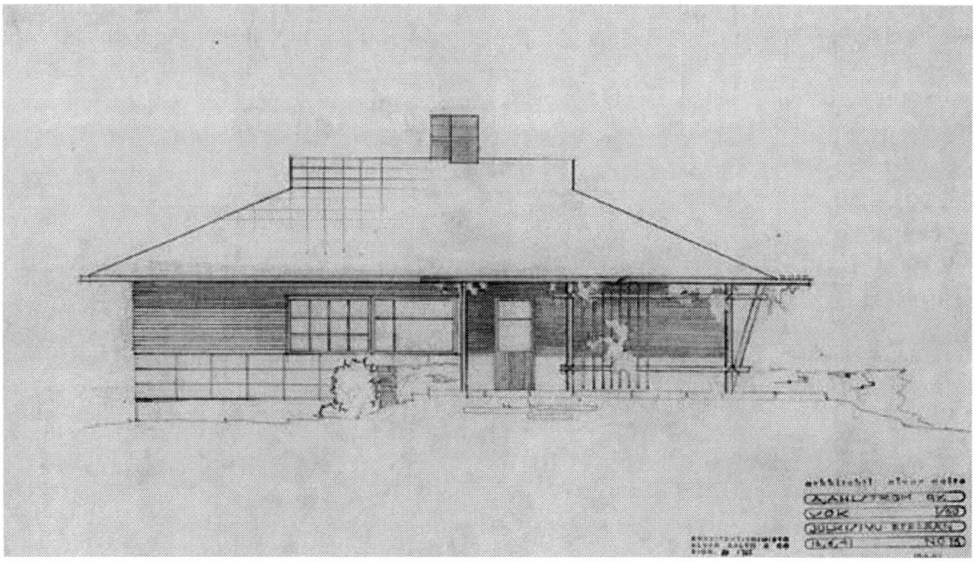

**Figure 6.** Alvar Aalto. VOK Standard Houses, for A. Ahlstrom Company, Pihlava, Pori, Finland, 1946. Source: [21]

The need to build new housing in Europe increased again after the First World War, which had severe consequences for the population. Interestingly, in the south-western countries of Europe in housing construction it is important to note the tendencies to turn to the typology of the traditional Mediterranean house. In Greece active construction work was carried out to restore 400,000 houses lost during the war. There was an increase in the number of economical multi-storey buildings, a characteristic feature of which was the presence of roof terraces and loggias-balconies. This feature is reflected in the flat roofs of modern buildings and in the five principles of Le Corbusier. The national features of the Greek house were used mainly in the construction of single-family country houses for the wealthy. Among them is the house of the architect Patroklos Karantinos, built for the artist Spyros Vassiliou in Athens (1958);

houses in the park areas of Athens Demetrios Pikionis, many country villas. A feature of traditional Greek housing, which belongs to the Mediterranean type, is the multilevel location of the house on a steep slope with the use of courtyards and terraces, asymmetry of spatial planning solutions, conciseness of forms, use as flat roofs - terraces (on the islands of Skyros, Andros, Mykonos) , and vaults or domes (on the island of Thira). During the reconstruction of the village on the island of Santorini (Thira), destroyed during the 1956 earthquake, architects P. Vasiliadis, A. Spanos and others applied the motives of folk architecture.

In the second half of XX century in Spain, in order to attract farmers to develop infertile lands, new settlements were built. In 1961, at the VI Biennale of Arts in Sao Paulo, Jose Luis Fernandez del Amo received a gold medal in the nomination "Urban Cluster Planning" for projects in the cities of San Isidro de Albatera (1953), Villalba de Calatrava (1955), Vegaviana (1956), El Realetgo (1957). In the architecture of these cities the architect combined the ideas of modern movement with the techniques of the Spanish traditional house. The architect used ceramic tiles, natural stones bleached with lime, which preserves the texture of the stone. This technique can be traced in the bare concrete surface of modernist buildings and in brutalism. Based on the prototype of the geometry of the Spanish folk dwelling, he designed a number of typical houses, compositionally close to a traditional home. Later in an article Jose Luis Fernandez del Amo wrote that while traveling to different parts of Spain, he learned a lot from anonymous architecture. Looking at the apartments, built on the basis of local traditions, the architect noticed the life wisdom of architects in understanding human needs. Del Amo also discovered the hidden techniques of spontaneous organization of villages and small towns, built without the participation of professional architects, which he repeatedly used in his practice [22]. The Spanish architect Alejandro de la Sota (1913-1996), like his compatriot Jose Luis Fernandez del Amo, took part in the state program of rural development. Between 1952 and 1963 he designed and built the city of Esquivel, located in the municipality of Alcalá del Río near Seville. The city was based on vernacular and Mediterranean architecture combined with the modern discourse of Modernism. The ideas of modernism manifested themselves in the strict geometry of the master plan, which was organized according to the radial-circular grid of streets. There is a traditional central city square and two public institutions (church and administrative center) are made by the architect outside the city. The settlement of Esquivel is designed for 161 families and consists of six types of houses. Each dwelling has traditional elements of Andalusian village buildings: the patio of a Spanish country house, high lime-whitewashed walls along the perimeter of the plot, small window openings, a roof with a small slope covered with tiles (Figure 7-9).

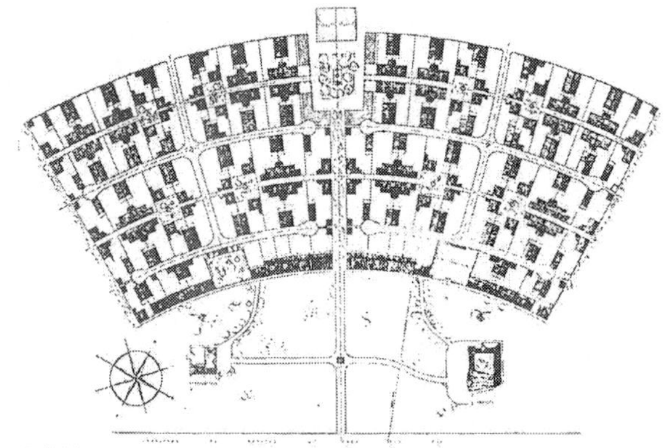

**Figure 7.** Alejandro de la Sota, Jose Luis Fernandez del Amo, master plan of Esquivel, Spain, 1954.

**Figure 8.** Alejandro de la Sota, Jose Luis Fernandez del Amo, photo of one of the streets of Esquivel, Spain, 1954. Source for figures 8, 9: [23]

**Figure 9.** Alejandro de la Sota, Jose Luis Fernandez del Amo, Esquivel, Spain, 1954. Source: [24]

Jose Luis Cert in 1949 in the village for workers in Chimbote (South America) used the Moorish-Spanish patio, which provided residents with isolated private living space [5, p. 16-17]. In Portugal after the overthrow of the dictatorship of Antonio Salazar and Marcel Caetano in 1975 the construction of social housing was organized under the state construction program SAAL (Serviço Ambulatório de Apoio Local). According to the program, late modernist architect Alvaro Siza designed three projects: neighbourhood of Bouca and neighbourhood of St. Victor in the city of Porto; suburban area of rural type Malageira (1977) near the city of Evora. Quinta da Malagueira, which consists of 1,200 one- or two-story buildings, was built over a period of 20 years on an area of 27 hectares. According to the project, the architect provided a courtyard in each apartment. Critics have drawn a strong parallel between Alvar Siza's social housing projects and the Dutch and German settlement projects (siedlungen) of the 1920s, as well as some of Adolf Loos's work. The use of flat roofs, white plaster on the exterior walls, a minimum of windows and doors, the lack of decor - these are the main common features.

Egyptian architect Hassan Fathy (1900-1989) and Indian Balkrishna Doshi (b. 1927), who received a European education, developed an innovative approach to building social housing in their homeland. It is important to note the interaction of a rational European approach with the Arab mentality of the East. Subsequently, this approach became widespread in social architecture. The design conditions for these architects were the lack of specialized professional builders, the shortage of building materials. For these reasons the architects used local materials and labor resources, used some elements of the building traditions of the region, which significantly reduced the construction budget and create comfortable living conditions in hot climates. Hassan Fathy applied these principles in the project of building the village of New Gourna (1945-1948) (Figure 10), Balkrishna Doshi in ATIRA & PRL houses (1957), Neighborhood in Indore (1983). The composition of the buildings of these complexes indicates the affinity with the traditional architecture of their countries.

**Figure 10.** Hassan Fathy, New Gourna Village, Egypt (1945-1948). Source: [25]

The famous project of the Peyrissac residence with agricultural territory in Cherchell, Algeria (1942), executed by Le Corbusier, in which the architect demonstrated sensitivity to the cultural and natural context. Le Corbusier turned to local folk architecture, as he mentioned in his book Oeuvre Complète (1945). The architect wrote that he designed the villa in such a way that a resident of Algeria could build a building of local stone by building familiar elements: columns, walls, vaults. Le Corbusier created a building from the basic forms of a Mediterranean house, which interpreted the ancient building tradition [1]. In this way, Le Corbusier translated the principle of working with historical non-professional prototypes into modern architecture.

The Indian architect Balkrishna Doshi (b. 1927), who worked closely with Le Corbusier and L. Kahn, used the experience of traditional architecture in the municipal housing complex in Aranya near Indore (1989), in particular, the features of Indian village planning, climatic techniques, adaptations of traditional buildings and cheap materials. This example demonstrates not only the development of traditional methods of housing construction in modern design, but also the close connection between Western and Eastern design methods in the twentieth century.

**In conclusion**, we note that the main prototype in the development of the principles of typification and standardization of social housing for modernist architects was traditional architecture. Modernist architects used in their decisions the spatial and functional organization of public housing, trying to combine the ideas of modern movement with traditional models and standards of the living environment. The use of traditional local materials and labour resources not only significantly saved the construction budget, but also prompted the decision of climatic adaptation of buildings.

The experience of folk architecture was broadcast through object-morphological and contextual channels of translation. The objects of translation in this case were spatial planning solutions and methods of adaptation to climatic and social conditions, borrowed from traditional architecture.

For architects of modernism the reproduction of the experience of traditional architecture has become a very spontaneous process of translating cultural skills into modern design. But this process has not been reflected in modern science.

It is also important to emphasize that thanks to the borrowing of traditional features, modern architecture became understandable to the consumer, and such morphological elements as a flat roof, white walls, simple rectangular columns and so on became a new and quite clear lexical code of modern architecture. If at first they were perceived as refined elegant and very exotic signs of new architecture, then later, as a result of repeated repetition of these forms, they became signs of architecture "modern movement" and created a widespread modern architectural language.

**References**

[1] Martynenko A S and Kudryashova I V 2016 Methods of broadcasting folk architecture in the architecture of modernism. Le Corbusier *Modern problems of architecture and urban planning,* **45** (Kyiv, KNUCA) pp 64 – 74 (in Russian)
[2] Wilk C 2008 *Modernism: Designing a New World: 1914-1939* (London: V&A Publishing) 448 p
[3] Dummett E 2005 Vernacular architecture, nature and the sacred Le Corbusier and the influence of the 'journey to the east' *Journeys of Discovery.* **4**. pp 1–13
[4] Passanti F 1997 The Vernacular, Modernism, and Le Corbusier *Journal of the Society of Architectural Historians.* **4.** pp 438–451. Available from Internet: http://www.jstor.org/stable/991313
[5] Giedion S 1984 *Space, Time and Architecture* (Moscow: Stroyizdat) 456 p (in Russian)
[6] Crane S 2011 Mediterranean Crossroads. Marseille and Modern Architecture (Univ Of Minnesota Press) 376 p
[7] Lejeune Jean-François 2010 Modern Architecture and the Mediterranean: Vernacular Dialogues and Contested Identities, ed J. Lejeune, M. Sabatino (Routledge) 320 p
[8] Lejeune J 2010 The modern and the Mediterranean in Spain: Sert, Coderch, Bohigas, De la Sota, Del Amo *Modern Architecture and the Mediterranean: Vernacular Dialogues and Contested Identities*, ed J. Lejeune, M. Sabatino (Routledge) pp 65–94
[9] Lyn Francis E 2010 Mediterranean Resonances in the Work of Erik Gunnar Asplund: Tradition, Color, and Surface *Modern Architecture and the Mediterranean: Vernacular Dialogues and Contested Identities,* ed J. Lejeune, M. Sabatino (Routledge) pp 213–230
[10] Avermaete Tom 2010 CIAM, Team X, and the Rediscovery of African Settlements: Between Dogon and Bidonville *Modern Architecture and the Mediterranean: Vernacular Dialogues and Contested Identities,* ed J. Lejeune, M. Sabatino (Routledge) pp 251-264
[11] Ikonnikov A B 1972 Gio Ponti *Masters of Architecture about Architecture* (Moscow: Art) p. 429–445 (in Russian)
[12] Muthesius H 2007 *The English House* ed Hermann Muthesius, Dennis Sharp (Frances Lincoln Publishers) 768 p
[13] Le Corbusier 1991 *Journey to the East* (Moscow: Stroyizdat) 120 p (in Russian)
[14] Le Corbusier 1976 Modulor. MOD 1. MOD 2 (Moscow: Stroyizdat) 238 c Available from Internet: http://corbusier.totalarch.com/mod_1/3 (in Russian)
[15] Cabanon de Le Corbusier, Roquebrune-Cap-Martin, France, 1951 Available from Internet: http://fondationlecorbusier.fr/corbuweb/morpheus.aspx?sysId=13&IrisObjectId=4659&sysLanguage=fr-fr&itemPos=8&itemCount=79&sysParentId=64&sysParentName=
[16] Sambricio C 2000 La normalizaciyn de la arquitectura vernacular: Un debate en la España de los veinte. Available from Internet: http://oa.upm.es/1555/
[17] Ikonnikov A B 1958 *Modern architecture of England. City planning and housing construction* (Leningrad: Gosstroyizdat) 207 p (in Russian)

[18] *Masters of Architecture about Architecture: Selected Excerpts from Letters, Articles, Speeches and Treatises* 1972 (Moscow: Art) 592 p (in Russian)
[19] Ikonnikov A B 1978 *Modern Swedish architecture* (Moscow: Stroyizdat) 151 p (in Russian)
[20] Treib M 2002 The Architecture of Landscape, 1940-1960 (Penn Studies in Landscape Architecture), ed Treib Marc (University of Pennsylvania Press) p 17
[21] Jetsonen Jari, Jetsonen Sirkkaliisa, Juhani Pallasmaa 2012 *Alvar Aalto Houses* (New Nork: Princeton Architectural Press) p 112
[22] Fernández del Amo J 1995 Delhacer de unos pueblos de colonización. Palabra y Obra: EscritosReunidos. (Madrid: COAM) 77 p
[23] Esquivel Available from Internet: https://archivo.alejandrodelasota.org/es/original/project/146 (in Spanish)
[24] Esquivel, 60 años después de los colonos Available from Internet: https://sevilla.abc.es/fotos-andalucia/20131021/esquivel-anos-despues-colonos-1511331669819.html#imagen1 (in Spanish)
[25] *Theatre, exterior façade New Gourna, Egypt* Archnet Available from Internet: https://archnet.org/media_contents/29892

IOP Conf. Series: Materials Science and Engineering **907** (2020) 012005 doi:10.1088/1757-899X/907/1/012005

# The architectural heritage of the second half of the 1940s - 1950s in the context of humanization processes and urban aesthetic comfort (on the example of Rivne city)

**M I Isupova**

Department of Architecture and Environmental Design, National University of Water and Environmental Engineering, Rivne, 33000, Ukraine

m.i.isupova@nuwm.edu.ua

**Abstract.** In recent decades, a lot of ecological, ergonomic, functional, planning and aesthetic problems have emerged in most cities of Ukraine and the world, the elimination of which involves the architectural environment humanization. The article is devoted to the study of the architectural heritage role in the context of the urban environment humanization processes and modern approaches to providing aesthetic comfort. On the example of Rivne city, the method of public opinion polls establishes specific features of the architecture, which, according to the residents of the city, provide aesthetic expressiveness of the architectural space.

## 1. Introduction
With the city's growth, negative and uncomfortable manifestations in the architectural environment accumulate and intensify. The humanization of the architectural environment as caring for residents and meeting their needs involves solving not only ecological, environmental, ergonomic, functional and planning problems that have emerged in recent decades in most cities of Ukraine and the world, but also achieving aesthetic and psycho-emotional comfort. The last two issues are quite complex in terms of their interdisciplinary status and the need for an individual approach in each case.

In addition to taking into account the general aesthetic categories of architecture and universal psychological aspects of human perception, the context of the urban environment also plays an important role: its history, "genius" of the place, regional features, traditions, mentality and preferences of residents, their understanding of beauty. Often, rapid urban changes are related to technological progress, social, economic and political transformations, and as well, as the factors mentioned above, they determine the context of the urban space.

The architectural environment sets the trajectory and movement within the living space, shaping the lifestyle. Architecture differs from other arts in that it is compulsory, and cannot be avoided, because architectural environment creates the immediate space of human daily being.

Therefore, the issue of humanization of the existing and projected architectural environment, interconnection and harmonious integration of new architectural objects into the urban ensembles that have developed historically, preservation and, if necessary, restoration of the historical buildings, become especially relevant.

## 2. The question of urban aesthetic comfort: the role of architectural heritage protection
It is worth noting that various aspects of humanization are closely linked: for example, improving the ecological situation by creating greenery at the same time improves the aesthetic image of the city, and

solving aesthetic problems in architecture improves the psychological state of residents. Therefore, except the utilitarian aspects of architectural environment humanization, the aesthetic component cannot be ignored.

"Reading" of architecture by people often takes place beyond awareness: city dwellers perceive the processes that take place inside or near buildings in their daily lives and usually don't notice surroundings, but "breathe them in". Although city dwellers take their urban surrounding for granted and rarely bring it to full consciousness, architecture can powerfully influence them [1]. The buildings themselves set the mood, acting as elements of the townscapes. However, it also turns architecture into a powerful tool that is able to manipulate people's mood and behavior.

Throughout human history this property of architecture has many examples of its usage by political regimes as propaganda. In totalitarian systems, such as Germany of Nazi period, architecture produced specific tools for "strictly planned political targets aimed to create desirable social respond" [2]. There are also reverse examples - it is well-known that abandoned areas with aesthetically indistinct buildings, or so-called slums, are characterized by an increase in the criminogenic situation.

Forms of human mentality and psychology are defined in the architectural environment. Buildings carry information about socially conditioned behavior and practical skills of people, about the relationship between them. The architectural environment embodies ethical and aesthetic ideals, connects different generations and eras, and is an important part of the collective memory of mankind.

C. Day notes that "old buildings are rarely just museums of a particular historical period. They have physical elements from many dates right up to the present; also, the imprint – both visibly and invisibly – of the many occupants, lifestyles and values that inhabited them" [1].

In general, architecture can be characterized as "the language of a long and deep collective mood". Architecture "embodies, consolidates, transmits people's emotions from generation to generation", acts as the "keeper of the organizing tradition" [3].

It is well-known that an aesthetically attractive urban environment is being provided when a harmonious sequence develops, "the new built upon the past" [1], and architecture objects are being built with taking into account the features of the existing historical and architectural heritage. Accordingly, harmony is lost when the past is destroyed or discarded.

Therefore, the continuity of the architectural tradition is an important criterion for ensuring city's residents psycho-emotional comfort. The preservation and maintenance of historic buildings in their proper form is an important task of the architectural humanization aesthetic aspect.

World and in particular European experience demonstrate a careful attitude to the historical and architectural heritage: buildings and their ensembles are protected and preserved in a form as close as possible to the original. Thus, the architectural canvases of cities do not lose their authenticity, and residents - a sense of rootedness, belonging to their history.

In parallel, there is an improvement of the existing urban environment with the use of techniques aimed at its humanization: "dissolution" of new objects in the historic environment (as an example - a cascade of escalators in Toledo, Spain), increasing walking areas in cities (in a lot of cities in European countries, such as the Netherlands, Denmark, France, Spain, Germany, Italy, Finland etc.) [4], improving the comfort of public spaces, etc.

Along with the composition, colors and other means of buildings artistic expression that affect human consciousness, an important role is played by architectural decoration. Over the last century, diametrically opposed views have been formed on the need to use architectural decoration systems and

IOP Conf. Series: Materials Science and Engineering **907** (2020) 012005 doi:10.1088/1757-899X/907/1/012005

buildings ornamentation. Thus, A. Loos, whose theory at one time had a significant impact on the formation of such architectural styles as constructivism and functionalism, treats ornamentation as a crime, considering its use a manifestation of savagery [5]. Another point of view about architectural decor emerged in the decade of postmodernism. For R. Venturi and D.S. Brown, architectural décor helps to integrate buildings within the urban realm, give them meaning in the eyes of the public [6]. Architects-postmodernists "tried to return to the architecture the lost fantasy and traditional figurative associations" using decorative elements and order forms [7].

Sometimes the use of certain decoration systems in architecture is dictated by the political regime - for example, the artistic image of Soviet architecture in the 1940s and 1950s was created under the influence of socialist ideologues. The language of the architectural style of neoclassicism was interpreted as the most universal, and therefore understandable to the wider public. In general, the universality of the neoclassicism architectural language is explained by the possibility of using similar signs in different cultural contexts to create metaphors that have different meanings. At the same time, there was a requirement for the "national form" and "socialist content" of the architecture in the 1940s and 1950s [8].

This made possible to use folk motifs allusions in buildings ornamentation and decorative architectural elements, and as a consequence - the development of the Ukrainian national style language in the postwar years' architecture. Therefore, even within a mentioned style, architects had some space for a creative research for the artistic expressiveness of the designed buildings, mainly by combining various elements of architectural decoration.

Typically, most buildings created under the influence of classical architectural styles or historicist trends in the XX century have facades rich in decorative elements that are an integral part of the historic architectural space of cities. However, the decor is one of the most vulnerable elements of the architecture, which is easily destroyed by climatic and human factors (Figure 1).

**Figure1**. Destruction of architectural decor threatens to lose the aesthetic appeal of historic buildings. Dwelling houses, Soborna Street, Rivne city. Fragments of facades. Source: photos by the author.

The loss of décor deprives buildings of their aesthetic expressiveness, and sometimes of architectural value in general, which leads to obvious negative effects in the context of the aesthetic comfort of the urban architectural environment.

Sometimes it is impossible to restore the lost decor due to the high cost of restoration work. However, in such cases, the use of modern technologies can significantly reduce the cost of restoring decorative elements of buildings without losing aesthetic quality. With the help of 3D scanning technology, it is possible to digitize the facade of an existing building and adjust the results, and with the help of 3D printing to create prototypes or even ready-made architectural details. First of all, this method is suitable for buildings whose decorative elements were made of concrete.

The one particular example of this approach is the restoration of a historic 1940s building in New York City on Fifth Avenue. Initially, the facade of the building was decorated with intricate ornamentation. Due to the high cost of facade restoration, the house was going to be demolished. New York architecture and engineering firm which focuses on technology-driven design and buildings restoration managed to save the building. They turned the project into a case study, exploring alternative ways to restore and maintain deteriorating ornamentation. To create the decorative element, a 3D form with a detailed pattern was printed. The final prototypes were manufactured using the printer for the casting molds and were fabricated out of concrete [9].

This method allows restoring the lost decorative elements of historic buildings, ensuring the preservation of the architectural environment aesthetic comfort and, obviously, has great potential for use around the world.

In Ukrainian cities, destruction of historical buildings decorative elements in the process of insulation of facades or installation on them advertising signs becomes a common problem. Such phenomena produce a deterioration in the aesthetic comfort of the urban architectural environment and cause negative emotions among city residents. Most of the buildings around the world built in the 1940s and 1950s now need major repairs to continue their existence in future, and it is important to ensure that the architectural decor of these buildings is properly preserved during the renovation process.

## 3. Analysis of architectural aesthetics in Rivne city

During the Second World War, the city of Rivne, like a lot of another cities in Ukraine, suffered significant destruction, especially in its central part, and in the immediate postwar years needed rebuilding.

Reconstruction of the 2nd half of the 1940s - 1950s reorganized the central space of the city and created new architectural objects (Theater Square ensemble, railway station, residential buildings on Soborna Street, etc.), stylistically belonging to Soviet neohistoricism, with facades decorated with order elements, ornamentation with plant motifs (often - with allusions to the Ukrainian folk style). Thus, the architectural objects of the postwar years are an important part of Rivne urban core and play a significant role in shaping city's image.

Usually, in the architecture of the 2nd half of the 1940s - 1950s, the effect of "greatness" that they sought to achieve was enhanced with the help of decoration systems. Also, the architecture of those years is characterized by the decision of the city center as a whole, compositionally and stylistically integrated ensemble. The architecture of Rivne's main highway, Soborna Street, is a vivid illustration of this approach. Together with the Theater Square, a kind of main festive avenue was formed, which became the compositional axis and core of the new center [10].

Most of the buildings of the 2nd half of the 1940s - 1950s retained their original appearance and did not require "decommunization" due to the lack of communist symbols in their architectural decoration. In the 2nd half of 20th - early 21st century other buildings were added in the central part of the town-planning ensemble, performed in more laconic and modern architectural styles, but the town-planning structure itself no longer underwent significant transformations.

IOP Conf. Series: Materials Science and Engineering **907** (2020) 012005 doi:10.1088/1757-899X/907/1/012005

As part of a study of the architecture aesthetics perception by the residents of Rivne city, the author conducted a poll about non-residential architectural objects. Poll participants were asked to choose from the list (Figure 2) the most aesthetically attractive buildings.

The poll showed the following results (Figure 3):

- 75.6% of forty respondents identified public buildings of the 2nd half of the 1940s - 1950s the most aesthetically pleasing;
- 11.1% preferred architectural objects of the 1960s and 1980s;
- 13.3% preferred architectural objects which were built during the years of Ukraine's Independence (1990s – beginning of XXI).

**Figure 2**. The list of buildings for aesthetic analysis in Rivne city: *a* - Architecture of the 1940s - 1950s. Building of railway station and Theater Square; *b* -Architecture of the 1960s - 1980s. Buildings of cinema and department store; *c* - Architecture of the 1990s –beginning of 21st century. Buildings of shopping malls. Source: photos by the author.

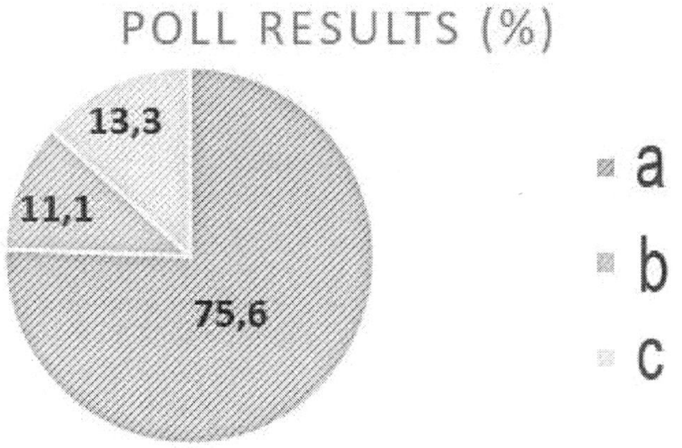


**Figure 3**. Buildings aesthetic analysis in Rivne city. Source: illustrations by the author

## 4. Conclusion

The humanization of the cities architectural environment is one of the most important tasks of the modern stage of human development. In addition to the utilitarian aspects of the humanization of the architectural environment, such as solving environmental, architectural and planning problems, the aesthetic component cannot be ignored.

Artistic, aesthetic, psychological properties of the architectural environment in each historical segment are formed by means of architectural expression, but it is also important to take into account the context of the environment, continuity of architectural tradition, harmonious integration of new buildings into the existing environment, protection of architectural monuments and their restoration. It provides aesthetic comfort of urban space, preserves the "genius" of the city.

City residents' points of view, especially on public buildings and spaces, should also play an important role in the humanization process. According to the poll, the majority of respondents - residents of Rivne identified richly decorated facades of buildings of the 2nd half of the 1940s-1950s most aesthetically appealing. According to these results, there is an unambiguous conclusion about the role of the architectural heritage of the 2nd half of the 1940s-1950s in providing architectural environment aesthetic comfort.

Preservation and restoration of the architectural decor of building complexes of the 1940s-1950s is an important issue in the context of architectural environment aestheticization and humanization in Rivne and other cities of Ukraine.

## References

[1] Day C 2004 *Places of the soul: architecture and environmental design as a healing art. Second edition* (Architectural Press) p 309
[2] Marek Antoszczyszyn 2017 IOP Conf. Ser.: Mater. Sci. Eng. 245 052062
[3] Bogdanov A 1911 *The cultural challenges of our time.* (Moskow: Izdanie S.Dorovatovskogo i A.Charushnikova) p 321 (in Russian)
[4] Shevchenko L 2019 European experience on humanization of the city's environment. *Scientific Achievements of the Present: Proceedings of IV International scientific conference* (Germany, Berlin: Tradition GmbH) pp 113-118 (in Ukrainian)
[5] Loos A 1997 *Ornament and Crime: Selected Essays* (California, Ariadne Press) p 204
[6] Moussavi F, Kubo M 2008 *The Function of ornament* (ACTAR, Harvard Graduate School of Design) p 192
[7] Nina Kazhar 2019 IOP Conf. Ser.: Mater. Sci. Eng. 471 082019

[8] Ikonnikov A 2001—2002 *Architecture of XX century. Utopia and reality. Book 1* (Moskow: Progress-Traditsiya) p 655 (in Russian)
[9] Ferello J 2018 *EDG creates customizable 3D-printed concrete molds* The Architects Newspaper, May 25, 2018, from https://archpaper.com/2018/05/edg-creates-customizable-3d-printed-concrete-molds/#gallery-0-slide-0
[10] Rudnytskyi A 1971 *Urban development of the USSR western regions and their socialist reconstruction* (Lviv) p 55 (in Ukrainian)

IOP Conf. Series: Materials Science and Engineering **907** (2020) 012006 doi:10.1088/1757-899X/907/1/012006

# Information Field of the Modern city silhouette

**L S Martyshova**

Department of Urban Planning, O. M. Beketov National University of Urban Economy in Kharkiv, Ukraine

larymarkaz@gmail.com

**Abstract.** A modern city in all its aspects is a product of diverse information and semantic features that have been expressed in different ways in different times. Time continuum of urban existence and globalization are leading to universalization of rules and methods, on the one hand, and to the destruction of the identity and integrity of the landscape and architectural urban environment, on the other. They are demolishing the system of conceptual references and transforming historic time-tested cultural codes. Today the system information balance is intensively evolving under the influence of time and objective information reality, shifting historically tested and conventional highlights. The information field of the city silhouette is considered as integrity, having a universal algorithm and code, whose revealing and defining promote the recognition and proper interpretation of the most typical features of the contemporary urban environment in the city silhouette structure. The use of the algorithm and code as design and analytical tools for city planning enables to create integrated structures and unify conceptual information on the city silhouette, which form the system of landmarks for human world perception.

## 1. Introduction

Information surrounds a person completely and continuously from the very birth. The newborn immediately begins exploring the available information field of environment along with solving different problems. To solve some of them it is enough to learn only environmental properties, objects and processes, while for the others, it is necessary to gain knowledge and create an integral world outlook. The modern world is a space of large-scale structures with multi-layer information, where technologies have become an instrument, while information has been used to influence people. Information field is a unified mechanism of context parameters impact (information selection, reception and transmission), based on the diverse perception of the environment that reflects the world outlook.

The first visual information that a person gets when approaching an urban area is its silhouette – a historic cumulative outline expression of city facilities in the skyline that is visible while approaching a city. The architectural silhouette provides a person with the initial landmarks in the city structure, while the skyline outline itself communicates essential information. The silhouette line is a kind of an electrocardiogram of the city life, especially at present, when its size is quite impressive. Therefore, the silhouette as a "boundary between the city "body" and the endless sky", conceptually contains two key elements, the *sky* and the *land,* their interaction and opposition [1, p.258]. Two spaces – the sky and the land – have long history of interaction, overlapping and reflection. The sky responds to the uplifts of the land, where natural landscape shapes and urban architectural shapes, being combined, form the interaction line – the city silhouette outline. These two elements influence each other, generate strong images, metaphors, symbols and signs, create meaningful structures, expressed in the encoded text of three-dimensional shapes. The landscape acts as a layer of information about given and existing natural forms, which have been inspiring mankind for centuries to create architectural forms – the models of universal landscapes (a "signified" in the sign structure). Architecture is primarily information about the reflected reality in the urban environment structure (a "signifier" in the sign structure). Therefore, both the sky and the land, as a system of objective emotional-and-imaginary oppositions, including tectonic, light, space-time, play an important part in various manifestations and relations. In the cultural tradition, the sky acts as a structure shaping thing generated by land (table1). The sky is a cover, protection, or a dome defending the land from adverse external factors and invasions. It provides a shade to the land, protects it from "inhuman" outer light.

The universal sky-land opposition is elaborated at the level of the earth surface shapes and is expressed as a geometric antithesis. The sky is conventionally represented by two images, having bright silhouette characteristics – mountains and domes. The mountain is considered to be a cosmic image of light and world integrity. Therewith, the dome symbolizes the way the sky limits the human inner world. The land is traditionally thought to be chthonian (dark and deep), related to the death realm. The loose mass of the mountain here brings about the associations of the heavy underworld and abyss. However, being "lit" with the skylight, inspired land becomes a favorable place for people to live. Thus, a city with its vertical structures and regularity is the most complete model of the Heavenly City. It combines the images of the sky and land, influenced by nature and perceived so that there is always interaction with nature and a divine element (table 1).

## 2. Silhouette

From time immemorial the man "has been invading the sky with vertical buildings, which were special to people". Religious and public facilities have always dominated the "city massive, and it is their interaction that visually shaped the silhouette, interpreted as an art object" [1, p.259-260]. Pyramids went into the sky with their waning steps or darted with the silhouette diagonal planes, belfries and spires of Christian cathedrals, prayer towers and mosques, the levels of Buddhist pagodas, "cur the sky open" with their sophisticated shapes, while palaces and temples were placed on high platforms. The silhouette spot is the most emotionally intense part of the visual information, whereas its morphological properties, related to the shape content trigger "the whole image recognizing process" [2, p.22].

The conventional human activities historically corresponded to the horizontal line, meaning that any vertical ones were efforts and challenges to the sky [1, p.258]. In landscape shapes the horizontal line usually corresponds to the endlessly "stretching" natural shapes of fields, seas, or steppes. In architectural shapes the horizontal line is easily found in the long lines of massive built-up areas and impressive public facilities. The vertical axis, as opposed to the horizontal one, in the landscape and architecture, is immediately identified by the viewer in the city silhouette as something unique, highlighted by nature or people and also conveying a special meaning (table 2). The equality of the "horizontal and vertical" lines exists only in the abstract system of Cartesian ordinates, since for the person the horizontal surface is not related to any difficulties in movement. Due to this reason, any horizontal misalignment is clearly recorded by our consciousness and remembered. This surface play creates the link between the internal and external (according to Heidegger, a "bend"), where "the land and world respond to each other by lifting an object as a product" [3, p.268]. This interaction outlines the object – its silhouette, freely accepting its "dents, planes and cuts, their correlation and alienation" [3, p.275].

The vector of architectural shape information, the architectural silhouette is "visually the most memorable, impressive property of the common space organization of the architectural form, reflecting significant topological features of the object content" [2, p.22]. Information qualities of the city silhouette are, firstly, defined by the mutual determination of the weight and space, because architecture becomes an art only if the synthesis of weight and space is "experienced as an artistic … a spiritual value". These are essentially rooted in the deep unconscious and organic instincts [4, p. 9].

**Table 1.** From the morphology to semantic-semiotic construction of the landscape-architectural composition

| Architectural object (morphological structure) SIGN – SIGNIFIER | Semantic-semiotic construction | |
|---|---|---|
| | MODEL, IMAGE | SIGNIFIED (meaning) |
| 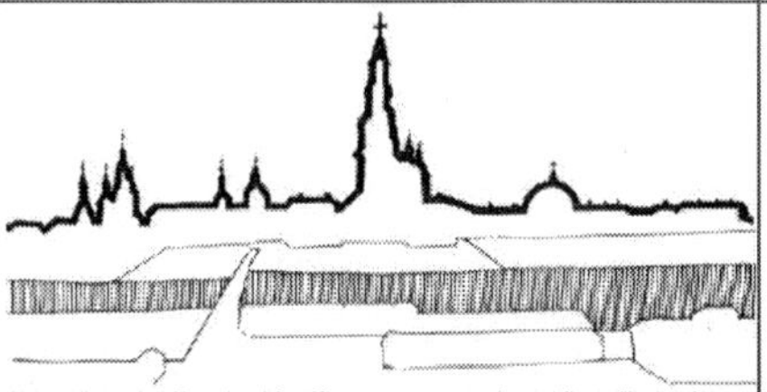 Pronounced vertical as a natural elevation vertex. | **«World tree»,** **«World axis»** | STATEMENT of STABILITY, CENTRALIZATION |
| 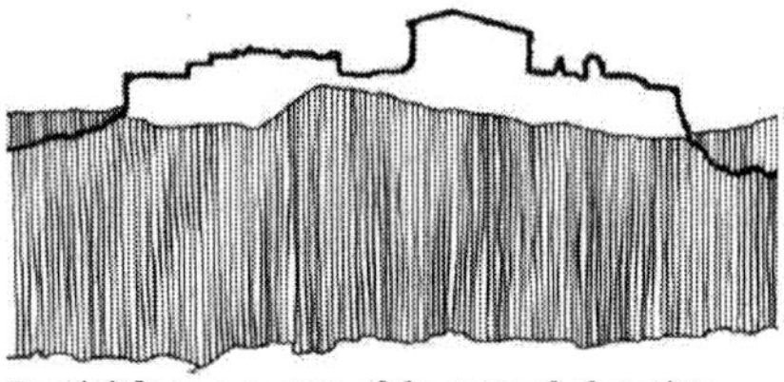 Spatial form, as a top of the natural elevation. |  Crown | SKY, POWER, DOMINANCE, CROWN |
| 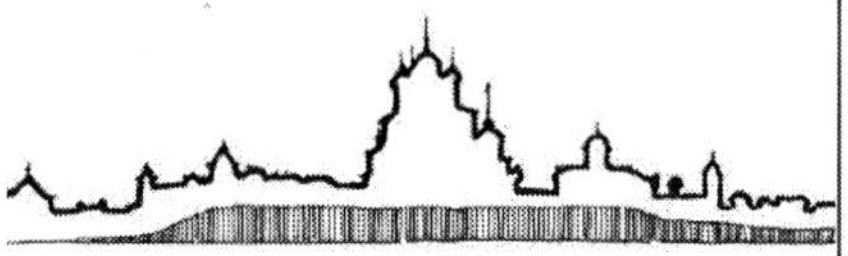 Dome or arch form as a repetition of natural form of an elevation (mountain or hill). |  Dome, arch, marquee as a cover | TOP, SKY, SKY DOME, SKY COVER |
| 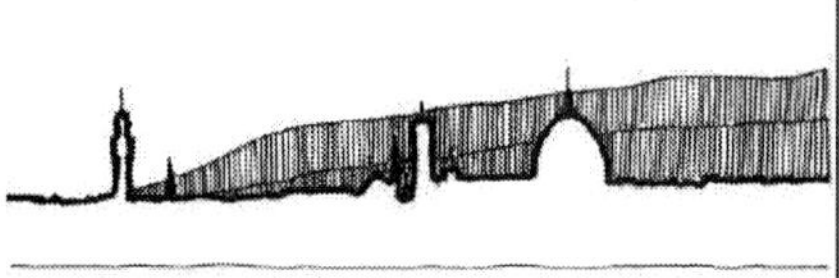 Dome form on the plane. | 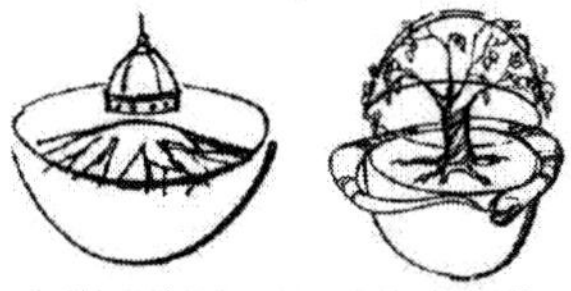 Arch, marquee, tent as a sacred center, opposed to the periphery | "SKY ON THE LAND", SACRED "MIDDLE" |
| 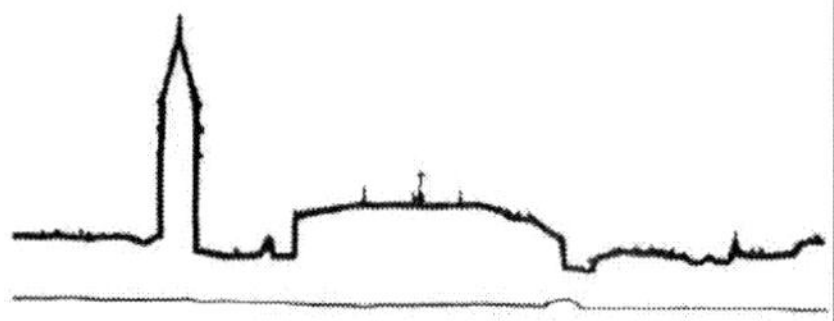 Active vertical, visible on the sky background. |  Column with a cloth thrown over it. Pillar, upon which the sky rests. «World axis». | PILLAR (SKY), POWER (SKY) |
| 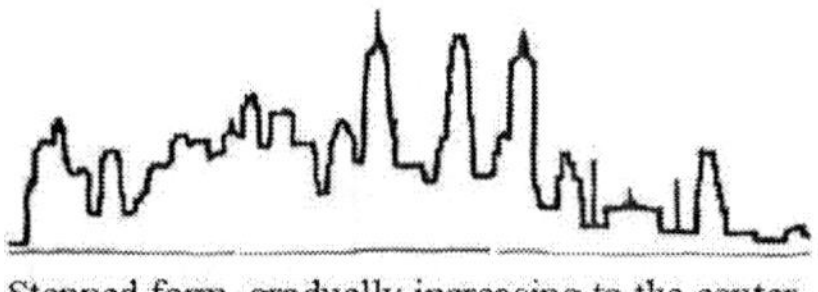 Stepped form, gradually increasing to the center. |  | COMMUNICATION (SKY AND LAND CONNECTION) |

Secondly, they are defined by the feeling of "marking and celebrating, since it is something long and stable, opposed to the temporary changeable flow…" [5, p. 10]. Thirdly, they are defined at the level of plastic world perception – anthropomorphic features of the otherness, since it is shaped in perception or creation. "As the cubic not-self is assessed in its relation to the human body, naturally, it is shown or created after its image and likeness and, in turn, the degree of anthropopathy is a criterion of value: the less organic the otherness is, the harder it is to experience it, live through it empathically. Still, it is perceived as morally valuable. In spiritual creativity a shapeless or sharp facet stereometric item can become a symbol of the matter hostile and rebellious to the spirit" [5, p. 10]. This position is directly related to natural (organic) properties of the architectural silhouette design, while the above features of the externally perceived architectural object encourage the formation of internally limited values. Contemporary architecture does not dominate any more in creating moral ideals and esthetic values; it is actively losing its in-depth semantics and content, generated by centennial traditions, science and culture. The phenomenon of the architectural and city planning information field as a system consists in conveying information from the author (architect, city planner) to the consumer (a citizen, a city dweller) via three-dimensional and silhouette forms. For instance, the dominated city silhouette form can convey information at a distance, in particular, of an internal space with important qualities that determine the city life and spirit.

**Table 2.** Stable interconnections of semantics landscape and semantics of the city's architectural silhouette

**Semantics of landscape form**

**Horizontal** surface of the earth – stability, peace, accessibility, lack of information.

**Vertical** – elevation, inaccessibility, correlation with the sky.

**Semantics of architectural silhouette**

**Vertical** – stopping the gaze, effort, "manifestation of human will", the connection of the sacred point of the city with Heaven.

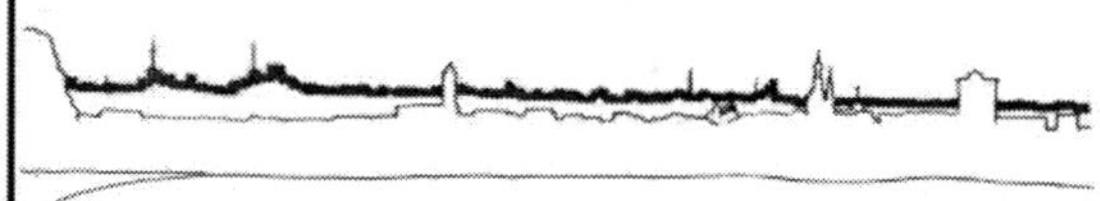

**Horizontal** – familiar, everyday world of residential areas.

Many **vertical** accents that form a picturesque silhouette, gives the human eye points that delay the gaze of a human being.

Differences between **horizontal** and **vertical** elements form heterogeneity, dissimilarity and rhythm disruption – unpredictability as a function of information content.
The presence of similarities, constancy, makes the silhouette predictable, and therefore less informative (according to Fomenko O. A.).

With time city silhouettes change, depending on numerous factors, including natural, social, economic, practical ones. In the ancient ad medieval times, for example, town silhouettes were normally practically oriented as they considered mainly their defense needs. That way a town or a fortress used to gain their natural silhouette, symbolic meaning as well as class, ideological and political content. It had to express power, strength, magnificence and inaccessibility from afar. It had to command respect and fear to external enemies and local dwellers. As a result, the urban silhouette was mainly formed by the shapes of the key public buildings, while a person, approaching the town or city, could easily recognize its facilities at a distance. During the following several centuries, the silhouette, especially that of larger cities, told about its size and financial standing, about the level of technologies and their development and thus the city showed its living standards from afar. It was due to those changes the silhouettes of modern cities were gradually complemented with large-scale palaces and administrative buildings, industrial facilities, office and shopping malls, and elite housing estates.

## 3. Information field

The information field or the *field* is used as a term in a number of sciences to describe the properties of the real space and real world and, it is usually related to the continuous or discrete totality of values, reflecting a properties or one of the properties of the environment. The unified information field system enables to combine various aspects and features, properties and concepts, while monitoring and analyzing their relations and interaction. The wide use of the concept in science and research allows us to use it in the information science, too, where architecture, as an interdisciplinary area, plays an important role in the visual information domain. The city silhouette information field is determined by the algorithm of the system elements interaction, where the system unites not only its structural elements but also various characteristics. The art of architecture and city planning is created in space and time, generating unique interdisciplinary knowledge products, complemented by the talents of architects. The formation of the algorithm as a tool to structure multidimensionality of contemporary information processes enables to create integrated systems of individual facilities, ensembles as parts of the multi-faceted urban design and eventually as the harmonious environment and holistic world image.

The history of architecture and city planning art shows that ancient cities normally kept their original easily recognizable images for centuries. "The law of the individual silhouette" of old cities appeared quite durable, as in the 19-20$^{th}$ centuries it was necessary only to comply with it, since the silhouette is the essential means of orienting in the city space and historical time [1, p.262]. That was once stated by N. Karamzin in his records. Nonetheless, rapid development and urban sprawl of the past two centuries disturbed the above compliance with the unwritten "law on the city spacial harmony and silhouette expressiveness" [6, p.237]. Those resulted in a new layer that complicated the information field of a modern city silhouette. Thus, what was once generated and repeatedly "overlayed" creatively complicated the city image and made it unique. City landscapes have become regular, logical and picturesque background for modern advanced urban architectural forms.

Information relations are vital for the information field. They reflect correlation between real objects and elements of the system. Information relations are a mandatory factor of people's knowledge of the world and their place in it. These relations may be expressed explicitly: in a hierarchy, the part and the whole, etc., or implicitly, for instance, in associations or images. Initially, the foundation of each city unique image is in the original combination of the elements of architectural and landscape environment. The organic connection between the city and natural environment has kept its top priority at all stages of the city planning development – from the ancient times to these days. The relief primarily determines the extent of its space expressiveness, since it is the base of the landscape and architectural urban space, while the picturesque silhouette, with numerous vertical highlights, provides people with the landmarks that are eye-catching and well-recognizable from any perspective (table 3).

## 4. Content (meaning)

An integral part of the city silhouette information field is its content, necessary for unambiguous interpretation of urban facilities and objects, as well as for their information clarity. The city silhouette information field is structurally complex and unites a variety of features from philosophical and functional to imaginary and psychological ones, which encourages the formation of internal incapsulated values. Semantic units create numerous structures, expressed with space stereometric means in city silhouette lines. In effect, the aim of creating the city silhouette as an object to be perceived is to inform of its content.

According to the information theory the content is to be understood as a major property of information. From the semantic load viewpoint, information depends on its purpose and target. The city or its fragment silhouette contains encoded morphological, semiotic, symbolic, metaphoric, semantic and other information on the functional and esthetic components of the urban architectural environment, and on the nature of the landscape as the city morphological basis. Primary data on a city is perceived by a person at a distance, from the silhouette line pattern. This information can also refer to the city life, city structural facilities – that is the first stage of cognition. By uniting with emotional and associative aspects, morphological characteristics become creative and imaginary, while the higher outline complexity generates an emotional and associative outburst, where most significance is paid to the very fragments of the silhouette that are hard to *read* in such multilayered and multifaceted text. In this case, architectural forms become more expressive and they are easier to be interpreted, especially if they are intensified with landscape shapes with similarities or differences (major types of interaction of three-dimensional shapes and the landscape).

### *4.1. Morphological structure*

The first stage of recognizing the silhouette, in particular, that of a city, is reading and analyzing the morphological structure of its composition. The next step is related to one of the main phenomena of the shape perception – the phenomenon of outlining a figure out of the background, since the whole perception field, primarily the eyesight, clearly differentiates various parts. A substantial figure or an object belongs to this part, while the other includes the background, where a "thing" is perceived and, unlike a figure, has the material property [8, p. 79]. The following step is in singling out the nature of correlation of the figure and background. It regards the architectural silhouette, the feature analysis of the natural countryside relief shape (a hill, plane etc.) and the form of an architectural object, as well as each further layer of the silhouette (table 3).

### *4.2. Sign structure*

The information value of an architectural facility is also determined by its sign structure. The sign structure is based on the sign, which is defined as a material object, having a meaning, and it is considered as a structure with double characteristics. These characteristics are formed by the opposition of the signified–signifier, where the signifier is the material value, where the culture and history are embodied, materialized, while the signified is the spiritual experience, giving the perspective for us to perceive, live through and understand the signifier. Thus, the sign is the result of the signified–signifier interaction, which reveals the cultural-historical – objective, factual and still experienced significance of the phenomenon [9]. Signs as the units of this structure are united and logically integrated to form a text – the material embodiment of the system. The system reflects esthetic properties – meaning and content, expressed through the elements of the silhouette composition. However, having identified the value as a symbol, and the signified as a thought or a reference, Ch. Ogden and A. Richards added the third referent element, representing an object (event) to refer to [2, p. 45]. Then the meaning is characterized by the relations between the sign (signifier) and the field of interpretation (signified), where interpretation is a way through a metaphor to an image. The metaphor, which can be based on the similarity of any qualities or properties (color, shape, volume, purpose, position etc.), enables to create full images based on bright and unexpected associations. Moreover, there is a definition of a sign not only as a material object that has its meaning, but also as a material object integrated with its meaning. Using this approach to the sign, it appears to be two-faceted, i.e. having form and content.

**Table 3.** Fundamental sign elements of the landscape-architectural city silhouette structure [7, p. 181]

| Landscape form as a sign | | | | Architectural equivalent of signified |
|---|---|---|---|---|
| SIGNIFIER | | SIGNIFIED | | |
| Landscape form | Silhouette shape | SKY | LAND | |
| Mountain  | Convex 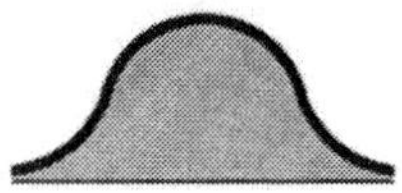 (upward) | Model of universe | | Signified as a model of universe |
| | | World mountain | | Marquee, hovel, tent, dome |
| Hollow  | Convex 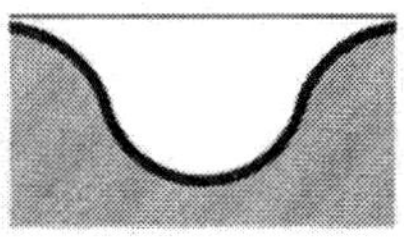 (downward) | Mountain – Sky | Cave in the mountain | Towers and pagodas, pyramids and ziggurats, stairs and steps, covers, marquees, canopies, tents |
| | | Entrance to the sky (mountain top) | Entrance to the underworld (piedmont) | |
| Plain  | Horizontal, at horizon level | Expanse, vastness, empty space | | Flat, horizontal forms |
| Field  | | Vastness and boundless expanse, bright open space | | |
| Forest  | Horizontal, above the horizon | | Wild form, danger | |
| Sea  | Horizontal, moving undulation | «Inverted» sky | Dept, going down | City |
| Steppe  | | Vastness | Vastness, oscillatory motion, rhythm, a sense of the infinite, referring to the «beginning» | |

When analyzing the sign structure of the city silhouette, it seems to be necessary to divide the sign structure into the typological elements according to the understanding of the composition as a single integrated unit to be perceived. This unit combines functional and esthetic aspects. Umberto Eco noted that "the very consideration of our relations with architecture proves that normally, dealing with it, we get involved in a communication act, which does not exclude functionality" [10, p. 204]; "...our approach to semiology allows us to see an architectural sign as a signifier, whose signified is its own functional purpose" [10, p. 211]. Yet, it is to be mentioned that the functional purpose of architecture may be studied not only from pragmatic perspective, but also from its artistic esthetic qualities. Consequently, from the communication viewpoint, the system of architectural signs must reflect both aspects.

The silhouette of a city or its ensemble may inform the viewer of the functional organization (the availability and location of the center and periphery, the availability of built-up areas in the top and bottom points of the relief etc.). The sign structure that reveals the correlation of the dominant and subordinate composition elements, the scale and rhythmical organization of the silhouette, and it reflects esthetic qualities as well. Logically and systematically united signs make up a text as the material side of the system. This text has a single meaning and can be considered as an indivisible signal (including an ensemble, a town etc.) with its specific features, recognizable to its recipients. The text must have its boundaries, the beginning and end, internal components and organization, providing it with integrity at the symbolic level and turning it into a literary text (as a total of phrases), which creates a meta-structure at the level of creative organization.

Silhouette forms with bright information qualities should include vertical and horizontal ones (either natural or man-made). Therefore, supposedly, silhouette drawings outlining the general landscape features have this important communicative feature: flat terrain (horizontal), highland (upward oriented) or lowland (deepened). The shapes, whose silhouettes are broken down or simplified have semiotic qualities. They also involve planned character of the viewed image, which forms a consistent layering of the silhouettes from different layouts and there emerges a difference between the silhouette of the object as visible on the sky background, on the distant landscape background or on the built-up area background. Another important sign factor is the degree of highlighting the city silhouette on the surrounding landscape, the opposition or, vice versa, fusion of anthropogenic and natural environment. All these sign forms are integrated thus forming the text, concept, referring to the essential concepts of semiotics, implying an invariant (stable) system of relations between structural levels and within the levels, the whole environment, the activity etc. [9, p. 63]. The semantic element of the textual (sign) structure is based on the signifier – signified relations. The concepts of "content" and "meaning" are tightly related though sometimes they are separated. According to G. Frege [11], meaning characterizes the relation between the sign and the object it denotes (e.g., words "house" and "home" linked with a place of residence), and also a reference to other signs. The meaning characterizes the relations between the sign, its meaning and the field of interpretation (naming). Accordingly, this concept is implicit, directly dependent on the knowledge on the object [12]. A strange thing may seem to be meaningless if there is no knowledge of its use, i.e. how to benefit from it or what it is for. The meaning may also include implied or expressed purpose of items, words, concepts or actions, given to them by a specific person or a community. The opposite of meaning is meaninglessness, i.e. the absence of a clear purpose.

### *4.3. Semiotic-semantic relations*

As regards semiotic-semantic relations among landscape and architectural city shapes and the principles of their compositional expression in the city silhouette, it is important to note U. Eco's statement that "the transfer of a feature to a different text is one of the essential means to create new meanings" [10, p.62]. Landscape forms underlying the city composition, as mentioned above, convey a traditional meaning, which is revealed through archetypal and mythopoetic codes. A mountain, a valley, a river, the land surface, a steppe or woods are themselves the sources of meaning generating images. Here, the meaning, understood as imaginary or real purpose, includes functional-pragmatic landscape uses and its esthetic-interpretational features.

According to the concepts of Gestalt psychology, the perception of the integral composition is related to the consciousness orientation on the perception of simple and regular structures. At the same time, the concept of simplicity is related rather to the organization of the complex material using the minimum number of systematized structural links than to the fact that the object has a limited number of features

[13]. However, it should be added that the creative aspect is inversely related to simplicity and features more complex perception, which can lead to the experience effect, according to the interpretation of the concepts of *disposition* and *composition* by L. Vygotsky. The simplest shapes, perceived by a person as two-dimensional (silhouette), are major geometric shapes (squares, circles, rectangles, triangle and a trapeze). In the architectural-landscape silhouette, a methodological way is a possibility to identify such elements as characteristic sings that are quite easy to read by an observer. These forms as signifiers may be interpreted imaginarily and symbolically as those meeting the major landscape opposition *sky-land* (signified). Nevertheless, their interpretation may be complemented with interconnections and relations between the forms themselves and their surroundings. For instance, we may identify structural options: whether the sky visually occupies more or less space than the land; the object silhouette is visible on the background of the sky or land, forest, sea etc. Therewith, the next step to take is to identify the features, characterizing their combinations and complementing the seeming simplicity. That kind of a complementary element includes, inter alia, a silhouette layout layering, blurred connections or the explicit opposition of the landscape and architectural forms. The identification of the relations among these forms appears significant, since landscape forms, as signs, convey their meanings (signifier – signified) to the architectural structure, while architectural forms with landscape features combine them with their own meaning. As a result, the information field of the city silhouette becomes a carrier of diverse meanings, strengthening (or mitigating) each other. Thus, it may be stated that the information load of the city silhouette is determined by the major morphological types: visual fixation spots, silhouette elements (landscape and architectural), typical linear shapes, systematization based on contrast or similarity, fixation of rhythmical and dynamic silhouette features, bringing the silhouette shape to a geometrically identifiable shape.

## 5. Conclusion

A modern city is full of diverse information, which complements and complicates its silhouette, most often reflecting neither its city planning structure nor its architectural image, neither its traditions nor functions, esthetics or other features, as it used to happen in the past. Beneficial space or silhouette locations in the urban structure are increasingly often occupied by insignificant public or historic legacy facilities, while the visual information elements (logos and street signs, landmarks and indicators, outdoor banners, communication towers etc.). This shift of the priorities entails not only adverse layering of different facilities in the city structure and silhouette, but also often to minimization of their significance, the loss of meaning and content or their complete destruction.

Architecture and city planning, as a system reflecting the environment integrity, encourage harmonization of the environment by the unity of the space principles and techniques as well as by the integrity of the information structure (algorithm) and communication language (code). The urban environment speaks its unique language, reflecting the dialogue in time using signs, symbols and meanings. The modern city and its silhouette are shown to the person as a work of art: here the nuclear historic center ensemble continues generating new structural elements. A new architectural layer of each period provides information in a new language, where the new contemporary language is the interpretation of the truly old, ancient, and historic archetype-based legacy. It is generated by the system of new relations in the information field, where signs and symbols are provided with fresh meanings. For instance, conventional harmony and logic op the historic city silhouette is destroyed, since the meaningful context of the modern environment does not correspond or come from history. That inevitably damages the information balances, ruins the urban environment integrity and the uniqueness of its code, generating a new reality with a new code and new algorithms. A person (both a maker and a consumer), who knows the code and the system (algorithm), can clearly read the multi-faceted structure of the modern city in its silhouette. Using conventional, historical categories, the city silhouette is a factor that facilitates person's orientation in the urban environment, forms the world outlook and establishes relations with the natural environment.

The silhouette of the modern city is influenced by globalization, information and other technologies development, which promote the emergence of new meanings. The synthesis of information and content parameters universalizes the diversity, removing uniqueness and actually leading to the fake meaning context, the code changes – the existing features are substituted with new ones. Certain meanings can be fully destroyed, while others are replaced with the opposite, thus distorting the conventional and understandable text. It requires creation of the new regularity (a new code) and a new language. There emerges imbalance not only in the semiotic, semantic but also generally in the global information field, generating a flow of images. These lose their meaningful elements and cannot be referred to a historic period or to the specific features of the landscape and culture, where the person loses the vital feeling of order and stability in the world around. It is based on the unity and systematization of landscape and architectural features that the city silhouette, which helps to build meaningful updated contexts, becomes unique. It contains combinations of codes, which algorithmically lead to diversity, and also stable images and meanings of the city structures in the information field of the city silhouette.

**References**

[1] Gutnov A E 1990 The World of Architecture (Moscow, Molodaya Gvardia)
[2] Fomenko O A 2000-2001 The Phenomenon of Architectural Silhouette. *Traditions and innovations in higher architecture and art education* Issue 6-1 KhDADM
[3] Podoroga V A 1993 *Landscape Metaphysics: communicative strategies in the philosophical culture of XIX– XX centuries* (Moscow, Nauka)
[4] Goldzamt E A 1985 *City planning culture of European socialistic countries* (Moscow, Stroyizdat)
[5] Gabrichevsky A G 1993 *Theory and History of Architecture: Selected papers* Kyiv
[6] Tonev L 1973 *The Composition of the Modern City* Sofia
[7] Martyshova L S 2014 *The composition and semantic role of landscape in the formation of the city silhouette*. Candidate thesis in architecture majoring 18.00.01 (Kharkiv, Kharkiv National Technical University of Construction and Architecture)
[8] Fomenko O A 2003 *The analysis and assessment methodology of the esthetic performance of morphological properties in architecture*. Doctorate thesis in architecture majoring 18.00.01 (Kharkiv, Kharkiv State Technical University of Construction and Architecture)
[9] Knabe G 2002 *Sign and its Properties*. Retrieved from https://art.1sept.ru/article.php?ID=200200401
[10] Eco U 1998 *The Absent Structure. Introduction to Semiological Research* (Saint-Petersburg, Petropolis)
[11] Frege G *Sense and meaning*. Retrieved from https://royallib.com/read/frege_gotlob/smisl_i_znachenie.html#0
[12] Gasparyan D E Language as Eigenform: Semiotics in the search of a meaning Vol. 34 No. 4 (2018) *Vestnik of Saint Petersburg University. Philosophy and Conflict studies. / Philosophy* https://philosophyjournal.spbu.ru/issue/view/287
[13] Rayevsky A A, Serebrennikova T A 2017 The Typology of information space in the context of organizing contemporary urban environment. *Fundamental Studies* № 4-1 P 64-69 Retrieved from http://www.fundamental-research.ru/ru/article/view?id=41436

# Mobile housing on the islands in post-industrial Dnieper

**V P Myronenko[1], T A Tsymbalova[2], O V Myronenko[3]**

[1]Department of design of architectural environment, Kharkiv National University of Civil Engineering and Architecture, Kharkiv, 61002, Ukraine
[2]Department of the architectural planning and town-planning, Dnieper State Academy of Building and Architecture, Dnieper, 49600, Ukraine
[3]Department of Architectural Design, Kharkiv National University of Civil Engineering and Architecture, Kharkiv, 61002, Ukraine

mironenkovp53@gmail.com

**Abstract.** Mobile housing has extensive experience in various regions of the world. The popularity of mobile housing is due to the practical advantages of their designs and universal application in various functional typological areas. The use of mobile residential structures in Ukraine has significant prospects for the development of the tourism business, in particular, with the recreational development of the islands. Analysis of the possibilities of using mobile housing in the natural-landscape potential of the islands of the post-industrial Dnieper. The study is based on an integrated approach – application of general scientific and applied methods: the study of literary sources regarding the historical experience of using mobile housing, historical and local history heritage, modern scientific developments and state regulatory documentation; photofixations, experimental modelling. The Dnieper islands constitute a significant part of the recreational and tourist potential in the structure of the complex landscape zone of the modern Dnieper city. The use of recreational mobile residential facilities with the possibility of their placement on the islands and in coastal waters is proposed. The use residential technologies for organizing recreational and tourist activities on the Dnieper islands seems promising, since it is based on the concept of an ecological approach to the recreational load on natural landscapes.

## 1. Introduction

Current socio-economic conditions in Ukraine, the level of provision population with capital housing, of different types and quality, promote the introduction of mobile residential technologies. It is advisable to apply the structural and technological advantages of mobile housing for the organization of temporary residence in many functional and typological directions, in particular, during the organization of recreational and tourist activities on natural landscapes.

The urgency of the work is connected with the problem of using the mobile form of recreational and tourist accommodation in the territories of the islands in modern Dnieper and improvement of the means of interaction of anthropogenic and natural objects.

An analysis of recent publications shows that a lot of research is being done around the world on of mobile architecture. Some theoretical developments of mobile architectural shaping are presented in scientific works: V. Vorobiev (transformation of mobile structures and emergence of kinematic architecture) [1]; D. Tkach (conceptual proposals for the formation of residential buildings using mobile objects built on the basis of conical surfaces) [2]. Topical issues related to regional problems of the use of existing coastal territories, creation of artificial land plots, development of tourist business were considered in their research: I. Cherniak, L, Yaruchyk (explored modern concepts of urban planning on the example of the Hague [3]; G. Klopko (a comprehensive strategy for the use of the coastal territories of Dnepropetrovsk) [4]; D. Semenov, S. Kaloshina (innovative methods of creating artificial island territories are presented) [5]; E. Bugriy (analysis of the problem of artificial islands in the context of accretion) [6]; I. Kovalchuk (the problem of optimization of tourist business due to the use of innovative technologies is investigated) [7].

The following study examines the possibilities of using a mobile form of housing for the organization of recreational and tourist activities in the island territories of modern Dnieper city.

## 2. Presenting main material

The use of housing that moves with people has been known since ancient times. Over the millennia, land and water mobile homes have been shaped depending on the degree of mobility and occupancy of individual ethnic groups. Most often, mobile homes were part of a marching property or road props and were made from lightweight materials - wood, wool, leather, fabric (the historical heritage of mobile architecture is represented by a very small number of archaeological evidences mainly fragmentary and interpretative) [8, 9].

The use of mobile housing has been inextricably linked to the history of the arid zone of the Ukrainian state since ancient times. Yurts, carts and tents of nomadic cattle ranchers were the main types of residential buildings of the steppe south until the 15th century (the safe mastering of the Wild Field by the settled population, the development of the urban economy and the active formation of the capital housing stock began only after the peace agreement with Turkey in 1774) [10].

Over the next two centuries, under the new conditions of rapid socio-economic development of the steppe region, the typology of mobile residential structures was largely confined to the use of temporary housing during construction.

In today's Ukraine, mobile housing technologies have great prospects, and in some cases, even some advantages over capital construction. For example, it is advisable to exploit the territories of the islands with mobile technologies in recreational and tourism activities, given the legal prohibition and restrictions on stationary construction, since mobile housing does not belong to capital real estate [11].

Research into the possibilities of using mobile homes in the islands, in the structure of complex development of recreational activity, can be carried out on the example of post-industrial Dnieper and its suburban areas.

The modern city of Dnieper (Dnepropetrovsk) is located on both banks of the Dnieper River at the confluence of the left tributary of Samara. The waters of the rivers divide the city into three unequal parts (the Dnieper River has a length of 31.5 km within the city; the Samara River - 9 km; inland water accounts for 18% of the city); numerous islands are located throughout the water area [12].

Unfortunately, the modern Dnieper islands are only a small part of the historical natural heritage left after the Dnieper HPP was flooded due to the construction of the GES in 1932. In particular, there were more than 120 islands with diverse landscapes between the cities of Dnieper and Zaporozhye - rocky, sandy, and covered with forests, shrubs, and grasses [13].

For the coastal population, the island territories were not only of great economic importance, the picturesque sceneries were fascinating and enriching spiritually. About one of the small islands - Lozovaty (an area of 2 ha, on which oaks, willows, vines grew), D. Yavornitsky wrote: “This is a wonderful island of all the islands that are between the Dnieper rapids. This is a wonderful picture painted by Nature herself. It makes one forget all that sad and helplessness that is in real life” [13].

All high rocky islands that never flooded during the spring flood were "museums" of antiquity of different eras- from the Stone Age to the Cossack era (after the construction of the Zaporozhye dam, 50 settlements were under water and a large amount of lands rich in unique historical and cultural heritage [13].

Modern islands constitute a valuable fund of general natural landscape potential Dnieper (a large island system has a conservation status within the Dnieper-Orel Nature Reserve) [12].

Directly, in the industrial centre, within the Dnieper water area, there are 24 islands and 2 peninsulas (19 islands are more than 1 km long and almost 300 m wide). The level use of island territories in the structure of urban environment is different. The largest anthropogenic load on natural landscapes has 5 islands:

- Monastyrsky - actively used as an urban recreational facility (the island is part of Shevchenko Park and connected to it by a bridge). But the quality of recreational services in a large area is low, because the island is crossed by arched supports of the Merefa-Kherson railway bridge (Figure 1).
- Stanovy - recreational and sports area in the structure of the rowing canal (the island is almost all restored to its new form after the flooding); connected with the mainland by a bridge.
- Feinberg - used to organize the infrastructure of a city cargo port; connected with the mainland by dam.
- Namystanka - the Kaydak bridge passes through the island (Figure 2).
- Kaminuvaty - the island has transmission line supports.

**Figure 1**. Monastyrsky Island, Dnieper, Ukraine.

**Figure 2.** Namystanka Island, Dnieper, Ukraine.

At one time, the problems of the use of islands located near urban residential areas were actively engaged in architectural workshops in Dnepropetrovsk. The most interesting, from the town-planning point of view, was considered the Green Island, located near the left bank of the Dnieper between the bridges Central and Amur. In one of the projects, the island is assigned the role of the main recreational facility of the Left Bank Community center (in the context of the structural planning concept of creating a large City Ensemble) [14]. There is currently no organized activity on Green Island (Figure 3).

**Figure 3.** Green Island, Dnieper, Ukraine.

Recreational use of the Dnieper islands has considerable historical experience. In the pre-revolutionary period and in the Soviet era, the islands were a popular holiday destination for citizens. Residents of Ekaterinoslav-Dnepropetrovsk massively rested on the beaches and forest clearings of the Green, Shevsky and Alekseevsky Islands (especially through water transport) in the summer.

It is advisable to carry out recreational activities on the islands of modern Dnieper, as part of the regional program of qualitative development of coastal territories of the rivers: the Dnieper, Samara, Orel, Kilchen, Sura, Voronoi, Osokorovka.

First of all, it is necessary to develop a comprehensive model of recreational and tourist development of coastal zones and water areas based on a study of the natural landscape fund and historical and cultural heritage. The Dnieper islands can be considered as separate attractive objects in the routes of historical and local lore (for example, a vivid "collection" of archaeological sites has been preserved on Kodachok Island).

The use of mobile homes in the islands could optimize the organization of household services in the general tourist-excursion structure and promote commercial interest in the tourism business (significant revenue side of the local budget) [7].

The following principles are proposed in the design of recreational mobile homes:

- Bionic shaping (based on the use of biomorphic shaping for the delicate artistic and aesthetic inclusion of anthropogenic objects in the natural environment.
- Ergonomics (comfortable design of life process based on taking into account sanitary-hygienic factors, as well as spatial and anthropological parameters).
- Environmental safety (the use of structures based on building materials of natural origin, including restorative natural materials of plant origin; introduction of energy systems based on renewable energy sources; waste disposal [16].

Accommodation of mobile recreational housing can be carried out both on the surface of the islands and on the water - off the coast.

It is advisable to use residential block-containers to set up mobile recreational complexes on land, fully equipped with furniture and engineering equipment (possible connection to centralized engineering networks and operation in offline mode); residential units are delivered to the island by floating means.

Installation of mobile residential units can be carried out directly into the prepared territory (crushed stone, sand or special plates are used to level the area). Spatial planning solutions can use modular elements and different blocking options (Figure 4).

**Figure 4.** Fragment of model of recreational residential complex.

Also, it is possible to use frame-awning and pneumatic structures, but their installation requires more time, and creating a comfortable design, arranging the necessary engineering and technical equipment is problematic.

Architectural organization of mobile housing on the water provides structural and technological solutions using floating foundations and lightweight frame structures for housing (at the same time, the houses on the water have a fixed location, that is, they are deprived of the possibility of free movement in space) [17].

Hydraulic structures, depending on the principle of floating base arrangement, can be of two main types:
- houses on pontoons;
- houseboat vessels - in conditions where the boats are not subject to further exploitation on the water (the boat is safe for housing).

The use of campers and trailers is not very desirable, because it's necessary to create a minimum vehicle load mode in the islands.

Effective recreational use of the islands requires comprehensive landscape surveys, first of all, definitions: highs and lowlands, the ratio of forest covered and open spaces, wetlands [18].

Since the forest plantations on the islands belong to the category of nature conservation (like all the forests of the Dnieper natural landscape), schemes of permissible loads on natural landscapes should be developed [12].

Of particular importance is the aesthetic appreciation of the artistic and spatial qualities of island landscapes: selection of scenic landscapes and viewpoints for visual perception of the coastal space, in particular, panoramic disclosures from the water area (this is an important factor that has a significant psychological effect on humans).

Based on a comprehensive analysis of island landscapes and historical and cultural monuments should be located:
- mobile housing;
- mobile social service infrastructure;
- elements of improvement (network of pedestrian paths and grounds: to view landscapes and expositions, sports, children's, for picnics).

But most importantly, in the process of recreational activity in island territories, it is to find a reasonable compromise between anthropogenic loading and the maximum conservation of natural landscapes.

In the present ecological environment of post-industrial Dnieper region, the problem of communicating with the environment is becoming more complex. The issue of preservation the Dnieper River itself is particularly acute now (the sanitary status of the Dnieper water area is currently unfavorable for human health) [12].

Protection of the Dnieper water-green landscapes, the possibility of their regulated recreational use is one of the means of formation environmental and social stability.

## 3. Conclusions
The need for recreational use of the Dnieper islands is due to their unique natural landscape potential and special urban value. Application of recreational mobile residential technologies in island territories, in the structure of a complex strategy for tourism business development on the Dnieper coasts, is appropriate, since it is based on the principles of maximum reservation of the ecological framework and the natural landscape heritage.

## References
[1] Vorobiev V V, Kozak Ya D 2010 Announcer of the Dnieper state academy of building and architecture **12** (Dnipropetrovsk: PSABA) pp 38-43
[2] Tkach D I 2011 System descriptive geometry (Dnepropetrovsk: PSABA) p 356
[3] Cherniak I, Yaruchyk L 2018 *World Science* **1** (Warsaw: SEC) pp 35-41
[4] Klopko G K, Kozienko E G 2012 Announcer of the Dnieper state academy of building and architecture **1-3** (Dnipropetrovsk: PSABA) pp 12-17
[5] Semenov D A, Kaloshina S V 2016 Announcer of the Perm national research polytechnic university **4** (Perm: PNRPU) pp 80-92
[6] Bugriy E V 2011 Scientific messages of the Sumy state pedagogical university are the name of A S Makarenko Geographical sciences **2** (Sumy: SSPU) pp 103-112
[7] Kovalchuk I E 2017 *Proceedings of the IV International Scientific and Practical Conference* (Lviv: Lviv Polytechnic) pp 692-693
[8] Andrianov B V 1985 *Nomadic population of the world* (Moscow: Science) p 282
[9] Maidar D, Purveyev D 1980 *From nomadic to mobile architecture* (Moscow: building publisher) p 216
[10] Starostin V S *The capital of the steppe region* 2004 (Dnepropetrovsk: Dniprobook) p 280
[11] *Land Code of Ukraine* The composition of the lands of the water fund 2001
[12] Shpilevsky I, Palecha Yu, and Shidlovskaya T 2019 *Making changes to the general development plan of the city of Dnipro* (Kiev: USRIUD) p 64
[13] Yavornytsky D I *Dnieper rapids* 1989 (Dnepropetrovsk: Beam) p 142

[14] *Architecture of Dnepropetrovsk* 2008 Edition of the Dnipropetrovsk regional organization of the NUA of Ukraine (Dnepropetrovsk: P. print) p 101
[15] Akinfiev E Ya 1889 *Vegetation of Yekaterinoslav at the end of the first century of its existence* (Ekaterinoslav: Printing house Pavlovsky) p 238
[16] Mironenko V P, Tsyimbalova T A 2018 *Proceedings of the IV International Scientific and Practical Conference* (Warsaw: SEC) pp 3-7
[17] Ekonomov I S 2010 *The principles of the formation of low-rise residential objects on the water* Abstract of the dis. of the cand. of architecture (Moskow: MARCHI) p 36
[18] Sycheva A V *Landscape architecture* (Moskow: Onyx - 21st Century) p 87

IOP Conf. Series: Materials Science and Engineering **907** (2020) 012008 doi:10.1088/1757-899X/907/1/012008

# The Basic Principles of Ecovillage Design

**O O Blahovestova[1], O O Pechertsev[1], S O Dansheva[2]**

[1]Department of Graphic, Kharkiv National University of Civil Engineering and Architecture, Sumska str. 40, 61002, Kharkiv, Ukraine.
[2]Department of Physics, Kharkiv National University of Civil Engineering and Architecture, Sumska str. 40, 61002, Kharkiv, Ukraine.

lena.pechertceva@gmail.com

**Abstract.** The article discusses the basic prerequisites for the emergence and development of ecological settlements, analyses the concepts of eco-settlement and sustainable development. As a result of the analysis of literature sources, the basic principles of eco-settlement's design were identified: the principle of ecological functioning, the principle of using traditional building materials, the technogenic principle and the principle of sustainable transport development.

## 1. Introduction

The emergence of increased interest in energy-efficient architecture is related to a number of factors. The dominant ones were the rapid technological progress of the 20th century, the expansion of technical capabilities and the introduction of new technologies, which ultimately led to the intensive natural resources depletion. All this entailed a violation of the natural balance in the natural environment. Thus, the question of a reasonable restriction of environmental management and environmentally prudent pressure on nature has come to the fore. One of the most effective methods for solving eco-problems was the emergence of energy-efficient buildings. Starting from the 60s, energy-efficient areas and eco-settlements have started appearing along with few energy-efficient buildings, especially in Europe, North America and Australia. It should be noted that most European countries are actively involved in the process of greening the cities, among these countries Sweden stands out (since 1980 about 20 eco-settlement projects have been implemented here), as well as the Netherlands, Denmark, Belgium and Norway. At the moment, in Ukraine and the CIS countries, environmental design is only limited to the creation of separate eco-housing objects, despite the fact that there is a fairly large number of unrealized projects of ecological settlements created by design centers specializing mainly in the field of environmental architecture.

## 2. The concept of eco-settlement

In order to understand the very essence of ecological settlements, it is necessary to turn to the interpretation of this term.

In the analysis of scientific works some variants of the term *"eco-settlement" (ecovillage)* definition were identified, the first of which was coined by R. Gilman in 1991 in the report of the Institute of Context for the Earth Trust, where he described it as "Human-scale, full-featured settlements in which human activities are harmlessly integrated into the natural world in a way that is supportive of healthy human development, and which can be successfully continued into the indefinite future [1]." According to Gilman, an eco-settlement, unlike traditional settlements, is a phenomenon of post-industrial society.

However, a more realistic explanation of this concept was suggested by Holick and Connelly: "a community of up to several hundred people that meet the material, economic, social, emotional, cultural, and spiritual needs of its inhabitants whilst being in harmony with its natural environment." This definition placed great emphasis on the human's spiritual needs among other facts but, in general, most of the ecovillagers do not share the same religious beliefs and quite often have different ones than other community members. Therefore, spirituality should not be considered as one of the core principles of this concept [2].

Another author, Albert Bates, gave an upgraded but at the same time a bit too detailed description of the ecovillage that could be used as a definition: "Ecovillages today are typically small communities with a tightly-knit social structure united by common ecological, social, or spiritual views. These communities may be urban or rural, high or low technologically, depending on circumstance and conviction. What they share is a deep respect for nature, with humans as an integral part of natural cycles. Ecovillages address social, environmental, and economic dimensions of sustainability in an integrated way, with human communities as part of, not apart from, balanced ecologies [3]"

Later, this definition was interpreted by eco-settlers themselves, whose interests were represented by the Global Ecovillage Network. In their view an ecovillage is an: "intentional, traditional; rural or urban community that is consciously designed through locally owned, participatory processes in all four dimensions of sustainability (social, cultural, ecological and economic) to regenerate their social and natural environments"[4].

*Sustainable development* is understood as a model for the use of resources, a model for interaction between people and nature, and a model for civilization development based on innovation, in which the vital needs of the current generation are achieved while preserving the environment, improving health, and without depriving future generations of such an opportunity.

At the same time, the problem of sustainability of natural-anthropogenic systems is solved in two ways.

**The problems of creating a favorable social microclimate:**

- Providing a variety of job opportunities that meet the aspirations and preferences of people.
- Meeting the increasing demand for housing, focusing on the construction of the particular ethnic requirements of people of different nationalities.
- Guaranteeing variety of high-quality services, including medical care and education.
- Expanding the range of leisure and recreation facilities.
- Interaction of settlements with the environment.

**The problems associated with urban ecology microclimate:**

- The use of energy-saving technologies and alternative / renewable energy sources, reducing the negative impact of energy production and usage on human health and the environment;
- Creation of favourable conditions for organizing and developing the private housing sector where appropriate, as well as identifying and enhancing its role for sustainable development of human settlements;
- Creation of convenient transport connections and the use of eco-transport;
- Development of energy-safe and inexpensive methods of construction, production and distribution of appropriate building materials, production of local building materials, based on the maximum use of local resources;
- Use of recycled materials and adoption of standards and regulations in construction, with the goal of enhancing the use of low-cost local building materials;
- Recycling wastes;
- Comfortable and healthy heating (cooling) systems using radiating surfaces that transfer heat directly to a person through waves, without preheating the air;

- Supply and exhaust ventilation, providing a constant flow of clean air without the effect of draft [5].

## 3. Basic principles of ecological functioning of eco-settlements

As a result of the analysis of the world's experience and theoretical studies of ecological settlements, the basic principles of ecological functioning and development of their living environment were formulated according to the concept of sustainable development:

- *The eco-equilibrium principle* – the principle of maintaining a state of the environment, which provides for its self-regulation, the reproduction of natural air, water, soil, plants and animals by taking measures aimed at stabilizing the anthropogenic load, compensating for adverse effects on the environment and observing conditions for the equal ecological balance of all elements of the system.
- *The energy saving principle* – orientation of design decisions on the use of engineering and technical systems that work on the basis of renewable energy sources: sunlight, geothermal sources, wind energy, tides, their rational location on the territory; the use of passive methods of energy saving and bioclimatic architecture techniques based on the maximum use of microclimate regulation methods, the use of effective heat-insulating materials, etc.[6]
- *The eco-functional principle* – the principle of planning the territory taking into account the natural and climatic features of the area, the choice of a favorable location for the site, a rational approach to the planning of the territory considering orientation to the cardinal directions, the establishment of landscaped areas in the "fabric" of settlements, minimizing the area of development.
- *The psychophysiological principle* – creation of a space favorable from the psychological point of view, guided by the methods of video ecology, taking into account the peculiarities of the human visual perception of the architectural environment and the basic principles of ergonomics.
- *The traditionalism principle* – construction from ecologically sound, traditional materials, intrinsic to the locality of construction, taking into consideration spiritual, cultural, national traditions and peculiarities of the development region on the design stage.
- *The social comfort principle* – conducting activities aimed at educating the population to the ecological principles and the concept of sustainable development, encouraging the population to have a healthy lifestyle, the developing ecotourism and permaculture.

## 4. The principle of using traditional and advanced building technologies in eco-settlements design

Traditional Ukrainian national architecture, which has been formed over the centuries and aimed at the rational use of local building materials and the construction of structures that would match the properties of these materials, has been reflected in nowadays construction. The technology of constructing clay buildings has been known for over six thousand years. However, despite the simplicity and cost-effectiveness of traditional forms of housing, along with environmental and progressive trends, there remains a lot of primitive, technically imperfect and conservative structures.

Several basic construction technologies were used for residential buildings: erection of a straw bale block house, straw bale cast (clay) house and mazanka. Mazanka is the warmest kind of house among the other ones built of clay. It is also the fastest in construction, but no less laborious. In Europe, mazanka was known even before the Middle Ages. Technologies of this type have been successfully used in construction so far, mainly in the Middle and Far East, in Africa, India and China, Germany and France (half-timbered houses), and even in Italy and Spain farm buildings are made in a similar way [7]. However, in order to understand the possibility of using such traditional technologies on the territory of Ukraine in modern realities, it is necessary to analyse the positive and negative features of this method of construction. The results are presented in Table 1.

**Table 1.** Analysis of positive and negative features of traditional Ukrainian houses.

| № | Negative features | Positive features |
|---|---|---|
| 1. | Insufficient thickness of enclosing structures, which leads to disruption of the thermal and sanitary-hygienic regime of the dwelling. | Use of local building materials: clay, wood, straw, reeds, etc. |
| 2. | Construction of a clay building is a very labour-intensive task and takes a long time to dry and shrink the wall. | Maximum simplicity of building structures. |
| 3. | Lack of thermal insulation of the floor due to lack of foundation in the usual sense. | Saving of wood achieved through the use of frame-type buildings and the active use of straw and reeds. |
| 4. | The occurrence of cracks due to the use of raw wood. | Environmental friendliness of building materials and construction technologies. |
| 5. | Increased fire risk of thatched and reed roofs. | Cost-effectiveness of traditional technologies. |
| 6. | The need for constant maintenance of thatched or reed roof. | Streamlined straw roofing helps wind protection. |
| 7. | Negative effect of rainfall on enclosing structures in case of insufficient roof overhang (less than 600-800 mm). | Decorative and picturesque home decoration emphasizes traditional Ukrainian vibrant color. |

Thus, taking into account all these shortcomings and paying attention to mistakes made in the construction of traditional dwellings, it is possible to use the experience of the previous generations' experience to build ecologically sound and comfortable housing in the design of ecological settlements. However, special attention should be paid to the advanced technologies that are actively evolving and fully consistent with the concept of sustainable development in the field of architecture and urban planning. Such technologies include the construction of straw blocks, geo-blocks, super-adobes, construction with the use of a wooden frame, construction from bamboo.

### *4.1. Straw blocks*

These are structures where the elements of the walls are straw bales stacked on top of one another or that filling a wooden frame. Such buildings have good thermotechnical characteristics due to the large thickness of the outer walls and the structure of tightly pressed straw. Their main advantages are low cost of construction, short construction time, environmental friendliness, operating economy, high fire safety (after finishing with plaster).

### *4.2. Geo-blocks*

The main feature is the use of local "geo" resources (earth blocks or natural stone) in the construction. For example, the presence of stone materials, especially in hot and arid climatic zones, allows to achieve optimal temperature comfort. In this case, the light openings in the building have minimal size to avoid overheating in the daytime or overcooling at night.

### *4.3. Super-adobe technology*

A distinctive feature of this technology is that the walls of the building are erected from simple sacks filled with soil. Unlike simple adobe structures, super-adobe homes can be used in flood-prone areas. Sacks with soil are traditionally used for flood control and dam construction, as well as for the construction of defense structures. Another specific feature of this technology is that it is much more difficult to erect straight walls rather than curved or domed. This is their advantage, as dome houses are very strong. Their shape balances the external load in all directions.

### *4.4. Wooden frame*

To date, the erection of wooden buildings involves both traditional methods of construction from wood and advanced technologies. For example, round sections of tree trunks can be used to save wood as the main wall material. This is considered a waste-free way of constructing walls.

Unlike straw and clay blocks, the use of wood as a constructive and enclosing material allows the creation of large-scale structures of different types. At the same time, the necessary level of comfort is achieved both due to the high thermotechnical characteristics of the wood itself, its aesthetic properties, and due to the spatial solution of the building.

### *4.5. Bamboo buildings*

Bamboo houses have been erected in regions with high seismic activity (mainly Southeast Asia) from time immemorial. The plant has tremendous flexibility, and in the case of an earthquake, tsunami or other natural disaster, the houses of this material are often simply bent, but not deformed like other structures. And even if the bamboo house falls when rocking, it will not cause death to people inside and near it because the material has a very light weight. In terms of environmental friendliness, bamboo is the leader among all kinds of wood known to the mankind. This plant grows so quickly that there is no need to use any chemical additives and pesticides when cultivating it. In addition, the bamboo stems contain unique components that have antiseptic and antibacterial properties [8, 9].

## 5. Technogenic principles of ecovillage design

To date, renewable energy sources are being actively used in the design of energy efficient districts and settlements. One of the major advantages of alternative energy is its environmental friendliness: the process of obtaining energy from renewable sources is not accompanied by the formation of environmental pollutants; does not lead to the destruction of natural landscapes; virtually eliminates hazards to biological substances; does not threaten ecological balance of ecosystems [10].

The following buildings can be distinguished by their orientation towards usage of different natural energy sources:

- solar-energy buildings (effectively using the energy of the sun);
- wind energy buildings;
- geo-, hydro- and aerothermal energy buildings;
- buildings with combined use of different natural energy sources [11].

Passive and active energy-saving technologies are also used when designing homes within eco-settlements.

Passive energy saving means the complex application of measures that help to keep heat, provide natural ventilation, cooling and insolation due to the natural properties of materials, structures and physical processes. Its essence is that the residential or non-residential premises have zero consumption of heat and electricity, i.e. there is an opportunity to control the heat loss and the excess heat indoors.

Passive or smart homes are self-sufficient in terms of heat and electricity generation through modern technologies that allow for absolute thermal insulation, by receiving electricity from solar or wind energy. The passive house also implements devices that allow efficient use of natural sunlight and produce fresh air inside the building, while maintaining the temperature balance in the premises. Passive home heating is due to the heat generated by the people living there, household appliances and alternative energy sources.

Active energy saving is defined as the product of sustainable change through the measurement, monitoring and management of energy use. In order to ensure active energy savings, it is necessary not only to install economical devices but also to control them only using the required amount of energy.

It is the control factor that is critical to maximizing efficiency. Currently, in the computer and telecommunications business there is a concept of "intelligent building". An intelligent building is a building with a single cable architecture that enables the circulation of the whole information flow:

telephony, LAN, video and other data down to the large life support and building management systems [12].

In Ukraine, it is more common to use passive energy saving systems through the use of energy-saving natural materials. Active energy saving systems (solar and wind energy, etc.) are more common in Europe and the Americas (Table 2).

**Table 2.** Energy saving systems in world ecovillages and ecovillages of Ukraine.

| № | Name and location | Year of foundation | Energy efficient technology |
|---|---|---|---|
| **1.** | Auroville, India. | 1968 | Utilization of solar and wind energy, technology of construction from the compressed soil. |
| **2.** | The Farm. Summertown, USA. | 1971 | Photoelectric elements, wind turbines, use of secondary resources, micro-hydroelectric power plant. |
| **3.** | Soldiers Grove. Clayton, Wisconsin, USA. | 1973 | Use of solar energy, biogas. |
| **4.** | Findhorn. Moray, Scotland, United Kingdom. | 1980 | Utilization of wind and solar energy, zero-emission construction. |
| **5.** | Okologisk, Landbisamfund. Denmark. | 1983 | Geothermal heat pump, restricting entry of cars into the territory. |
| **6.** | Crystal Waters, Australia. | 1985 | Passive solar heating, natural materials, solar panels. |
| **7.** | Ecolonia. Aalphen aan den Rijn, The Netherlands. | 1991 | Use of passive and active solar energy. |
| **8.** | Solar Village in Fefki. Lykovrissi, Greece. | 1991 | Active and passive solar systems, renewable heating systems. |
| **9.** | Anningerblick. Guntramsdorf, Austria. | 1992 | Passive solar energy, resource-saving building materials. |
| **10.** | Source Valley (Долина Джерел). Kyiv region, Ukraine. | 2003 | Construction from ecological materials. |
| **11.** | Rainbow (Радуга). Ukraine | 2010 | Use of solar energy, construction of ecological materials. |
| **12.** | Chmyrivka. Cherkasy region, Ukraine. | 2010 | Construction from natural materials. |
| **13.** | New Mayapur. Dnipropetrovsk region, Ukraine. | 2006 | Construction from natural materials. |
| **14.** | Green cliffs. Kyiv region, Ukraine. | 2012 | Straw construction. |

The *technogenic principle* also considers the eco-design of buildings as an integral part of eco-habitation in terms of use of various technologies in buildings, such as alternative energy sources (solar panels, wind generators, etc.), and their direct impact on the form and composition of structure.

In modern architecture, the introduction of wind turbines directly into the facade of the building is actively used. This was made possible after engineers created attractive, silent and compact windmills with sophisticated bending of the blades [13].

Also, a significant contribution to the formation of the three-dimensional and compositional structure of architectural structures was made by the placement of solar batteries and collectors on the facade and their introduction directly into it.

However, there is a need to differentiate the purpose of these technologies in the context of different architectural objects. There are two main principles:

- aesthetic purpose;
- practical purpose.

In this context, *aesthetic purpose* involves the use of alternative energy sources as decorative elements in most cases, even though these renewable energy sources do not cover the building's energy needs.

The *practical purpose*, as opposed to the aesthetic one, is to supply the building with necessary amounts of energy and, in some cases, even produce excess energy that could later be used to supply the neighbouring buildings [14,9].

## 6. The principle of sustainable transport developing

There is a number of transport types that can be referred to as the environmentally friendly means of transportation.

All eco-transport can be roughly divided into several types:
- electric;
- hydrogen;
- biofuel, independent of fossil fuels;
- bicycle transport, etc.

The term eco-transportation should be understood in the sense that it tends to minimize the impact of humans on the environment through vehicles' usage.

The main idea of eco-transport is to abandon internal combustion engines and to introduce electric vehicles. The main difference between electric cars and internal combustion engines is the ability to use different sources of energy: from burning fossil fuels at thermal power plants, using nuclear energy and renewable energy to various batteries. However, the main drawback of electric vehicles is the undeveloped gas station infrastructure. Electric cars can also use hybrid power supplies, where electricity is produced by combustion of diesel and gasoline [15].

However, electric vehicles are not limited to passenger cars. These are also *include* trains, airplanes, planes, seagoing vessels, motorcycles, scooters, etc.

Main characteristics of sustainable transport are listed below:

*in the field of natural environment:*
- limitation of emissions and wastes (polluting air, soil and water) to ensure the environment's ability to absorb / recycle / clean the reduced pollution;
- ensuring the functioning of vehicles at the expense of renewable or inexhaustible energy sources;
- utilization of natural resources used in vehicles and infrastructure (such as steel, plastic, etc.).

*in society:*
- ensuring the necessary level of accessibility for people and their property of this generation and all future generations;
- promotion of human health;
- assistance in maintaining a high standard of living;
- setting the noise penetration limit below the level accepted by society;
- security for people and their property.

*in the field of economy:*
- financial affordability for each generation;
- maximization of economic efficiency and minimization of economic costs in design and operation;
- support of a strong and diverse economy.

## 7. Conclusions

As a result of the analysis of the experience of designing eco-settlements abroad and in Ukraine, the basic principles of designing eco-settlements and their structural elements have been identified, taking into account the use of natural materials, energy efficient technologies and sustainable transport in their construction. It was found that it is possible to get a fairly economical and simple construction technique based on the existing one, considering all the shortcomings and using the positive features of traditional

Ukrainian architecture. With this goal in mind, it is possible to correct the basic errors by following steps: arrangement of high-quality tape foundation, protection of the adobe from the capillary suction of the plinth, the arrangement of a steep scaffold, the construction of walls with a thickness of not less than 500 mm, the attic insulation strengthening, the organization of a roof overhang of not less than 600-800 mm, and the employment of highest quality home maintenance techniques. The design of eco-habitats should also take into account the specific features of the local climate, the availability of certain local building materials, and subsequently choose technology that is appropriate in each particular case. At the same time, it is necessary to pay close attention to the advanced technologies which are actively developing and modernizing without losing their environmental friendliness.

The study also found that most Ukrainian eco-settlements significantly differ from the US and European eco-settlements, particularly by their inferior level of technological support, and are also characterized by a fairly low level of transportation support. These findings impose certain priorities for architects and urban planners of eco-settlements – namely, to take all the shortcomings into account and to correct deficiencies in order to bring eco-settlements in Ukraine to a higher level and ensure a decent standard of living for people.

**References**

[1] Ardzijauskaite V 2009 *Ecovillages: is it a way to reach environmental sustainability? Case studies in Denmark* (Denmark: Aalborg University) p 95
[2] Hollick M and Connelly M 1999 Learning from ecovillages worldwide *Communities* **104** 62-65
[3] Bates A 2003 Ecovillage Roots (and Branches) *Communities Magazine* **117** 10
[4] Grigor'ev V and Ogorodnikov I 2001 *Greening cities in the world, Russia, Siberia* (Novosibirsk: GPNTB SO RAN) p 152
[5] Maslov N 2003 *Urban planning ecology* (Moscow: FGUP Graduate School) p 285
[6] Mahlabani Y, Shahsavari F and Motevali Z 2016 Eco-village, amodel of sustainable architecture. *J Fundam Appl Sci* **8(3S)** 1835-1847
[7] Daniljuk A 1991 *Ukrainian hut* (Kiev: Scientific Thought) p 110
[8] Pecherceva O 2015 The traditional principles of structural elements disigning in the ecovillages, *Improving the organization of traffic and the transport of passengers and goods* (Minsk: BNTU) pp 46-53
[9] Blagovestova O and Pechercev O 2019 Using traditional and modern construction technologies in design of ecological settlements *Scientific Bulletin of Civil Engineering* **4** 5-10
[10] Selivanov N, Melua A and Zokoley S 1988 *Energy-active buildings* (Moscow: Stroyizdat) p 376
[11] Lapin J 2005 *Autonomous ecological houses* (Moscow: Algorithm) p 416
[12] Tetior A 2008 *Architectural and construction ecology* (Moscow: Academy) p 368
[13] Tabunshchikov Y, Brodach M and Shilkin N 2003 *Energy-efficient buildings* (Moscow: ABOK-PRESS) p 200
[14] Guk V and Pechertseva E 2014 Trends in the formation of energy-efficient eco-settlements, *Improving the organization of traffic and the transport of passengers and goods: a collection of scientific papers: based on the results of the annual International Scientific and Practical Conf.*, Minsk: BNTU) pp 41-47
[15] Kondrat'ev A 2012 The role of green transport in supporting sustainable urban development *Theory and practice of social development* **4** 342–44

# THESAURUS OF PARAMETRIC FORMATION PARADIGM ARCHITECTURAL SPACE

**D Sopov[1], E Protsenko[2], V Mironenko[1]**

[1]Department of Design of Architectural Environment, Kharkiv National University of Civil Engineering and Architecture, Kharkiv, Ukraine
[2]Department of Graphic, Kharkiv National University of Civil Engineering and Architecture, Kharkiv, Ukraine

dcopov93@gmail.com

**Abstract.** The realities of the existing ecological crisis raise the question of the need to create new methods in the approach to the design of an architectural habitat adapted to the surrounding nature. The paper discusses the principles and approaches of biomimetic design of the architectural environment, taking into account the ecological state of nature in Ukraine. At the same time, we are talking about the creation of a completely isolated, self-sufficient, self-contained ecosystem that will fully meet all the necessary requirements for the biological functioning of the human body and includes the natural environment with residential objects harmoniously embedded in it. The solution to these questions is possible with the integrated use of modern design methods in an adaptive architecture environment.

## 1 Introduction

The rapid development of technology leads to the inevitable change in the space around us. At the dawn of his existence, man sought to adapt to his environment, used natural shelters, existed in unity with nature. However, from the moment the first artificial dwelling was created, he became convinced of the possibility of protecting his existence by isolation from the environment. In our opinion, this was the source of the contradictions between man and nature, the depth and scale of which increased with the development of human civilization. In the middle of the twentieth century, these contradictions acquired a global, deadly nature for humanity [1, 2].

Destructive processes, already launched by nature on a planetary scale, began to cause concern not only among esotericists and members of religious sects but also among the governments of most countries of the world [3]. The increasing frequency, destructive power, and, most importantly, the factual impossibility of a long-term forecast of the occurrence and prevention of climatic collisions are causing increasing concern.

The central and eastern part of Ukraine is currently in illusory security, but the state of strategic waterways of the state, the critical pollution of small rivers (water resources), the ongoing process of burning and decay of peat deposits and the forest fires provoked by them are already causing great concern. Recent information about smog hanging over the capital of the state, the global coronavirus pandemic, the threat of activation of seismically dangerous zones, which are geographically tightly close to Ukraine, require a very serious and responsible attitude to the issue of environmental degradation of our habitat. Professional environmental analysts generally raise the question of the survival of the human race on our planet in the framework of the familiar and necessary conditions of existence [4, 5].

IOP Conf. Series: Materials Science and Engineering **907** (2020) 012009 doi:10.1088/1757-899X/907/1/012009

One of the important environmental problems is the rapid growth of urbanization and the global technicalization of society, which entails an increase in the imbalance between nature and the human environment. The adaptation of man to an artificial urban environment contributed to the transformation of his perception of nature itself as a resting place, and not a living environment. Figure 1 shows a diagram that characterizes the degree of environmental pollution of cities in Ukraine.

A distinctive feature of our time is the ever-growing dynamism of society, associated with the high pace of development of new technologies, expanding the scope of people's activities, increasing migration, population mobility, etc. According to sociologists, the structure of modern society has not finally formed and is in very complex dynamics, there is a fuzzy and vague social boundary, mobility of the vertical section of society when a person under the influence of various factors can change his social status at different periods of life.

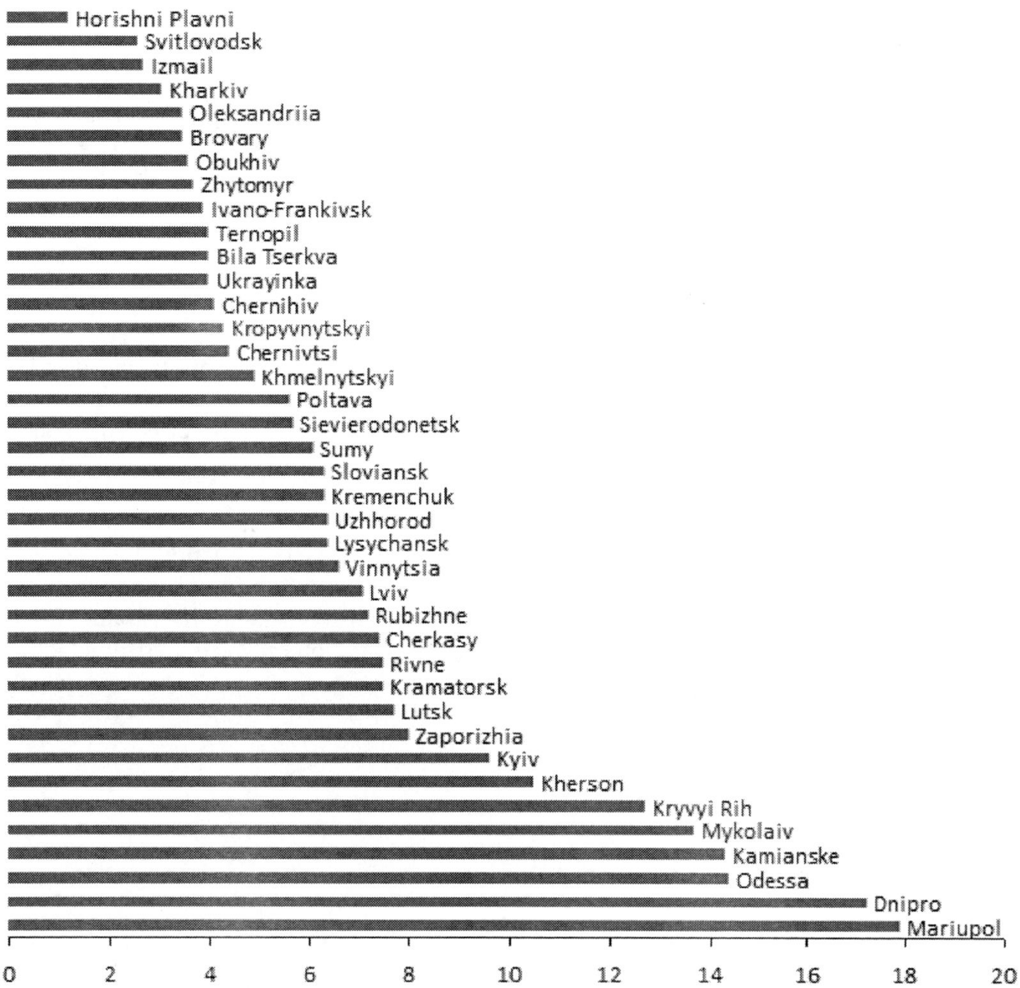


**Figure 1**. Indicators of environmental pollution of Ukrainian cities (according to the Central Geophysical Observatory) [6]

Since the issue of caring for the environment on a planetary scale looks like a utopia, a local solution to the problem of human survival in an already unfriendly environment is necessary.

**Prerequisites for the creation of ecosystems**

The realities of the existing environmental crisis raise the question of the need to create new methods in the approach to designing an architectural habitat adapted to the surrounding nature. Unfortunately, this new habitat can no longer be called simply the term "housing" in the usual sense of the word. It should be a completely isolated, self-sufficient, self-contained system that will fully comply with all the requirements for the biological functioning of the human body.

We are talking about creating a new ecosystem, which includes the surrounding natural environment with harmoniously embedded residential objects in it.

Such an ecosystem should be considered as a model of aggregates and relationships between its components:

- the functioning of all artificial objects of the system should be carried out in a closed cycle;
- creating a cycle of continuous wastewater treatment to reuse them as a renewable resource;
- disposal of solid waste using non-energy-intensive technologies and their reuse as raw materials for new building materials, fuels, fertilizers;
- providing water and food in the required amount utilizing mainly domestic resources;
- the energy supply of all technological processes due to alternative renewable energy sources.

At this stage of the creation of the project, in our opinion, the need arises to pose and resolve the following issues:

1. an adaptation of architectural objects to the environment due to the variability, flexibility, mobility of the architectural form, where continuous in time transformation becomes one of the types of their existence;
2. reliable life support due to the most economical consumption of available natural resources, environmentally friendly disposal of waste through their processing and reuse;
3. the correct choice of tools and methods at all stages of the design of such complex systems;
4. the selection of a reliable contractor for the implementation of the design and construction phases with the provision of their modern building materials.

The solution to these questions is possible with the integrated use of modern design methods in an adaptive architecture environment.

**Adaptive architecture as a means of ecosystem formation.** The search for an architectural form depends on the functional purpose of the structure and the area of the construction site. In our study, we focused on the natural environment.

In connection with the tendency of a steady increase in air temperature in Ukraine, the usual geographical boundaries of climatic zones are changing: the zones of arid territories from Odesa and Kherson regions are expanding towards the center of Ukraine (Fig. 2), the amount of precipitation in the Carpathian region and all western regions is growing (Fig. 3).

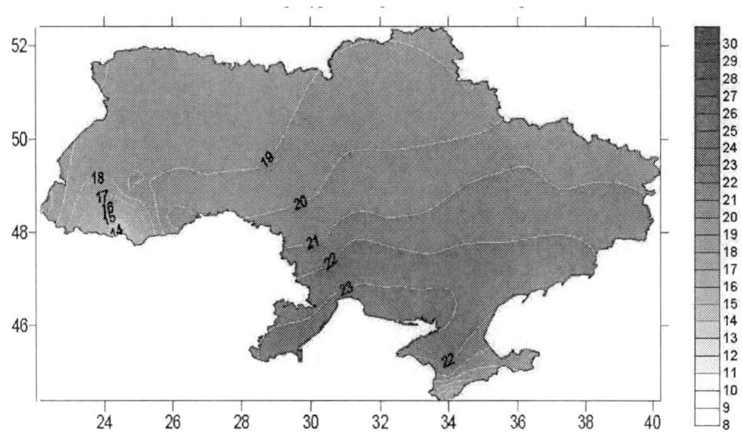


**Figure 2.** Air temperature in July 2019, Ukraine [6]

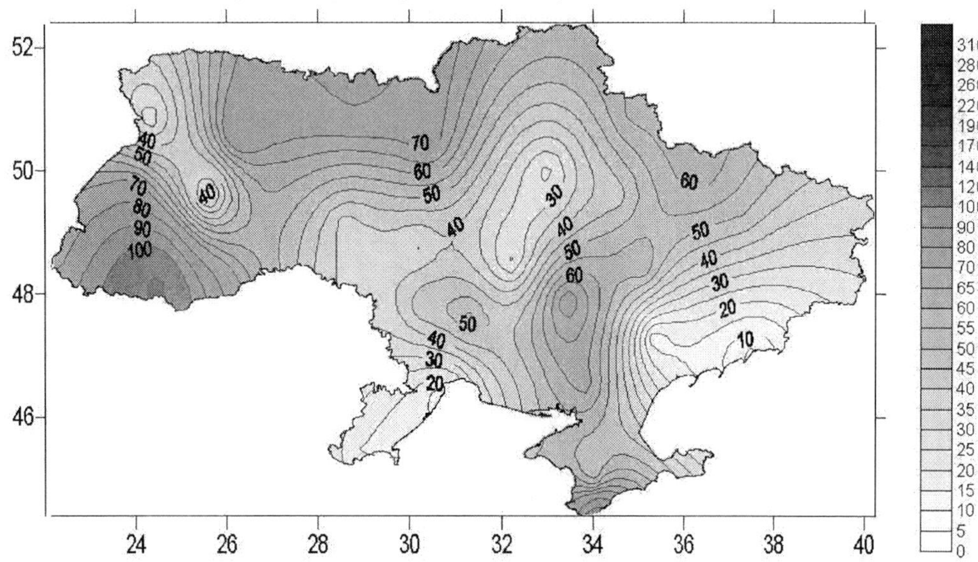


**Figure 3.** Amount of precipitation in July 2019, Ukraine [6]

One of the promising areas in the field of shaping in adaptive architecture is biomimetics - a method of creating architectural objects when borrowing ideas from wildlife. With a biomimetic approach, architects do not just copy natural images but carry out a detailed analysis of the principles of the living world and the harmonization of natural and architectural environments [7].

Inspiration in nature was found by many famous architects and designers - Alvar Aalto, Antonio Gaudi, Bruce Goff, Buckminster Fuller, Gustave Eiffel, Zaha Hadid, Imre Makovets, Rudolf Steiner, Santiago Calatrava, Hugo Hering, Fry Paul Otto, Frank Lloyd Raynite, and etc.

Architects have long used biomimetic principles in many structures. At the same time, they do not just copy the external form of natural analogs but take into account the nature of their structure, specific properties. Examples of this are shown in Fig. 4.

Biomimetic design principles can solve a number of urgent problems in architecture: the artificial habitat becomes more resistant to external influences, more comfortable and convenient for humans, and less harmful to the environment, more expressive. The formation and development of adaptive abilities of the artificial environment based on the study of ontogenesis and phylogenesis of living organisms occupy a special place when using biomimetic principles (Fig. 5).

The biomimetic concept of shaping is based on the possibility of practical associative application of the ideas of nature in architectural design to create effective ecosystems harmoniously integrated into the environment. Much attention is paid to the study of evolutionary processes during which living organisms adapt to changing environmental conditions. Since biomimetic design at different levels of the ecosystem takes into account organizational, production, and technological life support strategies, it refers to sustainable architecture.

| Construction | Natural analogue | The nature of the analogy | The nature of borrowing |
|---|---|---|---|
|  High-rise pipes |  Stalks of cereals | The sclerenchymal strands of the plant stem play the role of longitudinal reinforcement. The internodes of the stems are rings of rigidity. Oval vertical voids are located along the walls of the stem. | Subconscious: The complete analogy was determined after the pipes were created. |
|  Eiffel Tower | 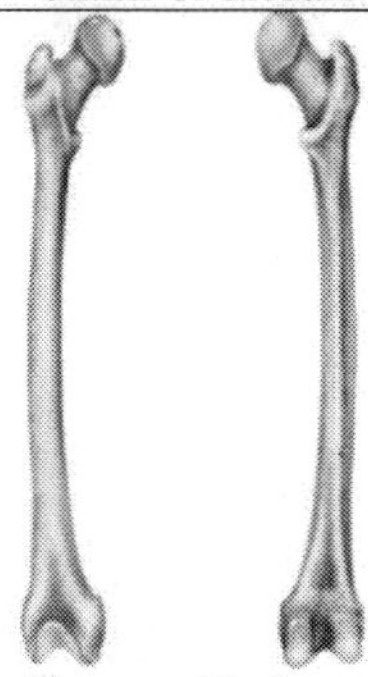 Human shin bone | Spongy tissue structure in the femoral head. The relationship between the formation of bone structure and external stress. | Conscious: Anatomy professor Hermann Von Meyer identified a network of ossicles with a strict geometric structure in the femoral head. Engineer Carl Cullman has theoretically substantiated load distribution. Natural load distribution using curved calipers was used by Eiffel. |
|  Funicular Nordpark, Innsbruck, Austria, arch. Zaha Hadid | 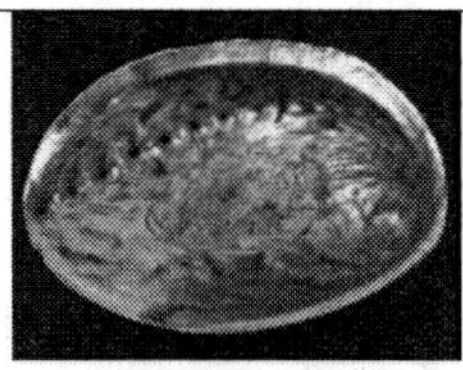 Sea clam shell | The layered structure of the shells ensures their resistance to external influences. | Conscious: The use of layered composite materials allows the use of complex architectural forms and ensures their structural reliability |
|  Vertical bionic city-tower arch. Maria Rosa Cervera and Javier Pioz |  Cypress | The shape and structure of the cypress crown, the structure of the root system. | Conscious: based on the growth and development pattern of cypress, the heat-regulating function of the skin for the outer cover. |

**Figure 4.** Examples of the use of biomimetic principles in architecture

If we consider the systematization of the borrowing of biomimetic principles, now it is worth highlighting the system developed by Janine Benius, an American researcher in the field of biomimetics, and supplemented by Pedersen Dawn. This scheme consists of two cross-systems, one of five types of borrowing - form, design, material, process, and function - and the other of three different levels of borrowing - the organism, behavior, and ecosystem. It turns out that this system allows you to comprehensively evaluate the borrowing of natural characteristics in an architectural project.

Types of borrowing reflect the elements of wildlife that are studied, analyzed, and used with the necessary characteristics. Borrowing levels reflect which particular object or group of objects has been studied. At the level of the organism, living organisms are studied, their structure, behavior, and vital functions. At the level of behavior, what has been done by a living organism, for example, to build a house or other structure, is investigated.

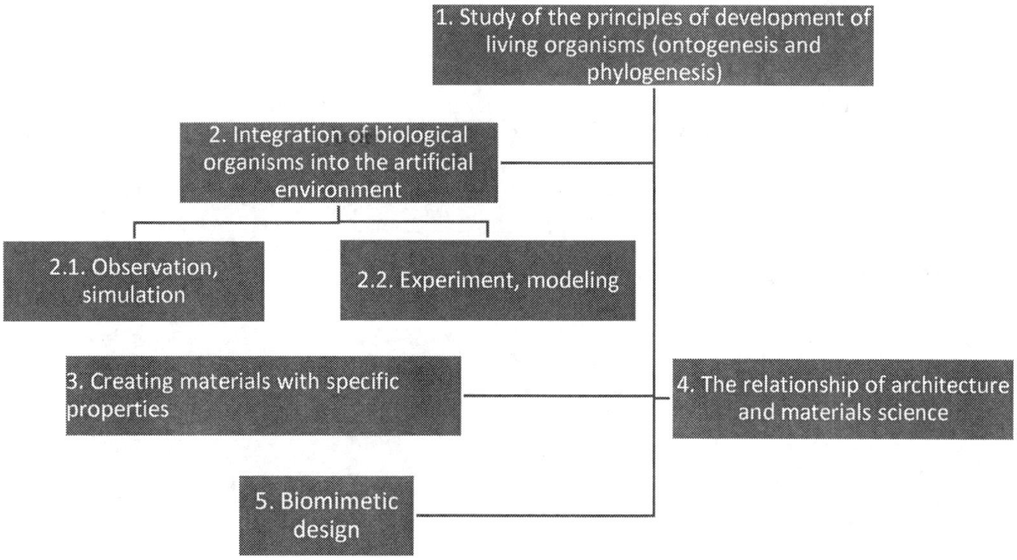


**Figure 5.** Biomimetic design principles

At this level, the shape of the natural object, the design features, the qualitative characteristics of the material, the principles of its functioning are also investigated. The third level, which involves the study of the ecosystem, explores complex natural systems, that is, the interaction of various living organisms and the environment itself as, on the one hand, the conditions for their interaction, and, on the other, its result. The third level is the most difficult to study and requires close cooperation of natural scientists and architects to achieve maximum effect. But at the same time, it is most effective in achieving a sustainable and comfortable architectural environment that can exist in harmony with nature.

**Biomimetic design solution analysis**

The artificial environment is becoming increasingly responsible for global environmental and social problems associated with the huge amount of waste, materials, and energy, as well as the greenhouse gas emissions associated with the habitats that people have created for themselves [8, 9]. It is becoming increasingly apparent that changes need to be made to how the created environment is created and maintained. Simulating the complex interactions between the living organisms that make up ecosystems is a robust perspective for human habitats

By the end of the 20th century, architectural composition finally got rid of the classic constraints of a global geometric prototype, symmetry and proportion. The compositional process can now be almost completely open to the final configuration of the project.

Computer technology has radically changed the nature of design, consistently shifting the emphasis from 2D drawings to 3D information models. Modern design includes comprehensive linking and comparison of information not only in the development and design of an architectural object, but also in the process of forming its original image.

The parametric approach requires new design thinking and deep knowledge of computer programs. This approach is necessary to create complex geometric and fundamentally new, unexpected shapes. The symbiosis of the architect's creative thinking and the capabilities of digital virtual reality modeling technologies allows us to take a fresh look at rethinking the role of form in architecture in the context of creating a comfortable and safe spatial environment.

A distinctive feature of the activities of researchers and practitioners in the framework of the parametric orientation is:

- the use of methods of interactive computer design and morphogenetic methods of modeling the architectural form;
- the application of complex computational geometry using advanced design techniques such as Tel-script or Rhino-script (Figure 6) and parametric modeling [10-12];
- creation of architectural forms capable of responding to changes in conditions without current human control, with the correction of structural and planning decisions.

**Figure 6.** Examples of creating complex geometric shapes using Rhino-script

Geometric modeling is the process of sequential creation of the structural geometry of an architectural object, based on the specification of an information array. Geometrization of an object, spatial design in three-dimensional modeling are necessary components for solving problems of parametric architecture, which can be used in the practice of designing various objects and which make full use of modern information technologies.

## Conclusions

The discourse proposed in this article is still a theoretical prerequisite for the creation of ecosystems based on the use of biomimetic design, which requires testing their ability to function in a sustainable and even restorative manner. The use of systemic biomimetics to simulate the functioning of ecosystems will allow for a deeper understanding of the nature of interaction in the "man-environment" system.

The main challenges faced by parametric architects are:

1. - lack of customer knowledge;
2. - insufficient funding;
3. - difficulties in production (many manufacturers are not ready to solve complex non-standard tasks)
4. - difficulties in mastering the software product.

Parametricism opens up wide opportunities for introducing innovative developments in architecture in related areas of knowledge, technical capabilities of new structures and materials, while it is necessary to work with complex information models of architectural objects based on computer technologies.

**References**

[1] Garnier L 2008 *Man and nature: making the relationship last* (UNESCO, Paris) p 142
[2] Mann D 2011 Natural world contradiction matrix: How biological systems resolve trade-offs and compromises *Procedia Engineering* **9** p 714–723
[3] Wuebbles D J, Fahey D W, *et all* 2017 *Executive summary* In: *Climate Science Special Report: Fourth National Climate Assessment* Volume **I** (U.S. Global Change Research Program, Washington, DC, USA) p 12-34 https://doi.org/10.7930/J0DJ5CTG
[4] Frick J 2017 On the survival of humanity *Canadian Journal of Philosophy* **47** (2–3) 344–367 http://dx.doi.org/10.1080/00455091.2017.1301764
[5] Arora N K 2018 Environmental Sustainability – necessary for survival *Environmental Sustainability* **1** (1–2) https://doi.org/10.1007/s42398-018-0013-3
[6] *Central Geophysical Observatory named after Borys Sreznevsky* 2019 URL: http://cgo-sreznevskyi.kyiv.ua
[7] Gruber P 2011 *Biomimetics in Architecture: Architecture of Life and Buildings* (Springer-Verlag/Wien, Austria) p 276 doi: 10.1007/978-3-642-11934-7_7
[8] Doughty M R C, Hammond G 2004 Sustainability and the built environment at and beyond the city scale *Building and Environment* **39**(10) 1223-1233 https://doi.org/10.1016/j.buildenv.2004.03.008
[9] Singh A, Nayyar N 2015 Biomimicry-An Alternative Solution to Sustainable Buildings *Journal of Civil Engineering and Environmental Technology* **2** (14) pp 96-101
[10] Stavric M, Marina O 2011 Parametric Modeling for Advanced Architecture *International Journal of Applied Mathematics and Informatics* **5** (1) pp 9-16
[11] Philip B, Williamson S, Woodbury R 2006 Parametric Modelling as a Design Representation in Architecture: A Process Account *Canadian Design Engineering Network Conference* (Toronto: July 2006)
[12] Schnabel M A 2007 Parametric Designing in Architecture. In: Dong A, Moere A V, Gero J S (eds) *Computer-Aided Architectural Design Futures (CAADFutures)* (Springer, Dordrecht) https://doi.org/10.1007/978-1-4020-6528-6_18

IOP Conf. Series: Materials Science and Engineering **907** (2020) 012010 doi:10.1088/1757-899X/907/1/012010

# Types of ecological relationships in the system "nature, artificial environment, humans". Peculiarities of modern eco interior development

**Svitlana Zymina**

Kyiv National University of Construction and Architecture, Architecture department, Department of the basics of architecture and architectural design, 31, Povitroflotsky Avenue, 03037, Kyiv, Ukraine

svetlanabzymina@gmail.com

**Abstract.** The article addresses the main ecological trends of the environmental protection in the field of architectural design and construction, which are united under the common name of eco-construction. The article analyses various types of environmental relationships according to the degree of harmfulness of their impact on the artificial environment, nature and humans because of their construction activities. A graphical model is proposed, which clearly demonstrates how ecological relations (harmless, conditionally harmful and harmful) interact with nature, humans and the artificial environment. The human being as a component in the chain of relationships is present in all these spheres; in the artificial environment factor - as motivation and the main driving force, and in the natural factor - as part of its creation, it always balances between nature and the artificial environment. The search for a solution to the global environmental problem has led to the emergence of such concepts as ecological interior and eco-style. Based on the author's classification, the peculiarities of the formation of a modern ecological interior are explained. The difference between the concepts of ecological interior and ecostyle is revealed, a new kind of ecostyle is proposed in the classification of eco-directions - pseudo-ecostyle, its features are considered.

## 1. Introduction

The issues of the relationship between humans and nature are the subject of study of such a scientific discipline as social ecology. Social ecology studies the interaction and relationships of human society with the natural environment and is based on the belief that almost all our modern environmental problems come from rooted social issues [1]. However, the qualitative difference between a human being and other creatures is that it is a human being who has these social problems. Moreover, only humans, besides a purely biological, direct metabolism and energy interchange with the environment, have a different type of exchange based on the results of labor activity and material production. The architectural field of activity is the main and the most extensive component of material production in the creation of an artificial human environment.

As a result of burning of all types of fuel (gas, petrol, firewood) the concentration of $CO_2$ is growing. Humans disrupt the carbon dioxide balance by mass deforestation, high industrial activity, increased transportation, urban growth. The Scripps Institution Oceanic Institute has published the latest $CO_2$ measurements taken daily since 1957 on the Mauna Loa volcano. Over the past 100 years, the concentration of $CO_2$ has increased by 40%, this is the highest level in 650 thousand years. To date, the level of $CO_2$ in the Earth's atmosphere has reached the highest levels [2]. The growth of $CO_2$ levels leads to global warming, since carbon dioxide, methane, water vapor - greenhouse gases, are the cause of an increase in the global temperature on the planet. This leads to the following: the melting of glaciers, an increase in the number of natural disasters (floods, hurricanes, typhoons, droughts), the extinction of the entire populations of insects, plants and animals. A human being is at the end of this chain and is directly connected with all these processes, continuing to excessively consume and increase in number. It is assumed that in the future the absolute amount of goods will decrease, and the number of buildings will double by the middle of the century, which will lead to the rapid growth of cities. Cities sprawl, capturing arable land, destroying forests and water sources, emitting even more $CO_2$ into the atmosphere.

To stop, or at least to slow down the population growth on the Earth is problematic, if not to say, impossible. The same goes for calls to reduce consumption. There is only one way out - to direct the efforts of mankind to the invention of technological innovations (the use of renewable energy sources and closed environmental production cycles).

## 2. The specifics of ecological relations between the artificial environment, humans and nature

The ecology of the Earth was taken notice of only in the middle of the last century, and visible efforts were beginning to be made only about 50 years ago. If we compare this half-century with the time span of human existence on the Earth, during which we mercilessly tormented her natural resources, it becomes clear that it is extremely difficult to overcome all the negative outcomes of human activity in the time allotted by nature. Fortunately, the humankind is already beginning to understand that taking steps in this area is vital.

In the contemporary world, scientists of all disciplines conduct researches on the problem of reducing the load of human activities from the natural ecosystem. Let us consider the main ecological areas of environmental protection in the field of architectural design and construction, combined under the general term ecological building, or sustainable building. Its goal is to reduce the level of energy and material resources consumption throughout the entire life cycle of the building: from choosing a site for design, construction, operation, to repair and demolition [3]. At the same time, it is necessary to maintain or even improve the quality of buildings and the comfort of the internal environment. Sustainable building extends and complements classical building design with the concepts of economy, usefulness, durability and comfort with a minimum load on the ecosystem. Among the techniques for designing and operating sustainable building, the following are the most common:

- subsurface buildings (energy costs for heating are reduced since the earth keeps more or lessstable temperature throughout the year);
- smooth, streamlined forms of construction (this ensures savings in operation due to a decreasein the building's resistance to wind load);
- the use of alternative and renewable energy sources (as opposed to oil, coal or gas, they are practically inexhaustible, accessible and do not harm the environment);
- landscaped roofs, facades and interiors (except for the original design, landscaping protects against overheating in summer, plants on the surface of walls and the roof of the house contribute to creating the most comfortable indoor microclimate in any weather, moisturize the air and clean it of harmful elements, improve the quality of the environment by filtering water and air);
- the availability of equipment for the collection, processing and secondary use of rainwater (water scarcity is a global problem for mankind, over 25% of the world population suffers from a shortage of drinking water [4];
- optimal thermal insulation and orientation (a well-developed system of insulation and orientation with respect to the cardinal directions of the house will save on heating in winter and spend less on air conditioning in summer);
- modern technologies that provide for the reduction of electricity consumption during the construction and operation of a building.

The listed techniques have their advantages and disadvantages. Innovative technologies in sustainable building are being developed and improved. We hope that soon the feasibility of energy saving techniques will increase without harming consumers and, most importantly, the environment.

Considering the problem of conservation of natural resources from the point of view of the relationship between humans, nature and the artificial environment, we conclude that the known methods of energy conservation have different effects on the nature and mankind. If we try to depict ecological relations and bring everything to a single model, we will get a graphic model of the relationship between nature and the artificial environment. (Figure 1). A human being as a component in the chain of relationships is in all of the listed points; on the part of the artificial environment - as motivation and the main driving force; on the part of the natural factor - as a piece of its creation.

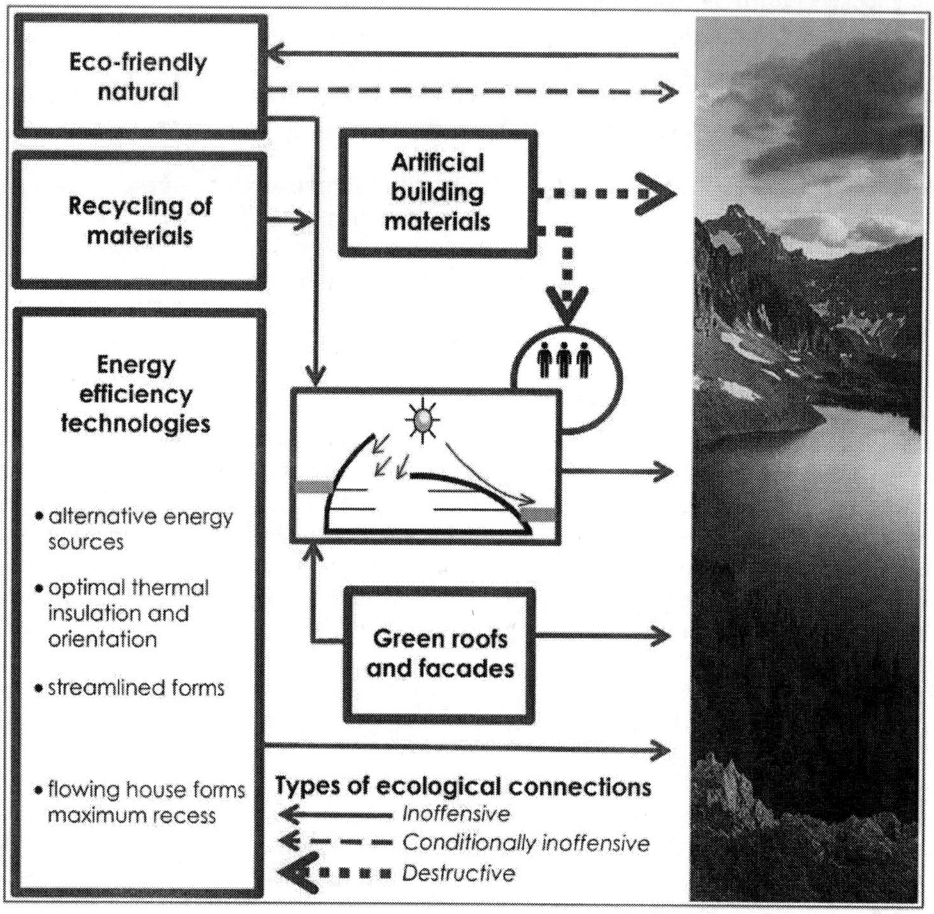


**Figure 1.** The model of ecological relationships between the artificial environment, humans and nature.

Thus, a human being balances between nature and the artificial environment created for herself. She cannot exist without them. Moreover, both environments must be loyal to her, that is, not to inflict harm by their impact. Hence the expression "environmentally friendly natural material". The term can be applied both to natural material and to a product from it. The oldest and the most widespread example of human interaction with nature is to take what it can give us (building material, water, air, etc.) and use it for our life. It would be ideal if, after utilization, nature would receive the same clean, non-polluted waste. With a reasonable balance of consumption, humanity would not have environmental problems. However, a couple of centuries were enough for a human being to disrupt the natural balance and face the threat of itsexistence. And yet, using sufficient and necessary efforts to clean up the waste of their life, humanity can return to nature at least conditionally harmless waste.

The term “conditionally harmless” waste should be understood as the waste that will not cause any significant harm to the nature, the nature will cope with it over time, using natural self-cleaning. The situation can be improved by reusing building materials. This will save both natural and energy resources.

The next block of energy efficiency techniques is the latest technologies that allow the conservation of natural non-renewable energy sources (gas, shale, peat). Among the list of the most common methods of energy efficiency are: alternative energy sources; modern construction technologies; optimal thermal insulation and orientation; smooth, streamlined forms of construction; maximum recess. The last three depend as much as possible on the will of the designer and will dictate the final architectural solution. Two more techniques from the list of “green technologies” should be considered separately. This is the maximum use of sunlight and landscaping of facades and roofs. Both techniques also actively influence architecture, but at the same time they have the maximum indicator of harmlessness of impact on the natural environment and even strengthen the positive natural factor. In addition to the decorative effect, plantings improve the indoor climate in any weather, increase air quality, saturating it with oxygen and purifying it from harmful elements. The undoubted advantages of vertical gardening are thermal insulation and sun protection, which positively affect the level of power consumption of the building, as well as a significant reduction in noise load [5].

Let us turn to the most controversial method - the use of artificial building materials and products. Artificial materials include brick, cement, reinforced concrete, glass, etc. They are obtained from natural and artificial raw materials, by-products of industry and agriculture using special technologies. During the production of artificial materials, a natural component is also involved, but in much smaller quantities, which means with less damage to the natural environment. It would seem that the ideal option was found, however, it should be remembered that some artificial materials can emit toxic substances during operation, and the production of such materials is based on the release of waste harmful to nature. Any effort to reduce the harmful factor leads to an increase in the cost of production, which is almost impossible in a market economy. In addition to the environmental awareness of the manufacturer, it is necessary to have strict legislation governing this area of production. Let us hope that in the nearest future the new technologies will appear that will make it possible to produce artificial materials without harmful effects on both humans and the environment.

## 3. Varieties of eco-directions in interior design

However, it should be noted that the harmless human impact on nature as a result of its construction activities is only part of the global environmental challenge. At the center of all efforts and thoughts is the human herself, with her desire not only to not harm the environment, but also to be in an environment that does not have a harmful effect on her health. Such an environment is the interior - the internal area of the building, the architectural and artistic design of which provides a person with aesthetic perception and favorable living conditions [6]. Interior components: the outer shell (floor, walls, ceiling) and subject content (furniture and equipment) ensure the implementation of the functional processes for which this interior space is created. The outer shell and subject content have the same ecological interconnections as in large construction. The search for a solution to the global environmental problem has led to the emergence of concepts such as ecological interior and eco-style. Ecostyle is one of the modern styles in the interior. It includes interiors where natural elements are visually present in one form or another - from natural components of the interior to graphic, outwardly naturalistic natural forms and objects (trees, rocks, sea). However, the concept of "ecological interior" is much broader than the concept of "eco-style". The proposed classification will help to understand the peculiarities of the formation of various environmental trends in interior design. (Figure 2).

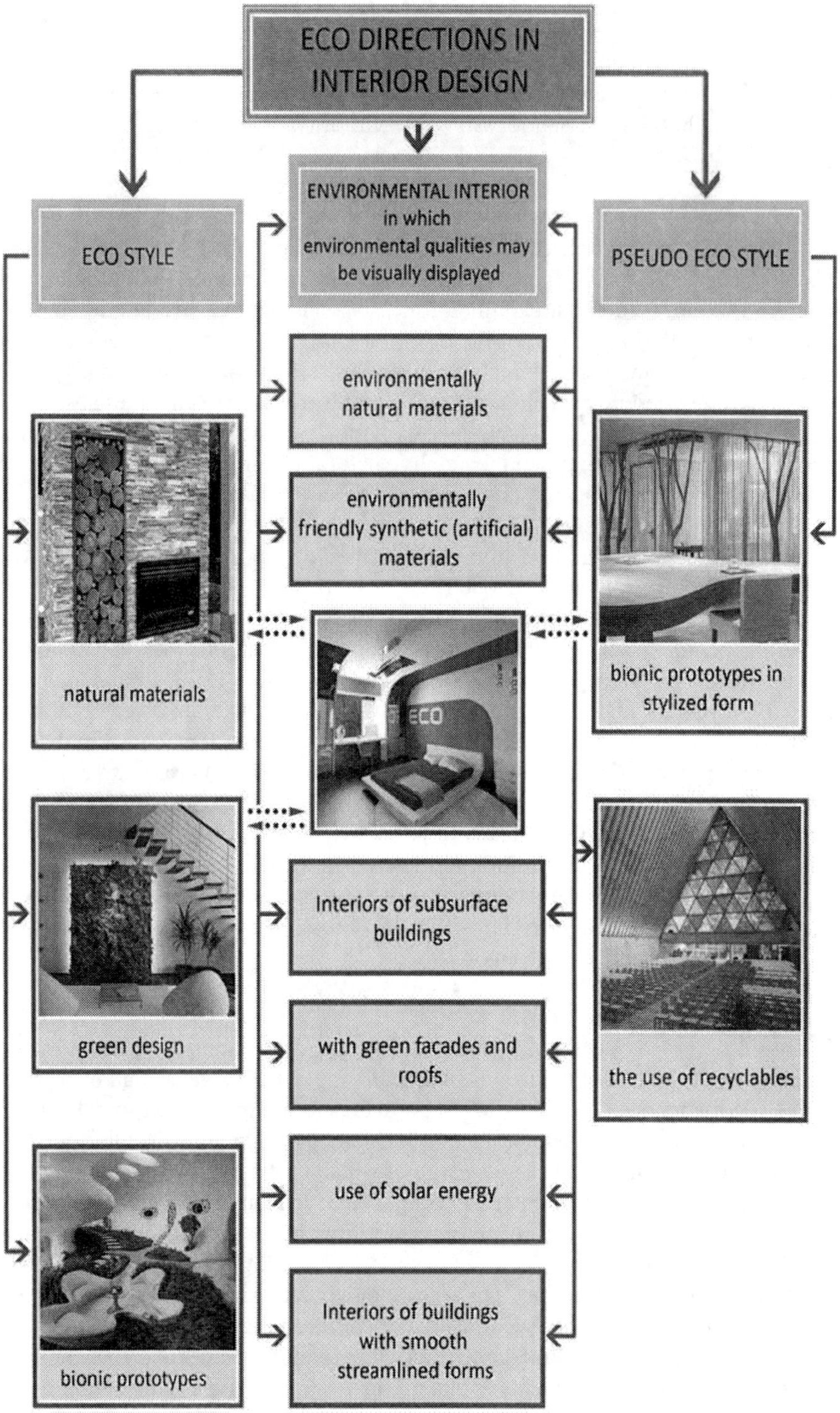


**Figure 2.** Classification of environmental trends in interior design.

Three major branches: eco-style, ecological interior and pseudo-eco-style share common ground and significant differences. Ecological interior is a trend and technology of designing. Essentially, any interior can be ecological if only environmentally friendly materials are used in it. Materials can be either natural or artificial materials (wood, clay, natural stone, brick, natural stone, glass, textiles, etc.), or synthetic, or artificial ones that do not harming human health. This is probably the main factor due to which the interior can be considered ecological. The ecological effect can be obtained by using the methods already mentioned in sustainable building (streamlined forms of construction, subsurface buildings, the availability of equipment for collecting, processing and recycling rainwater, the use of alternative energy sources). The mentioned methods have a different degree of manifestation in the interior. The interiors of buildings with streamlined shapes with their internal space repeat the geometry of the external. The interior space of subsurface buildings reflects with their geometry the technological feature of construction and non-traditional approaches to its lighting. The availability of equipment for the collection, processing and secondary use of rainwater and the use of alternative energy sources have no special effect on the formation of the interior.

The most striking and expressive ecological technique used by the branches of the "ecostyle" and "ecological interior" is the use of green spaces. With their help, you can not only transform the aesthetic appearance of any interior, but also beneficially affect the general condition of a person. In such a room, the microclimate improves, the air is cleared, it is easier to breathe, better to work, there is a feeling of peace of mind and balance. However, living plants put forward a number of requirements, such as: timely watering, lighting, temperature, the presence of a horizontal surface. Innovative technologies for the production of moss for the interior have нcome to the aid of designers. Artificially created moss can be placed vertically, and its diverse color scheme gives unlimited possibilities to designers. (Figure 3).

**Figure 3.** Examples of interiors using moss.

**Figure 4.** Examples of interiors in pseudo eco style.

The following, the most expressive technique when developing the interior in eco-style is the use of natural materials, notably, in their explicit, visual expression. These can be significant planes of textured unpainted wood, natural stone with an expressive untreated texture, or planes made of clay. However, it is visually impossible to assess the ecological purity of the material and even its genuineness. The texture of wood, clay, brick can be obtained on artificial materials as well. This aspect shows the main contradiction between the concepts of "eco-style" and "ecological interior".

There is another powerful ecological aspect where the interior can be considered ecological, although at first glance it does not always look like that. This aspect is the use of recycled materials. Nowadays innovative technologies for processing production waste and even the food industry are gaining momentum. More and more new, most unexpected offers appear on the market of this segment. The list of materials used for recycling is replenished every year with new, sometimes quite exotic ones (for example, fabrics imitating skin are made from the stomach of a cow). In recent decades, a whole list of architects working in this direction has appeared [7].

Another eco-principle is the use of bionic prototypes, when the forms imitate natural objects and have their own functional purpose. This principle does not provide for direct savings of material or natural resources. Rather, it aims to draw attention to the problem of ecology. The only condition for the manufacture of objects is the use of environmentally friendly, natural materials that do not harm human health. This principle has been named "organic functionalism." Its founder is the Finnish architect Alvar Aalto, whose fruit vase has glorified the style and spread the concept of "Finnish glass" to the whole world. A vase made in the form of curvature, repeating the pattern of the shores of the Finnish lake. The vase has become one of the most famous glass objects and is now in the collections of many museums in the world.

The imitation of natural forms in interior design when designing eco-style can be presented in two different versions. The first is the use of bionic prototypes in a accurate, realistic image. With this approach, the interior goes into the biostyle category. The second approach is the deliberate stylization of bionic prototypes using nature-specific colors and shades. This approach can be called a pseudo eco-style. Although, as mentioned earlier, such an interior can essentially be ecological when it is made from environmentally friendly materials. (Figure 4).

## 4. Conclusions

The above techniques for the design of the interior in eco-style do not provide for the completeness of possible options. However, the presented classification of eco-directions provides the necessary understanding of the structural relationships of various approaches and outlines some contradiction that arises between the concepts of "ecological interior" and "eco-style." The suggested model of ecological relations between nature, the artificial environment and humans gives a general picture of the mutual influence of these three components. Such an understanding of the situation facilitates the task of the world community in the struggle to improve the ecological state of our planet Earth.

## References:

[1] Haberl H, Fischer-Kowalski M, Krausmann F and Winiwarter V 2016 *Social Ecology: Society-Nature Relations across Time and Space* (Axel Springer SE)

[2] Information Agency LIGABusinessInform 2019 The concentration of carbon dioxide in the atmosphere broke the record. What does it mean? https://www.liga.net/society/cards/zemlya-na-poroge-katastrofy-chto-nado-znat-o-rekordnom-urovne-co2

[3] Ji Y and Plainiotis S 2006 *Design for Sustainability* (Beijing: China Architecture and Building Press)

[4] Hofste R W, Reig P and Schleifer L 2019 17 countries, home to one-quarter of the world's population, face extremely high water stress https://www.wri.org/blog/2019/08/17-countries-home-one-quarter-world-population-face-extremely-high-water-stress

[5] Kulik E and Yarosh T 2017 Pros and cons of vertical urban greening*: Proc. of the LX Student Int. Res. and Pract. Conf. Scien. community of students of the XXI century. Eng. Scien.* vol 12 (59) (Novosibirsk: ASW SibAK) pp 85-9

[6] Hnatuk L and Bakurova A 2014 Features of eco-design formation in office premises *Theory and practice of design* vol 6 (Kyiv: NAU) pp 43-51

[7] Zymina S and Mezhennaya N 2015 About some contradictions in the theory and practice of

eco-building and eco-design *Architec. Bull. of Kyiv National University of Construction and Architecture* vol 5 (Kyiv: KNUBA) pp 142 – 54

[8] Finnish Design Blog (2008) Alvar Aalto's Savoy vase continues to inspire designers https://web.archive.org/web/20120122010933/http://blog.finnishgifts.com/2008/07/25/alvar-aaltos-savoy-vase-continues-to-inspire-designers/

[9] Zymina S 2018 *Interior Styles* (Kyiv: Dovira)

[10] Karabanova T (2016) Shigeru Ban wins architecture's top award https://www.admagazine.ru/architecture/post-pochitaniya-sigeru-bana

[11] The World Commission on Environment and Development 1987 *Our Common Future* (Oxford: Oxford University Press) https://sustainabledevelopment.un.org/content/documents/5987our-common-future.pdf

IOP Conf. Series: Materials Science and Engineering **907** (2020) 012011 doi:10.1088/1757-899X/907/1/012011

# Schools of architecture as an instrument of architectural policy on the examples of European countries

**A O Yesipov**[1], **V P Myronenko**[1] and **N V Ivanova**[2]

[1] Department of Design of Architectural Environment, Kharkiv National University of Civil Engineering and Architecture, Sumska Str., 40, Kharkiv, 61002, Ukraine
[2] Department of Architectural Construction, Kharkiv National University of Civil Engineering and Architecture, Sumska Str., 40, Kharkiv, 61002, Ukraine

esipov1995@gmail.com

**Abstract.** The purpose of this article was to analyze the cultural policies in the field of architecture in various European countries and consider the principles of using the tools of schools of architecture as its element. Such issues are especially relevant for Ukraine, where the problem is the lack of regulated norms for Ukrainian architectural policy and specialized state institutions. The need to draw attention to the architectural education of Ukraine is due to its lack of actualization compared to advanced European countries and the depletion of the physical plant of universities. An analysis of foreign experience will make it possible to use the potential of universities to solve the problems facing the society.

## 1. Introduction

Architectural policy can be described as a set of principles, goals, objectives and forms of activities that guide authorities in the process of maintaining and developing the urban environment. It outlines the interests of a state with regard to architecture, contributes to the creation of state institutions and the introduction of tools to help achieve the desired results.

Nowadays, the goal of architectural policy in various cities and countries is to create conditions for the formation of an aesthetically beautiful, comfortable and diverse environment, providing prerequisites for a harmonious co-existence of citizens of different ages, backgrounds and wealth. The influence of urban environment on people comes to the fore in the discussions of politicians, architects and sociologists from different countries. Architecture, urban planning, landscape solutions determine the behavior, well-being and quality of life of all citizens. Understanding of this issue led to the fact that the relevant documents of regulated actions on the sphere of culture began to be developed.

## 2. The experience of the EU countries

The end of the XX and the beginning of the XXI century was a time for developing and passing of the cultural policies in the field of architecture at the state level around the world. Over the past twenty years such documents have appeared in many countries of Europe. Countries learn how to form architectural policy, share their experience with others. As a result, such process has a positive impact at the European level, making possible a closer rapprochement among member states and encouraging to improve the quality of architecture.

The EU resolution "On the quality of architecture in urban and rural environments" [2], accepted in 2001 under the chairmanship of France, establishes the following principles on which architectural policies of all EU countries should be based:

– improving the qualifications of customers and the level of public awareness in the field of architecture, urban planning and landscape design;
– the promotion of quality architecture through the construction of exemplary public buildings;
– stimulation of the international exchanges of knowledge and experience in the field of architecture with the help of student scholarships and grants, as well as hiring foreign professionals to work.

In the present situation, it should be noted a certain gap between cultural policy and the mechanism of its implementation. European countries have predominantly centralized models of the relationships between culture and state. First of all, due to the undeveloped multi-channel financing of cultural sphere, the formed state policy often appears not enough financially and materially (in legal and personnel relations) provided in its implementation at the municipal level. This requires a significant improvement of the "lower links" of the government and a direct interaction with the population, providing greater independence to regional and municipal cultural authorities [3]. They should have freedom of choice of methods and forms of work that provide the most effective ways of interacting with various organizations, institutions, faiths, commercial structures in solving cultural problems.

The author of the article assumes that schools of architecture, which distributed throughout Ukraine, working with the people and possessing the material and technical base as well as specialists, have huge potential, resources and the ability to be a key link between the authorities and the population in matters related to architectural policy. The integration into such projects will allow universities, staff members and students to increase their importance in society and to receive preferences and benefits from the state – it will help in promotion of architectural education.

It is proposed to consider the experience of European countries, as well as what resources and tools higher education have, and after this, to conduct a comparative analysis of some points that can be somehow implemented with the support of universities.

### *2.1 Belgium*

The Belgian program involves work in the educational field: it offers lectures handled by invited architects at universities, work in the field of continuing education, as well as internship programs and diploma certification in architecture in other EU member states. These goals are aimed at providing students with the opportunity to learn from practicing architects, sometimes world-class, as well as to increase the competitiveness of architects outside the country [4].

The Belgian concept involves taking advantage of the European Union to develop students' potential so then their schools of architecture will be competitive in the world. In this case, the measures are aimed at the long-term perspective and development of cooperation programs at the level of individual universities, but not affect usage of the material and technical resource base.

### *2.2 Croatia*

In Croatia, the idea of architecture is considered as the basis of national identity. In order to increase public awareness of the importance of a high-quality architectural environment, it is planned to introduce architecture into the National Program for Preschool and Compulsory General and Secondary Education, as well as to create a media platform [5].

The Croatian concept involves the use of scientific and pedagogical resources of university to increase awareness among the general public about architecture, to create a demand for the services of their graduates and for a high-quality architectural environment. Universities as state institutions, in most cases, can act as suppliers of architects and educators. The Croatian program does not give a comprehensive answer to the question: what opportunities should universities have in order to be included in architectural policy of the country? However, it can be said that the main ideas support the common policy of the European Union in this regard.

*2.3 Czech*

Czech policy is more detailed, here are some of the provisions that assume to engage architectural educational institutions [6]:

- to plan further training in cooperation with universities for civil servants in the form of courses aimed at presenting trends and best practices in the field of architecture;
- to expand the system of further education for architects, civil engineers and others involved;
- to revise educational programs for schools thus to include architecture, urbanism, spatial planning, landscaping and building culture in separate areas of education and relevant thematic sections. To pay attention to a person's attitude to his home and city, including caring for the environment where he lives. To use examples of good practice and architecture as educational material during training. To include practical exercises in education that teach how to improve an artificial environment. To ensure with the necessary teaching materials and professional support for teachers;
- to include training on topics related to architecture and building culture in primary and secondary schools in the teacher training system;
- to ensure background materials and initiate more opportunities in the public media to discuss topics related to architecture, urbanism, spatial planning, landscaping and building culture;
- to increase customer inquiries by using good practice examples. To support for sightseeing tours with architects or with their commenting on new buildings;
- to involve people (adults, children and especially schoolchildren) in projects that affect the public space.

The main directions of the Czech architectural policy generally assume cooperation with higher education institutions as suppliers of qualified personnel. Training courses aimed at raising the awareness of civil servants, the possibility of using university as a media platform, or working with people on projects that affect the public space – all of this assume using of the material and technical resource base. Such events can be held in universities as public hearings, seminars. It will increase the popularity of universities, as well as teachers will have an opportunity to earn extra money as consultants, guides or teachers while working with the general public.

*2.4 Denmark*

The basis of Danish schools of architecture is artistic, and institutions are not just seen as educational, but as cultural institutions with a deeper responsibility in their field of education. The programs need to be compared and their quality should comply with international standards to ensure Danish architectural studies are in line with best in the world. Higher education institutions in the field of architecture organize exhibitions, international lectures, conferences and other events throughout the year that are open to a wide audience.

Education and research in schools of architecture should take place in collaboration with the professional sector and be adjusted in accordance with its needs, and the sector should be involved in the quality development of universities. In order to provide the best conditions for further education and scientific research in collaboration with the sector, schools of architecture will create advisory councils that are closely related to school management. Council members should represent leading architectural firms and include international architects, and their responsibility is to provide advice on school development issue to Rectors and to minister of education.

Danish architectural education should be internationally competitive and engaging. For this reason, it is important that universities enter into strategic cooperation with leading international schools of architecture and contribute to enhance the international exchange. One way is to increase the supply of English-language programs that can attract foreign scholars, professors, exchange and scholarship students.

The constant development of new knowledge in the field of architecture is a decisive condition for qualitative progress in both education and the profession. Such development of knowledge can be successfully carried out in partnerships between universities and architectural firms. Universities will ensure it by accomplishing doctoral dissertations in collaboration with relevant partners and by increasing external research funding. By cooperating with this sector, schools of architecture must develop and adapt training courses to meet labor market needs.

One feature of the Danish experience is the various centers of architecture and construction, one of whose tasks is to build partnerships between schools of architecture and the commercial sector. This means that the centers work as the main operator in coordinating and implementing the new interagency architectural policy.

Center for Industrialised Architecture at The Royal Danish Academy of Fine Arts Schools of Architecture specializes in education and research in the field of industrial architecture and seeks to establish cooperation with parties in the construction sector. The subjects of the center's research include the deployment of the design process in industrial practice, building systems, mass customization in architecture and architectural potential of various building materials, construction technologies, industrial architecture as a lever for sustainable development, the importance of new digital technologies in the design process, etc.

The Danish Architecture Center is an information development center of architecture and construction, its task is to build partnerships that disseminate Danish architecture and construction. The activity is aimed at both the professionals and general public. The goal of the DAC is realized through exhibitions, meetings and conferences, educational services, Internet events and other information events. The goal is to contribute to the dissemination and development of Danish architecture at both national and international levels [7].

Thus, enhanced cooperation between universities and other components of the architectural sphere of culture, as well as the architectural sphere of industry, is a priority for Danish policy. This interaction should have positive results in terms of employing university graduates, as well as cooperation with other participants in the professional sector will allow the development of architectural education.

*2.5 Finland*

Finland has rich experience in promotion of the architectural policy: documents governing it have been passed since 1997. It's been 20 years since the first program was completed, the new one should be completed in October 2020.

The Ministry of Education explores possibilities for developing architectural education as part of voluntary adult education in order to improve the ability of citizens to make their own decisions on environmental changes. In the field of architectural education will be an opportunity to design and implement experimental buildings to demonstrate and to supplement teaching and research work [8].

Finland draws attention to educational work among the adult population, tools should be developed for this outside the higher education system, however, they must be based on the higher educational institutions, as well as additional opportunities should be provided for students of architecture schools.

*2.6 France*

The first document defining the vector of development in this direction – French Law on Architecture – was published in 1977. Without a doubt, France is a veteran in architectural policy. The cultural policy of France was considered by the author in a separate article [9], because thanks to her experience, the initiatives of France on schools of architecture should be considered with particular care.

In 2015, the Minister of Culture Fleur Pelleren, as part of its consideration of the draft law on creative freedom in the National Assembly [10], introduced the National Architecture Strategy [11]. The goal of the strategy is to change the attitude of the society towards architecture and to increase the role of architects in the community. Architecture remains as remotely understandable concept in the minds of many citizens, so it is necessary to develop awareness-raising activities aimed at the general public and work with stakeholders, especially local elected representatives. Engaging an architect is too often the exception rather than the rule. It is advisable to emphasize every architect's contribution with the help of new incentives. It is necessary to promote the know-how of architecture to owners and developers of buildings and to encourage the expansion of the sphere of influence of architects in new areas of construction. Supporting the export of architecture to the European and other international markets also provides an opportunity to develop. The export of architecture will not only adjust the trade balance of the sector, but also will demonstrate national architecture and young talents who possess all the skills which are necessary for more successful participation in the international competitions.

The strategy consists of six so-called "axes", each of them is aimed at the development of one of the areas of architecture culture. Fleur Pelleren is confident that the cultural, scientific, technical and architectural successes of initiatives is based on educational, research and business activities, and architecture schools are its epicenter [12]. The third axis of the national strategy is dedicated to them. Educational institutions of architecture with all their components: administrative, educational and scientific, as well as school students will be closely linked to the implementation of the strategy.

Public procurement should play an exemplary role for young architects who face difficulties in the early years of their careers. That is why the access for new firms of architecture to orders is facilitated.

Among other measures, there will be created the status of teacher-researcher. It should preserve academic and professional origin of the teaching profession and enrich their missions by expanding participation in teaching and research. Also, partnership research departments will be created in tandem with professional and economic entities to make institutions more recognizable.

A national assessment plan of the state licensed architect authorization will be launched to ensure proper project management on his behalf. It is especially important to know that the level of qualifications and competencies are necessary for recognition in the European Union.

Measures aimed at raising public awareness are already provided in "National Architecture Days", organized in October each year [13]. The Ministry of Culture welcomes cooperation with any partners who wish to participate in this event: professionals, local authorities, associations, public institutions. National Architecture Days are coordinated by the General Directorate of Heritage and held in regions by regional cultural management. They can expect on revitalization and greater mobilization of architecture networks at the national and regional levels: National Schools of Architecture (ENSA), National Council (CNOA) and Regional Councils of the Order of Architecture (CROA), House of Architecture, City of Architecture and Heritage (CAPA), Councils for architecture, urban planning and the environment (CAUE), a network of Cities and Lands of Art and History (VPAH), etc. [14].

The purpose of the Days of National Architecture is advancement of knowledge of the general public. They offer to reveal the riches of wonderful modern architecture, wherever it is, to show the history that the surrounding buildings tell us, to arouse interest and artistic feelings, to increase cultural, scientific awareness. These events are entrusted with the mission of informing the citizens that France requires architects to rethink future ways of housing, working, mobility, caring, education in a wider context of improving the architectural environment of cities. To achieve success in the ecological transition, come up with cities and territories of tomorrow. The program includes meetings and debates with figures in architecture, visits to construction sites and city walks, technical and art exhibitions or educational seminars. Visits to agencies of architecture, city routes, films and exhibitions are organized throughout France and help to interest the general public.

Many agencies of architecture have research practices that deserve improvement due to the presence of doctoral students who will base their work on the knowledge developed by the professional practice. The goal is for a quarter of post-graduate architects to be placed in professional structures by 2020.

"Cities and lands of art and history" is a French program in which communities commit themselves to develop action programs aimed at introducing young people to architecture and urban planning, promoting a heritage for a diverse audience, raising environmental awareness, training events for municipal staff, enrich knowledge through communication with artists and professionals, stimulate creativity by putting it into practice [15]. As of May 2019, the program included 195 cities and lands of art and history. The program is implemented at the expense of the Center for the Interpretation of Architecture and Heritage (CIAP), which is also supposed to be: a place for receiving visitors, a place of debate for public, a place for open events for tourists. The agreement provides for financial and technical support from the Ministry of Culture and Mass Media and includes an obligation for communities to use qualified personnel approved by the ministry (e.g. guide-lecturer and coordinator for heritage issues).

France has rich and wide experience in the development and the implementation of its architectural policy. More importantly, a special place is given to educational institutions of architecture, which play a key role in government strategy. Many of the objectives of raising awareness, conducting various events and interagency cooperation in architecture are impossible without qualified university staff. Attention is also drawn to specialists who graduate from schools of architecture since they are the driving force in the development of environment, design, architectural education of the country and others in the future.

*2.7 Germany*

It's certainly that the German population has a propensity for architecture: during studies almost every fifth person had a desire to get a job in the field of planning and construction. This interest makes it necessary to inform the majority population about the topic, and therefore, to develop appropriate projects for schools, for example. Against this background, architectural education in schools plays an increasingly important role. The corresponding projects have been developed by the twelve state architectural chambers and the Federal chamber of architects, nowadays it has already had a positive effect, for instance, the initiative "Architektur macht Schule" (Architecture in schools) [16].

But such experience from other European countries also show that a much wider integration of building culture topics into general education is possible, for example, in many countries it is included in the main curriculum on an ongoing basis. This is an important basis for awareness and active participation in the development of our artificial environment, and therefore Germany should be given even greater support in the implementation of it.

*2.8 Norway*

The Norwegian Center for Design and Architecture wants to contribute to the understanding, knowledge and usage of design and architecture in terms of business and community structures. A dedicated foundation will have to promote quality and innovation by using design and architecture to develop environment, products and services. This will stimulate innovation and rationalization, as well as increase value creation in Norwegian business and in the public sector [17].

Norwegian priorities are placed on the development of innovations in architecture and thereby popularization of architecture among the public. With this it is also planned to attract business investment in real estate and architecture. Students could help in the development of innovations in the field of architecture together with professionals, as a result such cooperation would benefit the development of professional skills of future specialists and would increase enrolment of the studied areas. Attracting business investment will help to develop higher educational institutions of architecture.

*2.9 Poland*

The opening of the National Institute of Architecture and Urban Planning in Poland aims: to create public awareness, to promote architectural and urban culture, to preserve the national architectural heritage of the XX century by research, documentation, popularization, exhibition, educational and publishing activities [18]. One of the objectives of the Institute is constant collaboration between architects, local authorities and users of public spaces in order: to improve the quality of environment, to find a balance in their development and to coordinate processes of their design and planning [19]. In the program of Polish architectural policy of 2018, there is a separate section devoted to architectural education, which shows the relevance of this topic and the work in this direction.

Polish both architectural and cultural policies have proved their worth on the example of preserving the architectural environment of the historical city centers in the post-war years. Such expertise, improvement in the work with society and cooperation with architects can lead to remarkable results. The architectural community of universities can find response and support in the community in their endeavors and gain the opportunity to influence municipal projects with their criticism.

*2.10 Portugal*

The priority areas in Portugal are informational, awareness-raising and educational measures.

Among educational measures, one can single out the stimulation of a sense of belonging, identity and responsibility among people in relation to the community and territory, increasing the interest and involvement of citizens and communities in the processes of participation, evaluation and decision making. It is supposed to inform the public about the importance of architecture, urban planning and landscape design to create a quality environment, as well as the dissemination of best practices in design, planning and preservation of the cultural landscape by organizing events, publications and exhibitions.

It is planned to integrate the theme "Architecture, city and landscape" into school curricula at various levels of non-specialized education, especially in curricula of primary and secondary schools, with the aim of promoting the construction culture. Training events will be organized in the field of: architecture and landscape, in particular on sustainable architecture and urban planning, architectural and urban preservation, reconstruction and restoration, landscape protection, management and planning.

The higher education sector is responsible for developing innovative conceptual and methodological foundations that can respond to the evolution of society and its values, to transferring such knowledge in an appropriate way not only to professionals, but also in forms that are understandable to the population, which contributes to the development of the territorial culture and promotes informed participation.

It is stated to promote the use of traditional knowledge, processes and methods for building construction and maintenance with the aim of preserving the heritage and passing it on to future generations. At the same time, the selection and the dissemination of examples of best practices and best experience will be carried out: research in the field of the sustainable construction, facilitation of relations with industry and the promotion of the technological innovations in construction, materials, energy production and consumption, and so forth [20].

All of the above requires the widespread involvement of both practitioners and theoreticians of architecture, thus enhanced cooperation and interaction is necessary between the scientific, technical and political communities, as well as between them and the population as a whole.

## 3. The results of the research

World experience clearly demonstrates that the demand for the profession of an architect cannot be overestimated since architecture is a cultural field, which the quality of people's life directly depends on. Therefore, it is necessary to pursue a consistent policy aimed at maintaining this area.

The architectural policy is developed both at national and city levels, however, the municipal authorities take responsibility for it and manage most basic tools. Having examined the European experience, the author of the article believes that schools of architecture in Ukraine could be exactly that key tool that is worth paying attention to. Architectural educational institutions have every opportunity to improve results in carrying out activities aimed at introducing a national cultural policy. In turn, universities and staff members can receive tax benefits or other privileges, as well as moral and legal support from the state.

Many of the tasks that the authorities of European countries assign to architectural communities should be carried out with the help of architecture schools at the planning stage. Some programs are not well-developed with regard to cooperation with universities, and some do not imply it at all.

The architectural policies of European countries cover a wide range of topics, but all of them have a similar focus. Among these areas we can distinguish:

- improving the business and social climate in the field of architecture and urban planning;
- development of infrastructure for professional activities;
- development of new technologies;
- increasing the professionalism of architects and stimulating interdisciplinary interaction;
- raising public awareness of architecture and urban development, involving local population in the development of design solutions;
- improvement of national architecture presence in the international arena;

- preservation and revitalization of cultural heritage;
- stimulation of a constructive dialogue between state, the professional community, business and public;
- development of architecture as a full-fledged mainstay of economy.

As for the tools of higher educational institutions, they can be divided into 2 groups. The first is scientific and pedagogical resource base (e.g. teachers and students), the second is material and technical resource base (e.g. premises and equipment of university). Among the general directions of European architectural policy, for example, engagement with scientific and pedagogical resource base includes grants and competitions for students of architecture, provides the opportunity for them to work together with professionals in other fields and learn from their experience, provides employment assistance to graduates, includes educational events for general public held by qualified teachers, etc. As concerns engagement with material and technical resource base, for example, it involves: the expansion of universities infrastructure, the use of universities premises for open seminars and lectures, excursions and other educational events, the investment attraction for universities, etc.

The above indicates that it is advisable for Ukraine:

- to develop and implement a more systematic and articulated architectural policy, as well as pay attention to the whole complex of higher architectural education after a thorough review of the EU countries experience;
- to draft legislation relevant to activity regulation of the state, municipal and university administration authorities to make successful realization of the established goals;
- to provide for the interaction of public with private institutions, which included in the architectural policy program implementation, and to create their rights and divide duties;
- to create benefit programmes for schools of architecture, as well as a set of principles and measures to support scientific and pedagogical, material and technical resource bases;
- to entitle the architectural community to determine the most effective tools for architectural policy implementation at both local and global levels.

In conclusion, I would like to emphasize the need to draw attention to the full range of the higher architectural education in Ukraine, especially schools of architecture. They have the potential to become a key link in the national architectural policy.

## References

[1] Council resolution of 12 February 2001 on architectural quality in urban and rural environments OJ C 73, 6.3.2001

[2] Borucka J, Macikowski B 2017 Teaching Architecture – Contemporary Challenges and Threats in the Complexity of Built Environment *IOP Conf. Ser.: Mater. Sci. Eng.* **245** 082058

[3] Ordre des Architectes 2016 *Guide de l'Architecte* 3rd edition (Bruxelles: Ordre des Architectesand Conseil francophone et germanophone) [in French]

[4] National guidelines for quality and culture of building – architectural policies of the republic of Croatia (2013–2020) URL https://mgipu.gov.hr/naslovna-blokovi-133/about-the-ministry-139/scope-of-the-ministry/housing-8652/apolitika-8754/8754

[5] Czech Ministry of Regional Development and Institute for Spatial Development 2015 *Architecture and Building Culture Policy of the Czech Republic* (Prague and Brno: Ministry of Regional Development and Institute for Spatial Development)

[6] Ministry of Culture 2007 *A Nation of Architecture, Denmark: Settings for Life and Growth: Danish Architectural Policy 2007* (Copenhagen: Ministry of Culture)

[7] Finnish Association of Architects and Arts Council of Finland 1999 *The Finnish Architectural Policy* ed H Korpelainen (Helsinki: Arts Council of Finland and Ministry of Education)

[8] Yesipov A O, Myronenko V P 2019 National strategy in the field of architecture as a part of cultural policy *Scientific Bulletin of Civil Engineering* **1(2)** 144–9 [in Russian]

[9] LOI n° 2016-925 du 7 juillet 2016 relative à la liberté de la création, à l'architecture et au patrimoine URL https://www.legifrance.gouv.fr/eli/loi/2016/7/7/MCCB1511777L/jo/texte

[10] Accroître le rôle des architectes URL http://www.culture.gouv.fr/Thematiques/Architecture/Strategie-nationale-pour-l-architecture
[11] Discours de Fleur Pellerin, ministre de la Culture et de la Communication, prononcé à l'occasion de la présentation de la Stratégie nationale pour l'architecture, à Paris, le 20 octobre 2015 URL https://www.culture.gouv.fr/Sites-thematiques/Architecture/Accroitre-le-role-des-architectes/Discours-de-Fleur-Pellerin-ministre-de-la-Culture-et-de-la-Communication-prononce-a-l-occasion-de-la-presentation-de-la-Strategie-nationale-pour
[12] Journées nationales de l'architecture URL https://journeesarchitecture.culture.gouv.fr/Journées nationales de l'architecture URL http://www.culture.gouv.fr/Thematiques/Architecture/Strategie-nationale-pour-l-architecture/Les-Journees-nationales-de-l-architecture
[13] Villes et Pays d'art et d'histoire URL https://www.culture.gouv.fr/Regions/Drac-Auvergne-Rhone-Alpes/Pole-action-culturelle-et-territoriale/Villes-et-Pays-d-art-et-d-histoire-Action-culturelle-patrimoniale
[14] Federal Foundation of Baukultur 2015 *Baukultur report 2014/15: built living spaces of the future– focus city* ed A Schmedding, D Michalski *et al* (Potsdam: Federal Foundation of Baukultur)
[15] Ministry of Culture and Church Affairs 2009 *Architecture.now: Norwegian architectural policy* (Oslo: Ministry of Culture and Church Affairs)
[16] Marczak P 2017 Shaping space programme as a tool for educating youth about architecture *IOP Conf. Ser.: Mater. Sci. Eng.* **245** 042075
[17] Andrzejewski P, Baranowski A, Buczek G, Chwalibóg A, Chwalibóg K, Śmiechowski D, Wojciechowski S and Wolski P 2018 *Polska polityka architektoniczna: polityka jakości krajobrazu, przestrzeni publicznej, architektury : rozwinięcie tematów* 3rd edition ed K Chwalibóg, I Hardej *et al* (Warszawa: Narodowe Centrum Kultury) [in Polish]
[18] Resolução do Conselho de Ministros 45/2015 2015 *Diário da República* **130** pp 4657–67 [in Portuguese]

IOP Conf. Series: Materials Science and Engineering **907** (2020) 012012 doi:10.1088/1757-899X/907/1/012012

# Multi-Level Viewing Points in the Urban Structure of a City

**V Kuzmych, Yu Petrovska**

Department of Design and Fundamentals of Architecture and Design, Lviv Polytechnic National University, Ukraine

yuliana.r.petrovska@lpnu.ua

**Abstract.** The authors of this research substantiate the statement that the proportional capture of the visual plane of the city panorama will be in constant dynamics in the context of both the planar expression and 3D observation of the volumetric solution of the spacial combination. A constant change of the viewing angle provides for the limitless quantity of viewing points concentrating on a cumulative picture of the angular perception. The article covers applied aspects of design and perception of the vertical and horizontal sector of capturing of the viewing panorama allowing obtaining multi-proportional correlations of the diagonal perspective and plane that in its turn impacts the general perception of the urban structure as a whole. The paper shows the importance of the spherical and angular range of the visual location that impact the psychological aspect of perception of the architecture as a whole, describes the dimensional and scale comparison of the main functional zones of a city, activation and concentration of accents of its mass with respect to the area of the general layout with addition of angular aspects of gravitational laws and identification of fundamentals of the linear and tonal perspective.

## Introduction

The significance of viewing points in the city planning is conditioned by enrichment and diversity of visual perception of the overall panorama of the city. Diversified unexpected perspective angles bring an intrinsic urbanized effect to the structure of the city. Each separate viewing point invites to contemplate inimitable perspectives and views and allows avoiding the monotony and sameness in perception of the visual space. The urban environment that is saturated with viewing points substantially enriches the urban structure of the city and opens opportunities for the aesthetic meaningfulness of the horizon. This paper looks into the issues of the visual value of viewing points and their angular panoramic perception.

## Analysis of recent publications on the subject

In their work, the researchers Kuzmych V., Petrovska Y. touch the issue of the harmony of a city image, design of urban planning elevations giving the possibility to obtain maximally harmonious correlations and proportions at the visual level by using software applications for development of the project documentation in the sphere of urban science (Kuzmych, Pavliv and Petrovska 2018). The researcher H. Koznarska describes the importance of viewing points for perception of the image of a city, identifies principles of their location and importance in perception of the image of a city (Koznarska, 2012). The analysis of the scenery opening of views of the landscape park in Kyiv by means of investigation of three pedestrian routes was covered by I.O. Sydorenko and V.V. Minder.

The researchers compiled schemes of the interchange of various types of spaces with identification of the amount of interchanges of the light and shade (Sydorenko, Minder, 2018). In his work, Ye. Vodzynskyi discussed the issue of the analysis of visual perception of architectural monuments, underlines the importance of quantitative assessment and described instruments of such assessment (Vodzynskyi, 2010). The researcher Donatella Scatena considers the perception as an important element in the process of formation of an architect or city planner. The action of perception is crucial for understanding and formation of an environment they work in. This ability is acquired by a specialist by way of study and research, experience and observation that allows him or her to shape architectural spaces in future (Scatena D. 2018). The researcher Dalia Dijokienė offers insight into the problem of the urban design as a component part of the art and architectural science and an indisputable direction of activity of a specialist in architecture. She assumes that cities play one of the most crucial roles in the state economy, while the urbanization offers a variety of prospects for a state (Dijokienė, D. 2013). At the same time, issues of the large-scale comparison of the main functional city zones with the total area and understanding of the panoramic layout in the urban science still remain unexplored. As well, the issue of formation of the architectural environment in conditions of a large city is still not covered.

**Author contributions**

This publication is written by two co-authors who expanded the existing materials of the research in the area of the urban science. Vasyl Kuzmych developed a compositional scheme of the interchange of the level of angular observations using the example of the observation wheel, as well as the scheme of spacial location of settlement objects. He covers the issue of the role of viewing points in multi-level angular perception of the urban environment. The researcher Yuliana Petrovska describes the notion of the panoramic layout on the urban science and the development of the architectural environment in conditions of a large city.

**Basic Theory Part**

The angular horizontal perception of a human takes place within the limits of the visual location that comprises 150 degrees. A human can examine the surrounding objects both in the horizontal and vertical planes. The best example of this is the observation wheel, where one can see a quick change of the above ground level within a short time and observation of panoramic views with account for viewing angles (Figure 1-2).

In the end, we obtain the ability of volumetric and spacial spherical perception of surrounding panoramas. However, the zone of the spherical perception is partially limited by the blind zone at the angle of 30 degrees. Thus, a human obtains the maximum visual information about the architecture within the observation range of 90 degrees. The sector of perception within the framework of the horizontal plane is more important for the human living. The aspect of the vertical angular perception impacting the locative function of the gravitation is also important. The mechanism of the spherical perception functions constantly and gives possibility of the spatial orientation. As a result of the acquired experience, the vertical angular aspect impacts the psychological perception of the panorama. A human organism feels the basic instinct of feeling of the foundation as a land plane in the balanced state. Basically, this feeling will depend on the dimension and area of the support foundation, where a human feel safe. Later on, humans adapted and mastered these factors on such construction elements as balconies and terraces. They are psychologically perceived by us as foundations of stability and balance. The angular perception of the space is no less important. In case of a change of the vertical perception angle, a feeling of stability can be retained irrespective of the level of observation subject to preservation of the body balance.

**Figure 1.** The London Eye from across the River Thames (Khamtran, 2009).

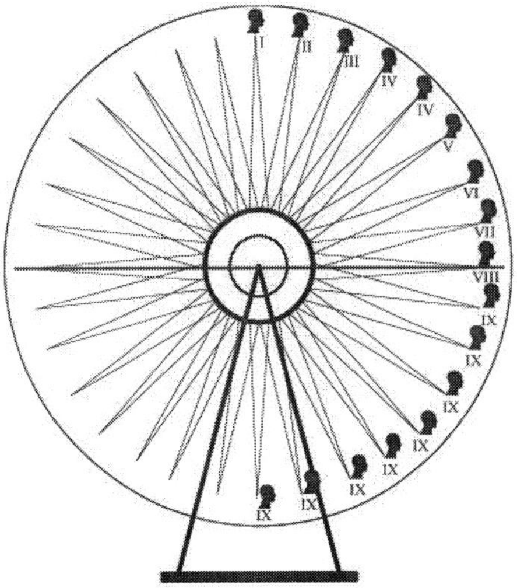


**Figure 2.** Compositional scheme of the change of the level of angular observations by the example of the observation wheel. I – top view, plan, master plan; II – bird's-eye view; III – panoramic view; IV – axonometric view; V – level view; VI – horizontal view; VII – ground-level view; VIII – zero-level view; IX – sub-level view (the scheme was developed by Kuzmych).

These parameters also include characteristics and features of types of the space. Increase of the area of the vertical capturing of the panorama by the human eye adds a certain monumentality, grandeur and sacral character to the layout. An important role in synchronization of the visual apparatus is played by the collaboration of cleido-mastoid neck muscles and extraocular muscles that keep the body in the state of equilibrium. Therefore, the level of the angular perception which differs from the usual one that retains the horizon line will be characterized by different psychological characteristics of perception of the environment. If we present it in the form of the received signal, the wider range of the tonal load of reflected flows will be perceived better. Functional and ergonomic factors are also appended by visual and psychological factors that jointly provide for the complex perception of the architecture as a whole. The perception of the architecture on the rugged relief has essential visual advantages over the plain territories. Enrichment of an object by deeply spatial surroundings gives it positive energetic advantages enlarging the range of perception.

Closeness of the space negatively impacts the human activity. In this case, the distance from the object or the point of perception of the environment also matters. The optimal structure of urban development must take into account not only the functional and planning structure, but also the visual and spatial one. The free scheme of scanning of spaces will always be the most comfortable option. Enclosed confined spaces of the short signal character are used as a mechanism of negative psychological pressure.

Analyzing various schemes, options of angular interchanges and panoramic views, we should mention that the composition of horizontal, vertical and diagonal character of perception envisages amplification of the diversity and development of building of architectural objects (Figure 3).

IOP Conf. Series: Materials Science and Engineering **907** (2020) 012012 doi:10.1088/1757-899X/907/1/012012

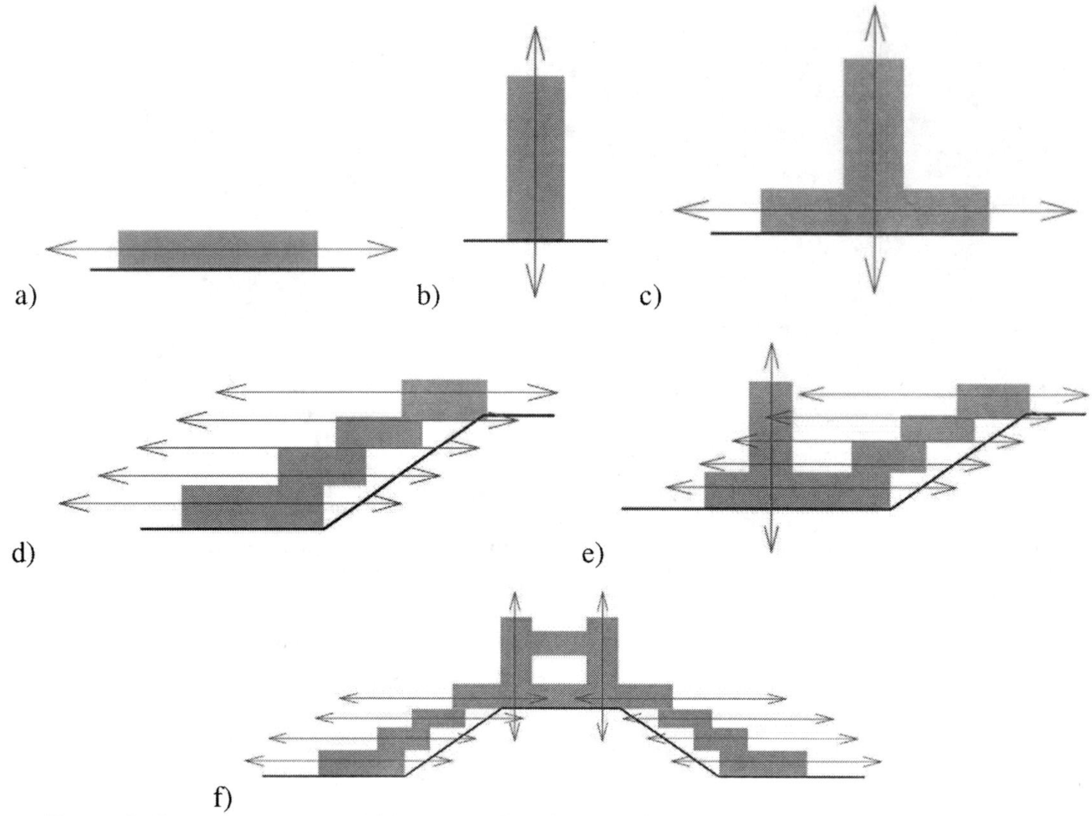


**Figure 3**. The scheme of spatial location of settlement objects: a) horizontal system; b) vertical system; c) horizontally vertical system$ d) horizontally terrace system e) vertically terrace system; f); ziggurat, horizontal and terrace system of settlement (the scheme was developed by V. Kuzmych).

Respectively, the diagonal development in the context of the visual perception is more expressive than the combination of horizontal and vertical ones. Going of visual stimuli of the human eye outside the horizon level gives the image a certain degree of monumentality and sacral character. This factor was taken into account in the process of construction of sacral or temple buildings. For intensification of effects of perception of these objects, the builders also used the dominant landscape, where the groundwork was laid. A typical example of this is represented by mounds, pyramids, ziggurats. Comparing objects in different systems of settlement or planning structures, we should note the fact that these options are characterized by various results of visual perception of the environment. In case of a horizontal low-rise housing system, the density and visual distances are relatively small. Such compositional approach was imposed, first of all, by the development of construction and climatic conditions. The modern construction technologies allow building of vertical systems of settlement giving the preference to large-distance perspectives. In case of a vertically horizontal option of construction, unequally distant perspectives are resolved depending on panoramic observations. In case of a horizontally terrace scheme of construction, the mechanism of visual perception is activated that is characterized by a broad range of the visual location.

Another option of the visual location is the scheme envisaging overlaying of multi-faceted, combined views and perspectives. A combined ziggurat and horizontal system of construction can be an example in this case. This scheme can be characterized by monumentality, grandeur and sacral character. For a long time, architects use this technique in pyramids, ziggurats and sacral buildings. This option of the urban development is rather popular and actively enriches the visual panorama of a city.

**Diversification of angles of perception of the visual panorama**

On the example of the observation wheel, a viewer can better understand the differentiation of viewing points in the process of perception of the urban environment. Typical characteristics of viewing points are their location above the ground, at the ground level or under the ground level as a plane. All these positions will have interim levels that will differ by the angular position relative to the ground level. The basic position of viewing points is characterized by the level of height at which a human visually perceives the surrounding environment. The conventional average level comprises 1 meter 60 centimeters. Such constant is used for the level of drafting of architectural plans. At this level, a viewer typically observes the environment. An important factor of this level is a natural feeling of gravitation ensured by the vestibular apparatus. We should draw pay attention to the angular feeling of the horizon line. According to the theory of evolutionary development, a human body clearly perceives and controls the verticality of its position. This ability is ensured by the gravitational attraction of the Earth. The next unchanging factor is the mechanism of perception of the horizontal axis expressed in the visual feeling of the horizon line. Therefore, the most important factor is the feeling of verticality and horizontality of the human body position. In case of a change of the height above ground, a viewer receives more axonometric or 3D picture as opposed to the frontal or facade projection. These axonometric views will differ from the height above ground, so architects introduced a notion of “bird’s-eye view”. Of course, the height of the bird’s-eye view is rather conventional. In the context energy viewing flows, the range of tonal correlations increases greatly thus enriching the overall perception of the situation. Since the axonometric perception from the bird’s-eye level is atypical and uncharacteristic for humans, in these cases the mechanism of visual therapeutic action is activated. For example, mountain trips and observation of the horizon from dominant surfaces are accompanied by certain positive signals based on non-standard positions and emotions. We recourse to certain abstraction for perception of the axonometric-angular view of the environment with conventional characteristics that partially ignore the feeling of gravitation. Wide panoramic and angular reach includes a large angle of the horizontal range. Layering of tonal layouts combines a summary panoramic view. Apart from that, sub-level types of the urbanized industry became widely popular owing to the current construction possibilities. Subways, underground parking lots, shopping and entertainment centers and other technological buildings are based on the principles of angular-and-altitude character. Looking at such objects, people must lift up their heads and change the viewing angle, thus creating non-standard visual signals that are not typical for the biologic nature of a human. The typical techniques of use of viewing points as tools of the architectural approach include the stylobate scheme of the viewing monumentality, creation of podiums, scenes, terraces, using semi-basement parts of buildings. Such techniques provide a possibility of higher angle of perception that improves visual qualities and adds monumentality. In many situations involving gatherings or concentrations of people, such as in concert, cinema or sports complexes, such tools provide a possibility of a targeted programmed observation. Elevations of surfaces of the angular perception ergonomically ensure uninterrupted high-quality perception of observed objects in exhibition complexes or lecture and audience halls. At the same time, these observed objects are placed above the horizon level.

**The role of viewing points in multi-level angular perception of urban environment**

Viewing points are inseparably connected with conditions of the gravitational stability. Standing of a human on different surface levels or artificial created constructions leads to different perceptions of the linear-and-outline contour and tonal-and-coloristic filling of the environment. Therefore, a change of the multi-level gravitational association leads to a change of the tonal perception of the environment. Accordingly, the set of short distance signals received at the level of the current area will be replaced by energy flows of long-distance perspectives of the open space. Such change of the height and angular observation of the panorama will provide for a change of energy viewing points, thus amplifying and enriching the general impression. In conditions of the existing urban structure, we can observe attempts of changing and amplifying of angular points and heights of the overall perception of the environment. Using the existing landscape dominants, architects try to maximally expand the horizon panorama. Development of technological and constructive possibilities of the construction industry provided visually new horizontal urban panoramas, thus enriching the energy chord of the surrounding environment. All this leads to increase of therapeutic energy flows. From the ancient times, people organized viewing points that were non-standard from the point of view of the angular perception, which received the heightened attention of viewers.

**Types of spaces and conditions of multi-level perception**

Creation of multi-functional spaces is inseparably connected with the filling and structure of a city. The aspiration to persevering of maximally open spaces remains an important factor in designing of architectural environments. Closed or partially open spaces have a less positive impact on the biological rhythm of perception of energy and visual communications. In these cases, a change of the angle of visual adaptation will be helpful. Adjustment of the visual perception and focusing at an object can improve the overall image of perception. Rotating adjustments of the sector of perception of objects neutralize the monotony of signals of visual location. In this case, a technique of using transparent external elevators can become useful from the point of view of constructive decisions. This will neutralize the feeling of oppression in closed spaces. The speed of external elevators will help to diversify perception images, enriching the overall viewing flow and avoid the monotony caused by being in a closed space. High-altitude options of using of external elevators neutralize the impression of the closed interior space with re-orientation at open exterior signals. The factor of the peripheral viewing angle will be relevant, working as a factor of visual limitation of contours of the perception image. The lesser distance to objects provides for the more sharp and accurate "readability" of the architecture as a whole.

**The notion of the panoramic layout in the urban science**

The panorama provides for the layout of city viewing points summarizing different locations of architectural objects in the space or in the environment. The maximum angle of horizontal capturing is caused by availability and grouping of certain urban functional zones or complexes. The composition of a panorama is formed at the account of the planning structure of a part of a city that is directly connected with the unity of the functional structure and the master plan. Very often, the structure includes administrative, residential, manufacturing and recreational subgroups united in a single function and task of the settlement structure. The basis of the visual presentation is concentrated on the functional zoning of a city thus expressing the capturing angle, range of observation and significance of a certain group or housing complex. To provide better and more substantial information on a part of the urban structure of a city, a focus point and a vertical angle of master plan observation is chosen and adjusted. Each specific planning of the structure requires the most appropriate viewing point in order to maximally express the compositional character and function of a city. The panorama can capture both a partially fragmentary range and a summary layout that captures the most typical planning and high-altitude urban structures. Therefore, an architect includes into the panoramic presentation the accentuated objects that should be emphasized by expressing the character and height of the buildings in a city or residential group. As a rule, such compositional decisions are subordinated to and represent the functional and planning structure of a city that express not only the horizontally planar structure of the master plan, but also the altitude-and-level characteristics of planning objects. Due to its orthogonal character, the panorama gives a feeling of deeply structural correlations of objects in the urban environment. In other words, it expresses the scale and dimensions of a city.

**Architectural environment in conditions of a large city or metropolis**

The high density of the urban structure and creation of artificial open spaces essentially change the aspect of observation and perception of the environment. In this case, a human intuitively tries to look up, thus changing the feeling of staying in the closed dense space to the feeling of liberation by entering into an open space. On densely built streets, where focusing of a look at a certain object takes place intuitively at the level of the psychological perception and obtaining of energy from the open space, a human eye looks for the energy of free flows. At the level of psychology, closed spaces and environments were always perceived with fear having a negative and depressive impact. Automatic search for larger distances is caused by the biological need to receive another range of frequency of energetic radiation inherent to humans.

**Conclusions**

In the current context, people become dependent on the density of construction caused by high prices for land and at the same time lost the battle at the visual and psychological level. Transfer from one closed space into another, more opened one, has become a characteristic feature of activity of the society. Attempts to partially solve the problem of space in the architecture began from using balconies, terraces, patios, penthouses, aerariums, and swimming pools. Analyzing the current architectural and construction industry, we can observe repeated transfers from one closed space to another. A window and a balcony are the best typical inventions. They give possibility to neutralize the feeling of the closed space by opening multi-distance views and ensuring the natural insolation. It's typical for people to perceive visual characteristics of the open space, therefore they feel attracted by the leisure in the open recreational space with its energetically positive aspects. Communication schemes of transfer from the place of living to the place of work, as well as the technology of manufacturing are basically associated with closed environments. The natural need in the open space was artificially replaced by the necessity of the closed environment. Transfer from the closed space of a living space is accompanied by the existing necessity of communication constructions, such as corridors, elevators, escalators, with transfer to closed spaces of transportation vehicles. The mirror-like nature of commuting ways similarly limits the ability of being in open natural environments. It is very important to have a green recreational zone of the urban forest type that fundamentally changes the energetic base of viewing points broadening flows of the positive visual character.

**References**

[1] Dijokienė D 2013 Significance of art and science of urban design in the study programme of architecture / Urbanistikos meno ir mokslo reikšmė architektūros studijų programoje. *Journal of Architecture and Urbanism (Electronic Materials)* **37** (4) pp 268-278 https://doi.org/10.3846/20297955.2013.866822
[2] Khamtran 2014 The London Eye from across the River Thames. [online] Wikipedia. (Last updated 4 April 2009) Available at: https://en.wikipedia.org/wiki/London_Eye#/media/File:London-Eye-2009.JPG [Accessed 15 March 2020]
[3] Koznarska H 2012 The value of publishing points for the perception of the image of the city (In Ukrainian). *Modern problems of architecture and urban planning* (Kyiv: Kyiv National University of Construction and Architecture) **30** pp 213-218
[4] Kuzmych V, Pavliv A & Petrovska J 2018 Angular harmony in architectural design of city sketch *Przestrzeń i Forma* (35) pp 107–112 DOI: 10.21005/pif.2018.35.C-01
[5] Scatena D 2018 Perceiving the void and the living being to build new environment friendly urban spaces. *Journal of Architecture and Urbanism (Electronic Materials)* **42** (1) https://doi.org/10.3846/jau.2018.2024
[6] Sydorenko I and Minder V 2018 Analysis of the system of species disclosure of park landscapes on the hill (In Ukrainian) *Problems of urban development (Electronic Materials)* **1** pp 111-120
[7] Vodzynskyy Ye 2010 Analysis of the specific disclosure of architectural monuments and its tools (In Ukrainian) *Ukrainian Academy of Arts, (Electronic Materials)* **17** pp 154

IOP Conf. Series: Materials Science and Engineering **907** (2020) 012013 doi:10.1088/1757-899X/907/1/012013

# Innovative Approaches in the Period of Mass Industrial Development (on the Example of Residential Areas of Kharkiv)

**O Buriak[1],O Vihdorovich[2], Y Gayevyi[3], A Golovchenko[1]**

[1] Department of the Fundamentals of Architecture, Kharkiv National University of Civil Engineering and Architecture, Ukraine
[2] Department of the Urban Planning and Urban Studies, Kharkiv National University of Civil Engineering and Architecture, Ukraine
[3] Department of Building Technology, Kharkiv National University of Civil Engineering and Architecture, Ukraine

abouryak@yahoo.com

**Abstract.** The article considers architectural, city planning, engineering and technological innovations in the building industry that were carried out in Kharkiv in the 1960-1980s and led to profound changes in the social organization of city life. During these years, Kharkiv confirmed its reputation of the scientific and technical capital of Ukraine in the field of architecture and city planning. The most important innovations that were worked out in Kharkiv as in one of the largest cities include the microdistrict system, a stepped public service system, and the focusing method in city planning. It was also innovative to create standard series of residential buildings for mass industrialized development at Kharkivproekt and Ukrmistobudproekt institutes. Technological innovations were created in high-tech house-building plants, organizational and managerial ones – in managerial combines capable of coordinating the entire project and investment cycle. Mass industrial residential development in the mid-1980s allowed liquidation of the housing crisis in Kharkiv. However, this period gave rise to a number of problems in functioning and maintenance of the huge residential areas created at that time. Analysis and objective assessment of the achievements and mistakes of the era of mass industrial construction is a task of current interest for Ukrainian architectural science.

## Introduction

In the 60s - 80s years of the past century, Kharkiv, as before in the 1920s and 1930s, was one of the leading experimental sites for urban planning, architecture and new building technologies. A large-scale program to provide the population with individual housing has opened a new stage in urban development in the former Soviet Union. Its implementation was expressed in numerous architectural and urban and engineering and technological innovations, encouraged the construction industry to form a new design typology, caused profound changes in the social organization of life of citizens. The role of the Kharkiv school of architecture and construction during this period was not limited to the implementation and testing of solutions created in central research and experimental design institutions. During these years, Kharkiv, confirming its reputation as the scientific, technical and educational capital of Ukraine, was a source of many significant innovations in architecture and urban development. Innovations in the design and implementation of mass industrial construction in Kharkiv covered numerous kinds of architectural, city planning, typological, organizational, managerial, engineering and technological innovations.

***The relevance*** of studying the period of mass industrial development today is evolving by the fact that large Ukrainian cities again, as half a century ago, are faced with a choice of strategies for spatial development, technological modernization and the beginnings of reorganization of urban life. Turning to their own experience from the middle of the twentieth century could clarify why and how in those years it was possible to organize and implement large-scale, truly innovative transformations in the huge and complex sphere of housing policy and urban development. The fact that the content of these transformations is relevant today not only in historical-theoretical, but also in practical and organizational terms, shows, in particular, the active and growing interest of Western researchers to the Soviet and, in particular, to the Kharkiv experience. Examples include research projects led by Theo Hauben (Rotterdam, the Netherlands) [1], Philipp Moiser (Berlin, Germany) [2], or Barbara Engel (Karlsruhe, Germany) [3]. This series can be extended.

In the history of national architecture, the representation of this most important stage in the development of post-war architecture still cannot find integrity. On the one hand, this period is quite thoroughly interpreted as a fundamental turn to solve the housing problem, to modernize the settlement system, social services, as well as all life support systems of the city and to realize the technological re-equipment of construction, and finally to return the architecture and urban planning in the USSR on the main path of world architecture development. On the other hand, the same era appeared as responsible for violating the traditional way of urban environment, for the loss of cities of individuality and human scale, for the dehumanization of the environment, – and all this combined with reproaches for secondary and non-independence, schematic urban thinking and primitive technical solutions.

***The purpose of this publication*** is to approach the formation of a holistic picture of changes in the architectural and urban planning sphere that occurred during the enrolment of mass industrial development, on the example of a large Ukrainian city, highlighting the innovative components in each of the main aspects of this picture – urban and planning, architectural and typological, engineering-technological and socio-functional.

Such a definition of the goal raises quite specific research objectives, most of which have been ignored by scientists for several decades, despite their obvious importance in gaining a systematic view of the recent history of Ukrainian city planning area. This is, firstly, a reproduction of a chronological sequence of events that determine the shift to industrialization of the city planning, and, secondly, closely related to the first task of identifying key figures – organizers, architects and engineers who had the greatest innovative contribution to creation of qualitatively new structures of research and experimental design as also of the building production systems.

This is followed by the task of identifying a system of innovations and, among them, those fundamental innovations that have allowed to create unprecedented in the history of construction technological processes of industrial production of residential buildings and organize the corresponding powerful chains of construction and installation works, engineering preparation of extensive territories for settlement zones development, reconstruction of a transport framework under many times increased intracity traffic, design and technological equipment of a complex technical system of transportation, installation and finishing of large-sized elements of panel buildings.

This set of tasks cannot be limited to the space of one city, even the largest - for the obvious reason that a number of systems that provided a revolution in urban planning, had a transregional or even all-Union scale – from the union-republican system of typical design to providing the building industry with assembly and transport machines and mechanisms. Even the more important seems to be the task of identifying the real sources of appearing of certain innovations and setting, where possible, the Ukrainian and local, Kharkiv priority in architectural, planning and engineering innovations.

It should be borne in mind that the scale of the phenomenon under study is extremely significant, and its structure and process are marked by a high degree of complexity. Therefore, the realistic goal for this publication is, as noted, to define an approach to constructing a general picture of that, without exaggeration, revolutionary rebuilding of all architectural, constructional, planning and socio-organizational activities which has been unfolded in the process of industrialization of mass housing development in the country. On the other hand, limiting the tasks of this article by Kharkiv is a conscious and well-considered decision that provides real opportunities to collect empirical material and access the necessary historical sources.

## 1. The period of mass industrial development in the architectural research

Period of the mass industrial development is widely reflected in the professional literature in the analytical and generalizing works of domestic authors. The works that resonate in the main content are devoted to the achievements of the period under study in scale of the former Soviet Union (Z. N. Yargina [4], Y. V. Kositsky, V. V. Vladimirov [5], V. A. Lavrov [6], V. T. Shimko [7], etc.), as well as the actual Ukrainian experience (M. M. Dyomin, G. I. Lavrik [8, 9], Y. P. Bocharov, G. Y. Filvarov [10], A. M. Pleshkanovska [11], N. M. Gabrel [12], Y. M. Bilokon' [13], V. I. Nudelman, Y. Y. Klyushnichenko, M. M. Kushnirenko [14]).

Almost all of these studies are characterized by a common planning and urban orientation. The predominance of this view of the epoch, figuratively speaking, "from a bird's eye", had its basis in the socio-political background in which the architectural theory existed and changed. It is easy to see that these texts almost without exception belong to the period of the so-called "perestroika" and the first post-Soviet decade, so they are based on methodological principles and a system of internal guild self-censorship, which has been developed in Brezhnev's time. The lack of research that would link all the significant levels of transformation that has taken place, including issues of economics, technology, research and development, social organization and stratification, etc., stems from the lack of "social order" to identify driving forces, goals, limitations and results of the implemented policy [15]. Maintaining the integrity of the process and the ability to regulate the situation should remain the party-government prerogative.

In this regard, a somewhat separate position is occupied by a textbook for architectural universities "Social Fundamentals of Architectural Design" (authors Z. N. Yargina and K. K. Khachatryants, published by Moscow Architectural Institute, 1990), which quite fully and interestingly reveals the systemic relationship between perspective architectural and city-planning research of that time and experimental-design developments. The nature of typical design, technology of zoning of typical projects and features of their introduction into the building industry are also discussed [16].

Quite naturally, the works devoted to a purely Kharkiv matters do not go beyond the discussion of urban composition (I. O. Alferov, V. L. Antonov, R. E. Lyubarsky [17]) and the historical evolution of urban development (Yu. M. Shkodovsky, I. M. Lavrentyev, O. Yu. Leibfreid, Y. Y. Polyakova [18]). At the same time, names of the real heroes of the era, Ukrainian architects and urban planners of this period – O. I. Zavarov, L. M. Tulpa [19; 20], E. G. Vainshtein, P. N. Nirinberg, L. D. Nivina, A. D. Konsulov and others remain known only to a few living contemporaries, witnesses and participants at the turning point of the 1950's and 1960's.

Over time, with the expansion of industrial development and the accumulation of both positive and critical experience of new urban planning practices, the view at the industrialization of construction has undergone a profound evolution. From an almost magical innovative means of transforming the living environment, the large building industry has gradually gained in the public eyes and in the professional reflexion an image of powerful and rigid, almost spontaneous force, which destroys the traditional urban lifestyle and dehumanizes the living environment. In the public consciousness the stable theme has gradually consolidated – the need to "humanize" the arrays of industrial housing, i.e. to return to the urban environment aesthetic quality and to overcome the socio-psychological alienation, which became to be increasingly associated with the effects of industrialization.

The economic and socio-political crisis of the late 1980s till mid-1990s, the cessation of governmental funding for housing developments, and the subsequent radical transformation of the economic system of Ukraine led to the final decline of the planning strategies and technologies of the large building industry. Under these circumstances, there was no incentive to invest in the study of the course and to record the creative results of the period of mass industrial development. The period, which until recently seemed a heroic era of domestic urban planning, was now perceived as a clear aesthetic

IOP Conf. Series: Materials Science and Engineering **907** (2020) 012013 doi:10.1088/1757-899X/907/1/012013

and socio-functional symbol of totalitarianism, which must be overcome and eradicated. The true significance of the architectural and urban heritage of the 1960s - 1980s (as well as their artistic and social heritage) began to loom in just two decades, closer to the mid-2000s, and primarily not in Ukraine but in Western and Central Europe, based on different, sometimes opposite life and professional intentions.

On the one hand, it is time for the creative vanguard of Western architecture to comprehend the experience of more and more frequent contacts with the former world from behind the Iron Curtain. These contacts eventually turned into an unformed feeling that in the recent wave of fascination with the avant-garde of the 1920s and condemnation of the eclecticism of the 1930s and 1940s, the Western analysts overlooked something big and important in the architecture of the East (and their admiration and repulsion were in fact prepared by pro-modern Soviet researchers and not without sophistication embedded in the minds of Western colleagues). Then it was time for young researchers from the Netherlands and Germany to discover for themselves and for us the ambiguous and therefore even more interesting achievements of the very recent socialist past [21]. A typical example is the book "Ideals in Concrete" [1], created as a result of a Dutch research expedition to study the environment of large areas of industrial development, which was conducted in a number of major cities in Ukraine and Russia. A comprehensive study of the experience of mass housing construction in the USSR is presented, for example, in the book "On the way to the typology of Soviet mass housing: a collection in the USSR 1955-1991" [2], which analyses the stages of mass housing development, probably for the first time on the basis of careful factual research.

Another impelling motivation to study the complex, heterogeneous and ambiguous heritage of the period under study arose in the context of the implementation in Germany of a number of large-scale government rehabilitation programs for similar housing estates inherited from the GDR. Since the early 2000s, the influential German magazine "Deutsche Bauzeitung" [22] has been systematically publishing research on the social issues of the architectural environment in the "Social City" series. In these studies the technique of architectural sociology is improved, conceptual principles of social rehabilitation of peripheral areas of big cities of the former GDR are developed, namely for areas of mass industrial building. These concepts also include accommodation programs for groups of residents with different social and property status, ways to intensify participation programs aimed at involving the population in improving their living environment, and so on.

Another, we can say, charitable purpose of these studies is to use the German experience of rehabilitation of industrial buildings for the possible deployment of similar programs in the former Soviet Union. It is also important to assess the practical results of the completed German programs, to find out the economic and social efficiency of the huge investments that were directed to them from the state budget of FRG [23]. Of fundamental interest is an international project initiated by the Karlsruhe Institute of Technology, entitled "Socialist City an Unloved Heritage?" Within the framework of the project, international research teams, in which scientists and specialists from Germany, Ukraine and the Russian Federation were represented, examined the results of rehabilitation projects in the areas of mass housing development in Berlin, Dresden and Halle-Neustadt. Following the interim results of the first phase of this research project, a number of interesting publications have already been published [3], the project will be continued in the near future.

## 2. The general innovative character of mass residential development in the 1960s – 1980s

The most important architectural and planning innovations that have been developed during the quarter of the century under study, especially in the largest cities of Ukraine, include the introduction of a microdistrict system with a tiered ("stepped") system of public service of the population. In the course of this implementation, in which Kharkiv played the role of one of the pioneers, fundamental shortcomings just in the functioning of the service system were revealed (Figure 1). Here Kharkiv city planners played a leading role in creating the method of so-called "focusing", which in the Ukrainian SSR was first used in Kharkiv. We should emphasize the crucial contribution of the architect and urban

IOP Conf. Series: Materials Science and Engineering **907** (2020) 012013 doi:10.1088/1757-899X/907/1/012013

planner Leonid Matviyovych Tyulpa [25], the main author of urban concepts and coordinator of design work on the most important residential areas of Pavlove Pole and Saltivka.

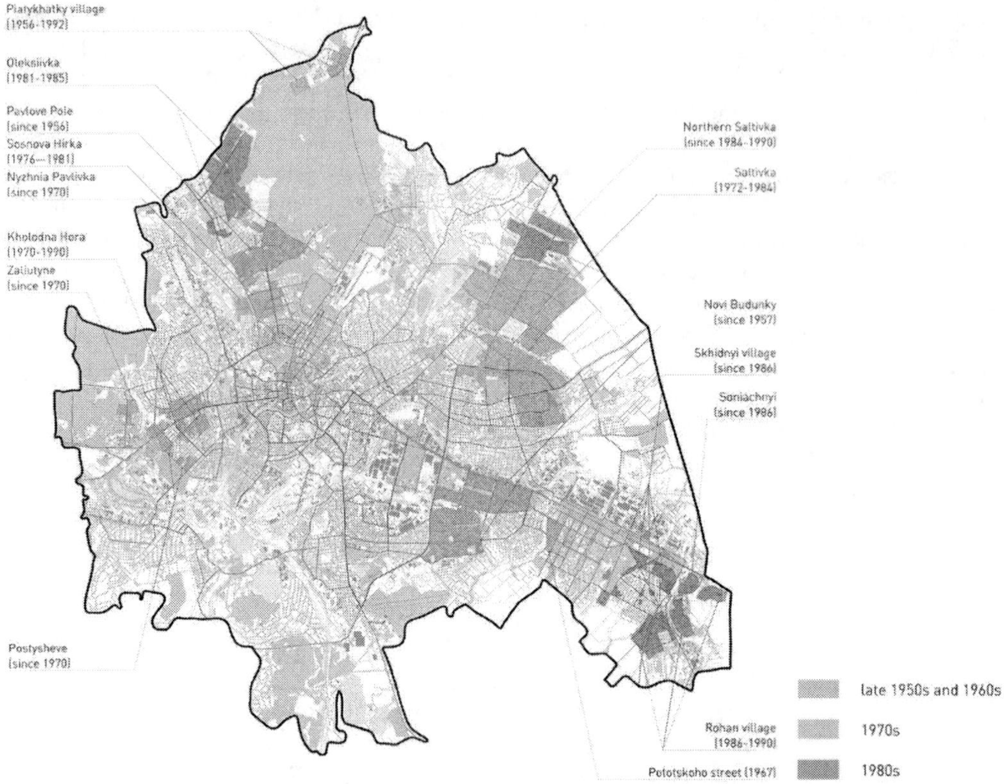


**Figure 1.** Development periods of industrial housing development in the city of Kharkiv (authors' scheme)

Fundamentally innovative was the activity of Soviet designers at the time in the field of creating typical series of residential buildings intended for mass industrial developments. In this direction, Kharkiv not only implemented and adapted to its technological capabilities a series developed in central and republican design institutions. On a par with them were innovative developments in a series of typical housing projects belonging to the "Kharkivproekt" and "Ukrmistobudproekt" institutes, with the interested participation of Kharkiv house-building plants (HBP).

Direct technological innovations in the period of industrial development were associated in Kharkiv with the creation of high-tech at the time house-building plants DSK-1 and DSK-2, the introduction of vibrating rolling conveyors production of large panels and the transition to fundamentally new flow building technologies. Due to this, the projects of standard series included original technical solutions for the serial production of prefabricated elements, innovations in housing planning decisions and in the structure of neighborhood development (Figure 2, [25]).

IOP Conf. Series: Materials Science and Engineering **907** (2020) 012013 doi:10.1088/1757-899X/907/1/012013

**Igor Alfiorov**
(1930-1978) architect,
Chief Architect of Kharkiv
(1965-1975).
Head of the development
of the masterplan for Kharkiv
(1967)

**Victor Antonov**
(1926-2015) Sc.D., architect,
professor, Chief Engineer
of the "Kharkivproject"
institute (1956-1967).
Head of the development
of the masterplan for Kharkiv
(1964)

**Yuri Vorobyov**
(1928-1999) Ph.D., engineer,
professor, Chief Engineer,
later Head of Kharkiv
House-Building Plant DSK-1
(1968-1973).
UkrSSR State Prize Laureate

**Oleksandr Gayovyi**
(1926-2008) Ph.D., engineer,
professor, Chief Engineer
of "Kharkivzhytlobud" Combine
(1966-1984). UkrSSR State Prize
Laureate,
Honored Builder of Ukraine

**Volodymyr Reusov**
(1925-2011) architect,
professor, Head of the
Building Department of the
Kharkiv Regional Party
Committee (1968-1985).
Honored Builder of Ukraine

**Leonid Tyulpa**
(1922-1994) architect,
urban planner,
author of the projects
of the largest Pavlove Pole
and Saltivka housing estates.
Honored architect of Ukraine

**Figure 2.** Architects and engineers, prominent organizers and active participants of mass housing industrial development in the city of Kharkiv

An important organizational and managerial innovation in Kharkiv, as in other major cities of Ukraine, was the creation of a unified system of building production in the structure of powerful managerial combines ("Kharkivzhitlobud" and "Kharkivprombud" of the Ministry of Industrial Construction of UkrSSR) that were able to effectively coordinate the project-investment cycle, including design and technological equipment, site preparation and implementation of construction on the scale of the metropolis. A significant innovation of the 1970s in the management and coordination of design

IOP Conf. Series: Materials Science and Engineering **907** (2020) 012013 doi:10.1088/1757-899X/907/1/012013

and construction was the transition to the creation of so-called City Planning Complexes as an enlarged basic unit of planning and implementation.

Innovative for the former USSR neighbourhood system was the culmination of a long and quite dramatic history of invention, multiple correction and implementation of this concept. This story spanned the entire first half of the twentieth century and unfolded in a number of developed countries in Europe and North America. It consisted of numerous and significant events in the social and architectural-urban thinking of the era and in construction practice. These include the well-known concept of the "Garden City" belonging to the Englishman Ebenezer Howard (1850-1928) [26], its modification in American housing in the 1920s, known as the program of "big quarters" [27], sociological analysis and critique of this practice by representatives of the Chicago School of Sociology [28], conceptual development and large-scale organizational, design and construction practice of Patrick Abercrombie (1879-1957), in particular in the project "Greater London" [29], and a significant number of repetitions, improvements and modifications of these samples almost all over the civilized world. But the completion and the most large-scale and consistent application of the microdistrict system was acquired in Soviet republics and in countries that aspired to a socialist orientation – on both sides of the "iron curtain". In a sense, these Soviet practices were the culmination and highest point of development of an entire era in world modernist urbanism [30].

## 3. New housing estates of Kharkov - experimental site of architectural and urban innovation

In the Ukrainian practice of industrial urbanism, a fundamental innovative step was taken in the development and implementation of the project for Pavlove Pole residential district in Kharkiv (since 1956, architects O. G. Krykin, L. M. Tyulpa and I. Ya. Feigin) – the first residential array on the free territory, designed for construction products of house-building plants [31]. The connection between the planning schemes in the design of the Pavlove Pole district and the nature of the construction industry in the form of newly established enterprises of conveyor production of residential and later social facilities was also innovative. In particular, the composition of the dotted buildings of the array was brought to life by the use as a basic planning unit of large-panel buildings – completed industrial products. Smaller architectural and planning innovations include the method of blocking buildings, which was first used also in Kharkiv, on Pavlove Pole (Figure 3, [32]).

**Figure 3.** Panorama of Pavlove Pole housing estate

Further search for more diverse opportunities to block houses prompted Kharkiv designers to improve the typology of large-panel buildings, stimulated the evolution of interior planning, development of a new typology of sections and stair-lift units, more flexible in terms of combinatorial urban planning capabilities. The use of various methods of blocking buildings marked later a departure from the original principle of the finished construction product like the whole house and provided a transition to more complex, variable and plastic ways of organizing the three-dimensional structure of the neighbourhood.

Detailed planning of residential areas using the method of "focusing" (Figure 4, [24]) for the first time in the USSR was developed in Kharkiv during the construction of Saltivka – the largest housing estate in Ukraine and in the former USSR (over 400 thousand inhabitants) (Figure 5, [32]). The method was based on the concentration of socio-cultural, commercial and consumer services in the vicinity of public transport stops and big connection hubs, with the appropriate calculation of accessibility radii. The "focusing" system made fundamental adjustments to the microdistrict neighbourhood system and practically eliminated the "first stage" of the service system within the microdistrict as a planning principle. Making these adjustments to the ideology and practice of microzoning was recognition of the principle of human mobility in the modern city and the actual rejection of the basic binding of the service system to the place of residence.

**Figure 4**. District no 602 in the structure of Saltivka housing estate. Project.

IOP Conf. Series: Materials Science and Engineering **907** (2020) 012013 doi:10.1088/1757-899X/907/1/012013

**Figure 5.** Microdistricts of Saltivka housing estate from a bird's flight

Typological innovations can be found in the development of Kharkiv's own series of industrial housing. These design developments of “Kharkivproekt” and “Ukrmistobudproekt” project bureaus coexisted with the optimization approach, which prevailed in the adaptation and partial improvements of the "all-Union" series, in particular in the work on the II-57 series, the basic for the DSK-1 house-building plant.

Technical and technological innovations are concentrated in the creation and cyclical modernization of the production of vibro-rolled panels, much more economical compared to frame-panel structures. Three BPS-6 vibro-rolling mills were installed at Kharkiv DSK-1 the first house-building plant in Ukraine. Undoubtedly innovative was the transition to the conveyor technology for the production of large panels and the corresponding current technology of installation of finished houses. Innovations in the technological processes of production and processing of assembly units were largely focused on finding ways to maximize the displacement of wet processes. Much effort, in particular, has been put into the implementation of sanitary cabins of full factory readiness.

Mass industrial housing construction allowed Kharkiv to eliminate the housing crisis around the mid-1980s. At this time, the city has reached a housing rate of 21.0 $m^2$ per capita. This figure, which is comparable to the European average, is still maintained in Kharkiv.

Innovative methods of designing and implementing mass construction were dictated by the shift of a unique historical situation, which stretched from the beginning of the Khrushchev’s "thaw" to the end of the period of "stagnation" in the late Brezhnev’s time, and, accordingly, by the action of a number of motivating factors. Among them, the leading role was played by the well-known and repeatedly mentioned by the bad word centralization and total nationalization of closely interconnected systems of the construction industry and typical architectural and urban planning. The privatization of the housing stock created in those years served as a kind of "airbag" for a large segment of the urban population in the following decades.

At the same time, the very specific nature of architectural and urban planning innovations of the period between the 1960s and 1980s led to a number of problems and difficulties associated with the today’s functioning and technical maintenance of the then created housing estates and complexes. Sober analysis and unbiased assessment of the achievements and failures of the era of mass industrial development is an important and relevant task for Ukrainian architectural science, especially against the background of the obvious shortage of new ideas and positive programs in today's urbanism.

## Conclusions

Mass industrial residential development of the late 1950s - early 1980s, which radically changed the face of modern Ukrainian city, only in recent years has become the subject of complex historical and architectural research. Kharkiv, the first capital of Ukraine, the second largest and most significance city in the country, represents a fertile field for such an analysis. Here remained crowded, Europe's largest arrays of large-panel buildings, not only fully functioning, but also continuously growing. Here is still available a historical evidence of how, over the course of a quarter of a century, a unique and highly productive design and production order has emerged, flourished and faded away, and in general ensured the successful implementation of a complex set of innovative and modernization measures.

The scope and specificity of the subject of these studies should be correlated with corresponding research approaches. The experience of the contact with the reality of the coup of the 1960s - 1970s allows researcher to outline their main characteristics:

Multidisciplinary, dictated by the close interconnection of purposeful changes carried out on different "floors" of urban development – from technical re-equipment and the creation of a new large building industry to a radical breakdown of architecture and urbanism thinking and to profound changes in the style of urban life.

Not only relevance, but also extreme urgency. Each year, it is becoming increasingly difficult to record the content and interconnection of events that are rapidly fading into the past, together with their participants and eyewitnesses, technical archives and huge factories. The urgency of a completely different plan is due to the fact that the gigantic volumes of large-panel development that have arisen with lightning speed now promise as quickly to enter the period of complete moral and physical depreciation. Kharkiv, like most large post-Soviet cities, is facing a period of forced intensive reconstruction of modernist sleeping areas.

The style of research needed in this context should be fundamentally different from the usual style of architectural and historical discourse. Its focus is not so much on architectural and urban morphology, not even on its functional patterns, but the meaning and structure of the transition situation, on the means to ensure a successful socio-technological breakthrough and ways of disposing of obsolete material objects and behavioral stereotypes.

Finally, the research of Kharkiv era of mass industrial development should solve the same problem with respect to the 60s - 80s as unfolded in recent years studies of the heroic era of the avant-garde of the 20s and 30s. The matter is to return to studies of Modern architecture the real historical scale of those architectural and urban innovations that took place in Ukraine in the amazing 20th century.

## References:

[1] Hauben T, Wagenaar C, Dings M and Linders J 2004 *Ideals in Concrete: Exploring Central and Eastern Europe* (Rotterdam: Fonds BKVB) p 179
[2] Moiser F and Zadorin D 2018 *Speaking of Soviet Typologies of Standardized Buildings. Industrial Multifamily Housing in USSR from 1955 to 1991* (Berlin: DOM publishers) p 447
[3] Engel B 2019 *Repetitive Residential Development in Socialist City. Heritage, Values and Possibilities* (Berlin: DOM publishers) p 240
[4] Yargina Z N 1984 *Town Planning Analysis* (Moscow: Stroyizdat) p 245
[5] Yargina Z N, Kositsky Y V and Vladimirov V V 1986 *Fundamentals of Town Planning: Textbook for Higher Education. Specialty «Architecture»* (Moscow: Stroyizdat) p 326
[6] Ikonnikov A V, Mikhalovsky E V and Lavrov V A 1973 *Historical and Architectural Heritage and Contemporary City* (Moscow: CSTI of Civil Engineering and Architecture) p 160
[7] Shimko V T 1990 *Architectural Development of Urban Realm: Architect Training Manual* (Moscow: Higher school 1990) 223
[8] Demin N M 1991 *Urban Layout Development Management* (Kiev: Budivelnyk) p 184
[9] Lavrik G I and Demin N M 1975 *Methodological Fundamentals of Areal Planning* (Moscow: Stroyizdat) p 98

[10] Bocharov Yu P and Filvarov G Y 1987 *Production and Spatial Organization of the Cities* (Moscow: Stroyizdat) p 256
[11] Pleshkanovs'ka A M 2005 *Functional and Planning Optimization of the Urban Areas Usage* (Kiev: Institute of Urban Studies) p 190
[12] Gabrel N M 1988 *Architectural Perfection of the Residential Development of the City Including Improvements in Maintenance of the Open Spaces. Thesis for a Candidate Degree in Law Sciences* (Moscow, Russia)
[13] Bilokon Yu M 2002 Functional and Structural Forms in Regional Planning. Scientific Publication (Kiev: Kiy) p 98
[14] Klyushnichenko Y Y 1999 *Socio-economic Fundamentals of Town Planning and Development* (Kiev: Ukrainian Academy of Architecture, MET of Town Planning) p 348
[15] Sazonov B V, Orlov M A and Fedoseeva I R 1970 Social Research in Architecture *USSR Architecture* **8**
[16] Yargina Z N and Khachatryants K K *1990 Social Fundamentals of Architectural Design. Textbook for Higher Education* (Moscow: Stroyizdat) p 242
[17] Alferov I A, Antonov V L and Lyubarsky R E 1977 *Formation of Urban Realm (in the Context of Kharkiv)* (Moscow: Stroyizdat) p 104
[18] Shkodovsky Yu M, Lavrentiev I N, Leibfreyd A Yu and Polyakova Yu Yu 2002 *Kharkiv Yesterday, Today, Tomorrow* (Kharkiv: Folio) p 206
[19] Tyulpa L 1963 *New Microdistricts in Kharkiv* **10** (Kiev: Building and Architecture)
[20] Tyulpa L 1964 *Saltivka District Planning* **3** (Kiev: Building and Architecture)
[21] Heatherley O 2015 *Landscapes of Communism* (London: Allen Lane) p 624
[22] Theresia-Heckemeier M 2001 *Social City* (Leinfelden-Echterdingen: Deutsche Bauzeitung) pp 61–65
[23] Meyerovitch M G and Frantseva Yu V 2017 *Problems of Complex Regeneration of Residential Area of the Regions of Large-panel Development. Possibility to Adapt German Experience to Social, Economical and Law Conditions of the Countries of CIS* (Irkutsk: News of Higher Educational Institutions. Investments. Development. Realty)
[24] Bouryak A, Antonenko N and Lavrentiev I 2017 *Leonid Tyulpa – the Architect of the Period of Mass Industrial Development* **issue 8** (Zaragoza: ZARCH Journal of Interdisciplinary Studies in Architecture and Urbanism) pp 154–169
[25] Source: prof. O. Buriak's personal archive
[26] Steuer M 2000 *A Hundred Years of Town Planning and the Influence of Ebenezer Howard The British* Journal of Sociology **51** pp 377–386
[27] Christensen C A 1986 The American Garden City and the New Towns Movement (Ann Arbor, MI: UMI Research Press) p 203
[28] Shonle R 1983 *The Chicago School of Sociology, 1918-1933* Urban Life **11** p 415
[29] Abercrombie P 1945 *Greater London Plan: 1944* (London: H.M. Stationery Office) p 220
[30] Belogolovsky V and Novikov F 2010 *Soviet Modernism 1955-1985* (Ekaterinburg: Tatlin) p 210
[31] Fomin V 1967 *Development of Kharkiv. Socio-economic and Cultural-household Issues of City Development* (Kharkiv: Prapor) p 31
[32] Stroyobzor. Portal of new buildings. 10.28.2017 How the sleeping areas of Kharkov were built [electronic resource] https: stroyobzor.uaassetsfiles_фото_www.спальные районы_стар.jpg)
[33] Photo by Vladimir Bysov; published with permission of the author

# Architectural and City-Planning Aspects of Innovation Hubs Formation in the Context of Creative Urban Regeneration (On the Case of Ukrainian Cities)

**K T Holubchak**[1], **O S Sleptsov**[2], **and R Tomlins**[3]

[1]Department of Architecture and Urban Planning, Ivano-Frankivsk National Technical University of Oil and Gas, Ivano-Frankivsk, Karpatska street, 15, 76019, Ukraine
[2]Department of Architecture Basis and Architectural Design, Kyiv National University of Construction and Architecture, Kyiv, Povitroflotsky Avenue, 31, 03680, Ukraine
[3]International Centre for Transformational Entrepreneurship (ICTE), Coventry University, Coventry, Gosford Street, CV1 5DL, United Kingdom

golubchak.kateryna@gmail.com

**Abstract.** Cities have become the key arenas of territorial competitiveness in the context of creative economy development and continuous innovation processes in all the spheres of social life. Functioning as a community of knowledge, innovation, creativity and learning, modern cities are becoming more dynamic, complex, diverse and open. In this context, new urban strategies and trends need to be taken into account. The concepts of "innovation hub" and "creative city" as a result of the process of creative urban regeneration are covered in the paper. The article states that innovative creative spaces provide unique opportunities to accelerate the generation and diffusion of innovation, especially through effective communication channels and a positive innovation ecosystem. The article covers the topic of architectural and functional aspects and design tendencies of innovative hubs in the structure of a creative city. The research material includes the architectural analysis of existing innovation hubs in Ukraine.

## 1. Introduction

In the face of globalization challenges, technological boom and the rise of creative economy and innovation, the competition increases not only at the international but also at the regional level [1]. Recent studies in the field of regionalism show that the region's competitiveness is based on its innovative ecosystem, integrating scientific, technological, investment and human potential on the basis of the network organization of innovation activities [1-7].

Today the dominant factors in the development of a modern post-industrial economy are human capital, knowledge, creative potential and the ability to generate ideas, and this has accordingly led to the transition of the global economic system to innovative tracks of economic development [8].

Therefore, the competitive development of countries at the present stage is based on the consolidation of innovative achievements, creative and intellectual potential and the formation of a number of strategic directions for the effective and rational use of this potential and the formation of creative urban centers (innovative hubs), which are the key elements of creative cities.

The American scientist Charles Landry, who is considered to be the author of the "creative city" concept, describes new urban planning strategies of transforming a city into a creative space. According to his theory, a creative city is a nucleus where "different types of cultural activity become an integral part of the economic and social life of the city, and include intellectual capital" [3]. Creativity is the key factor that enables modern city to respond to contemporary challenges and provide technological innovation. This will become more essential as we respond to the opportunities and challenges of Industry 4.0.[4].

Cities are becoming the nucleus of communication, creative collaboration, spaces for learning and creative expression despite the growth of virtual communication. The trend of dynamic growth in the number of innovative spaces and the creation of a network of "creative cities" worldwide confirms this development. However, the lack of theoretical basis and academic research on the impact of innovation and creative economy on the architectural and urban solutions of such spaces, as well as the lack of coverage of the phenomenon of the creative city and its importance for regional development in the case of Ukraine determine the relevance of the study.

As the global economy places the greater value on innovation, the essential role and value of innovation creative spaces will equally rise, becoming "a locus of experimentation in design and architecture" [9].

Thus, the purpose of the research is to outline the concept of a creative innovation hub and creative city, as well as to reveal the architectural and urban features of their formation.

To achieve this goal, the following tasks are formulated:

- to reveal the role and importance of creative industries in the territorial and socio-economic development of cities;
- to formulate a conceptual apparatus that defines the main terms of the study - creative city, innovative hub, creative cluster;
- to identify the architectural and urban planning features of the formation and functioning of innovative spaces in modern conditions of urban and regional development on the basis of analysis of global and Ukrainian experience;

The theoretical significance of this study is to develop strategic directions for the effective use of the region's creative potential and the organization of urban space by creating innovative creative spaces and further shaping creative cities.

## 2. Creative Economy as a prerequisite for the creative cities' emergence

### *2.1. Theoretical Discourses*

The national and regional level of innovative development and creative urban regeneration has attracted a lot of attention of foreign and Ukrainian scientists in recent decades. The theoretical basis of the research is formed on the basis of the theory of creative economy of J. Howkins [5], the theory of creative class of R. Florida [6] and the concept of creative city of Ch. Landry [3]. Based on the theories of these three authors, the urban creative economy can be described as a system consisting of three elements: the creative city, the creative class and the creative industries.

The creative economy aims to create an attractive investment climate, increase the region's competitiveness, promote the development of the social sphere, modernize the education sector etc. [1]. Within the creative economy, new economic models, new types of social relations and new cultural paradigms are emerging, as well as a new value, communication and instrumental sphere, in which consumption of goods and services fades into the background and the provision of opportunities for the creative expression and self-realization of each person comes first. The creative economy influences the image of cities and stimulates the urban community to a new, better life, to the interpenetration of different spheres of thinking and employment. Thus, cities become an immense space for communication, generating new ideas and economic growth [10].

A number of scientists from different fields such as S. Davymuka and L. Fedulova [1], A. Hautamäki [7], A. Scott [11] affirm that creativity and innovation, the generation of new knowledge are now the only ways to improve the competitiveness of each country and its regions.

Theoretical and practical issues of the development of creative economy in Ukraine, the concept of creativity in the context of territorial and socio-economic development of cities, as well as the parameters of creativity of territories and the possibilities of its assessment are covered in the works of Ukrainian scientists S. Davymuka and L. Fedulova [1], Y. Ushkarenko, A. Chmut, and K. Syniakova [2], I. Skavronskaya [12]. Challenges and prospects for the development of creative industries in Ukraine as well as an analysis of successful creative hubs were described in the research project conducted by C. Farinha [13].

However, the study of regional creativity, and in particular its architectural and urban planning aspects, cannot be considered complete because of the short study period and the pace of economic trends. In particular, it is rightly emphasized that the influence of individual factors on the formation of a creative environment in the region and the favorable conditions for the development of a creative economy may complicate the development of a regional policy of creative economy development [1]. Therefore, there is a need to investigate the relationship between innovative development and the creation of a creative environment in ensuring the socio-economic development of regions and updating their strategic orientations. In particular, there is a lack of theoretical base and academic research on the impact of innovation and creative economy on the design and architectural-urban solutions of innovative hubs. This is exacerbated by the lack of coverage of the phenomenon of the creative city, its importance for regional territorial and economic development and the possibility of its implementation in the conditions of Ukraine which determines the relevance of the study.

*2.2. The concept of "Creative City"*

Today, more and more European cities embrace a number of creative strategic approaches for economic development. This is confirmed by the trend of dynamic growth in the number of innovative spaces and the creation of a network of "creative cities" worldwide.

Bradford's paper 'Creative cities structured policy dialogue backgrounder' defines the key features distinguishing the creative city: "Creative cities are dynamic locales of experimentation and innovation, where new ideas flourish and people from all walks of life come together to make their communities better places to live, work, and play" [14]. The author finds that creative cities make a significant contribution to meeting national and regional policy goals of environmental sustainability, economic innovation and social inclusion. This gains additional relevance in current engagement with the Sustainable Development Goals [15].

Charles Landry [3] describes a new method of strategic urban planning in order to attract human creativity, imagination, talents and transform the city into an innovative creative environment. He defines creativity as the key enabler for a modern city to respond to contemporary challenges, provide technological innovation and rediscover its cultural heritage. A creative city is a place where "various cultural activities become an integral part of the economic and social life of the city and include intellectual capital in relation to products, processes and services" [3].

In the "Creative City" study the author defines the concept of urban circulation as a theoretical concept and a dynamic tool aimed at creating renewable urban energy that can govern the city. According to the researcher, creative urban spaces, wherein ideas are exchanged, generated and communicated, as well as the favourable atmosphere that promotes openness to new ideas are some of the factors that should be taken into account when identifying the creative potential of the city [3]. This creates a clear demand of urban design.

According to J. Verwijnen [16], "the notion of a creative city draws our attention to the fact that, beyond the traditional forms of urban renewal and regeneration, cities show a growing interest in creating centers that are saturated with an atmosphere of innovation and creativity" [16, 17].

Formation of creative cities is also a priority of the Government of Ukraine in "Strategy 2030". Namely, it is planned to realize positive dynamics in the formation of human capital, economic development and improvement of the quality of the urban environment through the development of the creative cluster [18, 24]. However, the Ukrainian experience of spatial-economic and urban development of most cities is characterized by imbalance, narrow profile orientation and the development of traditional industries. A promising way out of this state of affairs is to encourage and develop creative industries that have been recognized in foreign practice *for decades* as a powerful driving force for national and regional economic development and an effective tool for solving a number of key territorial and economic challenges within regional development. This position, along with the emergence of new forms of innovative creative spaces, namely innovation hubs, raises the question of exploring them in the context of architecture and urban planning – how and where they are most often formed, how they interact with existing buildings and environmental functions and what are their architectural peculiarities.

## 3. Innovation hubs as the key elements of a creative city's ecosystem.

### *3.1. Creative environment - the key component of the innovative ecosystem. Innovation hub concept*

The essential condition for creating a creative city is proper infrastructure whether intuitive and organic or structured and planned. One of the key tools for the realization of the concept of a creative city is the formation of creative clusters and innovative creative spaces. These are new structures represented in their wide typology - creative hubs, coworkings, fab-laboratories, business incubators, accelerators, which in their totality form a positive ecosystem for the development of innovation.

Creativity is an important characteristic of an environment conducive to innovation: the use of non-standard approaches to eliminating difficulties leads to the emergence of innovative results (qualitatively new technologies, products, etc.). S. Davymuka and L. Fedulova [1] note that creative activity acquires the characteristics of a self-developing process and initiates economic development, thus leading to the formation of innovative ecosystems. Generating and attracting creative ideas is the first prerequisite for a successful ecosystem to function: the existence of communities that create ideas and have the opportunity to come together and discuss them on an appropriate physical or virtual organizational platform, where the boundaries between the physical, digital, economic, social and cultural space blur [17]. Therefore, the need to provide a creative environment characterized by the presence of conditions that stimulate creative people to seek new, non-standard, more effective solutions to problems, and to implement measures that contribute to achieving their goals through rapid and successful adaptation to changing reality is becoming increasingly urgent.

Innovation hubs, being important centers of integration of education and art, research and business, generate a wide range of positive effects on the city's economy, its attractiveness and image, the creation of new products and innovative services, the support of talents and startups, fostering networks at regional and international levels and the provision of high-level educational opportunities which in general leads to the better quality of life. An inherent feature of innovation hubs is their multifunctionality and multidisciplinarity, which are supported by the lack of boundaries between social, economic, physical, digital and cultural space. In many cases, these spaces bring together creative class members, entrepreneurs and companies in one building, creating mixed spaces that lead to natural collisions and the sharing of ideas that foster innovation [6]. The main idea behind innovation hubs is to use technology, arts and science as the driving force for the revival and redevelopment of the city.

These are places where people can improve, learn and express themselves. But, of course, the greatest effect can be achieved when all these features are harmoniously combined physically and virtually through design and architecture. Therefore, the architectural analysis of the concept of innovation hub in the context of creative city regeneration requires careful research.

### *3.2. The architectural and design features of innovation hubs*

Different aspects of the influence of architectural organization of innovation hubs on the enhancement of creativity, innovations and overall development of creative economy are explored by I. Cunha and C. Selada [17], M. Osborne, [19], M. Meinel [20], A. Savanovic and M. Orel [19], B. Moriset [22], J. Matheson and G. Easson [23], K. Holubchak [24], V. Mulas [25], L. Narum [18].

The first recorded hub was founded in 2005 on the basis of an old warehouse in London, as a cross between a professional membership community, a learning lab and a business incubator [27]. The concept of the hub was architecturally designed with the aim to provide a collaborative space where people could meet, work and learn within a modular and flexible structure. The planning structure included a semi-open meeting room, co-working space with transformable leaf-shaped tables, kitchen and library [27].

Innovation hubs are rapidly becoming a worldwide phenomenon. In the last ten years, thousands of innovative spaces where people can design, test, scale and launch enterprising new ideas have popped up around the world. Nowadays, innovation hubs throughout the world designed to facilitate and foster collaboration are becoming powerful support environments for creative individuals who are creating collaborative communities and interconnected networks [21].

Positive developments in this area have also been felt in Ukrainian practice in recent years. Currently, there is a successful network of creative spaces on the territory of Ukraine, which are represented in their wide typology - from small coworking's and business incubators to huge creative centres. In a short span of time, they have gained popularity and have been fully implemented in urban life, and their numbers are steadily increasing. There are many successful examples of projects implemented in Ukraine, including: UNIT city (Kyiv), Creative Quarter (Lviv), Art Factory (Lviv), Innovation Center "Promprylad. Renovation" (Ivano-Frankivsk), Platforma Art Zavod (Kyiv), iHUB, the network of centres of innovation and entrepreneurship in the IT sphere (all over Ukraine) [13].

One of the most impressive in scale is UNIT City - the first innovation park in Ukraine (Figure 1) - a unique platform for business, education, cutting-edge technologies and creative economy development. The project is sponsored by Ukrainian architectural company TSEH, architectural firm MONO Architects and other architectural teams that have worked on individual objects, landscaping and interior design. Construction on 25 hectares of territory of the former motorcycle factory in Kyiv started in 2016 and now the business campuses, coworking, IT-school, a number of creative spaces for various events, cafes and gyms and a wide spectrum of entertainment facilities are functioning successfully, producing a wide range of economic and social values [28].

**Figure 1.** UNIT city innovation park project [28].

The project envisages the formation of a so-called "city in a city" – a full-fledged infrastructure for various purposes: innovative educational institutions, housing, offices, IT centres and accelerators, VR and AR laboratories, the medical centre, the kindergarten, production workshops and laboratories. The total area of UNIT City will reach the final stage of about 860,000 m² and it will become one of the largest in Europe, intended to integrate into the international innovation ecosystem [28]. All the buildings will be LEED-certified, which will confirm the environmental friendliness and energy efficiency of the innovation centre by one key criterion.

Innovation Hubs should be shaped by "the needs of local creative economy and are inserted in particular cultural context, supporting local creative industries and promoting their own values" [29].

In many cases, innovation hubs locate their functions inside abandoned buildings and former industrial sites, gradually renovating them. In this way these forgotten territories are inhabited and brought back to life making a huge contribution to urban development and renovation [13].

For example, the innovative centre "Promprylad. Renovation" in Ivano-Frankivsk (Ukraine) [30], formed on the basis of the former industrial manufacturing plant, is a connecting link between business and startups and makes an important contribution to the local economy by offering a platform for generating and testing new ideas leading to innovation, gradual job creation and urban regeneration (Figure 2).

The activity of the "Promprylad. Renovation" center focuses on the intersection of four areas of innovative development - informal education, contemporary art, urban planning and the new economy [30]. The building is currently undergoing revitalization and only the restored third "pilot" floor of the plant is fully opened, and has already confirmed the efficiency and the invaluable potential of the project. There is a wide range of amenities for the city's residents - educational and children's attractions, a conference room, coworking area, the workshop space, art gallery, offices of innovative enterprises and even the city's investment policy department. The Promprylad "city" is gradually becoming a hotbed of communication, creative collaboration, a space for learning and creative expression.

This project is a clear testament to how innovative centers and creative spaces are able to give new breath to undeveloped urban areas, industrial spaces and abandoned buildings. Therefore, this positive experience can become a successful model for implementation in other territories of Ukraine with unfavourable starting conditions, including those amongst industrial dereliction and outside the biggest Ukraine cities.

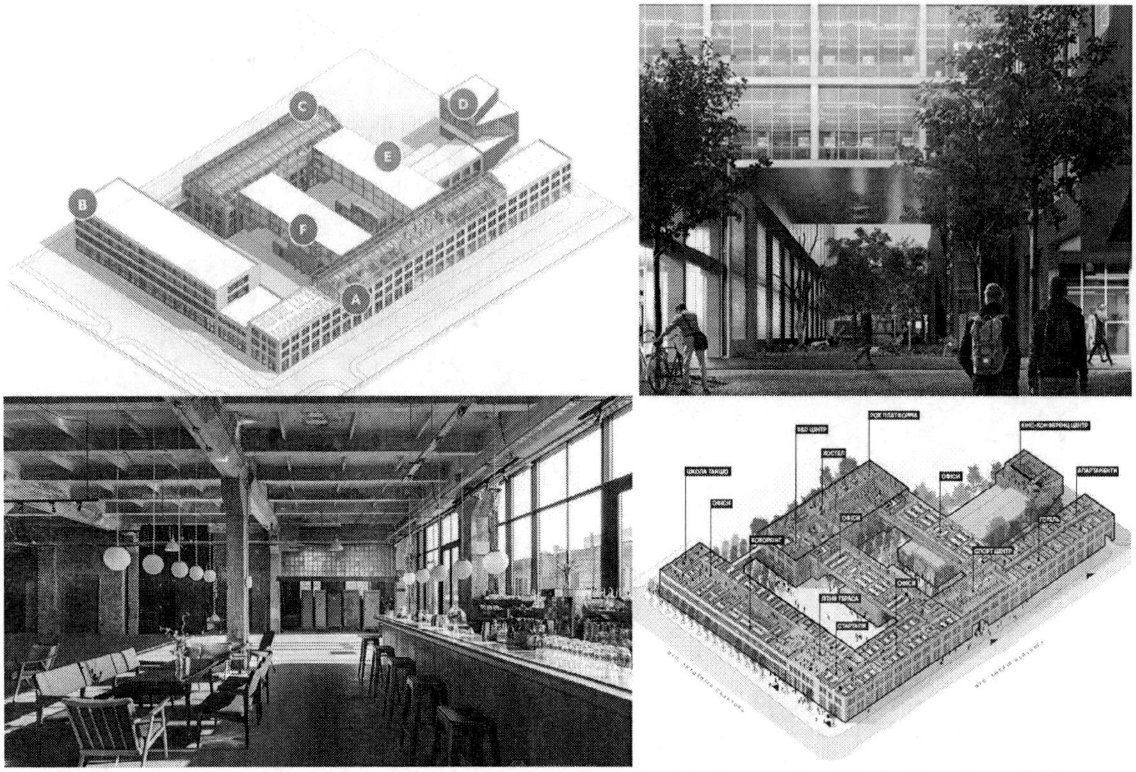


**Figure 2.** The innovative centre "Promprylad. Renovation" in Ivano-Frankivsk (Ukraine) [30].

The ideal architectural organization of a modern innovation hub from a social value perspective [31] aims to provide the community with a multifunctional collaborative environment with basic amenities such as open learning space, conference rooms and private offices workstations, lounge areas alongside self-service bars and a wide range of art and culture facilities such as exhibition halls and galleries or entertainment events' spaces [32]. Therefore, innovation hubs are perceived not only as interactive spaces, but more as "hybridized work and habitation ecosystems of inventive and productive communities" [21].

The analysis allows us to formulate the main trends of architectural and planning organization of innovative spaces:

– *Multifunctionality.*

The emergence of new forms of leisure, the need for communication, creative realization and the new demands of modern society require innovation hubs to be more diverse in their functional structure and list of premises. In addition to traditional planning elements, the planning solution of a modern innovation hub should include: lecture and multimedia classrooms, state-of-the-art technology rooms (AR and VR), children's rooms, creative studios and workshops, exhibition halls, coworkings, recreational greenhouses and atriums. Empowering the hub with new types of functional processes and activities will ensure the creation of a new image of the building as a socially important cultural center.

– *Flexibility and mobility of the functional structure.*

Flexible planning is an integral feature of the innovation hub, as it maximizes functional and psychological openness that promotes communication and interaction, allows users to easily adapt and navigate the services offered, freely choose the type of activity and be able to design the space around themselves.

In this case, the zoning of the space is not made by walls, but by visual distinction using color, floor level, ramps, lighting, furniture and transformed wall structures. Transformation of space and adaptation to specific activity is achieved due to modular equipment and furnishings, which are transformed into small areas for individual work or enlarged cells for co-working and group work according to the requirements of its users.

The planning solution should provide opportunities for adaptation to the dynamics of social processes, new trends and functions.

– *Creativity enhancing interior design*

Innovative creative spaces are aimed to foster idea generation, facilitate inspiration, skills' development, innovative thinking and communication. This is achieved by increasing the proportion of communicative space and space for creative realization, by filling the space with inspirational virtual and physical elements, like white boards for sketches, ideation cubes and others. The emergence of the latest design and digital technologies has greatly expanded the design space of architecture. The spatial structure of modern innovation hubs cannot be formed by physical material only, but it must incorporate a wide range of digital and light components, as well as new structural elements [24].

– *Inclusiveness and openness for everyone.*

The architectural planning of an innovation hub should ensure its accessibility and convenience for all age and social groups (children and pensioners, youth and disabled people), as well as for each individual user with their own needs and motivation, thus reducing social exclusion and preventing segregation [33]. Many of the innovation hubs are designed for the non-existent "average" user. However, it is important to take into account the needs of a wide range of users, the model and the specifics of their behaviour, thereby ensuring the high social performance of the building.

To solve this problem, local sociological surveys should be conducted and profiles of potential users and scenarios of their activity should be formed on the basis of them, alongside adopting key principles of accessible design.

– *Technology and interactivity.*

Technological innovations that permeate both the physical and virtual components of such hubs provide visitors with a wide range of opportunities for incubation of ideas, communication, learning, and information exchange [19]. The optimum functioning of an innovation hub is impossible without incorporating the innovative virtual, digital and information technologies and recent developments in the field of searching, storing and broadcasting information, multimedia stands, communication facilities and other high-tech equipment in its structure. The modern user requires a full range of the latest technological developments for a comfortable stay and full implementation of all requests, said developments being embedded in the structural elements of the building, furniture and the entire space of the hub.

– *The architectural expressiveness and aesthetics of the innovative space*

This feature implies the building's ability to meet the requirements of time and to reflect the aesthetic ideals of a society. World examples show original facade interpretations, the use of the latest structural technologies and building materials. The emotional and psychological expressiveness of innovation hubs becomes an instrument of its actualization and popularization among the population.

Facade and interior design should reflect the specifics and spirit of the place, and the building as a whole should position itself as an integral part of this place.

## 4. Conclusions

Creative cities and innovation centres that have been dynamically emerging throughout the country over the last few decades are called to underpin urban regeneration strategies in modern conditions, and are able to generate a wide range of positive effects, such as:

- significant economic growth and cultural development of the population;
- talent support and new jobs creation in the sphere of creative industries;

- production of innovative products and educational services;
- creative urban regeneration and branding of the city;
- promotion of Ukraine in the international arena.
- increase of investments and tourist attractiveness;
- development of continuous innovation.

However, it is important to place high demands on the architectural organization of innovation hubs in order to fully realize aforementioned positive effects. The article outlines the main trends that should be taken into account when designing an innovation hub, but it should be understood that there is no standard recipe on how to design the innovation hub correctly. The architectural organization of each individual project requires careful research and consideration of the cultural, economic, demographic characteristics of the local context. The architect needs to understand feelings, dreams, challenges of the communities this space is designed for. Thus, the architectural environment of the modern innovation space is a subject-space framework designed and co-designed to facilitate the implementation of a wide range of processes and activities: science and creativity, information exchange, communication, leisure, saturated with the processes of interpersonal communication.

**Acknowledgments**
The investigation is inspired and supported by the Creative Spark project, funded by the British Council.

**References**

[1] Davymuka S A and Fedulova L I 2017 *Creative Economy Sector: Experience and Directions of Development* (Lviv: Dolishniy Institute of Regional Research of NAS of Ukraine) p 103
[2] Ushkarenko Iu V, Chmut AV, and Syniakova K M 2018 Creative Economy: The Essence of The Concept and the Significance for Ukraine Under the European Integration, *Economy and Society* **18** 67-72 [in Ukrainian]
[3] Landry C 2012 *The Creative City: A Toolkit for Urban Innovators* 2nd Edition (London, UK: Earthscan) p 23
[4] Jackson D 2018 How Industry 4.0 Will Affect Entrepreneurs [Online]. Available: https://www.entrepreneur-resources.net/how-industry-4-0-will-affect-entrepreneurs Accessed April 5, 2020
[5] Howkins J 2013 *The Creative Economy. How People Make Money from Ideas* (London, UK: Penguin Books) p 27
[6] Florida R 2012 *The Rise of the Creative Class* 2nd ed. (New York, USA: Basic Books) 36.
[7] Hautamaki A 2010 *Creative economy and culture at the heart of innovation policy* (Helsinki, Finland: Helsinki University Print) p 18
[8] Fieraru V A 2017 Creative clusters within regional tourist and recreational complexes *Perspectives on science* **8 (95)** 74-77
[9] Wagner J and Watch D 2017 *Innovation Spaces: The New Design of Work* (Washington, DC: The Brookings Institution) 128
[10] Mengi O Durmaz Drinkwater S Öner A and Velibeyoğlu K 2017 Place management of a creative city: The case of Izmir. *International Journal of Knowledge-Based Development* **8 (3)** 271–291.
[11] Scott A J 2010 Cultural economy and the creative field of the city *Geografiska Annaler: Series B, Human Geography* **92 (2)** 115–130
[12] Skavronska I V 2017 Creative Industries in Ukraine: Analysis and Prospects of the Development, *Economics and Sociology* **10** 87-106
[13] Farinha C 2017 *Developing Cultural and Creative industries in Ukraine* (London, UK: British Council) p 38
[14] Bradford N 2004 *Creative cities structured policy dialogue backgrounder* (Ottawa: Canadian Policy Research Networks) 28
[15] Carpentier C-L 2020 Empowering MSMEs & Entrepreneurs to help us achieve the SDGs International Council for Small Business Webinar, March 31
[16] Verwijnen J 1999 The creative city's new field condition. Can urban innovation and creativity overcome bureaucracy and technocracy *Urban Development and the Information Society* 12–35.

[17] Cunha I V and Selada C 2009 Creative urban regeneration: the case of innovation hubs. *Int. J. Innovation and Regional Development* **1 (4)** 371–386
[18] The Strategy of development of the sphere of innovative activity for the period till 2030, 2019. [Online]. Available: https://zakon.rada.gov.ua/laws/show/526-2019-%D1%80?lang=uk. [in Ukrainian]. Accessed on: April 17, 2020
[19] Osborne M 2016 *Innovative Learning Environments* (Auckland: Core Education White Paper) p 46
[20] Meinel M Maier L Wagner T Voigt K I 2017 Designing Creativity-Enhancing Workspaces: A Critical Look at Empirical Evidence *Journal of Technology and Innovation Management* **1(1)** 1-12
[21] Savanovic A and Orel M 2018 *The Role of Creative Hubs in the Freelance Labour Market* (London: British Council) 27
[22] Moriset B 2013 Building new places of the creative economy. The rise of coworking spaces. 2013. [Online]. Available:. https://halshs.archives-ouvertes.fr/halshs-00914075 Accessed on: April 17, 2020
[23] Matheson J and Easson G 2015 *Creative Hub Toolkit* (London, UK: British Council)
[24] Holubchak K Chuchuk Yu Savchuk T Negrych I 2019 The phenomenon of creative economy in Ukraine and defining its place in the architectural space. *Advances in Economics, Business and Management Research,* **99** 73-78
[25] Mulas V Nedayvoda A and Zaatari G 2017 *Creative Community Spaces: Spaces that Are Transforming Cities into Innovation Hubs* (Washington, D.C: World Bank Group) p 75
[26] Narum L A 2013 *Guide Planning for Assessing 21st Century Spaces for 21st Century Learners* (Washington, D.C.: Learning Spaces Collaboratory)
[27] Bachman M 2014 How the Hub found its centre, Stanford Social Innovation Review, Available: http://www.ssireview.org/articles/entry/how_the_hub_found_its_center Accessed on: April 17, 2020
[28] Innovation Park UNIT City 2019 Available: https://ufuture.com/uk/nashi-kompaniyi/innovatsijnij-park-unit-city/ Accessed on: April 17, 2020
[29] British Council 2015 Creative Hub Toolkit, Available: http://creativeconomy.britishcouncil.org/blog/15/06/28/creative-hubkitmade-hubs-em erging-hubs/ Accessed on: April 17, 2020
[30] Promprylad. Renovation 2019 Available: https://warm.if.ua/projects/promprylad. Accessed on: April 17, 2020
[31] Tomlins R 2015 *Social Value Today: Current public and private thinking on Social Value.* (Coventry: HouseMark).
[32] Schuermann M 2014 *Coworking Space. A Potent Business Model for Plug'n Play and Indie Workers* (Berlin: Rocket Publishing)
[33] Hesemans H 2017 From Ceramic Centre to Creative Hub *The Maastricht Journal of Liberal Arts (MJLA)* **5** 18-25

IOP Conf. Series: Materials Science and Engineering **907** (2020) 012015 doi:10.1088/1757-899X/907/1/012015

# Innovative Component in the Creation of the Most Outstanding Greenhouses in the World of XIX - XXI Centuries. At the Junction of Architecture and Engineering

**L. Kachemtseva[1], N. Khoroian[1], O. Gella[1], S. Nikiforova[2]**

[1] Department of the Fundamentals of Architecture, Kharkiv National University of Civil Engineering and Architecture, Kharkiv, Ukraine
[2] Foreign Languages Department, Kharkiv National University of Civil Engineering and Architecture, Kharkiv, Ukraine

kachemtsevalubov@gmail.com

**Abstract.** The article examines the innovative contribution of architects and engineers to the design and construction of "translucent architecture". Examples of the most outstanding greenhouses since the middle of the XIX century to the beginning of the XXI century, as well as materials on projects, realizations and modern use of these constructions published in the public domain are considered. Particular attention is paid to the facts of professional success from the biographies of architects, engineers and entrepreneurs who created innovative buildings of "translucent architecture" during this period. It is concluded that the latest architectural and constructive solutions are connected with the paradigm shift in the XXI century, modern scientific achievements and high-tech engineering. The source of inspiration for the transfer of natural principles in architectural solutions are the factors of the latest innovative achievements of architecture and civil engineering.

## 1. Introduction

The term «innovation» was first introduced by an Austrian-American economist, Professor Josef Schumpeter, in the 1920s. According to his definition, innovation is the unification of all innovations available to the manufacturer – materials, production methods, opportunities to introduce products to the market and the creation of new markets [1]. Innovation is interpreted by Schumpeter as a new scientific and organizational combination of production factors, motivated by the entrepreneurial spirit of a special type of entrepreneur – "innovator". The key feature of innovation is novelty, but innovation is not just novelties, but the causative factors of production and production functions. The most important innovations are the introduction of new products and new production methods (commodity and technological). Basic innovations (large inventions) create the preconditions for the formation of new generations of equipment and technology, initiate secondary (small) innovations that develop the primary idea.

The term "innovation", born in economics, has become very popular. In authors' opinion, it can be used when considering the development of architectural and construction business in general as well as its individual segments. Professional publications traditionally emphasize historical, morphological and stylistic descriptions of objects. In the twentieth century, the publication of author's design concepts was added to this list. The innovative component is rarely the focus of architectural researchers.

IOP Conf. Series: Materials Science and Engineering **907** (2020) 012015 doi:10.1088/1757-899X/907/1/012015

**The purpose of this work** is to study the innovative contribution of architects and engineers to design and construction on the example of the most prominent greenhouses in the world since the mid-nineteenth century to this day.

## 2. Literature review

The work is based on the history of design and construction of greenhouses [2-8], published photographs, drawings and descriptions of famous architectural projects, their implementation [3; 9-14] and modern use [2; 10; 11]. Particular attention is paid to the facts from the biographies of architects, engineers and entrepreneurs – the creators of outstanding examples of "translucent architecture" [4; 7-9; 12] – and the role of innovative component in the history of their professional success.

## 3. Materials and methods of research

The main method of this research is a comprehensive and comparative study of various sources on the issue (drawings, photo materials of famous greenhouses, literary and biographical sources). The most iconic greenhouses of the late XIX – early XX centuries are analyzed from the position of identifying the innovative contribution of their creators to various fields of architecture, construction and engineering. As a result, the characteristic features of the work of architects were revealed, as well as a number of aspects of the activities of civil engineers and designers – from the sole creator to large design teams and engineering companies. It is noted that among the builders of greenhouses an important place is occupied by entrepreneurs manufacturing building structures.

## 4. Results

A notable example of innovation in architecture and construction business is the Crystal Palace by Joseph Paxton [2]. Under this name the exhibition pavilion that housed the first World's Fair entered the history of architecture. Joseph Paxton was not an architect by profession; he was a gardener and built greenhouses [3].

From Wikipedia it can be learnt that “Crystal Palace in London's Hyde Park was built in 1850-1851 of iron and glass for the 1851 World's Fair. Exhibition hall with an area of over 90,000 m$^2$, 564 m long and up to 33 m high was built under the direction of Joseph Paxton and accommodated up to 14,000 visitors. The palace was built in less than a year, thanks to the innovative idea of J. Paxton. It consisted in the use of modular iron and wooden structures. 5,000 people took part in the construction (no more than 2,000 people were on the site at a time). The glazing for the palace was made by the famous English company Chance Brothers. In total, more than 84,000 m$^2$ of glazing were produced” [4, 5].

The construction of the Crystal Palace went beyond the traditional construction practice of the time, which gave rise to new forms and aesthetic principles in architecture. With its new aesthetics, innovative design, organization and construction technology, the building belonged to the architecture of the future, the symbol of which it became. Design innovations have had an even stronger impact on the further development of construction equipment than external forms [6]. It can be stated that modern architecture "made of glass and metal" owes its origin to the Crystal Palace – and greenhouses.

Greenhouses are a special type of building that has certain requirements: year-round full illumination of the room at a sufficiently high height, lack of a large number of supports, and ideally – their complete absence, the possibility of ventilation, repair and replacement of light-transmitting coating, the ability to maintain temperature and humidity. When considering the experience of construction of these structures, it can be observed that the aforementioned requirements have become the basis for the search for new expressive forms and construction solutions, which continues today.

Until the middle of the XIX century, greenhouses or hothouses, as they were then called, were the subject of prestige for rich and influential people. The second half of the XIX century is the golden age of greenhouses. With the beginning of the widespread use of metal in construction, the construction of large-span structures became real. The largest greenhouses were built in England during the Victorian era.

One of the prominent and successful figures in the construction of greenhouses is Richard Turner (1798-1881). He was a manufacturer of iron products, one of the leading designers and builders of greenhouses in the Victorian era. Turner was a brilliant innovator who boldly expanded the construction capabilities of his time. The most famous of his surviving works is Palm House, the world's first building of such a scale with a wrought iron frame, not supported by columns (the so-called "Turner design construction") [7; 8]. The palm trees greenhouse built in the 1840s has become a hallmark of the Royal Kew Botanical Gardens. The greenhouse project for palm trees and other tropical and subtropical plants belongs to Demim Burton, and the realization is owned by Richard Turner. Before that, no one had ever built greenhouses of this size, so the architects used shipbuilding technology. The Palm House also looks like an inverted skeleton of a ship. The combination of technical and architectural skills has qualitatively changed the possibilities of botanical gardens.

One of Turner's first major buildings in London, Winter Garden in Regents Park, completed in 1846, no longer exists. And it was the first greenhouse in all of Europe open to the general public. It also became the prototype of a palm greenhouse in the gardens of Kew. The Winter Garden in Regents Park existed until 1932.

Turner worked extensively in Ireland, where he was born and grew up. One of Turner's most famous works, which has survived to this day, is the Curvilinear Range [7] in the Glasnevin Botanical Garden in Dublin. The garden was founded in 1795, it took several decades to form a collection, and then the question of building a permanent greenhouse for heat-loving plants arose. Work began in 1843. The Curvilinear Range of Glasshouses was built in the spirit of its time - from elegant metal structures with high vaults, a striking form with smooth curved lines. The large size of the greenhouse – more than 100 meters in length – did not prevent it from looking very light and almost airy. By 1995, when the botanical garden was celebrating its 200th anniversary, the Curvilinear Range of Glasshouses had been so carefully renovated that it received the Europa Nostra Prize from the European Union for its high skill in preserving architecture [8].

The first half of the XX century is a difficult period in human history. Two world wars, economic and industrial crises, revolutions, social upheavals... These were not the best times for such fragile buildings as greenhouses. At the same time there was the active development of science, technology, industry and bold social experiments, which are impossible without innovation in many areas, including architectural and construction business. The ideological and economic competition between the two systems, the capitalist and the socialist, which took place throughout most of the XX century, was also a stimulus for innovative discoveries and their implementation.

After the Second World War greenhouses were experiencing a renaissance; their construction uses technological achievements of previous stages and the products of latest advances in the construction industry. There were various constructive decisions in the practice of erecting greenhouses in the second half of XX century: two-sloped core designs of a covering with glazing, pyramidal translucent designs, arched and frame systems with total glazing, cable constructive systems with the suspended frame elements, mesh covers. The latter turned out to be the most potentially variable direction in the creation of engineeringly innovative and artistically expressive structures of "translucent architecture".

The first mesh covers and load-bearing structures based on them appeared during the height of the "first metal revolution". The inventor of this type of construction was the famous Russian engineer V. G. Shukhov (hanging mesh coatings at the industrial exhibition in Nizhny Novgorod, 1896) [9, 10]. It was he who first came up with the idea of using a joint static system of metal rods that intersect in two directions. In this design, the coating works as a whole, and all the rods carry approximately the same load, which allows to utilize rods of the same cross section. Later, architects of the high-tech movement – the famous Buckminster Fuller and Norman Foster – finally introduced mesh covers into modern construction practice. The study of the possibilities of mesh systems led to the appearance of the geodesic dome of B. Fuller (1950s), and then to the surfaces of double curvature in numerous variants [11].

In the middle of the XX century a new constructive discovery – a geodesic dome – made a great impression on the architectural community. A geodesic dome is a spherical structure assembled from tetrahedra, rods connected in a special way. In each of the nodes of the structure, three ribs of slightly different lengths converge, forming a geodesic structure, one of the strongest structures in the world. Richard Buckminster Fuller is considered to be the inventor of the geodesic dome. He had no higher education, and before doing his research, served as a naval officer and worked as a director of a construction company. He is called an architect, an engineer, an inventor. But he himself proclaimed the following: "In architecture, "form" is a noun; in industry “form” is a verb. I'm trying to connect them with each other" [12].

The first spherical grid was developed by engineer Walter Baurfeld in 1925. However, Fuller studied the shape of the sphere independently and in the 1950s patented the design of the geodesic dome and received several orders for its construction. This was the beginning of one of the most successful humanitarian design projects in the history of architecture of the twentieth century.

Geodesic domes brought Fuller international recognition. According to his project, a "golden dome" was built for the American National Exhibition in Moscow in 1959. In 1967, Fuller designed the US pavilion at the World's Fair in Montreal, Canada, which now houses the Biosphere Museum [13].

Fuller's innovative design has inspired designers around the world to create unique architectural objects. Geodesic domes have even appeared in futuristic designs of earthlings’ settlements on other planets.

Greenhouse designers also showed attention to Fuller's invention. In 1960, the Climatron was built on the grounds of the Missouri Botanical Garden in St. Louis, which became the world's first greenhouse made of plexiglass with a completely artificial climate. The diameter of the dome is 53 m, height is 21 m, the frame is made of aluminum pipes. It was designed by architects Murphy and Mack (T.C. Howard, Synergetics, Inc., Raleigh, N.C.) [14]. In 1961, the creators of the Climatron received the Reynolds Award for successful use of aluminum structures. In 1976, the greenhouse was named one of the 100 most significant architectural achievements in the history of the United States.

The world's most famous new dome-type structure is the Eden Project (2001), an indoor botanical garden in Cornwall, United Kingdom. The greenhouse complex is based on two huge geodesic domes, each of which consists of several translucent plastic spheres, divided into many hexagons and pentagons, which not only allow creating huge spherical surfaces, but also provide rigidity. The greenhouse complex is designed in the reclaimed areas of the quarry. Round domes of greenhouses are embedded into the existing landscape. Around the greenhouses a park area for walking and contemplation of the surrounding areas is organized. The idea belongs to Tim Smith, a project is designed by architect Nicholas Grimshaw and the engineering company Anthony Hunt & Partners. Project developer is David Langton. MERO company has designed and built greenhouses [15].

The architecture of the XXI century saw an emergence of new issues and guidelines related to the crisis at the turn of the XX-XXI centuries and a paradigm shift in modern architecture. Scientists distinguish "five metaconcepts: avant-garde architecture, architectural technicism, neohistoricism, megalandscape architecture and quasi-architecture" [16]. Along with the rapid development of civil engineering, the ecological ideological direction and the issue of sustainable development, as well as the issue of preserving life on the planet came to the fore in architecture and urban planning.

It is natural that due to the new understanding of the botanical garden and greenhouses as a public, multifunctional space in the XXI century, both an eco-botanical and at the same time technological complex, the creation of these objects is becoming an increasingly complex process. It no longer involves individual architects, engineers and entrepreneurs-innovators, but project groups, bureaus and companies, often several. The competitiveness of modern engineering solutions and the success of new buildings and complexes provide a new ideology, the innovative approaches to construction design.

In the Gardens by the Bay park project (2010) in Singapore a fantastic symbiosis of technology and natural diversity has been created. A unique garden of the future, eco-technological botanical complex offers solutions to many problems: from the preservation of natural biomes to the ecology of the modern metropolis. There are two giant domed greenhouses in the park complex - the Flower Dome and Cloud Forest. The greenhouses (covering about 1 hectare) developed by Wilkinson Eyre Architects are a clear example of the use of arch-frame structural system with a mesh cover. The design has no additional supports. Its base is a giant steel mesh, which carries 3332 glass panels and 28 metal arches mounted on the outside of the domes on top of the greenhouses. The design is actively aimed at minimizing environmental impact. Rainwater is collected from the surface of the greenhouse and circulates in the cooling system, which is connected to the so-called "supertrees". In "supertrees" water is used to cool hot air and circulating water [17].

According to the Plasma Studio project in 2013, a grand greenhouse was built in the Chinese city of Xian [18]. The glass building with a metal frame is partially sunk into the hill and resembles an exquisite crystal, shining in the sun with sharp edges. This impression is due to the faceted roof formed by triangular fractures that converge at different angles. Visitors enter the main hall of the building through a curved slot-tunnel that crosses the main facade like a crack. The interior of the greenhouse is divided into three main sections, dedicated to different climatic zones and landscaped with plants that are characteristic of these zones. The horseshoe-like shape of the object plan forms a large internal open-air space - it is the natural center of the complex, which connects other parts of the building in the open air.

In the XXI century the construction of individual greenhouse complexes has repeatedly become an event in the architectural world. A number of modern greenhouses with various innovative components, stunning design and architectural solutions are currently built around the world.

## 5. Conclusion

The flourishing of the most spectacular modern architectural and engineering ideas that we see today began long ago, during the formation of industrial society. This event has a distinct beginning, even a date – the First World Industrial Exhibition in London in 1851. It is marked by the construction of the Crystal Palace, the idea of which belonged to the innovator J. Paxton, and which together with the development of civil engineering, technology and materials, gave impetus to the birth of a relevant trend – called "translucent architecture".

Innovative developments in the field of architecture, engineering and construction technologies of past epochs were actively used in the XX century during the construction of greenhouse complexes. Along with the new interpretation of traditional structural schemes, innovative constructive solutions were invented, an example of them being a geodesic dome. With the use of the latest technologies, more and more opportunities emerged for the implementation of increasingly ambitious tasks that opened up new horizons for the development of architecture. Greenhouses with their requirements, special expressiveness and innovative methods of design solutions have repeatedly become symbols of the latest achievements in architecture and civil engineering of the last century.

In the XXI century the ecological ideological basis which directs possibilities of modern building technologies acquires urgency as the latest architectural trend. Today, the ideas of conservation of resources, sustainable development of settlements and the like give rise to new competition in the construction of unique translucent structures that become symbols of cities and countries worldwide. Avant-garde architectural designs of greenhouses, their chimeric bionic forms that appear more and more often, clearly show the possibilities of implementing innovative developments of modern engineering in the natural environment. Scientific achievements become a source of inspiration for the transfer of natural principles in the architectural solutions of modern greenhouses. Thanks to the combined efforts of today's innovators from various fields of science and construction, today we have the opportunity to implement projects based on the latest ideas, various technological innovations, unique and sometimes fantastic creative concepts.

## References:

[1] Shumpeter J 1982 *Theory of Economic Development: Studies of Entrepreneurial Profit, Capital, Credit, and the Business Cycle* (Moskva: Progress) p 456

[2] Official website of the Crystal Palace Foundation: 2020 *The Crystal Palace Foundation, specialists in the 1851 Great Exhibition, the Sydenham Crystal Palace and Crystal Palace Park* Accessed 11 May 2020 URL: http://www.crystalpalacefoundation.org.uk/

[3] The Great Exhibition of 1851 *Duke Magazine* Volume **92**, No.6, November-December 2006 Accessed 11 May 2020. URL: https://alumni.duke.edu/magazine/articles/great-exhibition-1851
[4] Chadwick G F 1961 *The Works of Sir Joseph Paxton* (London: Architectural Press) p 276
[5] Hix J 1974 *The Glass House* (Cambridge: The MIT Press) p 208
[6] Boriskina I V et al 2012 *Buildings and Structures with Translucent Facades and Roofs. Theoretical basis for designing translucent structures*, ed I V Boriskina (Sankt-Peterburg: Inzhenerno-informatsionnyi Tsentr Okonnykh Sistem) p 400
[7] Turner, Richard *Dictionary of Irish Architects* 1720-1940 Accessed 12 May 2020 URL: https://www.dia.ie/architects/view/5405/turner%2C+richard+%2A#tab_biography
[8] Malashkina O *Richard Turner - innovator of iron structures* June 14 2017 Berlogos Accessed May 11 2020 URL: http://www.berlogos.ru/article/richard-tyorner-novator-zheleznyh-konstrukcij/
[9] Grefe R 1995 *Shukhov V G (1853—1939). Die Kunst der Sparsamen Konstruktion* (Moskva: Mir) p 192
[10] Akopyan M and Vlasova E 2019 *Shukhov. Formula of Architecture* (Moskva: Kuchkovo pole Muzeon, Fond «Svyaz' Epokh») p 440
[11] R B Fuller and J Snyder 2010 *Ideas and integrities: a spontaneous autobiographical disclosure* (Baden: Lars Müller Publishers).
[12] Ananthasuresh G 2015 Buckminster fuller and his fabulous designs. *Resonance* **20**. 98-122. 10.1007/s12045-015-0159-6.
[13] Langdon D *AD Classics: Montreal Biosphere / Buckminster Fuller* 07 Oct 2018 *ArchDaily* Accessed May 11 2020 URL: https://www.archdaily.com/572135/ad-classics-montreal-biosphere-buckminster-fuller/
[14] Climatron Conservatory: History and Architecture. *Missouri Botanical Garden.* – Accessed May 11 2020 URL: http://www.mobot.org/hort/gardens/CLhistarchit.shtml
[15] Sochalin O *Paradise garden on the site of a clay quarry. The Eden project by Nicholas Grimshaw* 2007–2020 Architime.ru Accessed *May* 11 2020 URL: https://www.architime.ru/specarch/grimshaw/eden_project.htm#1.jpg
[16] Omuraliev D and Volichenko O The mainstream of the new architecture of the twenty-first century 2013-12-15 *Palmarium Academic Publishing* URL: https://www.ozon.ru/context/detail/id/31774495/
[17] Chirkov A Rukotvornyi oazis [Man-made oasis] July 12 2012 *Archi.ru* Accessed *May* 11 2020 URL: https://archi.ru/world/42527/rukotvornyi-oazis
[18] Xi'an Greenhouse / Plasma Studio 14 Aug 2013 *ArchDaily* Accessed *May* 11 2020 URL: https://www.archdaily.com/412656/xi-an-greenhouse-plasma-studio/

IOP Conf. Series: Materials Science and Engineering **907** (2020) 012016 doi:10.1088/1757-899X/907/1/012016

# Visual-communication environment of a modern city

**V P Dubinskiy [1] and A A Nesen[2]**

[1] Department of Architectural Design, Kharkiv National University of Civil Engineering and Architecture, Kharkiv, Ukraine
[2] Department of Architecture, O. M. Beketov National University of Urban Economy in Kharkiv, Kharkiv, Ukraine

vdubinskiy60@gmail.com

**Abstract.** The article aims at presenting an updated outlook on theoretical developments on the space of a modern city focusing on the formation of new dynamic environments, in particular the formation of a new object of the architectural space of a modern city – a visual-communication environment. The analysis of the research on the problem allowed authors to define conceptual issues of the subject of the research corresponding to modern world trends and scientific terminology. For the further emphasis of main subject of the research, the definition of the visual-communication environment of a modern city has been suggested. The importance of developing a comprehensive master plan of visual-communication environment of a modern city as the basis for possible solution of a number of problems of modern large cities has been determined. In addition, the definition and characteristics of the main structural parts of the comprehensive master plan of the visual-communication environment of a modern city (visual-communication frame, visual-communication outline) have been given. These categories allowed to reveal the nature of the visual-communication environment reflecting its compositional features.

*"The life of the village is narrative...*
*In the city the visual impressions change,*
*overlap, intersect, they are "cinematic".*
Ezra Pound [1]

## 1. The research of a modern city environment

The environment of a modern city is a space oversaturated with a variety of the latest innovative technologies and devices that are introduced into all spheres of human activities. The sphere of architecture is no exception: the architecture of the last decade of the twentieth century focuses on new super-power digital technologies and media.

At all stages of its development mankind tended to create the best means and ways of exchanging information for breaking down time, language, spatial and other barriers. It can be seen in the real world, when the steeple of the Catholic church is able to inform us that it is the place for meditation and religious meetings, bricks and ivy of the universities mean the years of education and scientific researches and the white smooth walls of the Greek temples reflect the birth of science, while the bright colors of the children's playgrounds denote games and joyfulness.

The form of media, their artistic and functional role in architecture, has been changing under the influence of society's needs. In the century of scientific and technical progress they became completely independent from the traditional architectural form, separating into an individual set of means of space systems organization. Now, they are on the level when it is possible to determine the tools and approaches applied in this sphere.

In the pursuit of satisfying practical needs, such as safety, convenience and comfort, as well as environmental friendliness and ergonomics of forms, the environment of a modern city demonstrates a complex urban network that is oversaturated with "bursts" of digital technologies. Such an environment manifests its architectural and design "interior" when interacting with a resident of a modern city in its daily routine. At this instant – the moment of interaction – chaotic "bursts" occur. First of all, it is a consequence of unawareness of theorists, designers, urbanists, project developers and residents of the city as a whole of the peculiarities of these systems, as well as restricted professional approach of some sciences to the study of this problem.

*1.1 Current situation and trends*

Synergetic, non-linear thinking and computerization of the general culture provoke re-thinking of the existing architectural space and determining the ways of its future development. According to Lev Manovich, "today we are witnessing the emergence of a new medium – meta-medium – a digital computer", which forms some kind of media systems, running through all spheres of life [2]. At present, there is no well-established general definition of these systems among the specialists of aforementioned fields; there is no derived general terminology. The analysis of the theoretical advances shows that initially media components were considered as an addition to the objects of an architectural design of a city. However, with the dynamic growth of cities and introduction of innovative technologies, they separated into an independent system, requiring a comprehensive consideration in real time. According to Norberg-Schulz, the main problem of the researchers studying urban space is considering this space a priori, though everyday life is to be the main task of architects and planners [3]. While studying such systems, first of all, it is necessary to accept their dynamism and flexibility, as in the result even an extensively designed city object faces physical changes at different levels throughout its existence.

Sometimes works researching this problem are called digital studies. Nevertheless, at present there is no single established academic term for such an interdisciplinary space. Such notions and terms as "information environment", "adaptive architecture", "intellectual architecture", "media city", "cyberspace" etc. are used.

The interpretation of "intellectual architecture" as a communication open system occurs for the first time in theoretical and practical research studies of the late 60s in Europe. British cyberneticist Gordon Pask suggested a model of an interactive environment including the real and the virtual, in which "computer systems, materials and space enter into a dialogue with their user having wide opportunities of studying him and adapting to his needs" in his work "Essays on Cybernetics" [4].

A significant contribution to the theory of urban planning was made by Kevin Lynch, who developed the theory of an urban form. He was the first to pay attention to the visual elements and cognitive ideas of the urban environment, the peculiarities of a city perception by its residents, which led him to drawing up the concepts of a mental image of a city and mental maps [5].

The concept of an architecture that responds to the changes in the environment easily, became the central subject of the leading trends after World War II. It aimed at creating a unique urban infrastructure, satisfying individual needs, and the architecture, which is constantly evolving socially, ecologically and technologically. French architect Ion Friedman, the founder and the leader of the Group of mobile architecture (GIAM), dedicated himself to the development of new "mobile" urban structures. His theoretical investigations crystallized into a complete project "A city in space", presupposing the creation of an enclosed, flexible space and temporary structures ("agglomerations"), able to rebuild itself if necessary. Friedman considered that the classical architecture underestimated the role of a resident and created a subordinate indifference. He believed in the main role of a resident in relations with an urban body – "The structures, forming a city, are to be skeletons, which are filled in accordance with the individual wishes of its residents" [6].

Nicholas Negroponte, an American information scientist, whose scientific work is related to an artificial intelligence, was the first to introduce a notion of an adaptive (sensitive) architecture. His works "The architectural machine" and "The machine of soft architecture" represent the first attempts of defining "sensitive architecture" as a product of natural integration of computer technologies into the space and structure of a modern city [7].

Australian media theorist Scott McQuire carried out a thorough cultural and historical analysis of the combinability of media technologies development with the progressive urbanization of the world. He paid special attention to the integration of media into modern urban life. This phenomenon is denoted by him as "relationship space", formed by the media which are becoming increasingly mobile, scalable and interactive [8].

According to their theory of tactical urbanism, architects Mike Lydon and Anthony Garcia believe that for progressing ahead even some prerequisites are enough. They consider that short-term creative projects make it possible to react to constantly changing living conditions and technologies [9].

In modern theoretic studies of such researchers as Akhmedova L.C. [10], Bilgin Altinöz [11], Dubinskiy V.P. [12], Esaulov G.V. [13], Kudaeva E.A. [14], Cihanger D. [15], Hudnik Sh. [16], Yarimbas D. [17] the focus is made on the emergence of new qualities and peculiarities of the urban environment. Various hypotheses on the formation of the environment of a modern city are put forward proposed but they are considered from the point of view of one of its functional characteristics and are not thoroughly investigated.

The analysis of the research works on the problem allowed the authors to single out a new special system – visual-communication environment, fulfilling a number of functions (aesthetic, communication, psycho-physiological, orientation, social, semiotic, ergonomic) encouraging qualitative perception of a modern city by its residents.

### *1.2 Problem formulation*

The environment of a modern city, influenced by the processes of globalization and mass consumption, is a kaleidoscope of various media components that are not subject to a single order. A modern city is no longer just a vacuum for living. The space is transformed every day – it becomes a layering of various structures and systems connected with several historical eras of a city transformation. New dynamic, innovative environments able to adapt and evolve are formed and each of them requires detailed investigation. The visual-communication environment is one of them.

A resident of a modern large city often falls into a dissociative network of media technologies, the components of which cause disorientation, loss of awareness of the current moment as it is. Widespread computerization engulfs all sections of a city. Furthermore, the authenticity of a city image, the idea of its subject-spatial environment and its unique historical heritage are lost. It becomes empty, vague, non-ergonomic, prone to gradual collapse. All these features lead to an unsatisfactory state of the visual-communication environment, dissatisfaction of the residents-users of this environment, in particular, and the dissatisfied city as a whole, making this issue an urgent and important subject for the research.

Innovative Internet technologies make us revise the most basic principles of designing architectural objects; correct, predict and clarify the changes in the environment of a modern city.

## 2. Visual-communication environment of a modern city

Due to the fact that the study touches upon the problems at the intersection of architecture, urban planning, cybernetics, semiology, sociology and psychology, their usage of similar research vocabulary became apparent, but the notions and terms in use are not always correlated. Thus, basic definitions were specified in the research process. It allowed the authors to derive a research vocabulary on the subject, corresponding to the current world scientific trends.

What is a visual-communication environment? What is its role in the formation of a modern city?

Visual-communication environment (VCE) is an independent functional object of an architectural design; a hierarchical system of a city, the forming components of which are anthropogenic and natural elements of an urban environment, equipped with visual and communication means.

Gradually, the elements of VCE have evolved from the primitive wall paintings and pictograms to modern innovative devices and full-fledged independent objects of a city environment: video walls, media facades, objects with systems of mobile reactive media, etc. Its functional potential has significantly expanded as well as its sphere of influence. Modern buildings are no longer monotonous static objects. These are full-fledged media components of VCE, able to read, analyze and respond to the surrounding vital processes. VCE of a modern city becomes "smarter" and more interactive.

In the course of the research it was revealed that VCE is a complex hierarchical system requiring a comprehensive investigation based on the principles of environmental and systematic approaches. VCE is directly related to the physical reality (for example, architectural buildings, natural and anthropogenic objects), as well as to people and their activities. Furthermore, there are less perceptible bonds describing the real state of VCE, such as socio-demographic situation, culture and general atmosphere.

It is known that any system is a set of interconnected components of various types, characterized by their unity. The first task of the analysis is to reveal the structure of the VCE system, its functions and features.

The main characteristic features of VCE are the following:

- a new form of communication based on the Internet;
- a new system of storing, moving and reading information in space;
- mimicry to the existing environment;
- a departure from stable geometric forms to smooth and flowing shapes;
- introduction of innovative technologies and devices;
- a new way of interaction with human sensations and feelings.

The main components of VCE are visual-communication frame and visual-communication outline, forming the skeleton and the body of VCE.

Visual-communication frame (VCF) is denotes the main means of revealing and forming the visual-communication environment of a modern city, which clearly identifies the hierarchy of compositional axes and visual-composition junctions; it forms the "skeleton" of the visual-communication environment, being a well-established transport and pedestrian infrastructure of a city, including city highways and district roads, pedestrian streets.

VCF is formed by a set of composite centers (nodes) and composite axes. In the hierarchical VCE system, composite nodes and axes are considered as additional (secondary) structure-forming elements.

Visual-communication outline (VCO) is a set of information elements, visual tools and objects, structured and subordinated to the general hierarchy of compositional axes and visual-composition junctions, creating the "body" of the visual-communication environment; it includes a diverse range of natural and man-made urban objects.

VCO is increased on VCF, and is a set of visual-communication zones with their visual-communication borders.

All of the aforementioned structural elements are parts of a comprehensive master plan of the visual-communication environment of a modern city.

The next step is to consider the purpose and features of the comprehensive master plan of the visual-communication environment of a modern city in detail.

Visual-communication master plan (VCMP) is a design document determining the fundamental model of the development of the visual-communication environment of a modern city, its layout, character, zoning and infrastructure not as a sum of separate parts but as a complex unity.

Designing of the visual-communication environment of a modern city, influenced by the interdependence of qualitative and quantitative changes, is carried out on the basis of this document. The main part of VCMP is a large-scale display of a city with a drawing of the designed object - the VCE system, plotted on it using a graphic overlay.

It is possible to solve a number of problems of modern large cities by following the regulations and restrictions of the comprehensive VCMP. Among such problems are: achieving the proper organization of information and orientation means in a city, preserving its historical heritage and emphasizing the authenticity of buildings, creation of a high-comfort and ecologically balanced environment for human life and activities, improving the aesthetic characteristics of objects, increasing safety, reducing time spending of people looking for something in a city. Certain regulation of advertising elements will allow to settle the problems of information pollution and media ecologization of a modern city environment.

Visual-communication environment is a sum of architectural functionality, visual epistemology and new interface operations. Since its components are introduced and implemented randomly now, the development of a comprehensive VCMP at the legislative level will allow to avoid further problems in the architectural design of modern cities.

It should be emphasized that the formation of VCE is to be carried out by taking into account the nature of the existing architectural and planning composition of a city consisting of anthropogenic and natural elements of the urban environment. The introduction of the VCE components can be a radical interference in a city and its architecture, but nevertheless one should take care of the architecture of individual buildings and take into account the peculiarities of the neighboring structures. The historical heritage of a city should be preserved, improved and complemented with the latest technologies. The interrelation of anthropogenic and natural components, improved and supplemented with innovative technologies, makes it possible to derive the following formula of VCE:

$$\mathrm{VCE} = \mathrm{InnT} \times (\mathrm{AE} + \mathrm{NE}),$$

where InnT – innovative technologies; AE – anthropogenic elements of the urban environment; NE – natural elements of the urban environment.

This formula shows that anthropogenic and natural elements are parts of a city. But their symbiosis with innovative technologies creates VCE of a modern city.

Since VCE is a dynamic, interactive system, it is influenced not only by qualitative and quantitative characteristics of a city but also by such constants as time (when/how long) and space (where/how). Due to built-in innovative technologies, it can change in space and time, adapt and transform depending on its environment.

$$\mathrm{T} > (\mathrm{VCE} < > \mathrm{R}) < \mathrm{S}$$

where VCE - the visual-communication environment of a modern city; T – time; S – space / environment; R – a resident of a modern city / a person-user.

Technologies development, computerization processes and architecture are in constant close interaction, influencing each other. A computer, as one of the components of VCE, and a person enter into progressively deeper relationship and the architecture acts as a platform, providing their interaction. Special attention should be paid to the important role of a resident-user in the system of complex relations between VCE and a modern city. Initially, a city forms the space and the environment, and influences the perception of a person. An architect transmits information to the resident through an architectural image, with any element of the subject-spatial environment participating in communication due to the data encoded in their form. But there is also a reverse effect coming from a resident, when he transmits specific information and data.

The narrative of the architecture on behalf of an architect may differ greatly from a resident's point of view. Moreover, each participant of this "conversation" may read a different plot ... that is why the perception of the same architectural object by a tourist and a student of architecture differs so much, though both of them are acceptable for architecture.

A resident of a modern city is not just an ordinary man, but also an active user of all interactive devices. In his relations with architecture a resident-user selects clues from the environment, as well as his memories and knowledge, which make him build a holistic comprehension of the environment. It is some kind of a test of his experience and immediate reaction. Thus, he can communicate with the environment he lives in, being an interface for transmitting information, but the environment can communicate with him as well. In this case, the aforementioned environment is VCE, responding to his requests, adapting and analyzing his characteristics and peculiarities.

Some design theorists believe that there is a constant reflexive conversation between a resident of a modern city and urban environment [18]. That is, people initially perceive the environment with already established preliminary comprehension, vague projection of the "complete product". But in reality, this projection is then re-projected, requiring dialectic refinement and re-determination of its parts. Hence, a constant dynamics of reprogramming of this image occurs, which makes VCE so inconstant and dynamic in its development but, at the same time, a unique environment of a modern city.

As previously noted, architecture turns from static to dynamic, establishing some kind of "receptive field" for interaction with a resident-user of a city, represented as VCE. It can read biometric data and adjust to a specific person or a group of people; remember his moods and requests; respond to the climatic conditions; constantly share information; be innovatively flexible. Buildings move, smell, talk and maintain themselves. However, proper dissemination of information among all residents is still burdensome and mutual dialogue is not always successful (that is, who, when and what data obtains?). Nevertheless, a modern city of the future turns to a smart cyber-space and an architect becomes a guide into this new space.

## 3. Conclusion

The study of an architectural environment of a modern city showed that city dwellers are in a continuous information flow. A city has passed from the static state of vacuum for living into the dynamic environment, which is in constant development and renovation. Although, according to their main purpose the buildings are considered to be an accommodation, they are presently on the way to media-computer renovation.

This leads to a progressive appearance of new environments, able to adapt, transform and pass from one state to another, regardless of their initial purpose. The visual-communication environment of a modern city (VCE) is one of such environments.

Because of the mass chaotic introduction of innovative technologies and devices of different forms and types, that have not been properly predicted, the environment of a modern city is now experiencing some of an information collapse. The chaotic use of the VCE components causes the pollution of an architectural image of a city. All these factors negatively affect a resident-user of a modern city (by disorientation, noise, information overload, abundant advertising, functional disparity of its urban elements, destruction of the stylistic unity of the architectural environment of a city, etc.).

Architecture has always been a reflection of the state of the world and a "testing ground" for the introduction of innovations, a stepping stone on the way to the future. However, currently architecture is characterized is a crucial period of its development due to the influence of several factors, changing both the fundamental nature of architecture and its role in everyday life. Said factors are the development of information technologies and widespread computerization, as well as a chain of global problems superimposed at one point.

Architects are faced with diverse tasks requiring immediate solutions and forecasts for the development of the future architecture and a city as a whole. Since each situation and context of the VCE components is unique, their careful consideration and analysis are required, taking into account the specific circumstances. All this takes architects to a completely new level of creative activities, focusing not only on the experience of previous centuries, but also on innovative technological progress.

VCE affects all aspects of human life, but its dominant sphere is the information component of a modern city. The solution of negative phenomena in this sphere can be in the so-called ecologization of the information environment. Although this problem does not have top priority in the world community now, it will be evident in a couple of years, as scientific and technical progress does not stay still.

Harmonization of a modern city life can also be achieved on the basis of a comprehensive master plan of the visual-communication environment. The creation of a coherent and unified system, where different layers do not contradict each other but join together in VCE, is the priority task. The city designed according to the requirements of the visual-communication master plan (VCMP) will stop being a projector with a large amount of visual load in the form of neon advertising, banners, storefronts and multi-colored light boxes. The city will become, first of all, a highly efficient comfortable environment, the utility coefficient of which will greatly increase due to the correct use and location of all structure-forming elements of VCE, subordinate to the unified system. Further development of a modern city's VCE is probably connected with the use of innovative technologies and the experiments in creating an environment capable of self-regulation, self-improvement and self-reproduction. It should be also noted that mutual co-existence of innovative technologies and the existing elements of the urban environment of anthropogenic and natural character is very important for improving the characteristics of VCE.

Considering VCE of a modern city, it can be concluded that it is a specific hierarchic system, penetrating the whole city and affecting all spheres of human life. Its formation and integration with the architectural environment occurs at different spatial levels and affects all areas of a city. It is also considered as an innovative component of a modern city, its unique digital code is used for transmitting all types of data and experience, and its language is built into the interfaces of innovative devices, building facades and street elements.

The essence of the above comes down to the fact that only the strict conformity to the developed comprehensive plan VCMP, based on innovative principles of its formation, will allow to create a quality visual-communication environment. Planning and forecasting of the development vectors of the visual-communication environment of a modern city makes it possible to provide comfort and safety, to avoid informational and media pollution. The criteria and indices of the comprehensive VCMP are the basis for planning activities, aimed at sustainable development of smart cities of the future, which is particularly urgent in the age of digital technologies.

**References**

[1] Pound E 1921 *Selected Poems* (London: Faber and Faber)
[2] Manovich L 2001 *The Language of New Media* (Massachusetts: The MIT Press)
[3] Norberg-Schulz C 1984 *Genius Loci: Towards a Phenomenology of Architecture* (New York: Rizzoli Press)
[4] Pask G 1961 *An Approach to Cybernetics* (London: Hutchinson & Co LTD)
[5] Lynch K 1960 *The Image of the City* (Massachusetts: The MIT Press)
[6] Friedman Y 1975 *Toward a Scientific Architecture* (Massachusetts: The MIT Press)
[7] Negroponte N 1975 *Soft Architecture Machines* (Massachusetts: The MIT Press)
[8] McQuire S 2008 *The media city* (London: SAGE Publications of London)
[9] Lydon M and Garcia A 2015 *Tactical urbanism: short-term action for long term change* (Washington: Islandpress)
[10] Akhmedova L S 2009 *Features of transformation of visual information and communicative field of the city* (Samara State Architectural and Construction University, Samara, Russia)
[11] Bilgin Altınöz A G 2010 Tarihi Dokuda 'Yeni'nin İnşası *Ege Mimarlık* pp 18-26
[12] Dubinskiy V P 2007 *Architectural and artistic principles of light and color environment formation in a modern city* (Kharkiv National University of Civil Engineering and Architecture, Kharkiv, Ukraine)
[13] Esaulov G V 2015 Information technologies in architectural and urban planning formation of living environment. *AMIT,* spec. vol. **15** pp 1-8
[14] Kudaeva E A 2016 Virtual communications in the city environment. *AMIT* **1** (34) pp 1-11
[15] Cihanger D 2018 Spaces by people: an urban design approach to everyday life *METU JFA,* vol. **36**, no. 2 pp 55-76
[16] Hudnik S 2003 Architecture and communication *UI,* vol. **14**, no.2 pp 104-108
[17] Yarimbas D 2018 Experiencing city by walking: Communications elements *ITU A|Z,* vol. **15**, no. 2 pp 19-30
[18] Snodgrass A B and Coyne R D 1997 Is Designing Hermeneutical? Architectural Theory Review *Journal of the Department of Architecture, The University of Sydney,* vol. **1**, no. 1 pp 65-97

IOP Conf. Series: Materials Science and Engineering **907** (2020) 012017 doi:10.1088/1757-899X/907/1/012017

# Problems of Architecture and Professional Development of an Architect beyond "Limits to Growth"

**O O Fomenko, S M Danylov and A M Izbash**

Department of Innovative Technologies of Design of Architectural Environment, Kharkiv National University of Civil Engineering and Architecture, Kharkiv, Ukraine

izbash280593@gmail.com

**Abstract.** The study examines the regional specifics of the formation of architectural specializations under the influence of social and resource crises experienced by Ukraine. Based on the work of D. Meadows and J. Forrester (system dynamics) a method was developed for the formation of "Knots of problems" based on "positive and negative feedback loops". Some "Knots of problems" and their influence on the formation of needs for architects and urbanists of the new generation were described on the example of Kharkiv. The identified needs are shaped by the acute socio-economic crisis experienced by the city. The authors argue that a focus on servicing territorial communities and developing strategies to enhance the role of society in creating an energy-efficient and environmentally positive architecture should be considered a prority in the work of architects of new formation. The research based on the identified socio-economic "Knot of Problems" describes its hermeneutic circle (a circle of interconnected systemic contradictions that prohibits the solution of identified problem). The analysis of the revealed "Circle" indicates that an achievement of a sustainable state by the cities of Ukraine is practically impossible without a significant increase in socialization in urban communities. The study focuses on the importance of universities and, in particular, architectural faculties, in the revival of territorial communities.

## 1. Introduction

Fifty years ago, an assistant at the Massachusetts Institute of Technology (MIT) Dennis Meadows was given an assignment by the Club of Rome to conduct an unusual study of the time [1]. The essence of an assignment was to compare some of the key trends in world development under the influence of the environmental, demographic, economic and resource spheres. March 13, 1972, in Washington, at the Smithsonian Institution, saw the first presentation of the collective work "Limits to Growth, a Report to the Club of Rome" [2, p. 208].

The authors of the book analyzed 12 calculated scenarios of a possible future of the world community in the period between 1970 and 2100. Some of the calculated scenarios ended with a sharp depopulation of the planet and the decline of material living standards. But alternative calculations of sustainable options for the future were presented, in which the population size and the satisfaction of its physical and spiritual needs were harmoniously combined. The relevance of the work of Meadows and his research team is confirmed by the fact that more than 30 million copies of the book in 30 languages have been purchased worldwide [3, pp. 59-63]. The book is still relevant and causes debate. After 1972, the following works were published: Donella Meadows "The Limits to Growth: 30 Years Later" in 2004 [4] and Jörgen Randers "2052: A Global Forecast for the Next Forty Years" in 2012 [5]. These works are a continuation of the work of Dennis Meadows and largely confirm and refine his calculations.

The national specifics and the events of spring 2020 are forcing us to consider a new perspective for the problems of the development of Ukrainian architecture and the profession of architect for the next fifty years. The paradox of current situation is that the estimated lifetime of the absolute majority of buildings being designed today goes beyond the forecasts of sustainable urban functioning in the way of life that has formed nowadays.

This article proposes a number of theses based on the ideology of the work of Meadows, specifically - the analysis of the conjugations of "positive and negative feedback loops" (Meadows [6]) linking architecture, ecology, socio-economic and geopolitical factors forming the life activity of Ukrainian cities.

## 2. The results of the study

According to the theory of modeling dynamic processes, any exponentially growing factor in the city system is involved in the positive feedback loop in one way or another [7]. The positive feedback loop is often called a "vicious circle." An example is the familiar "household price growth - housing demand growth" relationship: at certain time periods (for example, in 2002–2008, 2010–2014), prices per square meter of housing steadily increased in Kharkiv. This stimulated the interest of private investors in real estate, which triggered demand for housing and subsequently led to higher prices. As a result, the real estate market was overheating, creating a "soap bubble", which periodically "bursted". In 1-2 years, the "vicious circle" begins a new turn.

In the process of researching the problems of housing estate in Kharkiv, Ukraine, several groups of "Positive and negative feedback loops" that could potentially have a strong impact on the formation of the architectural component of the city were identified. This work proposes to identify such groups as "Knots of Problems". Together, these "Knots of Problems" create a unique map of forecasted events, specific only to Kharkiv and related to the adaptation of the architectural component of the city to the expected acute crises, including: potable water supply shortages, energy, technogenic, lack of engineering and technical staffing, food and environmental.

The uniqueness of the situation lies in the fact that the described problems are not critical at the theoretical level. However, in reality they get into the hermeneutic circle in following way:

1. Each of the "Loops" and "Knots" described further in the study severely restricts the city's access to social, technogenic and natural types of resources.

2. The shortage of social, technogenic and natural resources leads to the aggravation of corresponding crises.

3. To solve the emergent crises, the system needs social, technogenic and natural resources, access to which is restricted, as shown in paragraph 1. The circle is closed, creating a threat of the city's system going into a possible pre-catastrophic state, wherein the city cannot support the available number of its inhabitants.

According to authors, four among the many existing and predictable crises should be classified as basic. Their forthcoming nature makes it is necessary to adapt the architectural component of the city and partially change the current practice of preparing the next generation of specialists. These crises are listed below:

- Potable water supply shortages

Water management in the Kharkiv region nowadays is characterized by extensive water usage, formed during the Soviet era. The main problems are the non-separation of drinking and industrial water, leakage from the water supply network (according to the statistical reporting of 2-TP "Vodkhoz", measuring up to 33-40% of the supply [8, pp. 344-348]), poor metering and control of consumption associated with lowered water prices and the lack of interest of the public service in identifying water losses in network [9]. Kharkiv is one of the main consumers of clean potable water in the region. The stability of water supply is one of the basic factors for the stability of functioning of the entire city. Today, water management technologies are being actively developed around the world.This process was reflected in

IOP Conf. Series: Materials Science and Engineering **907** (2020) 012017 doi:10.1088/1757-899X/907/1/012017

the architecture by the points for voluntary certification of buildings accrued for water saving. It becomes an almost mandatory procedure for modern architects to develop competencies related to the implementation of rainwater collection and utilization systems in the design and construction practice, including the separation of sewage water into gray and black, the separation of tap water into drinking and technical, implementation of water treatment systems and saving systems [10]. Given the shortage of economic, natural and human resources, the problem of population's access to clean drinking water has a potential to become an acute crisis.

- Energy crisis

The city's existence and development is linked to the consumption of three basic resources: substance, energy and information. The consumption of these resources increases exponentially, which is accompanied by problems associated both with the shortage of resources and their mutual influence with the natural environment [11, pp. 2-14]. Since the First Industrial Revolution, the role of energy in meeting the vital needs of the city has been constantly growing. Architecture is one of the significant consumers of energy. Ukrainian cities are almost completely dependent on fossil fuels – such as gas, coal, oil and nuclear fuel. Compared to Europe, the share of alternative energy sources in the country's general market is unacceptably small and amounts to only about 1.2% of the total volume [12]. At the same time, a number of laws have been adopted in the country ("About energy efficiency of buildings", "About alternative energy sources" etc.) which are program documents for the energy and innovative development of Ukrainian architecture.

- Economic crisis

Economy is the main factor influencing the prospects for the modernization of the architectural component of the city. This thesis does not need interpretation. The unique problem of Kharkiv is well demonstrated by the data from the experiment described in the fifth section of the dissertation research of S.M. Danylov. The purpose of the experiment is to test the working capacity of the model complex for analyzing the city as a dynamic system developed in the dissertation research, in particular, the cognitive-research matrix. The object of study is obsolete buildings of the housing estate "Novi Doma". The objective of the study is to identify the "knots of problems" (sets of positive and negative feedback loops) associated with the adoption of the Law of Ukraine 2118-VIII "About energy efficiency of buildings" and the modernization of architectural objects in accordance to this law. Four possible scenarios of changing the socio-economic indicators of the housing estate were considered in the experiment depending on the selected estimate of modernization. The result of the study was the identification of problems associated with the insufficiently high financial situation of the population. The consequences of adopting an estimate of an option providing for the optimal modernization of buildings ($ 320 per m2) will be disastrous for their residents.

- Staffing crisis

Nowadays, the staffing crisis is not considered acute, or posing a threat to the stable functioning of the city. However, while considering this problem, it is necessary to take into account several loops of positive and negative feedback, specifically: qualification, demographic and emigration. The qualification "loops" arose as a result of the collapse of industry in the 1990s, when the reproduction cycle of engineering and labor in the construction industry was disrupted. By 2015-2020 there was an overlap of two "demographic pits" - the consequences of the Second World War and the 90s. In their totality, they caused a shortage of workers in middle and young age categories. Combined with the aging of population and a general decline in the birth rates, the demographic problem poses a serious threat to the stability of the city's life [13, p. 65]. In 2003, 2,887,900 people lived in the region of Kharkiv, 1,470,921 in the city itself. In 2020, the population of Kharkiv region and the city of Kharkiv totaled 2,658,461 people and 1,441,932 people correspondingly. Thus, the depopulation of the region amounted to 0.5% per year, and 0.15% of the city [14]. This fact is confirmed by the data of the Main Department of Statistics in the Kharkiv region. Over the same period, housing provision in Kharkiv increased from 18.2 to 24.1 $m^2$ per resident [15]. Labor emigration over the past five years has critically exacerbated the effects of qualification and demographic deficits. The shortage of qualified staff greatly inhibits the modernization of the construction industry, while the crises experienced by the country provoked a powerful outflow of talented youth from the country [16, p. 263].

The above crises are listed by the decreasing order of their threat to the stable functioning of Kharkiv. According to authors, the procedure for resolving these crises mirrors their order of severity. First step is the creation of a cluster of highly qualified staff, which would allow for stabilization of the economy. Subsequently, the economic stability and the availability of highly qualified staff will unlock the possibility of an effective reorganization of the energy sector, which will make it possible to stabilize the circulation of clean potable water between the city and surrounding territories.

This thesis is the result of many years of research into architectural innovation related to sustainable development. An architect who designs under the constraints imposed by the nZEB requirements (the nearly zero-energy buildings) is able to create complex objects that are significantly superior to buildings with traditional technical content. City of sustainable development, smart building / quarter / district / city, self-sufficient city, eco-city, cyber-city, CarboNZero, green standards – all of these terms are representing today's top technologies. Together, they create a special set of competencies that an architect must possess to design the entire life-cycle of an object from its design to disposal.

There are many precedents when an engineer who developed the technical components of a "zero" building became the head of the staff serving this building. The savings that an engineer brings in ensuring the maximum efficiency of the technologies involved significantly exceed the amount of his salary. This is exemplified in the Ciem building, Zaragoza. The building was commissioned by Zaragoza City Council as part of the Digital Mile project [17] (total area: 2 309 $m^2$; Cost / $m^2$: 2 283 €). The aim of the project is to demonstrate various solutions in the field of achieving the highest possible energy efficiency while maintaining user comfort. The main difference between CIEM compared to other similar buildings is the combination of bioclimatics (energy passivity) with energetically active systems (photovoltaic, wind, biofuel and geothermal systems). The combination of all this as well as the use of an intelligent control system, led to very satisfactory results. The building is also designed as a research laboratory, where a large amount of data is collected from sensors of all types. In addition to allowing the real-time determination of energy characteristics of a building, this provides an additional opportunity to optimize or improve building performance.

The combination of results of studies A) trends in the development of innovative technologies for sustainable development in architecture, and B) the architectural and urbanistic problems of the housing estate "Novi Doma", allowed to calculate the aggregated indicators of the city's needs for highly qualified specialists in the architectural and construction industry – namely, urbanists and engineers. The ideas about the circle of responsibilities and the set of competencies of the architect-urbanist, corresponding to the new needs of the city were formed[18].

The above calculations serve as the basis for forming the cycle planned by the authors of experiments. The expected results of the cycle are the formation of reasonable ideas about the nature of changes in the process of preparation of a new generation of architects able to work under conditions of many crises' exacerbation. The main difficulty in analyzing the prospects for the development of the architectural and urban areas in the practice and pedagogy of Kharkiv is the "Socio-economic knot of the problems of modernizing the architectural environment of the city".

A "knot" is formed by several tightly interconnected loops of positive and negative feedback. The following loops have the greatest impact on the problem: physical and moral obsolescence of buildings, degradation of the technical condition of utility networks; decrease in socialization of city residents; decrease in business activity; growth of corruption.

In the dissertation research of S.M. Danylov "Methodological foundations of modeling the city as a dynamic system" author proposes the concept of finding the optimal ratio of social, natural and technological resources necessary for the implementation of strategic and tactical tasks to modernize the architectural component of the city. The calculations of aforementioned work clearly signify that the social resource in the socio-economic conditions of Ukraine is the most important. At the same time, the parameters of socio-economic relations within the country indicate the fact that it is extremely difficult to use this resource.

A case in point is the negative experience of social relations obtained by architects and local authorities of Kharkiv in the process of reconstruction of a morally obsolete five-story residential building in Kharkiv ("Novi Doma" housing estate at Marshala Zhukova Ave., 21). The modernization of the building was completed by JSC Zhilstroy-1 in 2001 at the expense of investors and the municipality. This experience was considered as an experiment necessary to develop a socio-economic strategy for the renovation of obsolete city housing. The idea was to develop a project that was economically attractive to potential investors. The project's supposed payback was to be provided by the attic floor superstructure on five-story buildings with subsequent sale of new apartments.

The project by architect Nikonenko E.V. provided for the implementation of a complex of works on the overhaul of the building and the superstructure of the attic floor without resettling residents, increasing the energy efficiency of the building to class C. As a result of the work carried out, the costs of utilities for residents decreased by 56% on average. In the process of architectural supervision the complaints to law enforcement bodies were written against the architect, he was threatened with physical harm, some of the residents categorically refused to sign an agreement for repairs, etc. The result of the experiment was a complete loss of ethical and economic interest in this topic among architects, investors and the local government. Similar problems in the implementation of the "Regional programs of experimental reconstruction" were confronted by the organizers of experiments in different regions of Ukraine: 7 buildings in Luhansk region, 1 building in Odesa, 2 buildings in Poltava (for off-budget funds from investors), 1 in Kharkiv, 3 in Chernigiv, 6 in Kyiv and 2 in Sevastopol [19].

The hermeneutic circle of the problem in the formation of specialists professionally prepared to adapt the architectural component of Kharkiv to predicted crises is as follows: continued technical degradation of the architectural environment and engineering networks → it is difficult to solve the problem due to investments from the city budget, as Kharkiv is economically underdeveloped and significantly lags behind similar cities in the European Union → only social resources are abundant → there is an extremely high percentage of retirement age among highly qualified staff → the collapse of the country's economy leads to an extremely high motivation of young specialists to emigrate to Europe → a decrease in business activity due to fleeing of the nation's most active segment of the population leads to increased degradation of the architectural part of the city. The circle is closed.

According to authors, the breaking of the described hermeneutic circle lies in the development of scientific and practical activities of universities. Ukrainian universities have immense social reserves, their realisation is possible and, furthermore, vitally necessary. The above-mentioned cycle of planned experiments is dedicated to this problem, designed to identify optimal areas linking the educational, design, entrepreneurial activities of young professionals with the processes of revival of urban communities. Experience of Nikonenko E.V. convincingly showed how heavy the legacy of the USSR and the period of 90's is in the desocialization of city communities and the people mentality.

## 3. Conclusion

Dennis Meadows considers the ability of a region (or planet) to maintain the existing population density a result of the mutual influence of the biocapacity factors of the region, the availability of resources, the development of innovative technologies and the development of human potential. In the face of a decline in all four of these indicators, the survival of the city directly depends on the conscious and focused work of all its residents to correct the situation.

With a population of one and a half million, architectural and construction innovations become a weighty argument in solving the problems of adapting the city to escalating resource and environmental crises. Kharkiv and its region have a fairly high potential, allowing a positive assessment of the prospects of the city. In order to realize the existing potential, architects and urbanists of a new formation are needed, having competencies to transform the architectural environment of the city into a single, self-sufficient system.

Currently, the authors are working to determine the necessary architectural, construction and engineering competencies to form an architectural and urban specialization that meets the requirements of the city and its community. Scientific and practical directions are being developed in relation to the need for architects to work as leaders of multidisciplinary groups.

The main problem is the need to develop the principles of interaction between the architect and the territorial community in conditions of a depressive attitude of the population to any initiatives. This problem is evaluated and considered as a promising topic for a comprehensive study of urbanists of the following specialties: architects, sociologists, psychologists, managers, lawyers and city administrators.

## References

[1] Danylov-Danilyan V, Rafe I 2015 Beyond Growth. The computer model that stirred up the world *Science and life* **10**

[2] Meadows D L, Donella X, Renders J and Behrens III W 1991 *The limits to growth* – (Moscow: Moscow State University Press) p 208

[3] Norgard J S, Peet J and Ragnarsdottir K V 2010 *The History of The Limits to Growth* (The Solutions Journal) vol 1 chapter 2 pp 59–63

[4] Meadows D L, Donella H, Randers J 2004 *The Limits to Growth: The 30-Year Update* (White River Junction VT: Chelsea Green Publishing Co.)

[5] Randers J 2012 2052 - *A Global Forecast for the Next Forty Years*. (White River Junction: VT: Chelsea Green Publishing Co.)

[6] Meadows D L 1974 *The Dynamics to Growth in a Finite World* (Cambridge Mass: Wright-Alien press)

[7] Forrester J W 1961 *Industrial Dynamics* (Massachusetts: MIT Press)

[8] Serikova E N and Yakovlev V V 2011 *Communal services of cities* (Kharkiv: Kharkiv National Academy of Municipal Economy) pp 344-348

[9] Yakovlev V V, Lishchina V D, Babaev M V and Vasenko A G 2015 The source of water of Kharkiv and perspectives of the radial water intake *Journal of Taras Shevchenko National University of Kyiv*

[10] Minimum building water efficiency 2009 https://www.usgbc.org/credits/neighborhood-development-plan-neighborhood-development/v2009/gibp3

[11] Velikhov E P 2009 Energy evolution in the 21st century *Energy: economics, technology, ecology* **11** pp 2-14

[12] Voronovsky I 2019 *Development of alternative energy sources (Electronic Materials* vol **9**) (Tavria: Tavria State Agrotechnological University Press) chapter 1

[13] Ramazanov S 2008 *Demographic situation in Ukraine: causes and consequences (Electronic Materials* vol **6**: Historical Sciences) (Kyiv: Scientific chronicle of National Pedagogical Drahomanov University) p 65

[14] *Population Official site of the Main Directorate of Statistics of Kharkiv Region* URL: http://kh.ukrstat.gov.ua

[15] Makhonyuk O V 2019 *General trends of external migration in Ukraine. Migration in the conditions of transformation of regional labor markets of Ukraine: mechanisms of regulation* (Lviv: Dolishny Institute for Regional Studies of the National Academy of Sciences of Ukraine) p 263

[16] Manuel S I 2015 Ciem building *J. Construction21* URL: https://www.construction21.org/case-studies/es/ciem-building-zaragoza.html

[17] Danylov S M 2019 *Methodological bases of modeling the city as a dynamic system* (Kharkiv: Kharkiv National University of Civil Engineering and Architecture).

[18] Nekhoroshkov R B and Prasol V M 2009 Cost-effectiveness of the reconstruction of five-story buildings. Town-planning aspects of sustainable development of large cities *Digital repository of O. M. Beketov Kharkiv National University of Urban Economy*

# The problem of formation of innovative directions in the profession “Architect”

**Oksana Fomenko[1], Serhii Danylov[1], Oleg Sleptcov[2] and Andrii Izbash[1]**

[1] Department of Innovative Technologies of Design of Architectural Environment, Kharkiv National University of Civil Engineering and Architecture, Kharkiv, Ukraine
[2] Department of Fundamentals of Architecture and Architectural Design, Kiev National University of Construction and Architecture, Kiev, Ukraine

smd66smd66@gmail.com

**Abstract.** The paper investigates the phenomenon of the impact of innovation on the formation of competencies, specializations and specialities in the profession “Architect”. A research model is proposed that describes the impact on the profession of innovative, architectural and construction factors and the goal-setting factors of innovative knowledge and technologies. The features of the influence of changes in the socio-economic factor of the functioning of Kharkiv (Ukraine) on the formation of new urban areas in architecture are described. An experiment is described to create a cognitive model of the “Novi Doma” housing estate, Kharkiv, in which the authors of the study participated as expert architects. The deep socioeconomic crisis of the housing estate revealed and described in the experiment made it possible to draw conclusions about the parameters of a new scientific and educational direction in urban studies.

## 1. Relevance

Recently, the profession “Architect” has become a complex combination of many specialities and specializations. Over the past twenty years, under the influence of technological and information revolutions, a large number of innovations have appeared. Some of them, for example, computer literacy, have entered the core of the profession and have become practically mandatory for study by architects. Innovations gave rise to new areas of activity that were previously not inherent in the traditional circle of duties of an architect, which gradually begin to take shape in new specialities and specializations. A vivid example is the new specialization “architect-animator”. This is an unlicensed, yet already highly demanded type of innovative activity of the architect, associated with the animated presentation of architectural design concepts.

An interesting perspective on the phenomenon of the emergence of new specializations in architecture is provided by global statistics on intellectual property accounting. According to the WIPO (World Intellectual Property Organization) annual report on World Intellectual Property Indicators (WIPI), 3.3 million patent applications were filed worldwide in 2018, which shows a steady increase of 5.2% for the ninth consecutive year. Activity in the field of filing applications for registration of trademarks in the world has grown to 14.3 million, and in the field of industrial designs - up to 1.3 million per year.

According to the PCT (The International Patent System) for the period from 1995 to 2019, the number of patent applications in the following areas has increased: 1) Computer technology: 20 times; 2) Environmental technology: 4.5 times; 3) Civil Engineering: 5.3 times [1];

The above data indicate an "explosive" development of innovations that directly affect the formation of architecture and architectural specialities. The statistical approach allowed us to create a model that describes the impact of innovative knowledge and technologies on the formation of new competencies, specializations and specialities in architecture (Figure 1).

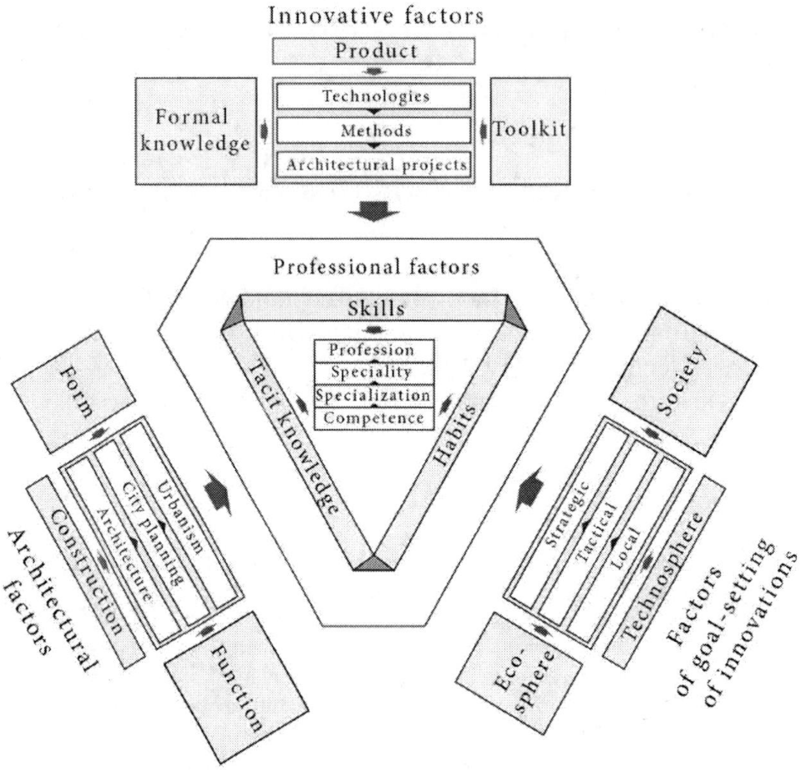


**Figure 1.** A model that describes factors that influence the formation of specializations and innovative competencies in architectural professions.

The model consists of four mutually related blocks: "Professional factors"; "Innovative factors"; "Factors of the goal-setting of innovations"; "Architectural factors". The model describes the formation of the first block - "Professional factors" under the influence of three others.

The first block - "Professional factors" is the object of study and consists of the following components: Competence, Specialization, Speciality, and Profession. The listed components are formed on the basis of skills, abilities and hidden knowledge. The term "Tacit knowledge" refers to the set of professional knowledge that is passed on from the master to the student and cannot be transmitted only in a verbal way [2]. For architecture, tacit knowledge is one of the significant factors in the formation of a specialist.

The second block - "Innovation factors" consists of the following innovations: technologies, methods, and architectural projects. Which, in turn, are formed under the influence of formal knowledge, tools and products.

The third block - "Factors of goal-setting of innovations" consists of the following hierarchical levels of goal-setting of innovations: local, tactical, strategic. The described levels are focused on achieving goals: environmental, technological and social, both individually and comprehensively.

The fourth block - "Architectural factors" describes the formation of architecture, urban planning and urbanism following the classical triad: form, function, construction.

The developed model made it possible to assess the prospects for the development of innovative areas in the profession and to develop several recommendations on the introduction of innovative training programs for Ukrainian universities.

In 2019, the authors of the study took part in an experiment related to the practical testing of the operability of the methods for modelling the city as a dynamic system developed at the Department of Innovative Technologies of Design of Architectural Environment (ITDAE) of Kharkiv National University of Civil Engineering and Architecture (KNUCEA). The authors' area of responsibility was the formation of multidisciplinary expert groups, as well as the study of social and legislative factors for the modernization of obsolete buildings of the "Novi Doma" residential complex, Kharkiv, Ukraine.

Social problems identified during the experiment became the reason for rethinking the developed model for the formation of architectural specialities. The specifics of the crises experienced by Ukrainian cities indicate that an exclusively technogenic approach is not able to effectively solve the challenges facing the country's architects. The problems identified an urgent need for the formation of a cluster of architects - urbanists capable of designing not only in the context of energy, economic and raw material crises, but also in solving acute social problems. Architects are closely connected with urban communities, living by their interests and able to gain the trust of these communities. This approach forms a cycle of tasks that were not previously solved in the national architectural paradigm.

This article describes the preconditions for the thesis about the need to develop a new direction in the architectural profession - the architect-urbanist, whose professional activity is focused on the formation of an architectural space based on the socio-economic interests of urban communities. The authors conducted research cycles in which the problem of a systemic contradiction of the investor interests and urban communities was identified. Moreover, as a rule, the interests of investors dominate, which periodically harms the city as a system and its brand. An architect - urbanist is able to identify the socio-economic resources of the community, develop on their basis strategic measures to transform the architectural component of the city and motivate the community to make the necessary changes.

## 2. Research methods

For the first time, the methodological foundations of the program complex for modelling the city as a dynamic system were developed at the ITDAE department of the Kharkiv National University of Civil Engineering and Architecture. They are based on a combination of two trends in modern science: optimizing the functioning of cities and managing large amounts of data about the city. This allows you to explore the dynamics of changes in the city over a certain historical time. The strength of these changes, their speed and acceleration determine the place and parameters of the existence of a particular phenomenon in the system. The totality and mutual influence of all dynamic processes in the system determine the parameters of the existence of the system itself.

The program complex consists of two hierarchically related, but autonomously organized parts: verbal-mathematical and imitation-cumulative. The cognitive-research matrix allows you to create information packages that verbally and statistically describe the principles of functioning of the subsystem "architecture", as part of the city system. The information storage model is responsible for the function of storing and processing information packets formed by the cognitive research matrix. Based on the accumulated data array, the information storage model displays and predicts in real-time the state of the city system in the Sustainability, Crisis and Catastrophe categories.

The specified model complex can be used as an architectural and urban toolkit for modelling, monitoring and strategic planning of the functioning and development of the city as a dynamic system. This will potentially allow solving a set of problems associated with adapting the architecture of post-industrial cities to modern conditions. The complex nature of this program allows you to take into account the goals of both the "architecture" subsystem and the city system as a whole, and find a coordinated, compromise solution for its optimal functioning and development.

The research put forward the thesis that Kharkiv needs a cluster of architects who are able at the system level to develop strategic solutions for transforming the architectural component of the city in the face of many systemic crises.

The basis for this thesis was the discussion at the interdepartmental seminar at the Academic Council D 64.056.02 of Kharkiv National University of Civil Engineering and Architecture. At the seminar, the report of Danylov S. M. was heard on the results of dissertation research for the degree of Doctor of Architecture "Methodological basis of city modelling as a dynamic system" [3]. The fifth section of the thesis is devoted to: "An experiment to test the performance of a cognitive-research matrix as part of a model complex for analyzing and modelling a city using the example of the "Novi Doma" housing estate in Kharkiv". The object of the experiment: housing estate "Novi Doma". Subject: impact of the law of Ukraine "About the energy efficiency of buildings" adopted in 2017 on the dynamics of socio-economic indicators of the "Novi Doma" housing estate.

## 3. The results of the experiment

During the experiment, more than 350 information blocks were formed and parametrically connected, describing: depreciation of engineering networks and buildings of the district; increased energy consumption of buildings; increase in the welfare of residents of the area; activity ACAB (Association of co-owners of an apartment building) on the rehabilitation of buildings; ageing residents of the area; labour emigration; marginalization of the population; the social activity of the population of the district; increase in deductions for capital repairs of buildings, etc.

A comparison of the graphs constructed on the basis of the information blocks made it possible to identify several crises, the dynamic indicators of which are close to the transition to the pre-catastrophic state. The study revealed a deep social crisis, which is the main factor hindering almost all the measures being developed to increase the stability of the region.

As one of the analytical methods developed in the dissertation research, the concept of the "Site of Problems" was proposed.

The "Knot of Problems" describes conflicts of centripetal and centrifugal efforts that arise in the city system in the processes of functioning. Centripetal efforts arise in response to changes in the conditions of the city's functioning and are aimed at preserving it as a system. Centrifugal centred on mutations in urban space and aimed at adapting the system to new challenges. As a result of changes in external and internal conditions, the goals of centrifugal and centripetal efforts contradict, provoking how the transition of the functioning processes of the city system from resistance to crisis and in the opposite direction.

The authors of the study took part in the above experiment as experts, forming the "Legislative knot of problems", which affects the prospects for the modernization of residential buildings in the "Novi Doma" district of Kharkiv. According to the tasks set for Fomenko O. O and Izbash A. M., the "Knot of problems" was formed on the basis of the following features: A) on a hierarchical basis - international agreements in which Ukraine, national legislation and regional acts participate; B) By scope - legislative regulation of the main subsystems of the city of the Ecosphere, Technosphere, Population; C) In the direction of the efforts - centripetal, centrifugal. The study identified 12 legislative acts related to the socio-economic field of implementation of the Law of Ukraine "About the energy efficiency of buildings" (Table 1).

The structure of the formed "Legislative knot of problems" made it possible to determine legislative restrictions and stimulating aspects that ensure the processes of energy modernization of residential buildings in the district and city. The picture below (Figure 2) shows an experiment on graphically displaying the "Legislative knot of problems" indicating the following factors provoked by the analyzed laws in the city system: A) Centripetal efforts. In this case, laws aimed at maintaining the integrity and stability of the city system are considered. For example, the law "About atmospheric protection", which sets limits on the emissions of harmful substances by architectural objects into the atmosphere during their life cycle; B) Centrifugal efforts. They are aimed at adapting the architectural environment of the city to the new realities of its functioning.

IOP Conf. Series: Materials Science and Engineering **907** (2020) 012018 doi:10.1088/1757-899X/907/1/012018

**Table 1.** The laws and articles of these laws that form the "Legislative knot of socio-economic problems of the "Novi Doma" housing estate" related to the modernization of the housing stock of the district following the requirements of the Law of Ukraine "About the energy efficiency of buildings".

| № | The legislative aspect | Comments |
|---|---|---|
| 1 | "About protection of atmospheric air " | Section IV. Article 23. Conditions for the Design, Construction and Reconstruction of Enterprises and Other Objects that Affect or May Affect the Condition of Atmospheric Air |
| 2 | "About the protection of the environment" | Section XI. Article 51. Ecological Requirements for the Placement, Design, Construction, Reconstruction, Commissioning and Operation of Enterprises, Structures and Other Objects |
| 3 | "About alternative energy sources " | Section II. Article 9. Stimulation of electricity production from alternative energy sources |
| 4 | "About the complex reconstruction of quarters (residential districts) of outdated housing stock" | Article 9. Identified sources of financing of measures for reconstruction of residential quarters of cities. |
| 5 | About Energy Saving | Article 14. Areas of Use of Energy Saving Funds |
| 6 | About Local Self-Government in Ukraine | Article 26. The exclusive competence of village, settlement, city councils |
| 7 | "About the energy efficiency of buildings" | Article 6. Minimum requirements for energy efficiency of buildings |
| 8 | "About the peculiarities of exercising property rights in an apartment building" | Article 12. Costs for managing an apartment building |
| 9 | "About the association of co-owners of an apartment building" | Article 11. Keeping the residential complex on balance<br>Article 12. Management of indivisible and common property of a residential complex |
| 10 | About the Energy Efficiency Fund | Article 15. Functions of the Fund |
| 11 | "Program of partial reimbursement to Kharkiv individuals for interest on energy-saving loans for 2017-2020" | Article 4. Financial support of the Program |
| 12 | Energy Saving and Energy Efficiency Program of Kharkiv Housing Fund for 2018-2022. Kharkiv city 2017 | Article 1. The current state of the housing stock in Kharkiv |

For example, the Law on Energy Efficiency, putting forward requirements to reduce buildings energy consumption due to their modernization; C) Sustainability. A chain of legislative ties that have a positive response in the city system. For example, the National Law "About Energy Saving" has led to the emergence of the National Law "About the Energy Efficiency Fund", which is the basis for the creation of the regional regulation " The program of partial interest rate reimbursement to individuals of Kharkiv on energy saving loans for 2017-2020 years". Taken together, laws create positive efforts to enhance the city's system; D) Contradictions. In this case, the logical contradictions of the Laws are not meant, but the contradictions arising between the subsystems of the city in the processes of their functioning. In the formed "Knot", as a potential contradiction, insufficient financing for the modernization of the housing stock of the district from the means of city and state budgets is reflected; E) The crisis. In this case, the potential possibility of provoking a socio-economic crisis by law is displayed. For example, the National Law “About Energy Efficiency”, through the Law “About features of realization of the property right in the apartment building”, affects the welfare of residents of apartment buildings, obliging them to certain expenses to bring their property into line with the law.

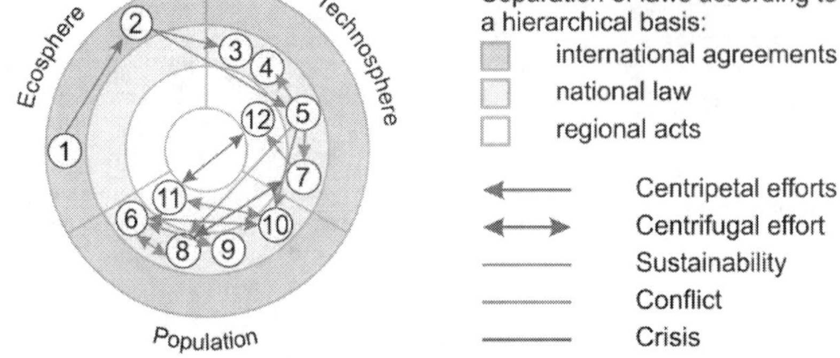


**Figure 2.** The legislative knot of problems of modernization of obsolete housing in the “Novi Doma” district of Kharkiv.

Thanks to the analysis, it became possible to assess the share of financial participation of the state, investors and residents of the district in the process of modernization of obsolete housing. According to aggregated indicators, the share participation of residents of the “Novi Doma” district in financing measures for the modernization of housing, the owners of which they are, is defined as 70%. Following the order of the Ministry of Regional Development, Construction, Housing and Communal Services of Ukraine, № 172 dated 07.11.2018 “About approval of the Procedure for certification of energy efficiency and the form of energy certificate”, the concept of “Reference building” was used in the experiment.

The study of international experience in the modernization of obsolete residential facilities has identified four prototypes that have become the basis of the “Reference Buildings”. The selected prototypes made it possible to determine the types, nature and cost of modernization of buildings in the range from 160 to 320 US dollars per m2 (Table 2) and through them, the average financial burden per inhabitant.

**Table 2.** Types of reference buildings

| Energetic building class, cost modernization m² | Reference Building/ Address / Architect | | Energetic building class, cost modernization m² | Reference Building/ Address / Architect | |
|---|---|---|---|---|---|
| 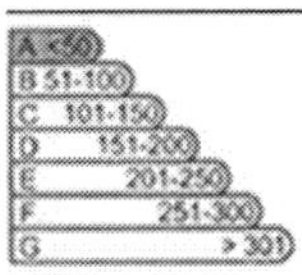 | LWG Leinefelde<br>Germany<br>Stefan Forster<br>Architekten<br>350 USD/m2 |  | 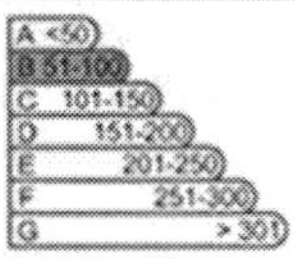 | Tomsk, Russia<br>Komsomolsky ave,71<br>LLC SV-Trade<br>205 USD/ m2 |  |
| 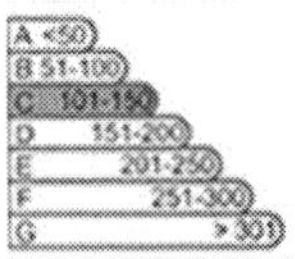 | Kaliningrad, Russia<br>Leninsky ave., 2-4<br>Arthur Samits and<br>Ilya Kiselev<br>280 USD/m2 |  | 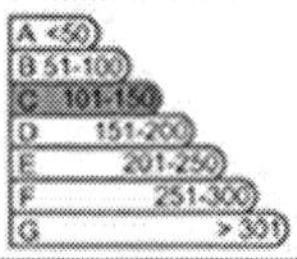 | Kharkov, Ukraine<br>ave. Marshal Zhukov, 21<br>Nikonenko Eugene<br>160 USD/ m2 |  |

The obtained results made it possible to calculate four possible options for the socio-economic consequences of the modernization of buildings for the population of the “Novi Doma” housing estate. At a round table devoted to the end of the first stage of the experiment, a discussion was held by experts of the Legislative, Social, Technogenic and Economic “Knots of Problems”.

Through the efforts of all experts who took part in the experiment, the Comprehensive Problem Knot was compiled. The “Knot” was compiled by comparing the dynamics of the base growth [4] rate graphs indicators are shown in Table 3.

**Table 3.** Graphs of the growth rates of indicators included in the "Integrated Problem Knot".

| Indicator | Basic growth rate relative to 1960 | Indicator | Basic growth rate relative to 1960 | Indicator | Basic growth rate relative to 1960 |
|---|---|---|---|---|---|
| 1 Population density | 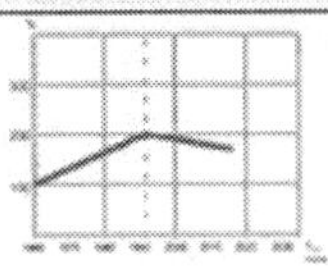 | 2 Infrastructure facilities | 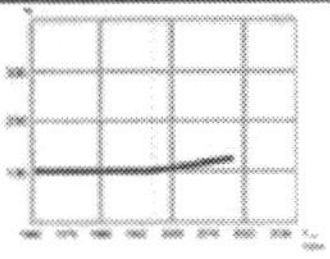 | 3 Purchasing power | 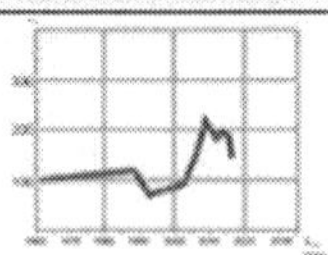 |
| 4 The ratio of young to old | 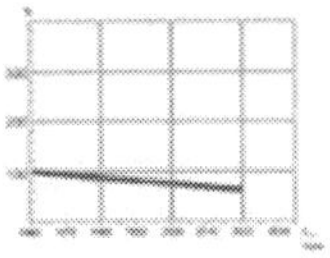 | 5 Housing Aging | 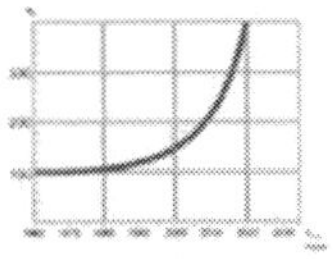 | 6 Branding the district | 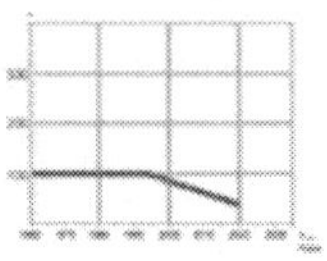 |
| 7 Modernization of buildings | 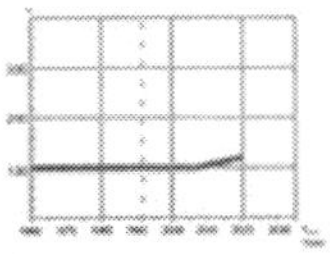 | 8 Electricity prices | 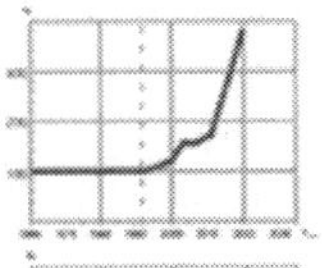 | 9 Gas prices | 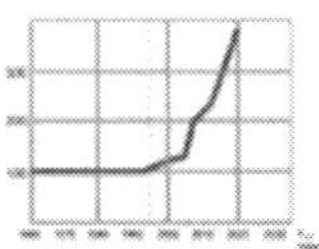 |
| 10 Communal payments | 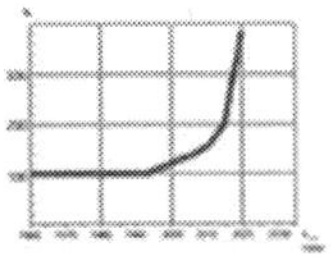 | 11 Food prices | 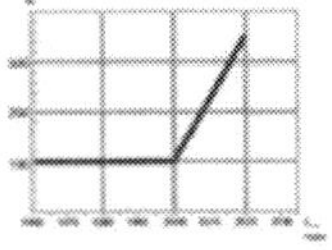 | 12 Deductions for repairs | 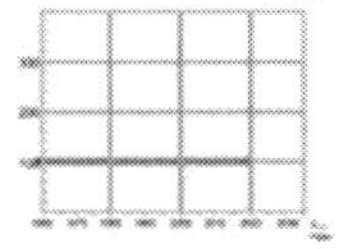 |
| 13 Lonely people | 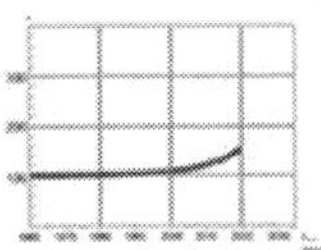 | 14 Socialization | 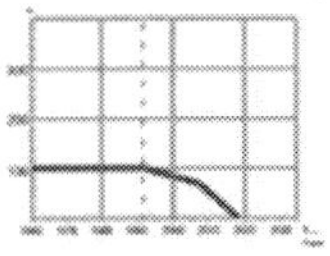 | | |

## 4. Conclusions from the experiment

The law "On energy efficiency" concerning the obsolete housing stock of the "Novi Doma" district of Kharkiv will not be implemented. The following factors indicate this: According to the data of the Main Department of Statistics in the Kharkiv region in 2015 the provision of housing for Kharkiv residents averaged 24.1 m2 [5]. From which it follows that the average cost per 1 inhabitant for modernization of housing amounted to about 3856 US dollars following the fourth minimum indicator of the cost of m2 of modernization of the "Reference Building". In 2020, the average loan rate for "Warm loans" at the six largest banks in Ukraine amounted to 18% per annum. In accordance with the calculation of the loan calculator minus a 40% share of municipal and state financing, the repayment of loans amounted to 96 monthly payments of 1575 UAH. following the exchange rate on 02/08/2020. The average living wage (including utility bills) is 2270 UAH/ month per person [6]. The average pension is 3019 UAH. (January 2018). The statistical report "The distribution of the population in terms of average total income equivalent (2010-2018)" Indicates that 33.7% of the city's population has an income of less than 3360 UAH. This means that at least a third of the district's residents are physically unable to take part in financing the modernization of their housing (Table 3. paragraphs 3; 8; 9; 10; 11).

In the last decade, a decrease in the perception of the district's brand associated with the following factors has been identified: 64% of the district's buildings are morally obsolete (table 3. paragraph 5); 90% of engineering networks have expired [7]. These factors were reflected in the ranking of Kharkiv districts in terms of the cost of apartments the housing estate has entered the top three districts with the worst rates [8]. As a reaction, according to the results of surveys of real estate agencies, from 25 to 30 residents with an income above the average moved from the estate area every month to more prestigious areas. These data indicate a gradual deterioration in the housing estate situation. Also, almost complete lack of motivation for prompt action to solve the problem of obsolete housing stock among local authorities and investors.

The housing stock of Kharkiv includes 9363 houses with a total area of 41 771.3 thousand m2. Most of the housing in the city was built on the projects of mass series in the 60-80s of the last century. The pace of residential buildings obsolescence exceeds the pace of new housing construction, overhaul, modernization and reconstruction combined.

Almost the entire housing stock is characterized by the problem of technical and moral ageing of the engineering equipment of houses - heating systems, hot and cold water supply, sewage, sanitation, electricity, lighting, ventilation, elevator facilities. Hydro and thermal insulation of roofs and walling are worn out, which causes them to leak and freeze. Depreciation or absence of thermal insulation of the heating and hot water pipelines lead to significant losses of thermal energy in the houses [9].

In March 2019, the authors of this study received consultations on the prospects for the renovation of obsolete housing in the city at the city's chief architect Chechelnitsky S.G. and Nikonenko E.V. - the chief architect of the project for the modernization of a five-story residential building in Kharkiv, at the address 21, Ave. Marshala Zhukova ("Novi Doma" housing estate). A complex of design works was carried out on the building's overhaul, attic superstructure, without resettling residents, increasing the building's energy efficiency to class C. The negative experience described by the architect in the process of architectural supervision of the reconstruction of the house gave the study authors a new perspective on the problem of modernizing obsolete housing stock.

The modernization of the building at 21, Ave. Marshala Zhukova was completed by JSC "Zhilstroy-1" in 2001 at the expense of investors and the municipality. Repair and construction work was carried out without the resettlement of residents. This experience was considered as an experiment necessary to develop a socio-economic strategy for the renovation of obsolete housing in the city. The idea was to develop a project that was economically attractive to potential investors. The project payback was supposed due to the attic floor superstructure on five-story buildings with the subsequent sale of new apartments.

The number of efforts to establish constructive interaction with the residents of the house made by the project organizers forced them to reconsider their attitude to this area of economic activity. Architect Nikonenko E.V. describes the social problem of home modernization as an extremely negative experience that he would not like to repeat.

Thus, the basis of the crisis revealed in the experiment lies in acute socio-economic problems blocking all attempts by interested parties to solve the problem of progressive degradation of the technical condition of the housing stock of the "Novi Doma" district. The severity of social problems led to a decrease in the motivation of investors and local governments to almost zero. In Kharkiv over the past 20 years, the authors of the article revealed only one case of the modernization of the building described above. Delaying the solution to this call can lead to irreversible consequences and depopulation of the housing estate.

The facts cited in the conclusions of the experiment prompted the determination of paragraph 14 "Socialization" (table 3) as a crisis threatening the transition of the dynamics of the functioning of the region to a pre-catastrophic state. A study of the causes of the identified crisis showed that similar problems are characteristic of almost all cities of the country.

The study revealed critically low socialization of the population of the district and city. The conducted express survey gave the following results, 100 surveyed residents of the district do not trust: the government - 89%; Banking system - 72%; Local governments - 63%; Residents of the district - 54%; Residents of the house - 49%; Investors - 86%.

One of the results of the experiment was an awareness of the responsibility of architects for the development of new strategic approaches to solving the problem of the obsolete real estate fund of the city. Given the acute shortage of resources, traditional architectural and town-planning approaches to this problem are practically powerless. The main problem associated with the prospects for resolving the identified crisis is the lack of urban communities. The effectiveness of public and private investment programs in the current conditions does not solve the problem. A promising way out of their designated crisis seems to be: A) The revival of territorial communities, presumably based on the ACAB (Association of Co-Owners of an Apartment Building); B) Nurturing a cluster of young, motivated professionals capable of developing effective step-by-step strategies for resolving problems facing communities;

C) Development of methods for overcoming crises due to the rational use of social, technological and natural resources of the city region.

These theses indicate the need to develop innovative, experimental competencies in the speciality "architectural urbanism".

## 5. Aspects of the formation of new specializations in architecture.

Analysis and development of strategic models for solving the problems identified in the experiment allowed us to draw the following conclusions: 1) In the city of Kharkiv, from 15 to 25% of the population lives in buildings whose estimated life is over; 2) The level of socio-economic development of Kharkiv does not allow us to hope for a quick and effective solution to the problems of old housing; 3) In the city there is practically no cluster of young managers, scientists and practitioners capable of solving the problems facing the city at the level of strategic management;

A scientific and practical response to the identified challenge was the dissertation research of graduate student Izbash A.M. led by doc. arch. prof. Fomenko O.O. (authors of the article): "The role of innovative technologies in expanding the core of the profession "architect". This study develops a model for the development of new architectural competencies under the influence of innovations that have come to architecture and construction.

The point of this work is an innovative approach, considering the cluster of architectural urbanists as a compensation buffer in resolving social, technological and environmental systemic conflicts that arise in the formation of the architectural environment of the city and affect the principles of functioning of urban space.

The new strategy for the stable functioning of the city in the conditions of an acute shortage of resources requires not only appropriate innovations, but there are also problems that make it necessary to pay special attention to the idea of educating a cluster of urbanists of a new formation for Kharkiv, among them: 1) Depopulation of the city. Over the past 30 years, the population of Kharkiv has declined from 1,623,000 to 1,441,932 people [10] ; 2) Low purchasing power per person - € 1 830 [13]; 4) Stagnation of the city's economy; 5) The marginalization of the population; 6) Housing stock obsolescence;7) Degradation of utility networks, etc;

With a high degree of probability it can be argued that traditional architectural approaches based on the active exploitation of raw materials and economic resources will be ineffective. The urgent need to transform the urban space (more than 400,000 people live in buildings whose estimated operating life has been completed or will be completed in the next 10 years), together with the stagnation of the economy and lack of resources, creates a very different "map" of the architect's activities;

## 6. Conclusions.

The theses presented in the article indicate the need to consider innovative educational processes as part of the practical activities of architects aimed at adapting the city to new challenges. To do this, it is proposed to develop the following scientific and educational areas: A) Development of a strategy for interaction between the university and local authorities; B) Implementation of the principles of regional environmental management as the basis for the sustainability of the city. This indicates the orientation of the city's economy to the exploitation of the region's social, natural and technological resources, and a decrease in dependence on energy and raw materials exports; C) Development of strategies for the participation of the university in the processes of the revival of territorial communities. For Ukrainian cities, the theme of the revival of urban communities is one of the most important factors in increasing their sustainability. During the existence of the USSR, urban communities were almost destroyed. In this case, the term "urban community" means a compactly living community of people united basing on a common ethical, economic, cultural and other interests. Conscious of the community of these interests and defending them;

Calculations of the city's potential need for urban architects made it possible to determine the following ratio: 1 specialist per 3500-5000 inhabitants. For the city of Kharkiv and the region, this means jobs for 350-600 architects-urbanists. For the universities of the city, the formed need makes it possible to open a new scientific and educational field with the prospect of the annual graduation and further employment of 15-25 specialists.

The authors of the article are convinced that the identified problems and crises associated with the architectural component of the city of Kharkiv are characteristic of many cities of the post-Soviet space. The development of scientific, educational and urban strategies for resolving the challenges facing the city are indispensable for its survival. The nature of the revealed crises and their development scenarios indicate the fact that the European experience is in many ways not applicable to local socio-economic realities. For this reason, so much attention is paid to the problems of the revival of urban communities, as a non-alternative factor in the renovation of the city and its architecture.

One of the most important results of the experiment was the awareness of the role and responsibility of architects for the revival of Ukrainian urban communities. Architects have a unique set of competencies that allow them to form multidisciplinary teams, technically accurately calculate the necessary and permissible changes in the architectural fabric of the city and convey the results of their research in aesthetically and emotionally verified forms of architectural visualizations. In practice, today, there are no other professions capable of combining such a wide range of humanitarian and exact sciences.

## References

[1] *World Intellectual Property Indicators: Filings for Patents, Trademarks, Industrial Designs Reach Record Heights in 2018* 2019 (Geneva: WIPO) URL: https://www.wipo.int/pres-room/en/articles/2019/article_0012.html
[2] Polanyi M and Sen A 2009 *The Tacit Dimension* (Chicago: University of Chicago Press)
[3] Danylov S 2019 *Methodological basis of city modeling as a dynamic system* (Kharkiv: Kharkiv national university of civil engineering and architecture) URL: http://kstuca.khar-kov.ua/ndial/nauka/208-specializovana-vchena-rada-d-6405602.html
[4] Farris P W, Neil T Bendle T N, Pfeifer P E and Reibstein J D 2010 *Marketing Metrics: The Definitive Guide to Measuring Marketing Performance* (Upper Saddle River, New Jersey: Pearson Education)
[5] *Housing Fund of Kharkiv Oblast* 2018 (The head office of statistics in the Kharkiv region) URL: http://kh.ukrstat.gov.ua/zhytlovyi-fond-1995-2012
[6] Koshanskaya I and Pisarenko M *Minimum wage and cost of living: how the amounts will change from January 1* (News Channel 24) URL: https://24tv.ua/en/prozhitochnyj_minimum_2020_minimalnaja_zarplata_2020_ukraina_summa_vyplat_n1220160
[7] Kobzar N 2020 *Kharkiv heating systems are 90% worn out – Andreev* (Kharkiv: Status quo). URL: https://www.sq.com.ua/rus/news/novosti/17.01.2020/harkovskie_teploseti_iznoshenu_na_90proc_andreev
[8] Karpushenko A 2019 *Rating of Kharkiv districts in terms of the cost of apartments* (Kharkiv: StroyObzor) Build a review URL: https://stroyobzor.ua/news/reyting-rayonov-harkova-po-stoimosti-kvartir-fevral-2019-go.html
[9] Kharkiv City Council 2017 *About the introduction of a minimum fee to the Program for energy efficiency of the Kharkiv housing fund for 2015-2020* URL: https://doc.city.kharkov.ua/uk/pro-file/document/view/id/678786
[10] *Map of the Month: GfK Purchasing Power Europe* 2019 (Bruchsal: GfK) URL: https://www.gfk.com/insights/news/map-of-the-month-gfk-purchasing-power-europe-2019/

IOP Conf. Series: Materials Science and Engineering **907** (2020) 012019 doi:10.1088/1757-899X/907/1/012019

# Innovations in the Architectural Environment: the Impact of Society on the Positive Perception and Conflict of the Incomprehensible

**Nataliia Mezhenna and Daria Filippova**

Department of Fundamentals of Architecture and Architectural Design, Kyiv National University of Construction and Architecture, 31, Povitroflotsky Avenue, Kyiv, 03037, Ukraine

archiknuba@gmail.com

**Abstract.** The article is focused on issue of human perception of new objects of architecture and design, overcoming people's commitment to the old and familiar. New ideas often face a negative reaction from society. Modern architecture, any projects of modern style, cause not only curiosity in its viewers but irritation, stress, negativity, psychological aversion as well. Original and common information of perception of the architectural environment is characterized by a dialectical unity. With complete originality there is a conflict of the incomprehensible. With complete commonness – the conflict of oversaturation. Unapprehensiveness generates the transfer of functions, the transfer of an object to another functional group. The more developed the society and the specific observer are, the easier it is to perceive new forms and technologies for them. There is a larger amplitude scale at the stage of information filtering. The development of society brings perception to a new stage, new forms and styles transform into typical ones, a psychological adaptation occurs. Crisis situations in a country and the general perception of life as critical, non-positive notion forms a negative reaction to new proposals. The situation is similar to "Seligman's dogs", the so-called learned helplessness. The psychological impossibility of non-action is the impossibility of thought, the unwillingness to look for analogies and images that would lead to metaphorical matches. Often the perception of a new object can be influenced in advance by technologies and mass media. In order to resist these technologies, to have his own, personal opinion, a person needs a high level of knowledge, world outlook, synthesis of information about the object. The formation of new human values, raising living standards lead to an adequate perception of the new architectural environment.

## 1. Introduction

A person lives in a constantly changing world that contains immense amount of diverse information, both necessary and unnecessary for their adequate existence. Only one thing remains in a human life mostly unchanged – architecture. Architecture embodies cognitive, semantic, emotional needs, speaks about the past and the future, and reflects certain values. The semantic and visual diversity of the architectural environment is an important basis for human contact with the outside world. Thus, the willingness to accept or not to accept the innovations in the architectural environment directly depends on human values, both personal and "imposed" by society, laws, local preferences and so on. Moreover, no matter how properly the architectural environment is designed according to all building codes, modern trends and ergonomic indicators, it can still be rejected by an individual or a society. Having analyzed the existing architectural objects of Kyiv (as well as almost all major cities in the world), we can say with confidence that architects mostly do not consider the most basic psychological principles of human perception of space. Whereupon, the architecture and its environment, its saturation with visual elements contributes to the mental state of a person, which, in turn, has a great influence on human psychosomatics, behavior, social development and more.

## 2. Perception of new in a sustainable architectural environment

The main parameters that affect the emotional perception of the architectural environment are:

- saturation with stress factors, environmental factors, anthropogenic factors, oversaturation with stimuli, etc.;
- social orientation of the environment, participation in the development or limiting of social interaction;
- orientation factor, "easy use" of the city by its residents and guests in accordance with their ideas and values;
- fullness of emotions, aesthetics of the subject-spatial environment;
- cultural and educational factor, the city as an intellectual, cultural and visual environment;
- the opportunity to participate in solving issues – how residents perceive specific decisions on the development of their environment and how fully they participate in making such decisions.

It is the last parameter that is very often rejected in our society. Foreign experience shows that any city-planning project of a town or district should begin with a dialogue between the project architect, builder, designer and ecopsychologist, as well as with the search for subject-spatial solutions that will favor positive regulation of human behavior through emotions.

In some European countries, architects in their pre-project work often use the empirical method in identifying the specifics and preferences of people who will be in direct contact with thi environment daily. For example, the yard of the house was landscaped and all ergonomic parameters, safe pavement, aesthetically pleasing content had been taken into account, but people living in this house try to be in this area as rarely as possible, avoiding it. So why does a person reject this healthy zone? The reason is very simple - most people who live here do not feel comfortable in areas where trees are over a meter high, they would be more comfortable if shrubs or shrub-type trees were planted instead. Thus, when designing this yard area, it would be desirable to conduct a survey of residents about their preferences, and then the architects would design an environment that residents would accept for its psychological comfort [1].

## 3. Conflict of the incomprehensible

The perception of modern interventions in the existing architectural environment (both positive and negative) is associated with resistance, overcoming the commitment of people to the old and familiar. Progressive architecture is implemented in the life and consciousness of our society and country not only physically but also morally, so it can be noted that this is a manifestation of some resistance, a crisis in man – the psychological perception of new information in the format of architectural space. It should be noted that today there is a lot of verbal violence in speeches, comments to publications, on television and on all social networks: strange invectives, attempts to offend, provocative statements, even death threats. Almost any new idea, direction, proposal, even positive a priori, very often faces a negative reaction from society. Furthermore, the more atypical a new work is, the more negativity it can induce. People subconsciously (or consciously) look for disadvantages first, underestimate the value of the idea, look for (and are sure to find) something bad.

One of the most famous historical examples of this attitude to the bright innovations in architecture was the Eiffel Tower in Paris. At first, the French perceived it very negatively, but over time it became one of the main symbols of the city. The famous Sydney Opera House was also received negatively during its long-term construction and at the start of its operation. Now, speaking about Australia, this distinct object cannot go unnoticed. These architectural works have even become the main visual symbols of their countries. There are many more examples in history when an architect tries to design a work of their time, even look ahead a little stylistically, but the understanding of the quality and necessity of this work by contemporaries comes a little later. Therefore, authors will consider the perception of new architecture and design in urban environments, why new works are perceived neutrally or more optimistically by the population in some countries, while in others they face complete rejection, or even provoke aggressive reaction.

This reaction extends almost entirely to modern architecture, especially in historic city centers, the reconstruction of historic buildings, any projects of modern style. The person unequivocally perceives a situation of "the-new-in-historical" as stressful, atypical and non-positive [1,3]. A new construction presenting itself as a historical building, though a fake one, a conscious compilation of architectural styles of antiquity is perceived in a calmer way, almost imperceptibly. This architecture is often considered "bad taste" by professionals, while the viewers on average like it. Alternatively, they do not notice anything at all, do not feel any changes in the environment.

At the same time, we can say that anything new, unknown, unfamiliar and unrecognizable causes not only curiosity in its viewers, but irritation, stress and negativity as well. An unfamiliar silhouette, the novel shapes, unusual designs are perceived as stressful, induce a psychological rejection. In order to perceive the new adequately, it is necessary to overcome the internal resistance. Connecting progressive architecture and its implementation with the life and consciousness of our society and country not only physically but also morally, we can note another manifestation of the crisis of society – the psychology of perception of new information in the format of architectural space. However, the somewhat provocative modern stylistics, which have become commonplace in Europe, America and some Eastern countries, often provoke lively debates in Ukraine, which are far from positive (Figure 1).

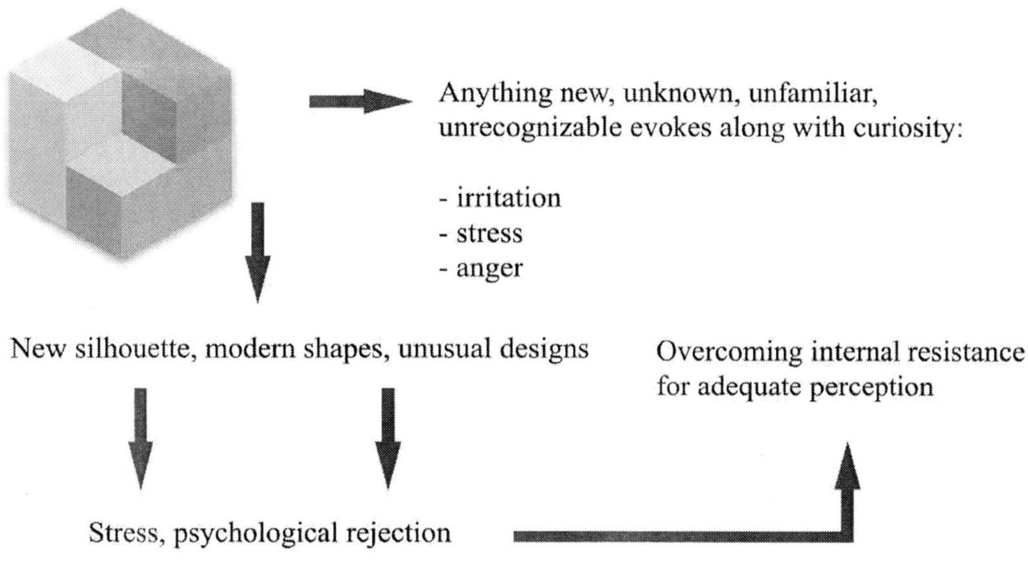


**Figure 1.** The model of adequate perception of a new object

The average observer does not possess a professional understanding of architecture and makes his opinion about what he sees based on his own values, which were formed by certain factors. The basis of human perception of a new object of the environment is physical phenomena that allow analyzing information with the help of sensory organs. Each person has his own perception of the environment, which is influenced by his age, education, life experience, religious views, fashion, personal preferences, professional qualities, country of birth and residence as well as many other reasons that shape the personality of a particular observer. Therefore, the question arises: is it possible to determine the composition of emotional and aesthetic factors, their nature and form with sufficient quality [2]? Thus, an individual perceives and analyzes the subject environment not only as a separate set of sensations and emotions, but also as a certain integral image of individual elements. Through the unification and generalization of specific details, a person's consciousness establishes a general image or content, subconsciously comparing it with similar objects and ideas. The consciousness of an individual also has the property of assigning different qualities to the objects of the environment, at the same time endowing it with positive or negative features. Dependence of the perception on the personality of the subject, the peculiarities of their psychodynamics, experience, volume and content of memory is defined as apperception. Even the apperception of a particular person can change under the influence of changes in his age, world outlook, preferences, raising of cultural level and knowledge. It follows that the more educated the society, the higher its general level, the easier and more optimistic the innovations will be perceived.

Why are the first impression, the first human emotions from what is seen so important? Drawing a parallel with the animal world, we can say that a person often acts in accordance with emotions or even solely on instincts. For example, after receiving a dose of negativity, during an argument or a push, a person begins to shout back under the influence of his emotions and instincts, behaves aggressively morally and physically. This is the reason why so often people in a state of affect do terrible things that they regret later, that is, act only as animals that are unable to analyze and process information. Even people who are engaged in self-development, psychology, yoga (or other psychological practices), who train their emotions and learn to control them, cannot always follow how their body unconsciously activates certain reactions. What can be said then about the majority of the population, who do not even think about what an emotion is, where and why it arises and what to do with it, how to manage it? In addition, even though the perception is a controlled process in its part, the other part, the subconscious perception, is an unknown, unconscious, unexplored region of the human self. Innovations in the architectural environment stimulate a person to certain actions and reactions, to interaction in a society, to building a dialogue.

U. Pavlov derived the sequence of pyramidal arrangement of levels of perception on the model of the subjective world of a person. This sequence corresponds to the sequence of formation of social and individual experience of a human being. For their effective existence and functioning in a complex social environment, people had to develop the intellectual abilities to effectively communicate, learn, and, most importantly – to understand not only the actions but also the thoughts and desires of their fellow tribesmen. A human being is adapted to abrupt changes in the environment and its functional systems are represented by five other levels of perception in addition to the lower two. All major functional systems of animals are represented only by the two lower models of environmental perception. At the same time, the above-mentioned formed systems are basic and necessary for coexistence in natural conditions for animals. Therefore, basic levels of perception are important for both the human and animal worlds. An architect who has the knowledge and the means to influence human behavior and perception of the world has a powerful weapon that can either "protect" and develop society or cripple it (Figure 2).

It can be observed that some new buildings do not cause any reaction in society, and the appearance of others causes considerable public attention, more often negative than positive. Why does a person like or dislike a new building, the reconstruction of a historic building in a modern contrasting style? A striking example is the Podil Theater in the historical environment of the Andriyivsky Descent in Kyiv, which has invoked completely opposite opinions of citizens, most of whom have condemned the implementation of this object for one reason or another.

What exactly does this building represent? The Podil Theater is a challenge of a new time for the paradigm of today. This architectural product has violated the boundaries of people's usual perception of architecture, especially sharply seen against the background of the historic buildings of the Andriyivsky Descent, which is also one of the hallmarks of the city. Color, shape, style of the building look completely different, there are no usual outlines, no features of the classical style, a bold coloristic scheme of the facades. But can we say that "unlike others" means worse? "Unlike others" may mean a more progressive, modern, an element of the architecture of the future, utilizing a new approach, a new style etc.

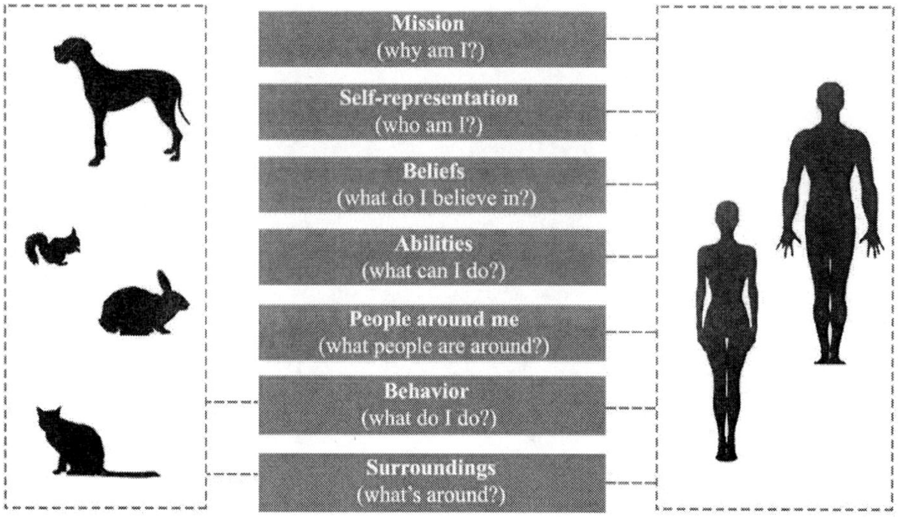


**Figure 2.** Levels of perception on the model of the subjective world of a person

In the process of observing the object, a person perceives images, which are special cases of signals-codes, which, on the one hand, are symbols of their time, the period when the building was created, and on the other - can be perceived by future generations by taking into account the knowledge and emotions pertaining to this period of history. For example, the allegories on which the medieval and classical art of painting, mosaics and reliefs in the temples and palaces of Europe is based, are not always clear to modern viewer at first sight. Their perception requires special knowledge and intellectual level. An ignorant person will see just a ram, a snake, an eagle or a flower, not realizing the biblical images behind these symbols, and not understanding the deep allegorical meaning presented by the author. Meanwhile these allegories were clear for their contemporaries, perceived and read, apparently, at first sight.

The same thing happens with the perception of spatial objects. An architectural or design work reflects the artistic and aesthetic mastery of reality, which a person learns based on their spiritual strength and the activation of life-affirming symbols. A person perceives a new object via the following main stages:

1 - designation - "Something new!";
2 - question - "What is it?";
3 - review (outside, from afar, up close, inside) - "How does it look?" (visuals, touch, taste, sound, etc.);
4 - search for identifiers, signs, symbols and other known meanings - "What does it looks like?";
5 - identification, recognition - "It's it!";
6 - doubts, the right to make a mistake, the possibility of variability - "Maybe something else?";
7 – matching the expectation - "yes or no?";
8 - conclusion about the quality, the result of perception - "Positive/negative/neutral" (Figure 3).

Thus, an object is perceived most positively when it is quickly recognizable, but not commonplace. Original and commonplace information in architecture, subject design and architectural environment design is characterized by a dialectical unity. Therefore, in the absence of preliminary knowledge by the observer, with the complete originality there is a conflict of misunderstanding at its sign system. With the complete banality of information there is a conflict of oversaturation. In both cases, the observer's perception of the object is blunted. The person simply does not notice the object or, alternatively, does not want to make an effort for the perception of it due to irritation by its incomprehensibility. People perceive familiar forms faster and better than unfamiliar ones. With great originality or incomprehensibility there is a conflict of the incomprehensible [3].

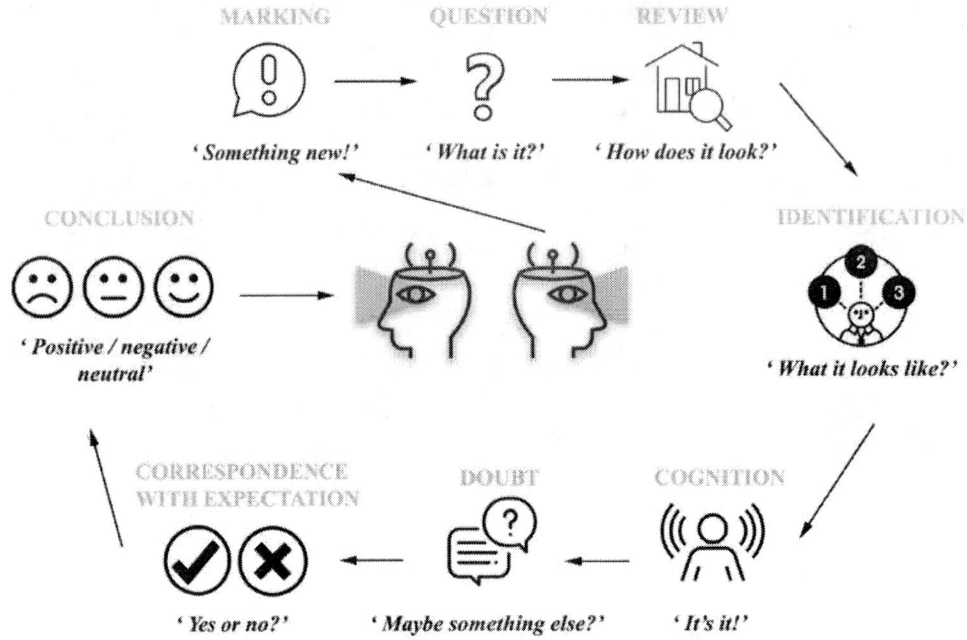


**Figure 3**. Stages of perception of a new object by a person

The result of the complete final perception of the object is highly dependent on the criterion of novelty for a given observer. If the subject does not fall into any category of the imagination, it either causes excessive attention, or may go completely unnoticed. Another option for objects with complete novelty is the transfer of functions. The mobile phone can be perceived as an ornament, or a tool for digging the ground by the inhabitants of the wild tribe... That is, ignorance creates a transfer of functions and the transfer of the object to other functional group than its inherent one. There are many graphic parodies of well-known architectural works that were not perceived well by contemporaries, in addition to verbal expressions.

## 4. Psychology of apperception of the individual and society

I. I. Seredyuk called the perception of architecture, formed by the combined experience of past generations, a relevant vision. Because of genetic perception, our vision does not always respond immediately to current needs, being influenced by outdated patterns of the past. Therefore, a new, relevant perception must fight the genetic, which at the same time serves as the basis of its development [4].

The famous Japanese architect Kenzo Tange formulated his attitude to traditions: "in order to turn a tradition into something creative, it must be denied and, in a sense, destroyed. It must not be canonized, but debunked "[5]. This means a new reincarnation of history, a new view of it through the prism of modernity. By accepting this maxim, we can talk about the adaptation of both the new in the historical space and the historical, which acquires new qualities in the modern environment [6].

The more developed the society in general and the specific observer, as an individual, the easier it is for them to perceive the new, to take on new forms and technologies. This is because of the existence of a larger amplitude scale at the stage of filtering information, finding matches, the subconscious listing of known images; a person with a broader mind is better prepared to perceive a new image. The development of society brings perception to a new stage. When new forms, constructions and styles become typical, occur more often, it leads to psychological habituation and cognition of forms. The situation ceases to be critical and stressful, a psychological adaptation occurs. Brand new, atypical, original, first becomes familiar and typical. The observer ceases to wonder what she did not like before, becomes accustomed and begins to perceive it as normal, and later – as positive. Over time, the atypical becomes typical, the original – commonplace.

It is important to note the fact that the general state of society always has a psychological effect on the individual and shapes their perception of the environment as a whole, as well as its individual elements. The tensions and crises in a country, the general perception of this part of life as critical and non-positive forms a person's negative reaction to positive things, such as new proposals, modern architecture, the latest works of design. And the more atypical they are, the worse the reaction. This situation is similar to the situation with "Seligman's dogs", described as learned helplessness, wherein they cannot control negative, unacceptable events.

By definition, learned helplessness is a condition of a person or an animal when the individual does not try to improve their condition (does not try to avoid negative stimuli or get positive ones), although they have such an opportunity. It usually appears after several unsuccessful attempts to influence the negative circumstances of the environment (or avoid them) and is characterized by passivity, refusal to act, unwillingness to change the hostile environment or avoid it, even when the opportunity arises. According to several studies, it is accompanied for humans by a loss of sense of freedom and control, disbelief in the possibility of change and in their own strength, melancholy, depression and even the acceleration of death [7]. The phenomenon was discovered by American psychologist Martin Seligman in 1967 [8,9].

It can be stated that for a person who is under pressure of circumstances, or society that is in a critical period, in a critical, depressed state, there is a psychological impossibility of new, independent thoughts, actions and movements. Apathy and depression are defined not only by the inability to act, but also by the negativism of the psychological reaction, unwillingness to open to the new. Impossibility of non-action is the impossibility of thought, unwillingness to look for analogies and images, to compare signs and symbols, which would lead to at least some metaphorical matches and, consequently – recognition and psychological perception of the modern new image. To some extent, such reactions can now be observed in Ukrainian society, which on the one hand is in a state of development, and on the other hand is under great pressure by the crisis conditions of present day. It can be asserted that all societies of different countries experience a somewhat slow reaction to the new, modern and atypical in difficult crisis conditions. At such times, the creation of commonplace works and the return to the images of the past occur, classical styles are often reinterpreted, and the copying and stylization of images is abundant (Figure 4).

Often, a positive or negative perception of a new object can be psychologically prepared by society, technology, mass media, if someone needs such a reaction. A person subconsciously prepares themselves for perception, ready in advance for a programmed reaction; they will look for (and always find) exactly those qualities (positive or negative) for which they were psychologically prepared. The observer arrives at this thought, expresses it, not even realizing that it is not their personal choice, that they have been carefully prepared and programmed for it. In order to oppose these technologies, to have an independent personal opinion about the newly created object, a fairly high level of knowledge, world outlook and the synthesis of information about the object are paramount.

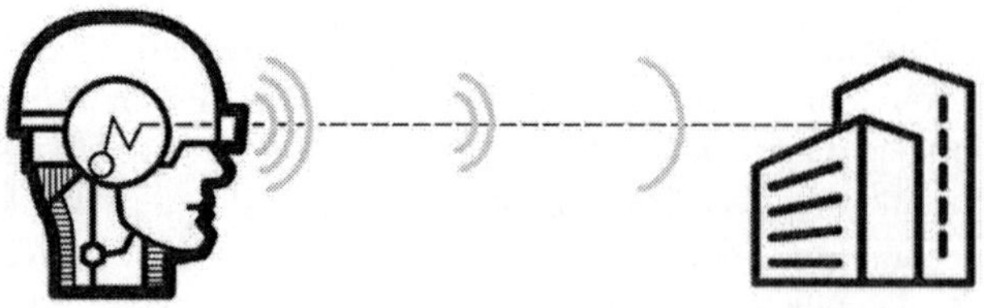


**Figure 4.** Principles of the formation of architectural style's perception for a person

## 5. Conclusions

In conclusion, the importance of the comprehensive development of society for the best movement forward, for a full understanding of the prospects of new and modern solutions in various fields of art needs to be emphasized. The formation of new human values will lead to a change in architectural demand, which, in turn, will affect the quality of the architectural environment. The execution of this algorithm of actions will increase the living standards of the population and its psychological stability.

## References

[1] Mezhenna N 2012 Integration of object of historical urban environment in composition of contemporary building *Scientific enquiry in the contemporary world: Theoretical basics and innovative approach* **4** pp 3-5

[2] Strautmanis I 1971 Arhitektoniskas vides emocionali estetiska uztvere *Arhitektura un pilsetbuvnieciba Latvijas PSR* **II** p 29

[3] Mezhenna N Y 2019 Positive perception and conflict of the incomprehensible - psychology of apperception of personality and society *Architectural Bulletin of KNUBA: Nauk.-vyrob. Collection* (Kyiv: KNUBA) **20** 200-210 (in Ukrainian)

[4] Sereduk I I 1979 *Perception of the architectural environment* (Lviv: Higher school: Lviv University Publishing House) (in Russian)

[5] Tange K 1959 An approach to Tradition *The Japan Architect* **1-2** 55-58

[6] Shevtsova G V 2011 Architectural Tradition as a Vector of Japanese Megacities' Development *Contemporary Architecture of the World: Collection of Scientific Articles of the Russian Academy of Architecture and Construction Sciences, Research Institute of Theory and History of Architecture and Urban Planning* **1** 291-300 (in Russian)

[7] *Learned helplessness* URL: https://en.wikipedia.org/wiki/Learned_helplessness

[8] Seligman, M E P 1975 *Helplessness: On Depression, Development, and Death* (San Francisco: W H Freeman)

[9] Seligman, M E P 1972 Learned helplessness *Annual Review of Medicine* **23 (1)** 407–412

IOP Conf. Series: Materials Science and Engineering **907** (2020) 012020 doi:10.1088/1757-899X/907/1/012020

# Symbolism and functional organization of architecture of the Orthodox religious and educational complex

**V Khaidukov, O Sleptsov**

Department of Fundamentals of Architecture and Architectural Design, Kyiv National University of Construction and Architecture, 31 Povitroflotsky Avenue, Kyiv, 03680, Ukraine

draw.vlad@gmail.com

**Abstract:** The focus of the article is a scientific investigation of the symbolism, functional organization and peculiarities of the formation of the Orthodox religious and educational complex and its components: the spiritual and educational centre and the monastery complex. The topic of the article is devoted to the study of cruciform structures and the actual scientific and applied problem of forming the conceptual foundations of religious and educational complexes. A place of worship is one of the few visually recognizable types in the entire architectural typology. It is the symbolism, which is one of the foundations of sacred architecture, that formed the visual expressiveness of the structures. The cross, as one of the main symbols of Christianity, influenced the formation of temple architecture. In general, the evolution of the symbol of the cross in architecture has created three basic types of structures: cruciform, cruciform transitional and circular. The above structures directly affect the formation of the architectural space of the sacral structures. On the basis of the established principles and techniques of construction of architectural and spatial solutions of buildings and structures of Orthodox architecture and the proposed method of architectural organization of Orthodox centres and complexes, it becomes possible to develop rational architectural and construction solutions.

## 1. Introduction

It is well known that spiritual and educational activities play one of the most important roles in forming the cultural centre of the community. It is the basis of many major functions of any sacred structure, such as the monastery complex. Therefore, these activities previously indirectly influenced the socio-economic development and economic life of the state [1]. Temples and monasteries are one of the ways of cultural and spiritual development of the population. Unfortunately, the buildings in the monastery complex do not always meet the modern requirements of society. If the development of the temple, as canonical architecture, is on-going, the accompanying buildings often remain traditional in a functional sense.

Influence on the formation of monastic complexes was mainly due to socio-cultural processes. In the dark periods of history around the complex were growing up the walls, turning it into a fortress. The inclusion of the monastery in the structure of the city caused the destruction of fortifications, except when the monastery itself was not a catalyst for the development of settlements around it. The most important effect is the separation of religion from the state, during which many buildings of the complex have lost their historical function and in need of refurbishment and the creation of new types of architectural spaces.

After analysing a separate building in the monastic complex identified the problem of the conversion of the greater part of those premises for modern requirements. This is true for educational institutions (theological seminaries and academies) who need language laboratories, etc.; libraries that require electronic equipment; venues that require the use of furniture that can be transformed into a larger, mass homes that require a separate bathroom for each living cell, in fact, turning into a number of cells, with the prospect of development in the form of their completion, under the joint shell as well as the need gets spread arhontariks (from Greek αρχονταρικόν) where conducted spiritual conversations with pilgrims; Bishop's house, which requires changes in administrative functions, with possible space for a temporary stay of the Bishop.

Turning to history, architecture is based on a symbolic understanding of space. The proportions in the buildings were determined by the symbolic values of the forms. Using certain laws of geometry, the building turns out to be a sacred force. In ancient Greek and Roman temples, direct relationships were established between architectural proportions and cosmic models, and architects emphasized the idea of spiritual ascent. Similarly, Coptic temples [2] and temples of the Balkan countries [3] were formed in the same way.

The cruciform structure is directly connected with the sacral structures which are the centre of any religious and educational complex.

The modern requirements that are given by the society require the formation of a special area around the sacral structures, whose main function should be educational, religious and educational activities. This is very important in the countryside, where there used to be an enlightenment house at church and school.

Defining the Orthodox SEC (spiritual and education centre) is quite difficult because of the current content formation. The problem was investigated in the scientific works of PhD in Theology Priest O A Mumrikov [4], PhD in Pedagogic sciences S I Abramov [5], PhD in Pedagogic sciences T A Stanovskaya [6].

One of the forms of culture is leisure. Recently, the experience gained in pedagogy over the period of its development defines leisure activities in the following forms: holidays, entertainment, recreation, creative activity, self-education.

In modern conditions, the study of moral and ethical norms in Orthodoxy, in addition to traditional forms of educational activity, such as Sunday schools, has a practical manifestation of leisure activities in order to consolidate the obtained theoretical knowledge. It is suggested that an example of a leisure institution that would integrate all five of the above forms is a spiritual education centre.

Therefore, the article is devoted to the formation of the spiritual and educational centre as a type of spiritual institution whose main activity is leisure. Also, the topic of the article is devoted to the study the cross structures that may influence the formation of the Orthodox religious and educational complex.

## 2. Analysis of recent research and publications

The article analysed the scientific works of Sleptsov O S [7], Bulichova T V [8], Chen L Y [9], Stepanova I V [10], Gutnov A E, Lezhava I G [11] and other. The last two scholars have devoted their work to urban planning.

Thus, the religious and educational complex, modelled on the heritage of the NER group ("New Element of Settlement"), also uses a similar concept for relatively independent living in any locality.

One of the trends in her scientific work Bulycheva T V studied different principles of function, functional process and zoning of sacred complexes. Stepanova I V summarized the following functional factors of formation of modern Orthodox temple complexes:

- clear zoning and consolidation of functions;
- creation of multifunctional centres and complexes.

The latter in turn was organized with the principle of blocking and cooperatives, as well as the versatility of the interior space.

It is established that the principle of multifunctionality as an example of the Autonomous existence of the complex is also the basis of the monastery. Clear zoning helps to speed up functional processes that take place in it.

In his book "Architectural design and reconstruction of Orthodox churches" Sleptsov O S has identified additional functions: museum, memorial, charitable, ethnographic and others. Also, the pilgrim road was added as one of the main object of the complex. It was also determined zoning. The complex consists of nine areas: input; temple, memorial, office, household, educational, charity; sporting, recreational and economic.

Stepanova I V in her scientific work described the need to develop a more flexible space-planning structure that plays the decisive role in the formation of architectural ensembles.

Chen L Y in 2015, has considered the issues of historical designs of the Ukrainian monasteries of the XV – XVIII century. Analysis was by visual perception. You have defined the following composite solution:

- the temple forms an introverted space of the monastery complex and bell tower extroverted;
- Church and bell tower form the extroverted space;
- Church and bell tower form introverted and extroverted space.

Sleptsov O S found that the form of the monastery dominant contains the modern architectural and artistic features: imagery, economy and great acoustics.

The best of the forms was chosen as the rotunda.

Cruciform structures are indirectly linked to scientific research in various fields of knowledge. The terminological analysis of scientific publications provides a list of authors and terms related to the aforementioned structures. In Architecture: Kryvoruchko Y I ("Sacred") [12], Pavlov N L ("Natural rhythm") [13], Volegova A A ("Macrocosm, Microcosm") [14]. In Religious Studies and Philosophy: John of Kronstadt ("The Cross of the Primary") [15], Ostashova N V ("Symbol") [16]. In Art Studies: Zolotarchuk N I ("Processional Cross") [17], Kuzenko P Y ("Sign") [18], Bogomolets O V ("Solar Sign") [19]. In History and Archaeology: Kupriyanova O I ("Stavrography") [20].

The scholars identified above, have studied the problem and obtained the result in one way or another related to the religious and educational complex. This is due to the fact that the complex combines both the sacral complexes and the centres connected with the spiritual development of man.

## 3. Results

The concept of the spiritual and educational centre (SEC) is to supplement the liturgical, social and cultural functions. The SEC must meet all the requirements of leisure activities.

The three main functions of the Orthodox SEC are following: axiological; educational; family, creative and / or cultural and leisure activities. SEC specialization depends on the predominance of the any type of leisure. On the basis of the main functions, was formed the main zones of the Orthodox SEC: entrance; cultural and educational; cultural and leisure; administrative; office and household.

The features of cruciform structures consisting of directions and intersections have been established. The importance of architectural and spatial organization in the urban and architectural context is determined. A religious and educational complex consisting of a temple, monastery or temple complex, and a spiritual and educational centre may, by its architectural and planning decision, include a cruciform structure. Also, in terms of planning, the religious and educational complex is approaching urban planning. Therefore, its combination with the cruciform structure forms a centric, relatively timely accessible and dualistic development: at the same time compact and with a comprehensive perspective of development planning organization of the complex.

The following scientific studies of the above complex were tested by the method of experimental design. The religious and educational complex project consists of a monastery complex and a spiritual and educational centre. The main results of the study were presented at the XXVIII International Review-Competition of graduation projects of graduates of architectural and artistic specialities of higher education institutions of Ukraine, which took place in Rivne.

It is established that in the monastery complex there were gradual changes in the main and secondary functions, which directly influenced the functional process of the complex as a whole. The accompanying functional process of the monastery complex was constantly transformed, but, owing to the essence of monasticism, remained the basis of the autonomous system. Therefore, the perception of the monastery as an architectural ensemble has always been a major issue at the design level.

All functional processes were intended to ensure the relative independence of the complex from the environment, regardless of whether it was architectural or natural. It was the territorial location that influenced the premises program. Modern requirements have to approval the main and secondary functions of the monastery complex, more detailed rules of design, revision of the functional processes that take place in it and attention to the achievements in other types of buildings and structures.

It is determined that the typical planning decision, in addition to the natural specifics of the monastery complex, prevents the actual individuality of the planning decisions as a whole. With regard to the space-planning decisions of the monastery complex as a grouping of the main and secondary premises into a single composition, the presence of such decisions is not possible due to the ensemble perception of the monastery as such. Therefore, it would be more acceptable to disclose problems through composite solutions. Compositional decisions emphasize the ensemble only in terms of changes in the dominance of the bell tower and the cathedral. These types of structures are always a compositional centre, and the accompanying structures are a functional need of the monastery complex. The typological planning decision is not always successful in the context of the monks' stay or visit to the laity.

It was found that the national influence on historicism did not bypass the principle of formation of the monastery complex. In a broad sense, the imitation of historical styles is sometimes referred to as religious cannons. From an architectural point of view, refinement is needed. Religious cannons have almost no effect on form, but are of great importance for the planning of sacred structures and the orientation of their individual elements around the world. Therefore, having determined the essence of the formation of the monastery complex, it is necessary to outline its problems. Over the centuries, elements of architecture have created unique shapes that are recognizable to any person and not confused with any type of building. New forms are not always accepted by society, perhaps because of fears of loss of visual difference among other buildings. Although moulding experiments have to take place because of the need for new construction. In foreign practice this need is not so necessary. Preserving historical monuments is more important than building new monastery complexes. Therefore, many questions in the formulation were raised at the beginning of the restoration of sacred construction in the last century.

On the one hand, the national influence on historicism did not bypass the principle of formation of the monastery complex. Thus appeared unique structures were created, for example, in the Ukrainian Baroque. It is especially organically organized around any monastery ensemble.

It is established that the construction of the monastery complex is one of the longest processes, but under the current conditions the question of its continued existence and active participation in the social life of the surrounding settlements remains open. Moreover, the need for educational activities in sparsely populated areas is not addressed by the construction of a cultural or educational home. Therefore, it was decided to expand the architectural typology of the religious and educational complex (REC). It has two main components: the monastery complex as a religious centre and the spiritual and educational centre as the centre of enlightenment. Therefore, such a REC is an adjoining centre and complex. In addition to the main, it includes minor complexes, such as the spectacular. Their number is not regulated and depends on the type of organization of the complex.

The analysis of typological planning decisions on the example of a generalized scheme of the monastery complex allows grouping all functional processes into seven components: the entrance group, the sacral, housing, socio-cultural, economic-production, social and household, educational.

The main function is divided into: religious, educational, uniting. The secondary function is divided into: agricultural, industrial, tourist.

It is determined that the form in the world of Orthodox monastery complexes is generally associated with national style and Revivalism. Of course, there are unique structures that are most dependent on unusual terrain. So it was decided to create a form that combines tradition and novelty. Using the method of exclusion for superfluous architectural elements, the form was created as the Christian symbol and was supplemented by a traditional architectural element, such as an arch. The ensemble uses a radial system with a dominant in the centre with parametric development. Thus, the cathedral remains the dominant of the monastery complex. If the bell tower is dominant, it is located separately from the monastery complex, but on an elevated relief, like the cathedral. Usually, the bell tower is a chapel, and a recreation centre and entrance group are organized around it.

During the analysis of the monastery complex, the main and secondary functions of the religious and educational complex were formed. The main functions are following: religious, enlightenment, educational.

The secondary functions include production, tourist, cultural and leisure.

The typological scheme of the complex consists of the centre, complex and related types of buildings and structures: spiritual and educational centre, monastery and entertaining complex, entrance group and cumulative square, two entrance gates and transport stops with parking lots, baptistery, bell tower with chapel. It is also necessary to create two recreational areas (Figure 1).

It is established that the planning of the complex in terms of approaching urban development.

Artistic concept has been created. It is based on a site that is already positioned as a sacral territory, it is in pristine condition. Therefore, it was decided to create a monastery complex that would restore the historic function of the territory. This complex will become the centre of spiritual life with all the necessary functions. It will include temple, enlightenment, living quarters, recreation area, catering, exhibition halls and assembly halls, convention rooms, parking.

Cruciform, transitional and circular structures were used at different planning scales. According to the provided structures, the concept of religious and educational complex was formed (Figure 2).

**Figure 1.** Computer generated bird point of view of the religious and educational complex.

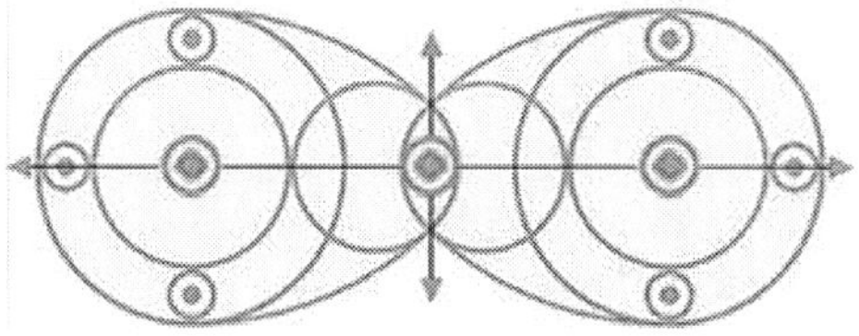

**Figure 2.** Scheme of the religious and educational complex with cruciform and circular structures.

In order to put this idea into practice, the author interprets the typological planning scheme through the master plan into the planning scheme of the monastery complex. Conventionally, the concept is considered in terms of three aspects: functional, social, symbolic.

The value of the first aspect is the reproduction of the educational activities of the monastery complexes.

The second aspect is related to the need of society for spiritual development.

The third aspect is the connection of symbols and their religious and Christian explanations in terms of the symbolism of numbers, etc.

The symbols were divided into two categories: numeric and figurative. The first one is the number 8 and 10. In Christianity, eight means restoration and rebirth. Eight resurrections: three in the Old Testament, three in the Gospels, and two in Acts 9:20. The number eight in the Bible is the number of completeness and is associated with a new beginning - the beginning of a new life, a new era (2 Peter 2:5). The value of the number ten among Christians symbolizes the Ten Commandments of the Decalogue (Exodus 20:2-17), (Deuteronomy 5:6-21), the parable of the ten lamps, virgins (Matthew 25:1–13) and talents (Matthew 25:24–25). The second category is the image of a butterfly. Mainly the interpretation of this symbol related to the belief of Christians. In Christianity, the stages of butterfly development are embodied in life, death, and resurrection. Butterfly is sometimes depicted in the hand of a baby Christ. In accordance with the above aspects, the monastery complex and the spiritual and educational centre at it were designed. The religious and educational complex contains the symbol of butterfly (Figure 2).

The functional aspect is manifested in the spiritual and educational centre. The social aspect is related to the choice of terrain around the complex. The symbolic aspect became the basis for the organization of space at the urban and architectural levels.

It is recommended to use a cruciform transitional structure at the urban level but a circular structure should be used for individual architectural objects.

It is stated that the monastery complex has 5 main functional zones: Main temple area, living area, socio-cultural, educational function, economic production, social and household services,

Features of the planning solution are:

- at the level of transport links: intercepting parking, environmental friendliness, harmonious integration into the existing transportation system;
- at the level of the master plan: symbolism, parametric horizontal and all-round development.

The creative search for the dominant form of the monastery complex consisted of four variants. The best option is similar to a rotunda, bionic with a comprehensive facade inspection and has the shape of a Christian symbol with traditional tectonic architectural elements – arches.

It is determined that the result of creative search for the shape of buildings and structures should be based on a circular structure, taking into account the cruciform structure of the REC.

Features of the organization of architectural space are based on the interconnection and orientation of the dominant: the cathedral temple and the bell tower. Based on scientific research, it was decided to create an introverted-extroverted architectural space.

It is stated that the main function of the religious and educational complex is divided into: religious, enlightenment and educational. The secondary function of the complex is: production, tourist and cultural and leisure. Functional processes that take place in the complex, in addition to the processes of the monastery and spiritual and educational centre, include the processes of the spectacular complex, baptismal and pilgrimage paths.

The planning features of the religious education complex include: comprehensive development and symbolic placement on the master plan.

It is determined that a harmonious architectural space is created on two levels: enlarged (macro); medium (micro and meso).

The latest technologies allow you to create a flowing space between the interior and exterior, between the architectural environment and the architectural space (as defined by Y A Rochnyak [21]). Such space creates a visual and functional connection between the two levels.

At the enlarged level, space harmonization takes place through the visual and functional equilibrium of buildings or structures and the environment. On an average level, the harmonious architectural space of a particular typology is linked to its expressive concept, main features and functional zoning.

Aspects of the harmonious architectural space of the SEC are as follows:

- imagery – the visual and informational emphasis of the institution; occupies the facade of the SEC: it is completely, partially, part of the ensemble;
- recognizability – a visual and cultural sign of the institution;
- emotional comfort – harmonious perception by all organs of human feeling;
- self-understanding – the visual and functional communication of the institution; clear visual representation of functional zones;
- security and accessibility – the ability to freely get into all the functional areas of the institution;
- contextual – the spatial and informational content of the institution; preserving valuable architectural space and the environment, including historical and architectural monuments.

It is established that the cruciform structure has the following features: centrality and dualistic architectural and planning decisions. The architectural space acquires mono- and multifunctional properties. The vertical direction is connected to the dynamics, the way to the desired. The horizontal direction acquires a static meaning, namely stability. Additional directions, such as diagonal ones, are used to make changes to the structure while maintaining centrality. The use of these directions forms a transitional structure from which a new centric, for example, circular can be formed. Each type of structure is a reflection of the symbolism of Christian temples (Figure 3).

The architectural image is also a reflection of the symbolism in sacred architecture. An architectural image is formed at the intersection of a structure, regardless of the number of directions. Each direction is in a separate plane. The direction of the architectural image together with the vertical and horizontal directions in each plane forms a projection of the architectural image. These projections form an architectural image. Each structure is different in number of directions. The directions are as follows. The vertical direction is connected to the dynamics, the way to the desired. The horizontal direction becomes static, namely stability. Additional directions, such as diagonal ones, are used to make changes to the structure while maintaining centrality. The architectural image is a symbol in the cruciform and circular structures. In each structure, it looks different, because as the number of values increases, the type of structure changes (Figure 3). A circular structure was laid in the temple. Egg is stated as the architectural image. The total number of main facets of the temple reflect the symbolic meaning of the number eight.

**Figure 3.** Main temple of the complex

Temples in the form of a cross meant that the Cross of Christ is the foundation of the Church and the ark of salvation for people. Temples in the form of an eight-pointed star marked themselves The Star of Bethlehem and the Church as a guiding star to the Saviour research in the life of the future (eighth) century. Round temples mean the universality and eternity of the Church [22]. Thus, cruciform structure symbolizes the ark in the meaning of salvation, cruciform transitional structure symbolizes the Star of Bethlehem in the meaning of the way to salvation and circular structure symbolizes the universal in the meaning of eternity of Church.

## 4. Conclusions

It is determined that the existing perspective and need of modern society in the spiritual and educational centres is not reflected in reality due to insufficient study of the problems of organization of their architectural and planning structure and finally uncertain activity of the SEC, which actualizes the continuation of scientific researches of their architecture.

The Orthodox Spiritual and Educational Centre complements the sacral core with additional architectural spaces – leisure, recreation, rehabilitation, training, creativity, transforming the spiritual and educational centre into a religious and educational complex.

The features of cruciform structures are established. Their relationship with the design and perception of architectural space, such as sacral structures, is determined.

The symbolism of the religious and educational complex is formed by Christian interpretations of numbers and signs, cruciform, transitional, circular structures and architectural image.

Thus, the formation of an architectural image is possible by filling it with different meanings. For example, a characteristic feature of a sign is uniqueness, a symbol is ambiguity. The projection of an image is a sign and the image is a symbol formed from these signs.

As a result of the experimental design of establishments of this type, the analysis of the normative base for the design of sacral and public buildings of different specializations and scientific researches, the functional zoning of educational establishments was determined and certain functional groups of premises were provided.

**References**

[1] Kirichevsky P 2013 *Ukraine of the twentieth century: culture, ideology, politics* **18** 219.
[2] Guirguis M N, Dewidar K M, Kamel S M, Iscandar M F 2018 *Ain Shams Eng. J.* **9** (4) 3093- 3101
[3] Ilvitskaya S V 2005 *Patterns of the formation of the architecture of Orthodox monastery complexes: the example of the Balkan countries* (Moscow: Russian State Library) p 432.
[4] Mumrikov O A 2010 *St. Tikhon's University Review.* IV. Pedagogy. Psychology **3** (18) 13-20.
[5] Abramov S I 2010 *Spiritual and moral education of schoolchildren in the system of additional education in a small town* (Moscow: Russian State Library) p 234.
[6] Stanovskaya T A 2017 *Teaching Orthodox culture in Russian schools in the second half of the XIX – early XX centuries* (Moscow: Russian State Library) p 198.
[7] Sleptsov O S 2014 *Architectural design and reconstruction of Orthodox temples: A textbook for universities* (Kyiv: A+S) p 272.
[8] Bulychova T V 1994 *Architecture of modern Orthodox spiritual centers in Ukraine* (Kyiv: Kyiv National University of Construction and Architecture Press) p 17.
[9] Chen L Y 2015 *Bulletin of Lviv Polytechnic National University.* Architecture **816** 213-218.
[10] Stepanova I V 2013 *Sučasnì problemi arhìtekturi ta mìstobuduvannâ* **33** 478-485.
[11] Gutnov A, Lezhava I, Baburov A, Dumenton G, Sadovsky S and Kharitonova Z 1966 *New element of settlement: on the way to a new city* (Moscow: Stroyizdat) p 127 (in Russian).
[12] Kryvoruchko Y I 2018 *The phenomenon of the sacred in the development of cities and territories: the experience of Ukraine* (Kyiv: Vernadsky National Library) p 778.
[13] Pavlov N L 2017 *Architecture and Modern Information Technologies* **4** (41) 68-87.
[14] Volegova A A 2007 *Semiotic aspects of cosmogony and cosmology in monuments of architecture and urban planning* (Moscow: Russian State Library) p 294.
[15] John of Kronstadt 2018 *About the Cross of Christ* (Directmedia) p 184.
[16] Ostashova N V 2004 *The symbol of the cross in the history of culture* (Moscow: Russian State Library) p 175.
[17] Zolotarchuk N I 2019 *Bulletin of KNUKiM.* Series in Arts **41** 216-225.
[18] Kuzenko P Y 2007 *Art culture. Actual problems* **4** 609-617.
[19] Bogomolets O V 2015 *Current problems of philosophy and sociology* **8** 14-17.
[20] Kupriyanova O I 2012 *Vlast* **2** 53-56.
[21] Rochnyak Y A 2001 *Bulletin of Lviv Polytechnic National University.* Architecture **429** 84-87.
[22] Smirnova I 2006 *The Secret History of the Cross* (Moscow "EXMO") p 236.

# Conceptual modeling to control the quality parameters of the BIM model

**Olga Kysil, Sergii Mikhalchenko, Valery Tovbych**

Department of Information Technologies in Architecture, Kyiv National University of Construction and Architecture (KNUCA), Povitroflotskyi Avenue 31, Kyiv, 03037, Ukraine

solomyab@gmail.com

**Abstract**. This study pays attention to the quantitative and qualitative analysis of the project in the BIM (building information model) form, as well as the relevance to this scientific topic. The definition of ETR (Electronic Terms of Reference) is given. The working hypothesis of the research is described, which consists of several main stages. The novelty of the application of quantitative and qualitative analysis to BIM is emphasized and the method of its automation is described. An example of a typical object is chosen, on which the practical stage of assembling the ETR is carried out and an algorithm for quantitative and qualitative analysis on this basis is developed.

## Introduction

The purpose of this article is to describe the general method of qualimetric assessment of the quality of the BIM model and general algorithms of the system and integrated approaches in this direction. The general study of the conceptual modeling of the future properties of the BIM project is based on a fairly wide range of works of modern scientists. Typically, the range of interests includes the business effect of BIM technology, analysis of energy efficiency models, improving the management of the design process, and the entire life cycle of the facility. Attention is paid to communication between project participants, data protection, classification of elements, etc. [1, 2, 3].

The authors of this article took a decision to focus on the implementation of theoretical research, which at the time of its appearance (the '80s, '90s) had no prospects for implementation due to insufficient hardware capacity. Despite this, interactive procedures for making design decisions were studied, even aesthetic issues, which are extremely hard to formalize [4]. The mathematical model of the system approach required a lot of time and miscalculations, the complex one mainly used the anthropo-dependent expert method. Thus, the topic of quality in architecture does not have currently implemented electronic control tools and is a poorly structured multicriterial task. In addition, the existing correlations between impact factors require the development of a complex design decision-making apparatus.

## Relevance of the study

One of the conclusions of research on the methodology of optimization of the architectural environment was the statement about its complex organizational structure, which is based on various criteria [5]. Multicriterial is a feature of modern architecture. Accordingly, the popular view of the construction site through the monetary dimension does not always justify itself. The emergence of many energy efficiency and green building certification systems in the last decade underscores this. Finding a balance between influencing factors is an urgent scientific problem.

The founder and ideologist of qualimetric analysis is G.G. Asgaldov [6, 7]. Also, a significant contribution to the development of architectural qualimetry was made by D.N. Yablonsky, G.D. Yablonska [17], etc. Researches of the system approach of an estimation of quality in architecture were engaged in: G.I. Lavrik [8], M.M. Demin. Systematicity in architecture in general is a topical issue [9], which became one of the starting points for the study of the BIM model as a complex system [15].

Modern scientists are searching for the optimal mathematical model to determine the optimal performance of complex structures, which are architectural objects. The method of analysis of hierarchies (MAI) is considered in modern research as one of the most effective methods of decision-making. It allows you to clearly and rationally structure a complex problem of decision-making in the form of hierarchy, compare and quantify alternative solutions, taking into account not only tangible or intangible quantitative qualities of objects but also subjective qualities or processes of a spiritual nature [10]. In addition, some research presents the algorithm of the procedure for determining the weight of property indices of urban quality of life on the basis of qualimetric method expert assessments. An important basis for the formation of ETR as a qualimetric standard is a theoretical algorithm related to the development of a universal method of quantitative assessment of urban quality of life, which allows to assess any object [11].

The study focuses on the possibility of applying a BIM model of qualimetric quality assessment based on two approaches: integrated and systematic. But in modern sources there is no transitional link in the application of these methods in BIM technology. No systematic analysis of the BIM model and its input and output data as an architectural and construction system. There are no methods and means of assessing the quality of the BIM model as a product of architectural and construction design, which must be implemented with maximum aesthetic, economic, technological and environmental effects.

Approval of the concept of BIM implementation in Ukraine at the state level at the end of 2019 made it possible to further work on proposals for changes to DBN 2.2.3 for the implementation of BIM in Ukrainian regulations. In line with this study - ISO 22263: 2008, dedicated to quality management. In addition, amendments were made to the Law of Ukraine "On Building Standards" to improve rationing in construction: added "parametric rationing method in construction - a way to establish requirements for the object of rationing in construction, which provides for the definition of objectives and / or safety parameters, functionality and quality of the object of rationing in construction (criteria, requirements for performance and / or its indicators.) However, it should be noted that according to the analysis of modern research, determining the quality of the model is quite relevant not an issue for the global design community. To date, comprehensive tool of its kind for BIM technology to designers available. The use of survey results can be widely represented in various aspects of construction, architectural and economic fields of activity:

- A tool for improving the public procurement system. It is from this sector that the implementation of BIM at the state level in many developed countries began. Optimizing transparent decision-making helps to improve the economic performance of the construction industry and the return on investment in the public construction sector.
- Application of decision-making meanwhile the public hearings. The clarity of the method helps to compare the performance of different projects and make appropriate independent decisions.
- A method of improving the assessment of private investment risks, which is carried out through the modeling of quality indicators according to the criteria important for a particular business plan.
- Quality control of the project team for both its leader and the investor. Transparency in the adherence to the schedule and the quality of the work corresponding to the ETR improve the economic effect of cooperation for both parties.

– Grounds for a legal assessment of actions to change the terms of reference in the design process, which entail iterative revisions of the project, and, accordingly, the financial obligations of the parties.

**Scientific hypothesis**
The algorithm for obtaining scientific results consists of 3 stages:

Analytical stage:
- Study of existing methods of multicriterial analysis and optimization of indicators to improve the result;
- Systematization of modern criteria for assessing the quality of architectural and construction object;
- Conducting business analytics of existing BIM software to identify existing automatic means of obtaining indicators from the model;

Methodical and theoretical stage:
- Definition of the main groups of quality assessment criteria;
- Description of the BIM model as a system with the interaction of subsystems and over systems;
- Defining the paradigm of the Electronic Terms of Reference as a control system for the input database of the model;
- Formation of the property tree. Development of mathematical and analytical apparatus to obtain research results.

Practical stage:
- Block diagram of creating a template of the Electronic Terms of Reference;
- Block diagram of the software module to determine quality indicators, their aggregation, and weight distribution;
- Technical description of the software implementation of the mathematical and analytical apparatus of the study.

The object of research is the BIM model as a system of heterogeneous data types at the input and output.

The subject of the study is a multicriterial analysis of the BIM model.

The purpose of the study is to develop a methodology for applying automatic multicriterial analysis to the product of construction information modeling to manage the quality of the modeling product.

**Scientific novelty**
The study for the first time describes the analysis of quantitative and qualitative design characteristics of the BIM model. It predicts the economic, aesthetic, environmental and technological properties of the project.

Methods are implemented using the:
1) definition of indicators of 4 main categories of assessment: economic, technical, aesthetic, ecological;
2) obtaining consolidated quality indicators of different categories from the BIM model;
3) comparison of the reference indicators for electronic terms of reference;
4) automation of multicriterial comparison of model indicators with reference values.

It should be noted that there are now several documents governing the quality of certain aspects of BIM processes and information product. These are Employer's Information Requirement (EIR), BIM Management Plan (BMP), BIM Execution Plan (BEP). But all these documents do not have the requirements of the needed quality of architecture, which reproduces the model.

The paper concludes on the possibility of taking ETR as a qualimetric standard, provided that it determines the values of weights and formulates on this basis the reference indicators of properties. The conclusion is based on the existing modern theses about the numerical values of the characteristics of the reference sample. They are set as accepted reference values, which are used as an agreed reference for comparison and which are defined as:

a) theoretical or established value based on scientific principles;
b) assigned or certified value based on experimental data from certain national or international organizations;
c) agreed (based on consensus) or certified value, based on joint experimental work carried out by a scientific or engineering team [12].

**Significance of the study**

The practical significance has been studied on the example of designing the residential area of a functional-spatial military cluster using BIM technology. The development of a software tool based on the developed method of quantitative and qualitative analysis of the BIM model will allow managing the qualitative characteristics of the project according to existing regulatory parameters and customer-specified conditions at the initial stage of the life cycle of the object - its project. The tool described in the study improves both the economic, technical, environmental and social efficiency of the product of architectural and construction design.

The application of the method of quantitative and qualitative analysis of the BIM model is appropriate for the introduction of standardization, in particular in the field of design of special structures. The Technical Committee for Standardization TK-320 "Special Purpose Buildings" has expressed its intention to use this method after its final development.

In addition, the application of this method makes it possible to more accurately determine the material consumption and total cost of construction. This saves significant funds and prevents corruption in tenders for the design and construction of buildings and structures to state order. The practical significance of the project exists both at the national level, from the regulation of public procurement and public hearings, and in the private construction sector: to assess business risks and mutual control of the customer and the contractor.

It should be noted that an integral part of the ETR is the regulatory framework, which corresponds to the type of design object. In the example presented in the article, these are military standards that always provide for action within the described framework [13]. It is the initial constraints in the data fields of the ETR table. But according to the modern method of parametric examination of the building, which is gradually being replaced by fixed State Building Standards, it is such a flexible system of setting initial parameters as ETR that allows to combine the wishes of the customer with constant building standards. Besides, the parametric method of rationing is practiced in countries with already implemented at the state level BIM technology.

**Method of qualimetric analysis of BIM model**

This method in the study is developed on two bases: integrated and systemic approaches. Both options are implemented in the process of forming the ETR according to the request of the customer or investor and the building codes in force in the country of construction. To further assess the design qualities of the model, ETR is taken as a qualimetric standard. It can also be assumed that ETR is a conceptual BIM model [14], which precedes the construction of the design BIM model for further work with related specialists at other stages of the life cycle.

In previous theses, the study defines the following groups of quality criteria: technology, economics, ecology, esthetics. To form such groups of criteria, an analysis of international certification systems DGNB, LEED, BREAM, which reflects the current indicators of environmental quality and sustainable development of architectural objects [15].

At this stage, the study tested the formation of ETR for BIM model of a typical residential military facility. The reason for choosing such an object is its typological and functional features against the background of a clear structure [16]. The composition of the terms of reference in this case has a basic core, which is a description of the constant prerequisites, and additional options that change according to age and status factors. In the process of research, a tree of properties of the residential area was constructed, the data of which were transferred to the ETR spreadsheet. An assumption was made about the similarity of the structural structure of the table data and the property tree, which was confirmed during the work. Given the traditional application of the expert method of decision-making, there is a certain problem of automating such a process. In the course of the work, a hypothesis arose as to the possibility of providing expert roles to the architect, project manager and customer, who fill in the ETR table. Also, the "expert" regulator of parameters is a database of standards, which is automatically connected to the table at the time of specifying the typology of the object. It determines the possible valid value ranges in the data fields of the ETR table.

## The results of the study and its prospects

The application of the qualimetric method of assessing the quality of the BIM model by an integrated approach takes more hardware resources and time due to a large branched system of indicators for four groups of criteria, each of which has its subgroups. However, the ETR formed by this method allows to describe in detail the needs of the customer and it is easier to harmonize them with the standards.

The compilation of ETR according to the system approach is a fast method of conceptual modeling of business effects from the project. Aggregation of indicators helps to adjust the project more easily to achieve the optimal picture.

In the example given in this article, it is advisable to use a systematic approach to determining the quality of the project. This is especially convenient in the case of developing a typical model. For further application of the model for a specific design task, it is possible to supplement the ETR and recalculate the indicators by a comprehensive method.

During the project, the so-called self-learning BIM model is created. In fact, it can be used as a conceptual template for objects of a similar typology. In the process of developing design solutions and comparing their success with ETR, machine learning can occur [14]. The prospect of such software is difficult to overestimate in the context of the versatility and complexity of modern architecture.

## Conclusion

According to the experience of architects involved in BIM technology design, a significant problem is outdated design standards and the application of the rules of paper documentation to modern projects. Design models have in electronic form the whole range of relevant information. But now there is a conflict between new technologies and generally accepted design processes. The terms of reference are an element of such a conflict, as it is one of the initial defining documents for the project. Besides, to achieve maximum economic and social effects, the architect must be able to monitor quality indicators at any stage of design.

The study described in the article is part of a project to develop software tools for all stakeholders of the BIM process. Previous surveys have depicted the high level of interest to such a software shown by the professionals.

## References

[1] Olawumi T O and Chan D W M and Wong J K W 2017 *Evolution in the intellectual structure of BIM* (Taylor and Francis Ltd.) p 1060–1081

[2] Ashworth S, Tucker M and Druhmann C 2016 The role of FM in preparing a BIM strategy and employer's information requirements (EIR) to align with client asset management strategy *15th EuroFM Research Symposium, Milan.*

[3] Succar B and Poirier E 2020 *Lifecycle information transformation and exchange for delivering and managing digital and physical assets* (Automation in Construction) vol **112**.

[4] Tovbych V 1986 *Interactive graphic modeling of architectural environment taking into account features of its visual perception* (Kiev, Kyiv National University of Construction and Architecture the dissertation)

[5] Myronenko V 1999 *Methodological bases of optimization of the architectural environment. Author's abstract. dissertation of doctor of architecture* (Kharkiv State Technical University of Construction and Architecture)

[6] Azgaldov G and Kostin A 2012 *Quality - as the most important object of quantitative analysis*, (Barcelona, Jan. 2012. Accessed)

[7] Azgaldov G 1996 Construction of a tree of indicators of properties of object *Standards and quality* Vol. **11**

[8] Lavryk G 2002 *Fundamentals of systems analysis in architectural research and design* (Kyiv, KNUBA, Ukrainian Academy of Architecture) p 140

[9] Panova L 2010 *Systematics of the architectural environment: monograph*, (Kharkiv, Kharkiv National Academy of Urban Economy) p 235

[10] Badyul M and Kramarenko V 2013 Application of the method of analysis of hierarchies in design and construction *Construction. Materials science. Mechanical engineering. Series: Energy, ecology, computer technology in construction* vol **70**

[11] Lisnychenko S 2015 *General principles for determining the importance of indicators of urban quality of life* (Kyiv, Online).

[12] Motalo A and Stadnyk B and Motalo C 2020 Analysis of methods and types of measurements in qualimetry *Scientific journals and conferences, ISTCMTM* Vol **78**

[13] *Base camp facilities standards* 2004 URL: https://docplayer.net/29960552-Base-camp-facilities-standards.html

[14] Kysil O and Levchenko O 2018 *Software tools for вім analysis and neural networks of artificial intelligence on its basis* (Academia.edu, Collective monograph) Vol. **2**

[15] Kysil O 2020 Systematisation of BIM data for project quality management *International Scientific Conference Topical issues of modern design*

[16] Mykhalchenko S 2016 Cluster approach to the formation and development of special areas in terms of approximation to the standards of developed countries *Urban Planning and Spatial Planning* Vol. **59** p 336

[17] Yablonska G 2013 *Economic efficiency and quality of urban and architectural solutions* (Kyiv, Kyiv National University of Construction and Architecture) p 120

[18] Knysh V and Yablonska G 2017 Methodical improvement of design as a strategic and tactical means of materializing architecture *Urban Planning and Spatial Planning* vol **65**

IOP Conf. Series: Materials Science and Engineering **907** (2020) 012022 doi:10.1088/1757-899X/907/1/012022

# The conception and development tendencies of the helicourts network design in the world

**N H Semyroz[1], S S Kysil[2], O S Sleptsov[1], O O Safronova[4] and T V Bulhakova [5]**

[1] Department of Fundamentals of Architecture and Architectural Design, Kyiv National University of Construction and Architecture, Kyiv, Ukraine
[2] Department of Interior Design and Furniture, Kyiv National University of Technologies and Design, Kyiv, UA

kysil.ss@knutd.edu.ua

**Abstract.** The article analyses the origin and development of helicourts [1,2]. This is the new type of constructions which located on top of the roofs of skyscrapers and designed for users of air transport. Historical analysis was performed on the topic of cities design within the vertical plane, including the idea of traffic separation. The modern tendencies of how this issue is resolved were analyzed on the example of São Paulo and Kyiv. It is established, that in modern metropolitan areas there is ongoing new technological revolution, which aims at a radical change of transportation mode in cities, leading to the emergence of the modern aircraft. The main principles of formation of helicourt design decisions within the structure of high-rise buildings in metropolitan areas are: structurality, dominance and form. It has been determined, that the preservation of historical development, natural areas, and improvement of both transport connectivity and the environment where human activity is concentrated in cities can be solved by the introduction of a complex of structures that serves helicopter aviation, and that are located within the structure of multi-storey buildings.

## 1. Introduction

Growth of urbanization level, automobilization, increase of urban population on the planet, increase of urban development density as well as of buildings height encourage the creation of a transport network with infrastructure located on the roofs of high-rise buildings and skyscrapers. The intensity of population movement within the urban structure is increasing and the use of different modes of transportation is increasing [3,4]. Theoretical methods of research confirm the increasing role of small aircraft such as helicopter transport as a promising type of a passenger transport for major cities [5]. One way to create a convenient, safe, comfortable connection in cities is to introduce infrastructure for helicopter aircraft, such as helicopters (Figure 1).

During the process of research, the scientific and theoretical novelty of the proposed hypothesis was determined. The key element is the vertical stratification of the pedestrian transport system. The practicality of this hypothesis consists in the improvement of the living conditions of society, especially in metropolitan areas. This hypothesis allows for a fundamental change of the urban situation in cities. For the people and their lives, it is advisable to keep the natural environment at the level of pedestrian traffic, and to move all the implementations of transport needs above their heads. In the future, managed and unmanaged transport routes and related infrastructure will be located at the first transport level, and at the higher levels, there will be different types of aviation and space transport [6, 7].

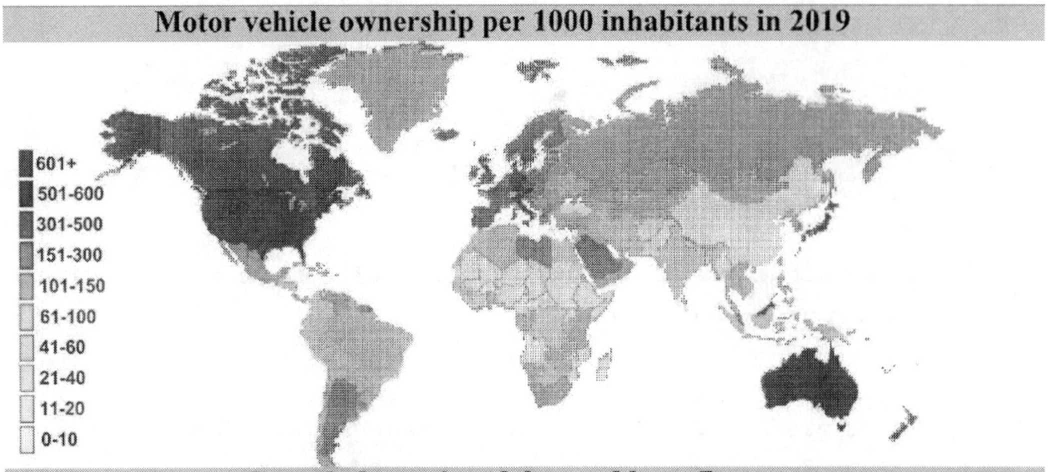


Growth dynamics of the world car fleet

| Year | 1900 | 1914 | 1921 | 1940 | 1950 | 1960 | 1970 | 1980 | 1990 | 2000 | 2019 |
|---|---|---|---|---|---|---|---|---|---|---|---|
| Cars, millions | 0,011 | 1,826 | 10,927 | 46,057 | 70,388 | 126 | 230 | 362 | 550 | 700 | >1 one billion |

**Figure 1.** World level of motorization in 2019

*The purpose of this study* is to analyze the tendencies and principles of how building complexes for helicopter aviation, that are a part of a larger the structure (high-rise buildings), are organized.

Analysis of the recent research of the XXI century has shown that scientist in their scientific works consider in detail the role of various types of urban transportation in the everyday life of society. Foreign studies, related to the construction of heliports, helipads, and helistops are represented by the work of such American researchers as Michael J. O'Donnell, John Leverton, Leopold A. Goldschmidt, Nicholas H. Ludlow, Alex de Voogt [8-12].

Tendencies of how architecture is integrated in urban environment and how architectural space is designed for helicopters were studied by such famous architects as: Oscar Niemeyer, Henry N. Cobb, Robert Moses, Amnon Niv, Philip Johnson, Toyo Ito, Rem Koolhaas, Zaha Hadid and others [13,14,15,16].

In Ukraine transport situation was the topic of studies by: V. Zahreba, E. A. Reitsen, A. K. Starynkevych, G.I.Filvarov, V. I. Chekmarev [17]; in Russia – S. A. Waksman and others. Studies related to the design of rooftop heliports were made by: Y. V. Verjuzhsky in his work «Design of airports», V. M. Pershakov, A. O. Belyatynsky, T.V. Blyzniuk in the monograph «Vertodroms» [18, 19].

## 2. Methodological approach

The methodology of the research consists of complex theoretical, empirical, retrospective and factor analyzes. The theoretical method of research is to predict the development of helicopter aviation as a type of urban passenger transport in couple with its infrastructure network. In this article, the theoretical method of developing a metropolitan transport scheme is based on Kyiv as an example. Last time transport mobility of Kyiv's population was comprehensively investigated in 1986, when the Master Plan was created. Afterwards, the basic indicators have been recalculated using different methodologies, but it should be noted, that during the last decades in Kyiv the traffic flows and conditions of how it is served have changed radically; mobility of the population of the city and its suburbs has considerably increased [20].

The empirical research in this paper was conducted to develop and make integrated decisions on the rational and safe use of various modes of urban transport and their interaction to meet the diverse transportation needs of the population. The empirical methods of the general methodology in this study include compositional aesthetic evaluation, comparison method and modeling. Composite aesthetic evaluation includes selective examination and photo-fixation, visual inspection and architectural and structural measurement of the main parameters of helicourt components as architectural objects in the structure of public and residential buildings. The method of comparison made it possible to determine the complete set of domestic and foreign databases regarding design solutions for these objects. The peculiarities of three-dimensional spatial formation of helicourts were revealed and as a result, a comparative analysis of the domestic and foreign experience of designing and construction of helicourts was made. The last stage of empirical research was modeling. This covers the creation of creative experimental models and basic functional-planning and architectural design solutions of helicourts.

A retrospective analysis of the study revealed the dependence of the evolutionary formation of the helicopter transport from the graphic, illustrative as well as research samples.

Based on factor analysis, there was studied influence of external and internal factors on the architectural and planning organization of helicourts in residential and public buildings. External factors include: socio-economic, natural, climatic, urban, cultural and historical. Internal are: functional planning, structural-technological, ecological and artistic-aesthetic.

## 3. Definition of the term «helicourt»

Growth of the motorization level, insufficient number of parking spaces, low traffic capacity of streets, poor quality of road surface leads to a crisis of transport situation in major cities of the world. There are traffic jams and environmental imbalances in cities [21]. Helicopter transport is still far from being widespread, but the volume of transportation by helicopter is constantly increasing, and there is every reason to believe that it will take one of the leading places in the urban transport system. At the same time, an unmanned aerial vehicle – a drone, which may be used in urban conditions, – will replace the helicopter in the near future.

In turn, the accelerating pace of life both in Ukraine and in the world requires the introduction of new high-speed modes of transport, namely, – air transport. In many countries in the metropolitan areas, there are complexes of structures that are located on the roofs of multi-storey buildings for servicing aircraft. In Ukraine, such complexes just started being born. At the same time, arises the necessity to regulate their location in cities [22].

An analysis of the different interpretations of aviation terms made it possible to propose a new term to further refine the object of study. It makes sense to name the above-described structure a «helicourt». Helicourt is a complex of structures that service air transport passengers, that is located exclusively on the part crowning a high-rise building or is a part of the traffic interchange platform and has a special equipped ground intended for takeoff and landing of air transport [23]. Usually, these are helicopters (Figure 2).

## 4. Historical analysis of design experience

«The future will free us from our constant commitment to the land» – it was said by modernist architects at the beginning of the 20th century and they were absolutely convinced that the traditional street is outdated. They imagined all sorts of alternatives: intricate multi-level cities with underground tunnels, and all kinds of streets in the sky. The idea of traffic flows separation originated in the 1820s and was widely cited in literature and illustrations, and in the 20s of the 20th century, overpasses and tunnels appeared in major American cities.

In 1904, French architect Tony Garnier showed a plan for the Industrial City with 35,000 inhabitants (Figure 3). His city was organized based on the principle of spatial separation of functions and it is one of the earliest examples of vertical zoning. On the Master Plan of the exemplary industrial city from the Tony Garnier's book, «Une Cité Industrielle», there are residential, recreational and industrial areas [24]. In 1927, Ludwig Hilberseimer designed his high-rise project, High-rise City (Hochhausstadt, Figure 4), where the planar buildings stood side by side, and the transport and pedestrians moved along elevated from the road level sidewalks.

IOP Conf. Series: Materials Science and Engineering **907** (2020) 012022 doi:10.1088/1757-899X/907/1/012022

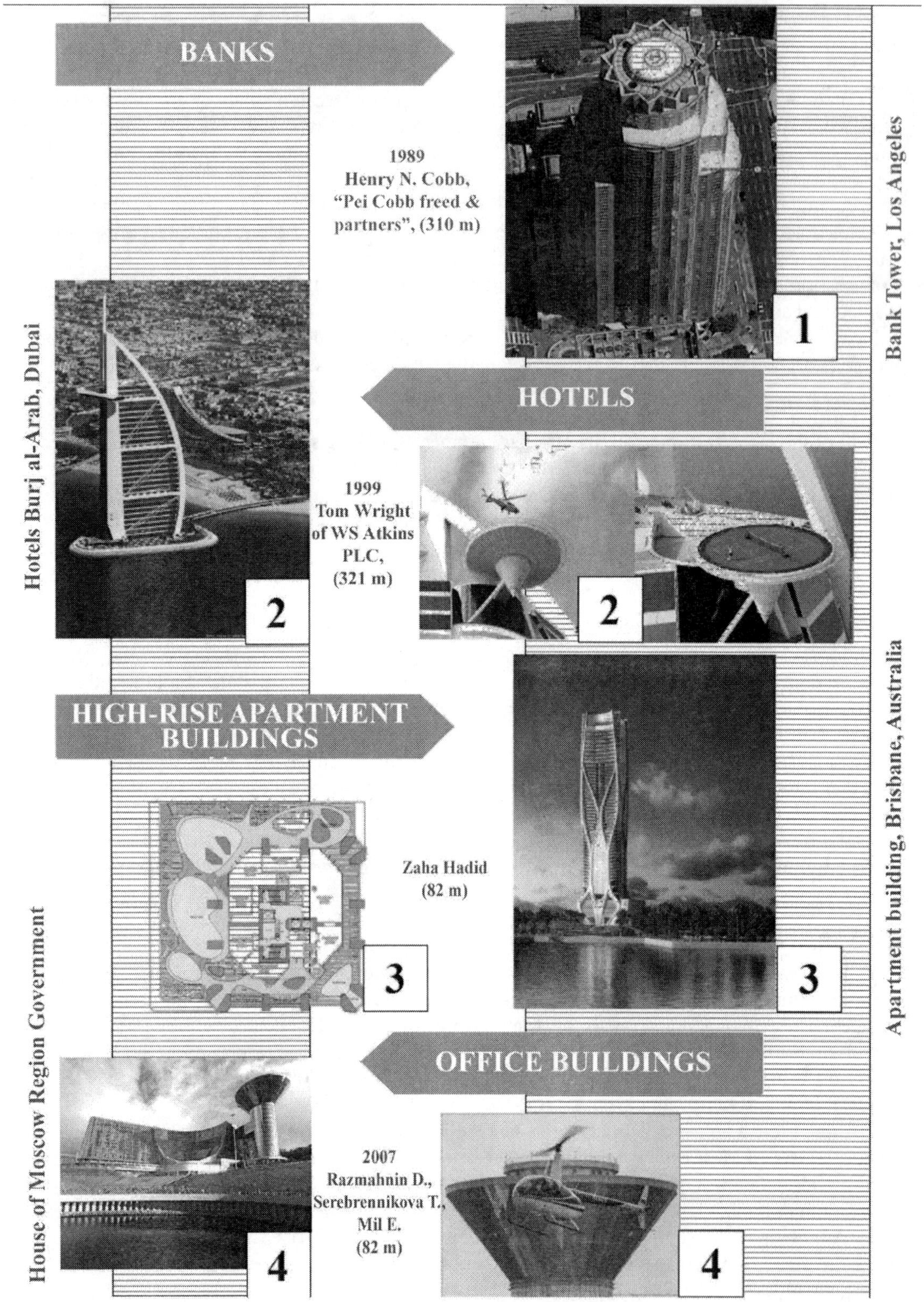


**Figure 2.** Experience in building helicourts in the structure of residential or public high-rise buildings

IOP Conf. Series: Materials Science and Engineering **907** (2020) 012022 doi:10.1088/1757-899X/907/1/012022

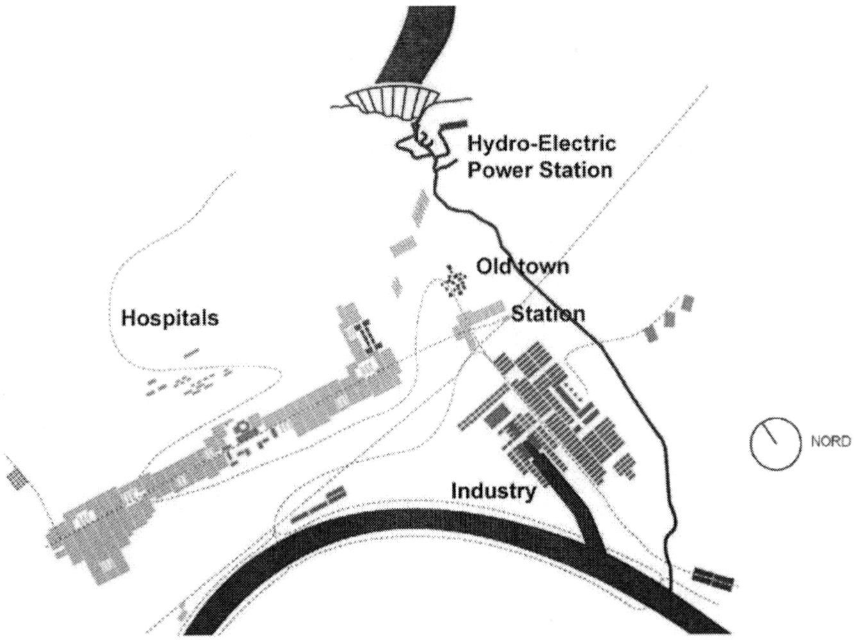


**Figure 3.** Plan of Industrial city, Tony Garnier, 1904 – 1917

**Figure 4.** High-rise City (Hochhausstadt). Perspective View: North-South Street. View of multi-storey buildings

The key ideas for urban development were the ones from Le Corbusier: zoning principles, the combination of high-rise residential development with landscaping and the differentiation of transport infrastructure. A complete embodiment of the architecture and modernist mid-twentieth-century urban development has been found in the new capital city of Brasilia, built according to the master plan of Lúcio Costa [25, 26].

## 5. Location of helicourts in the structure of cities

These works formulate the foundations of a modern urban city: transport interchanges, the elimination of the traditional street, reinforced concrete tower buildings, the use of functional zoning. Improvement

IOP Conf. Series: Materials Science and Engineering **907** (2020) 012022 doi:10.1088/1757-899X/907/1/012022

of the urban form scoped on the vertical dimension was the idea of many architects, especially considering the fact that air travel was taken as something special. In modern metropolitan areas, the street has ceased to be a place for pedestrians; it became predominantly focused on the intense automotive traffic. There is a new technological revolution happening which aims at a radical change of vehicles in the cities, at the emergence of the newest aircraft: drones, airplanes, airbikes [27].

Increasing traffic congestion and the onset of economic upturn in Brazil have affected the business aviation industry's upswing. Sao Paulo – the largest city in the country has the world's second largest helicopter park, second only to New York today. The rapid development of the helicopter industry in São Paulo is explained by the search for personal safety and convenience. 80% of all flights are business aviation, including police, radios, and television helicopters and a small fleet of «Robinson» helicopters used to search for stolen vehicles.

On June 1, 2012, a record was made in South America's largest city: streets were blocked by 295-km traffic jams. According to the data from the Association of Brazilian Helicopters Pilots ABRAPHE (Associação Brasileirade Pilotosde Helicóptero), by the end of 2011, between 480 and 490 helicopters were registered in the metropolis. The most popular models are: «Eurocopter Astar», «Agusta-109» and «Sikorsky S-76», which are used for both business aviation and oil platform service. There are many «flying» aerotaxis operating in São Paulo, especially on the most important air route Rio de Janeiro – São Paulo. Out of the 260 helipads, about 200 are located on the roofs of houses [22].

The activity of the modern metropolis aims at performing various functions: economic, administrative, scientific, cultural, and wellness-related. The increasing intensity of urban population requires a modern, high-speed transport connection.

The city of Kyiv already ranks 33rd on the European part of the Eurasian continent in terms of population concentration and economic activity. At the same time, 16 helipads are planned to be built in Kyiv [20]. They are necessary, first and foremost, for the prompt transportation of seriously ill patients and the provision of emergency medical care (Figure 5).

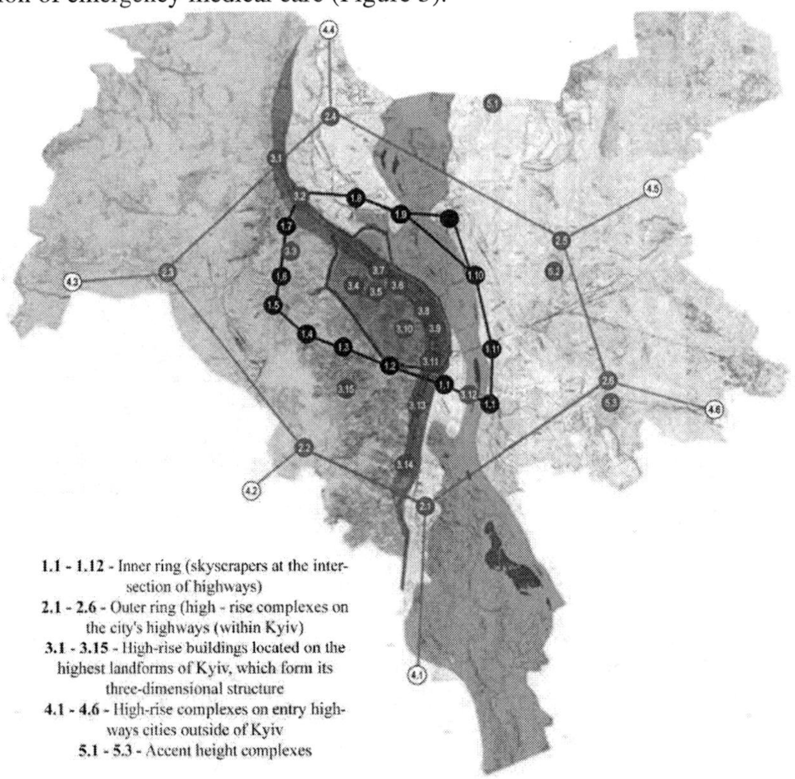


**Figure 5.** Scheme of placement of high-rise buildings and complexes according to the relief map of the city in the long-term development of Kyiv

Besides helicopters and helipads intended for use by public institutions, according to the «Concept of placement of high-rise buildings and structures in Kyiv until 2020», it states that Kyiv will have high-rise buildings built, and they will need helipads for rescuing people in case of fire. Functional purpose of those high-rise complexes – creation of business, financial, hotel, cultural, educational and sports centers serving the whole city – implementing day-to-day functionality of the city as a capital. On the rooftops of high-rise residential buildings, there are platforms provisioned for landing the emergency cabin of fire helicopters [28, 23].

## 6. Universal principles for the design of helicourts

The overview of foreign and domestic experience shows that in metropolitans today it is necessary to apply universal principles for the design and construction of helicourts. It makes sense to split these principles in three directions, according to the classical triad in architecture – functions, constructions, and forms. Functionality covers the principle of structurality. Constructiveness covers the principle of dominance. In addition, form design covers the principle of form. The principle of technological determinism unites these principles [23].

## Conclusions

The activity of the modern largest city is aimed at performing various functions: economic, administrative, scientific, cultural and wellness. The increasing intensity of urban population requires a modern, high-speed transport connection.

In this study it was found, that up to this date the complex of buildings for helicopter aviation, located in the structure of residential and public buildings, has not been sufficiently studied. In Ukraine, as well as in other metropolitan areas of the world, high-rise construction is rapidly being implemented and is growing at a rapid pace; buildings and residential complexes are being constructed, aviation is being developed.

In turn, project activity should outmarch, anticipate and plan the interconnectivity of scientific and technological progress, the economy with the demands and needs of society to ensure a comfortable life. Moreover, it is with the development of aviation transport, especially in the major cities, that it is possible to: resolve urban air transport corridors; deployment, location, parking, refueling and repair of aircraft, including sanitary aviation. Also important is the protection against air pollution from emissions into atmosphere, and noise pollution by the aircraft.

Further research is needed to study the stages of development of a new type of air transport for urban planning, the economy, and the safety of using passenger drone above the city.

## References

[1] Semyroz N H 2016 *Current state of design and construction of helicorts* Inventor's certificate № 66640 (Kyiv: in Ukrainian)
[2] Semyroz N H, Sleptsov O S 2017 *Project proposals «Helicort»* Inventor's certificate № 70133 (Kyiv: in Ukrainian)
[3] Kysil S S 2013 Construction issues of multi-storey car parks in large cities of Ukraine *Mistobuduvannia ta terytorialne planuvannia* **50** (Kyiv: KNUBA) pp 275-280 in Ukrainian
[4] Kysil S S 2014 Directions of building design for multi-storey car parks in major cities *Arkhitektura i sovremennye informatcionnye tekhnologii (AMIT)*, (Moscow: Moscow Architectural Institute). URL: http://www.marhi.ru/AMIT/2014/3kvart14/kisil/abstract.php in Rusian
[5] Semyroz N H 2013 Helicorts in the cities *Suchasni problemy arkhitektury ta mistobuduvannia* **34** (Kyiv: KNUBA) pp 560-568 in Ukrainian.
[6] Semyroz N H 2014 A study of the historical development of helicopter transport and a retrospective of the development of helicopter industry *Arkhitekturnyi visnyk KNUBA* **3** (Kyiv: KNUBA) pp 91-97 in Ukrainian
[7] Frost J B 1996. *British helipads.* (Chester, UK: Appledore Publications)

[8] Michael J O`donnel 2012 *Aviation Circular No: 150 /5390-2C* (initiated by: AAS-110 C. Heliport design) p 188
[9] De Voogt A J 2006 Hospital helipad architecture: ideals of safety and design. InCor Wagenaar (ed.) *Hospital Architecture* (AZG: Groningen) pp 141–157
[10] De Voogt A J 2007 *Helidrome Architecture*. (Rotterdam: 010 Publishers)
[11] ICAO 1995. *Heliport manual*. (ICAO Publications Montreal, Canada)
[12] Timko T & Ištoka Otković I 2017 Comparison of different heliport solutions for the clinical hospital center in Osijek. *Elektronički časopis građevinskog fakulteta Osijek* **8** pp 58-67 doi: 10.13167/2017.14.7.
[13] Niemeyer O 1956 *Notes on Brazilian Architecture* In S. Papadaki, O. Niemeyer: Works in Progress (New York: Reinhold) pp 11–14
[14] Koolhaas R 2002 *The Great Leap Forward*. Harvard Design School Project on the City. (New York: Taschen)
[15] Urban F 2016. Robert Moses: The master builder of New York City *Planning Perspectives* **31**: 4 659-661 doi: 10.1080/02665433.2016.1203114
[16] Noever P, Haslinger R & Himmelblau C 1991 *Architecture in Transition: Between Deconstruction and New Modernism* (Munich)
[17] Reitcen E A & Kaddakh Kh 2000 Modelling traffic flows in cities *Bezpeka dorozhnoho rukhu Ukrainy* **1**(6) (Kyiv: NDTsBDR) pp 41-46 in Ukrainian
[18] Veriuzhskyi Yu, Rodchenko O V & Hyrych V Yu 2011 *Airport Design* (Kyiv; NAU) p 257 in Ukrainian
[19] Semyroz N H, Pershakov V M, Bieliatynskyi A O & Blyzniuk T V 2014 *Heliports* (Kyiv: NVF «Slavutych-Delfin») p 345 in Ukrainian
[20] *The development strategy of the city of Kyiv until 2025*. Approved by the decision of the Kyiv city council №824/7060 of December 15, 2011, a new edition of 2016 (in Ukrainian). https://doi.org/10.1787/9789264301436-en
[21] Kysil S S 2016 *The principles of the architectural and planning organization of multi-storey car parks (for example, the largest cities of Ukraine)*: Abstract of the dissertation of the candidate of architecture (Kyiv National University of Construction and Architecture, Kyiv, Ukraine) p 21
[22] De Vasconcellos E A 2005 Urban Change, Mobility and Transport in São Paulo *Three Decades, Three Cities, Transport Policy* **12** (2) pp 91–104
[23] Semyroz N H 2017 *The principles of the architectural and planning organization of helicorts:* Abstract of the dissertation of the candidate of architecture (Lviv Polytechnic National University, Lviv, Ukraine) p 24
[24] *Le Musée Urbain Tony Garnier* 2019 (Lyon, France)
[25] Le Corbusier 1929 *The City of Tomorrow and its Planning* (London: John Rodker)
[26] Le Corbusier 1971 *Looking at City Planning* (New York: Grossman)
[27] Bulhakova T V 2010 The problem of compositional analysis of the urban environmen. *Suchasni problemy arkhitektury ta mistobuduvannia* **23** (Kyiv: KNUBA) pp 52-60
[28] Sleptsov O S 2010 *The architecture of civil structures: Industrialization* (Kyiv, A+S)

IOP Conf. Series: Materials Science and Engineering **907** (2020) 012023 doi:10.1088/1757-899X/907/1/012023

# Innovative concepts in the Kharkiv projects of J. Steinberg of the capital period

**O V Maimeskul, A S Mamon and M V Rusanova**

Department of the Fundamentals of Architecture, Kharkiv National University of Civil Engineering and Architecture, 61002, Kharkiv, str. Sumska, 40, Ukraine

maryrusanova19@gmail.com

**Abstract.** The innovative concepts of Jacob Steinberg, one of the leaders of constructivism architecture, the head of the Society of Modern Architects of Ukraine (OSAU) are considered. The projects of J. Steinberg by the purity of functionalist literacy are comparable to the projects of leading OSA masters. Among Steinberg's Kharkiv projects - recognized as masterpieces of constructivism - the Builders' Club, the building of the Central Committee of the CP(b)U, a complex of the Kharkiv Civil Engineering Institute on Shatilovka, a residential complex on the Pushkinskyi entrance. The unique Steinberg's objects, that were revolutionary for their time, weren't taken under protection and today they have been lost, rebuilt, or are in ultimate limit state. His legacy is still waiting for his collectors, organizers and researchers. Comprehensive research is needed, on the basis of which it will be possible to develop a strategy for documenting and preserving the Kharkiv objects of an outstanding Ukrainian architect. This applies to the possibility of complete restoration of the houses at the Pushkinskyi entrance, removal of distortions introduced into the building of the Builders' Club during the post-war restructuring, graphic reconstruction of the lost Central Committee building and irreversibly rebuilt complex of ex. KhICE. The report discusses the prospects for the formation of citywide and national programs for research, documentation and preservation of the architectural heritage of interwar modernism.

## Introduction

Jacob Steinberg (1896-1982) - an outstanding local architect-constructivist, the author of a number of original projects and implementations in Kharkov, Kiev and other cities of the former Soviet Union.

Jacob was born in Kiev and received here an architectural education in the Kiev Art Institute in 1925. The commitment of the young architect to the ideas of the avant-garde is evidenced by the fact that in the 1923/1924 academic year he refused to do a course project "in the proposed styles" [1]. The administration of the institute regarded this as a rebellion and Steinberg was threatened with expulsion, but the conflict was settled and students received the unspoken right to carry out their projects without reference to historical styles. The level of ambitions of the young architect is approving by the fact that as a graduation project, he (in collaboration with I. Malozemov) completed a competition project for the House of State Industry in Kharkiv (Gosprom). His project under the motto "Screw" participated in the competition along with the projects of 17 other contestants and received one of the prizes [1]. In this first major project of

Steinberg, the author's handwriting was not yet fully disclosed. The building aims towards the classic composition of volumes, and the pitched roofs of the two central towers, although they enliven the silhouette, makes the building look a bit archaic.

The grandiose construction site of the Gosprom built under the project of Leningraders S.S. Serafimova, S.M. Kravets and M.D. Felger, became for Steinberg the first responsible place of work. Here he became the foreman of one of the construction sites, and later went to work in the design office. Since it turned out that the metal large-format windows that were provided for by the project could not be made at Kharkov factories, all the wooden windows of the building were done according to Steinberg's project. Traditional designs of wooden frames did not match the avant-garde image of the building. Based on the sketches of Steinberg, several variants of wooden frames were done, which were subsequently installed in the window openings of the already built southern wing of the building [2]. Thus, the most successful option was chosen.

Due to the acute shortage of qualified personnel of builders and architects, the design teams established at large construction sites had particular value. One of such groups of specialists who worked on the construction of the Gosprom industry headed by Steinberg became the basis of the architectural sector of the Industroy or, in other words, the All-Ukrainian Design and Construction Society "Construction Industry" (Industroy). Steinberg worked from 1928 to 1934 as the chief architect of this organization. Mostly, the company was engaged in the construction of industrial buildings of various profiles. Among them, Steinberg personally creates a project of sugar factory at Vesely Podil station near Kharkiv.

Industroy also took an active part in architectural competitions. Five years after the competition for the Gosprom building, Steinberg took part in the competition for the project of the Theater of Mass Musical Action in Kharkiv. 145 projects from different countries was submitted to this international competition. The project, carried out by Steinberg in collaboration with Yu. P. Afanasyev, V. P. Kostenko, M. L. Movshovichem, R. M. Friedman was submitted to the competition on behalf of the Glavproekt. The consultant of the project team was the artist of the theater V. G. Meller [3]. Unfortunately, the competitive projects of the Theater of Mass Musical Action were lost during the Second World War. After the creation in 1934 of the organization Tsivilbud on the basis of the Industroy, Steinberg headed their architectural department and led it until 1936. After that, until the Second World War, he led the design teams of various organizations. J. A. Steinberg conducted a public activity by leading the Association of Modern Architects of Ukraine in 1928-1929 [4].

In the evacuation, Steinberg worked as the head of the design bureau of the factory No. 60 in the city of Frunze (modern Bishkek). In 1944, Steinberg had returned to Ukraine and developed a competitive project of reconstruction for str. Khreshchatyk in Kiev (not implemented).

## 1. The Builders' Club on Sq. Rudnev

One of masterpiece of Jacob Steinberg in Kharkiv - is the Builders' Club on Sq. Rudnev (now the square of the Heroes of Heaven hundreds, 1927-1930, in collaboration with I. I. Malosemoff and I. F. Milinis). This project (Figure 1) is one of the examples of "pure" orthodox constructivism [5]. It is no coincidence that he was among the few Kharkiv projects that were published on the constructivist magazine CA (Modern Architecture), the front-liner at that time [6], along with such projects as the theater of mass musical action (brothers A. A. and V. A. Vesnin) and project of the post office (P.A. Golosov). S. O. Khan-Magomedov included the Builders Club among the hundred masterpieces of the Soviet architectural avant-garde [7].

In the process of constructing the Gosprom, J. Steinberg took part in the creation of a number of collective competitive and custom projects. Among them, the frankly constructivist project of the Builders' club is of the greatest interest. In the same 1927, the young architect took part in two competitions that were highly resonant - the project of the Government House in Kharkiv and the project of the facade of the station in Kiev [1]. In the project of the club Steinberg sought to reveal the image through new architectural forms close to the machine industry, showing the capabilities of advanced construction equipment. The author's innovative approach is reflected in the expressive dynamic composition and the entire volumetric-spatial structure of the building. The facade of the club resembles the Villa Savoy in Poissy, France (architect Le Corbusier) with its horizontal volume and perspective view of the courtyard, but with the addition of a transverse protruding element of a rounded glazed staircase (Figure 2). The inner open courtyard-lobby plays the role of the spatial core of the composition, from which you can get into the various functional parts of the building, the communication of which is also carried out on the second floor. The club scene has two portals: one of which worked in the winter, the other (with an amphitheater designed for 1200 people) in the summer. All this created an original modern architectural image. Since the mid. 1920s the architecture of club buildings was hotly discussed in a professional environment, which led to a gradual shift in the focus of club work from a spectacular theater and cinema zone to a coterie one. This was reflected in design tasks: the using of a small hall becomes more relevant, the club necessarily includes audiences, workshop and exhibition rooms, the number of study rooms, studios are significantly increased, the area of libraries are expanded, the area for working with children, as well as the sports part. In addition, there was a change in the perception of the image of the club, which now had to reflect the ideology of the dictatorship of the proletariat, the revolutionary ideas of the time. In the project of the Builders Club, young authors managed to meet new needs and win the competition. The premises of their club included: an auditorium with 1200 seats with a stage, a reading and lecture hall with 150 seats each, a library with 35,000 volumes, a gym, a canteen, an audience of 200 seats, a number of rooms for clubs, entertainment and recreation, as well as nursery. J. Steinberg successfully grouped the rooms, dividing them functionally into 2 blocks: "noisy" in the left and "quiet" in the right wing of the building. The total capacity of the club was: in the winter - 2500, in the summer - 400 people [8].

**Figure 1.** Club of builders (1930) [7].

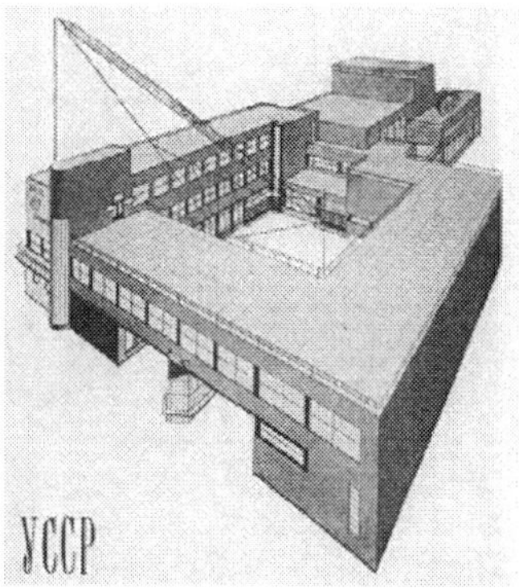


**Figure 2.** The Builders' club, J. A. Steinberg, I. I. Malozemov, I. F. Milinis [6].

During the Second World War, the building of the Builders Club was badly damaged and then underwent two reconstructions: in the 1950s and mid-1970s. The utilitarian approach that prevailed in architecture at

IOP Conf. Series: Materials Science and Engineering **907** (2020) 012023 doi:10.1088/1757-899X/907/1/012023

that time led to the disruption of the internal structure and facades of the building. During the reconstruction, the courtyard was redeveloped into a covered hall, the third floor was built on the two-story part of the building, the windows were changed to traditional, the flat roof was turned into a pitched roof, the modernist concrete facade was tiled with ceramic tiles and the inscription with the club name on the facade was removed. As a result, the building has changed beyond recognition, likened to the faceless boxes of the 1960s. it ceased to exist as an example of pure constructivism (Figure 3).

**Figure 3.** Club builders. Today's state (author 's photo).

## 2. The Residential buildings on the Pushkinskyi entrance

Residential buildings on the Pushkinskyi entrance 7, 7A and 8 (1929–1930, co-authored with R. Fridman, engineer V. Trubins) are examples of "academic" constructivism [5]. In the flat facades of 4-story residential buildings author alternates horizontal rows of windows and flat open balconies of apartments, the graphicness of which resembles a hostel of the Bauhaus complex in Dessau, Germany (architect V. Gropius), with vertical glazing of stairwells (Figure 4). The accent is enhanced by the protruding parapet in the central part of house number 8, which at the same time fix the symmetry of the entire complex of residential buildings. The purity of the solution for the composition of the U-shaped residential building with the courtyard (Figure 5), which is revealed gradually in steps to the road, can be compared with the organization of the side buildings of the Project House of the Government Complex on Dzerzhinsky Square, (architect S. S. Serafimov, M. A. Zandberg- Serafimov), built in the same period.

**Figure 4.** Residential complex at Pushkinskiy entrance, 7, 7A [9].

**Figure 5.** Residential complex at Pushkinskiy entrance, 8 [9].

An unique approach developed for the construction of houses on the Pushkinskyi entrance and described by Steinberg himself in the magazine "Budivnitsvo" (Building): "work was oriented on three such indicators: speed of execution; the use of new designs and a new methods; cheapness" [9]. As a result, the production plan was drawn up with the intention of sequentially carrying out work in such a way that immediately after the construction of the first floor it is cover, allowing, in parallel with the construction of the walls of the second floor, to begin internal work on the first. Thus, during the construction of the fourth last floor, then all floors had already been made intermediate ceiling, black floors, covering, partitions and partial plaster. Following the installation of partitions, central heating, plumbing and sewage were arranged. This innovative approach allowed the construction of the house #8 in 3 months and 3 weeks (the foundation of the house was laid on July 17, 1929, and on November 1, 1929 the first inhabitants moved into the house and by November 10 the house was completely ready to move in). In his article, Steinberg also concluded: “This method of performing work must be introduced at all construction sites, instead the old method, in which internal work does not begin before the building is covered with a roof ...” [9].

When designing the house number 8 they chosen the rectangular configuration of the object as is more economical. The house is composed of four sections, and accordingly, has four stairwells with two exits (to the courtyard and to the street). A total of 32 apartments have been designed. A typical section is two apartments: one- and three-room, serviced by one stairwell. The layout of the building is asymmetric: the two sections located to the right of the central axis are arranged in a different way: one section is two two-room apartments, and the second section is three- and four-room apartments. The house was indebted to its future residents for such “atypicality”: four-room apartments were built for department heads and chief engineers of the Civilbud (Civil construction) Society. When designing the living space, English principles were adopted: the apartments did not have walk-through rooms, most of the walls were freed for furniture. To the great regret of the architects, the rooms were not equipped with standard furniture, however, they hoped that this was a matter of the near future and would soon become a part of the architectural project, citing another capital as an example - “Moscow is already making such attempts in the residential construction of “Mosrada” in this construction season ".

U-shaped residential building, composed of 8 residential sections of two types with a large courtyard and a round fountain in the center, has two addresses: Pushkinskyi entrance 7 and 7a. Each wing of the house had its own customer: the first was built for Giprokoks, the second - for Building Trust 86.

Now, unfortunately, residential buildings on the Pushkinskyi entrance, 7, 7A and 8 are quickly destroyed, most of the original windows were replaced with modern metal-plastic ones, the balconies are glazed, built-in shops with entrance structures attached to them appeared on the ground floors. In the current state, these striking examples of residential complexes of the first wave of modernism cannot even get into the register of monuments of local importance.

## 3. The Kharkiv Institute of Civil Engineering

Jacob taught at a number of construction and art educational institutions in Kharkiv and Kiev, in particular, at the Kharkiv Art Institute and at the Kharkiv Institute of Civil Engineering (KhICE) created in 1930. This year was a turning point for the structure of higher education in the USSR: by a government decree, many faculties of multidisciplinary universities were transformed into separate higher educational institutions. KhICE was created on the basis of the architecture faculty of the Kharkov Art Institute and the construction faculty of the Kharkov Technological Institute (now the National Technical University “KhPI”).

Immediately after the creation of KhICE, the construction of a new institute complex was started. The development of the project was entrusted to J. A. Steinberg, at that time an assistant professor of the Faculty of Architecture at KhICE. For construction, a site of 6.7 hectares was allocated in the area of Shatilov cottages. At that time, this area was built up with estates and dachas and was subject to redevelopment in accordance with the general plan of the "socialist reconstruction of Kharkiv", as the capital of Soviet Ukraine. The site was located half a kilometer from the new administrative center of the Ukrainian Republic - former Dzerzhinsky Square, at the beginning of the new central avenue, on both sides of which the complex of educational, administrative and residential buildings of the institute was to be located. On the right side of the avenue (when viewed from the side of the city center), educational buildings and laboratories were designed. On the opposite side, student dormitories, professors' houses, a sports stadium and a canteen, which were the second stage of construction, were planned.

In the absence of practical experience in the construction of large university complexes, J. A. Steinberg (co-authors of the project, architects R. Fridman and A. Zaslavsky) applied a new spatial planning concept to the project, which envisaged the creation of a single complex consisting of separate buildings interconnected by warm transitions. Such an organization was called upon to ensure the autonomous existence of all departments of the university and, at the same time, provide the opportunity for quick communication of each of the departments with a common library, assembly hall and other common areas.

The planning structure of each of the educational buildings was also innovative. The project used a combination of corridor and non-corridor planning schemes [10]. The first floor of each of the four-story blocks was designed in transit, it was occupied by vestibules, wardrobes and stairs leading to the upper floors. These functional areas were planned in such a way that noise from students moving along the main transit routes did not interfere with studying in laboratories located on the second floor. The second floor had a corridor layout, the third, on which there were classrooms and rooms for general classes, was corridor-less. On the last, fourth floor there were drawing rooms. Each building had its own nodes of vertical communications, which provided a significant degree of autonomy of the buildings.

**Figure *6*.** The first building of KhICE, 1932 (photo from author 's archive).

Pre-war photographs give an idea of the external appearance of the first KhICE building (Figure 6). According to the principles of functionalism, the facades of the institute reveal its internal structure: the windows of the third and fourth floors are larger than the windows of the first floors, since more sunlight was needed for the rooms located on them. The characteristic plastic accents were glazed volumes of vertical communications that protruded strongly from the plane of the facade, inside of which there were ramps popular in constructivism architecture.

The construction of the first KhICE buildings was completed in 1932 [10]. The design and construction of subsequent buildings took place during the period of obsolescence of modernist architecture and the imposition of the official concept of “development of cultural heritage”. Therefore, the stylistic changes were made to the project.

The KhICE complex was badly damaged during the Second World War. The surviving photographs clearly show the destruction: all the windows were lost (through the empty openings of the staircase of Building A, the ramp replacing the staircases is clearly visible), however, the general configuration of the buildings designed by Steinberg has been preserved. In the post-war years, the building was transferred to the Mining Institute and was restored according to the project of N. M. Podgorny, who was then head of the department of descriptive geometry at this institute. Podgorny used in the project two partially surviving

IOP Conf. Series: Materials Science and Engineering **907** (2020) 012023 doi:10.1088/1757-899X/907/1/012023

buildings from the complex designed by Steinberg, however, the appearance of the buildings was fundamentally changed: the main entrances received solemn classical porticoes. The construction of a building with a massive tower, on the corner of Lenin Avenue (now Nauki Avenue) and Serpova Street, marked the final rejection of Steinberg's spatial concept. However, in the rebuilt building, some elements remind of the pre-war KhICE, for example, semi-cylindrical ledges of stairwells, which are not at all organic for the neoclassical image of the Mining Institute.

## 4. The Central Committee building

The greatest interest from the number of Kharkov projects J. A. Steinberg represents the overbuilding of the building of the former Provincial Zemstvo for the Central Committee of the CP (b) U. In 1925, the Central Committee of the Communist Party (Bolsheviks) occupied two adjacent buildings on K. Liebknecht Street (today Sumska Street), previously owned by the Provincial Zemstvo. One of the buildings was built in 1900 according to the project of arch. A.B. Minkus, the second was attached close to it in 1914 (architect V.V. Velichko). The facades of both buildings were made in the Neo-Renaissance style. Since these buildings could not accommodate the government apparatus of the Ukrainian Republic, J. A. Steinberg was commissioned to complete the design of the new Central Committee building. The unique concept of his project, which is undoubtedly innovative for its time, the author outlined in the article "Nadbudova budynku CC KP(b)U" [Overbuilding of the Central Committee building of the CP(b)U] at the "Budivnitstvo" [Construction] magazine [11]. In this article, the architect not only formulates the idea of the project, but also describes the specific difficulties that had to be encountered during the design.

**Figure 7.** The building of the Central Committee KP(b)U, 1932 (photo from author 's archive).

**Figure 8.** The building of the Central Committee KP(b)U, Steinberg's project.

Instead of demolishing buildings, Steinberg proposed combining them, adding three floors to this. At the same time, the historical facades were preserved in their original form with the complex decor of the Ionic order, and at the same time they were supplemented by a large-scale overbuilding, made in the methods of modernist architecture (Figure 7, 8). The architect argued his idea: "A new house on the street, as a rule, appears between neighbors, each built in its own style ... As a result of the proximity of the horizontal buildings, a kind of dynamic street ensemble arises. Those laws that operate in the horizontal direction should work in the vertical direction: just as the ensemble is formed horizontally, it is formed in the vertical direction ..."

It should be noted that such ideas were unusual for the architecture of early functionalism, which, as a rule, focused on the construction of new buildings, and preserving the heritage of previous eras was not considered an important task. The idea of vertical zoning of the city was widespread among the supporters of functionalist urban planning, especially among those architects who advocated the idea of compact development of the city. However, the ideas of Steinberg in this case are different in meaning from the ideas of the "vertical city" by Antonio St. Elia and Ludwig Gilbersimer. If the concepts of the latter dealt with the vertical separation of functional zones or traffic flows: warehouses and transport at the lower levels, housing and administrative buildings - at the upper, then Steinberg alternates along the vertical axis of the historical "layers" of the city. The author himself emphasized that his project was completed "in four dimensions", where the fourth dimension meant time [11]. In the same article, the author reflects on the possibilities of a holistic reconstruction of the historical centers of cities, when "... the overbuilding of buildings along the street is planned in such a way that horizontal belts of overbuilding floors appear under the whole block, under which the gothic and Empire style of the old buildings will remain, and the street, preserving them the ensemble will take on a new, predetermined by plan, appearance". Steinberg interpreted the three- and four-story overbuilding above historic buildings as "a modern, functionally faithful entablature" [11]. Such an architectural "collage" is more like postmodernism than early functionalism.

However, the design of the overbuilding of two historic buildings lurked many other difficulties. The existing entrance group of buildings of the former Zemstvo (vestibules, wardrobes) could not meet the requirements of the design assignment, according to which the total volume of the complex was to increase 2.5 times. The main staircase in one of the buildings was in the center and was lit only by the upper light, which made it impossible to operate it in a five-story building. In addition, an elevator was required. All this led to the creation of a new entrance group, which took the corner of the current streets of Sumy and Svobody. Due to this, on the side facade from the side of Svobody Street, the two-story facade of the 1900 building was surrounded by a modern, modernist facade on the right and left, and above it there are four new floors, which were distinguished by long ribbon windows.

From the courtyard to the building was added a hall for general meetings, designed for 650 seats in the hall and 70 seats on the podium, with a room for stenographers to work. The hall was illuminated with the help of a shed ceiling, the windows of which were not visible to those sitting in the hall, which created a soft and diffused light.

In general, the idea of building the building of the Central Committee of the Communist Party of Ukraine took into account the specific circumstances of the place to a greater extent than was based on the ideas of modernist architecture. After 50 years, Charles Jenks will call this approach "adhokism" and include it in the list of basic techniques of postmodernism [12]. This suggests that the concepts of Jacob Steinberg were innovative for his time. In his projects, he combined the actual achievements of his contemporaries with his own view of the problems of architecture and urban planning.

The overbuilding of the Central Committee building was basically completed in 1931, but the rush during construction work led to the collapse of the plaster of the upper floors during early frosts. Restoration work lasted until 1932. At the same time, it was decided to remove the classic decor from the facades of historic buildings. Thus, the finely traced ionic columns on the facade of the building erected by V.V. Velichko. The decoration was replaced by a series of uniformly rectangular blades in plan that visually supported the overhanging modernist part of the building. Of course, such a change in the facades led to a significant simplification of the appearance of the Central Committee building, and the loss of expressiveness of the author's intention.

During the hostilities, the Central Committee building was partially destroyed by a direct hit by an aerial bomb and after the war it was demolished. In 1954, a new house was built on this site for the regional party committee under the project of V.M. Orekhova and V.P. Kostenko. In terms of volume, the new building corresponded to its predecessor, however, it was dramatically different stylistically: the facade was made in the style of the solemn neoclassicism characteristic of the first post-war years. The massive columns of the modernized five-story warrant have nothing to do with the J. A. Steinberg project Steinberg. Today, this building is located on the Kharkiv Regional State Administration.

**Conclusions**

Jakob Steinberg, one of the leaders of the Constructivism architecture, in 1928-1929 he was the head of the Society of Modern Architects of Ukraine (OSAU). Steinberg's unique projects carried out in Kharkov in 1920-1930 reveal him as a professional and a leader in the era of modernism, ahead of the thinking of contemporaries. His work, imbued with the spirit of innovation in everything - in architectural and planning, structurally and technical and purely compositional, artistic aspects is underestimated and insufficiently studied. His revolutionary architectural heritage, which deserves to be protected, is today have been lost, rebuilt, or are in ultimate limit state.

There is good reason to consider it necessary to create a systematic program for the study of the architectural heritage of J. A. Steinberg, which could be the beginning of a joint company to preserve the heritage of modernist architecture of the city and country, and which will include:

- historical-architectural and historical-cultural studies and documentation of all the famous projects of the author in Kharkiv in accordance with modern standards developed by ICOMOS and DOCOMOMO;
- graphic reconstruction and creation of detailed 3D models of these objects;
- publication of J. A. Steinberg's monograph, as well as the creation of an album of his works by local specialists directly related to the architect or covering his practice in their research, with the possible involvement of a Kiev colleagues;
- conducting public lectures, studying the Steinberg's heritage as part of the curriculum of architectural schools in Kharkiv and Ukraine;
- determination of the strategy for conservation of objects on the basis of research. It seems possible to fully restoration of the houses at the Pushkinskyi entrance, removal of distortions introduced into the building of the Builders' Club during the post-war restructuring, graphic reconstruction of the lost Central Committee building and of the irreversibly rebuilt complex of ex. KhICE;
- definition of scenarios for the development of the city, in which there is a need to restore the heritage of modernist architecture (architectural tourism);
- inclusion of buildings in the registers of monuments of the appropriate level (local, state significance)

Examples of successful research and/or restoration of objects of modernist architecture of individual authors is the work of foundations of such masters as Le Corbusier (created in 1968), Mies van der Rohe (1983), Oscar Niemeyer (1988) or the Bauhaus Dessau Foundation (1996) that dedicated to a separate complex, as well as attracting the world community through a series of conferences dedicated to the legacy of modernist architecture in cities such as Gdynia or Tel Aviv. For his long practice, J. A. Steinberg participated in the same architectural competitions in which Le Corbusier and other Western masters participated. His objects were declared as the pure instance of the constructivism architecture. His articles were published in such magazines as "SA"(Modern architecture), "Nova Generatsiya" (New generation), "Budivelnik" (Builder) and others, in which articles of famous domestic and western architects were published. Nevertheless, Ukrainian science has far from fulfilled its duty to the master. Its architectural, scientific, pedagogical and journalistic heritage is still awaiting its collectors, systematizers and researchers, as well as bold programs for the restoration and reconstruction of its architectural objects.

**References**

[1] Khan-Magomedov S O 1996 *Architecture of the Soviet avant-garde* **1** (Moscow: Stroyizdat) 709
[2] Zvonitsky E M and Leibfreyd A Yu 1992 *Gosprom* (Moscow: Stroyizdat) 80

IOP Conf. Series: Materials Science and Engineering **907** (2020) 012023 doi:10.1088/1757-899X/907/1/012023

[3] Kazus I A 2009 *Soviet architecture of the 1920s: design organization* (Moscow: Progress-Tradition) 464
[4] Bachinskaya L G 2010 My teachers *Nowadays problems and architecture and public education* **25** 63-77
[5] Bouryak A P and Kraizer I I 2007 Stylistic attribution of the Dzerzhinsky square (presently Freedom square in Kharkov) *Heritage at risk. Special. The soviet heritage and European modernism* (Berlin) 80-85
[6] Steinberg J A, Malosemoff I I and Milinis I F 1927 Builders' Club, Kharkiv *Sovremennaya arkhitektura* **3** 91
[7] Khan-Magomedov S O 2005 *One hundred masterpieces of Soviet architectural avant-garde* (Moscow: Stroyizdat) 456
[8] Steinberg J A 1928 Recipe for "citizen happiness" *Nova Generatsiya* **2** 127-29
[9] Steinberg J A 1930 From the best construction experience of 1930 *Budivnitstvo* **2**
[10] Smolenskaya S O 2017 *The architecture of avant-garde modernism in Ukraine: genesis and heritage* (text of a doctoral dissertation) Kharkiv National University of Civil Engineering and Architecture Ukraine
[11] Steinberg J A 1931 Overbuilding of the Central Committee building of the CP(b)U *Budivnitstvo* **9** 33-36
[12] Jencks Ch and Silver N 1972 *Adhocism: the Ca se for Improvisation* (New York: Doubleday) 248

# Reactive powder concrete incorporating metakaolin and fly ash for monumental architectural objects

**L Dvorkin[1], V Zhitkovsky[1], N Lushnikova[2] and M Fursovych[3]**

[1] Department of Building Products Technology and Materials Science, National University of Water and Environmental Engineering, Rivne, Soborna st. 11, Ukraine, 33028
[2] Department of Architecture and Environmental Design, National University of Water and Environmental Engineering, Rivne, Soborna st. 11, Ukraine, 33028
[3] Department of Automobile Roads, Bases and Foundations, National University of Water and Environmental Engineering, Rivne, Soborna st. 11, Ukraine, 33028

E-mail: v.v.zhitkovsky@nuwm.edu.ua

**Abstract.** The paper gives coverage on the properties of reactive powder concrete, incorporating metakaolin and fly ash and polyacrylate superplasticizer. There are given results of comparative research of reactive powder concrete (RPC) containing silica fume. There are also considered the impact of curing conditions on the strength of RPC specimens. The combination of fly ash and metakaolin is the effective composite mineral additive that allows to obtain RPC without silica fume. The optimal dosage of metakaolin in composite admixture from the point of maximum flexural and compressive strength was about 10% by cement weight. The curing of the RPC specimens in the hot water at $t=80^0C$ increases compressive strength of concrete at 20-25%. Adding the combination of metakaolin and fly ash with permits to achieve RPC high values of mechanical properties compatible to RPC containing silica fume. High fluidity of fresh concrete along with high performance of hardened concrete allows to apply RPC for manufacturing architectural elements of non-linear geometry, thin structures, and other.

## 1. Introduction

Reactive powder concrete (RPC) is the most effective type of fine-grained concrete, which has increased homogeneity, strength and deformability is developed in France in the 1990s [1]. For such concrete, the typical compressive strength is within the range from 150 to 200 MPa [1, 2]. According to the data of researches [3, 4], such concrete can achieve strength up to 800 MPa under special conditions. As it is shown by the many researches [1-3], RPC is also characterized by a high crack resistance, which is characterized by the higher ratio of flexural strength to compressive strength.

Significant progress in RPC technology at the present stage has become possible mostly due to the development of production and use of chemical and mineral admixtures, in particular superplasticizers and silica fume [1, 5]. Fine powders such as ground quartz and silica fume with a particle size 0.1-600μm are applied for RPC manufacturing [6-8]. The grain distribution of different powders is optimized to achieve the increased density of the concrete matrix [7, 9, 10]. High-range water reducers permit to reduce the water-cement ratio (W/C) from 0.4… 0.5 to 0.2 [9].

The effectiveness of the RPC is confirmed by data on its use for different types of specific structures. RPC is applied in construction of roads [11, 12] and bridges [6, 7, 8, 9], as well as fortification structures and many other projects [4, 9, 13]. Due to its increased strength, durability and radiation resistance, it can be used as a reliable material for containers of radioactive wastes from nuclear power plants [14]. It is also used for thermal protection of building structures, as it provides better fire and heat resistance than ordinary high-strength concrete. High energy absorption provides greater structural reliability of RPC in earthquakes [15].

Despite the fact that the production cost of RPC is generally higher than that for ordinary concrete, there are some economic advantages in its application. Ultra-high mechanical characteristics of RPC provide reduction in thickness of concrete elements, which leads to materials savings and the cost of manufacturing [16]. RPC has advantages as an architectural concrete, especially at the construction of monumental structures and elements of non-linear geometry with high requirements for physical and mechanical properties. Due to the high workability, RPC improves shaping properties, which allows to make both thin-walled structures and to create a variety of complex architectural forms with high surface quality. Dispersed reinforcement of short steel fibers can replace completely or partially traditional reinforcing elements. Technology of RPC involves the introduction of high-range water reducers and highly active dispersed mineral admixtures, mainly silica fume [5]. Under modern circumstances, silica fume is a scarce component and its application makes some difficulties with its transportation and dosing. There are known the research results of influence of alternative mineral admixtures on properties of special types of concrete: UHPC incorporating very fine fly ash or metakaolin in replacement of silica fume [17], RPC containing fly as mineral admixture [18], fly ash and slag as alternative silica containing mineral admixtures for RPC [19, 20]. The research data on determination the curing conditions on RPC with alternative mineral admixtures are given in [21, 22, 23]. However, there is a limited data on the application of fly ash in combination with metakaolin in RPC and the impact of terms of curing on mechanical properties of such concrete.

There are research results describing the application of alccofine – a new generation micro fine supplementary cementitious material for HSC and RPC. According to the data its effectiveness of alccofine on workability and compressive strength of RPC is higher than silica fume [24], as well as split tensile strength [25]. However, this admixture is rather expensive and currently unavailable in Ukraine.

## 2. Aim and scope of the research

The aim of the research was to estimate the possibility of partial or complete replacement of silica fume as part of the RPC with more available mineral admixtures. The influence of fly ash and metakaolin as local mineral admixtures, on mechanical properties of RPC under different terms of hardening was under investigation.

As fly ash has low pozzolanic activity and low water demand, whereas metakaolin has high pozzolanic activity and high water demand, it was suggested that the combination of these admixtures will have a synergetic effect on properties of RPC. There had been determined the terms of thermal treatment of RPC (type of thermal treatment and temperature of isothermal heating). The comparative study of influence of silica fume on compressive and flexural strength of RPC was considered.

## 3. Materials and methods

Portlan Cement CEM I 42,5R, fine quartz sand with grain size 0.16-1.25 mm were used. Polyacrylate superplasticizer Mapei Dynamon SP3 was added in all the batches of the experiment. As mineral admixtures metakaolin, fly ash and silica fume were applied (see table. 1).

W/C ratio of fresh concrete was determined by the required fluidity at Suttard viscometer equal to diameter of flow spread 25…30cm. For each batch cube specimens with 10 cm edge and prisms 4×4×16 cm in size were prepared.

**Table 1.** Mineral admixtures used in the research.

| Mineral admixture | Specific surface area (SSA), $cm^2/g$ | Manufacturer |
|---|---|---|
| Metakaolin (MK) | 10,125 | LLC "Western Kaolin Company" (Ukraine) |
| Fly ash (FA) | 2,527 | Burshtyn coal-fired power plant (Ukraine) |
| SikaFume-HR/TU | 23,158 | Sika |

The compressive and flexural strength at the age of 1, 7, 28 days of hardening under normal conditions and under steam curing and hot water saturation was determined. For the selection of effective compositions of reactive powder concrete, $C_f$ coefficient was calculated. It characterizes the specific consumption of cement per compressive strength unit. Y. Bazhenov and V. Kalashnikov used this coefficient [26] to determine the effectiveness of the compositions of reactive powder concrete

$$C_f = \frac{C}{f_{cm}}\ ,\ kg/MPa \qquad (1)$$

where $C$ is the cement content per 1 $m^3$ of concrete, kg; $f_c$ - compressive strength of concrete, MPa.

## 4. Results and Discussion

### *4.1. Determination the impact of type and dosage of mineral admixtures on strength and $C_f$ coefficient*

At the first stage of the research there was considered the influence of the type of mineral admixtures at equal dosages on compressive and flexural strength and effectiveness coefficient of reactive powder concrete.

Concrete compositions and results of the experiment are given in table 2 and figure 1. As can be seen from the figure 1b and table 2, specimens of RPC made of high-flowable mixture with a maximum content of Sika Fume-HR/TU 360 kg/$m^3$ have compressive strength 162.4 MPa at the age of 28 days of normal hardening. The results meet the known data [4, 30]. Partial or complete replacement of silica fume with other investigated admixtures makes it possible to obtain concrete with strength values in the range of 102-133 MPa. Such values of strength, while ensuring high fluidity of fresh concrete, indicate that the RPC can be characterized as material with high workability and high performance under conditions of various loads occurring at performance of the building structures.

The flexural strength for all the specimens varied within the range from 18.3 to 32.2 MPa (figure 1a).

As is known [28], the ratio of flexural strength to compressive strength characterizes the resistance of concrete to the formation of cracks and its deformability. For ordinary concretes, the ratio $f_t$ to $f_c$ is usually within the range of 0.12 - 0.15 and for reactive powder concrete it is equal to 0.16 - 0.2. Therefore, the increased crack resistance and deformability of RPC is a determining property that allows to recommend this concrete for structures operating under conditions of high dynamic loads.

In the case of the fly ash use, the maximum strength was achieved by increasing the content of mineral additives up to 360 kg /$m^3$ at a later age. As is known [29], due to the high consumption of cement, a significant amount of calcium hydroxide is released in hardened cement matrix of RPC. Mineral admixtures interact actively with it to form water insoluble compounds. Fly ash, in its turn, exhibits some plasticizing effect due to the peculiar shape of the grains [30]. It allows to keep low value of W/C ratio and dense structure of cement stone even at high cement consumption.

The use of highly active admixtures of silica fume and metakaolin proved to be the most effective from the point of compressive strength values. The addition of metakaolin leads to maximum compressive strength value 133.4 MPa and showed slightly less efficiency than silica fume (162.4 MPa). The kinetics of the strength of the RPC (figure 1) corresponds to the known data [27].

**Table 2.** Impact of mineral admixtures on the flexural and compressive strength of reactive concrete.

| No. | Cement (kg/m³) | Quartz sand 0.16… 1.25 mm (kg/m³) | Mineral admix-ture, (kg/m³) | Water, (l/m³) | W/C | Flow spread dia-meter, cm | Concrete strength (MPa) / Effectiveness parameter (kg/MPa) | | | | | |
|---|---|---|---|---|---|---|---|---|---|---|---|---|
| | | | | | | | $f_{c.tf}^1$ | $f_{cm}^1$ / $C_{f1}$ | $f_{c.tf}^7$ | $f_{cm}^7$ / $C_{f7}$ | $f_{c.tf}^{28}$ | $f_{cm}^{28}$ / $C_{f28}$ |
| **SikaFume-HR/TU (SF) (SSA=23,158cm²/g)** | | | | | | | | | | | | |
| **1** | 1080 | 1200 | 120 | 240 | 0.22 | 30 | 10.1 | 47 / 22.9 | 25.6 | 95 / 11.4 | 29.7 | 143 / 7.6 |
| **2** | 840 | 1200 | 360 | 240 | 0,29 | 12 | 12.7 | 43,3 / 19.4 | 28.2 | 125.5 / 6.7 | 32.2 | 162.4 / 5.2 |
| **Fly Ash (FA) (SSA=2,527 cm²/g)** | | | | | | | | | | | | |
| **3** | 1080 | 1200 | 120 | 240 | 0,22 | 30 | 6.46 | 31.5 / 34.2 | 19.22 | 70.3 / 13.7 | 23.1 | 100.9 / 9,5 |
| **4** | 840 | 1200 | 360 | 240 | 0,29 | 30 | 4.76 | 22.8 / 36.8 | 15.22 | 68.5 / 12.2 | 18.3 | 114.5 / 7.34 |
| **Metakaolin (MK) (SSA=10,125 cm²/g)** | | | | | | | | | | | | |
| **5** | 1080 | 1200 | 120 | 240 | 0,22 | 30 | 9.8 | 48.1 / 22.5 | 22.7 | 103.1 / 10.5 | 26.6 | 133.4 / 8.1 |
| **6** | 840 | 1200 | 360 | 270 | 0,33 | 10 | 7.8 | 27.2 / 30.9 | 17.2 | 78.5 / 10.7 | 20.1 | 102.4 / 8.2 |

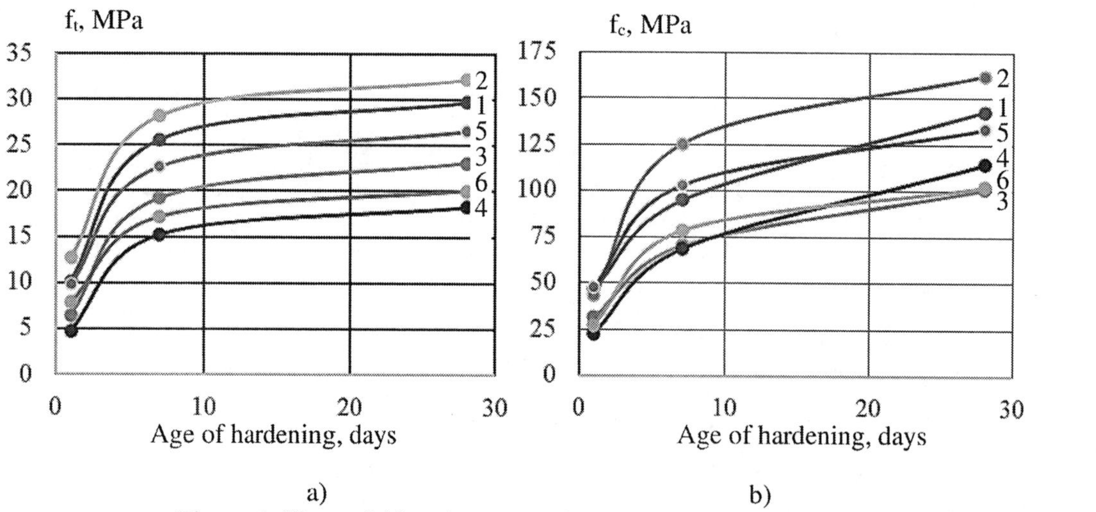


**Figure 1.** Flexural (a) and compressive strength (b) kinetics of RPC depending on the type and dosage of mineral admixture (# of batches according to table 2)

Specimens containing silica fume (360 kg/m³) and metakaolin (120 kg/m³) had maximum rate of hardening. It should be mentioned that the specimens reached values of compressive strength after 1 day of hardening from 23 to 48 MPa. This strength can be considered sufficient for high-speed construction.

The nature of the effect of different types of mineral admixtures on the flexural strength of RPC (table 2) is the same as for the compressive strength. Composites containing silica fume, have the maximum flexural strength from 29 to 32 MPa, composites containing metakaolin and fly ash has slightly lower values (26 - 27 MPa for metakaolin).

The maximum efficiency of the metakaolin is observed at its consumption of about 10% by binder weight, this result confirms the data obtained by us earlier in the study of high-strength concrete [28]. A further increase in the amount of metakaolin leads to the increase in water consumption of the concrete mixture and an increase in W/C, while the strength of concrete decreases (figure 1).

The compositions containing the raised amount of Sika Fume-HR/TU (table 1) are considered the most effective from the position of the 28-day strength per kilogram of cement. RPC with the admixtures of fly ash and metakaolin in higher quantities in terms of efficiency of cement are equal to RPC with lower content of silica fume (120 kg/m$^3$). Thus, considering the established effectiveness of fly ash and metakaolin for RPC, as well as the difference in mechanism of their action on the properties of fresh and hardened concrete, we can expect their mutual effect on the properties of RPC.

*4.2. Determination the impact of metakaolin dosage in composite fly-ash -metakaolin admixture*

At the second stage there was determined the effect of combination of fly ash and metakaolin on RPC strength properties. When performing a series of studies, the composition of RPC was as follows: cement - 840 kg/m$^3$, mineral admixture - 360 kg/m$^3$, quartz sand 0.16-1.25 mm - 1,200 kg/m$^3$, the consumption of superplasticizer polyacrylate type Dynamon SP-3 - 2% by the weight of the binder (cement + mineral admixture). Water consumption was determined to provide flow-spread diameter 25... 30 cm at the Suttard viscometer. As a mineral admixture, fly ash was used; in addition, metakaolin additives in the amount of 5, 10 and 15% by weight of cement were added to the concrete.

The obtained results (table 3, figure 2) indicate that additional introduction of metakaolin leads to increasing RPC strength: compressive strength of concrete increases up to 122... 126 MPa. However, increasing in metakaolin content by more than 10% by weight of cement is inefficient due to a significant increase in the viscosity of the mixture and subsequent increase in water consumption. The positive effect of fly ash on the flowability of fresh concrete compensates to some extent with the increase in viscosity inherent in metakaolin [27].

The effectiveness of the compositions of mineral admixtures from the standpoint of flexural strength was similar. The maximum values of $f_t$ (23-25 MPa) are achieved when combining 10% of metakaolin by weight of cement with fly ash.

**Table 3.** Research results of metakaolin and other mineral admixtures on strength properties of reactive powder concrete.

| No | Content of MTK (%) by cement mass | W/C | Flow spread diameter (cm) | Concrete strength (MPa) depending on the age of hardening | | | | | |
|---|---|---|---|---|---|---|---|---|---|
| | | | | $f_{c.tf}{}^{1}$ | $f_{cm}{}^{1}$ | $f_{c.tf}{}^{7}$ | $f_{cm}{}^{7}$ | $f_{c.tf}{}^{28}$ | $f_{cm}{}^{28}$ |
| **1** | 5 | 0.29 | 30 | 7.21 | 21.0 | 17.09 | 75.8 | 22,77 | 115.6 |
| **2** | 10 | 0.29 | 30 | 7.65 | 29.2 | 18.87 | 79.9 | 25.28 | 126.2 |
| **3** | 15 | 0.29 | 30 | 7.12 | 20.1 | 16.91 | 73.1 | 21.81 | 111.0 |

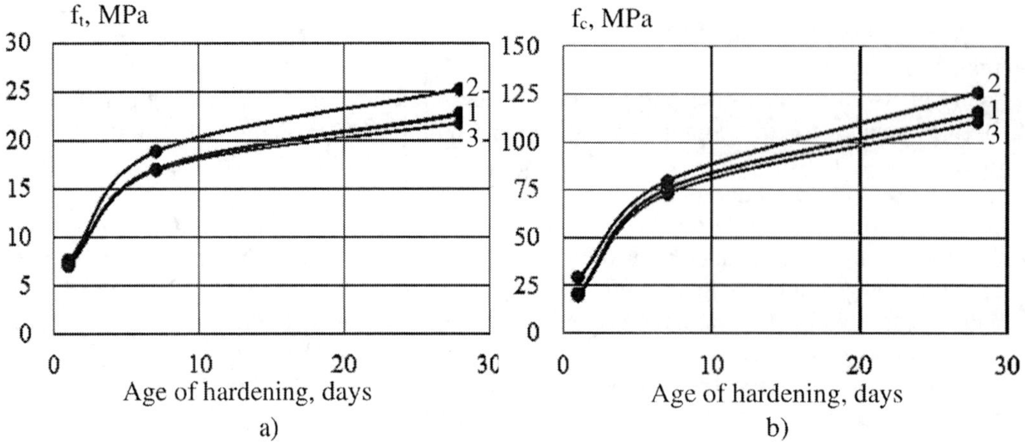


**Figure 2.** Flexural (a) and compressive strength (b) kinetics of RPC depending on the metakaolin content (# of batches according to table 3)

*Influence of curing mode on the mechanical properties of RPC*

In order to determine the effect of heat treatment on the curing process and the strength of RPC containing fly ash and metakaolin, the third stage of experiments was conducted. The content of metakaolin was 10% by weight of binder. Steam curing was carried out in the laboratory steaming chamber according to the following mode: preliminary aging - 2 hours; temperature rise at a rate of 25°C/h; isothermal heating - 8 hours, cooling under normal conditions. Compressive strength of the samples was determined after 2 h of steam curing and at the age of 7 and 28 days. Along with steam curing there are research results, where maximum hydration and pozzolanic reactivity in RPC was obtained at hardening the specimens in the hot water [31]. To determine the curing efficiency of the RPC compositions, the specimens were cured in water at a temperature of 50°C and 80°C. The results of determination the strength are given in table 4 and figure 3.

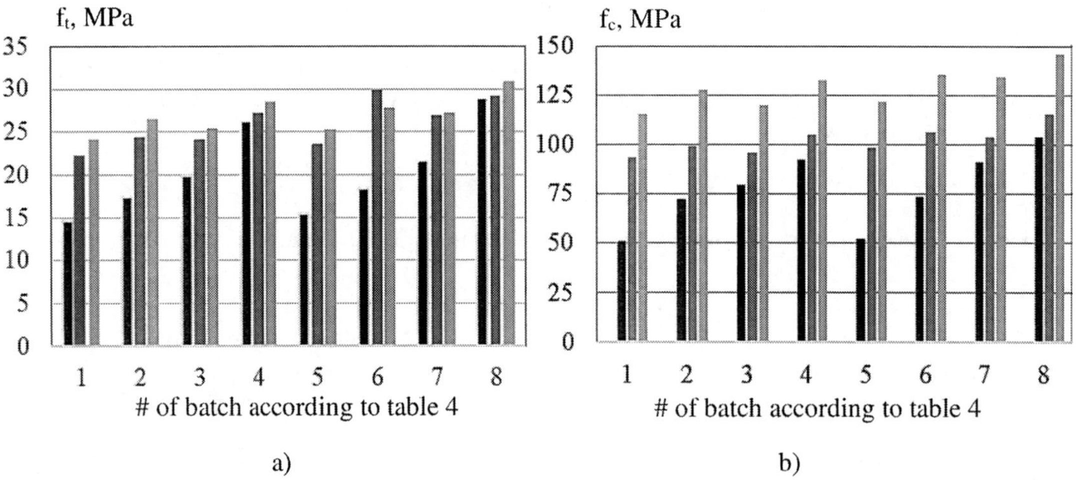


**Figure 3.** Flexural (a) and compressive strength (b) of RPC depending on the curing conditions and composition of mineral admixture (# of batches according to table 4)

after steam curing · at 7 days of normal hardening
at 28 days of normal hardening

**Table 4.** Impact of steam curing on properties of RPC.

| No. | Type of mineral admixture (kg/m³) | W/C | Flow spread diameter (cm) | Concrete strength (MPa) depending on the age of hardening | | | | | |
|---|---|---|---|---|---|---|---|---|---|
| | | | | $f_{c.tf}^{SC}$ | $f_{c.tm}^{SC}$ | $f_{c.tf}^{7}$ | $f_{c.tm}^{7}$ | $f_{c.tf}^{28}$ | $f_{c.tm}^{28}$ |
| **Steam curing** | | | | | | | | | |
| **Isothermal heating, t=50⁰C** | | | | | | | | | |
| **1** | Fly ash (FA) | 0.29 | 30 | 14.6 | 51.2 | 22.5 | 93.7 | 24.1 | 115.6 |
| **2** | Fly ash (FA)+ metakaolin (MTK) | 0.29 | 30 | 17.4 | 72.4 | 24.40 | 99.2 | 26.5 | 127.6 |
| **Isothermal heating, t=80⁰C** | | | | | | | | | |
| **3** | Fly ash (FA) | 0.29 | 30 | 19.9 | 79.7 | 24.1 | 96.1 | 25.4 | 120.0 |
| **4** | Fly ash (FA)+ metakaolin (MTK) | 0.29 | 30 | 26.2 | 92.6 | 27.2 | 105.3 | 28.5 | 132.6 |
| **Water curing** | | | | | | | | | |
| **Isothermal curing, t=50⁰C** | | | | | | | | | |
| **5** | Fly ash (FA) | 0.29 | 30 | 15,4 | 52,2 | 23,5 | 98,4 | 25.2 | 121.7 |
| **6** | Fly ash (FA)+ metakaolin (MTK) | 0.29 | 30 | 18,4 | 73,6 | 29,9 | 106,2 | 27.7 | 135.8 |
| **Isothermal curing, t=80⁰C** | | | | | | | | | |
| **7** | Fly ash (FA) | 0.29 | 30 | 21.6 | 91.0 | 26.9 | 103.9 | 27.2 | 134.4 |
| **8** | Fly ash (FA)+ metakaolin (MTK) | 0.29 | 30 | 28.9 | 103.9 | 29.1 | 115.7 | 30.8 | 145.9 |

Steam curing of RPC at a temperature of 50°C causes an increase in compressive strength for 1 day on average by 2.5… 3 times, for 28 days - by 1.5… 2% comparing to normal hardening. Increasing the temperature of isothermal heating clearly causes an increase in the strength of the RPC. Compared to the specimens hardened under normal conditions, steam curing at 80 °C causes an increase in strength for 1 day in 3.5… 4 times, for 28 days - by 4… 5%. As is known [32], heat and moisture treatment significantly activate pozzolanic activity of various mineral admixtures. The obtained results show that the use of heat steaming significantly increases the strength caused by the addition of fly ash (figure 3). At 50-80 °C temperature range, heat treatment accelerates the interaction of calcium hydroxide with amorphous silica of fly ash. The positive effect of steaming is also reflected in the flexural strength (figure 3a). The maximum values of the strength for RPC specimens were obtained after curing in water with a temperature of 80° C (table 4). RPC containing composite mineral admixture containing metakaolin 10% and fly ash (10%) reached a compressive strength of 145 MPa at the age of 28 days of hardening, flexural strength - 30.8 MPa, which is 15… 17% higher than the strength of concrete of the same composition hardened under normal conditions.

## Conclusions

The results of determination the influence of different types of mineral admixtures (highly reactive silica fume and metakaolin and low-reactive fly ash) and their compositions on the compressive and flexural strength of RPC and kinetics of strength growth prove the possibility to obtain concrete with compressive strength values in the range of 100 - 133 MPa.

It is established that the combination of metakaolin with fly ash is an effective composite mineral admixture that allows to replace silica fume in RPC. The compressive strength of such a composition under normal curing conditions is – 124 - 126 MPa, and after keeping in water at a temperature of 80 °C it reaches 145 MPa. The synergistic effect of this composite admixture is explained by the positive effect of spherical ash-removal particles on the workability of concrete mixtures, that to some extent compensates for the increase in viscosity caused by adding metakaolin. High flowability of fresh concrete along with high mechanical properties permit to apply RPC as architectural concrete, in particular for large-scale objects.

## References

[1] Al-Azzawi A A, Abdulsattar R, Al-Shaarbaf I 2018 *A* State-of-the-Art Review on Reactive Powder Concrete Slabs *Int. J. of App. Eng. Res.* **13** (1) pp 761-68

[2] Bhusari J, Gumaste K S 2017 Characterization of Reactive Powder Concrete for its mechanical properties *Int. J. of Civil Engineering and Technology* **8** (5) pp 8-13

[3] Reda M M, Shrive N G, Gillott J E, 1999 Microstructural investigation of innovative UHPC *Cem. Concr. Res.* **29** pp 323-329

[4] Richard P and Cheyrezy M 1995 *Composition of reactive powder concrete, Cem. and Concr. Res.* **25**(7) pp 1501-1511

[5] Chan Y-W, Chu S-H 2004 Effect of silica fume on steel fiber bond characteristics in reactive powder concrete *Cem. and Concr. Res.* **34** pp 1167-1172

[6] Shaheen E and Shrive N G 2006 Optimization of Mechanical Properties and Durability of Reactive Powder Concrete *ACI Mater. J.*, Vol 103, No **6**, pp 444-451

[7] Richard P and Cheyrezy M H 1994 Reactive Powder Concretes with High Ductility and 200-800MPa Compressive Strength *SCI SP* **144** pp 507-518

[8] Rebentrost M and Cavill B. 2006 Reactive Powder Concrete Bridge *AustRoads Conference* pp 1- 11

[9] Cyr M F and Shah S P 2002 Advances in concrete technology *Adv. in Build. Tech.* Vol. **1** pp 17-27

[10] Bonneau O, Lachemi M, Dallaire E, Dugat J, and Aitcin P C 1997 Mechanical Properties and Durability of Two Industrial Reactive Powder Concretes *ACI Materials Journal* Vol. 94 No. **4** pp 286-290

[11] Hattatoglu F, Bakis A 2017 Usability of ignimbrite powder in reactive powder concrete road pavement *Road Materials and Pavement Design* 18:**6** pp 1448-1459

[12] Cwirzen A, Penttala V and Vornanen C 2008 Reactive powder based concretes: Mechanical properties, durability and hybrid use with OPC Cem. and Concr. Res. Vol. **38** pp. 1217-1226

[13] Blais P Y and Couture M 1999 Precast, Prestressed Pedestrian Bridge – World's First Reactive Powder Concrete Structure *PCI Journal* May -June pp 60–71

[14] Gökçe H S, Yalçınkaya Ç, Tuyan M 2018 Optimization of reactive powder concrete by means of barite aggregate for both neutrons and gamma rays *Constr. and Build. Mat.* Vol. **189** pp 470-477

[15] Lee N P and Chisholm D H 2005 Study Report Reactive Powder Concrete *BRANZ* Vol. **146** pp 1- 29

[16] Dowd W 1999 Reactive Powder Concrete – Ultra-High Performance Cement Based composite, NOVA award nomination, *Construction Innovation Forum* (Electronic Materials)

[17] Ultra High Performance Concrete (UHPC*)* 2004 *Proceedings of the International Symposium on Ultra High Performance Concrete* Kassel, Germany September 13-15, ed. by M Schmidt, E Fehling, C Geisenhanslüke, Kassel University Press

[18] Kakad P R, Gaikwad G B, Hetkale R R, Kolekar D S, Paul Y 2015 Reactive Powder Concrete Using Fly Ash, *Int. J. of Eng. Trends and Tech.* (IJETT), Vol. **22**(8) pp 380-383

[19] Yazıcı H, Yiğiter H, Karabulut A S, Baradan B 2008 Utilization of fly ash and ground granulated blast furnace slag as an alternative silica source in reactive powder concrete *Fuel* Vol. **87** (12) pp 2401-2407

[20] Peng Y, Hu S and Ding Q 2010 Preparation of reactive powder concrete using fly ash and steel slag powder *J. Wuhan Univ. Technol.-Mat. Sci. Edit.* 25 349–354

[21] Yazıcı H, Yardımcı M Y, Aydın S, Karabulut A S 2009 Mechanical properties of reactive powder concrete containing mineral admixtures under different curing regimes, *Constr. and Build. Mat.* Vol. **23**(3) pp 1223-1231

[22] So H, Jang H, Khulgadai J et al. 2015 Mechanical properties and microstructure of reactive powder concrete using ternary pozzolanic materials at elevated temperature. *KSCE J Civ Eng* **19**, 1050–1057

[23] De Hong W, Ju Y Z, and Zheng W Z 2014 Mechanical Properties of Reactive Powder Concrete Containing Fly Ash under Different Curing Regimes *Appl. Mech. and Mat.* **597** pp 320–323

[24] Patel D, Patel I, Shah J 2018 A Study on Properties of RPC With Various Silica Fume And Alccofine Content *JETIR* Vol. **5**(11) pp 589-595

[25] Kumar S, Acharya G, Mhamai S R K 2015 Reactive Powder Concrete with mineral admixtures Vol **2**(6) 1749-1757

[26] Bazhenov Y M, Demyanova V S, Kalashnikov V I 2006 *Modified high-quality concrete.* (Moscow, Association of Civil Engineering Universities) p 368

[27] Collepardi M, Collepardi S, Troli R and Coppola L 2003 Innovative Concretes (SCC, HPC and RPC) in the Field of Architectural, Civil and Environmental Engineering *Proceedings of the Workshop on New Technologies and Material in Civil Engineering, Milan* pp 1-8

[28] Dvorkin L I, Lushnikova N V, Runova R F, Troyan V V 2007 *Metakaolin in construction mortars and concrete* (Kyiv, KNUCA Publishing House) p 214

[29] Ng K M, Tam C M and Tam V W Y 2010 Studying the production process and mechanical properties of reactive powder concrete: A Hong Kong study *Magazine of Concr. Res.* **62**(9)

[30] Dvorkin L I, Solomatov V I, Vyrovoy V N 1991 *Cement concrete with mineral fillers* (Kiev Budivelnyk) p 136

[31] Morsy M S, Galal A F and Abo-El-Enein S A 1998 Effect of temperatures on phase composition and microstructure of artificial pozzolana-cement pastes containing burnt kaolinite clay *Cem. and Concr. Res.* Vol. **28** pp 157–160

[32] Dvorkin L I 1988 Effect of active admixtures in plasticized cement-based concrete *Proceedings of Universities. Civil Engineering and Architecture* **9** pp. 53-57

IOP Conf. Series: Materials Science and Engineering **907** (2020) 012025 doi:10.1088/1757-899X/907/1/012025

# Design features of low-rise buildings walls erected from permanent formwork ComBlock units

**K V Spirande[1], Yu V Bondarenko[1], M V Iakymenko[1], R M Shemet[1] and V V Bielka[2]**

[1]Department of reinforced concrete and stone structures, Kharkiv National University of Civil Engineering and Architecture, 40 Sumska Str., 61002 Kharkiv, Ukraine
[2]Building and Production Company AVcom Ltd., 17 Pashchenkivska Str., 61017 Kharkiv, Ukraine

karina5516@ukr.net

**Abstract.** The study focuses on the design of load-bearing monolithic concrete and reinforced concrete walls of low-rise buildings erected from permanent formwork ComBlock units for construction, in areas without episodic influences and analysis of their work. While developing the recommendations the main attention was paid to working out design solutions of load-bearing monolithic walls with minimal reinforcement consumption. The guidelines contain general provisions for the design and calculation of load-bearing vertical elements of a house; numerous studies of the stress-strain state of monolithic vertical load-bearing elements of low-rise buildings with different structural schemes.

## 1. Introduction

Nowadays, a tendency of expanding construction of individual low-rise housing takes place in many countries including Ukraine. Structural elements of monolithic concrete and reinforced concrete are often used in their design and construction. Buildings with energy-saving exterior walls have become of great demand. Modern construction technologies are aimed at simplifying and accelerating the construction process.

One of the possible ways to solve these problems is to use permanent formwork with high thermal insulation properties for the construction of such buildings. Kharkiv building and production company AVcom is engaged in the construction of such houses with ultra-low energy consumption with the aid of the permanent formwork ComBlock.

The increase of economic efficiency of these structures is related to the reducing of their metal content.

The load-bearing vertical structures of low-rise buildings with a rigid structural scheme feature relatively small vertical stresses in the concrete walls provided tensile forces. They do not exceed the tensile strength of concrete. Hence, the installation of assumed reinforcement is not necessary. Furthermore, the objective of minimization of used constructive reinforcement becomes of current relevance.

The problem of developing design solutions and methods of calculating the load-bearing structures of the above-mentioned buildings with minimum consumption of steel reinforcement has been solved within this study.

## 2. Analysis of recent research and publications

Permanent formwork is special units fabricated from different materials. This technology cardinally simplifies the process of monolithic construction, and the building itself does not require additional insulation. There are several known variants of this type of formwork, which use slabs of fibre mat, wood fibre concrete, cement-bonded chipboards or other materials, in which cement is comprised as a component.

Energy-efficient wood fibre concrete blocks of permanent formwork have become widespread, namely: BRISOLIT (Germany, 1934), DURISOL (Switzerland, 1932), TECOLIT (Russia, 2014).Wood chips are used for their manufacture. They are mixed in certain proportions with cement and mineral additives, and expanded polystyrene or PIR (polyisocyanurate) [1, 2, 3]. In 1956, the family company VELOX WERK patented a similar technology of monolithic construction in permanent formwork of VELOX chipboards [1] in Austria. The system fully corresponds to all modern standards, consumer requirements and has a fairly high competitiveness.

In the sixties of the last century, the Austrian engineer proposed to make a permanent formwork from expanded polystyrene in the form of hollow blocks. It is possible to make a wall like from children's bricks, and then to fill in internal cavities with concrete. After the concrete hardens, a wall in the form of a sandwich consisting of concrete, dressed in a heat-insulating "shirt" of expanded polystyrene, is obtained. Such a wall requires additional insulation no longer. Since the invention of expanded polystyrene blocks of permanent formwork have constantly been modernized, introduced into production by various companies in many countries: PLASTBAU (Switzerland) [1]; Ecostone (Finland); IntegraSpec (Canada) [4]; Isodome (Poland) [5, 6]; Plastbau and Thermoblock (Ukraine) and others [7, 8].

Insulating concrete formwork industry continues to strive for improvement and creates exceptionally energy-efficient, firm and environmentally friendly structures due to many years' experience and innovation.

The analysis of the existing similar permanent formworks and technologies has proved that there are no unified regulatory documents for designing houses with their aid. Therefore, it became necessary to develop methods of calculating and designing the walls of low-rise buildings erected from blocks of permanent formwork ComBlock.

## 3. Results of theoretical studies

Kharkiv building and production company AVcom is engaged in the manufacture of permanent formwork ComBlock units with adjustable width of the concrete core of the wall (Figure 1). This unit consists of a panel of external insulation (thickness from 50 to 300 mm) and an internal gypsum panel (which does not require further plastering) (Figure 1, a) or a removable inner panel of the formwork (Figure 1, 2).

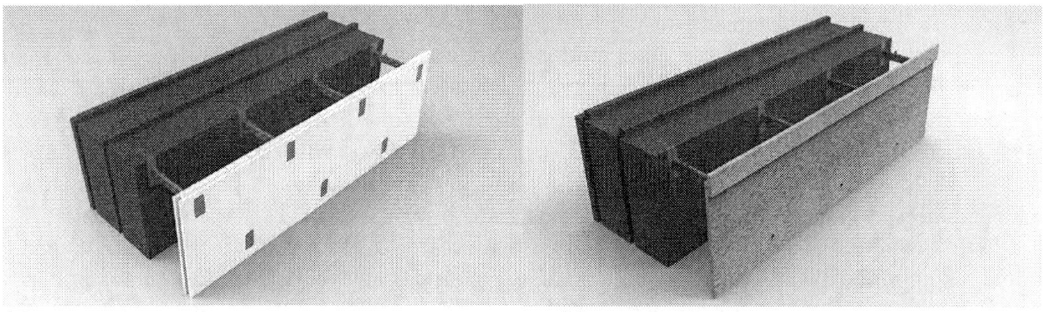

(a) (b)

**Figure 1**. ComBlock with gypsum panel (a) and with removable formwork panel (b)

The panels are interconnected by adjustable plastic screeds (jumpers), which allow changing the thickness of the concrete core of the wall in the range from 100 to 430 mm [9]. Walls built with ComBlock have significant advantages over walls made of other materials, for example, cheaper than brick - by 30%, aerated concrete - by 10%; they are more environmentally friendly and fireproof than walls made of thermo blocks.

After the abolition of state building codes DBN V.2.6-6-95 in 2015, which specified the standards for the design, construction and operation of the construction system "Plastbau" (similar to ComBlock), no regulatory framework exists currently in Ukraine. The need of developing guidelines for the design of buildings using permanent formwork ComBlock is especially important. The need is relevant under the conditions of the introduction of new regulations DBN V.2.6-98:2009 (National Building Standards) and DSTU B V.2.6-156:2010 (National Standards of Ukraine) for the design of heavy concrete and reinforced concrete structures developed on the basis of Eurocode 2 EN 1992-1-1:2004 in our country.

The authors performed numerous studies of the stress-strain state of load-bearing vertical elements using permanent formwork ComBlock units. There were developed recommendations for the design of load-bearing concrete and reinforced concrete structures of frameless (wall) and frame-wall monolithic residential buildings up to 5 floors for construction in areas with normal soil conditions in non-seismic areas.

### *3.1. Basic design requirements*

It is known that the load-bearing structures are calculated for the first and second groups of limit states, taking into account the most unfavourable combination of loads. As a rule, the deformation (stiffness) characteristics of reinforced concrete elements are taken on the basis of methods that consider the nonlinear operation of structural materials, when determining the forces in the load-bearing elements of structures and horizontal displacements of the system.

The values of nonlinear stiffness of reinforced concrete elements are taken depending on the stage of design, calculation requirements and the nature of the stress-strain state of the element. The calculation of the structural system is performed for all successive stages of construction (in case of a significant change in the design situation) and for the stage of operation. At the first stage of calculation of the structural system, whereas reinforcement of concrete elements is unknown, the nonlinear operation of the elements should be estimated by reducing their stiffness by means of hypothetical generalized coefficients or using a "fictitious" modulus of concrete deformation. At the following stages of calculation of the structural system, when the installation of reinforced concrete elements is known, the specified values of elements stiffness are introduced into the calculation. These values are determined taking into account reinforcement, crack formation and development of inelastic deformations in concrete and reinforcement according to DBN V.2.6-98:2009 and DSTU B V.2.6-156:2010.

At the first stage of calculation, it is allowed to take approximate values of the elements stiffness in order to estimate the forces in the elements of the structural system. It is taken into account that the distribution of forces in the elements of structural systems depends not on the value but mainly on the stiffness ratio of these elements. It is recommended to take the specified values of stiffness in accordance with DSTU B V.2.6-156:2010 for more accurate assessment of the distribution of forces in the elements of the structural system. A significant reduction in stiffness in the bending plate elements (because of the possible formation of cracks) should be considered in comparison with the eccentrically compressed elements. At the first approximation, it is advisable to take the modulus of elasticity of concrete $E_c$, which is equal to its calculated value in accordance with table 3.1 of DSTU B V.2.6-156:2010.

At the following stages of calculation at known reinforcement, the specified stiffness of elements is accepted taking into account reinforcement, existence of cracks and inelastic deformations in concrete and reinforcement. The values are defined according to the operating regulatory documents.

In addition, it is recommended to accept rational constructive parameters of walls, which are established on the basis of the technical and economic analysis when designing. For low-rise buildings with permanent formwork ComBlock the advisable thickness of monolithic walls should not be less than 150 mm, concrete class - not less than C15/16, the amount of reinforcement - in accordance with DSTU B V.2.6-156:2010.

*3.2. Calculation and construction features of buildings with bearing concrete walls*

The calculation of bearing capacity of concrete elements under the action of longitudinal compressive force and bending moment should be performed for sections that are normal to their longitudinal axis.

In general case, the calculation of the reinforced concrete structure is performed on the basis of diagrams of concrete and reinforcement deformation and nonlinear deformation model. It is allowed to perform the calculation of concrete elements by the simplified method. In this case, the ultimate deformations of stretched concrete are used as a criterion for failure. The use of simplified diagrams of "stress-strain" dependence is possible if they are equivalent or more conservative than those given in paragraph 3.1.4.1 DBN V.2.6-98:2009.

Concrete elements are calculated according to the ultimate forces, which are determined on the basis of the following preconditions: the sections after deformation remain flat; equations describing a complete diagram of concrete deformation are considered valid. It is allowed to use a two-line diagram of concrete deformation with a limited length of the horizontal branch; the largest elongation of the extreme stretched fibre equals – $2f_{ctm}/E_{ck}$.

The influence of longitudinal bending and random eccentricities should be taken into account, when calculating the eccentrically compressed concrete elements.

As for the walls made in compliance with the corresponding construction and care regulations, temperature and shrinkage deformations can be neglected. In the absence of measures to prevent tensile fracture of cross section, it is necessary to limit the maximum eccentricity of the force application in the section in order to avoid the formation of large cracks.

Numerous researches of a stress-strain condition of bearing and self-bearing walls from permanent formwork ComBlock units of various constructive schemes have been performed on the basis of the offered various architectural and design solutions of buildings. Gradual calculation of load-bearing walls (columns) of buildings has been done using software complexes LIRA-SAPR and MONOMAKH-SAPR.

*3.3. Numerical studies of stress-strain state of walls of low-rise buildings erected from blocks of permanent formwork ComBlock*

Estimated schemes of frameless buildings are classified [10]: by the design nature of spatial work- in to one-, two- and three-dimensional; by the type of unknowns - into discrete, discrete-continuum and continuum; by the type of construction, which is the basis of the calculation scheme – into rod, plate, combined ones.

The forces acting in the wall structures and floors should be determined on the basis of spatial calculation schemes. In this regard, the nature of external loads application, the characteristics of calculated system, and the required accuracy of the calculation ought to be taken into account. Spatial calculation schemes allow to determine the forces in the structures and their displacement from external loads in any direction. The three-dimensional calculation scheme takes into account the peculiarities of the interaction of load-bearing structures most accurately, but the calculation based on it is the most complex.

Discrete computational models based on mathematical and geometric sampling of spatial structures calculated by the finite element method (FEM) became the most widespread. Sampling parameters and types of finite elements are accepted with regarding the requirements for the accuracy of the model reproduction of the real structure.

The spatial structural system is a statically indeterminate system. Sampling of structural schemes is advisable to perform with the aid of shell and rod finite elements used in the calculation complex.

When constructing a finite-element calculation model, the dimensions and configuration of finite elements are assigned on the basis of the possibilities of the specific calculation programs used. They are accepted so that the required accuracy in determining forces applied to the length of columns and area of floor slabs, foundations and walls should be provided. Here with, the total number of finite elements in the calculation scheme, which affects the duration of the calculation, is taken into account.

The calculation of structural systems by the finite element method is generally performed using special certified computer systems. The requirements of DBN V.2.6-98:2009 and DSTU B V.2.6-156:2010 as for physical and geometric nonlinearity and second-order influences are implemented in these systems. Such a system is LIRA-SAPR. To simplify and accelerate the creation of the calculation scheme, it can be used the software package MONOMAKH-SAPR. The automated breakdown into finite elements is carried out in this software package when choosing the size and configuration of the elements (Figure 2).

(a) (b)

**Figure 2.** General view of spatial calculation models in PC MONOMAKH-SAPR (a) and calculation scheme of walls in finite elements (b) in PC LIRA-SAPR

At the first stage of calculation, it is recommended to take the modulus of elasticity of concrete $E_{cd}$, which is equal to its calculated value in accordance with table 3.1 DBN V.2.6-98:2009 in order to estimate the forces in the elements of the structural system.

Generally, the calculation of the load-bearing capacity of the walls can be performed as flat selected elements on the combined applying of normal and shear forces, bending moments, torques, and transverse forces applied on the sides of a flat selected element. The forces are obtained from the calculation of a structural force by means of the finite element method according to DSTU-N B V.2.6-205:2015.

The calculation from the surface of the wall is performed similarly to the calculation of flat floor slabs, determining the values of the ultimate bending moments taking into account the influence of normal forces.

According to the results of the calculation, an image of the deformed scheme is formed (Figure 3, a). The construction of is fields of displacements (Figure 3, b) and stresses for wall elements is performed.

At the next stages of calculation, the specified stiffness of plates is accepted with reinforcement, existence of cracks and inelastic deformations in concrete and reinforcement in mind, provided the known reinforcement. All the parameters are defined according to the statutory regulatory documents.

In this case, the calculation of the walls will be carried out by dividing the flat element into separate layers (upper, middle and lower) with the definition of the main stresses on each layer separately and the identified elements, in which cracks were formed.

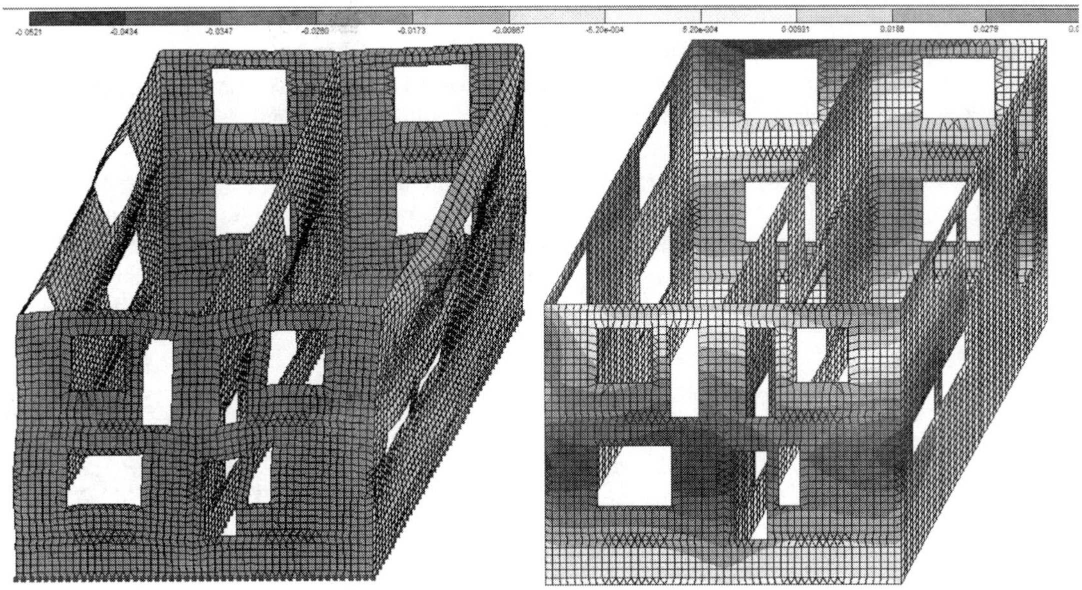

(a) (b)

**Figure 3**. Results of deformations calculation: a - deformed scheme; b - isofields of displacements

The calculation with account of the inelastic deformations and real diagrams of concrete deformation (considering the physical nonlinearity) will allow to determine the location and direction of cracks formation, and perform, if situation so requires, the necessary reinforcement (Figure 4). The diagram of the concrete state can be specified in the form of complete or simplified one according to the class of concrete in compliance with item 3.1.4 DBN V.2.6-98:2009.

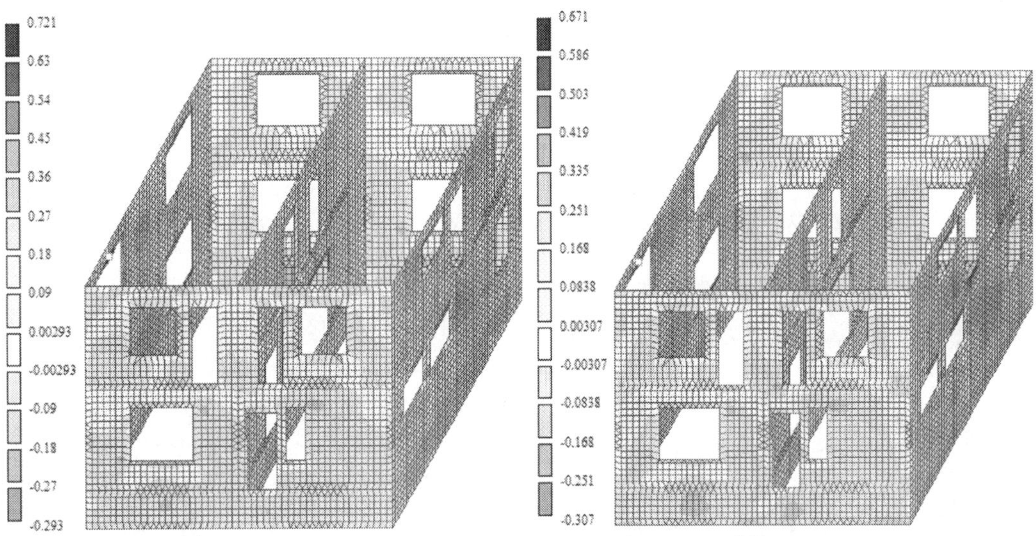


(a) (b)

**Figure 4.** Results of physically nonlinear calculation on the isofields of main stresses: a - the upper layer; b - the bottom layer

After determining the forces in the cross sections of the walls, it is necessary to check the bearing capacity of dangerous concrete sections according to the method described in paragraph 2.2.1

DBN V.2.6-98:2009. If the load-bearing capacity of the section is not provided, it is necessary to continue the calculation in the PC LIRA-SAPR in ARM-SAPR with the determination of the number of longitudinal and transverse reinforcement of the walls.

There is a redistribution of forces in monolithic reinforced concrete. The height of the jumper has an overlimited height from the top of the opening to the bottom of the next slot located above it.The reinforcement of the wall can be taken into account thereby.

The calculation of jumpers can be performed as for a wall beam (Figure 5, a) or a beam with clamped supports (Figure 5, b). If there is a small distance from the top of the slot to the bottom of the floor, and the wall is more like a jumper in this place, then it should be reinforced as a jumper (Figure 6). If there is a significant distance from the top of the opening to the top of the wall, the opening must be framed with clamps. When reinforcing the jumper its cross-sectional height can be considered of 300 mm or less.

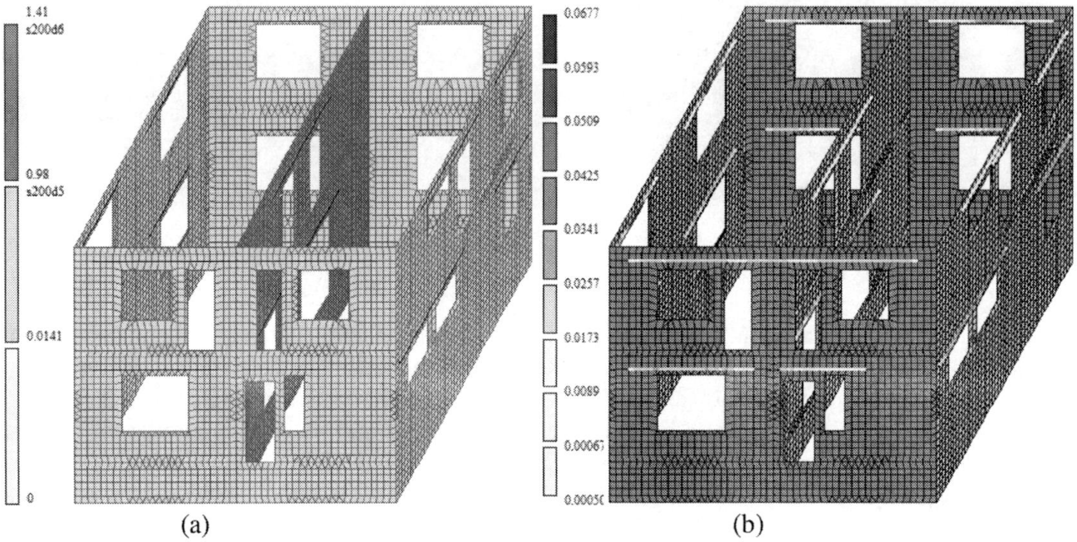


**Figure 5.** The cross-section area of reinforcement: a - by 1 LM near one surface on the X axis (horizontal); b - percentage of reinforcement

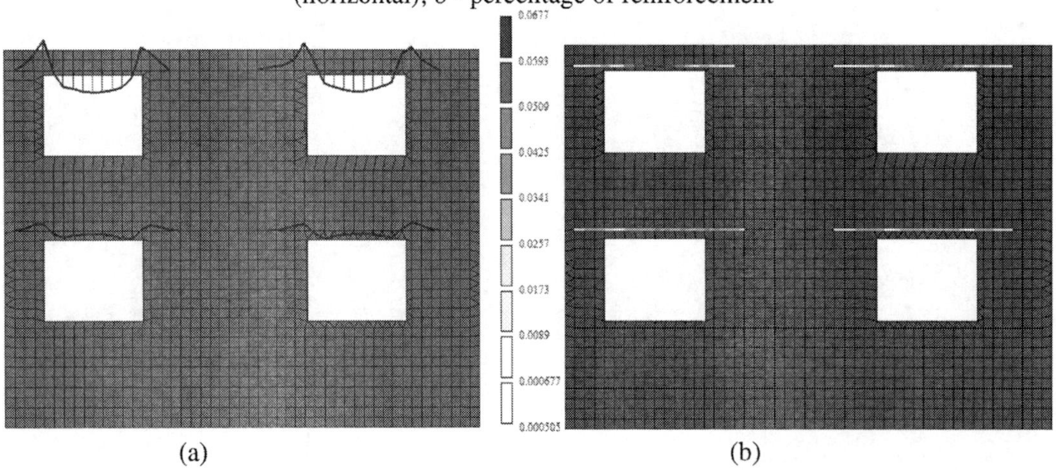


**Figure 6.** Calculation of jumpers: a - nature of bending moments diagram; b - percentage of reinforcement

For monolithic concrete walls, a calculation scheme with hidden columns and crossbars can also be used (Figure 7), where monolithic belts under the floor slabs are applied as crossbars.

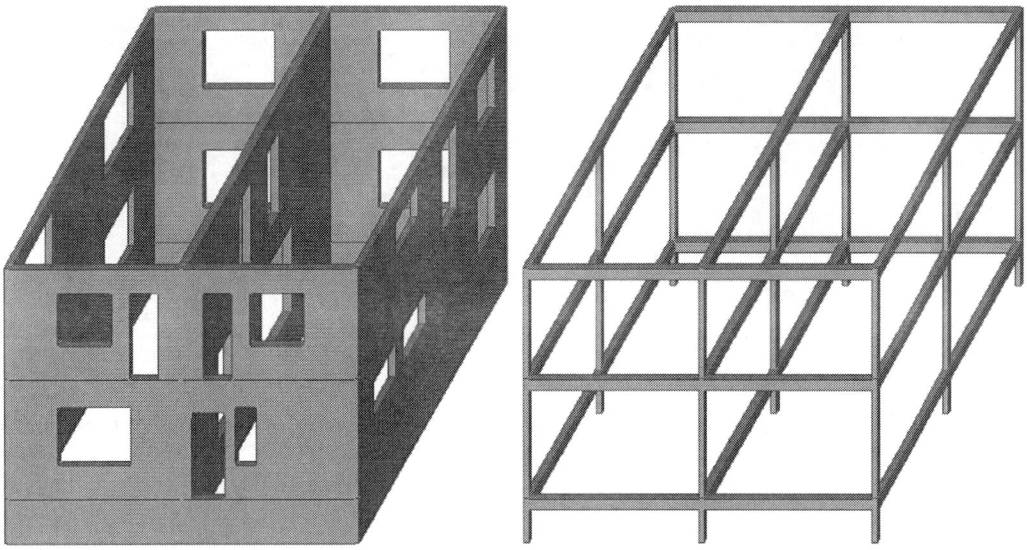

(a) (b)

**Figure 7**. Estimated model of load-bearing walls with a hidden frame

The calculation outcomes indicate that the stress-strain state of the load-bearing walls does not change significantly compared to the option without hidden rod elements (Figures 8 and 9), and the presence of hidden elements will cause the increase of the reinforcement cost.

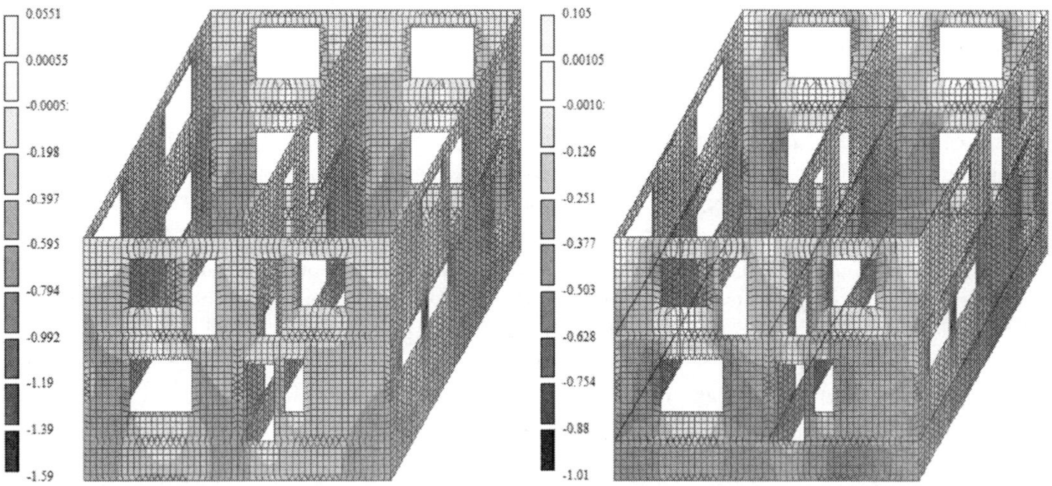


(a) (b)

**Figure 8.** Stress isofields $N_y$ (MPa) in the walls of frameless schemes without hidden (a) and with hidden (b) columns and crossbars

The findings of the analysis of the carried-out numerous researches of the stress-strain condition of bearing and self-bearing walls from permanent formwork ComBlock units for buildings of various constructive schemes allow to come to the following conclusions.

IOP Conf. Series: Materials Science and Engineering **907** (2020) 012025 doi:10.1088/1757-899X/907/1/012025

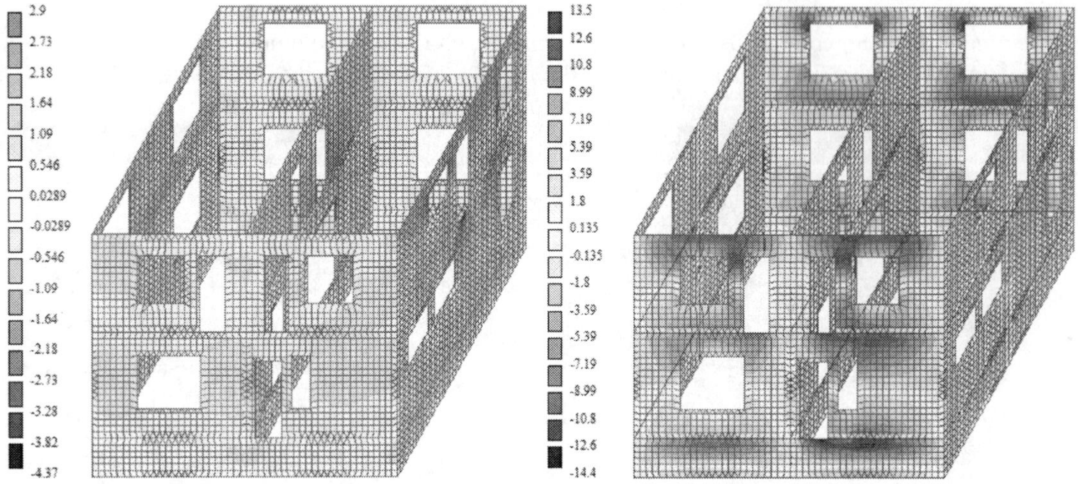


(a) (b)

**Figure 9.** Isofields of bending moments $M_x$ (kPa): a - frameless without hidden columns and crossbars; b - frameless with hidden columns and crossbars

It has been proved worthwhile to perform the calculation of low-rise buildings using permanent formwork ComBlock under construction in areas with normal soil conditions in non-seismic areas without hidden columns and crossbars in the following sequence:

1. The constructive system and the corresponding calculation scheme of the building are accepted on the basis of the offered architectural and design solutions.

2. The gradual calculation of the load-bearing walls (columns) of the building is performed using the appropriate software packages (for example, LIRA-SAPR or MONOMAKH-SAPR).

3. The received forces in walls (columns) are analysed and two possible algorithms of the further calculation are considered depending on the received results:

- in case of the absence of stresses in the structural elements, which exceed the calculated tensile strength of concrete, i.e. $\sigma < f_{ctd}$, testing design calculation of sections both for structures of non-reinforced and low-reinforced concrete is performed in accordance with the requirements of section 12 of DBN V.2.6-98:2009, section 9 of DSTU B V.2.6-156:2010. Structural reinforcement of load-bearing elements of the building is developed, including areas above the slots in the walls (jumpers) and joints of walls with floors and coverings corresponding to the requirements of state building codes and design solutions proposed by the authors.
- if stresses take place in the structural elements that exceed the design characteristics of concrete tensile strength, i.e. $\sigma \geq f_{ctd}$, the calculation of reinforcement of load-bearing walls and columns of the building is performed (continued) by means of software packages. Reinforcement of bearing elements of the building is developed on the basis of the specified calculation.

The outcome analysis of the performed numerical researches testifies that it is necessary to carry out the specified sequence when designing each object separately within designing bearing walls of low-rise buildings with the use of permanent formwork.

## 4. Conclusions

The obtained research findings revealed that it is possible to minimize the consumption of reinforcing steel in monolithic concrete walls made with permanent formwork ComBlock units.

Design (working) reinforcement in load-bearing vertical elements of such buildings should be installed only in cases where stresses occur in structural elements that exceed the calculated tensile strength of concrete.

For the most frequently used architectural-design and structural solutions of low-rise buildings,

IOP Conf. Series: Materials Science and Engineering **907** (2020) 012025 doi:10.1088/1757-899X/907/1/012025

which are erected in areas with normal soil conditions in non-seismic areas, their vertical reinforcement is not required due to the absence or presence of insignificant tensile forces in the walls. In this case, they can be made of non-reinforced or low-reinforced concrete with structural reinforcement, including joints between walls themselves, walls with floors and coverings and areas above the slots in the walls (jumpers).

Herewith, it is important and necessary to perform the proposed above sequence of calculations and design for load-bearing walls of such low-rise buildings. That allows to minimize the cost of reinforcement for each object individually, taking into account it's architectural and design features and construction conditions as well.

The proposed calculation method can be used in the design of low-rise buildings with load-bearing monolithic concrete and reinforced concrete walls with other technologies of their construction.

## References

[1] Badin GM and Sychev S A 2013 *Modern Technologies of Construction and Reconstruction of Buildings* (St. Petersburg: BHV-Peterburg) 288

[2] Mackova D and Spisakova M 2015 Comparative Analysis of Houses Built from Insulating Concrete Formwork – case Study *SSP - Journal of Civil Engineering* **10**(2) 139-146

[3] STO 12601302–002–2016 *Design of Buildings and Structures Using the Tecolit Housing System (for Usual Conditions and Construction of Sites with Seismicity up to 7 Scales on Scale MSK-64, included)* (Chelyabinsk)

[4] *IntegraSpec ICF Building System. Engineering Guide – Details and Tables* 2019 (Kingston: Integraspec) 36

[5] Guidelines for the calculation and construction of walls in the system "Izodom 2000 Poland" 2011 *Information book "Izodom 2000 Poland"* **2** 45

[6] Fedyuk R S 2013 Cast-in-Situ Reinforced Concrete Walling Using Permanent Formwork of Polystyrene Foam *Vestnik of Irkutsk State Technical University* **10**(81) 185-190

[7] Taran V A 2013 Self-Supporting Walls Made of Polystyrene Foam Concrete in Concrete Forms *Modern Industrial and Civil Construction* **9**(1) 49-58

[8] Meneylyuk A I, Cherepaschuk L A and Oliynyk N V 2018 Analysis of New Constructive Solutions of Energy Efficient Heating Constructions *Young Scientist* **1**(53) 435-438

[9] *Album of technical solutions. Permanent formwork "ComBlock"* 2019 (Kharkiv: Construction and Production Company AVcom) 24

[10] *Manual for the Design of Residential Buildings* Issue 3 Constructions of residential buildings (to SNiP 2.08.01-85) 1989 (Moscow: Stroyizdat) 304

# Peculiarities of erecting walls for houses and buildings by technology of monolithic construction using permanent formwork unit

**S V Butnik[1], V A Viatkin[1], M N Dzhalalov[1], I V Hovorukha[1] and V V Bielka[2]**

[1] Department of Building Technology, Kharkiv National University of Civil Engineering and Architecture, Sumska St. 40, 61002 Kharkiv, Ukraine
[2] Limited Liability Company «BUILDING and PRODUCTION COMPANY «AVcom», Pashchenkovskaya St. 17, 61017 Kharkiv, Ukraine

s.butnik@ukr.net

**Abstract.** The article deals with constructive and organizational-technological solutions when performing works on construction of external and internal bearing walls with the use of combined permanent formwork. The technology allows to reduce the construction time by combining in the system of the retained formwork unit three technological processes, which are performed simultaneously by one group of contractors - erection, insulation and wall decoration. Depending on the thickness of the insulation in such houses and buildings, it is possible to achieve close to zero energy consumption.

## 1. Introduction

Saving energy and heating costs is an issue facing any community. In Europe, energy efficiency issues are covered in Directives 2002/91 / EC, 2010/31 / EC and EPBD (COM / 2016/0765). Ukraine is moving somewhat slowly, but still in the direction of legally approved principles of energy efficiency. The Law of Ukraine "On Energy Efficiency of Buildings" stipulates that in order to gradually increase the energy efficiency of buildings, a national plan for increasing the number of buildings with close to zero energy consumption is approved. In the near future, we will have to use less energy for heating than we have used so far [1]. So, the main characteristic of the buildings of the future is low energy consumption [2].

The practice of modern construction of energy-efficient comfortable houses on the Ukrainian market is represented by various materials and technologies. The main ones are aerated concrete, bricks, houses and buildings erected by monolithic technology in permanent formwork. Permanent formwork, in turn, is represented by different options. But such important parameters of the future house (building) as: energy saving, tightness, duration of construction, durability, environmental friendliness, cost, fire danger, etc. always come to the fore [3-8]. All these requirements are met by the combined permanent ComBlock formwork that is confirmed by two patents and positive responses of customers.

The object of research is houses and buildings erected with the use of permanent formwork ComBlock of the AVcom trademark.

The subject of research is the technological regulations for construction of walls according to the technology of monolithic construction to ensure energy efficiency of the building.

The purpose of research is organizational and technological solutions in the construction of external and internal load-bearing walls with simultaneous use of insulation and decoration to ensure energy efficiency of buildings.

## 2. Materials and Methods.

The construction of external walls using this technology consists of the following layers:

- insulation with a thickness of 50 to 300 mm (outside);
- concrete core with a thickness of 100 mm (according to calculations);
- gypsum panel 30 mm thick (inside).

The construction of internal load-bearing walls consists of two gypsum panels 30 mm thick and a concrete core 100 mm thick (according to calculations).

The components of the permanent formwork block (insulation and gypsum panel) are interconnected by plastic adjustable jumpers, which fix the desired thickness of the concrete core with an adjustment step of 25 mm. In addition to the connecting function, jumpers are used to fix the reinforcement in the design position (Figure 1). The material of the jumper is corrosion resistant.

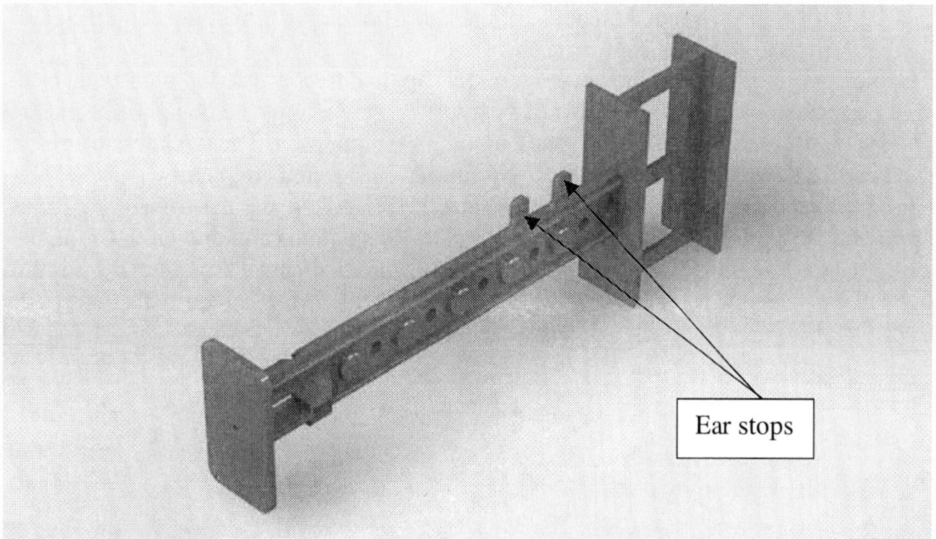


**Figure 1.** Adjustable plastic jumper in assembled form.

The permanent formwork filled with concrete forms a monolithic wall of estimated thickness depending on design loading. As insulation it is possible to use Neopor® - with density of 15 kg / m$^3$, (EPS70), expanded polystyrene M25 with density not less than 15 kg / m$^3$ (DSTU B EN 13163: 2012 Building heat-insulating materials - Products made from foam polystyrene (EPS). Technical specifications (EN 13163: 2008, IDT)) and mineral wool - with density of 135-145 kg / m$^3$. On the inside of the formwork are gypsum panels that do not require further finishing (plastering, linings).

The block of permanent formwork can be ordinary and angular.

Ordinary Comblocks. The ComBlock system with a gypsum panel (Figure 2), used for construction of external bearing walls, consists of insulation panel which is from 50 to 300 mm thick and the gypsum panel being 30 mm thick, and they are connected between themselves by plastic adjustable jumpers (7 pieces per one form).

One ordinary module of the ComBlock system is 1000 mm long, 300 mm high, the thickness of module depends on the designed thickness of insulation and the concrete core of wall. The thickness of the concrete core of wall can be from 100 mm to 430 mm. No further plastering is required.

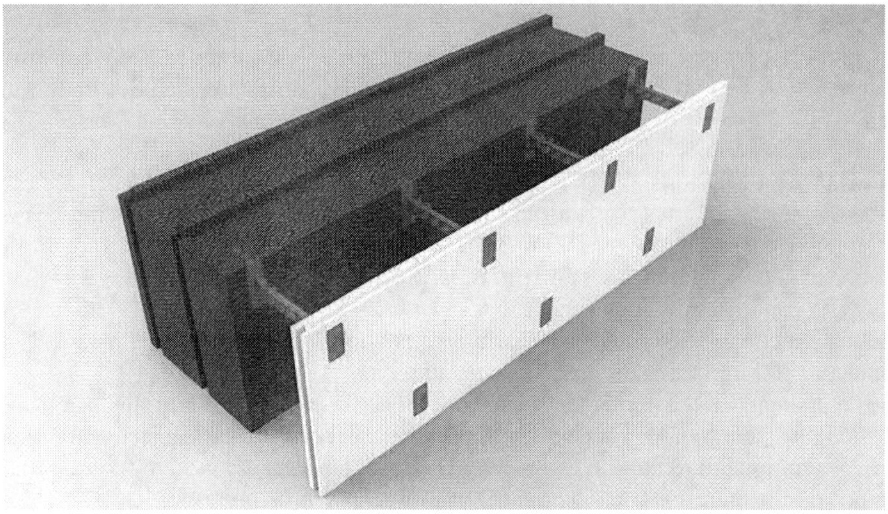

**Figure 2.** Scheme of the ComBlock system with gypsum panel.

The block of internal bearing wall (Figure 3) is applied to construction of internal bearing walls. The form of permanent formwork consists of two gypsum panels connected by plastic adjustable jumpers (7 pieces per form). One block is 1000 mm long, and 300 mm high. The thickness of the concrete core of the wall can be from 100 mm to 430 mm. No further plastering is required.

Corner Blocks. The internal side length of one module of the angular ComBlock for plaster panels is 250 mm and 500 mm. The length of the outer sides (insulation) and the width of the block depend on the designed thickness of insulation and the concrete core of the wall. The thickness of the concrete core of the wall can be from 100 mm to 430 mm. No further plastering is required. (Figure 4).

**Figure 3.** Scheme of the inner load-bearing wall Comblock.

IOP Conf. Series: Materials Science and Engineering **907** (2020) 012026 doi:10.1088/1757-899X/907/1/012026

**Figure 4.** General view of the corner ComBlock with gypsum panel.

The ends of the outer and inner panels of the Block are made with spikes and grooves, which provides density when combined, prevents the flow of concrete mix through the joints, and prevents the formation of cold bridges.

## 3. Approbation of the Research

Technological regulations for the construction of buildings and structures using permanent formwork ComBlock are to assemble blocks of permanent formwork; to install the whole course of them into the design position; to install reinforcement; and to fill it with concrete.

Installation of the course of permanent formwork blocks begins with mounting of corner blocks being leads for the process (Figure 5). During installation, the vertical and horizontal position of gypsum panels and insulation ones, as well as the horizontal position of gypsum panels and insulation ones relative to each other are checked.

**Figure 5.** Installed corner block with plasterboard.

When installing the next courses of Comblock, a 250 mm bond is required according to the technology. Corner blocks are left and right, made asymmetrically (inner plasterboard is shaped like the capital letter "L" and has dimensions of 250x500 mm).

After mounting of angular blocks, installation of ordinary blocks (1000 mm long), with controlling their position on cords, is started. And, thus, the first course of blocks of external and also internal walls is completed (Figure 6).

After installation of the course of formwork blocks, vertical and horizontal fittings should be installed (Figure 7).

Mounting of horizontal fittings begins with the installation of working rods.

Reinforcement with vertical rods is carried out taking into account the spatial location of the reinforcement in the wall structure. Vertical working rods are installed with the connection of the lower ends to the free length of reinforcement bars and combining them with clamps. Next, from the bottom to the top other clamps are put in, fixing the vertical rods.

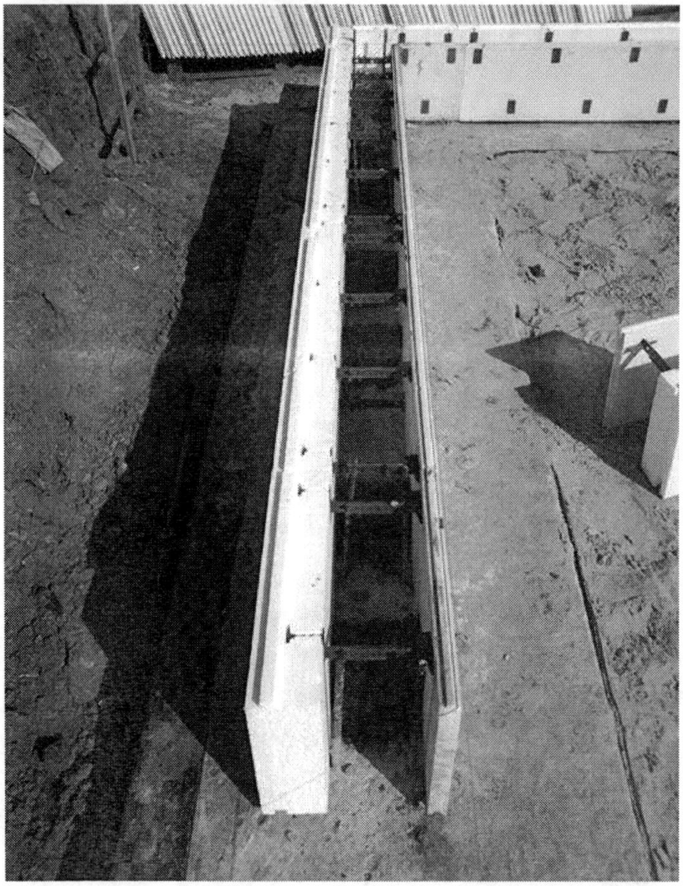

**Figure 6.** The first course of blocks of external walls

After installation of the course of ComBlocks, and horizontal and vertical armature, laying of concrete mix starts according to the project.

The concrete mixture is prepared in a concrete mixer in the immediate vicinity of the site where it is laid.

When laying the concrete mixture, it is pierced or compacted with a low-power vibrator to avoid violating the integrity of the blocks of permanent formwork.

**Figure 7.** Vertical reinforcement installed.

Particular attention in the system of fixed formwork Comblock is paid to ensure tightness when installing door and window openings.

In places of execution of window or doorway openings the ordinary block is cut to the size required by the project. In the ends of the concrete core of wall on the side of the opening, formwork of insulation, or template-formwork is installed, which is fastened with wooden struts (Figure 8).

**Figure 8.** Arrangement of door and window openings.

Installation of the window block can be carried out in two options:

• fastening of the window block with external exposing of thermoinsulation;

• fastening of the window block without external exposing of thermoinsulation.

For "mounting with external exposing" in the outer layer of insulation in the lower part of the window opening a recess is cut, in which trioterm insulation (or its analogue, which is dense, with a bulk density of 200 kg / m3) is inserted, unfastened and concreted in the design position, which is the basis for installation of the window unit in the area (body) of the insulation. Trioterm is additionally fixed on metal brackets which are monolithic in concrete walls that provides reliable fastening of the window block.

In the case of "mounting without external exposing ", the window block rests on the concrete part of the wall. In this case, you need to lay insulation at the ends, as well as at the bottom and top of the window unit (if it is designed so).

Conformity of thermal and sound insulation characteristics of enclosing structures made with the use of technology of fixed formwork Comblock, allows to achieve resistance to heat transfer much higher (Table 1) than the requirements of regulatory documents of Ukraine [2].

**Table 1.** Dependence of heat transfer resistance on insulation thickness in Comblock system.

| Thickness of insulation, mm | 100 | 150 | 200 | 300 |
|---|---|---|---|---|
| Thermal transfer resistance R, $(m^2 \cdot K)/W$ | 3.48 | 5.04 | 6.6 | 9.73 |

The use of permanent formwork in the construction of walls speeds up and simplifies construction by combining several processes in one technological cycle (load-bearing wall with a designed heat transfer resistance is erected in one technological cycle). The design of the permanent formwork provides tightness, and concrete in it, after setting and gaining strength, becomes a structural part of the finished wall.

The floor from ComBlock is erected in 10-12 days (walls + insulation + plaster) by a team of 4 people, while brick walls are erected in 25-30 days without insulation and plastering.

Due to such thermal insulation, this construction system allows to save energy resources significantly when heating buildings and houses, which becomes especially important when energy prices are constantly rising.

## 4. Conclusions.

Construction with the permanent formwork of the Comblock has a number of advantages:
- the insulation is only exposed to the outside;
- a significant reduction in heat loss during the operation of buildings and buildings;
- reduction of construction time due to uniting three technological processes in the block system of permanent formwork performed simultaneously by one team of contractors - erection, insulation and wall finishing;
- possibility of application of the ComBlock permanent formwork technology in seismic areas.
- the ComBlock system makes it possible to obtain high indices of heat transfer resistance with minimum expenses of the customer.

## References

[1] Law of Ukraine “On energy efficiency of buildings” URL: https://zakon.rada.gov.ua/laws/show/2118-19

[2] State building regulations. 2017 *B.2.6-31-2016. Thermal insulation of buildings* (Kyiv: Ministry of Construction of Ukraine) p 37

[3] Savyovsky V V and Dzhalalov M N 2009 Influence of construction conditions on the effectiveness of building insulation *Scientific Bulletin of Civil Engineering* **52** 288–292

[4] Savyovsky V V, Dzhalalov M N, Savyovsky V V and Mulyar A N 2010 Energy audit and thermal modernization of buildings. *Construction of Ukraine* **6** 3–7

[5] Savyovsky V V and Dzhalalov M N 2011 Investigation of the influence of technological features of the thermal insulation device on the energy efficiency of external building envelopes *Scientific Bulletin of Civil Engineering* **66** 116–122

[6] Dzhalalov M.N 2015 Analysis of the condition of structural elements of buildings of housing fund and methods for eliminating defects with application of energy-saving technologies *Bulletin of Civil Engineering* **3** 28–32
[7] Kolomiets I V and Dzhalalov M N 2017 Investigation of the factors affecting the technical and economic indicators of the device for the thermal insulation of the external walls of buildings *Book of abstract the VI International scientific and technical conference of young researchers and students "Current issues in modern technologies"* (Tern., 16-17 November 2017), vol. **I** pp 107
[8] Dzhalalov M N, Butnik S V, Kolomiiets Y V and Hovorukha I V 2019 Technological characteristics of determining the energy efficiency of enclosing structures of the building exemplified by a general education school. *Scientific Bulletin of Civil Engineering* **2(96)** 212–221

IOP Conf. Series: Materials Science and Engineering **907** (2020) 012027 doi:10.1088/1757-899X/907/1/012027

# Role of potential-determining ions and water molecules in the structure formation of cement compositions

**A Plugin[1], T Kostyuk[2], D Bondarenko[3], O Pluhin[1], V Arutyunov[4]**

[1]Building Materials and Structures Department, Ukrainian State University of Railway Transport, Feuerbach sq. 7, 61050 Kharkiv, Ukraine
[2]Building Materials and Products Department, Kharkiv National University of Civil Engineering and Architecture, Sumska str. 40, 61002 Kharkiv, Ukraine
[3]Computer Science and Information Technologies Department, Kharkiv National University of Civil Engineering and Architecture, Sumska str. 40, 61002 Kharkiv, Ukraine
[4] JSC "Trest Zhilstroy", Alchevskikh str. 43, 61002 Kharkiv, Ukraine

aaplugin@gmail.com

**Abstract**. The paper is concerned with dealing with an urgent problem of present-day construction materials science and engineering such as increasing the strength of cement stone when using finely dispersed fillers. It has been found that the physical and mechanical properties of Portland cement-based materials largely depend on the strength of the cement matrix modified by finely dispersed fillers, in particular, on the electric surface properties of all components and pore walls (equilibrium electric surface potential and double electric layer with potential-determining ions and counter-ions and at the interface between the solid phase and water). Studies were conducted to determine the values of equilibrium electric surface potentials of perlite and glass hollow microspheres. The system-forming hierarchical structure of water acting as a matrix or a crystal-forming platform for the synthesis of spatial structures during the formation of a crystalline cage on the surface of micro-fillers was substantiated. The assembly elements of these systematized structures were investigated using electron microscopy. It should be noted that when water films are supersaturated with ions during cement hydration, crystals nucleate and grow on the surface of micro-fillers, which leads to self-compaction and hardening of the cement matrix at the microlevel.

## Introduction

The works of the school of thought led by A.N. Plugin [1, 2] demonstrate that the hardening of cement binders depends on electro-heterogeneous interactions between the particles of cement hydration products, mineral additives and aggregates with unlike surface charges. In this context, the interaction is carried out through thin, monomolecular, in many cases, water interlayers that remain between the particles during the hydration hardening.

Currently, cement compositions keep on improving to obtain new properties by introducing various dispersed components such as perlite sand, fiberglass, glass microspheres, etc. [3-6]. Their electric surface properties, the nature of the distribution of water on their surfaces, and their role in the structure formation and formation of the properties of cement compositions remain poorly understood.

IOP Conf. Series: Materials Science and Engineering **907** (2020) 012027 doi:10.1088/1757-899X/907/1/012027

**The purpose of the research** is to clarify the electric surface properties of new disperse components of cement compositions, the laws of distribution of potential-determining ions and water over the surfaces of disperse particles, and their effect on the structure formation.

**Literature Data Analysis**

A specific feature of the electric surface properties of disperse materials is the discreteness of the surface charge, due to the fact that the double electric layer is formed by potential-determining ions on the surface of the solid phase, which are point charges located at a considerable distance from each other and which are essentially active centers on the surfaces of the disperse structure [1, 2].

According to the papers [1, 2], all structural elements of the concrete mixture and concrete have marked electric surface properties in the form of a surface charge, which is described by the density $q^0$, $C/m^2$, surface (electric surface) potential $\psi^0$, $V$, and equilibrium electric surface potential $\psi^0_p$, V, and a double electric layer (DEL) at the interface of the solid phase and the water. The structure of DEL, including ions that are specifically adsorbed, is shown in figure 1.

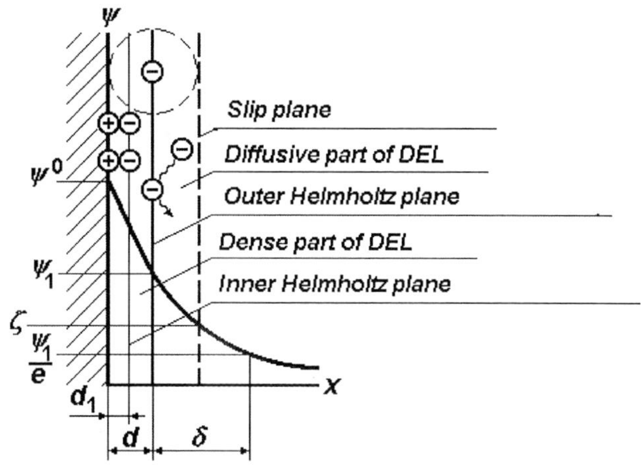


**Figure 1.** The structure of DEL [2]: $\psi^0$: electric surface (surface) potential; $\psi_1$: potential at the border between the dense and the diffusive parts of DEL; $\zeta$: electrokinetic potential; $\delta$: the thickness of the diffusive part of DEL.

In the concrete mixture, the concentration of electrolyte, mainly $Ca(OH)_2$, quite quickly becomes significant (0.02 mol/L, and even higher at the beginning of hardening owing to the supersaturation of the liquid phase); therefore, all water physically and chemically binds to the surface of the structural elements and becomes part of DEL. The action in such a physically and chemically bound water layer within DEL is determined by the electric field that is created by the potential-determining ions of the particles, that is, the electric surface potential $\psi^0$, and the high polarity of the water, which transfers the electric field to a distance of 1 μm or more [1, 2, 7 ].

$$\delta_{ac} = \frac{В/\rho^в}{Ц \cdot s^ч} = \frac{Ц \cdot В/Ц}{\rho^в \cdot Ц \cdot s^ч} = \frac{В/Ц}{\rho^в \cdot s^ч} = \left[\frac{1}{(кг/м^3)\cdot(m^2/кг)} = m\right] \quad (1)$$

Having a huge specific surface area due to the most finely dispersed component, hydrosilicate gel, cement stone is a highly concentrated disperse system with a specific surface area of 250÷700 $m^2/g$ = 250000÷7000000 $cm^2/g$.

As a result, the thickness of the water layer on the gel particles, and, consequently, other particles of cement stone, ranges from 0.63 nm $6.310^{-10}$ m, which corresponds to the size of the water associate [8], to 10 nm $1 \cdot 10^{-8}$ m, which corresponds to the size of the pores in the particles and globules of the hydrosilicate gel [2].

Consequently, for cement stone with a low W/C ratio DEL is flat, with no diffusive part, which allows considering the interactions between the particles and the surfaces using the disjoining pressure equations for a flat DEL [2, 9]. In this case, the interaction is determined by the potential difference between the layer of potential-determining ions (PDI) and the layer of counter-ions (CI), which is named the electric surface potential in the papers [1, 2]. The absolute value of the electric surface potential $\psi^0_{ep}$ is determined using the experimental calculation method by the value of standard electrode potentials $E^0$ for simple substances [2].

The electric surface potential of a complex substance under standard conditions is determined by the equation:

$$\psi^0_{ep complexsubs\tan ce} = \frac{\sum \psi^0_{ep.i} \cdot \nu_i}{\sum \nu_i} \quad (2)$$

and the equilibrium electric surface potential (at different pH levels) $\psi^P_{ep}$ of complex substances is determined by the equation:

$$\psi^P_{ep} = \psi^0_{ep complexsubs\tan ce} + 0.059pH \quad (3$$

**Findings of the research**

Equilibrium electric surface potentials of expanded perlite sand, fiberglass and glass microspheres were obtained based on the average stoichiometric composition and electric surface potentials of oxides.

The surface charge of glass (glass fiber and glass microspheres) was determined based on its average stoichiometric composition of $Na_2O{\cdot}CaO{\cdot}6SiO_2$ and the electric surface potentials of the oxides according to the method [2]:

Absolute electric surface potential:

$$\psi_{ep}{}^0 = (2.207 + 1.37 + 6 \times 0.56)/8 = +0.217 \text{ V} \quad (4)$$

Equilibrium electric surface potential at pH = 12 (pore electrolyte of cement stone):

$$\psi_{ep}{}^P = \psi_{ep}{}^0 - 0.059\text{pH} = 0.217 - 0.059 \times 12 = -0.499 \text{ V} \quad (5)$$

where 2.207; 1.37; 0.56 are the absolute electric surface potentials of $Na_2O$, $CaO$ и $SiO$, respectively.

The calculation of the electric surface potentials of perlite is shown in Table 1. As one can see, the grains of expanded perlite both in water (pH~7) and in compositions based on Portland cement (pH~12) have a negative electric surface potential and, consequently, a surface charge.

**Table 1**. Calculation data for the electric surface potential of perlite

| Oxide | $\psi^P_{ep}$, V | $x$, % | $x \cdot \psi^0$ | $\bar{\psi}^P_{ep\ perlite}$, B | $\psi^P_{ep,perlite}$, V, at pH | |
|---|---|---|---|---|---|---|
| | | | | | 7 | 12 |
| $SiO_2$ | −0.55 | 72 | −39.6 | | | |
| $Al_2O_3$ | 0.33 | 15 | 4.98 | | | |
| $Fe_2O_3$ | −0.32 | 1.4 | −0.442 | | | |
| $CaO$ | 1.38 | 2.2 | 3.036 | −0.19 | −0.64 | −0.90 |
| $MgO$ | 1.13 | 0.36 | 0.405 | | | |
| $K_2O$ | 2.35 | 2.85 | 6.707 | | | |
| $Na_2O$ | 2.21 | 2.85 | 6.308 | | | |
| Σ | | 96.66 | −18.61 | | | |

Thus, in the environment of cement paste and stone, fiberglass, glass microspheres, perlite have negative surface potential values; fiberglass and microspheres: $\psi_{ep}{}^P = -0.499$ V, and perlite:

$\psi_{ef}{}^{P} = -0.90V$. Therefore, they can be a substrate for crystallization of hydration products with a positive surface charge, which corresponds to the diagram shown in figure 2.

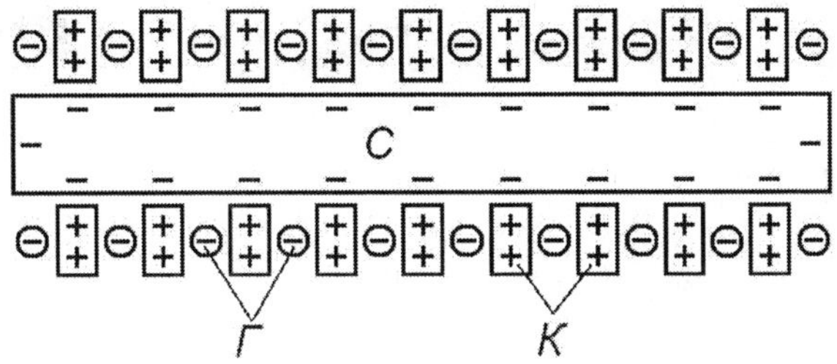


**Figure 2.** Estimated structure of cement hydration products around fiberglass:
C - fiberglass; K - crystalline hydrates of portlandite, hydroaluminates and hydrated calcium sulfoaluminates; Г- particles of hydrosilicate gel.

Silicones used in cement mixtures are, as a rule, organosiloxane polymers (polyorganosiloxanes) with the structural formula $[R_2SiO]_n$, where R is an organic group (methyl, ethyl or phenyl), such as the simplest representative of silicones, polydimethylsiloxane:

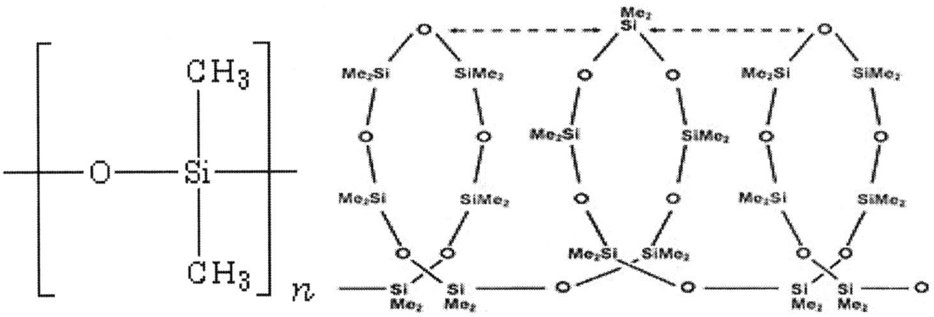


**Figure 3.** The structural formula of polydimethylsiloxane and the spiral structure from it due to the dipole moment.

These groups can have a dipole moment relative to the siloxane chain and consequently enable the silicone film to have hydrophilic properties and an electric surface potential, that is, the ability to serve as a substrate for cement hydration products (by analogy with [6, 10]).

The dipole moment is determined by the structure of the siloxane group, which results in the polarization of the bond between silicon and hydrocarbon. The extremely flexible siloxane chain Si-C and Si-O reacts to the dipole moment and "tries" to arrange the siloxane units in such a way that the positively charged silicon of one unit is located opposite the negatively charged oxygen atom of the other unit, figure 3.6 (a). As a result, a singular spiral structure arises [11]. In this case, hydrocarbon radicals are directed outward, and hydrophilic associations are drawn inward. This structure allows water structures to be organically integrated into the lattice metrics. In the volume of water, the main structural element is also formed as a non-linear element, but twisted into a spiral. Moreover, the whole variety of combinatorial water structures on the surface of the polysiloxane chain will be a variety of the same spiral cycles of different symmetries [11].

In 1993, the American chemist Kenneth D. Jordan gave his suggestions of stable "quanta of water", which consist of 6 molecules [12]. According to the ideas of Dr. Head-Gordon's group (2002), water molecules form "true bricks" of water through hydrogen bonds, which are topological chains and rings of many molecules, figs. 4 to 5.

IOP Conf. Series: Materials Science and Engineering **907** (2020) 012027 doi:10.1088/1757-899X/907/1/012027

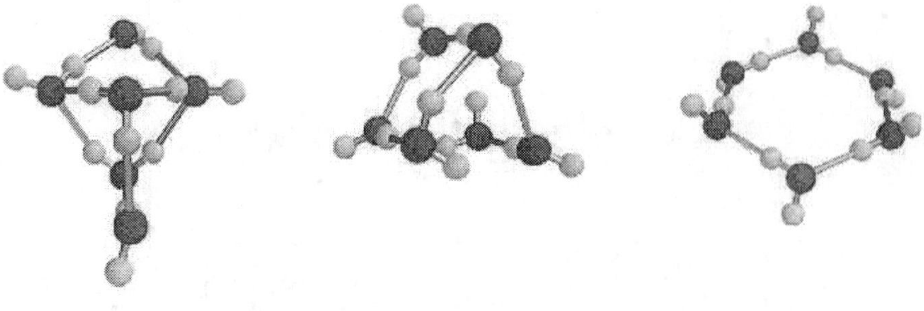

**Figure 4.** Models of water associates

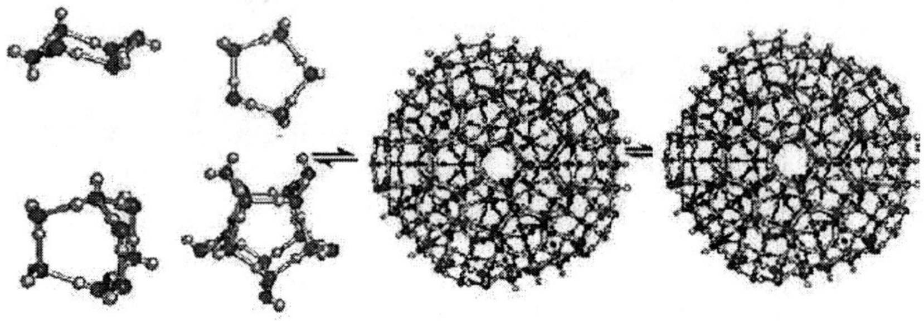

**Figure 5.** Associations of water clusters [62]

The ideas about the structure of water, especially those related to the formation of large clusters, figure 5, are ambiguous. It is widely acknowledged that there is a common opinion on the existence of a quasistructure of water and initial associates of 5 water molecules on average. It is this structure that determines the interaction between water and the walls of capillaries. This arises from the ideas of A.N. Plugin on the dielectric permittivity of water [2], where the taking into account of the associates of 5 molecules of water with unidirectional dipole moments allowed deriving an equation for the dielectric permittivity of water, the calculation of which, unlike all existing equations, gave the value of $\varepsilon = 80$. In addition, it was shown that the dielectric permittivity of water at a charged surface varies from 2.16 to 80 at a distance to 1 μm from this surface, which corresponds to the propagation of an electric field of point charges through polar liquids according to G.I. Distler [7].

From this perspective, the following pattern of the behavior of water on the surface of the structural elements of concrete with charged surfaces seems most realistic. The structure of large water associates and the bonds between them are likely to be disrupted under the influence of the electric field of the surface and, more precisely, the PDI. Closely at the surface, the polar molecules of water will be adsorbed on the PDI. In this case, the adsorption of molecular associates of 5 molecules of water with unidirectional dipole moments should be most probable. In the intervals between the PDI, single unassociated molecules will be adsorbed, which will make it possible to obtain the most dense structure of cement stone, figs. 6 to 8.

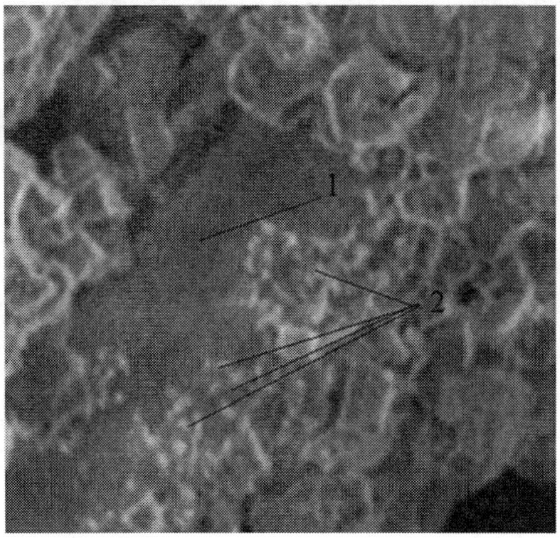


**Figure 6.** The nucleation of the crystalline phase on fiberglass: 1) fiberglass; 2) nucleation of crystalline hydrates on the associates and individual water molecules.

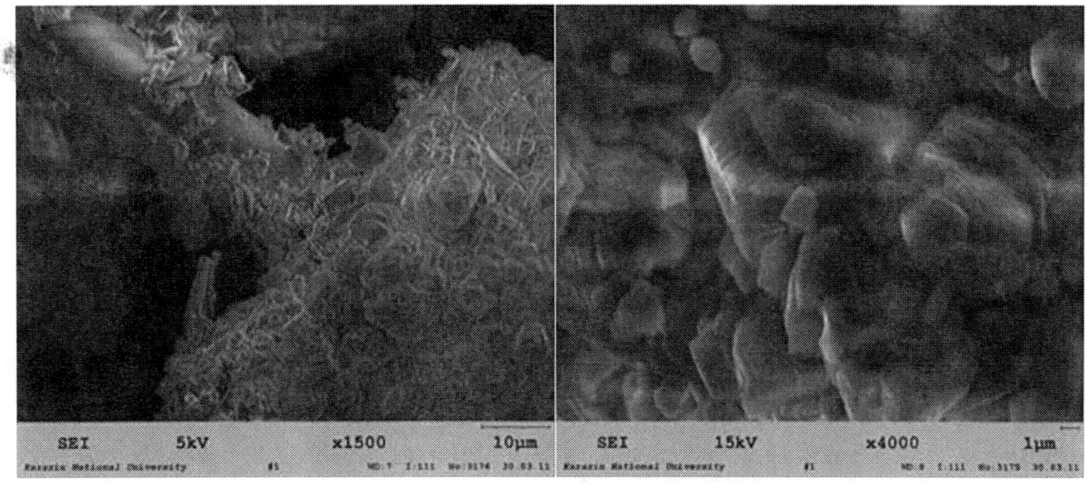


**Figure 7.** Formations peculiar to portlandite and calcite form a dense structure on fiberglass.

The system-forming hierarchical structure of water act as a matrix or "scaffold" for the self-organization of spatial structures during the formation of a crystalline cage on the surface of a polyester fiber coated with a silicone film. The assembly elements of these systematized structures are visible in the electronic image, Figure 8a. When water films are supersaturated with ions during cement hydration, crystals grow on the surface of the polyester fiber, figure 8, b.

IOP Conf. Series: Materials Science and Engineering **907** (2020) 012027 doi:10.1088/1757-899X/907/1/012027

a b

**Figure 8.** A through pore of perlite (a) and of a hollow glass microsphere (b), densely filled with cement hydration products; individual particles close to cubic shape: calcite from 0.6 microns in size.

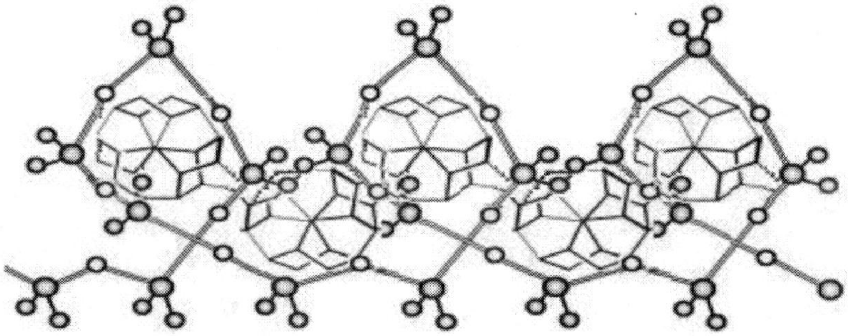

a)

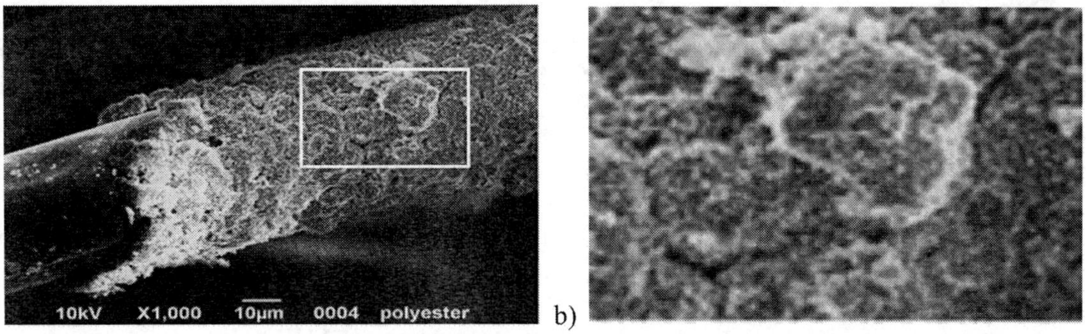


b)

c)

**Figure 9.** Formation of a crystalline cage on a silicone film of a polyester fiber:
A - A structural model of a siloxane chain with an integrated spiral-shaped structural element of water;
b - The growth of crystalline structures on the surface of a polyester fiber treated with a silicone sizing agent; c - A fragment of crystalline structures on a polyester fiber when magnified up to 3000x.

**Findings**

Therefore, it has been found that in a cement stone environment perlite, fiberglass and glass microspheres have a negative electric surface potential (charge) and can be used as a substrate for crystallization of hydration products with a positive surface charge.

The dipole moments and hydrogen bonds of water molecules promote the formation of the "structure" of water from molecular associates, consisting of about 5 molecules with dipole moments oriented in the same direction, and single water molecules.

From this perspective, the following pattern of the distribution of water by the charged surfaces of the structural elements of concrete seems most realistic. The structure of large water associates and the bonds between them are likely to be disrupted under the influence of the electric field of the surface and, more precisely, its potential-determining ions. Directly at the surface, the associates of 5 molecules of water are adsorbed mainly on the potential-determining ions, and in the intervals between the PDI, single unassociated molecules are adsorbed, which ensures the most dense structure of cement hydration products.

**References**

[1] Plugin A N, Vdovenko N V, Biryukov A I, Ovcharenko F D. 1988 On the mechanism of the emergence of the electric surface potential of various substances in aqueous solutions. *Proceedings of the USSR Academy of Sciences* Issue **3** Vol. 298 pp 656-661.

[2] Plugin A N, Plugin A A, Trikoz L V, Kaganovsky A S, Plugin Al A 2011 *Fundamentals of the theory of hardening, strength, deterioration and durability of Portland cement, concrete and structures made of them.* Vol. **1** Colloid chemistry and physicochemical mechanics of cement concrete. Edited by A N Plugin (Kyiv: Naukova Dumka) p 331

[3] Bondarenko D A 2010 *A dry mortar for obtaining thermal insulation materials with low vapor permeability.* Dissertation for the degree of Candidate of technical sciences. (Kharkiv: Kharkiv State Academy of Railway Transport) p 198

[4] Salia M G 2012 *Cement-based waterproofing coating with improved crack resistance.* Dissertation for the degree of Candidate of technical sciences. (Kharkiv: Kharkiv State Academy of Railway Transport) p 153

[5] Plahotnikov K V 2019 *Mineral thin-layer thermal insulation coating with improved physical and mechanical properties* Dissertation for the degree of Candidate of technical sciences (Kharkiv: Kharkiv National University of Civil Engineering and Architecture) p 158

[6] Distler G I 1972 *The electrical structure of actual surfaces of solids and the formation of boundary layers with special properties that ensure the transmission of the long-range influence of solids* (Moscow: Nauka) p 245

[7] Schukin E D, Pertsov F V, Amelina E A 2006 *Colloid chemistry* (Moscow: Higher school) p 444

[8] Lobyshev V I, Solovey A B, Boulyenkov N A 2004 Computer-assisted modular design of parametric water structures. *Biophysics* Vol. **48** No. 6

[9] Schukin E D, Pertsov F V, Amelina E A 2006 *Colloid chemistry* (Moscow: Vysshaya Shkola) p 444

[10] Plugin A A, Salia M G, Kostyuk T A 2011 Isotropic microreinforcement of cement stone with hydration products to increase the physical and mechanical properties of waterproofing coatings. *Visnyk of NTU "KhPI": Proceedings of works. Thematic issue "Chemistry, chemical technology and ecology"* **50** pp 97–103

[11] Levitsky M M 2009 *The many faces of siloxanes* Available at : http://him.1september.ru/2003/45/1.htm

[12] *The formation of water clusters* Available at : https://www.o8ode.ru/

IOP Conf. Series: Materials Science and Engineering **907** (2020) 012028 doi:10.1088/1757-899X/907/1/012028

# Architectural construction using technologies based on knowledge about the flow of Bingham plastic fluids in various pipelines

**A A Zadorozhnyi**

Department of Construction, Track and Handling Machines, Ukrainian State University of Railway Transport, Feuerbach sq., 7, Kharkov, Ukraine, 61050

zsnj1971@ukr.net

**Abstract.** Currently, such types of engineering work as construction of classic buildings and structures, as well as erection of new ones that have an irregular shape, i.e. curved, arched, elliptical, round and similar surfaces, can be performed using complex technological equipment, such as combined external or internal formwork, the use of shotcrete technology together with the off-formwork construction technologies (3D printers). One of the prerequisites for uniform transportation of media by a concrete pump through pipelines, concrete conveying pipes, and most importantly, their homogeneous distribution by an extruder during layering, is a study based on mathematical modeling and comparative analysis of mathematical models of the flow of Shvedov-Bingham fluids in pipelines of various cross-sections, determining the performance, optimal rheological properties of mortars and concrete liquids. This will allow further control of their fluidity and plastic viscosity during the extrusion process.

**Introduction.** The study of the flow of Bingham fluids fed with concrete pumps and mortar pumps through pipelines of various cross-sections to the extruder of a 3D printer is currently relevant. As the viscous-plastic fluids move in the pipeline, the change in the elastic limit $\tau_y$ arising between the inner surface of the pipe of a radius $R_{mp}$and the concrete mixture will be determined by the pressure difference $\Delta P$ on the pipeline section $\Delta L$ under consideration, created by the piston of the concrete pump or mortar pump. The flow of such concrete mixtures in the pipeline is characterized by properties that must be taken into account in the study of the movement processes and creation of working models for concrete mixtures.

**Review of recent research and publications.** Processes occurring in non-Newtonian fluids, such as concrete mixtures, are considered as those in a Shvedov-Bingham medium. The concrete mixture is a multicomponent medium that is described by the rheological laws of the flow of Bingham fluids [1, 5, 9- 11].

In-depth research of the processes of flow of Bingham fluids and the resulting mathematical models was carried out by such authors as W.L. Wilkinson, A. V. Gnoevoy, V. Prager, B. M. Smolsky, Z. P. Shulman, V. V. Gorislavets, S. S. Kutateladze, and others. The proposed models are widely used for rigid viscous-plastic media, which include concrete mixtures, and generalize most of the classical equations of the rheological state of fluids [2-4].

**Definition of the purpose and objectives of the study.** The purpose of the study is to identify inaccuracies in determining the elastic limit of $\tau_y$ in the experimental plotting of the flow curve of Bingham fluids that affect the flow characteristics of concrete pipelines, which, in turn, will allow a reasonable approach to the calculations and designing of new concrete pumps and mortar pumps [8], which are part of the technological lines for the non-formwork construction of irregular architectural structures [6, 7].

The objective of this work is to determine the productivity $Q$ (relative flow rate) and the speed of the central core $\Delta U_p$ depending on the parameters of the concrete mixture and the radius of the pipeline $R$.

**The main part of the study.** To calculate the flow rate of visco-plastic fluids and to determine the flow rate $Q$ of Bingham liquids moving through pipes of circular cross-section, the regularity shown in (Fig. 1) is generally used.

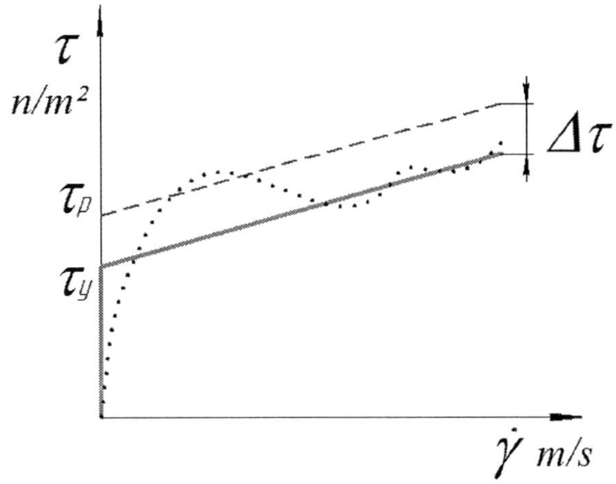


**Figure 1.** One-dimensional flow of viscous-plastic Bingham fluids (media)

Figure 1 shows the approximate position of the points of experimental values for determining the dependence of the shear stress $\tau$ on the shear rate of fluid layers $\dot{\gamma}$.

In this case, the elastic limit is determined with an accuracy $\Delta\tau$, , i.e. $\tau_y \pm \Delta\tau$.

For further convenience, we take $\tau_y$ as its possible minimum value, $\tau_{ymin}$ then $\tau_p = \tau_y + \tau$.

Coefficient of plastic viscosity $\mu_p = tg\alpha$.

In our case, it is reasonable to trace the influence of some accuracy$\Delta\tau$ on the above-mentioned flow characteristics of Bingham fluids.

Suppose the inner radius of the pipe is $R$.

The distribution of flow rates of the fluid over the radius can be found from the rheological equation, which for Bingham fluids is written as follows:

$$\tau - (\tau_y + \Delta\tau) = \mu_p \cdot \dot{\gamma}. \quad (1)$$

where $\tau$ is the shear stress; $\tau_y$ is the static elastic limit of the Bingham fluid; $\Delta\tau$is a specified accuracy; $\mu_p$ is the coefficient of plastic viscosity; $\dot{\gamma}$ is the rate of shear of the fluid layers.

Since the shear rate of the liquid layers$\dot{\gamma} = \frac{du}{dr}$ (where $r$ is the radius of the liquid flow layer), then we find:

$$\frac{du}{dr} = \frac{\tau - \tau_p}{\mu_p}. \quad (2)$$

IOP Conf. Series: Materials Science and Engineering **907** (2020) 012028 doi:10.1088/1757-899X/907/1/012028

Provided the fluid adheres to the walls of the pipe ($r$=$R$, $u$=0), after integrating this equation with $r$, the following flow rate equation is obtained$u(r)$:

$$u(r) = \frac{(R-r)\cdot\tau_w}{\mu_\rho}\cdot\left(\frac{R+r}{2R} - \frac{\tau_\rho}{\tau_w}\right). \tag{3}$$

where $\tau_w = \frac{R\cdot\Delta P}{2\Delta L}$is the shearing stress on the pipe walls; $\Delta P$ is differential end pressure at section $\Delta L$ [1].

As it is known, the central part of the flow (a cylinder with a radius $r_p$) of the Bingham fluid moves as a solid [1].

Since the shearing stresses$\tau_w$ are distributed along the radius of the pipe $r$ according to the linear law $\tau = \tau_w \frac{r}{R}$, then at the boundary of the plastic flow zone $\tau_0 = \frac{\tau_\rho\cdot R}{\tau_w}$. If we denote $\frac{\tau_\rho}{\tau_w} = \tau_{\rho 0}$, and $\frac{\tau_y}{\tau_w} = \tau_0$,, then:

$$r_\rho = R\cdot\tau_{\rho 0}, \tag{4}$$

where, $\tau_{\rho 0}$ are dimensionless quantities.

The flow rate at the boundary of the plastic zone (or, the same, the flow rate of the $u_\rho$central core) is found from equation (1.3), after the expression for $r_\rho$ is substituted (4).

$$u_\rho = \frac{R\cdot\tau_w}{2\mu_\rho}\cdot\left(1 - \tau_{\rho 0}\right). \tag{5}$$

where, $u_\rho$is the flow rate of the central core at the boundary of the plastic flow zone.

Let us write the flow rate equation $u(r)$ (1.3) in dimensionless form, for which the following flow rate profile designations are introduced$U(\rho)$:

$$\frac{r}{R} = \rho,\ \tau_{01} = \frac{\tau_\rho}{\tau_w} = \rho,\quad U(\rho) = \frac{u(r)\mu_\rho}{R\tau_w}.$$

The following expression for the flow rate profile is obtained $U(\rho)$:

$$U(\rho) = 0{,}5(1-\rho^2) - \tau_{01}(1-\rho). \tag{6}$$

Fig. 2 shows graphs of flow rate profiles of Bingham fluids$U(\rho)$ for various values $\tau_{01}$.

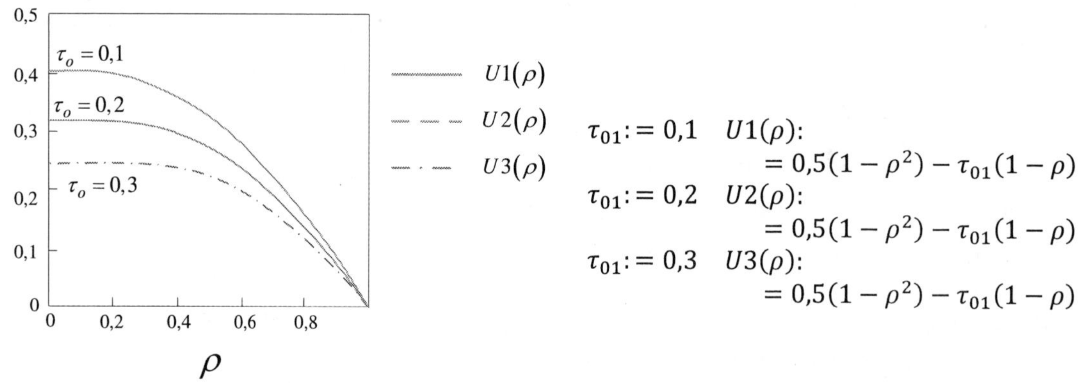


**Figure 2.** The flow rate profiles $U(\rho)$ for different values of $\tau_{01}$

Based on the above, an expression for the flow rate of the fluid is found:

$$Q = \pi \cdot r_p^2 \cdot u_p + \int_{r_p}^{R} 2\pi \cdot r \cdot u(r)\, dr = \frac{\pi \cdot R^3 \cdot \tau_w}{\mu_p} \cdot \left(\frac{\tau_{\rho 0}^2}{2} + \frac{\tau_{\rho 0}^4}{2} - \tau_{\rho 0}^3\right) +$$

$$+ \frac{\pi \cdot R^3 \cdot \tau_w}{\mu_p} \cdot \left(\frac{1}{4} - \frac{\tau_{\rho 0}^3}{3} - \frac{\tau_{\rho 0}^2}{2} + \tau_{\rho 0}^3 - \frac{5}{12} \cdot \tau_{\rho 0}^4\right) = \frac{\pi \cdot R^3 \cdot \tau_w}{\mu_p} \cdot \left(\frac{1}{4} - \frac{\tau_{\rho 0}}{3} + \frac{\tau_{\rho 0}^4}{12}\right). \quad (7)$$

The resulting expression is consistent with the Buckingham formula [1]. Let us consider how much the flow formula changes when the elastic limit$\tau_y$ changes by $\Delta\tau$.

In addition to denotation $\tau_{\rho 0}$and $\tau_0$, we introduce a dimensionless value $\Delta\tau_0 = \frac{\Delta\tau}{\tau_w} = k$.

The flow rate at the elastic limit$\tau_y$ is denoted $Q^/$, and at $\tau_0 - Q$.. Respectively, for the relative increment of the flow rate, the dependence (7) is valid.

$$\Delta Q = \frac{Q - Q^/}{Q^/} = \frac{Q}{Q^/} - 1. \quad (8)$$

The formula (7) includes the value$\tau_{\rho 0}^4$ .

$$\tau_{\rho 0}^4 = \tau_0^4 \cdot (1 + k)^4 = \tau_0^4 \cdot (1 + 4k + 6k^2 + 4k^3 + k^4) .$$

Since $k << 1$, $k^3$and $k^4$ are dropped in this expression.

On the basis of the above, here we get the expression $\frac{\tau_{\rho 0}^4}{12} \approx \frac{\tau_{\rho 0}^4}{12} \cdot (1 + 4k + 6k^2)$ .Then for the total flow$Q$ we find the following values:

$$Q = \frac{\pi \cdot R^3 \cdot \tau_w}{\mu_p} \cdot \left[\frac{1}{4} - \frac{\tau_0}{3} - \frac{\tau_0 \cdot (1+K)}{3} + \frac{\tau_0^4}{12} \cdot (1 + 4k + 6k^2)\right]. \quad (9)$$

According to the above, at the elastic limit$\tau_y$ for $Q^/$when $k = 0$the flow rate expression is found:

$$Q^/ = \frac{\pi \cdot R^3 \cdot \tau_w}{\mu_p} \cdot \left(\frac{1}{4} - \frac{\tau_0}{3} + \frac{\tau_0^4}{12}\right). \quad (10)$$

Based on the above, the formula for the relative change in flow (formula 1.8) will have the following form:

$$Q = \frac{\frac{1}{4} - \frac{\tau_0 \cdot (1+k)}{3} \frac{\tau_0}{12} \cdot (1+4k+6k^2)}{\frac{1}{4} - \frac{\tau_0}{3} + \frac{\tau_0^4}{12}} - 1 = \frac{\tau_0 \cdot K \cdot [-4 + 2\tau_0^2 \cdot (2+3k)]}{3 - 4\tau_0 + \tau_0^4}. \quad (11)$$

If we take into account that $\Delta\tau = 0.1\tau_y$, i.e. $K = 0.1$and $\tau_0 = 0.2$then we get $\Delta Q = -0.036$.

Thus, with an increase of $\tau_y$ by 10%, the flow is reduced by 3.5%. In addition, $\Delta Q$ depends on $\tau_0 = \frac{\tau_y}{\tau_w}$ in the non-linear way.

The radius of the central core increases by $k$:

$$\Delta r_p = \frac{r_p^/ - r_p}{r_p} = \frac{\tau_0 \cdot (1+K) \cdot R - \tau_0 \cdot R}{\tau_0 \cdot R} = k. \quad (12)$$

If we take that $k = 0.1$or 10% then the flow rate of the central core of the Bingham fluid flow will change as follows:

$$\Delta U_\rho = \frac{U_\rho^/ - U_\rho}{\mu_\rho} = \frac{U_\rho^/}{U_\rho} - 1 = \frac{(1+\tau_{\rho 0})^2}{(1+\tau_0)^2} - 1 = \frac{2\tau_0 \cdot k \cdot (-1+\tau_0+0.5\tau_0 \cdot k)}{(1+\tau_0)^2} = \frac{0.04}{0.64} \cdot (-1 + 0.21) = -0.049. \quad (13)$$

I.e. the flow rate $U(\rho)$is reduced by 5%. All changes depend on the relationship $\tau_0 = \frac{\tau_y}{\tau_w}$.

**Conclusions from the theoretical research and prospects, further development in the field.** For pipelines of a radius $R$ in which concrete and mortar mixtures are moving, the dependences of the flow rate profiles of the central core of Bingham fluids$U(\rho)$ are found for different values$\tau_{01}$. For a given accuracy$\Delta\tau$ and $k = 0.1$or 10%the flow rate decreases by 5%. The dependences of the relative change in the flow $Q$are obtained in the process of comparing the Shvedov-Bingham and Buckingham dependences. When $\tau_y$increases by 10%, the flow rate decreases by 3.5%. In addition, $\Delta Q$ depends on the ratio $\tau_0 = \frac{\tau_y}{\tau_w}$ in a non-linear way.

## References

[1] Wilkinson W L 1964 *Non-Newton fluids* (Moscow: Mir) p 216

[2] Gnoyevoy A V 2004 *Basics of theory of flow of Bingham media* (Moscow: FIZMATLIT) p 272

[3] Smolskiy B M, Shulman Z P, Gorislavets V M 1970 *Rheodynamics and heat transfer of non-linear viscous-plastic materials* (Minsk: Nauka I tekhnika) p 240

[4] Prager W 1963 *Introduction into mechanics of continuous media* (Moscow: Inostrannaya literatura) p 406

[5] Weipert D 1993 *Rheologie der Lebensmittel* (Hamburg: Behrs) p 620

[6] Bugayevskiy S, Babaev V, Shmukler V, Gerasymenko V, Zadorozhny A, Viselga G 2017 Principles of Design and Formation of a Self-Supporting Skeleton in Constructing Buildings of Complex Geometric Shapes with the Use of Shotcrete Technology *The 13th International Conference Mechatronic Systems and Materials (MSM-2017)* (Vilnius, Lithuania) pp 28

[7] Bugaevskyi S O, Zadorozhnyi A O, Kovrevskiy A P, Stefan O N, Bugaevskyi M S 2017 Study of properties of self-compacting concrete in time *Road transport. Collection of proceedings* (Kharkiv: KHNADU) pp 116-124

[8] Remarchuk N, Zadorozhnyi A, Chmuzh Y 2017 Methodology of Creating the Hydraulic Actuator at the Stage of its Design *Eastern-European journal of enterprise technologies* No. 2/7 **86** pp 42- 50.

[9] Zadorozhnyi A A, Kovrevski A P 2017 Analysis of the flow of Bingham fluids though circular pipelines *Collection of research papers of UkrDUZT* Issue **168** pp 44-49

[10] Zadorozhnyi A A, Kovrevski A P, Chovnyuk Y V, Remarchuk N P 2018 Features of the flow of liquids of variable viscosity by the pipeline of a various form of transverse section. *Collection of scientific works of the international scientific conference „Technology and transport infrastructure"* (Kharkiv: UkrDUZT) pp 24-25

[11] Zadorozhnyi A, Kovrevski A, Chovnyuk Y, Remarchuk N 2018 Flow of a Bingham Fluid Through Circular Pipes with Variable Viscosity Coefficient Along the Pipe Length *International Journal of Engineering and Technology* Vol **7** No 4.3 p 100-104

IOP Conf. Series: Materials Science and Engineering **907** (2020) 012029 doi:10.1088/1757-899X/907/1/012029

# Influence of genesis of mineral components of asphalt concrete for its durability in aggressive environments

**Sergii Yefremov**

Department of Technology of Road-Building Materials and Chemistry, Kharkiv National Automobile and Highway University, 25, Yaroslava Mudrogo str., Kharkiv 61002, Ukraine

svefr@ukr.net

**Abstract**. The results of determining the durability of asphalt concrete are presented by the method developed at the Department of Technology of Road-Building Materials and Chemistry of KhNADU. The features of the resistance of asphalt concrete on various rocks by the influence of liquid aggressive environments are shown. The coefficients of stability of asphalt concrete to the action of liquid aggressive media on asphalt concrete from mixtures of monomineral compositions made from various rocks are determined. Thus, despite the difference in the origin of the rocks that make up the asphalt concrete, a long service life of road coatings based on them can be ensured, which in turn can ensure high profitability of using these coatings even on roads with heavy traffic and difficult conditions exploitation.

## 1. Introduction

The wide variety of crystalline rocks of the deposits of Ukraine gives the possibility of a large number of options for selection and use in a wide range of mineral materials of different origin as constituents of asphalt concrete mixtures. And the study of this possibility in terms of its impact on the durability of asphalt concrete is becoming more and more urgent every year. Particularly urgent is the study of the durability of asphalt concrete with the simultaneous action of stress and a liquid aggressive environment. The use of local rocks, such as basalts, quartzites, limestones, sandstones, or other geological sites, whose deposits are much closer to the construction site than granite materials used for road construction, can significantly reduce the financial cost of transportation, and in some cases also for extraction, without losing the normative quality of asphalt concrete coating in which these mineral components can be used. In the first place, it is necessary to rely on the regulatory requirements for asphalt concrete, in National Standard of Ukraine DSTU B V.2.7-119: 2011 [1].

The main operational indicator of asphalt concrete, i.e. the material most used in road construction, is its durability. In the regulatory requirements for asphalt concrete [1] analog of its durability is the coefficient of long-term water resistance, which is determined by the degree of bending strength of its samples under the action of water during the normative time compared to samples aged in air at temperature (20±1) °C [2]. The constituents of this indicator are the strength of the bending specimens, which have been tested in a water-saturated and non-saturated state. But these strength indicators very roughly reflect the effect of the aquatic environment on the durability of asphalt concrete. A more sensitive indicator for predicting the durability of asphalt concrete may be the coefficient of resistance of asphalt concrete to the influence of liquid environments ($K_{l.e.}$) on its durability, which was used in previous works [3-6]. The components of this indicator were the time before the destruction of asphalt

IOP Conf. Series: Materials Science and Engineering **907** (2020) 012029 doi:10.1088/1757-899X/907/1/012029

concrete samples, which were tested for bending in the aquatic environment ($t_{l.e.}$) and the time before the destruction of asphalt concrete samples, which were tested for bending in air ($t_{a.}$) and which were used as the ratio (1) of the first to second:

$$K_{l.e.} = \frac{t_{l.e.}}{t_{a.}} \quad (1)$$

Measurements of time before the destruction of asphalt concrete beams were carried out in seconds and the results ranged from 10 s to $10^5$ s. Such a large range demonstrates the greater sensitivity method of measurement the time before the destruction to that of other methods. The numerical value of the coefficient of resistance of asphalt concrete to the influence of liquid environment is less than 1, since $t_{l.e.}$ always below $t_{a.}$ as a result of the adverse effect of the aquatic environment.

Mineral components of asphalt mixes in different territorial regions of Ukraine may be different in origin and mineral composition, but the main thing is so that they do not reduce the inter-service life of asphalt concrete pavement of roads in which they are used. To avoid reducing the durability of asphalt concrete, it was hypothesized that rocks, most of which include minerals of the silicate and carbonate class, can be used as components of asphalt mixtures [7]. And these rocks are present throughout the massif of the Ukrainian Crystal Shield, the Carpathians and the Crimea.

## 2. Geological survey of rock-forming minerals of Ukraine

Along with granites, as traditional mineral materials used in the production of asphalt mixtures, the northwestern and central parts of the Ukrainian Shield are composed of such rocks as basalts, gabbro, diorites, diabases, pegmatites, amphibolites, migmatites, labradorites, syenites, quartzites, gneisses, shales and marbles [8]. The south-eastern and Azov parts of the Ukrainian Shield are mostly represented by such rocks as basalts, gabbro and sandstones [8, 9]. The Carpathian region consists mostly of limestone, but there are granites, diorites, gneisses, andesites and sandstones [8]. The Crimean region is mostly represented by limestones and shale rocks of various origins [8].

Such a wide range of these rocks by origin causes a significant amount of rock-forming minerals [10-16]. But in turn, the properties of minerals depend on the chemical composition, building of the crystal lattice and many of their physical properties [11, 12, 17-21]. Depending on these properties, the structure and texture of the asphalt concrete is formed, which includes selected mineral components and in which the durability of asphalt concrete pavement of roads depends.

The choice of mineral components was based on three principles: first - rocks must have a brand beyond the strength of not less than M800; secondly, most rock-forming minerals must be of silicate or carbonate origin; third, the deposits of the selected rocks should be represented by the largest number in the respective region. Effective study of the durability of asphalt concrete was based on the initial physical and mechanical properties, chemical composition and the influence of the origin of selected rocks as components of the asphalt mixture, the interaction with organic petroleum binders [9-11, 22-27]. Based on this knowledge, four types of rocks were accepted for testing: basalt (Rafalivske deposit of Rivne region), limestone (Old Crimean deposit of the Autonomous Republic of Crimea), sandstone (Vilyanivske deposit of Luhansk region) and granite (Redutskoe deposit of Poltava region). selected as a comparative reference rock, which is traditionally used for the production of asphalt mixtures.

For all mineral materials accepted for testing, as potential components of the asphalt concrete mixture, the main physical and mechanical properties were determined, which are presented in table 1. The mineralogical composition of basalt consists of: silica (40÷52 %); microliths of plagioclase; clinopyroxene (augite); titanomagnetite; hornblende and as an igneous rock belongs to the class of "basic". The chemical composition of basalt includes the following compounds: $SiO_2$; $K_2O$; $Al_2O_3$; $Fe_2O_3$ $TiO_2$; $FeO$; $MnO$; $MgO$; $CaO$; $P_2O_5$ [10, 11], this causes their effect on the unsatisfactory interaction with organic petroleum binders and the adhesion of bitumen to the mineral surface of acid rocks is significantly reduced [23-27].

**Table 1.** The main physical and mechanical properties of the original rocks.

| Indicator | Test results | | | | Requirements of DSTU B V.2.7-119: 2011 for asphalt concrete type "V" I brand |
|---|---|---|---|---|---|
| | granite | basalt | limestone | sandstone | |
| True density, kg/m$^3$ | 2720 | 2870 | 2850 | 2690 | - |
| Average density, kg/m$^3$ | 2700 | 2850 | 2690 | 2470 | - |
| Porosity, % | 0.7 | 0.7 | 3.3 | 8.2 | - |
| Grade of initial rock by crushability | M1200 | M1200 | M800 | M800 | not less than M800 |

Dense marble-like limestone has the following mineralogical composition: calcite; dolomite. The chemical formula of limestone includes the following compounds: $CaCO_3$; $MgCO_3$ [10, 11]. The interaction of these compounds has a satisfactory effect on the adhesion of bitumen to the mineral surface of alkaline rocks [23-27].

The mineralogical composition of sandstones as sedimentary rock fragments is due to the largest amount of quartz with some inclusions of feldspars and mica. The sandstone tested had iron oxide cementation and, accordingly, the chemical formula included the following compounds: $SiO_2$ and $Fe_2O_3$ [10, 11]. As is known from many sources [23-27], such a set of chemical compounds is not very conducive to good adhesion of bitumen to the mineral surface of acid rocks.

Comparative (reference) granite material, as an acidic igneous rock, in mineral composition was represented by: quartz (>50 %); plagioclase; feldspars; mica. Chemical composition of granite: $SiO_2$; $K_2O$; $Al_2O_3$; $Fe_2O_3$ $TiO_2$; $FeO$ [10, 11]. Since granite is an acid rock, the adhesion of bitumen to its mineral surface is not very satisfactory.

## 3. Materials

Samples of all rocks accepted for testing were crushed and scattered into fractions starting from 10÷15 mm and ending with a fraction <0,071 mm. Subsequently, asphalt concrete mixtures were formed in fractions and each of them included only one rock, ie all mixtures were single-rock. The particle size distribution for all mixtures was the same, which is presented in table 2.

**Table 2.** Granulometric composition of asphalt concrete mixtures on granite, basalt, limestone and sandstone stone materials

| Leftovers | Content by weight, % of mineral grains larger than this size, mm: | | | | | | | | | Bitumen content, % |
|---|---|---|---|---|---|---|---|---|---|---|
| | 10 | 5 | 2.5 | 1.25 | 0.63 | 0.315 | 0.14 | 0.071 | <0.071 | |
| Partial | 10 | 15 | 12 | 18 | 6 | 11 | 10 | 3 | 15 | 6,0 |
| Full | 10 | 25 | 37 | 55 | 61 | 72 | 82 | 85 | 100 | |
| Requirements to grading (full leftovers) DSTU B V.2.7-119 V type | 0 ÷ 10 | 25 ÷ 35 | 36 ÷ 49 | 48 ÷ 63 | 60 ÷ 75 | 71 ÷ 84 | 80 ÷ 89 | 84 ÷ 91 | 100 | 6,0 ÷ 7,5 |

BND 60/90 bitumen was used as an organic petroleum binder for testing. The content of bitumen in all asphalt mixtures was 6.0 %, which corresponded to the optimal content of bitumen in the asphalt mixture, which consisted of one rock using granite.

To determine the structuring properties of the asphalt-binding component of asphalt concrete, several tests of mineral powders from the previously listed rocks were performed, the results of which are presented in Table 3. According to the results of tests of powders of different genesis, obvious facts attract attention: the highest bitumen content of limestone mineral powder is most likely due to the most developed surface of grains and probably corresponds to its largest specific surface area, which allows optimal packaging of mineral particles completely covered with bitumen; the minimum porosity at consolidation allows to reach the minimum water saturation that in turn allows to increase coefficient of long water resistance as a guarantee of durability.

**Table 3.** Normative properties of mineral powders of source rocks.

| Indicator | Test results | | | | Requirements of DSTU B V.2.7-121-2003 |
|---|---|---|---|---|---|
| | granite | basalt | limestone | sandstone | |
| Grain content, % by weight: | | | | | not less than |
| smaller than 1.25 mm | 100 | 100 | 100 | 100 | 100 |
| smaller than 0.071 mm | 73 | 72 | 73 | 74 | 70 |
| Porosity when compacted, % | 29.0 | 31.9 | 27.5 | 23.8 | not more than 35 |
| Swelling of the mixture of powder and bitumen, % by volume | 2.05 | 2.03 | 2.41 | 2.12 | not more than 2.5 |
| Bitumen capacity, g | 51.2 | 52.4 | 59.8 | 58.7 | not more than 65 |
| Average density of asphalt binder, kg/m$^3$ | 2140 | 2230 | 2160 | 2090 | - |

Cylindrical specimens with a diameter of 71.4 mm were made of asphalt concrete mixtures, which consisted of one rock, which were tested in order to obtain their physical and mechanical properties. The test results are shown in table 4.

**Table 4.** Physical and mechanical properties of asphalt concrete on granite, basalt, limestone and sandstone mineral materials.

| Indicator | | Test results | | | | Requirements of DSTU B V.2.7-119: 2011 for asphalt concrete type "V" I brand |
|---|---|---|---|---|---|---|
| | | granite | basalt | limestone | sandstone | |
| Average density, kg/m$^3$ | | 2325 | 2453 | 2394 | 2298 | - |
| Water saturation, % | | 3,0 | 2,5 | 1,9 | 3,0 | not more than 3.0 |
| Swelling, % | | 0 | 0 | 0 | 0 | - |
| Limit compressive strength, MPa, at temperature, °C: | 0 | 9,6 | 10,8 | 10,7 | 9,2 | not more than 12.0 |
| | 20 | 4,8 | 5,5 | 5,8 | 5,7 | not less than 2.7 |
| | 50 | 1,8 | 1,9 | 2,6 | 2,1 | not less than 1.3 |
| Coefficient of water resistance at long water saturation (on the 15th day) | | 0,90 | 0,92 | 0,93 | 0,90 | not less than 0,90/0,86[a] |

[a] The numerator contains the requirements for the coefficient of long-term water resistance for the upper layer of the coating, the denominator - for the lower layer of the coating.

## 4. Durability test results and their analysis

To determine the durability of asphalt concrete, which consisted of one rock, 4×4×16 cm beam specimens were made of single-rock asphalt concrete mixture, which were tested for bending in order to obtain time to their destruction at different stresses with simultaneous action of different of liquid aggressive environments. The test temperature was maintained close to 21 °C. Asphalt concrete beam samples were tested on a press, as shown in Figure 1, in the stress range from 0.75 MPa to 2.50 MPa. The obtained results of durability are presented graphically in Figures 2 and 3, as well as in the form of indicators of resistance of asphalt concrete against the action of liquid aggressive media in table 5.

**Figure 1.** Scheme of the test lever press to determine long-term tensile strength in bending: 1 - four-pressure press lever; 2 - counterweight; 3 - transmitting rod; 4 - limit switch; 5 - asphalt concrete beam; 6 - lower supporting cylinders ∅ 10 mm; 7 - loading area; 8 - upper cylinders of the loading platform ∅ 20 mm.

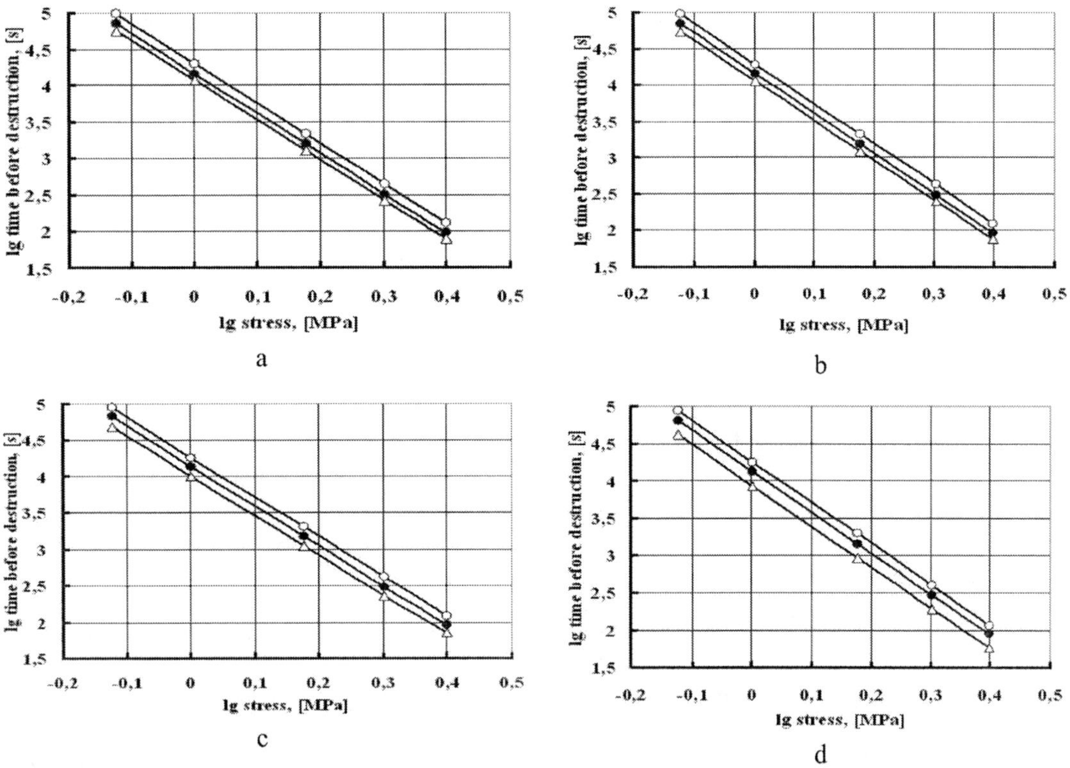


**Figure 2.** Dependence of the time before the destruction of asphalt concrete (type “V”) on limestone (a), granite (b), basalt (c) and sandstone (d) mineral material with 6.0 % BND 60/90 bitumen on stress, at a temperature of 21 °C in environments: ○ – air; ● – distilled water; △– 5 % aqueous solution of NaCl.

IOP Conf. Series: Materials Science and Engineering **907** (2020) 012029 doi:10.1088/1757-899X/907/1/012029

**Table 5.** Resistance of asphalt concrete on various mineral components to action of liquid aggressive environments.

| Indicator | Test results | | | |
|---|---|---|---|---|
| | granite | basalt | limestone | sandstone |
| Time before destruction in 5 % aqueous NaCl solution at a stress of 0.75 MPa, s | 54002 | 49011 | 56175 | 41320 |
| Time before destruction in distilled water at a stress of 0.75 MPa, s | 69671 | 67712 | 71567 | 64010 |
| Time before destruction in air at a stress of 0.75 MPa, s | 93335 | 90625 | 95625 | 88007 |
| Coefficient of resistance of asphalt concrete ($K_{l.e.}$) to the influence of 5% aqueous NaCl solution at a stress of 0.75 MPa | 0,57 | 0,54 | 0,59 | 0,47 |
| Coefficient of resistance of asphalt concrete ($K_{l.e.}$) to the influence of distilled water at a stress of 0.75 MPa | 0,75 | 0,75 | 0,75 | 0,73 |
| Time before destruction in 5 % aqueous NaCl solution at a stress of 2.51 MPa, s | 77 | 70 | 80 | 59 |
| Time before destruction in distilled water at a stress of 2.51 MPa, s | 93 | 90 | 96 | 88 |
| Time before destruction in air at a stress of 2.51 MPa, s | 124 | 120 | 128 | 117 |
| Coefficient of resistance of asphalt concrete ($K_{l.e.}$) to the influence of 5% aqueous NaCl solution at a stress of 2.51 MPa | 0,62 | 0,58 | 0,63 | 0,50 |
| Coefficient of resistance of asphalt concrete ($K_{l.e.}$) to the influence of distilled water at a stress of 2.51 MPa | 0,75 | 0,75 | 0,75 | 0,75 |

All time dependences presented in Figure 2 for asphalt concrete on different mineral components are located in the same way - above all are the dependences of the time to failure of asphalt concrete tested in air, in the middle, tested in distilled water, and the lowest dependences relate to the test 5 % aqueous NaCl solution. This location corresponds to well-known ideas, because the operation of asphalt pavements in natural conditions under the action of air depends only on the value of the applied voltage, as well as its durability. However, as climatic conditions change, the action of stress can be accompanied by the action of liquid media, such as rain or melt water and chloride solutions, which are used to combat winter slipperiness. But some differences distinguish these dependencies. Closest to each other are these three dependences for the limestone filler, which indicates the best interaction of mineral components with bitumen, and this corresponds to well-known ideas and previously performed research [23-27].

The resistance of asphalt concrete on the limestone components of the aggressive action of liquid environment is the highest, as evidenced by the coefficient of resistance of asphalt concrete to liquid environment ($K_{l.e.}$), which was at a stress of 0.75 MPa for 5% aqueous salt solution 0.59 (table 5). The duration of the aggressive environment at the lowest stress of 0.75 MPa was more than 11 hours, which allowed for such a long time to penetrate the aggressive environment between the bituminous film and the mineral surface and contribute to the destruction of asphalt concrete, so $K_{l.e.}$ for all single-breed asphalt concrete samples the smallest. The action of the aqueous medium is less aggressive, which is also confirmed by the values of $K_{l.e.}$, which for almost all samples was 0.75 at all stresses. However, if we consider the location of the lines of dependence of durability of all monomineral asphalt concrete samples accepted for testing in Figure 3, they are very close to each other, and some even merge, which indicates the overall high stability of asphalt concrete on all accepted rocks.

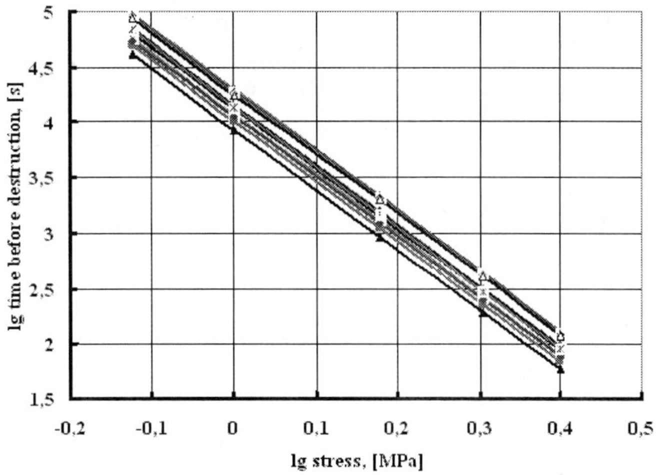


**Figure 3.** Dependence of the time before the destruction of asphalt concrete (type "V") on various mineral materials with 6.0 % BND 60/90 bitumen on stress, at a temperature of 21 °C in environment: ○ – limestone air; ● – limestone distilled water; × – limestone 5 % aqueous solution of NaCl; + – granite air; - – granite distilled water; ◆ – granite 5 % aqueous solution of NaCl; ◇ – basalt air; ✱ – basalt distilled water; ■ – basalt 5% aqueous solution of NaCl; △ – sandstone air; □ – sandstone distilled water; ▲ – sandstone 5 % aqueous solution of NaCl.

## 5. Conclusions

Analyzing the test results, we can assume that for all asphalt concretes accepted for testing with the use of mineral materials of different rocks in general, satisfactory durability indicators were obtained, indicating a sufficiently high resistance to liquid aggressive media. However, it should be noted that asphalt concrete, in which the limestone mineral component was used, may have the longest service life in road surface, because its coefficient of resistance of asphalt concrete to the influence of liquid media ($K_{l.e.}$) is the highest at all stress levels. This is explained and corresponds to the well-known laws [23÷27] of better interaction of alkaline mineral surfaces with bitumen in comparison with acidic ones. The predominant surface interaction with alkaline mineral surfaces allows to obtain the best bitumen coating of limestone grains, which in turn allowed to obtain the highest durability, given that the tensile strength of the original rock in terms of fragmentation is not the highest.

Against the background of high normative test results of asphalt concrete on limestone, other results are also quite high and they all meet the requirements of DSTU B V.2.7-119:2011 [1] for asphalt concrete type "V" I brand, as well as it is characteristic and for durability. Compared with the obtained normative coefficient of water resistance at long water saturation on the 15th day ($K^{w}_{l}$) for asphalt concrete on granite, the same indicator for asphalt concrete on basalt is higher, which is explained by higher density and strength of the original basalt rock in turn, it was possible to obtain a higher density and strength of asphalt concrete on its basis, but the coefficient of resistance of asphalt concrete on basalt to the influence of liquid media ($K_{l.e.}$) is lower, because its mineral composition contains more augite, hornblende and Labrador than granite, which adversely affects its durability. In general, it can be argued that the results allow the use of dense basalt, dense marble-like limestone and fine-grained sandstone with iron oxide cementation for the production of asphalt mixtures to replace granite rocks.

## References

[1] National Standart Ukraine DSTU B V.2.7-119:2011 Asphalt mixtures and road and airfield asphalt. Specifications
[2] National Standart Ukraine DSTU B V.2.7-319:2016 Asphalt mixtures and road and airfield asphalt. Test methods.

[3] Yefremov S V 2009 The role of the temperature factor in the process of destruction of asphalt concrete under the combined influence of mechanical loads and aggressive environments *News of DonNABA* (Makiyivka: DonNABA Press) **75** pp 20-26.
[4] Yefremov S V, Hamad R A, Zolotarev V A, 2010 Influence of aqueous surfactant solutions on the adhesion of bitumen to the substrate and the durability of asphalt concrete *Proc. of the Int. Science-and-Technology Conf. "Project the exploitation of non-rigid road roads"* (Kharkiv: KhNADU Press) pp 189-196.
[5] Yefremov S V 2013 Influence of structure-forming components on durability of asphalt concretes in aggressive environments as a factor of safety of highways *Bulletin of KhNADU* (Kharkiv: KhNADU Press) **61-62** pp 239-242.
[6] Yefremov S V 2017 Influence of temperature on durability in aggressive environments of asphalt concrete on granite and limestone crushed stone *Bulletin of KhNADU* (Kharkiv: KhNADU Press) **79** pp 123-127.
[7] Yefremov S V 2014 Influence of the origin of mineral components of asphalt concrete on its physical and mechanical properties *Proc. of the Int. Scientific-Practical Conf. "Improvement of structural, technological and operational performance of roads and artificial structures on them in student research and young scientists"* (Kharkiv: KhNADU Press) pp 239-242.
[8] Kulemsa V G, Feshchenko N I, Feshchenko N M 1975 *Reference manual on mineral building materials* (Kiyiv: Budivelnik) p 360.
[9] Pigulevsky P I 2011 The material composition of the upper mantle of the southeastern part of the Ukrainian shield according to geological and geophysical data Scientific *Bulletin of the National Mining University* (Dnipropetrovsk: RIK NMU of Ukraine) **1** pp 5-10.
[10] Levinson-Lessing F Yu 1955 *Petrography* (Moscow: Publishing House of the Academy of Sciences of the USSR) p 447.
[11] Lazarenko E K 1951 *Mineralogy course* (Kiev: State Publishing House of Technical Literature of Ukraine) p 688.
[12] West Terry R 1995 *Geology applied to engineering* Prenttice Hall Englewood Cliffs, NJ.
[13] Barth T F W 1962 *Theoretical Petrology 2nd ed.* John Wiley & Sons, New York.
[14] Ernst W G 1969 *Earth Materials* Prenttice Hall Englewood Cliffs, NJ.
[15] Mason B 1960 *Principles of Geochemistry 2nd ed.* John Wiley & Sons, New York.
[16] Williams H, Turner F J, Gilbert C M 1954 *Petrography* W.H. Freeman and Co., San Francisco.
[17] Berry L G, Mason B, Dietrich R V 1983 *Mineralogy 2nd ed.* W.H. Freeman and Co., New York.
[18] Herman C, Zim H S, Shaffer P R 1963 *Rocks and Minerals* Golden Press, New York.
[19] Klein C, Hurlbut C S Jr 1985 *Manual of Mineralogy 20th ed.* John Wiley & Sons, New York.
[20] Sinkankas, J 1966 *Mineralogy* A First Course, D. Van Nostrand Co., Princeton, NJ.
[21] Wieser M E, Holden N, Coplen T B, Böhlke J K, Brand W A, Bièvre P De, Gröning M, Loss R D, Juris Meija, Takafumi Hirata, Prohaska T, Schoenberg R, O'Connor G, Walczyk T, Yoneda Sh, Xiang-Kun Zhu 2013 *Atomic weights of the elements 2011* (*IUPAC Technical Report*) Pure and Applied Chemistry **85** № 5.
[22] Zaritsky P V, Tykhonenko D G, Gorin M O, Andreev V V, Degtyarev V V 2012 *Geology with the basics of mineralogy* (Kharkiv: Maidan) p 584.
[23] Kolbanovskaya A S 1955 *Coupling of bitumen with a mineral surface Study of hydro- and heat-insulating materials and structures* (Moscow: Publishing House for Construction and Architecture) pp 66-70.
[24] Lysikhina A I, Sitskaya R M, Avlasova N M, Yastrebova L N 1952 *On the stability of bitumen and their interaction with mineral materials* (Moscow: Dorizdat) p 175.
[25] Kolbanovskaya A S, Mikhailov V V 1973 *Road bitumen* (Moscow: Transport) p 259.
[26] Rudenskaya I M, Rudensky A V 1984 *Organic binders for road construction* (Moscow: Transport) p 229.
[27] Lysikhina A I 1962 *Pavement and foundation with the use of bitumen and tar* (Moscow: Avtotransizdat) p 360.

# Nanomodified cement composites for thin walled architectural structures

**Andrii Plugin[1], Kateryna Pushkarova[2] and Maryna Sukhanevych[2]**

[1]Department of Building Materials and Structures, Ukrainian State University of Railway Transport, Feuerbach Square, 61500, Kharkiv, Ukraine
[2]Department of Building Materials, Kyiv National University of Construction and Architecture, Povitroflotsky Avenue 31, 03037, Kiyv, Ukraine

m.suhanevich@ukr.net

**Abstract.** The article presents the studies of the processes of structure formation in the system "cement-plasticizer - crude carbon nanotubes", established the relationship with the parameters of the structure, physical-mechanical and operational properties of composite materials from the standpoint of physico-chemical mechanics of dispersed systems and shows the influence of processes on the physical-mechanical properties of cement composites. The plastic strength and physical and mechanical properties of nanomodified cement mortars and fine-grained concretes have been studied and it has been shown that the use of the developed composite materials should be used to obtain fine-grained concretes for thin-walled architectural structures.

## 1. Introduction

Recently, in the field of production of building materials are becoming increasingly popular nanotechnologies that provide the ability to control the processes of structure formation at the nanoscale as a result, aimed at changing the composition and structure of tumors, and hence the performance of artificial stone [1, 2].

Various technological techniques, which can be attributed to nanotechnology, allowed to significantly improve the physical and mechanical properties of various nanosized additives to modify metal, ceramic, glass matrices [3], compositions based on Portland cement [4]. The use of artificially synthesized carbon-containing nanoparticles - fullerenes, astralenes, single- and multilayer nanotubes, as additives in the composition based on Portland cement has increased strength, impact resistance, reduced deformability of artificial stone, etc. [9, 10]. The greatest increase in physical and mechanical properties was provided by carbon nanotubes [14], especially added in the form of dispersions in plasticizers [15] and at certain optimal dosages of nanotubes [16]. In [15] there was a much greater plasticizing effect from the joint use of a plasticizing additive with nanotubes than from the plasticizing additive alone. Unfortunately, the use of nano-additives in cement concrete is limited due to their high cost. Reducing the cost in this case can be achieved through the use of crude carbon nanotubes.

Investigating the reasons for the influence of nanoadditives, in particular, nanotubes, on the properties of compositions based on Portland cement [5-10], gypsum [11], various authors note their influence on the structure, hydration processes over time, the formation of distribution boundaries phase etc. Researchers [6] explain the improvement of the properties of artificial stone due to its micro reinforcement due to the creation of conditions for the formation of elongated crystalline new formation, which grow and intertwine, forming a spatial grid, reinforcing artificial stone. In [12] it was found that nanomodification of cement stone to a greater extent affects the structure of hydrosilicate phases, increasing their density, and nanoparticles with high reactivity, large surface area and significant surface energy play the role of crystallization centers, nano reinforcement or nanofiller porosity of composites [13]. The better ability to modify slag-containing cements in [16] is also explained by the peculiarity of the morphology of new formation, in particular by the fact that in the presence of blast furnace granular slag aluminosilicates or natural aluminosilicates structures of lower basic hydrosilicates of elongated shape are formed, for which nanotubes are substrates. However, such explanations of the reasons for the improvement of the physical and mechanical properties of cement composites by nanotubes are a statement of their influence on the morphology of artificial stone and do not reveal the real mechanism of this improvement.

Although many researchers consider nanotechnology to be something new, in the technology of mineral binders and concretes, such methods and means, such as plasticizing building mixtures by adsorption of surfactant molecules, have been known since the first half of the twentieth century. In the scientific community it is believed that the hardening of mineral binders is the process of dissolving particles of matter, the formation of nanosized particles of hydrates and their aggregation into a strong stone. Therefore, it is natural to say that the scientific basis of construction nanotechnology began to create in the XIX century, V. Michaelis and A. Le Chatelier (colloid-chemical and crystal-chemical theory of hardening of Portland cement, respectively), as well as in the XVIII century. A. Lavoisier, who was the first to establish the laws of gypsum hardening [17].

However, to a qualitatively new level of nanotechnology of building materials in the middle of the twentieth century brought out the work of P.O. Rebinder, who developed colloid chemistry, outlined in its composition the physical-chemical mechanics of dispersed systems, developed on its basis a modern theory of hardening of Portland cement (hydration structure). It is the laws and methods of physical and chemical mechanics that allowed A.M. Plugin to supplement Rebinder's theory with electro heterogeneous mechanism of structure formation, to explain and make regulated water-tight permeability and excessive creep of concrete, to explain the mechanism of plasticizing action by adsorption of anionic surfactants (surfactants) on positively alluvial products permeable action on the basis of Portland cement in concrete [17].

Based on the above, the aim is to establish the mechanisms of influence of carbon nanotubes in combination with plasticizers, on the structure and properties of compositions based on Portland cement in order to obtain fine-grained concrete for thin-walled architectural structures.

## 2. Raw materials and test methods

Theoretical studies were performed by analyzing the compliance of the previously defined processes of structure formation and properties of Portland cement-based compositions with carbon nanotube additives and plasticizing additives to the known laws of colloid chemistry and physical-chemical mechanics of dispersed systems.

Portland cement CEMII/A-S-32,5, plasticizing additives such as: lignosulfonate with esters of polycarboxylate (SikaPlast 520), naphthalene formaldehyde, melamine formaldehyde (Muraplast FK-98), polycarboxylate were used for experimental researches. As a nanoadditive, crude multilayer carbon nanotubes coated with a silicon airgel with interplanar distances of 0.34-0.36 nm and a particle size of 60-200 nm were used.

Aqueous dispersions were made from carbon nanotubes and plasticizers using a cavitation mixer, in which the content of nanotubes and plasticizers was taken as 1% by weight of cement.

Determination of the features of the structure of artificial stone and identification of its elements was carried out by physical-chemical methods using a scanning electron microscope REMMA-102-02.

Physical-mechanical and operational properties of cement compositions and fine-grained concrete were determined in accordance with current regulations.

## 3. Results and discussion

Comparison of electron micrographs of treated (CNT) and untreated carbon nanotubes (UCNT) (Fig. 1) and studies of crude tubes showed that the diameter of the tubes is 10–20 nm, and crude tubes contain a significant number of gel formations (Fig. 1, b), which consist of an aerosil residue in the amount of 6–20% and graphite in dimensions comparable to the transverse dimensions of nanotubes of 10–20 nm.

*a)*

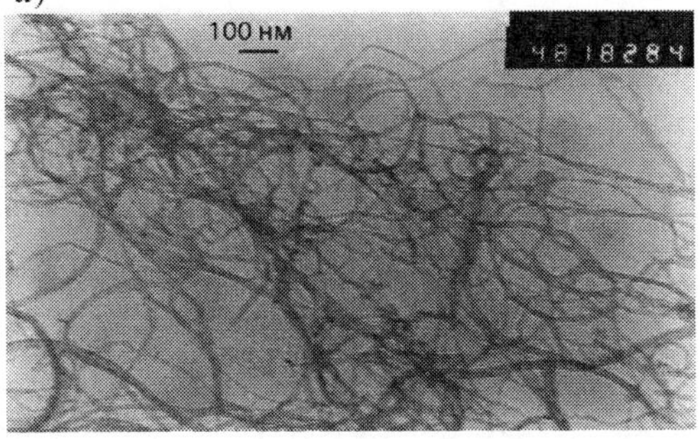


*b)*

**Figure 1.** Electronic micrographs of carbon nanotubes:
a - treated, transmission electron microscope; b - untreated, scanning electron microscope, x2500General view of the model center and layout of thermocouples:1 - container; 2 - supports; 3 - a stack for burning stack; 4 – thermocouples.

As shown in [15], in the case of modification of cement systems by nanotubes, the largest increase in physical and mechanical properties provided their introduction in the form of dispersions in solutions of plasticizers [15]. Plasticizing additives in cement concretes are anionic surfactants, the molecules of which are diphilic and consist of a hydrophobic hydrocarbon radical R and hydrophilic functional groups, which dissociate in aqueous solution, resulting in the formation of macroanions $R\text{–}(SO_3^-)_n$ or in polycarboxylate $R\text{–}(COO^-)_n$ (Figs. 2, 3, a). Carbon nanotubes (Fig. 1, 3, b) are hydrophobic, so they do not acquire a surface charge in the aqueous medium. In an aqueous solution, according to the rule of polarity alignment, surfactant molecules are adsorbed on the surfaces of nanotubes by their hydrophobic parts, forming nanoparticles that have a negative surface charge (Fig. 3, c). In areas of the surface of nanotubes covered with amorphous silica, the adsorption of surfactants does not occur.

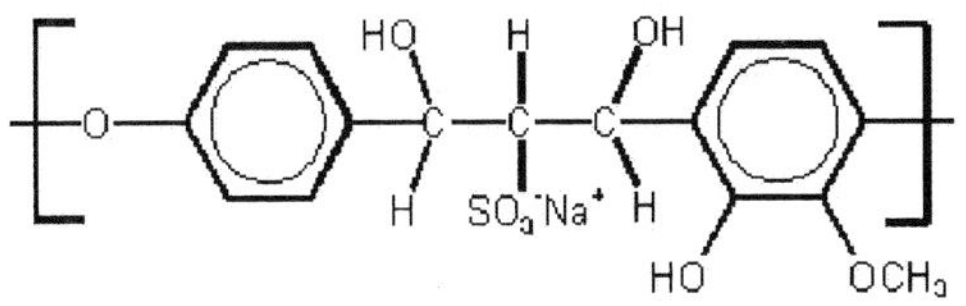

*a)*

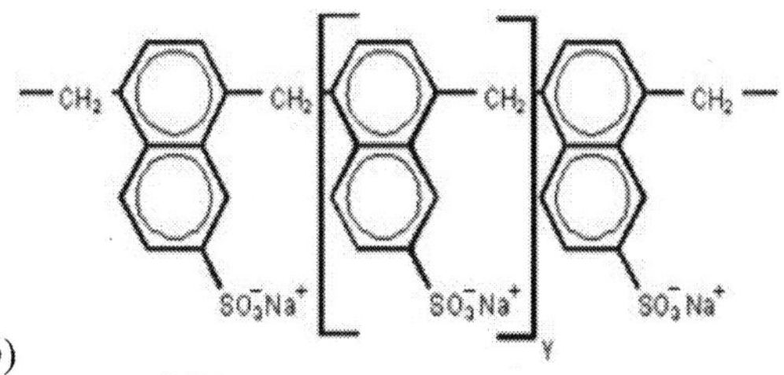

*b)*

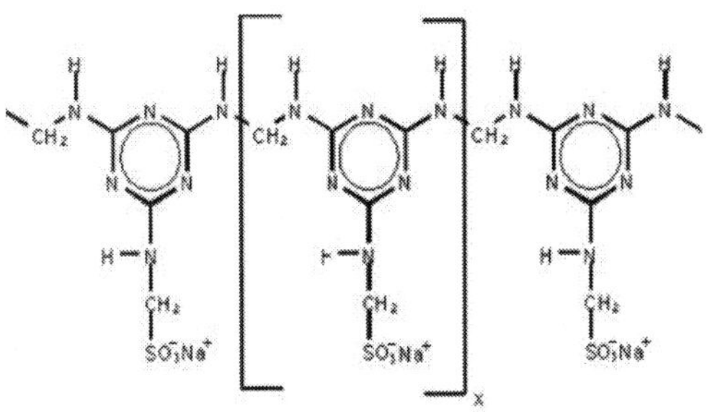

*c)*

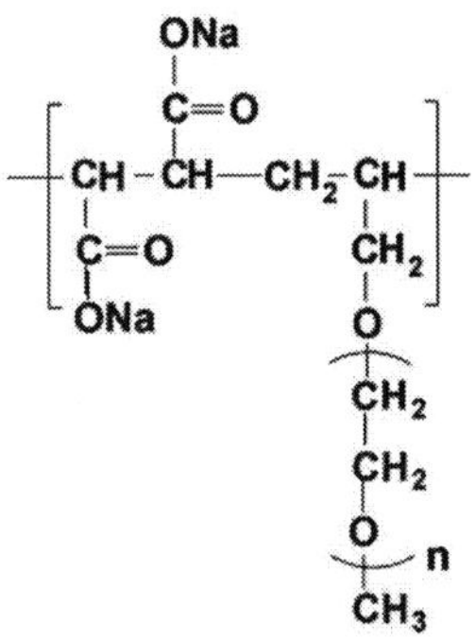

*d)*

**Figure 2.** Molecular structure of surfactants used as plasticizers, lignosulfonate (a), naphthalene formaldehyde (b), melamine formaldehyde (c) and polycarboxylate (d) type

IOP Conf. Series: Materials Science and Engineering **907** (2020) 012030 doi:10.1088/1757-899X/907/1/012030

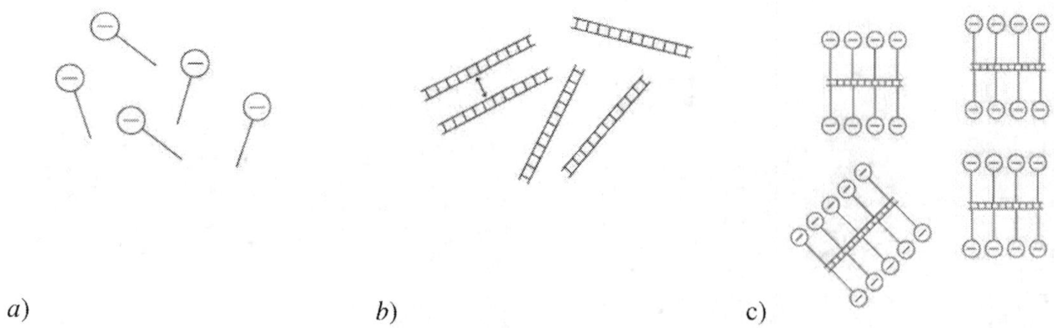


**Figure 3.** Schematic images of anionic surfactants (a), carbon nanotubes (b), particles - aggregates of surfactant molecules with carbon nanotubes (c)

The fractional composition and average size of these particles, as well as the stability of dispersions of them in a surfactant solution of lignosulfonate with polycarboxylate esters, naphthalene formaldehyde, melamine formaldehyde, polycarboxylate type were studied. It was found that the particle sizes of dispersions are for surfactants of the type: lignosulfonate with polycarboxylate esters, melamine-formaldehyde - 5–1000 nm, on average 200 nm; naphthalene-formaldehyde, polycarboxylate - 30–1500 nm, on average 300 nm. The minimum average particle size was observed for the content of nanotubes in the dispersion of 0.5–1%, which obviously provides the limiting monomolecular adsorption of surfactants on their surface. The obtained dispersions are stable over time, sedimentation and aggregation of nanoparticles were not observed.

It is known that the high shear resistance and viscosity of cement paste, mortar and concrete mixture without plasticizer additives due to the fact that on the surface of Portland cement particles in the aqueous medium are formed areas with negative surface charge ($C_3S$, $C_2S$), which predominate and with positive surface charge ($C_3A$, $C_4AF$) [18–20]. After mixing with water between oppositely charged areas, electrostatic attraction forces appear, which together with the always adhering molecular forces cause high shear resistance and viscosity of the entire dispersed system (Fig. 4, a).

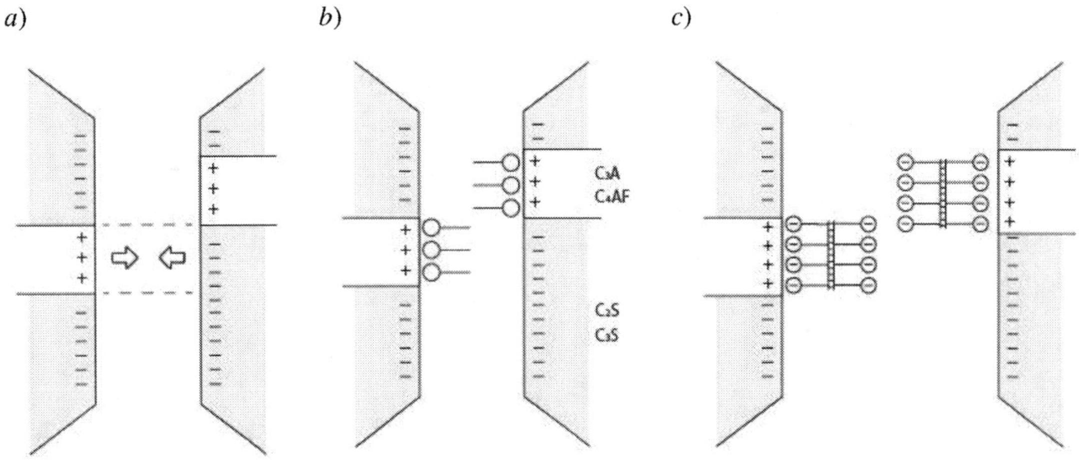


**Figure 4.** Schemes of interaction between Portland cement particles: a - without plasticizing additives; b - with the addition of anionic surfactants; c - with the addition of anionic surfactants modified by nanotubes.

The mechanism of action of plasticizing additives is the adsorption of anionic surfactants on positively charged areas of the surface of Portland cement particles $C_3A$, $C_4AF$ (Fig. 4, b). Due to this adsorption, depending on the number of anionic functional groups, neutralization or even recharging of

$C_3A$ and $C_4AF$ regions occurs, which eliminates electrostatic attraction from interparticle interactions, significantly reducing the shear resistance and viscosity of the system.

Nanoparticles - surfactant aggregates with carbon nanotubes have a significant number of anionic groups oriented in all directions along the tube (Fig. 3, c). Adsorbed on sites $C_3A$, $C_4AF$, they do not neutralize their positive surface charge, but recharge it to an intense negative, causing growth in the interparticle interactions of the repulsive component. This causes a greater plasticizing effect of the combination of surfactants with carbon nanotubes than from surfactants alone.

Based on this structure of nanoparticles - surfactant units with nanotubes, their surface charge is negative both in areas with adsorbed surfactant molecules and in areas covered with amorphous silica. Therefore, they can be substrates for crystal hydrates that have a positive surface charge - portlandite $Ca(OH)_2$, hydroaluminates and hydrosulfoaluminates of calcium.

Experimental verification (Fig. 5) of the proposed mechanism of plasticizing and structure-forming action of nanoparticles - surfactant aggregates with carbon nanotubes showed that the greatest effect on reducing the shear resistance and viscosity of cement paste over time is observed in the use of plasticizing additives lignosulfonate with polystyrene esters or melamine-formaldehyde (Muraplast FK-98) type and UCNT in the amount of 0.05% [11,12]. It was found, that the presence of a plasticizer in slag-containing cement paste significantly reduces its plastic strength and prolongs the induction period, with the greatest elongation (about 3 hours) occurs with the introduction of additives polycarboxylate type MC Power Flow 3100 and lignosulfonate with polycarboxylate esters.

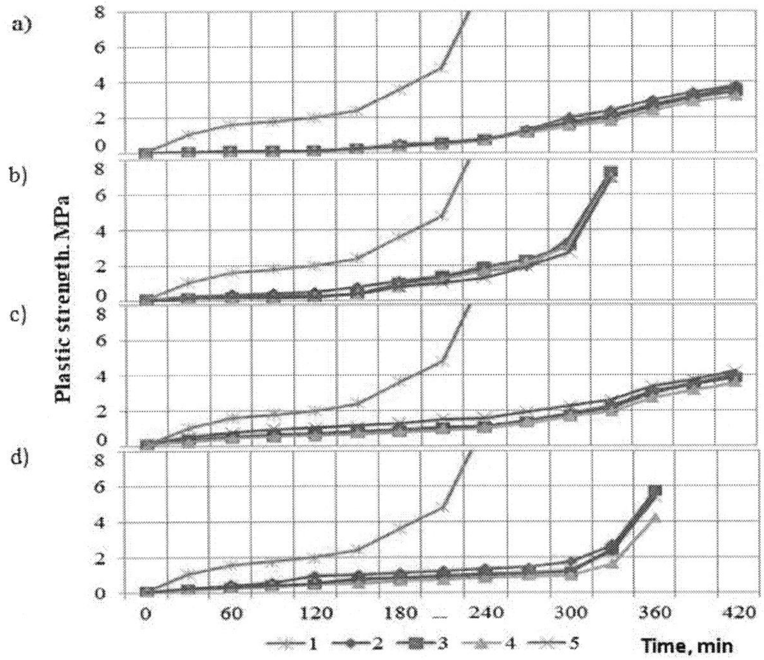


**Figure 5.** Change in plastic strength over time of compositions based on slag-containing cement (1), with the addition of plasticizer (2) and modified dispersions containing nanoadditives (3) in plasticizers of different types: a) PC (Power Flow 3100), b) NF (СП-1), c) LST + EP (Sika-Plast-520), d) MF (Muraplast FK-98): 1 wt.% UCNT.

The study of the influence of the structure-forming action of nanoparticles - surfactant aggregates with carbon nanotubes showed that the greatest effect on increasing the strength of artificial stone is observed in the case of plasticizing additives lignosulfonate with polycarboxylate esters (SikaPlast- 520) or melamine formaldehyde (Muraplast FK-98) 05%. The compressive strength of slag-containing cement stone due to the introduction of nanoscale additive increased by 21–39% (Fig. 6, a) bending strength - by 13–24% (Fig. 6, b).

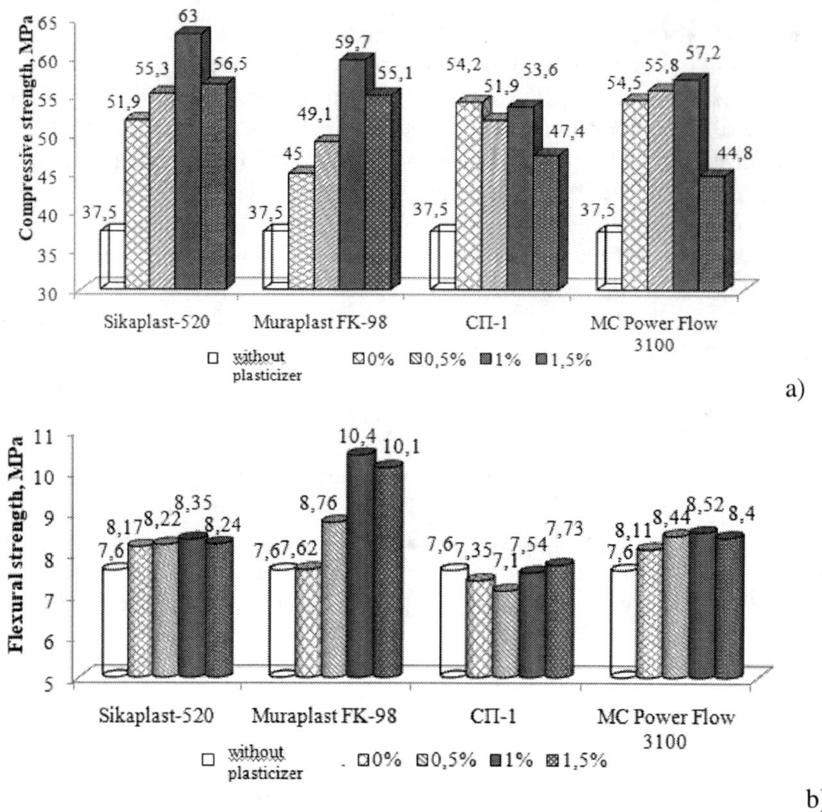


**Figure 6.** Kinetics of change in strength for 28 days of hardening under compression (a) and bending (b) of samples of slag-containing cement mortar modified by dispersions of nanotubes in plasticizers of different types with different content of nanotubes

In addition, the positive effect of nanomodification on the pore structure of the cement matrix in the direction of formation of micro- and nanopores of close diameter has been established, and their ordering is also observed. It is shown that the structure is dominated by closed pores, which increases the water resistance and corrosion resistance of the composite material. The effect can be explained in terms of the energy interaction between the components of the system, as carbon nanoobjects create a large number of uncompensated surface charges, and the formation of electrokinetic barriers to the transfer of aggressive substances through the body of artificial stone.

The introduction of UCNT dispersions leads to structural changes in the cement-sand mortar with the formation of a dense shell on the surface of the solid phases of cement grains and quartz sand. The number of contact interactions in the structured boundary layers increases and spatial frame cells are formed in the cement matrix with dense layers of crystal hydrates, which leads to a general compaction of the matrix.

The study of the microstructure of nanomodified artificial stone shows that there is the formation of lamellar calcium hydrosilicates of hexagonal shape, which increase in layers, forming a fairly dense and impermeable structure (Fig. 7). Such new formation in the hydration products of the sample with a pure plasticizer are completely absent. Over time, the presence of nanotubes contributes to the complete recrystallization of needle crystals into plates, which are located on top of each other and in shape resemble a crystal lattice of graphite. This phenomenon can be explained by the epitaxial growth of calcium hydrosilicates due to the formation of crystals on the surface of carbon nanotubes, which are formed from several rolled into a tube hexagonal graphite planes.

**Figure 7.** Photographs of the microstructure of artificial stone based on: slag-containing cement with the addition of 1 wt.% dispersion of crude carbon nanotubes in plasticizer Muraplast FK-98 after curing for 7 (a, c) and 90 days (b, d), respectively, at an increase of x3000 (a, b), x1000 (c), x2500 (d)

It is confirmed that the composition of the cement matrix plays a decisive role in the composite's acquisition of special properties, and the combination of slag-containing cement with bentonite in the presence of NVNT in melamine-formaldehyde plasticizer gives the best results for all studied performance properties. Cement composites are obtained, which are characterized by compressive strength of 35...49 MPa, flexural strength of 12...17 MPa, water absorption coefficient of 0.018...0.021 kg/m$^2$·√h, water resistance of 0.74...0.87 MPa, corrosion resistance in sulfate medium 0, 73... 0.92, frost resistance up to 200 cycles. Studies have also shown high crack resistance of nanomodified cement composites, which is 1.5...1.7 times higher than similar to the unmodified system.

A composite material has been obtained, which can be used as a cement matrix in the creation of fine-grained and coarse-grained concrete, special-purpose solutions, including waterproofing, repair, protective coatings for the restoration of reinforced concrete structures.

The most effective use of nanomodified cement composite material is planned as a matrix for obtaining fine-grained concrete for thin-walled architectural structures for landscape design, working under the influence of dynamic loads and aggressive influences of water and sulfate environments, elements of enclosing structures river banks, etc.

## 4. Conclusion

1. It is established, that for the modification of composite materials based on Portland cement in order to improve their physical and mechanical characteristics, it is advisable to use as an additive an aqueous dispersion of crude carbon nanotubes and anionic surfactants - plasticizers for concrete. Part of the surface of nanotubes is covered with amorphous silica, the rest is a hydrophobic carbon surface.

2. In an aqueous dispersion, carbon nanotubes and molecules of anionic surfactants form nanoparticles-aggregates with a negative surface charge, in which according to the rule of polarity alignment surfactant molecules are adsorbed by hydrophobic hydrocarbon radicals on carbon surfaces of nanotubes, hydrophilic anionic functional groups - outward in all directions.

3. The plasticizing effect of co-administered anionic surfactants and carbon nanotubes is greater than the plasticizing effect of surfactants. This increase is due to complete recharging of positively charged parts of the surface of Portland cement particles due to adsorption of these nanoparticles-aggregates instead of neutralization due to adsorption of surfactants only and complete replacement in contacts

between cement particles in the test of electrostatic attraction between oppositely charged areas of electrostatic surfaces.

4. Nanoparticles-aggregates of surfactants and nanotubes cause a structuring effect on cement stone, playing the role of negatively charged substrates for crystal hydrates of portlandite and calcium hydrosulfoaluminates with a positive surface charge. Together they form in the cement matrix a reinforcing spatial frame structure with dense layers of crystal hydrates, which causes its overall compaction and strengthening. Amorphous silica contained in crude nanotubes is involved in the hydration of Portland cement with the formation of additional low-base calcium hydrosilicates with a negative surface charge, which build up in layers on lamellar calcium hydrosilicates and form a fairly dense and impermeable structure of cement.

**References**

[1] Vauchsky M 2005 Directed formation of an ordered supramolecular crystal hydrate structure of hydrated mineral binders *Bulletin of civil engineers* **2** 44-47

[2] Ponomarev A 2009 High-quality concrete. Analysis of possibilities and practice of using methods of nanotechnology *Construction Engineering Magazine* **6** 25-33

[3] Gusev B 2009 Problems of creating nanomaterials and the development of nanotechnologies in construction *Nanotechnologies in construction: a scientific Internet journal* **2** 5–10

[4] Korolev E 2014 Nanotechnologies in building materials science. Analysis of the state and achievements. Ways of development *Construction materials* **11** 47–78

[5] Ydovich M, Ponomarev A, Gareev S 2008 Surface-active properties of nanomodificated plasticizer *Construction matersals* **3** 2-3

[6] Perfilov A, Alatortseva U, Dmitruk M, Perfilov V 2009 Application of modifying nanoadditives for increased strength of fiber concrete *Proceedings of the universities. Construction* **8** 17-20

[7] Sukhanevych M 2017 Principles of composite construction of waterproofing materials on the basis of cements modified with micro- and nanoadditives *Building materials, products and sanitary equipment* **5.8** 149-157.

[8] Pushkarova K, Sukhanevych M, Marsikh A 2016 Using of untreated carbon nanotubes in cement composition *Materials Science Forum* **865** 6-11

[9] Sanitsky M, Marushchak U, Mazurak T 2016 Nanomodified Portland cement compositions with high strength at an early age *Building materials, products and sanitary ware* **57** 147–154

[10] Tolmachev S, Belichenko O 2011 Properties of fine cement concretes with carbonaceous nanoparticles *Proceedings of the 4-th International Conference on Concrete Repair* 313 – 324

[11] Derevianko V, Chumak A, Vaganov V 2014 Influence of nanoparticles on the processes of hydration of semi-aqueous gypsum *Building Materials* **7** 22 - 24

[12] Pushkarova K, Sukhanevych M, Martich A 2014 Features of the processes of structure formation of waterproofing coatings modified by nanotubes *Building materials, products and sanitary ware* **53** 8-13

[13] Pushkarova K, Sukhanevych M, Martich A 2014 Use of complex additives containing carbon nanotubes and plasticizers for modification of cement compositions *Building materials, products and sanitary ware* **51** 14-20

[14] Pushkarova K, Dvorkin L, Gradoboev O, Zaichenko M, Kaganovsky O, Plugin A,Tymoshenko S, Shabanova G 2014 Energy-saving mineral binders and composite building materials based on them 272

[15] Plugin A N, Plugin A A, Trikoz L 2011 *Fundamentals of the theory of hardening, strength, destruction and durability of Portland cement, concrete and structures made of them. Colloid chemistry and physicochemical mechanics of cement concrete* **1** 331

[16] Pluhin A, Plugin O, Borziak O, Kaliuzhna O 2020 The Mechanism of a Penetrative Action for Portland Cement-Based Waterproofing Compositions *Proceedings of CEE Lecture Notes in Civil Engineering* **47**

[17] Pluhin O, Plugin A, Plugin D, Borziak O, Dudin O 2017 The effect of structural characteristics on electrical and physical properties of electrically conductive compositions based on mineral binders *Matec Web of Conference* **116** 9

IOP Conf. Series: Materials Science and Engineering **907** (2020) 012031 doi:10.1088/1757-899X/907/1/012031

# Radiation protective properties of fine-grained concretes and their radiation resistance

**D V Anopko, O A Honchar, M O Kochevykh and L O Kushnierova**

Department of Building Materials, Kyiv National University of Construction and Architecture, Povitroflotsky Avenue 31, 03680, Kyiv, Ukraine

anopko.dv @knuba.edu.ua

**Abstract.** Studies have shown that the thickness of the half layer of the gamma radiation flux attenuation with E = 662 keV is 3,5-5,2 cm (depending on the concrete density average) and for a neutron flux with E = 2,5 MeV is 5-6 cm. The radiation resistance of concrete under the influence of gamma radiation large doses, their activity and radionuclide composition after irradiation with neutrons was investigated. To study the behaviour of concrete under the influence of gamma radiation, two series of samples were made. One series was the control, and the second was exposed to gamma radiation. The temperature of the samples during testing did not exceed 40 °C, the dose of gamma radiation was $10^9$ rad. Its value corresponds to the dose that concrete can receive when it comes into contact with highly radioactive wastes from the "Shelter" facility for 300 years. The characteristic of an industrial gamma-ray plant is radiation energy of 1,25 MeV and a dose rate of 2 Mrad/h. Using this setup allows you to reach a dose of $10^9$ rad in less than a month, and $10^8$ rad in 4 - 5 days. Concrete that reached the age of 28 days and stored under normal conditions was exposed to gamma radiation. Analysing the obtained data of physical and mechanical tests of radiation-protective compositions, it can be noted that large doses of gamma radiation practically do not affect the strength of the developed material.

## 1. Introduction

Many years of world experience in the selection of materials for radiation protection shows that concrete is the main material. It can combine heavy components to protect against gamma radiation and light components to absorb neutron fluxes. This fact, as well as manufacturability, low price and the possibility of using local materials, have determined the widespread use of concrete for protection against radioactive radiation.

The nature of the interaction of γ- and neutron radiation with concrete is different. Electromagnetic γ-radiation can interact with electrons of an atom, nuclear particles (nucleons), with an electric field that surrounds charged particles (nuclei and electrons), and with a meson field that surrounds nucleons [1-3]. Any of these types of interaction of γ-quanta can lead to their complete absorption, elastic or inelastic scattering.

As a result of the interaction of photons with matter, one part of their energy is converted into the kinetic energy of charged particles, and the second part is converted into the energy of secondary photon radiation. The probability of the interaction of γ-radiation with a substance increases with its density, with an increase in the atomic number; therefore, it is advisable to use denser and heavier materials to absorb it [1-4].

The energy perceived by an atom from a neutron that bombards it is expended on its displacement, excitation or ionization. As a result, it changes the material initial microstructure. Local changes (defects) in the crystal lattice and microstructure of the material, which were formed under the influence of irradiation, lead to a profound change in the crystal, molecular and supramolecular structure. This is accompanied by a change in all the properties of the material and is the reason for the decrease in its strength and durability.

## 2. Materials and methods of research

Radiation-protective properties to the action of γ-radiation and neutrons were determined by comparing the intensity of flows through the material under study and the standard. Concrete plates with a size of 100×100×15 mm were used as samples. The $^{137}$Cs isotope was used as a γ-radiation source, the spectrum of which is predominantly a line with the energy of E = 662 keV. The diameter of the radiation source is 6 mm and the thickness is 8 mm. An MKS-01R radiometer with a BDKB-01R detector was used to study the γ-radiation dose rate. A $^{252}$Cf neutron source with average neutron energy of E = 2.5 MeV was used to study the radiation-protective properties of concrete in a neutron field. An MKS-01R radiometer with a BDKN-03 detector was used to record the intensity of fast and intermediate neutrons.

Concrete cubes of 20×20×20 mm in size were used to study the formation of secondary γ-radiation with an energy of up to 7,7 MeV. Samples were irradiated in the active zone of the WWRM nuclear reactor in a "wet" channel with the flux density of $5{,}1 \cdot 10^{13}$ n/cm$^2$ for 275 hours. The temperature of the samples did not exceed 50 °C. During the test the flux density was $5{,}05 \cdot 10^{19}$ n/cm$^2$.

To assess the radiation resistance of the composites, the samples were subjected to long-term γ-radiation at the Belgorod-Dniester OJSC "Hemoplast" with an industrial certified γ-installation for sterilizing products with emission energy of 1,25 MeV and a dose rate of 2 Mrad/h. The temperature of the samples did not exceed 40 °C during the test. The radiation dose of concrete received from γ-radiation was received $10^9$ rad. Its value corresponds to the dose that concrete can receive when it comes into contact with highly radioactive waste from the Chernobyl sarcophagus over 300 years. The radiation resistance of concrete was evaluated by reducing the tensile strength in bending and compressive strength on samples with dimensions of 40x40x160 mm.

## 3. Research results

*3.1. Study of radiation-protective concrete properties from the γ-radiation action.* The research was conducted at the Scientific Center of the Institute of Nuclear Research of the National Academy of Sciences of Ukraine. The geometry of the experiments is shown in Figure 1.

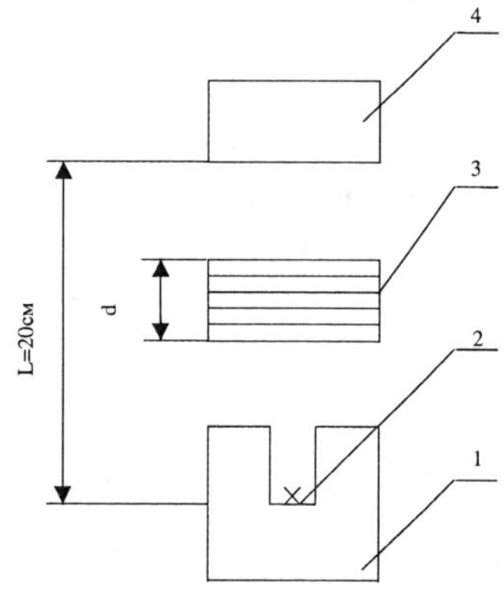


**Figure 1.** The geometry of the samples and equipment arrangement: 1 - container; 2 - γ-radiation source; 3 - test samples; 4 - detector.

To characterize the degree of absorption of ionizing radiation, such indicators were used [1]:

1. Half-value thickness (HVT) - is the thickness of the protective material layer necessary to halve the radiation intensity.

2. Relaxation length - the thickness of the material layer, which reduces the flux density of particles or photons (quanta) by *e* times (2,73).

3. Linear attenuation coefficient $\mu$ ($cm^{-1}$) - is a function of the photon radiation energy and the substance type: $\mu = d\varphi_\omega / (\varphi_\omega dx)$. It characterizes the relative change in the photon radiation energy flux density $d\varphi_\omega / \varphi_\omega$ when passing through a layer of matter at $dx = 1$.

4. Mass Attenuation coefficient $\mu_n$ ($cm^2/g$) - the ratio of the linear attenuation coefficient $\mu$ to the density of the substance $\rho$ ($g/cm^3$) depends on the atomic weight of the element: $\mu_n = \mu / \rho$.

The compositions of the samples are given in the table 1.

**Table 1.** The average density and composition of radiation-protective concrete.

| № | The composition, mass% | Ratio binder / aggregate | Average density, kg/m$^3$ |
|---|---|---|---|
| 1 | Modified binder - 75<br>Dispersed iron (0,08-0,16мм) - 25 | 3:1 | 2350 |
| 2 | Modified binder - 65<br>Dispersed iron (0,08-0,16мм) - 35 | 2:1 | 2520 |
| 3 | Modified binder - 50<br>Dispersed iron (0,08-0,16мм) - 50 | 1:1 | 2860 |
| 4 | Modified binder - 35<br>Dispersed iron (0,08-0,16мм) - 65 | 1:2 | 3640 |
| 5 | Modified binder - 35<br>Dispersed iron (0,8-2,0 мм) - 65 | 1:2 | 3900 |
| 6 | Modified binder - 25<br>Dispersed iron (0,8-2,0 мм) - 75 | 1:3 | 5130 |

The distance between the γ-radiation source and the detector was constant and equal to 20 cm. In studies, the dose rate $I_0$ was first measured at a distance of 20 cm (when the studied samples are absent between the detector and the radiation source). The dose rate was $I_0 = 24,2$ mSv/h. One, then two, three, four and five plates of each series were installed in series' first. The dose rate was measured for each thickness obtained. For samples of each thickness, 6 measurements were carried out at an exposure of 10 C. Based on the measurement results, the average power value for each thickness was found. The measurement results are shown in table 2.

**Table 2.** The results of measurements of radiation protective properties of concrete in the field of γ-radiation.

| The concrete composition according to the table 1 | Linear attenuation coefficient, μ, sm$^{-1}$ | Half-value thickness, cm | Radiation Power Relaxation Length, cm | Mass Attenuation coefficient $\mu_n$, cm$^2$/g |
|---|---|---|---|---|
| 1 | 0,1276 | 5,43 | 7,84 | 0,043 |
| 2 | 0,1317 | 5,26 | 7,59 | 0,045 |
| 3 | 0,1320 | 5,25 | 7,58 | 0,043 |
| 4 | 0,1320 | 5,25 | 7,58 | 0,042 |
| 5 | 0,1558 | 4,45 | 6,42 | 0,040 |
| 6 | 0,2167 | 3,20 | 4,61 | 0,042 |

IOP Conf. Series: Materials Science and Engineering **907** (2020) 012031 doi:10.1088/1757-899X/907/1/012031

The measurement results showed that the attenuation of the radiation dose rate for the chosen experiment geometry for the studied series of samples is well described by the exponential law [1]:

$$I = I_0 e^{-\mu X} \tag{1}$$

$I$ – the dose rate at a distance of 20 cm when the studied samples are between the detector and the radiation source; $\mu$ – linear attenuation coefficient ($cm^{-1}$); $X$ – thickness of samples, cm.

The attenuation of the γ-radiation dose rate is also confirmed by the fact that the dependence $ln[I_o/I(X)]$ is well described by a linear function

$$B(X) = \mu(X) \tag{2}$$

For example, Figure 2 shows the dependence of $ln[I_o/I(X)]$ on thickness $x$ for concrete compositions 5 and 6. It can be seen from this figure that the experimental values of $ln[I_o/I(X)]$ are well described by line dependence $B(X) = \mu X$. The slope of the straight lines for concrete compounds 5 and 6 is different. The tangent of their slope determines the linear attenuation coefficient $\mu$.

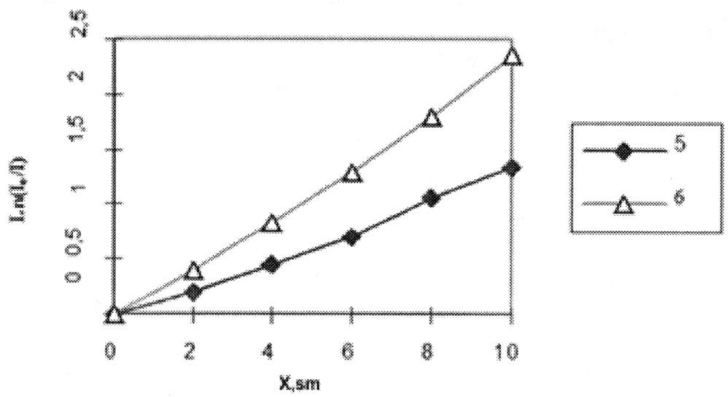


**Figure 2.** The dependence of $ln(I_o/I)$ on the thickness of the sample.
Legend: 5, 6 - concrete composition numbers.

The linear attenuation coefficient μ for each series of concrete compositions was calculated by the formula $[I_o/I(X)]$

$$\mu = \sum_{i=1}^{5} YiXi \Big/ \sum_{i=1}^{5} XiXi, \quad Yi = ln[I_o/I(X)] \tag{3}$$

In the calculations according to formula (3), the average value of the plate thickness along its diagonal was used. The calculations showed that concretes of compositions 5 and 6 have the highest linear attenuation coefficient of γ-radiation. This is due to higher rates of their density. The data obtained correspond to known patterns [1, 2]. From the data obtained it follows that concretes of 5 and 6 compositions are most effective as radiation protective materials from the action of γ- radiation.

*3.2. The study of radiation protective concrete properties under the neutron radiation influence.*

We studied the attenuation of the intensity of the flux of fast and intermediate neutrons as they pass through a set of concrete plates in an amount of one to five pieces for each composition. The neutron source was in a paraffin container at a depth of 10 cm from the surface of the container opening. The distance between the detector surface and the hole was remained unchanged (10 cm). The thickness of the concrete between the detector and the radiation source was increased by setting concrete plates in the gap between the detector and the radiation source. The results showed that for the chosen geometry of the experiments, the dependence of the neutron flux intensity on the thickness of the shield is well described by the exponential dependence [1]:

$$\varphi = \varphi_0 e^{-X/L} \tag{4}$$

$\varphi_0$ – the intensity of the flow of fast and intermediate neutrons without plates of concrete between the neutron source and the detector; $L$ - the length of neutron relaxation in the medium, cm; $X$ - thickness of a set of plates of concrete, cm.

To obtain the necessary statistics for each thickness, we measured the neutron flux intensity 6 times with an exposure time of 10s. Then found their average value. To calculate $L$, the average value of the plate thickness along its diagonal was used. In the experiments, $\varphi_0$ = 62,4 n/cm$^2$ s. The experimental values for different series of concrete are given in table 3.

Since the dependence $ln[\varphi_0/\varphi(X)]$ is well described by the linear function $B = X/L$, using the least-squares method, the relaxation length L can be calculated by the formula:

$$L = \sum_{i=1}^{5} (XiXi) / \sum_{i=1}^{5} YiXi \ , \tag{5}$$

Studies of radiation-protective properties of the proposed concrete formulations have shown that they have high protective indicators. Thus, the Half-value thickness for neutron radiation is 4,7-6,2 cm. The relaxation length of the neutron radiation power for fast and intermediate neutrons in the studied concrete samples is within the measurement accuracy and weakly depends on the composition.

**Table 3.** The results of measurements of radiation protective properties of concrete in a neutron radiation field.

| The composition of concrete according to the table 1 | Half-value thickness, cm | The relaxation length of the power of neutron radiation, cm |
|---|---|---|
| 1 | 4,64 | 6,69 |
| 2 | 5,20 | 7,46 |
| 3 | 6,17 | 8,91 |
| 4 | 5,07 | 7,32 |
| 5 | 5,58 | 8,05 |
| 6 | 5,01 | 7,23 |

Thus, the tested concrete compositions according to the results of studies shown in tables 2 and 3, effectively protect against gamma - and neutron radiation. This is evidenced by the insignificant thicknesses of the layers of half-value thickness of neutron and gamma radiation.

*3.3. Study of the radionuclide composition and activity of radiation-protective concrete.*

From literary sources, it is known [2, 3] that during irradiation of heavy concrete based on metal aggregate under the influence of neutron radiation, secondary $\gamma$-radiation with an energy of up to 7,7 MeV is formed. Therefore, it is necessary to study the radionuclide composition and activity of radiation-ionizing fine-grained concrete after their irradiation with a neutron flux.

Studies were performed on concrete compositions 3, 4 (table 1). For this, cubic samples of 20×20×20 mm in size were made, three samples of each composition. After radiation, for the decomposition of short-lived radionuclides, concrete samples were kept in “hot chambers” for 168 hours. After that, the samples were removed from special ampoules and their radionuclide composition and activity were examined. The exposure dose rate of $\gamma$-radiation was determined at a distance of 10 cm from the samples. The measurement was performed using a germanium-lithium gamma spectrometer with a NUC 8092 analyzer system and subsequent processing of gamma spectra with a special AK program. The measurement results with an error of ± 10% are shown in table 4.

As follows from the data given in table 4, the greatest contribution to the activity of concrete samples is made by the radionuclides $^{51}Cr$ and $^{59}Fe$, which are formed in the reactor core from the reaction $(n,\gamma)$. For example, a $^{51}Cr$ radionuclide ($T_{1/2}$= 27,7 days) is converted to $^{51}V$ via electron capture. This transformation is accompanied by the emission of γ-quanta with energy of 0,320 MeV, x-ray radiation and electron radiation. Receipt scheme $^{51}Cr$: $^{50}Cr + (n,\gamma) \rightarrow {}^{51}Cr \rightarrow {}^{51}V$.

The calculated specific activity of $^{51}Cr$, which is formed in the reactor core at a flux density of $5{,}1 \cdot 10^3$ n/cm$^2$, per 1 g of natural Cr is $1{,}06 \cdot 10^{11}$ Bq/g. This is 3-4 orders of magnitude more than concrete.

A similar pattern of the formation of the $^{59}Fe$ radionuclide, which is formed from natural $^{58}Fe$ by (n,γ) reaction: $^{58}Fe + (n,\gamma) \rightarrow {}^{59}Fe \rightarrow {}^{59}Co$.

The specific activity of $^{59}Fe$ is $1{,}2 \cdot 10^{10}$ Bq/g, which is 3 orders of magnitude more than that of concrete.

**Table 4.** Radionuclide composition and activity of concrete

| The composition according to table 1 | Concrete sample | Radionuclide | Half-life, days | Specific activity Bq/g (Ki/g) | Exposure dose rate γ-radiation, mR/h |
|---|---|---|---|---|---|
| 3 | 1 | $^{51}Cr$ | 27,7 | $1{,}07 \cdot 10^7 (2{,}88 \cdot 10^{-3})$ | 8,0 |
| | | $^{54}Mn$ | 312 | $3{,}95 \cdot 10^6 (1{,}06 \cdot 10^{-4})$ | |
| | | $^{59}Fe$ | 44,6 | $8{,}99 \cdot 10^7 (2{,}43 \cdot 10^{-3})$ | |
| | | $^{60}Co$ | 5,272 (years) | $6{,}79 \cdot 10^5 (1{,}83 \cdot 10^{-5})$ | |
| | 2 | $^{51}Cr$ | 27,7 | $1{,}02 \cdot 10^8 (2{,}76 \cdot 10^{-3})$ | 7,0 |
| | | $^{54}Mn$ | 312 | $3{,}85 \cdot 10^6 (1{,}04 \cdot 10^{-4})$ | |
| | | $^{59}Fe$ | 44,6 | $8{,}89 \cdot 10^7 (2{,}40 \cdot 10^{-3})$ | |
| | | $^{60}Co$ | 5,272 (years) | $6{,}72 \cdot 10^5 (1{,}82 \cdot 10^{-5})$ | |
| | 3 | $^{51}Cr$ | 27,7 | $1{,}05 \cdot 10^8 (2{,}8 \cdot 10^{-3})$ | 9,0 |
| | | $^{54}Mn$ | 312 | $3{,}9 \cdot 10^6 (1{,}05 \cdot 10^{-4})$ | |
| | | $^{59}Fe$ | 44,6 | $8{,}96 \cdot 10^7 (2{,}42 \cdot 10^{-3})$ | |
| | | $^{60}Co$ | 5,272 (years) | $6{,}77 \cdot 10^5 (1{,}82 \cdot 10^{-3})$ | |
| 4 | 1 | $^{51}Cr$ | 27,7 | $1{,}48 \cdot 10^8 (4{,}1 \cdot 10^{-3})$ | 12,0 |
| | | $^{54}Mn$ | 312 | $4{,}03 \cdot 10^6 (1{,}09 \cdot 10^{-4})$ | |
| | | $^{59}Fe$ | 44,6 | $8{,}45 \cdot 10^7 (2{,}28 \cdot 10^{-3})$ | |
| | | $^{60}Co$ | 5,272 (years) | $6{,}75 \cdot 10^5 (1{,}82 \cdot 10^{-3})$ | |
| | 2 | $^{51}Cr$ | 27,7 | $1{,}46 \cdot 10^8 (3{,}93 \cdot 10^{-3})$ | 12,3 |
| | | $^{54}Mn$ | 312 | $4{,}0 \cdot 10^6 (1{,}08 \cdot 10^{-4})$ | |
| | | $^{59}Fe$ | 44,6 | $8{,}41 \cdot 10^7 (2{,}27 \cdot 10^{-3})$ | |
| | | $^{60}Co$ | 5,272 (years) | $6{,}82 \cdot 10^5 (1{,}82 \cdot 10^{-3})$ | |
| | 3 | $^{51}Cr$ | 27,7 | $1{,}4 \cdot 10^8 (3{,}78 \cdot 10^{-3})$ | 12,0 |
| | | $^{54}Mn$ | 312 | $3{,}91 \cdot 10^6 (1{,}06 \cdot 10^{-4})$ | |
| | | $^{59}Fe$ | 44,6 | $8{,}26 \cdot 10^7 (2{,}23 \cdot 10^{-3})$ | |
| | | $^{60}Co$ | 5,272 (years) | $6{,}79 \cdot 10^5 (1{,}82 \cdot 10^{-3})$ | |

*3.4. Study of the radiation resistance of concrete in the gamma radiation field*

The durability of radiation-protective concrete is determined not only by their high ability to absorb ionizing radiation, but also possess high radiation resistance in the field of ionizing radiation. Radiation resistance is the ability of a material to maintain its properties after irradiation [2, 3]. From literary sources, it is known [3, 4] that concretes based on Portland cement and based on slag-alkali cement during prolonged exposure to γ-rays have a fairly high radiation resistance. The radiation resistance of the studied non-shrinking concrete, which contains a significant amount of highly basic calcium hydrated sulfoferrite with prolonged exposure to γ-radiation, is not known. Therefore, special studies were conducted on the behavior of these composites under the action of ionizing radiation.

The study of the influence of prolonged exposure to γ-radiation on the properties of the material was carried out on the concrete, the compositions of which are given in table 1.

Two groups of concrete samples were used: one - was not exposed to γ-radiation, the second - was exposed to γ-radiation.

Concrete, which reached the age of 28 days and stored under normal conditions, was exposed to γ radiation. This setup allows getting the required dose in a short period of time - $10^9$ rad less than a month, and $10^8$ rad - in 4-5 days. This makes it possible to study the structure of the material without taking into account the significant influence of cement stone hydration processes on it.

The impact on the material of such a significant energy impulse violates its thermodynamic equilibrium. This creates the conditions for the appearance of significant fluctuation processes in it, directed at the formation of structure regularity violations. To reduce the intensity of fluctuations, it is advisable to create composites with a minimum value of their anisotropy of properties, which is controlled by the size of the aggregate and the degree of cement expansion [5]. This was realized during the creation of these composites and described in [6, 7]. The results of mechanical and physico-mechanical tests of concrete are given in tables 5. Tests on the radiation resistance of concrete under the influence of gamma radiation showed that they do not lose compressive strength when absorbed at a dose of $10^8$ and $10^9$ rad (table 5). These results are consistent with known data [2, 4].

The results obtained indicate that, as a result of radiation, the compressive strength of the samples not only did not decrease but also increased. This can be explained by the fact that the intensification of the concrete structure formation process by the influence of gamma-ray, and the resulting structural disturbances, are directed at increasing their compressive strength.

The emerging structural disturbances can be characterized as self-organizing, which contributes to an increase in the strength of the material [5].

**Table 5.** Change in compressive strength and bending strength of concrete under prolonged exposure to gamma radiation

| The composition of concrete according to the table 1 | Compressive strength, MPa | | | | Bending strength, MPa | | | |
|---|---|---|---|---|---|---|---|---|
| | Control samples, age, days | | Samples that have absorbed the dose, rad | | Control samples, age, days | | Samples that have absorbed the dose, rad | |
| | 28 | 56 | $10^8$, age 32 days | $10^9$ age 56 days | 28 | 56 | $10^8$, age 32 days | $10^9$ age 56 days |
| 1 | 55 | 55 | 55 | 56 | 1,9 | 2,0 | 1,5 | 2,0 |
| 2 | 51 | 52 | 51 | 52 | 1,9 | 2,1 | 2,0 | 2,4 |
| 3 | 51 | 51 | 52 | 52 | 1,6 | 1,85 | 1,85 | 2,3 |
| 4 | 53 | 54 | 54 | 55 | 1,2 | 1,3 | 1,3 | 1,4 |
| 5 | 48 | 49 | 48 | 49 | 1,2 | 1,2 | 1,2 | 1,2 |
| 6 | 64 | 66 | 67 | 70 | 0,93 | 0,94 | 1,00 | 0,76 |

Analyzing the data obtained, it can be noted that using the proposed methods for regulating the composition and technology of fine-grained concrete (changing the phase composition of the binder hydration products towards the formation of highly basic hydrosulfoferrites and calcium sulfoaluminoferrites due to the introduction of silica fume, iron oxide additives, and metallic iron, using aggregates of different dispersion) managing the formation of self-organizing structures in cement and obtaining a high-quality artificial stone.

So, gamma irradiation of samples with a dose of $10^8$ rad leads to a transformation of the structure of cement and concrete based on it (compositions 1, 2, table 5), which is accompanied by a temporary decrease in the strength during bending of concrete. Further irradiation with a simultaneous restructuring of the structure causes an increase in the strength of the material during bending. The introduction of the proposed addition of silica fume with iron oxide and pure metal eliminates the negative effect of gamma radiation on the strength of concrete; there is even an increase in bending strength over the entire radiation interval.

It should also be noted that the dispersion and amount of aggregate affect the resistance of concrete to gamma radiation. The use of dispersed aggregate provides an increase in the strength of concrete in bending (composition 4, table 5). An increase in the dispersion of the aggregate (0,8 - 2,0 mm) is accompanied by a decrease in the growth of concrete strength during irradiation (composition 5, table 5). With an increase in the amount of aggregate (composition 6, table 5), there is a significant increase in compressive strength and a slight decrease in concrete strength in bending after exposure to gamma radiation with a dose of $10^9$ rad, which does not exceed allowable values. The obtained values of concrete bending strength will increase in time as a result of the binder hydration process, and the compressive strength of concrete will also increase after the action of $\gamma$-radiation.

## 4. Conclusion

The developed composites have half-value for the $\gamma$-radiation of the $^{137}$Cs isotope with energy E = 662 keV – 3,20…5,29 cm, and for neutron radiation $^{252}$Cf with energy E = 2,5 MeV – 5,01…6,17 cm. The developed concrete compositions have a small activity (3-4 orders of magnitude lower) compared to natural chromium and iron if the latter are exposed to neutron radiation. The influence on the material of such a significant energy impact removes it from thermodynamic equilibrium, which creates the conditions for the occurrence of significant fluctuation processes in it, aimed at creating violations of the regularity of the structure. To reduce the intensity of fluctuations, it is advisable to create composites with a minimum value of the anisotropy of their properties, which in the developed materials is controlled by the dispersion of the aggregate and the degree of expansion of cement. This confirms the promise of using the proposed fine-grained concrete as radiation-protective materials.

It was established that the studied composites have significant radiation resistance in the fields of gamma radiation at a dose of up to 1000 Mrad. They are characterized by an increase in compressive strength and stable values of tensile strength in bending. Based on the data obtained, it is possible to develop concrete compositions for industrial production and use. It is also of interest to study the properties of these composites at the meso- and microlevels after prolonged irradiation with neutrons and gamma rays, and in the future, use of the theoretical method for modeling the transfer of neutron flux in various media [8, 9]. Simulation of the passage of neutrons through concrete layers of various thicknesses, in particular, a neutron flux with an energy of up to 14 MeV, will allow one to calculate the protective properties of the studied composites for plasma focus units, thermonuclear reactors, and fast neutron reactors [10 - 11], since experimental studies with sources of such high power is associated with a number of technical problems.

## References

[1] Egorov Yu A 1983 *Fundamentals of radiation safety of nuclear power plants* (Moscow: Energoatomizdat) vol **2** p 330
[2] Dubrovskiy V B 1977 *Radiation resistance of building materials* (Moscow: Stroyizdat) p 278
[3] Dubrovskiy V B and Ablevich Z 1983 *Construction Materials. Building materials and structures for protection against ionizing radiation* (Moscow: Stroyizdat) p 240
[4] Petrova T M, Komohov P G et al 1997 Radiation-resistant concrete based on slag-alkali binders *Cement* vol **1** pp 33-35
[5] Sheynich L A and Pushkareva E K 2009 *Self-organization processes of the structure of building composites* (K: Gamma-print) p 153
[6] Sheynich L A and Anopko D V 2011 *Radiation-protective non-shrinking fine-grained concrete on cast-iron aggregates* (K: Gamma-print) p 198

IOP Conf. Series: Materials Science and Engineering **907** (2020) 012031 doi:10.1088/1757-899X/907/1/012031

[7] Anopko D V, Gonchar O A, Kobyilyanskiy I Z 2019 Composite radiation protective cement with enhanced thermal and crack resistance *Scientific Bulletin of Civil Engineering* (Kharkiv: KhNUBA) vol **98** pp 231-240

[8] Wagner J C, Peplow D E, Mosher S W, Evans T M 2011 Review of Hybrid Deterministic/Monte Carlo Radiation Transport Methods, Codes, and Applications at Oak Ridge National Laboratory *Progress in Nuclear Science and Technology* vol **2** pp 808-814

[9] Romanenko I M, Golyuk M I, Nosovskiy A V, Gulik V 2018 Investigation of a New Composite Material Based on Ultra Heavy Concrete and Basalt Fiber for Radiation Protection from Gamma Radiation *Yaderna ta radiatsiyna bezpeka* vol 1(**77**) pp 52-58

[10] Gulik V and Tkaczyk A H 2014 Cost optimization of ADS design: Comparative study of externally driven heterogeneous and homogeneous two-zone subcritical reactor systems *Nuclear Engineering and Design* Issue **270** pp 133-142

[11] Sharifi Sh, Bagheri R, Shirmardi S P 2013 Comparison of shielding properties for ordinary, barite, serpentine and steel-magnetite concretes using MCNP-4C code and available experimental results *Annals of Nuclear Energy* Issue **53** pp 529-534

IOP Conf. Series: Materials Science and Engineering **907** (2020) 012032 doi:10.1088/1757-899X/907/1/012032

# Modification of the structure of the stabilized basalt fiber

**V I Gots[1], P P Palchyk[1], S A Tymoshenko[1], S P Palchyk[2]**

[1]Department of Technology of Building Constructions and Products, Kyiv National University of Construction and Architecture, Povitroflotsky Avenue 31, Kyiv, Ukraine

[2]Department of Cybersecurity and Computer Engineering, Kyiv National University of Construction and Architecture, Povitroflotskyi Avenue 31, Kyiv, Ukraine

PPPalchik@ukr.net

**Abstract.** The study of thermophysical, physical-mechanical and acoustic characteristics of coarse continuous basalt fiber in the conditions of influence of operational factors and corrosion environment is carried out. It is shown that obtaining barrier materials with high thermophysical and acoustic properties is possible by creating a porous system in the material with a predetermined textural characteristic of pores of different class, purpose and combination. The influence of the porous system of coarse continuous basalt fiber on the operational properties of flexible heat and sound insulation products and materials has been studied. It is determined that the performance characteristics of flexible thermal insulation materials are in direct proportion to the degree of porosity of the constituent materials and their structural characteristics. Shown that the increase in the operating temperature of these materials, sound and thermal resistance is determined by the chemical and mineralogical composition of the source materials and their structural characteristics. The influence of the mechanism of selective leaching of coarse continuous basalt fibers on their textural characteristics and the zone of operating temperatures is studied.

**Introduction.** The requirements of regulatory documents on thermal protection of buildings and structures have led to a steady trend to create new materials that ensure the sustainable operation of structure and equipment in conditions of intense exposure to destructive factors of external influence. Such materials are subject to a set of (sometimes mutually exclusive) requirements that must combine: physical and mechanical, chemical, hygienic, thermophysical, decorative and other properties of materials. In this regard, there was a need to create materials. Which at low average density, high strength is able to withstand significant temperature loads, the impact of extreme natural and operational factors.

The development of modern technology necessitates the creation of materials that are characterized by low average density, able to withstand the effects of high temperatures and aggressive environments. One such material is mineral fibers based on erupted rocks. The raw materials for their production are non-deficient, widespread in Ukraine the main rocks of magmatic origin - basalts and their structural analogues - diabases, amphiboles and others. The relatively low melting point makes it possible to use these rocks for the production of mineral fibers in a simplified scheme, directly with the loading of rock into the smelting unit, which ensures high efficiency of the product, stability of process parameters and no harmful load on the environment.

The development of construction production to increase the requirements for structural and thermophysical characteristics of buildings requires the creation of materials that in their properties should significantly exceed existing ones.

In this regard, there is a need to create products that have a low average density, high strength characteristics that can withstand temperature loads, the impact of aggressive environments.

One of the materials that meet the above requirements are concretes based on continuous basalt fiber. In addition to high technical performance, their lower cost should be taken into account, compared to other types of materials for reinforcement. on elasticity it prevails over steel.

The main difference of basalt concrete is its high strength at all types of stress and the ability to withstand large deformations. The relative deformation of basalt concrete, without the formation of cracks can reach 0.7....0.9%, which is much higher than that for conventional concrete.

Basalt fibers of the VRV and STV type from the basalt of the Usachkovsky, Talne, and Yakova Dolyna deposits were used for research.

When processing basalt fiber of a certain chemical and mineral composition with solutions of certain chemical reagents, it is possible to obtain porous siliceous fibers with high technical characteristics and performance properties. In terms of chemical composition, they consist in the vast majority of modifications of silica. Therefore, the question of durability and stability in aggressive environments of porous basalt fiber, given that the wall thickness of its framework is extremely small and can reach only a few dozen angstroms, is of paramount importance and requires detailed research.

The paper considers one of these aspects, namely the stability of the modified basalt fibre in complex chemical solutions that simulate the composition of the pore liquid in real conditions of solidification of the material.

Basalt fibers, in contrast to glass fibers have a high tendency to crystallization, higher temperature resistance and chemical resistance. Depending on the temperature and heating time, the content of oxides in them changes significantly. Thus, when heated above 600 $^0$C, the beginning of the transition of FeO to $Fe_2O_3$ is observed. When heating basalt fibers to 450 ... 500 $^0$C there is a slight decrease in mass, which is due to the loss of chemically bound water, and with subsequent heating - an increase in mass, which is associated with the addition of oxygen in the oxidation of ferrous iron to trivalent. The chemical resistance of heat-treated fibers to water and alkali is practically unchanged, as these fibers in the initial state have a high resistance in these media. Acid resistance of heat-treated basalt fibers increases with increasing temperature and reaches its maximum at 800 $^0$C. There is an increase in corrosion resistance of heat-treated basalt fiber in an alkaline environment.

As can be seen from Fig. 1. the decrease in the strength of basalt fibers after heat treatment is insignificant. On average, the decrease in the strength of basalt fiber is in the range of 16 ... 24% and in any case has a much smaller value compared to glass fibers.

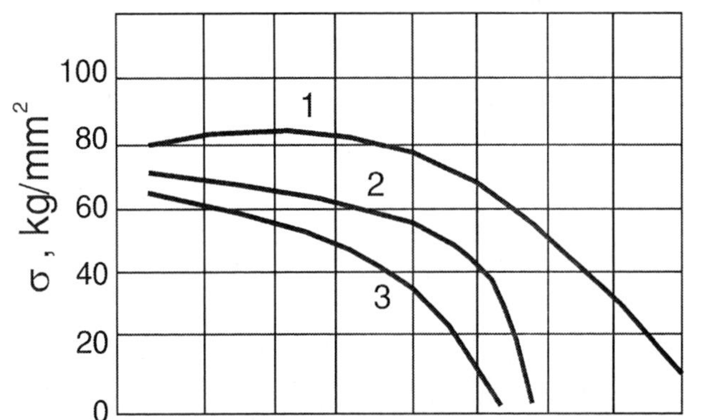


**Figure 1.** Dependence of strength of mineral fibers on heat treatment temperature.
1- basalt fiber; 2 - glass fiber; 3 - mineral wool

The study of the chemical stability of different types of basalts - vitreous, recrystallized, original, porcelain, it was found that vitreous basalt, which has not been subjected to heat treatment above the transition temperature, is destroyed by acid solutions. After leaching of the glass basalt, a pure porous silica skeleton of the fiber remains.

Thus, the leaching of all types of basalt fibers can be explained by the structural ratio of different phases and the position of a number of oxides. which are in the siliceous skeleton of the fiber. These phenomena are confirmed by theoretical statements about the leading role of the method and the degree of supercooling of the starting material and the state and temperature at which the fibers of a certain chemical and mineralogical composition were fixed.

**An overview of the problem status**

The aim of the research was to obtain porous coarse continuous basalt fibers with elevated operating temperature.

An important area in the production of mineral fibers is the use of rocks as a one-component raw material. In the technological process of obtaining the working melt it is possible to exclude the stages of adjustment of mineral and chemical composition of raw materials, silicate formation and lighting, which significantly increase the complexity of production and, consequently, lead to a significant increase in the cost of finished products. It is known that basalts, diabases, amphibolites, andesitic, diabase and pyroxene porphyrites, andesitobasalts and others can be used as one-component raw materials for the production of mineral fibers. The fibers obtained from this raw material are not only not inferior to traditional glass and mineral fibers, but in a number of properties far outweigh them. In the table 1.

**Table 1.** The main technical characteristics of mineral fiber-based various materials

| № | Fiber material | Density, $kg/m^3$ | Tensile strength, MPa $10^3$ | Young modulus, MPa $10^3$ | Elongation at tension, MPa $10^3$ |
|---|---|---|---|---|---|
| 1 | Steel fiber | 7.8 | 0.8… 2.15 | 200 | 3… 4 |
| 2 | Asbestos | 2.6 | 0.91… 3.1 | 68 | 0.6 |
| 3 | Fiberglass | 2.6 | 1.05… 3.85 | 70… 80 | 1.5… 3.5 |
| 4 | Carbon | 2.0 | 2.0 | 245 | 1.0 |
| 5 | Basalt | 1.1… 1.6 | 4.0… 12.0 | 184… 210 | 1.2… 1.8 |

This is the basis for the intensive development of the production of heat and sound insulation products using mineral fibers. Table 2 shows the comparative characteristics of the main technical characteristics of thermal insulation materials using mineral fibers.

**Table 2.** Comparative characteristics of the main technical indicators of thermal insulation materials based on mineral fibers

| № | Indicators | BTSH (10mm) | HSP (20mm) | BTSH (30mm) | SHAON (15mm) | SHAON (15mm) |
|---|---|---|---|---|---|---|
| 1 | Operating temperature, $^0C$ | 700 | 700 | 1150 | 400 | 400 |
| 2 | Thermal conductivity, VT/mK | 0.055 | 0.055 | 0.055 | - | - |
| 3 | Linear density, tex | 25 | 60 | 100 | 56-79 | 200-240 |
| 4 | Sorption hydration | 2 | 2 | 16 | - | - |
| 5 | Density, $kg/m^3$ | 450 | 200 | 190 | - | - |

As can be seen from the data given in table 1.2 heat-insulating products (TZIV), for the production of which basalt fibers were used have an unconditional advantage over other types of TZIV based on mineral fibers and, moreover, based on fibers made of polymeric materials. But at the same time as meeting the growing needs of industry in the production of thermal insulation materials come to the fore requirements to improve their performance, including such as operating temperature, thermal conductivity, density and others. It should be noted that basalt fibers and materials based on them have high thermal and insulating properties, as well as structural properties. In terms of temperature resistance, basalt fibers and products based on them significantly outperform similar indicators of heat and sound insulation products based on glass and mineral fibers.

Analyzing the structure, mineral and chemical composition of the fibers, it is possible to conclude that there are common features in all thermal and sound insulation materials, which impose certain restrictions on improving their performance. Thus, a common feature of all heat and sound insulation materials is the use for their production of fibers of dense structure with mineralogical and chemical composition, which is similar to the raw material, which predetermines the honey performance. Therefore, one of the ways to increase the performance of products based on inorganic mineral fibers is to modify the structure of the fiber, as well as adjust its chemical and mineralogical composition in the direction of increasing the operating temperature and thermal insulation properties of the fibers. In this direction, the possibility of replacing fibers with a dense structure in thermal insulation materials with basalt fibers with a three-dimensional microporous structure and with an adjusted mineral and chemical composition was considered. The research was carried out on thickened continuous fibers made from basalt raw materials of the Usachkivske, Yanova Dolyna (Rivne region) and Khayla-Choprak (Donetsk region) deposits. The chemical composition of basalt raw materials of these deposits is given in table 3.

**Table 3.** Chemical composition of basalt raw materials of different fields (% wt)

| № | Field | $SO_2$ | $TiO_2$ | $Al_2O_3$ | $Fe_2O_3$ | FeO | MgO | CaO | $Na_2O$ | $K_2O$ |
|---|---|---|---|---|---|---|---|---|---|---|
| 1 | Kryvyi Rih | 54.6 | 0.2 | 13.3 | 1.9 | 9.0 | 5.56 | 4.44 | 2.09 | 1.31 |
| 2 | Berestovetske | 49.03 | 2.85 | 12.59 | 3.88 | 10.15 | 5.47 | 9.54 | 2.34 | 0.66 |
| 3 | Selchivske | 55.6 | 0.66 | 19.4 | 7.54 | 7.54 | 3.84 | 7.91 | 1.86 | 1.56 |
| 4 | Yanova Dolina | 48.8 | 2.75 | 15.0 | 8.47 | 6.39 | 5.13 | 8.34 | 1.5 | 0.75 |

**Results**

As can be seen from the results of the experiments shown in tables 1-3 the lowest value of the viscosity is a solution of basalt charge of basalts with a high content of iron compounds (in oxide and oxide form), manganese compounds and alkali metals. These materials include basalts of Hein-Chokhrak field of Donetsk region and vice versa basalts of field Yanov Dolina of Rivne region, which contain high content of compounds of aluminum, titanium and calcium show the highest values of melt viscosity. Accordingly, the melting temperatures of the basalt charge of these deposits are inversely related (1390$^0$C for Hein-Chokhrak field of Donetsk region and 1540$^0$C for the field of Yanova Dolina, Rivne region). The temperatures of the upper and lower limits of crystallization of melts of basalts of different deposits are in the range of 1165… 1295$^0$C and also depend on their chemical composition and structural characteristics of the rock. X-ray phase analysis of samples of basalt melt (melting point 1540$^0$C, holding time 0.5 hours), showed the presence of augite and magnetite.

Experiments to determine the cooling rate of basaltic melt of different deposits Fig. 1. showed that melts with a high content of iron and alkali metal compounds (curves 2, 3.) have a significantly lower cooling rate compared with solutions (curve 1) in which the content of iron and alkali metal compounds is limited to 8…11% and 0, respectively 8…1.4%.

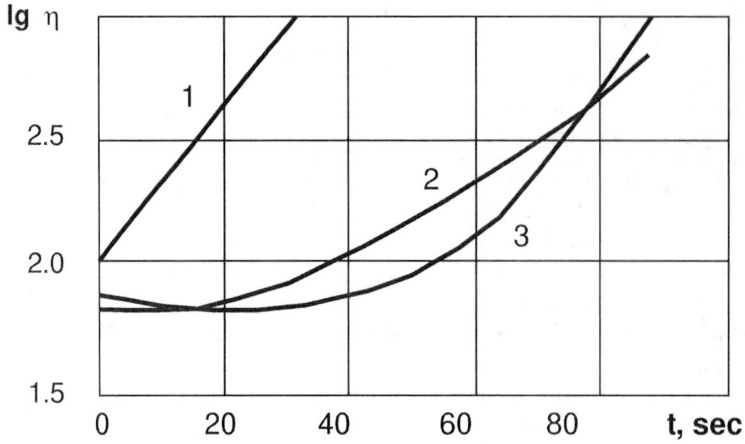


**Figure 2.** The rate of cooling of the melt of basalt rocks and their structural analogues
1. Melts with a low content of compounds of iron and alkali metals;
2. Melts with high content of iron compounds;
3. Melts with high content of iron and alkali metal compounds

These features of the cooling rate of the basalt melt of different deposits can be explained by a decrease in the temperature transparency of melts with a high content of iron and alkali metal compounds. Thus, the most important properties of basalt melts, which provide the possibility of forming a continuous coarse fiber to obtain the specified characteristics are the viscosity of the melt and its crystallization properties.

Determination of the stability of the obtained continuous coarse basalt fiber in a corrosive environment was determined by the method of its treatment with solutions of different chemical nature at a temperature of $95 \pm 5^0C$.

## Conclusions

According to the obtained results, porous basalt fiber with the texture of the specified parameters can be obtained using basalts of certain deposits. Technological processing of basalt fiber, in this case, is carried out with the help of reagents and with parameters that ensure the formation of a silica skeleton of a certain structure. This provides a porous quartzite fiber with high operating temperatures, which significantly exceeds the operating temperature of basalt fibers.

## References

[1] Iler R K 1979 *The Chemistry of Silica* (New York: John Wiley & Sons) p 896
[2] Fournier R O 1985 Behavior of silica in hydrothermal solutions *Review in Economic Geology* **2** 45-60 https://doi.org/10.5382/Rev.02.03
[3] Allègre C-J, Michard G 1974 *Introduction to Geochemistry* (Springer, Dordrecht) https://doi.org/10.1007/978-94-010-2261-3
[4] Abrahamyan AV 1963 Study of the process of leaching of vitreous basalts *Glass and ceramics* 7
[5] Christensen V, Jensen S, Guldberg M & Kamstrup Ole 1994 Effect of Chemical Composition of Man-Made Vitreous Fibers on the Rate of Dissolution in Vitro at Different pHs *Environmental health perspectives* **102** Suppl 5 83-86 doi: 10.1289/ehp.94102s583.

IOP Conf. Series: Materials Science and Engineering **907** (2020) 012033 doi:10.1088/1757-899X/907/1/012033

# Influence of whitening additives on the properties of decorative slag-alkaline cements

**V I Gots, A G Gelevera, O N Petropavlovsky, N V Rogozina and V V Smeshko**

Research Institute of Binders and Materials named after V D Glukhovsky, Kiev National University of Construction and Architecture, 03037 Povitroflotskyi Avenue, 31, Kyiv, Ukraine

a-gelevera@ukr.net

**Abstract**. Traditional decorative cements are made on the basis of white cements and the demand for them is constantly growing. But such cements have all the disadvantages of clinker cements, namely, high energy intensity and high price. An alternative may be slag-alkali cements having higher physical and mechanical characteristics and lower price. Therefore, these studies were aimed at choosing such additives that would have the maximum whitening properties and would enhance the stabilization of bleaching processes during hydration and hardening of slag in the presence of alkalis, to determine the simultaneous effect of three additives on the whiteness and physical-mechanical characteristics, as well as to determine the optimal the content of these additives in slag-alkali decorative compositions. In this work, as a result of the studies performed, the possibility of obtaining effective slag-alkali white cements as the basis for the production of decorative concrete with acceptable economic characteristics is proved. The influence of each of the bleaches on such characteristics of decorative slag-alkali cements as strength, setting time, whiteness is established. Methods for optimizing compositions are identified.

## 1. Introduction

Decorative cements are used to increase the architectural expressiveness of buildings and perform finishing work. Demand for them and requirements for their operational characteristics are constantly growing [1]. Usually they are made on the basis of white cements [2]. However, such decorative cements have all the disadvantages of Portland cement, including high energy demand and high cost. They are not always able to provide a decorative ecological and comfortable coating with improved performance [3-5].

The use of decorative alkaline activated cements, in particular slag-alkali cements, can be an important alternative in the production of white and non-ferrous cements. In addition, they are able to demonstrate a number of special properties цвета color stability, durability, high strength and adhesion, etc. [6-12]

The problem associated with the use of slag-alkali cements as a basis for decorative cements with high whiteness (70...95%) is the instability of the chemical composition of slags and the acquisition of green color in the process of hydration of slag-alkali cement due to the oxidation processes of sulfide compounds present in slag [13].

In previous studies in this direction [14-17], insufficient attention was paid to the influence of individual additives and their interaction on the decorative and physic-mechanical properties of slag-alkali cements.

Therefore, these studies were aimed at choosing such additives that would have the maximum whitening properties and enhance stabilization of bleaching processes during hydration and hardening of slag in the presence of alkalis, to determine the simultaneous effect of three additives on the whiteness and physic-mechanical characteristics, as well as to determine the optimal the content of these additives in slag-alkali decorative compositions.

## 2. Raw materials and research methods

In the studies, two blast furnaces granular slags were used, which significantly differed in the content of iron oxides. The chemical composition of the slag is presented in Table 1.

**Table 1.** Composition of blast furnace granulated slag.

| Slag | The content of oxides, % of the mass | | | | | | | | | | $M_b$ |
|---|---|---|---|---|---|---|---|---|---|---|---|
| | $SiO_2$ | $Al_2O_3$ | CaO | MgO | FeO | $SO_3$ | MnO | $TiO_2$ | roasting losses | Σ | |
| Dnieper | 37,90 | 6,85 | 45,35 | 5,21 | 0,31 | 2,6 | 0,11 | 0,35 | 1,34 | 100,02 | 1,13 |
| Kryvyi Rih | 39,2 | 5,1 | 47,14 | 4,42 | 1,65 | 1,72 | 0,16 | – | 0,6 | 99,99 | 1,16 |

To compare the influence of grinding conditions on the whiteness of cement, slag was ground both in a conventional ferrous metal mill with steel grinding bodies, and in a mill with alubite (high alumina) grinding bodies and a lining. The degree of grinding of the slag was 4650...4770 $cm^2/g$ according to Blaine.

Non-hygroscopic sodium metasilicate ($Na_2O{\cdot}SiO_2{\cdot}5H_2O$) in powder form was used as the alkaline component of slag-alkali cements.

$TiO_2$, $CaCO_3$ (90%) and kaolin of class KH 84 (84%) were used as whitening additives.

To produce 4×4×16 cm beam samples of 1:3 composition from a mixture of normal consistency, standard sand of the Gusarovskoye field, Kharkov region, was used.

All components were mixed in standard Hobart mixers.

Technological and physic-mechanical properties of slag-alkali cements were determined in accordance with state standards in force in Ukraine.

The whiteness of the samples was determined using an NS810 spectrophotometer with a wavelength range of 400...700 nm. For comparison, we used a sample of Ral 9016 with a brightness of L = 98,85. Samples from cinder-base alkali cement test of normal density were studied after 28 days of hardening.

To optimize the composition of decorative cements, ease of processing the obtained data and create a mathematical model of the experiment in the form of regression equations, we used the mathematical design of the experiment in the form of a full factorial experiment $3^3$, which traditionally has proven itself in scientific research.

## 3. Research Results and Discussion

### *3.1. Study of basic physical-mechanical characteristics and whiteness*

First, studies were carried out on the effect of bleaching additives on the degree of whiteness of the compositions and on their physical and mechanical properties using a full factorial experiment. Factors, the boundaries of their variation, and the experimental matrix with response functions are presented in Table 2 and Table 3.

The Dnieper slag, ground in an ordinary steel mill and therefore partially contaminated with metal grinding media, and the Dnieper slag, ground in a mill with alubite grinding media and a lining, were used.

**Table 2.** Source data

| № | Factors | Units | Codes | Factor Variation Levels | | |
|---|---|---|---|---|---|---|
| | | | | −1 | 0 | +1 |
| 1 | $TiO_2$ content | % | X1 | 0 | 4 | 8 |
| 2 | Kaolin content | % | X2 | 0 | 7 | 14 |
| 3 | $CaCO_3$ content | % | X3 | 0 | 12 | 24 |

**Table 3.** Full factorial experiment matrix.

| № | Plan matrix in codes | | | Plan matrix in physical terms | | | Strength $R_{str.}$, MPa, after, days | | | Setting start time, min. | Degree of whiteness, % | |
|---|---|---|---|---|---|---|---|---|---|---|---|---|
| | X1 | X2 | X3 | $TiO_2$ % | kaolin % | $CaCO_3$ % | 2 | 7 | 28 | | milling in a steel mill | milling in alubite mill |
| 1 | 1 | 1 | 1 | 8 | 14 | 24 | 25,2 | 30,4 | 38,6 | 33 | 88,4 | 94,4 |
| 2 | −1 | 1 | 1 | 0 | 14 | 24 | 31,5 | 38,8 | 46,2 | 36 | 74,8 | 79,8 |
| 3 | 1 | −1 | 1 | 8 | 0 | 24 | 29,7 | 40,2 | 46,4 | 38 | 83,5 | 89,1 |
| 4 | −1 | −1 | 1 | 0 | 0 | 24 | 35,4 | 46,8 | 56,0 | 37 | 66,0 | 70,5 |
| 5 | 1 | 1 | −1 | 8 | 14 | 0 | 32,5 | 41,0 | 52,2 | 57 | 83,0 | 88,6 |
| 6 | −1 | 1 | −1 | 0 | 14 | 0 | 36,1 | 43,9 | 49,9 | 46 | 63,2 | 67,5 |
| 7 | 1 | −1 | −1 | 8 | 0 | 0 | 35,9 | 46,2 | 57,0 | 69 | 83,7 | 89,4 |
| 8 | −1 | −1 | −1 | 0 | 0 | 0 | 39,0 | 47,4 | 56,7 | 54 | 60,3 | 64,3 |
| 9 | 1 | 0 | 0 | 8 | 7 | 12 | 31,4 | 39,5 | 48,5 | 42 | 82,2 | 87,9 |
| 10 | −1 | 0 | 0 | 0 | 7 | 12 | 36,1 | 44,2 | 52,1 | 36 | 63,7 | 68,1 |
| 11 | 0 | 1 | 0 | 4 | 14 | 12 | 31,8 | 39,2 | 45,0 | 40 | 81,4 | 87,1 |
| 12 | 0 | −1 | 0 | 4 | 0 | 12 | 35,5 | 45,8 | 52,3 | 47 | 77,5 | 82,8 |
| 13 | 0 | 0 | 1 | 4 | 7 | 24 | 30,1 | 39,4 | 47,5 | 34 | 81,2 | 86,9 |
| 14 | 0 | 0 | −1 | 4 | 7 | 0 | 35,5 | 45,0 | 54,7 | 54 | 75,5 | 80,8 |
| 15 | 0 | 0 | 0 | 4 | 7 | 12 | 33,5 | 42,4 | 49,8 | 40 | 77,7 | 83,2 |

Note. The amount of metasilicate in the compositions – 10%, slag – Dnieper.

As a result of processing the data in Table 3, we obtained regression equations for three responses – strength after 28 days ($R_{28}$), the start time of setting ($\tau$) and degree of whiteness (W) when grinding slag in a mill with alubite grinding media:

$$R_{28} = 49{,}85-1{,}81\cdot X_1-3{,}65\cdot X_2-3{,}57\cdot X_3+0{,}47\cdot X_{11}-1{,}23\cdot X_{22}+1{,}27\cdot X_{33}+0{,}5\cdot X_{12}-2{,}48\cdot X_{13}-0{,}75\cdot X_{23}$$
$$\tau = 40{,}17+3{,}1\cdot X_1-3{,}3\cdot X_2-10{,}3\cdot X_3-1{,}28\cdot X_{11}+3{,}72\cdot X_{22}+3{,}72\cdot X_{33}-1{,}13\cdot X_{12}-3{,}63\cdot X_{13}+1{,}88\cdot X_{23}$$
$$W = 83{,}16+9{,}92\cdot X_1+2{,}11\cdot X_2+3{,}01\cdot X_3-5{,}16\cdot X_{11}+1{,}79\cdot X_{22}+0{,}69\cdot X_{33}-1{,}0\cdot X_{12}-1{,}63\cdot X_{13}+1{,}53\cdot X_{23}$$

The strength of the samples after 28 days was in the range 38,6...57,0 MPa, the onset of setting – 33...69 minutes, the degree of whiteness – 64,3...94,4%.

Diagrams of strength, setting start time and whiteness are presented in Fig. 1...Fig. 3. For the convenience of their construction, each 3-factor equation was divided into three pseudo 2-factor equations with X3 ($CaCO_3$) – 0%, 12% and 24%

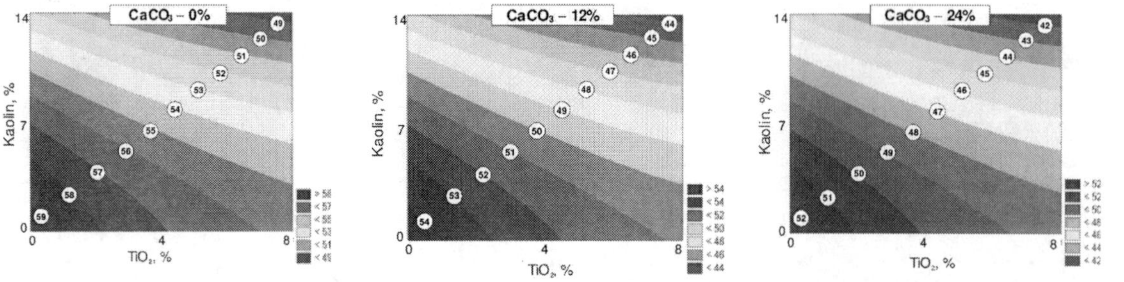


**Figure 1.** The isolines of the strength of slag-alkali compositions after 28 days

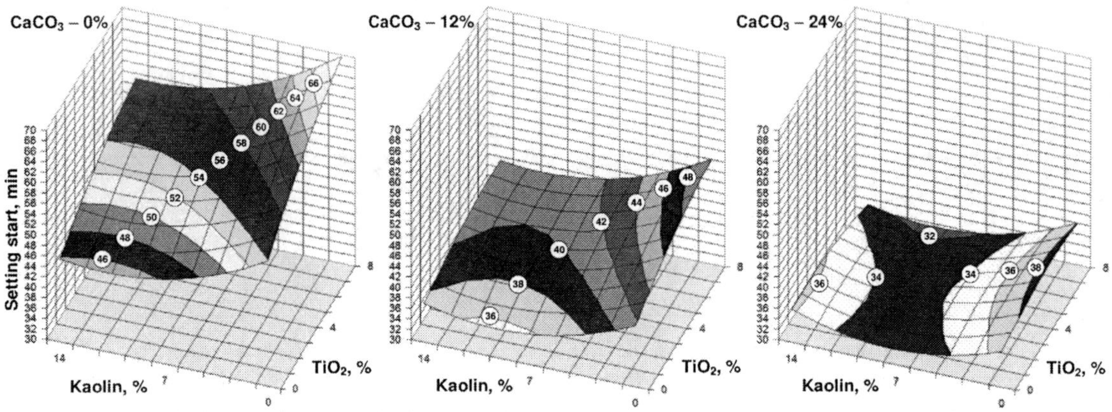


**Figure 2.** Isosurfaces and isolines of the start time of setting slag-alkali compositions.

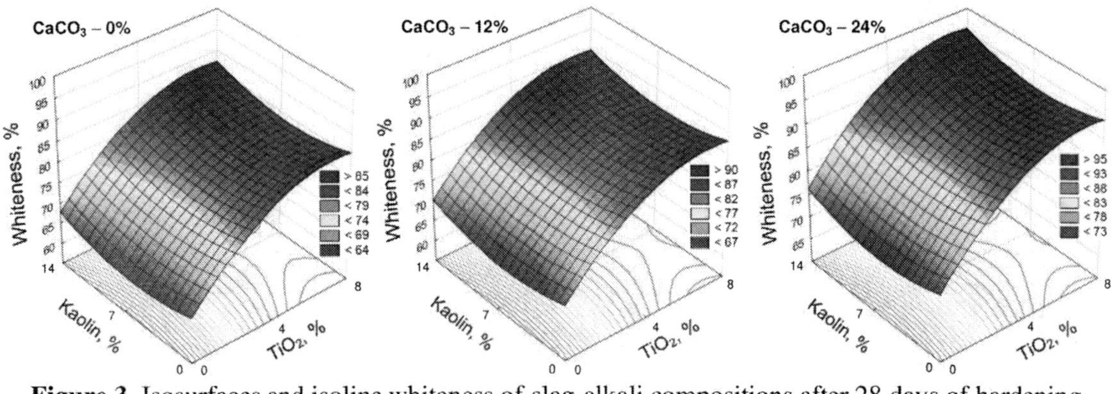


**Figure 3.** Isosurfaces and isoline whiteness of slag-alkali compositions after 28 days of hardening.

Slag-alkali decorative cements had good dynamics of hardening in the early stages and after 2 days showed a strength of 25,2...39,0 MPa.

Depending on the degree of whiteness, white clinker cement is divided into three grades: 1 (premium) – 80...95%, 2 grade – 75%, 3 grade – 68%. Therefore, based on these requirements, most of the slag-alkaline compositions, even after grinding the slag in a steel mill, can be attributed to the decorative, as they had a degree of whiteness of – 70%.

As a result of the regression analysis of the obtained equations, the significance of each of the coefficients of the equation (the so-called Pareto effects) is determined, that is, the influence of each of the factors and their interaction on the response functions. The results are presented in Fig. 4.

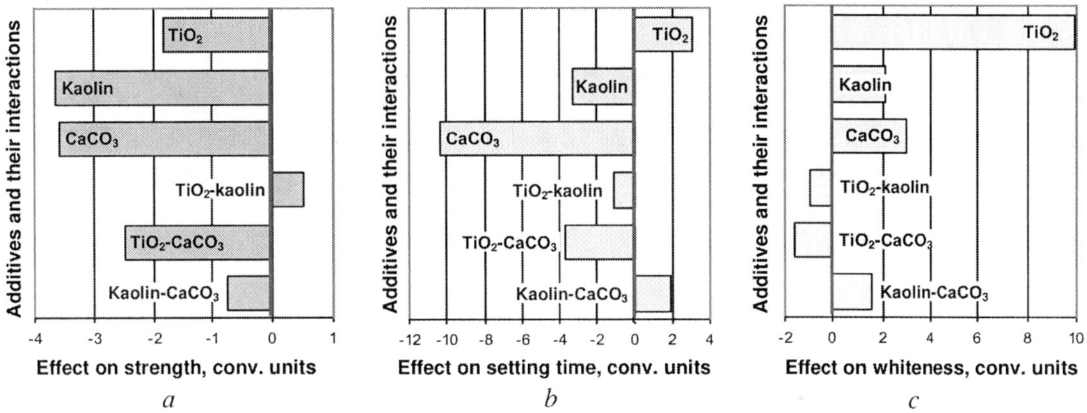


**Figure 4**. The degree of influence of whitening additives on the properties of slag-alkali decorative cements: *a* – on compressive strength; *b* – on setting start; *c* – on whiteness.

As can be seen from Fig. 4, *a*, all individual whitening additives do not contribute to the increase in strength. The greatest negative effect on the strength of slag-alkali cement compositions has kaolin, slightly less than – $CaCO_3$ and the smallest – $TiO_2$. The interaction of factors X1 and X2 ($TiO_2$ and kaolin) contributes to a slight increase in strength.

If kaolin and $CaCO_3$ shorten the setting time, then $TiO_2$ (X1) and the composition of kaolin with $CaCO_3$ (interaction of factors X2 and X3) somewhat contribute to their increase (Fig. 3, *b*).

From the point of view of increasing whiteness, the most positive effect is exerted by $TiO_2$, then $CaCO_3$ and the least influence is влияние kaolin (Fig. 3, *c*). The interaction of factors X1 and X2 ($TiO_2$ and kaolin) and X1 and X3 ($TiO_2$ and $CaCO_3$) somewhat impair whiteness.

### *3.2. The effect of iron oxides on whiteness*

Checking the effect of iron oxides was carried out as follows. As a control composition, composition N° 4 (see table 3) based on the Dnieper slag was adopted, where the amount of FeO was 0,35% (Table 1). For comparison, the same composition N° 4 was used, but based on Kryvyi Rih slag, where the amount of FeO was 1,65% (Table 1). The third point was composition N° 4 on Kryvyi Rih slag, where additional FeO was added to the total amount of 3,3%. The research results are presented in Fig. 5. The mathematical dependence of whiteness on the FeO content was expressed by the regression equation shown in the graph.

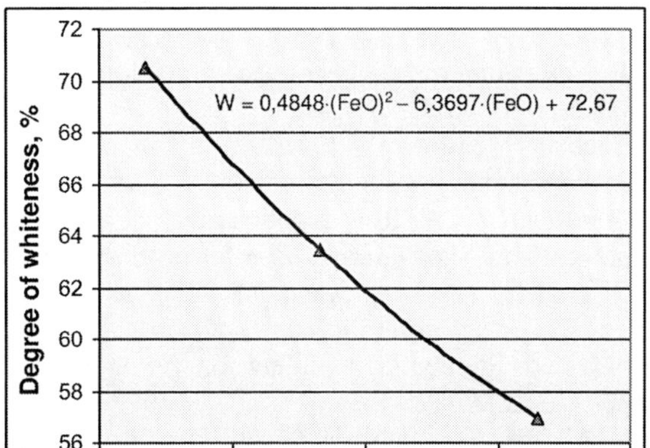


**Figure 5**. Effect of FeO additive on the whiteness of slag alkaline cement.

### *3.3. The effect of sulphide compounds of slag's*

In the initial stages of hardening, slag-alkaline compositions may acquire a blue-green color. In the literature [13], this is associated with the formation of compounds of the type $FeSO_4{\cdot}7H_2O$ and this was the basis for concerns about the possibility of obtaining white decorative cements on their basis. But if slag-alkali concrete is exposed to air, then its surface quickly and substantially "bleaches", obviously, due to carbonization [18]. The introduction of whitening additives in its composition completely eliminates the risks of painting the surface in undesirable colors and shades.

### *3.4. Economic aspect.*

If we compare the cost of bleaches, then it roughly corresponds in such a way – "$CaCO_3$ : kaolin : $TiO_2$" as 1:6:12. That is, the most expensive component is $TiO_2$ and therefore systems with a significant $TiO_2$ content are economically inefficient.

Analyzing Table 3 and evaluating the effectiveness of the compositions from the point of view of their economy, it can be said that slag-alkali compositions make it possible to significantly reduce or completely abandon the use of $TiO_2$. So, for example, in compositions N° 2 and N° 4, $TiO_2$ is generally absent, but due to only kaolin and $CaCO_3$, a whiteness of 70,5...79,8% and a strength of 46,2...56,0 MPa are ensured. And in warehouses N° 11...15 with a quantity of $TiO_2$ 4%, a degree of whiteness of 80,8...87,1% and a strength of 45...54,7 MPa are ensured. Therefore, titanium dioxide can be used to trim compositions, and as the main bleaches use cheaper kaolin and $CaCO_3$.

## 4. Conclusions

Thus, as a result of the studies, the possibility of obtaining effective slag-alkali white cements, as the basis for the production of decorative concrete based on them, with acceptable economic characteristics, has been proved. The influence of each of the bleaches is established, the role of FeO on the deterioration of the whiteness of decorative compositions is indicated. Methods for optimizing compositions are identified. The high activity of slag-alkali cements allows the introduction of bleaching additives in the amount of 25% or more without significant deterioration of physical and mechanical characteristics.

Further work will be directed to deeper physical and chemical studies, the study of the catalytic effect of $TiO_2$ (both conventional and anatase) on the restoration of whiteness during operation, the effect of the dispersion of bleaches on decorative properties, the establishment of the interaction of bleaches with dyes in highly alkaline environments, the study of white stability and non-ferrous slag-alkali cement to the effects of UV and heat-moisture treatment of concrete and to perform long-term tests to establish the durability and operational characteristics of white and colored slag-alkali cements.

## References

[1] Semenov V M 2010 Paintwork materials for the protection of concrete and cement surfaces. *Paint and varnish industry* **11** pp 23-27

[2] White cement [Electronic resource] *Portal of Budpostach* Access mode: https://pp-budpostach.com.ua/a118411-belyj-tsement.html

[3] Berezina L A and Abdulberov Z A 2009 Application of kaolin by IMERYS Minerals in the manufacture of paints and varnishes *Paintwork materials and their use* **1-2** pp 61-65

[4] Levashova Yu S, Kosenko N O and Lebedeva O S 2019 Investigation of the parameters of the microclimate of large crowds *Scientific Bulletin of Construction* **93(3)** (Kharkiv) pp 217-221

[5] Petrenko G S 2010 Modern pigments and fillers for the paint and varnish industry *Paint and varnish industry* **9** pp 30-36

[6] Krivenko P V 1994 Alkaline Cements: *materials First Intern. Conf. "Alkaline Cements and Concretes"* (Kyiv) pp 111-129

[7] Krivenko P V, Petropavlovsky O N, Gelevera A G, Vozniuk G V and Pushkar V I 2009 Commercially-produced alkaline cements and their efficiency *Scientific-Technical Collection*

*"Resource-Saving Materials, Structures, Buildings and Structures"* **18** (Rivne) pp 64-71

[8] Krivenko P 2014 Alkali-Activated Materials – 55 Years of Experience *Proc. of the International Conference on Alkali-Activated Cement and Concretes* (Chongqing, China) pp 1-5

[9] Shi C, Krivenko P V, Della Roy 2012 *Alkaline activated cements and concretes* (in Chinese, Authorized translation from English): Monograph (Taylor & Francis) p 326

[10] Fernández-Jiménez A, Garcia-Lodeiro I, Maltseva O and Palomo A 2019 Hydration mechanisms of hybrid cements as a function of the way of addition of chemicals. *Journal of the American Ceramic Society* **102 (1)** pp 427-436

[11] Krivenko P, Petropavlovsky O, Kovalchuk O, Pasko A and Lapovska S 2018 Design of the composition of alkali activated Portland cement using mineral additives of technogenic origin *Eastern-European Journal of Enterprise Technologies* 4/6 (**94**) pp 6-15

[12] Krivenko P V, Petropavlovsky O N, Gots V I and Rostovskaya G S 2009 Alkali activation of Composite Cement *Ibausil. Internationale Baustofftagung* **1** (Weimar) pp 445–456

[13] Chaouche M, Gao X X, Cyr M, Cotte M and Frouin L 2017 On the origin of the blue/green color of blast-furnace slag-based materials: Sulfur K-edge XANES investigation. *J Am Ceram Soc.* **100** pp 1707-16 doi:10.1111/ jace.14670

[14] Krivenko P V, Kovalchuk A Yu and Ostrovskaja L M 2011 Studying of possibility of increase of slag-alkali cements whiteness degree. *Collection "Building materials, products and technical equipment"* **41** (Kyiv, Research Institute of Building Materials and Products) pp 10-14

[15] Krivenko P, Petropavlvskyy O, Puskar V and Ostrovska L 2011 Decorative alkaline cements *IV Intern. Symp: Non-Traditional cement & Concrete* (Brno) pp 257- 265

[16] Kryvenko P, Sanytsky M, Kropyvnytska T, Kotiv R 2014 Decorative multi-component Alkali Activated Cements for restoration and finishing works *Advanced Materials Research* **897** Trans Tech Publications, Switzerland pp 45-48 doi:10.4028/www.scientific.net/AMR.897.45

[17] Krivenko P V and Kovalchuk A Yu 2019 Management of the decorative properties of alkali cements *Scientific Bulletin of Civil Engineering* **2(95)** (Kharkiv) pp 280-285

[18] Kovalchuk O Yu 2014 Research of carbonization of alkaline concretes *Collection "Building materials, products and sanitary ware"* **52** (Kyiv, Research Institute of Building Materials and Products) pp 19-24

IOP Conf. Series: Materials Science and Engineering **907** (2020) 012034 doi:10.1088/1757-899X/907/1/012034

# Resource saving and energy saving at the simultaneous production of two types of cement

**A A Riazanov[1], V I Vinnichenko[2], A N Riazanov[1], R Z Rakhimov[3] and I V Nedoseko[1]**

[1] Department of Building Constructions, Ufa State Petroleum Technological University, Ufa, Russian Federation
[2] Department of Mechanization of Construction Processes, Kharkiv National University of Civil Engineering and Architecture, Kharkiv, Ukraine
[3] Department of Building Constructions, Kazan State University of Architecture and Engineering, Kazan, Russian Federation

vvinnichenko@ukr.net

**Abstract**. Modern trends in the development of the global cement industry are to reduce the energy intensity of production and reduce the environmental burden on the environment. These goals can be achieved by using fuel-containing man-made waste in place of natural raw materials and fuel. Great decrease of fuel costs by 20-40 % when making Portland cement is shown to be achieved by combining calcination of both Portland cement clinker and carbonate-silica mix in one heat and technological cycle to get cements with low-temperature calcination. Extra heat effect is achieved by adding combustible industrial wastes, like ashes from power plants or coal waste, to the raw mix to get cements with low-temperature calcination. Further modification of lime-silica cements obtained by Portland cement clinker within 20-40 % range by weights is shown to get mixed product with strength indices in sand mortar corresponding to 0,6-0,8 grade strength of original Portland cement. Combined method for making two kinds of binders is shown to reduce carbon dioxide by 22-60 % calculated as the final modified product as compared with traditional making of Portland cement clinker. When substituting 50 % of clinker part by cements with low temperature calcination with certain composition, grade strength of the mixed product is similar to Portland cement indicators.

## 1. Introduction

Currently production of Portland cement has exceeded 4bln tone per year and keeps growing. In the nearest future Portland cement remains the only binder due to wide occurrence of mineral resources on the Earth applied for making it and accessibility of technology. New trends for the development of the world cement industry are reduction of both the energy intensity of Portland cement production and environmental impact. Emission of greenhouse gases as carbon dioxide under dissociation of both calcium and magnesium carbonate inside the cement kiln thereby making about 500kg per 1 ton of clinker are considered to be the most negative side effects of production. It is not impossible to greatly reduce $CO_2$ emissions directly when producing Portland cement because raw material compositions applied for clinker making contain much carbonate. However, it may be done by adding various mineral admixtures at clinker grinding thereby reducing specific yield of carbon dioxide per unit of the final mixed product.

## 2. Combined method in producing Portland cement

### *2.1. Ways to reduce fuel costs*

Energy consumption under production of Portland cement is 60-70 % of cost thereof, with one part thereof being to electric energy, while the other three parts – to heat energy [1]. Potential ways to reduce energy costs under production of Portland clinker are:

- application of energy-efficient technologies and equipment;
- gas recovery and recirculation;
- application of material resources reducing heat loss affected by reactions;
- utilization of wastes containing fuel components in compositions thereof;
- reducing of heat loss in equipment.

The most efficient ways to save energy without reequipping cement plants are heat recovery of waste gases from kilns as well as utilization of wastes containing fuel components in compositions thereof [2]. Ashes from thermal power stations and coal beneficiation wastes are regarded as large wide-spread man-made fuel-containing mineral wastes. Large-scale utilization of such wastes may enable to solve serious environmental problems faced by many industrial regions of our planet [3, 4].

Direct application of fuel-containing wastes instead of part of argillous components for producing Portland clinker is not quite efficient, as it enables to substitute no more than 10-15 % of natural argillous raw material that negligibly reducing fuel capacity of high-temperature process for clinker making [5].

### *2.2. Binders after low-temperature calcination*

Production of mixed binders made from energy-intensive local cements modified by clinker adding under grinding may prove to be viable solution in the current conditions. This category includes binders for calcination obtained by heat treatment before sintering of the raw charge consisting of carbonate rock and argillous component [6-11]. The calcination temperature therewith is 900-1100 °C while the final product after fine grinding is a hydraulic binder identical to roman cement by its main properties and compound composition obtained from natural carbonate raw materials with argillous admixtures of 20-25 % by weight [12]. Natural carbonate-argillous mixture ratio as well as hydraulic activity of such binder is defined by hydraulic index value m to be within 1.1-1.7 range [13]. However, CaCO3 in the mixture before calcination will be 70-75 % by weight. Therefore, carbon dioxide yield will also be similar to values of Portland cement production. Lower energy intensity will be the technological advantage therewith due to reducing the calcination temperature by 350-550 °C. Such binders alone have relatively weak ultimate strength to be 5-15 MPa in sand mortar, and are differed by deferred kinetics of setting and hardening under normal conditions. However, if modified by active mineral admixtures under grinding their strength becomes 25 MPa and approaches the values of average grade cement. Significant effect (up to 35 MPa) is achieved under integrated application of both mineral and chemical admixtures [14]. The idea itself for reasonable receiving of both intermediate product as clinker and final product as modified (mixed) binder is to be changed if simultaneously try to significantly reduce energy intensity of Portland cement and carbon dioxide yield at clinker calcination. We may consider therewith both Portland cement modification by adding the binder that is less energy-intensive and more eco-friendly by $CO_2$ emissions, and vice versa, modification of more energy-efficient lime-silica cement obtained at low temperature calcination by adding Portland cement clinker at grinding. There are several ways to increase energy efficiency, with the first one being the reduction of clinker share in the mixed binder and increase of clinkerless share therein. Energy intensity of the clinkerless share may be decreased by reducing carbonate content in the mixture. However, it is not possible for roman cement made from natural mix by availability of the hydraulic index in the permissible range. Application of binders for low temperature calcination with far less carbonate content in the raw mix is considered to be effective. A V Volzhensky et al [15,16] have proposed the way to get low-temperature cement by simultaneous calcination at 900-1100 °C of either limestone or dolomite with fly ash of thermal power stations or waste coal. Full spectrum of binders after low temperature calcination with strength of compression of 5-20 MPa in sand mortar and mechanism of hardening

IOP Conf. Series: Materials Science and Engineering **907** (2020) 012034 doi:10.1088/1757-899X/907/1/012034

typical to lime-pozzolanic cement of mixed grinding is supposed to be obtained by this technology. Principal technological advantage of the indicated cements is that the most energy-consuming dissociation process of $CaCO_3$ and $MgCO_3$ is actually provided by the heat generated at simultaneous calcination of fly ash or waste coal. Carbonate content in the mix is much lower that with Portland cement and is 25-50 %. Therewith ash or waste coal content in the mix is 50-75 % by weight thereby providing quite high heat capacity of the raw mix in the indicated range of initial components ratio that made 3245-6160 kJ/kg for the mixes being analyzed. Lime-ash, lime-glynite, dolomite-glynite and dolomite-ash cements obtained after calcination refer to hydraulic binders. Their modification by adding Portland cement clinker enables to regulate strength properties in wide range 'Figure 1' thereby greatly widening areas for applying modified cement [17]. Besides, carbon dioxide yield expressed as final mixed product is reduced by 22-60 % per 1 ton of modified cement depending on components ratio in the raw mix and in the final product 'Figure 3'.

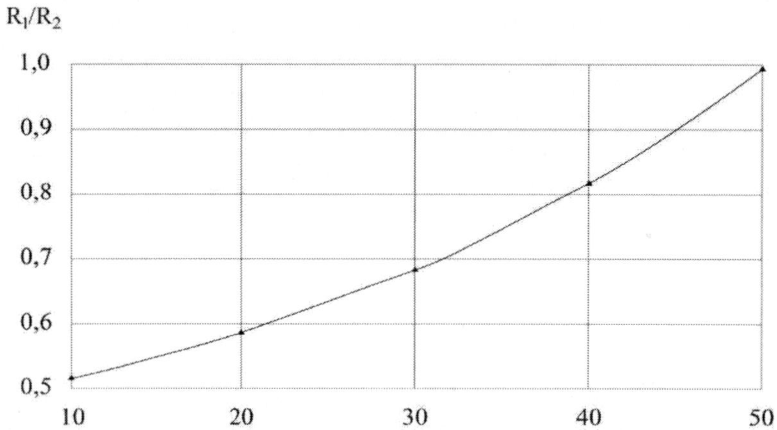


**Figure 1.** Relation between average relative strength of modified binder samples and Portland cement clinker content

- $R_1$ – compression strength of samples with modified binder;
- $R_2$ – compression strength of samples with Portland cement.

Energy efficiency ratio $K_e$ 'equation 1' is proposed to be used for evaluation of heating properties of two-component fuel-containing mixture and calculation of components ratio therein:

$$K_e = \frac{m^w \cdot Q_i^w}{Q_i^o} \tag{1}$$

where, $Q_i^o$ - required specific fuel costs for calcination the raw mix, kJ/kg; $Q_i^w$-heat capacity of combustible waste, kJ/kg; $m^w$- specific content of combustible waste in the raw mix, kg/kg.

To burn the raw mix without process fuel ($K_e$=1), specific content of waste in 1 kg of two-component raw mix will be 'equation 2':

$$m^w = \frac{Q_i^0}{Q_i^w} \tag{2}$$

The required reference fuel consumption (kg) for calcination of 1 t of raw mix (at $K_e$<1) will be 'equation 3':

$$B_f = \frac{1000 \cdot Q_i^0}{29308}(1 - K_e) \tag{3}$$

therewith, decrease in process fuel consumption for calcination 1t of raw mix will be:

$$E_f = K_e \cdot 100 \tag{4}$$

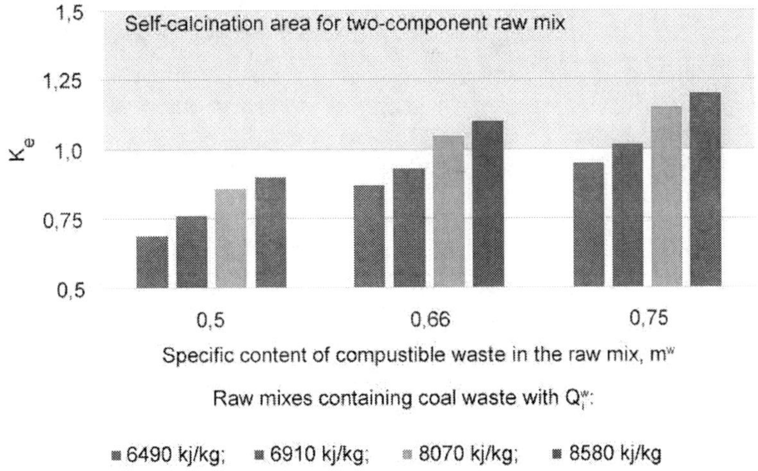


a

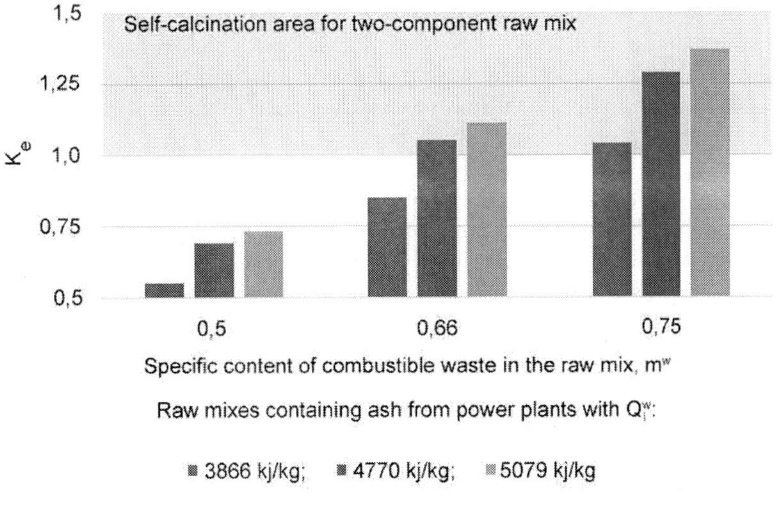


b

**Figure 2.** Energy cost for calcinations of fuel-consuming raw mixes at production of low-temperature cements

The analysis of the obtained calculated relations indicated in 'Figure 2 (a, b)' has revealed that $K_e$ = 0.55-1.37 for the laboratory compositions at heat capacity of fly ash at 3860-5070 kJ/kg and its composition in two-component mixture amounting to 50-75 %. We have $K_t$ = 0.69-1.2 for raw mixes with waste coal at their heat capacity of 6490-8580 kJ/kg. The process fuel therewith under the established heating process for calcinating raw mixes with values of $K_t \geq 1$ is not required.

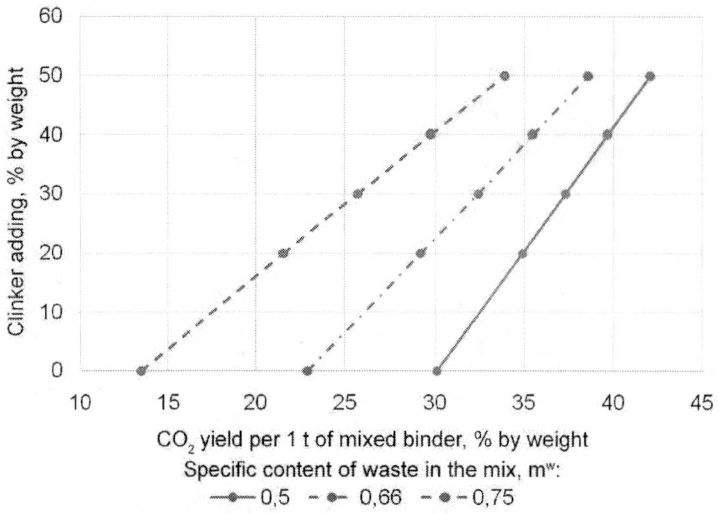


**Figure 3.** Properties of modified cements by specific yield of $CO_2$

Possibility of self-calcination of two-component carbonate-ash mixture at 1100 °C was confirmed by production of pilot batch of 60 tons of lime-ash cement by Voroshilovgrad Cellular-Concrete Plant in 1989 (now, Ukraine). The 30m rotary lime kiln was used for calcination of carbonate-ash mixture. Fly ash content in the mixture was 66 % by weight. The pilot batch of autoclave gas-concrete items was made from lime-ash cement obtained with no process fuel. Later this plant released the pilot batch of lime-brick cement made from gravitational waste coal.

*2.3. Combined heating process*

The authors calculated the combined method for producing modified Portland cement while simultaneously firing two raw mixes in two rotary kilns. In a low-temperature furnace, the heat of the gases that leave the high-temperature furnace is utilized. In a high temperature furnace, the firing temperature is 1300-1500 °C. And in a low-temperature furnace, the firing temperature is 1000-1100°C.

Firing products are Portland cement clinker (high temperature furnace) and calcareous or dolomite clinker (low temperature furnace). In the future, two types of clinker are crushed together.

The heat gas flow of the high temperature furnace is directed to a low temperature furnace, where the final product, for example, is dolomite clinker. Thermal power represents the difference between the required thermal power for firing dolomite clinker and the thermal power of the gas stream that leaves the high temperature furnace.

Thermal power is the amount of heat that must be supplied to the furnace per unit time to ensure the quality of the final product.

Thermal power is defined by equation (5):

$$N = P \cdot q \tag{5}$$

where, P·- kiln capacity, kg of clinker per sec., q – specific consumption of heat energy for making 1kg of clinker, kJ/kg.

The energy indices of combined firing are calculated for the Kramatorsk cement plant. The furnace produces clinker by wet production method. The furnace capacity is 16 tons of clinker per hour. The temperature of the gases leaving the Portland cement kiln is 500 °C.

Capacity of heat flow of gases is defined by equation (6):

$$N_d = V \cdot c \cdot t \tag{6}$$

where, V – gas volume exhausted from dolomite kiln, $m^3/s$, c – gas heat capacity, $kJ/m^3$ K, t – gas temperature.

Decrease of specific fuel consumption will be 'as in equation (7)':

$$B_f = \frac{N_d}{Q} \tag{7}$$

where, Q – calorific value of standard fuel - 29 308 kJ / kg.

Costs for process fuel in the dolomite furnace can also be reduced through organ-containing waste. In Ukraine, one of the variants of such waste is coal enrichment waste.

The thermal capacity of the dolomite furnace is calculated for the variant: gases from the Portland cement kiln are sent to the furnace and there are wastes containing organic matter in the feed mix. Calculation results are given in the table 1.

**Table 1.** Energy indicators of cement rotating kiln at combined procedure for making two kinds of binders

| The thermal energy that a dolomite furnace, kW | Thermal energy of gases Portland cement kiln, kW | Thermal energy released by waste combustion, kW | Thermal energy of combustion of high-quality fuel, kW |
|---|---|---|---|
| 8932 | 4652 | 4280 | - |

More efficiency in the combined production of binders by extra reducing heat loss may be achieved at simultaneous calcination of two raw mixes in two-section rotary kiln.

## 3. Conclusions

Combined method for making two kinds of binders is shown to reduce carbon dioxide by 22-60 % calculated as the final modified product as compared with traditional making of Portland cement clinker.

The proposed combined procedure for making Portland cement clinker and cements with low-temperature calcination will enable to greatly expand possibilities to utilize large volumes of industrial wastes, like ashes from power plants and coal waste.

## References

[1] Vinnichenko V and Ryazanov A 2018 Energy Efficiency of Binder Application in Concrete *International Journal of Engineering & Technology* **7** No 4.3 335-338

[2] Rakhimov R Z, Magdeev U Kh and Yarmakovsky V N 2009 Ecology, scientific achievements and innovations in production of construction materials under and with industrial raw materials *Construction materials* No **12** 2-5

[3] Vinnichenko V I and Ryazanov A N 2017 Ecological indices of manufacture of Portland cement clinker and production of the dolomite clinker *6th International Scientific Conference "Reliabiality and Durability of Railway Transport Engineering Structures and Buildings"*(MATEC Web of Conferences 116, Kharkiv, Ukraine) pp 75-79

[4] Shelikhov N S, Rakhimov R Z and Senyushkina O L 2004 Technological aspects for producing low-grade hydraulic binders based on resource and energy saving *The Eights Academic Reading of Russian Academy of Architecture and Construction Sciences. Current situation and prospects for the development of construction engineering* (Samara) pp 592-595

[5] Luginina I G, Ibatullina L Kh et al 1983 Application of coal waste in cement production *Cement* No 11 **6**

[6] Shelikhov N S and Rahimov R Z 2008 Hydraulic lime and roman cement from mineral raw material of Tatarstan *Non-Traditional Cement and Concrete. III International Symposium* (Brno) pp 712-718

[7] Hughes D C, Jaglin D, Kozlowski R and Mucha D 2009 Roman cements - Belite cements calcined at low temperature *Cement. Concrete* No **39** 77-89

[8] Tislova R, Kozlowska A, Kozlowski R and Hughes D 2009 Porosity and specific surface area of Roman cement pastes *Cement and Concrete Research* **39** Iss 10 950-6
[9] Hughesa D C, Sugdena D B, Jaglina D and Mucha D 2008 Calcination of Roman cement: A pilot study using cement- stones from Whitby *Construction and Building Materials* No 22 1446-1455
[10] Shirin-Zade I N 2010 Structure of argillous-composite materials *Construction materials* No **3** 33- 34
[11] Barbane I, Vitynya I and Lindynya L 2013 Research of chemical and compound composition for roman cement synthesized from Latvian clay and dolomite *Construction materials* No **1** 40-43
[12] Volzhensky A V 1986 *Mineral binding materials* (Moscow: Stroyizdat Publ) p 130
[13] Shelikhov N S, Sagdiev R R, Rakhimov R Z and Stoyanov O V 2013 Roman cement of low-temperature calcination *Bulletin of Kazan Technological University* **16** No 19 pp 62-66
[14] Sagdiev R R, Shelikhov N S, Rakhimov R Z and Stoyanov O V 2013 Effect of technological production conditions and admixtures on properties of composite carbonate-argillous binder *Bulletin of Kazan Technological University* **16** No 5 pp 110-113
[15] Volzhensky A V, Ryazanov A N, Chistov Yu D and Karpova T A 1989 Fuel-saving technology for making lime-ash cement *Construction materials* No **5** 9-10.
[16] Volzhensky A V, Chistov Yu D, Ryazanov A N, Karpova T A, Pogostnov A P, Zubov Yu V and Maletin V V 1987 The way for making lime cement *Patent SU A1 No.1700909, MPK C04B 2/10*
[17] Riazanov A N, Vinnichenko V I, Riazanova V A, Riazanov A A and Nedoseko I V 2018 Structure and Properties of Lime-Ash Cement and its Modification *Construction materials* No. **1-2**. 18-23

IOP Conf. Series: Materials Science and Engineering **907** (2020) 012035 doi:10.1088/1757-899X/907/1/012035

# Innovative building materials in creation an architectural environment

**V A Abyzov[1], K K Pushkarova[2], M O Kochevykh[2], O A Honchar[2] and N L Bazeliuk[1]**

[1]Department of Interior Design and Furniture, Kyiv National University of Technology and Design, Nemyrovycha-Danchenka Street, 2, 01011, Kyiv, Ukraine
[2]Department of Building Materials, Kyiv National University of Construction and Architecture, Povitroflotsky Avenue, 31, 03680, Kyiv, Ukraine

pushkarova56@gmail.com

**Abstract.** Progressive building materials and technologies can ensure not only the durability of buildings and structures that are operated in difficult conditions and the consumption of a minimum amount of energy with little impact on the environment, but also will contribute to the creation of an effective and harmonious architectural environment in accordance with the requirements of sustainable development. An important role in solving these problems is assigned to recycled (reusable materials); traditional natural and local building materials that due to technical progress have gained new prospects for use in modern environmental and energy-efficient structures and buildings ("old-new" natural materials); nanomaterials and nanotechnologies, the use of which will not only improve the quality and properties of materials, and, accordingly, the environment, but also create completely new materials and architectural solutions with a set of properties. The purpose of this study is generalization, systematization and examination of these innovative materials and products.

## Introduction

At the present stage of society's development, the goal of international humanitarian action and one of its socio-economic objectives is to ensure a "Sustainable Development of the Architectural Environment". It is currently one of the European Union's regulatory requirements. The need to ensure an ecological balance between meeting the modern needs of mankind and protecting the interests of future generations, including their need for a safe and healthy environment, defines current global construction trends, which indicate that the most promising direction is the development of urban environment based on energy-saving technologies, as well as environmentally friendly building materials.

The construction industry is the biggest sector of the economy as regards management and in terms of raw materials flow. Most capital, both the financial and the natural, is invested in building structures.

Leadership in Energy and Environmental Design (LEED) is a set of rating systems for the design, construction, operation, and maintenance of green buildings, which was developed by the U.S. Green Building Council. Another certificate system that confirms the sustainability of buildings is the British BREEAM (Building Research Establishment Environmental Assessment Method) for buildings and large-scale developments. Currently, World Green Building Council is conducting research on the effects of green buildings on the health and productivity of their users and is working with World Bank to promote Green Buildings in Emerging Markets through EDGE (Excellence in Design for Greater Efficiencies) Market Transformation Program and certification [1].

It is well known that green building is based on two basic principles. First, at all stages of project implementation the negative environmental impact should be minimal. Second, maximum comfort and safety should be created for those who work or reside in the building. These aspects are the focus of the certification systems mentioned above. "Green" building as a trend is becoming increasingly popular. More and more green projects of various functional buildings are being implemented every day.

In addition to environmental problems it is also important to solve the problem of energy conservation. To solve these issues, it is necessary to analyze carefully building materials and choose those that will contribute to the efficient use of energy, greater comfort and lower costs taking into account the life cycle of the materials, which includes its production, use and disposal (recycling).

Building materials 'creation (taking into account all the technical advantages that the scientific base of modern materials science gives) must consider the economic, environmental, social and ergonomic aspects in the life activity of mankind and production, including: ensuring the minimum energy costs at all stages of production and use; possibility of material renovation, maximum re-use of material (recycling), preservation of environment. In this regard, a new term has emerged - "Eco-friendly materials". It is applied to materials and structures that promote a healthy lifestyle and the conservation of natural resources.

In building, environmentally-friendly materials (also known as green building materials) are those in which, for their production, placing and maintenance, actions of low environmental impact have been performed. They have to be durable, reusable, include recyclable materials in their composition and have to be from resources of the area where the building activity will take place –they have to be local materials. These materials also have to be natural (soil, adobe, wood, cork, bamboo, straw, sawdust, etc.) and must not be spoilt by cold, heat or humidity.

The idea behind sustainable architecture is to build in a way that reduces harmful impact on the environment. The leading components of a sustainable building architecture are the application of sustainable building materials such as organic compounds or recycled materials and the use of environmentally friendly methods of waste management. The key role is played here *innovative materials*, including traditional ("old - new") building materials made on the basis of modern technologies.

There are different views and opinions in the modern scientific literature regarding the typology of modern innovative building materials. The most complete classification of innovative materials is presented in the book "Smart Materials in Architecture, Interior Architecture and Design" [2]. *Along with this, the purpose of this study* is to get acquainted with such innovative materials and products as: recycled (reusable materials); traditional natural and local building materials that due to technical progress have gained new prospects for use in modern environmental and energy-efficient structures and buildings ("old-new" natural materials); nanomaterials.

## Recycled materials

The reuse of building materials and industrial waste is currently under development. With the improvement of processing technologies for various types of materials suitable for re-production and waste there is an intensive expansion and use of them in world practice. The use of industrial waste is particular relevance in conditions with limited sources of supply of material and financial resources.

In the recent past, construction sites that had to be demolished were destroyed in such a way - they were blown up, and then the landed mass was taken out. As a result, there were huge blockages of concrete, metal, glass, which was not easy to disassemble. Recycling allows to reuse construction waste without harming the environment. Throughout the world, recycling of construction waste is a very profitable industry. The volume of construction waste is increasing every year and, according to the participants of this promising market, the main problem is not transportation, but recycling and, importantly, ecological disposal of construction waste. Due to improvements in technology and legislation, countries such as Denmark, Netherlands, Sweden, where more than 90% of waste is currently being recycled, have been able to reach a very high level of construction waste recycling.Thus, construction based on recycled materials can be an effective means of saving money and protecting the environment. Against this background, let's consider a good examples of recycling various building materials.

***Recycling of rocks.*** The mining, quarrying and processing of rocks in quarries and stone works generates a lot of waste, the amount of which can reach up to 80% of the volume of rocks under development. From an economic and environmental point of view, it is advisable to use this waste for the production of rubble, sand, artificial blocks and decorative plates [3, 4]. In the manufacture of granite-crushed stone, a large amount of siliceous waste in the form of pulp (when washing crushed stone) or screenings with grain sizes up to 5 mm is formed. This waste can be used in the production of silicate bricks, cellular concrete, as well as decorative extruded products. Crushed stone and sand with high decorative properties are used for the manufacture of artificial blocks and slabs.

***Recycling of ceramic materials.*** The problem of ceramic waste recovery is very acute for manufacturers of ceramic products. The most advantageous solution, given the high cost of disposing of such waste and, consequently, causing damage to the environment, is the possibility of returning it to the production process using special technology that should take into account economic (cost of raw materials), technological (adding chamotte to ceramic masses without extension grinding cycle) and environmental aspects. Disposal of obsolete ceramic bricks is possible as a thinner additive in the composition of the ceramic mass to obtain a ceramic shard for construction purposes and for the partial replacement of natural sand in the production of concrete for small-sized road paving elements. The fired waste of both glazed and non-glazed products can be returned to the production cycle using a line whose main equipment is grinding equipment for preliminary crushing of ceramic waste whose dimensions can reach 300 mm and a hammer mill, which ensures crushing to a powder state with the formation high content of fine fraction.

***Recycling of glass and mineral melts products.*** The problem of processing for broken glass waste has now become one of the most pressing, because the production of glass products requires huge energy costs and glass is one of the most difficult to recycle waste. Glass is not subject to destruction under the influence of water, atmosphere, solar radiation, frost. In addition, glass is a corrosion-resistant material that does not break down under the influence of an overwhelming amount of strong and weak organic, mineral and bio acids, salts, as well as fungi and bacteria. Therefore, if organic waste (paper, food waste, etc.) is completely decomposed after 1...3 years, polymeric materials - after 5...20 years, then glass, like steel, can survive for tens or even hundreds of years without much damage [4].Glass waste is divided into returnable (own) and secondary (purchased), which is due to two sources of their formation (production and consumption). The returnable glass waste is fully consistent in chemical composition with the glass produced on this furnace, in connection with this glass factories use almost all of their own glass waste, with the exception of the broken reinforced glass, triplex, mirrors and high-quality glassware made of colourless glass. Secondary glass waste is formed in the sphere of consumption (food industry, trade, house-building plants, construction) and, as a rule, in chemical composition does not correspond to the glass of the furnace into which it is loaded. Currently, secondary glass wastes are used in the manufacture of glass containers, glass mosaic and facing tiles, roofing materials, floor tiles, artificial slate, marble, staple fiberglass, etc. The processing of glass bottles (specifically, defective and broken bottle) directly at the glass factory itself is an example of rational waste management.

***Secondary processing of metals.*** Ferrous metals include iron and its alloys, which make up more than 90% of the total global metal production. A large share in the total volume of solid waste belongs to metal waste. Secondary metal resources are made up of scrap (43%) and waste (57%) [3]. Utilization of ferrous metal is an important aspect for the preservation of the environment and the conservation of natural resources. For example, the steel cladding and wood used in the shed project are completely recycled and reused (Fig. 1, a). Scrap is sorted both by chemical composition and by the overall dimensions of products or structures. If the work is carried out with large metal structures, then loading equipment is used for sorting. Sorting by chemical composition is carried out taking into account the metal quality indicator, as well as by its type. In addition, when sorting, separation can take place upon

IOP Conf. Series: Materials Science and Engineering **907** (2020) 012035 doi:10.1088/1757-899X/907/1/012035

the fact that the alloying and carbon components are contained in the scrap metal. In small enterprises, sorting is often done manually, right at the acceptance stage. If the pieces of metal are small in size and weight, then they are pressed into briquettes. Also sheets of metal are subjected to grinding in shredders (cutting sheets into fragments).

***Recycling of concrete.*** Concrete scrap is formed during the dismantling of structures, slabs of temporary roads, piles, during the testing of structures and the accumulation of substandard reinforced concrete products, which occur during the replacement of housing stock and in cases of natural disasters. Crushed concrete stone is used as a large aggregate in various branches of construction industry. The embodiment of concrete processing products as one of the most common building materials in the world is the Hanil Visitors Centre & Guest House project (Fig. 1, b).The facade is made with both gabion walls, as well as concrete cast with fabric.This saves raw materials and improves the environment.

a) Shed No. 8841North Cll Loma Linda, Tucson, AZ, United States US, Tucson
https://architizer.com/blog/inspiration/stories/recycled/

b) Hanil Visitors Center & Guest House, BCHO Architects. Maepo, South Korea
https://architizer.com/blog/ inspiration/stories/ recycled

c) "Circular Pavilion", Encore Heureux architect cooperative, Paris, France
https://www.bmiaa.com/circular-pavilion-building-with-recycled-materials-in-paris/

d) Pavilion EcoARK, Taipei, Taiwan
https://inhabitat.com/amazing-plastic-bottle-architecture-withstands-earthquakes-in-taipei/ecoark-recycled-plastic-bottle-building-in-taipei-7/

**Figure 1.** Examples of using recycled materials in architectural environment

Upon receipt of crushed stone from concrete, fuel consumption is 8 times less than during its extraction in natural conditions, and the cost price of concrete on secondary crushed stone is reduced by 25% [5]. The most promising is the mechanical method of activation of crushed stone from concrete. With mechanical activation methods, self-grinding by mixing crushed stone in mixers or grinding in ball mills with metal balls is provided. The efficiency of using crushed stone from concrete increases if it replaces the aggregate from natural stone material in the manufacture of precast concrete structures directly at the enterprise that processes substandard products. Since the material is the result of waste processing, the initial stage of production is the collection of construction waste. The production process consists of the following steps: selecting the right starting material, evaluating its suitability, collecting garbage and transporting it to a recycling point, crushing concrete (asphalt concrete, brick) using hydraulic shears /hydraulic hammers, crushing pieces in a crusher, sifting crushed stone: identifying glass and metal residues fractionation using screens.

***Recycling of wood.*** Wood is one of the most important materials of human life. The use of wood in the production of building materials, furniture manufacturing and in other sectors of economic activity is often uneconomical and is accompanied by the formation of a significant amount of waste (35 ... 45%), which can be recycled. Therefore, the processing of wood waste is one of the pressing issues of economical production [6]. The resources of the integrated use of wood include all lumpy waste from the woodworking industry, coniferous and deciduous low-grade wood (firewood), small-sized wood from thinning, bulk waste (sawdust) and bark, as well as logging waste (branches, branches, stumps), and even used wood panels (Fig. 1, a). The “Circular Pavilion” building follows the principles of circular economy, where one person’s waste is another person’s resource (Fig. 1, c). Thus, the façade is made of 180 wooden doors recuperated from the rehabilitation of an apartment building; the stone wool comes from the disassembly of a supermarket roof; the floor is composed of used wooden exhibition panels; the furniture is collected from the different junkyards in Paris; the lighting comes from obsolete street lights; the windows and the wooden strips are taken from Parisian building sites surp. Stemwood, which is the main object of forest exploitation, accounts for 70% of the total mass of the tree, the bark is about 9%, twigs 8, stumps and roots 13%. Thus, wood waste is represented by scraps of wood, roots, stumps, bark, wood chips, twigs, shavings and sawdust. Wood debris is produced in large quantities at enterprises engaged in woodworking, sawmilling, plywood, furniture, etc.

***Recycling of polymer materials.*** Plastics are used relatively little as secondary raw materials. This is explained by the variety of types of plastics, which in addition to polymers contain a large number of other components, which complicates the sorting and processing of plastic waste, especially household waste. In all civilized countries society is responsible for the proper disposal of plastic waste. The most environmentally friendly solution for the disposal of polymer waste lies in its recycling. Indeed, it is in this way that natural resources are saved and the environmental impact is minimized. The need for polymer processing is because for the complete decomposition of plastic products it will take about 150 years. Decaying, plastic releases toxic substances that adversely affect human health. Plastic processing is necessary for improving the ecology of our planet and the economical use of its resources; reduction of waste in landfills; creating raw materials for its recycling; creating fuel material [3]. The main directions of plastic waste disposal: landfill; processing using factory technologies; co-incineration of plastic waste with municipal waste; pyrolysis and separate burning in special furnaces; the use of plastic waste as a finished material for other technological processes. The processing of plastic waste by factory technologies can be considered as the optimal method of their use for obtaining building materials, including plastic blocks, bricks, profiles. Plastic bottle architecture is fantastic at turning a problem into an eco-friendly opportunity. The amazing EcoARK in Taipei (Taiwan) is one such example (Fig. 1, d). Built from 1.5 million recycled plastic bottles, this massive pavilion is surprisingly strong enough to withstand the forces of nature - including fires and earthquakes.

**"Old-new" natural materials**

In recent years, an ecological trend is developing in construction, which aims to use natural materials. These materials can be of both plant and animal origin. This eco-building model does not require high-energy expenditure for production and contributes to the development of energy-saving investments that meet current technical requirements. These types of products are perceived as healthy and cheap, in many cases locally available. In addition, the above material solutions can have a significant impact in modern construction due to the increase in prices of traditional construction products and due to energy savings during construction and when using the investment. Building materials of natural origin can be used as materials for thermal insulation, e.g. straw, sheep's wool or cellulose, and can be used as construction or finishing materials, such as plywood, wood fiber materials with external gypsum or wooden board or clay plaster with straw [7]. Examples of materials of natural origin are: dried and compacted earth, wood, plywood, wood or hemp fiber insulation boards, straw, sheep's wool, cellulose, chipboard with an external gypsum board or wooden etc.

One of the oldest building materials is ***land.*** In each region and in the history of each country, we can find monumental buildings, as well as rural buildings made from land mixed with other natural materials. Buildings from the ground are seen as healthy, friendly to people and the environment [8]. Common soil products are unfired, dried bricks and blocks or walls made of rammed in formwork. Due to the potentially low production impacts, rammed earth has recently become a highlight amongst eco-friendly and sustainable architect for a vernacular green building material for its "Eco Houses".

The next group of products is building materials made of ***fibers natural, such as flax, hemp or wood.*** These products are characterized by good heat and sound insulation. Their disadvantage is their flammability, which is why they must be protected with fireproofing impregnates. They are also not resistant to rodents and show poor resistance to moisture. They are often used with "housing", i.e. with gypsum boards, plasterboard or also wood-like ones [7].

***Straw*** under the right conditions of collection is a very good heat-insulating material that can be used in construction. High heat engineering properties are possessed by wheat, rye and barley straw. A house that is built from straw or straw blocks is called an ecohouse (ecological house) and building houses from straw is called green building (Fig. 2.a). Every year a huge amount of straw is produced in the world, most of which remains in the fields or burned. This raw material can be used in construction, which would solve many social and economic problems in various countries of the world [9]. Straw is a renewable raw material that is cheap, easy to turn into building material, and also easy to dispose of after many years of use: burn or leave for decay in the open. Two types of straw blocks are used for construction: the first – from pressed dry straw, and the second - pressed straw blocks treated (coated) with clay mortar.

The technology of building houses from **reeds** has been developed. It provides the buildings' construction based on wooden frame panels with reed stems; the reed cavity is covered with clay. Hemp fiber lay between them so that there are no bridges of cold. Then the panels are installed, plastered with clay and lime, and brick partitions are made inside. Reed mats are made on special machines - reeds are evenly placed between the wooden racks of the machine and pressed. The advantages of a house made of reeds are: environmental friendliness, resistance to decay, seismic resistance and ability to withstand strong winds, water resistance, low thermal conductivity (thermal conductivity of reeds is 4 times lower than that of wood; 7 times lower than that of brick), high noise insulation, does not require additional ventilation. Yet, with all the obvious advantages of a reed house, there are some points that need to be considered during construction. Bulrush is a natural material. Reed slabs and reed roofs are cellulose-containing building materials; therefore, they are fire hazardous and need to be treated with protective agents - flame retardant. This will not only create refractoriness, but also give water-repellent properties and provide bio-protection.

Sun-dried ***clay*** is one of the oldest building materials that has been used for more than 10.000 years to build houses, mosques, churches, palaces and cities. Currently one-third of the global population lives in this vernacular architecture made with the man-made stone-like construction material. This building material, commonly made from a mix of clay and organic materials, is resisting the long span of time and weathering (Fig 2, a). Used correctly, the clay making technology has many benefits: the technique used is very simple with zero maintenance cost, local availability and affordability, and has a good thermal and acoustic insulation. Because of this, earthen architecture has many advantages from a sustain ability perspective. The use of this eco-friendly construction material nowadays could be a viable alternative to solve our 21st Century big concerns of energy and climate change. For this reason, it is important to understand this material characterization, diagnosis and suitability in construction, and to understand how can it be a good construction material for a particular site [10]. Modern clay architecture uses clay tiles, bricks, plasters, beaten clay, ready to use mortar and other clay brick products made for different uses. The creation of expressive and interesting architectural forms using clay bricks is also possible by combining it with modern materials like glass or clean plastered surfaces that create contrasting solutions One of the good examples of contemporary clay brick architecture is the famous Indian Institute of Management – Ahmedabad building.

a) A cottage built using straw blocks
https://www.google.com/url?sa=i&source=images&cd=&cad=rja&uact=8&ved=2ahUKEwiE5JDqgsrmAhVtposKHaebDfoQjRx6BAgBEAQ&url=http%3A%2F%2Fposelok.com.ru%2Ffoto-doma-iz-gliny-i-solomy.html&psig=AOvVaw25R5qZfZHOydFr_lOvO5AY&ust=1577130324176303

b) Load-bearing structures made of bamboo
https://www.lifegate.com/people/lifestyle/bamboo-construction

**Figure 2.** Examples of using "Old-new" natural materials in architectural environment

***Bamboo*** as a construction material is traditionally associated with the cultures of Central and South America, Asia and the South Pacific. It can be used to build entire homes and other structures that are earthquake-resistant and can survive strong winds and storms. Only a few of thousands of known bamboo species can be used for construction purposes. Bamboo, which may be cut and laminated into sheets and planks, is an ideal construction material as it is durable, sustainable, and environmentally friendly. It is as strong as steel in terms of tension, and because it is cylindrical and hollow, it is stronger than concrete in terms of compression, thereby giving it a high strength-to-weight ratio. It also weighs much less than traditional building materials such as wood, concrete, and steel. As a flexible material, bamboo may be curved or flattened by the application of heat and pressure (Fig. 2, b). It is also naturally water-resistant, so unlike wood, it does not have the tendency to warp due to heat or dampness. Bamboo has poor fire-resistance and limited durability when exposed to UV rays and humidity. This can cause excess swelling and the lack of humidity can cause shrinking and/or cracking [11]. At the same time, the use of this original material in Europe has been gaining ground recently. Bamboo is used as a facing and finishing material in the exterior decoration of various buildings, as well as decorative material in interiors.

## Nanomaterials

An analysis of current trends in the introduction of new building technologies and materials in economically developed countries of the world suggests that the basis for the dynamic implementation in practice for the next 10...20 years will be materials and technologies obtained on the basis of achievements and developments in the field of nanotechnology [12]. The decisive factor is that the very nanoscale of the system components results in new functionalities and properties for improving products or developing new products and applications."

Today in the world with the help of nanotechnology on an industrial scale, cement, ceramics, metal alloys, plastics, paints and other materials with unique properties are produced. Nanomaterials are used to improve thermal properties, increase the efficiency of energy transfer, lighting, heating [12].The use of nanomaterials in construction is important not only to improve the properties of materials, but also from the point of view of solving the problems of energy conservation and ecology.

The work of scientists in the field of nanotechnology in construction was reflected in the production of materials such as high-strength concrete, high-strength steel, structural composites, nanocoatings, innovative films, nanocomposite pipes, fiberglass composite reinforcement. Energy-efficient mineral-based heat-insulating materials with low thermal conductivity, sorption humidity and increased noise absorption, nanoscale organic-mineral modifiers for road concrete are being developed; nanometric metal-mineral biocidal additives for paints and varnishes, mortars and concrete, working in conditions of biological aggression; high-strength concrete with a low average density; nanometric compensators of internal stresses, etc.

Nanomodified concrete (including lightweight), taking into account its cost, is effectively used in the construction of high-rise structures; in bridge construction, road works and in the construction of hydraulic structures; in seismically active territories, regardless of climatic zoning. Structures do not require waterproofing, are characterized by high crack resistance. Also, such concrete can be used in interior design. For example, a high workability index allows you to create products by casting or other methods. Such concrete is used for the manufacture of original interior items, as well as countertops, sinks and other products.

Self-sealing multilayer composite nanomodified coatings with improved waterproofing and anticorrosion properties can be used for surfaces that work in an aggressive environment, as well as for decorative finishes, for example, for floor coverings. The use of composite reinforcement in concrete allows to increase such properties as corrosion resistance, durability, and also helps to reduce the weight of concrete structures. In addition, surface modification of such reinforcement using carbon nanocomponents increases its adhesion to concrete, increasing the strength and crack resistance of the structure at the macro level.

Thus, through the use of nanomaterials, one can easily control the structure, density and other characteristics of concrete. With the advent of nano-concrete, architects and designers have new opportunities for creating highly complex structures that allow the realization of the most daring architectural solutions [12]. In addition to the use of nanotechnology, all buildings of unusual shapes are modeled using programs in 3-d graphics. An example of such a building constructed using IT technology is the Guangzhou Opera House.

A great achievement in the field of nanocoatings was the imitation of the effect of lotus petals, which are completely invulnerable to water. "Lotus-Effect" is one of the best-known means of designing surfaces with nanomaterials. As a result, the building of the Bolshoi National Theater (architect Paul André) appeared in Beijing, its huge egg-shaped dome made of glass with nanocoating based on the $TiO_2$ catalyst, due to which the dome is not subject to contamination and wetting by precipitation and is capable of self-cleaning (Fig. 3).

Translucent nanocoatings have the ability to accumulate solar energy, so they can be used on windows and facades of buildings not only to create the effect of solar panels, which reduces energy costs, but also gives the facade a stylish look.

Nanoglass is obtained by applying a thin layer of In-$SnO_2$ metal oxides by pyrolysis in the process of producing float glass. The heat transfer coefficient of such glass is reduced by 70...80%, and the thermal conductivity of the glass with its use is 2...2.5 times. If a special composition with titanium oxide nanoparticles is sprayed onto the surface of the still-float glass, this coating ensures the neutralization of organic compounds on the glass surface and complete hydrophilization, which contributes to self-cleaning of the glass.

A product known as Nanogel, a form of aerogel, not only provides high performance thermal insulation but also effective sound insulation. It is almost as if it derives from another star, or at least fresh from the NASA laboratories. The latter is not so far from the truth, as this is where aerogels were first developed for various applications designed for outer space. In reality aerogel is relatively banal: it is simply an ultralight aerated foam that consists almost 100% of nothing other than air (the exact figure varies between 95% and 99.9%).

**Figure 3.** $TiO_2$-based nanocoating glass on the facade, National Theater, Beijing, PRC.
https://www.google.com/url?sa=i&source=images&cd=&ved=2ahUKEwiAurnYisrmAhXwmIsKHa5GBzsQjRx6BAgBEAQ&url=http%3A%2F%2Fdoramakun.ru%2Finteresting%2Fasian-facts%2FChina-Bolshoj-Nacionalnyj-teatr-v-Pekine.html&psig=AOvVaw2nJAuvQM0LHx5sJyyp92Ro&ust=1577132479317254

The remaining foam material is a glasslike material, silicon dioxide, also known as silica. In addition to its thermal insulating properties, aerogel also acts as a sound insulator according to the same basic principle. The air molecules immovably trapped in the nanopores of the aerogel stop sound waves from passing through the material.

The prospects for the further development of building nanomaterials are as follows: the foundations of buildings with a self-regulating system for compensating soil shrinkage; load-bearing structures of buildings, monitoring their own stress-strain state; building envelopes and roofs that accumulate the energy of the sun; coatings that respond to the psychophysical state of people, photocatalytic coatings. All these characteristics should become the basis of the modern "smart home" of the new generation.

The use of nanomaterials in construction: allows the use of new architectural solutions; reduces construction costs and the pace of construction of buildings; improves the quality of structures and its operational characteristics; contributes to the preservation of the environment; ensures compliance with safety standards and requirements; allows designers to adapt buildings to biologically similar forms, creating a model of architecture that fully interacts with the climatic, chemical, kinetic and social aspects of life, reducing the ecological footprint of modern society in an urban environment.

## Conclusion

The need to ensure an ecological balance between meeting the modern needs of mankind and protecting the interests of future generations determines the current world construction trends, which indicate that the most promising direction is the development of architectural environment based on energy-saving technologies also environmentally friendly building materials. It is shown that all modern building materials should meet the requirements of the concept of sustainable development and green building. The modern problems of environmental protection are connected with the problem of energy saving. To solve these issues, it is necessary to analyze carefully building materials and choose those that will contribute to the efficient use of energy, greater comfort and lower costs taking into account the life cycle of the materials, which includes its production, use and disposal (recycling).

The leading components of a sustainable building architecture are the application of sustainable building materials such as organic compounds or recycled materials and the use of environmentally friendly methods of waste management. The key role is played here innovative materials, including traditional ("old - new") building materials made on the basis of modern technologies. These types of products are perceived as healthy and cheap. In addition, the above material solutions can have a significant impact in modern construction due to energy savings during construction and when using the investment.

The use of nanomaterials in construction is important not only to improve the properties of materials, but also from the point of view of solving the problems of energy conservation and ecology. The use of nanomaterials in construction: allows the use of new architectural solutions; promotes environmental conservation; ensures compliance with safety standards and requirements; allows designers to adapt buildings to biologically similar forms, creating a model of architecture that fully integrates with the climatic, chemical, kinetic and social aspects of life.

**References**

[1] *EDGE Buildings, Build and Brand Green* URL: http://www.edgebuildings.com
[2] Ritter A 2007 *Smart Materials in Architecture, Interior Architecture and Design* (Birkhauser – Publishers for architecture, Basel-Boston-Berlin) p 191
[3] Kryvenko P V, Pushkarova K K, Baranovsky V B, Kochevykh M O 2012 *Building materials science: Textbook* (Kyiv: Lira-K) p 624
[4] Janiak J 2017*Recykling i upcylking w służbie architektury. Estetykaobiektów z materiałów wtórnych (*Przegląd budowlany vol 1) pp 27-30
[5] Krivenko P V, Pushkarova K K and Kochevych M O 2001 *Aggregates for concrete: Textbook* (Kyiv: FADA Ltd) p 399
[6] Dvorkin L I and Pashkov I A 1989 *Building Materials from Industrial Waste* (Kyiv: Higher School) p 208
[7] Dachowski R, Kamionka, L W & Gałek K 2019 Unconventional buildings materials in biplot context, *Scientific Review Engineering and Environmental Sciences* vol **28** (2) pp 268-277
[8] Wasilewska A and Pietruszka B 2017 *Materiały naturalne w ekobudownictwie* (Przegląd budowlany) vol **10** pp 50-53
[9] Wasilewska A and Pietruszka B 2016 *Słoma w ekobudownictwie* (Ekologia a Budownictwo) p 353
[10] Ben Guida D 2016 Sun-dried Clay for Sustainable Constructions. *International Journal of Applied Engineering Research* **11** No 6 pp 4628-4633
[11] Nurdiah Esti 2016 The Potential of Bamboo as Building Material in Organic Shaped Buildings *Procedia - Social and Behavioral Sciences* **216** pp 30-38 doi: 10.1016/j.sbspro.2015.12.004.
[12] Leydecker S 2008 *Nano Materials in Architecture, Interior Architecture and Design* (Birkhauser, Basel-Boston-Berlin) p 192

IOP Conf. Series: Materials Science and Engineering **907** (2020) 012036 doi:10.1088/1757-899X/907/1/012036

# Processes of structure formation and neoformation of basalt fiber in an alkaline environment

**O Yu Berdnyk, O V Lastivka, A A Maystrenko and N O Amelina**

Department Technology of building constructions and wares, Kyiv National University of Construction and Architecture, Povitroflotskyj avenue, 31, 03037, Kyiv, Ukraine

kseniareznik87@gmail.com

**Abstract.** Several approaches can be used to determine the effect of an alkaline environment on glass materials (basalt fibers). One of the most common approaches for the determination of the alkali resistance of a substance is to measure the relative mass loss after the interaction of the object of research with the alkaline environment. According to the practice guidelines, the approach involves determining the alkali resistance of a glass by measuring the mass loss per unit area of the glass under test after boiling it in an alkaline solution. But the disadvantage of this approach is it can't be applied to basalt or another multicomponent fiberglass as it doesn't introduce different speeds of fiber dissolution.

## 1. Introduction
There are many ways to determine the effect of an alkaline environment on glass materials (e.g. basalt fibers). One of the most common ways to determine the alkali resistance of the substance is to measure the relative mass loss after the interaction of the study object with the alkaline environment. This approach has a significant advantage over the measurement of the relative mass loss due to the increase of the sample surface; therefore, the shape of the sample has no particular importance.

## 2. Analysis of recent research and publications
In recent years, research on the effect of an alkaline environment on the structure of the basalt fiber, as well as problems of obtaining and use of silicate melts of various functional purposes became the subject of the works by Jigiris, Makhova, and Aslanova.

## 3. The purpose of this work
The purpose and objective of the present research are to study the morphology of the basalt fiber surface, its elemental and phase composition before and after its interaction with an alkaline environment. These objectives allow identifying the details of the fiber etching process in an aggressive environment and the impact of this process on the coating process.

## 4. Materials and methods of research
Basalt fibers and alkaline solutions were used as raw materials. The first group includes a basalt fiber that completely dissolves in etchant solution; second group covers the basalt fiber that partly dissolves with formation of basalt fiber fragments without the change of its structural properties; third group includes the basalt fiber that keeps the initial form with formation of surface micropore structure [1].

Thus, divergence in interaction of basalt fibers with etchant solution became a pre-condition for the necessity of research (table 1).

**Table 1.** Chemical composition of basalts of deposits

| Name of deposit | $SiO_2$ | $Al_2O_3$ | $Fe_2O_3$ | FeO | MgO | CaO | $TiO_2$ | $R_2O$ |
|---|---|---|---|---|---|---|---|---|
| basalt | 50.6 | 16.0 | 14.8 | 8.1 | 5.1 | 9.8 | 0.9 | 3.2 |

## 5. Research results

The morphology of the basalt fiber surface involves significant changes after 8 days of holding the fiber in an alkaline solution. Shows basalt fiber data of the scanning electron microscope (SEM), made at different intervals of time in Figure 1.

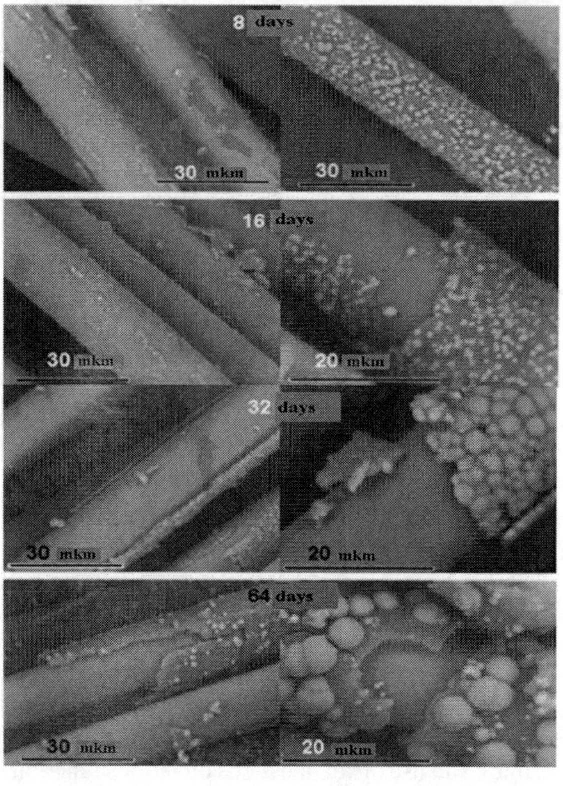


**Figure 1.** Uncoated basalt fiber after etching in the sodium hydroxide solution.

With an increasing holding period (16 days or more), it can be observed that the corrosion layer consists of two layers [2, 3]. The first one is on the surface and is tightly adjacent to the fiber and consists of particles of hexagonal lamellar shape directed perpendicular to the surface of the basalt fiber Figure 2.

The second – an external layer – consists of spherical particles. With the increasing etching period, there is an increase in the size and number of spherical particles, as well as an increase in the thickness of the corrosion shell. After 64 days of holding the basalt fiber in an alkaline solution, the number of fibers with peeling corrosion significantly increases. The shell thickness makes 650-750 nm after 16 days of etching and increases up to 1.5-2 μm after 64 days of etching. Changes in the elemental composition of the fiber surface, depending on the holding time in alkaline solution, were studied by elemental microanalysis [4]. The selection criteria were the integrity of the "corrosion shell – fiber" pair and the morphological features of the corrosion shell for each sample. The results of the EDS analysis of fiber surface studies after 8 days of etching are shown in Figure 3.

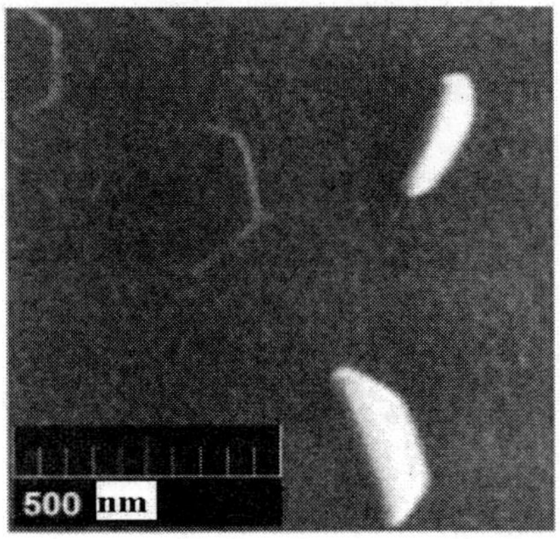


**Figure 2.** Hexagonal lamellar particles on the basalt fiber after etching in a sodium hydroxide solution

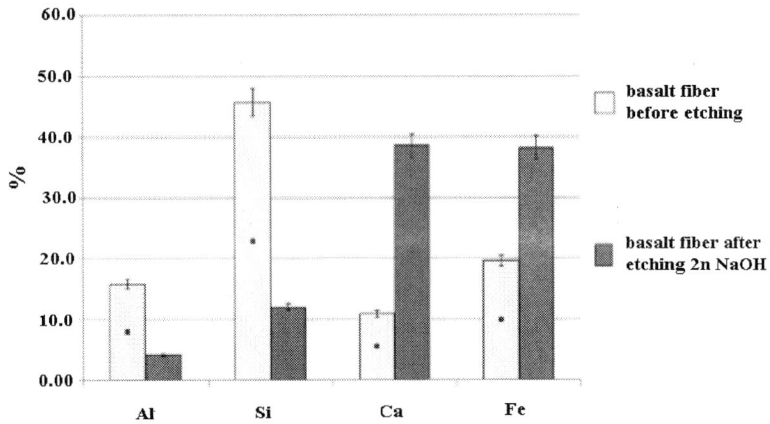


**Figure 3.** Results of the EDS analysis of the uncoated basalt fiber before and after etching (8 days)

According to the results of studies, it can be concluded that the concentration, the main components of basalt fibers, silicon, and aluminum - are decreasing, and the concentration of iron and calcium - are increasing. The concentrations of MgO and $TiO_2$ are too low to be detected by the EDS analysis [5].

To determine the phase composition of basalt fiber before and after etching, the methods of IR spectroscopy and X-ray phase analysis were used.

The IR spectra of the initial fibers and fibers after etching in an alkaline medium are shown in the Figure 4.

The bands attribution is shown in Table 2. There is a group of very intense bands in the range 1000-1200 $cm^{-1}$.

According to the data [6,7] the bands can be assigned to the valence fluctuations of the Al-O-Si composition. In the region of 1400-1700 cm-1, there is a group of bands of medium intensity. These bands can be attributed to the valence fluctuations of the MeOH bonds (under Me are meant the metals that make up the fiber). In the infrared spectrum of the etched basalt fiber, there is a change in the intensity of the bands, namely, the intensity in the region of 1000-1200 cm-1 decreases. It can be assumed that the decrease in the intensity of the bands in the region of 1000-1200 cm-1 is due to the dissolution of the aluminosilicate composition in the surface area of the fiber, whereas the increase of the intensity of the bands in the region of 1400-1700 cm-1 is associated with the formation of metal

IOP Conf. Series: Materials Science and Engineering **907** (2020) 012036 doi:10.1088/1757-899X/907/1/012036

hydroxides on the fiber surface. A very intense band of ~ 1450 cm-1 is present in the IR spectrum of the corrosion shell, which is exfoliated. It can be attributed to the valence vibrations of the carbonate group. The X-ray phase analysis confirms the presence of sodium carbonates and iron bicarbonates in the corrosion layer Figure 5.

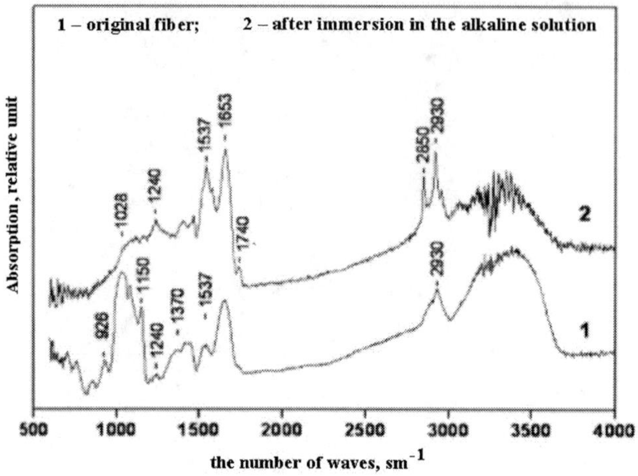


**Figure 4.** IR spectra of the basalt fiber: 1 – original fiber; 2 – after immersion in the alkaline solution

**Table 2.** Results of the basalt fiber IR spectra studies

| Name of deposit | $SiO_2$ | $Al_2O_3$ | $Fe_2O_3$ | FeO | MgO | CaO | $TiO_2$ | $R_2O$ |
|---|---|---|---|---|---|---|---|---|
| basalt | 50.6 | 16.0 | 14.8 | 8.1 | 5.1 | 9.8 | 0.9 | 3.2 |

*Corrosion in a saturated $Ca(OH)_2$ solution*

After etching of the original basalt fiber in a saturated $Ca(OH)_2$ solution during 16 days, two types of morphological features can be observed on the fiber surface, namely, not tightly placed crystals with the length of 2-3 μm Figure 6. According to the SEM analysis, both neoformations and their clusters consist of tightly placed elongated plate crystals [8].

The EDS analysis of different basalt fiber surface areas with no clusters shows no significant changes in the elemental composition of the surface compared to the one of the not etched fiber Figure 7.

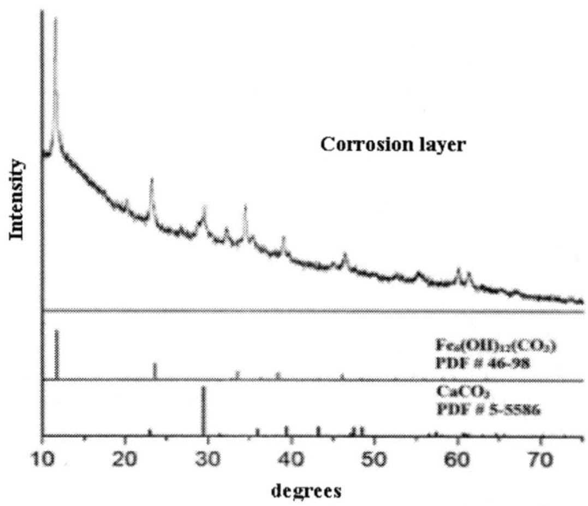


**Figure 5.** X-ray pattern of the corrosion layer

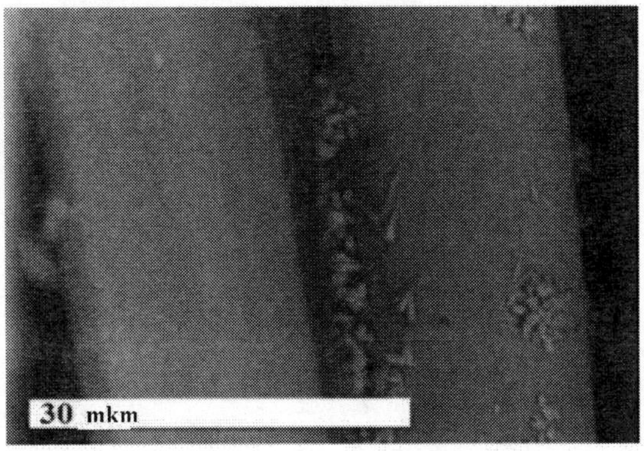


**Figure 6.** An SEM image of the basal fiber after 16-days etching in a saturated $Ca(OH)_2$ solution

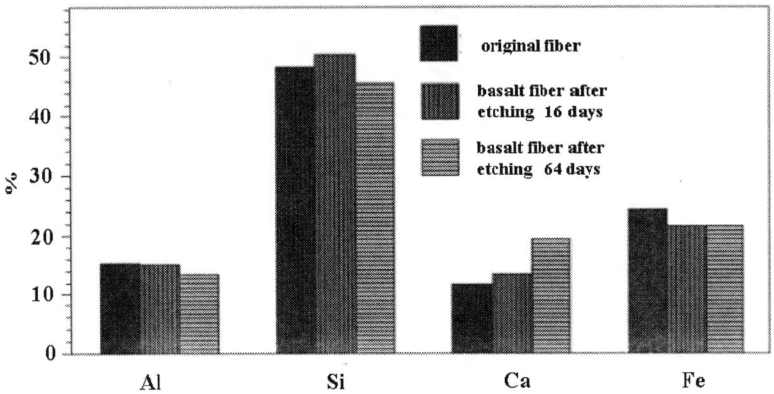


**Figure 7.** Normalized EDS analysis data obtained from those areas of the basalt fiber surface on which no aggregates of needle-like crystals were observed after immersion of the fiber into the saturated $Ca(OH)_2$ solution for 16 and 64 days

It should be noted, sufficient changes in the elemental composition of the original basalt fiber could be observed after its immersion into the NaOH solution for 16 days. Thus, the basalt fiber undergoes more changes after its etching in the NaOH solution, than in the solution of Ca hydroxide due to the lower concentration of the latter [9, 10].

The SEM images of the basalt fiber after 64 days of etching in a saturated $Ca(OH)_2$ solution are shown in Figure 8.

The picture shows the number of neoformations increases with the increasing duration of etching. A larger number of round-shaped particles cover nearly the entire fiber surface (Figure 8 a).

Koga observed a similar morphology for calcite formed during the reaction of $Ca(OH)_2$ with $CO_2$ under conditions of 100% humidity. Along with the round-shaped particles, there are plate crystals of hexagonal shape (Figure 8 b) and flower-like clusters present on the surface. The clusters cover a large part of the fiber surface (Figure 8 c). At higher magnification, it can be seen that these clusters are polycrystalline (Figure 8 d). According to the X-ray phase analysis data, the phase composition of the corrosion shell corresponds to calcite (Figure 9).

The method of X-ray phase analysis discovered no Fe-containing phases. However, some of the morphological figures, namely plate hexagonal crystals which are typical for hydroxycarbonate of iron formed at high pH, may indicate the formation of this phase [10]. Perhaps the quantity of the Fe-containing phase on the etched basalt fiber surface is below the visibility limit of the X-ray phase analysis method.

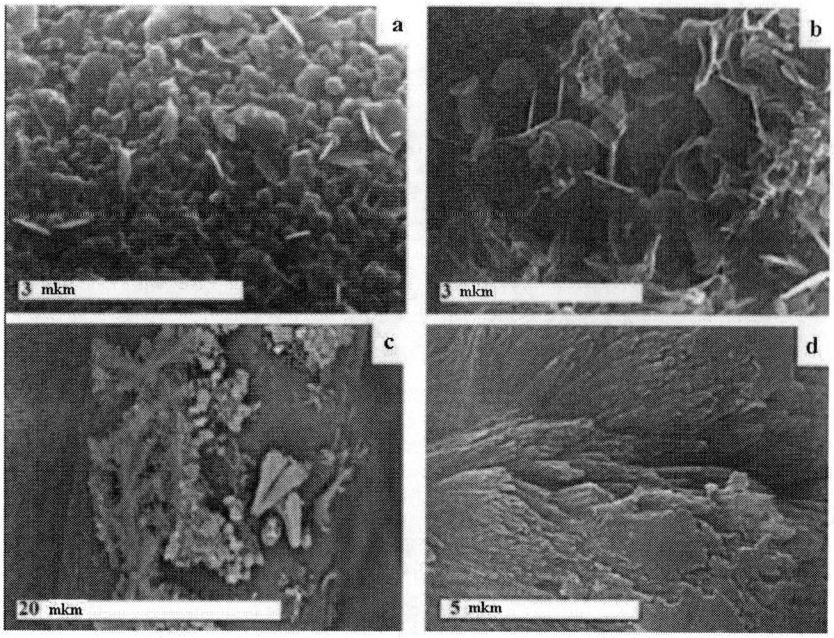


**Figure 8.** SEM images of the basalt fiber after 64 days of etching in a saturated solution of $Ca(OH)_2$

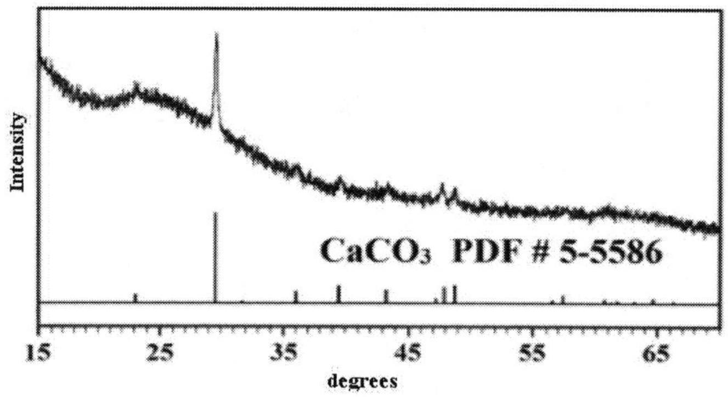


**Figure 9.** X-ray pattern of the basalt fiber after 64 days of etching in a saturated $Ca(OH)_2$ solution

The long lasting 64-day etching of basalt fibers in the $Ca(OH)_2$ solution is followed by the change in elemental composition (Figure 8.), morphology and phase composition (Figure 9). As it can be seen from Figure 8 the content of Al and Si on the basalt fiber surface decreases and on the opposite, the content Ca increases. More than 70% (of the mass) Ca are noticed in the neoformations on the surface after immersion of the fiber in a $Ca(OH)_2$ solution. These data together with the data of the X-ray phase analysis prove that a neoformation is a calcium-containing phase like for example $CaCO_3$ calcite.

The origin of calcite on the surface can be explained by the fact that the gaseous $CO_2$ slowly enters the solution in a saturated $Ca(OH)_2$ solution and reacts with it, forming $CaCO_3$. Then $CaCO_3$ deposits in the form of crystals on the surface of the basalt fiber, possibly in the places with no defects. The recovery of Ca in the solution takes place due to the dissolution of a new portion of basalt fibers in a $Ca(OH)_2$ solution. Further increase in the onset of calcite occurs in the process of dissolution and deposition. The SEM analysis showed that the binding of the corrosion shell to the fiber core decreases with increasing etching time. As a result, the corrosion membrane is torn away from the fiber. The basic patterns are revealed during the long-term etching of the fiber in a solution of $Ca(OH)_2$, similar to those that were established for etching of basalt fiber in a solution of NaOH. The corrosive shell does not prevent the alkali from penetrating the surface of the basalt fiber, first because of poor bonding to the surface of the fiber and deflection, and second, because of the porosity of the corrosion shell.

## 6. Conclusions and prospects for further studies

Based on the data of SEM/EDS, X-ray phase analysis and IR-spectroscopy we can conclude that the corrosion shell, formed on the original basalt fiber after its etching in an alkaline solution of NaOH, consists of plate hexagonal crystals of $Fe_6 (OH)_{12}CO_3$ ferrous bicarbonate directed perpendicular to the surface of the fiber and outer layer, made of globular particles of calcium carbonate.

## References

[1] Bartenev G M 1980 High-temperature relaxation mechanisms in inorganic glasses *Journal of Non-Crystalline Solids* 285-298

[2] Hoto R 2011 Optimization of mechanical properties of basalt woven *Proceeding of the 181 CCM, Korea*

[3] Gots V., Palchik P, Berdnyk O 2018 Investigation of properties of modified basalt fibres *Scientific journal "EUREKA: Physical Sciences and Engineering"* Tallinn 43-48

[4] Mathis D, Blanchet P, Landry V, Lagière Ph 2019 Thermal characterization of bio-based phase changing materials in decorative wood-based panels for thermal energy storage *Green Energy & Environment* vol **4**(1) pp 56-65

[5] Makhova M F 1990 Relationship between the viscosity of melts and the composition of rocks upon receipt of glass fibers *Glass and Ceramics* 19-21.

[6] Novitsky A G 2003 Basalt raw materials. Technology of choice for the production of fibers for various purposes *Ukrainian chemical industry* 47-52

[7] Tsapko Yu, Tsapko A and Bondarenko O 2019 Effect of a flame-retardant coating on the burning parameters of wood samples *Eastern-European Journal of Enterprise Technologies* vol 2/10 (**98**) pp 49-54 doi: 10.15587/1729-4061.2019.16359.

[8] Tsapko Yu, Bondarenko O, Tsapko A 2019 Research of the efficiency of the firefighting roof composition for cane *Materials Science Forum* vol **968** pp 61-67 doi: 10.4028/www.scientific.net/MSF.968.6.

[9] Han B 1981 Crystallization processes in petrurgic melts and glasses (*Czechoslovakia, Marianske Lazne*) 28-41.

[10] Erdoğan Y 2016 Production of an insulation material from carpet and boron wastes *Bulletin of the Mineral Research and Exploration* vol **152** pp 197-202.

[11] Berdnyk O Yu, Gots V I, Rogozina N O, Maystrenko A A 2019 Production of modified basalt fibre for heat-insulating products manufacturing *IOP Conference Series: Materials Science and Engineering* **708** 012082

IOP Conf. Series: Materials Science and Engineering **907** (2020) 012037 doi:10.1088/1757-899X/907/1/012037

# Temperature-time monitoring of concrete hardening of a wind farm foundation

**O V Kabus[1], O OKulaienko[2], A V Ryabushko[2], V P Sopov[1] and O V Usherov-Marshak[1]**

[1] Department of Physical and Chemical Mechanics and Technologies of Building Materials and Products, Kharkiv National University of Civil Engineering and Architecture, Sumska str. 40, 61002 Kharkiv, Ukraine
[2] Department of Automation of Production Processes, Kharkiv National University of Civil Engineering and Architecture, Sumska str. 40, 61002 Kharkiv, Ukraine

[1]E-mail: calorimetry_centr@ukr.net

**Abstract** The foundations of wind farms are massive structures that require temperature control of concrete hardening during their construction. This is due to the need to ensure crack resistance and specific operating conditions of engineering networks located in the body of the structure. The work describes a comprehensive study of the foundations construction on the Zaporizhzhya wind farm. The study included calorimetric analysis using adiabatic calorimetry and on-site temperature monitoring of the object using a continuous temperature measurement system. Laboratory studies allowed verify the composition of concrete proposed by the concrete manufacturer and develop recommendations for concreting. The data obtained showed that the hardening conditions approach adiabatic ones with a foundation slab thickness of more than 2.5 m. This leads to a temperature gradient of more than 25 ° C between the internal and external parts of the structure. The adopted recipe and technological solutions have ensured the achievement of acceptable conditions for the foundation slab construction in the winter.

## 1. Introduction

Wind power as an alternative renewable source of electricity in Ukraine is going through a period of active development. Structurally, a wind power station (wind farm) is a group of wind generators, which consist of an electric generator driven by wind energy, a mast and a foundation. The architectural expressiveness of the landscape with the working wind farms is characteristic.

Despite the apparent structural simplicity, the concrete foundation of the wind generator is a complex structure of massive reinforced concrete, which is designed to guarantee the long-term stability of the wind farm under the influence of the environment, including wind loads, climatic and corrosive effects. The massive foundation in this context requires special measures in its manufacture [1].

The concept of massive concrete is clearly defined by the normative document of the American Institute of Concrete ACI 207.1R-05 [2] as "any volume of concrete with dimensions large enough to make decisions to minimize cracking due to heat generation during cement hydration and changes in concrete volume". The heat-stressed state of the hardening mass is the object of comprehensive research and a number of regulatory documents aimed at obtaining defect-free durable concrete [3-7].

## 2. Purpose of the study

Verification of recipe and technological solutions for the reinforced concrete foundations construction of the first stage of the construction of a 500 MW wind farm in Priazovsky and Melitopol districts of Zaporizhzhya region. Construction is developed by the company EuroCape Ukraine I. The number of foundations is 27. It is planned to install up to 167 wind generators.

## 3. Research methodology

The study was performed by two methods:
- calorimetric analysis of the additives effect on the hardening of cement and concrete by determining the heat release parameters accompanying the hydration of cement and exothermy of concrete [8];
- continuous temperature-time monitoring of concrete hardening at the facility to obtain temperature values at different points and sections of the foundation.

## 4. The study performance

### *4.1. General technological information*

The reinforced concrete construction of the foundation slab has a rounded shape. The lower part of the base is in the form of a cylinder with a diameter of 19 m and a height of 1.1 m. The middle part is a truncated cone with a height of 1.4 m, which goes into a cylindrical sole with a diameter of 6.2 m and a height of 0.7 m for fastening the mast of the wind generator. In the center of a heavily reinforced foundation, a system of plastic cable channels is installed on the frame (Fig. 1, a). The temperature of their operation should not exceed 70 $^0$C.

a) b)

**Figure 1.** The foundation plate of the wind farm
a) the cable channels location; b) foundation concreting in progress.

Concrete volume – 585 m$^3$. Concrete class C35 / 45.
Concreting conditions are winter, at a temperature of -5 $^0$C ... + 5 $^0$C.
The initial temperature of the concrete mixture is 15 $^0$C.
Slump class – S4 (16-20 cm).
Concrete mixture supply – concrete pump (Fig. 1, b).
Concreting time – 24 hours.
Materials and compositions provided by the concrete manufacturer (table 1).
Characteristics of materials:
- cement – Portland cement CEM II/A-S 32.5-LH;
- crushed stone – granite, fractions 5-10 and 10-20;

IOP Conf. Series: Materials Science and Engineering **907** (2020) 012037 doi:10.1088/1757-899X/907/1/012037

- sand – quartz, particle size modulus – 1.5;
- additives: polycarboxylate type superplasticizer with an admixture of lignosulfonates, an antifrosty accelerating additive and a hardening retardant.

**Table 1.** The compositions of concrete

| Components | #1 | #2 |
|---|---|---|
| Cement, kg | 560 | 560 |
| Coarse aggregate 5-10, kg | 325 | 325 |
| Coarse aggregate 10-20, kg | 835 | 835 |
| Sand, kg | 550 | 550 |
| Water, kg | 190 | 185 |
| Superplasticizer,% | 0.8 | 0.8 |
| Antifrosty additive,% | 0,5 | - |
| Hardening retarder,% | - | 0,3 |

*4.2. Calorimetric analysis*

The study was performed at the KNUCEA calorimetric center. To assess the maximum temperature of concrete heating, the ability to control the intensity and completeness of cement hydration with the addition of chemical additives, the adiabatic calorimetry method was used [9-10]. Figure 2 shows the temperature dependences of the hardening of concrete samples for 72 hours.

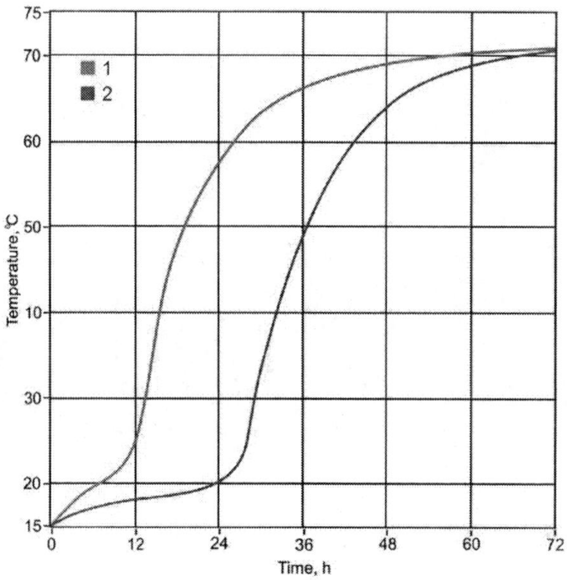


**Figure 2.** Temperature of concrete samples in an adiabatic calorimeter

The results of measurements of the concrete hardening temperature indicate an initial delay in hydration for 9 hours for composition 1 and 24 hours for composition 2. After that, in both cases there is an intensive temperature increase for 24 hours. A significant difference in the indicators of the initial temperature of both compounds is leveled by 72 hours. The absolute value of the temperature during this period was 70.5 °C, and its excess relative to the initial value was 55.5 °C. This indicates the possibility of a significant increase in the temperature of concrete and the achievement of a critically acceptable value of 70 $^0$C during hardening of concrete in the body. The adding of the hardening retarder to reduce the heating temperature is inefficient and can lead to a prolonged initial deceleration at low temperatures with freezing of the liquid phase [11]. It is concluded that it is necessary to change the initial temperature of the concrete mixture from 15 °C to 6-8 °C. For work in the summer period, this composition cannot be used or it is required to organize layer-by-layer concreting [1].

*4.3. Temperature-time monitoring of concrete hardening in the foundation structure*

A measurement system with a special monitoring features was developed. There are eight temperature sensors (Pt100 resistance temperature transducer) connected to a multi-channel temperature meter (Fig. 3). The measured data is automatically transferred to a computer.

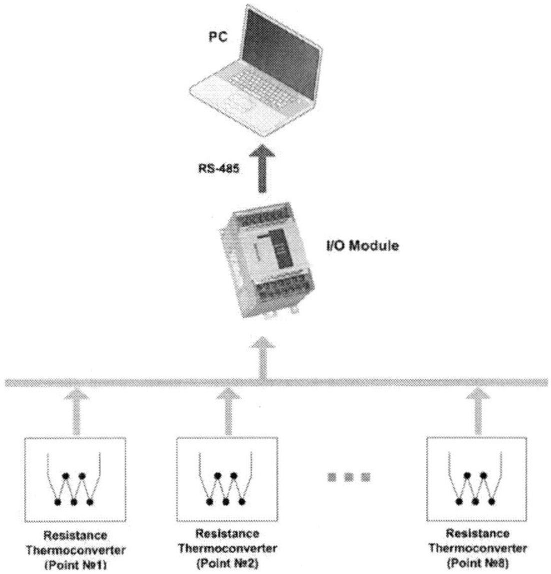


**Figure 3.** Measurement system structure

Temperature sensors were installed at the characteristic points of the foundation plate (Fig. 4). Sensors 1-3 were located in the center of the structure with a maximum height of 3.2 m (section I-I), sensors 4 and 5 – at the point of transition from the basement pedestal to the base with a section height of 2.5 m (section II-II). Sensors 6-7 – were located at a distance of 60 cm from the edge of the plate with a section height of 1.2 m (section III-III). Sensor 8 was used to measure the ambient temperature under the tent, which was installed above the slab after concreting. The results of monitoring the hardening temperature of the foundation slab after concreting are presented in Fig. 5.

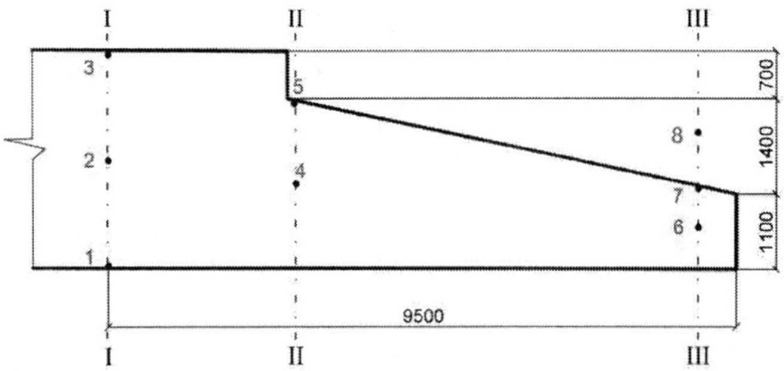


**Figure 4.** Sensor layout

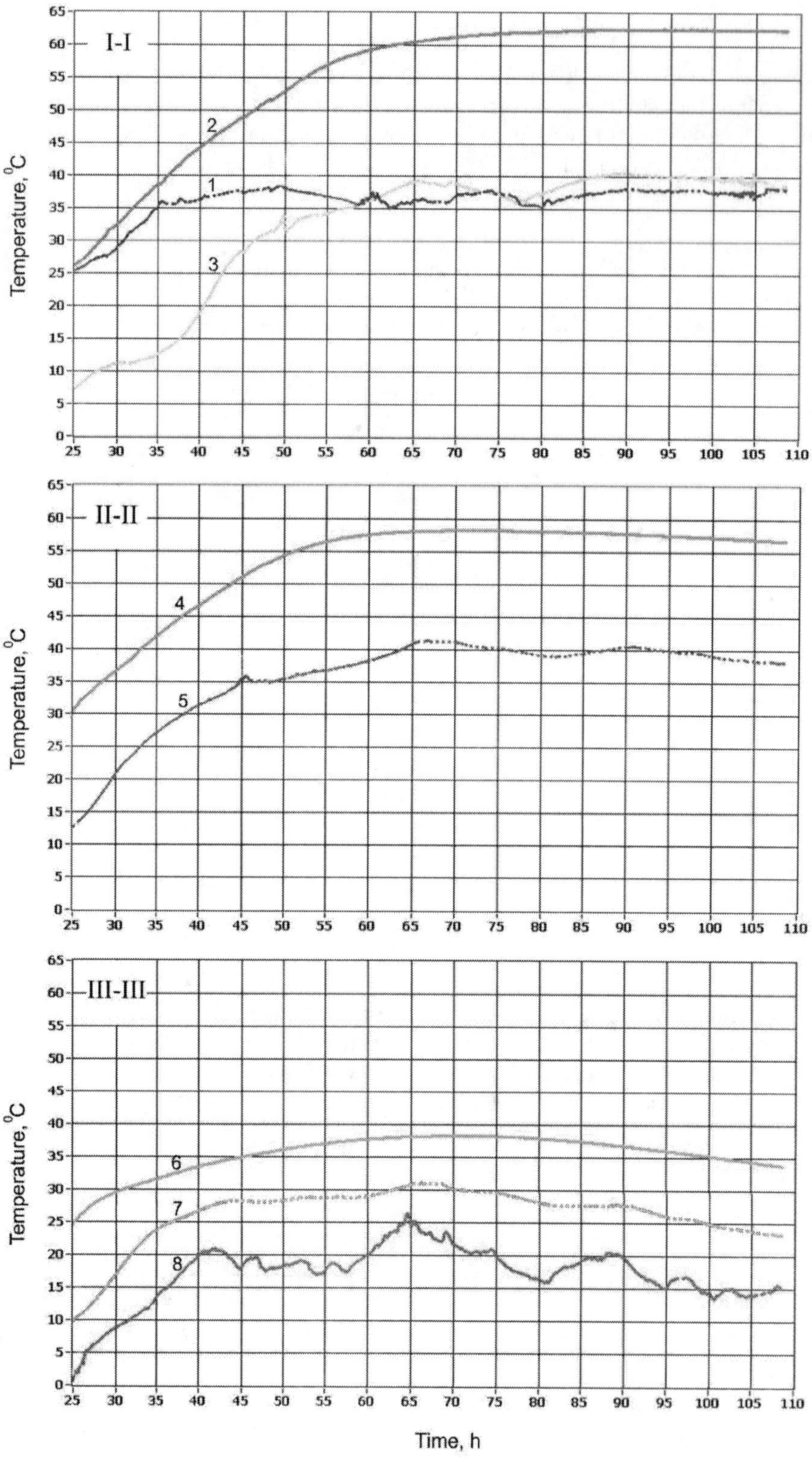


**Figure 5.** The hardening temperature of the foundation plate

The maximum temperature of 62.5 °C is observed at point 2 of section I-I, where the hardening conditions approach adiabatic ones. The initial temperature was 7.5 °C, and the temperature increase was 55 °C. Due to the significant time interval of concreting of the lower and upper parts of the foundation, there is a significant lag in the rate of temperature rise over 60 hours, after which the temperature values are equalized. The temperature gradient between the inner (2) and outer (3) parts of the foundation is significant. Its maximum value of 27 °C was observed after 38 hours of hardening. Then this value decreased to 20-25 °C.

In section II-II, the maximum temperature was 58 °C, and the temperature gradient was 16-18 °C. The minimum rate of temperature rise was recorded in section III-III, where the maximum was 38 °C, and the gradient was from 15 to 8-10 °C. The installation of the tent over the foundation creates favorable conditions for hardening to reduce temperature gradients. At the same time, the temperature of concrete rises by 15-20 °C compared with the temperature of the outside air.

Due to significant temperature gradients, for further equable cooling condition of the foundation structure, sheltering of the foundation surface was organized with a 10 cm thick insulator. The cooling rate did not exceed 1-2 °C per day until the temperature in the center of the foundation dropped to 30 °C. The data obtained made it possible to establish the nature of the development of temperature hardening conditions in different sections and to develop organizational and technological solutions for concreting subsequent foundations.

## 5. Conclusion

The temperature regime of concrete hardening is one of the key factors in the construction of wind farm foundations. The adopted recipe and technological solutions ensure the achievement of acceptable conditions for the construction of the foundation slab in the winter. However, it is advisable to continue research on improving the concrete recipe to reduce temperature gradients in the central part of the foundation of the wind farm.

## References

[1] Yuwen J. Honggang L. Actual temperature evolution of thick raft concrete foundations and cracking risk analysis. *Advances in Materials Science and Engineering.* V.2019, 2019. 7029671.

[2] ACI 207.1R-05:2012. Guide to Mass Concrete. *Reported by ACI Committee 207.* 2005. 30 p.

[3] Kiernozycki W. Betonowe konstrykcjt masyvne. *PC, Krakow.* 2013. 320 s.

[4] Han S. Assessment of curing schemes for effectively controlling thermal behavior of mass concrete foundation at early ages. *Construction and Building Materials.* 2020. 230.117004.

[5] Troyan V.V. Technological bases of increase and forecasting of durability of concretes of massive constructions. *K.: «Interservice».* 2017. 238 p.

[6] Trapeznikov, L.P. Temperature crack resistance in massive concrete. *M.: Energoatomizdat.* 1986. 272 p.

[7] Semenov K.V., Barabanschikov Y.G. Crack resistance of massive concrete foundation slabs and its provision during the construction period in winter. *Construction of Unique Buildings and Structures*, 2014. №2 (17). p. 125-135.

[8] Usherov-Marshak A.V., Kabus A.V. Functional kinetic analysis of the effect of admixtures on cement hardening. *Inorganic Materials.* 2016. V.52 (4). pp. 435-439.

[9] RILEM TC 119-TCE. Avoidance of thermal cracking in concrete at early ages. *Materials and Structures.* 1997. V.30. pp. 451-464.

[10] EN 12390-15:2019 Testing hardened concrete. Adiabatic method for the determination of heat released by concrete during its hardening process.

[11] Usherov-Marshak A., Zlatkovski O., Sopov V. Calorimetric study of frost attack during cement hardening. *Journal of Thermal Analysis and Calorimetry.* V.68 (1). 2002. p. 223-230.

IOP Conf. Series: Materials Science and Engineering **907** (2020) 012038 doi:10.1088/1757-899X/907/1/012038

# Application of the easy concentration effect in concrete technology

**Alexandra Shishkina, Alexander Shishkin**

Department of Technology of Building Products, Materials and Structures, Kryvyi Rih National University, Kryvyi Rih, Ukraine

5691180@gmail.com

**Abstract.** The results of studies of the influence of nano-modifiers on the appearance of dimmers of colloidal surfactants on the compressive strength of concrete and the rate of its formation are presented. It has been shown that colloidal aliphatic surfactant increases the strength of concrete. Moreover, the joint introduction of a colloidal aliphatic surfactant and traditional hydrophilic surfactants leads to an even greater increase in the strength of concrete. A significant increase in concrete strength occurs at an early age in concrete.

## 1. The essence of the applied scientific problem

The essence of the applied scientific problem, which is considered in the paper, is to develop the scientific basis of nanomodification of fine-grained concretes by introducing into their composition filled micelles of colloidal hydrophobic aliphatic surfactants in low concentrations.

## 2. Actuality of theme

Certain types of concrete and reinforced concrete materials, products and structures, namely construction solutions for surface decoration and repair of building structures, thin-walled and thin-walled products, reinforced concrete, including monolithic, building structures, it is advisable to make of fine-grained concrete.

Every year in the world practice of concrete and reinforced concrete production the production of high-quality concretes of fine-grained concretes is growing rapidly and has become an objective reality, caused by considerable savings of material and energy resources. Significant scientific advances in the creation of superplasticized binders, microdisperse mixtures with reactive powders from high-strength rocks have allowed water-reducing effects to be reduced to 60% by the use of superplasticizers of oligomeric composition and hyperplasticizers of polymeric composition.

Typical nanowires in Portland cement systems are $SiO_2$ micro- and silica nanocrystals, $TiO_2$, $Al_2O_3$, $Fe_2O_3$, $CaCO_3$, shungite, taurite and carbon nanomaterials (carbon nanotubes and nanofibers) [1,2].

Research S M Tolmachov, O A Belichenko [3] showed that with the introduction of a small amount of CNTs (0.0045 and 0.0225%), the compressive and flexural strength increases by 1.2 and 1.5 times, respectively.

In the writings of M V Sukhanevich [4] showed that the introduction of multilayer carbon nanotubes together with plasticizer C-3 in cement mortar leads to increased flexural strength and reduced porosity. Moreover, the pore distribution becomes more uniform [5].

In the implementation of complex modification of the superplasticizer and active silicon dioxide is achieved by providing water-reducing properties and is the binding of calcium hydroxide with hydrosilicates, and the process occurs without increasing the solid phase, ie without the formation of internal stresses. The overall porosity of the structure is reduced by the replacement of the mechanically weak calcium hydroxide phase by the stronger hydrosilicate phase [5].

These concretes, regardless of their scope, must have certain general properties. Namely, sufficient durability and high speed of formation of its structure that will shorten the time of production of such concrete products and structures or shorten the construction time.

One of the disadvantages of many cement-water systems is their relatively low rate of structure formation and, therefore, its basic properties. Yes, for the reaction of hydration of cement minerals, the usual conversion time is several weeks or even months, so finding effective methods of accelerating these reactions is an urgent task.

In addition, the use of some surfactants may make them incompatible with cement [6]. Therefore, studying the effects of chemicals in low doses is of great interest [7, 8].

At this time, a steady trend has been the use of various types of catalysis to accelerate virtually all reactions used in organic chemistry. Even those transformations that were previously carried out without the use of any catalysts are now involved in a range of catalytic processes, reflecting the general general direction [9, 10]. The interest in solutions of colloidal surfactants arises from their general ability to structure water molecules, and to promote peptization of binder particles, that is, to act as a catalyst in chemical reactions [11, 12]. These substances activate the water, which eliminates their negative impact on the processes of hydration of cement. Therefore, determining the effect of low dose surfactants is relevant.

The novelty of the work is to summarize the experience of using filled micelles in concrete technology and to determine that their most rational creation on the basis of colloidal surfactants of aliphatic type at low concentrations.

## 3. Materials used in the research and methods of their implementation

The investigated design projects were best performed by Portland Cement 42.5 of PJSC Heidelberg Cement. (Kryvyi Rih, Ukraine). River sand was used as a small concrete filler. As the surfactant forming the hydrophobic micelles (MPAR) used sodium oleate, as plasticizers used: superplasticizer Master Silk, as well as hyperplasticizer company Srase "Adva 151". The plasticizers were dissolved in water to a concentration of 0.001. The plasticizer aqueous solution in the amount calculated according to the design of the experiment was added to the container with a metered amount of water for mixing the concrete. The compositions of the tested concrete mixtures were determined in accordance with the requirements of the rules.

## 4. Research results

Due to the fact that a special role in the problems of forming the optimal structure of cement composites is given to mixing water with cement, as the main component. Water nanostructuring, which leads to changes in its parameters, is a way to improve the rheological characteristics of cement dough and the physical and mechanical properties of nanomodified cement stone and concrete made on its basis. This occurs through the activation of the liquid phase and the physicochemical processes that form the structures of the composites. Well-known experimental data indicate that aqueous systems in which solutes are in low (picomolar) and ultra-low (picomolar) concentrations have a number of special unclear properties. One of the features observed in the field of low or ultra-low concentrations of soluble substances is the nonlinear dependence of the physicochemical properties of the solution on the concentration.

Indirect characteristics of the degree of structuring of water, namely the size of the solvation shell around nanomodifiers - MPARs, may be electrical conductivity, pH of the suspension and change of the meniscus in the capillary. Measurement of these indicators allows us to judge the comparative size of the solvation shell and, accordingly, the size of these molecules that are adsorbed on it.

These data allow to state that the introduction of nanomodifier - MPAR dimers in water leads to the formation of a volumetric fractal grid due to changes in the water's own structure, which are manifested in changes in its physicochemical properties.

Based on the study of the physicochemical properties of aqueous suspensions with different concentrations of dimmers, the interval of a nanomodifier in which a volumetric fractal mesh arises that is capable of altering the electrochemical properties of miscible water is determined. As a result of researches the mechanism of nanostructuring of water is formulated by sorption of hydroxyl groups by the surface of the nanomodifier. This revealed a poly-extreme relationship between the concentration of the MPAR nanomodifier dimmers and the properties of the suspension. The results of complex studies revealed that the optimum concentration of nanomodulator dimmers - MPAR in mixing water is in the range of $10^{-3}$ ... $10^{-6}$ %. In the case of mixing water structured with nano-modifiers - MPAR dimmers, the dispersion system "Portland cement-water" should demonstrate the properties of water as a structured system, namely: providing a high degree and speed of hydration of dry matter (rapid dissolution of inorganic salts) at a deeper and more uniform the bonding of the filler particles with cement. The electrokinetic potential of formed surfaces and bulk agglomerates of hydrate complexes of solids charge increases. The degree of topography of the active centres of the surface of the particles of the dispersed phase increases.

According to the results of the integrated analysis of dispersed components (Portland cement), which was carried out by sieve analysis, it was found that there was a significant reduction in the particle size of Portland cement in the dispersion system "Portland cement - water - MPAR" due to the action of water structured by MPAR dimmers. This is indicated by the increase in dissolution of Portland cement particles when mixed with water, structured with MPAR dimmers. The results of the experiments showed that the introduction of a low concentration of MPAR in the cement-water dispersion system under the experimental conditions leads to an increase in the amount of bound water in concrete and a significant increase in the rate of formation of its structure throughout the study period.

At a certain interval of change of concentration of MPAR ($10^{-3}$...$10^{-6}$ %) the specific electrical conductivity of the system practically does not change (Figure 1).

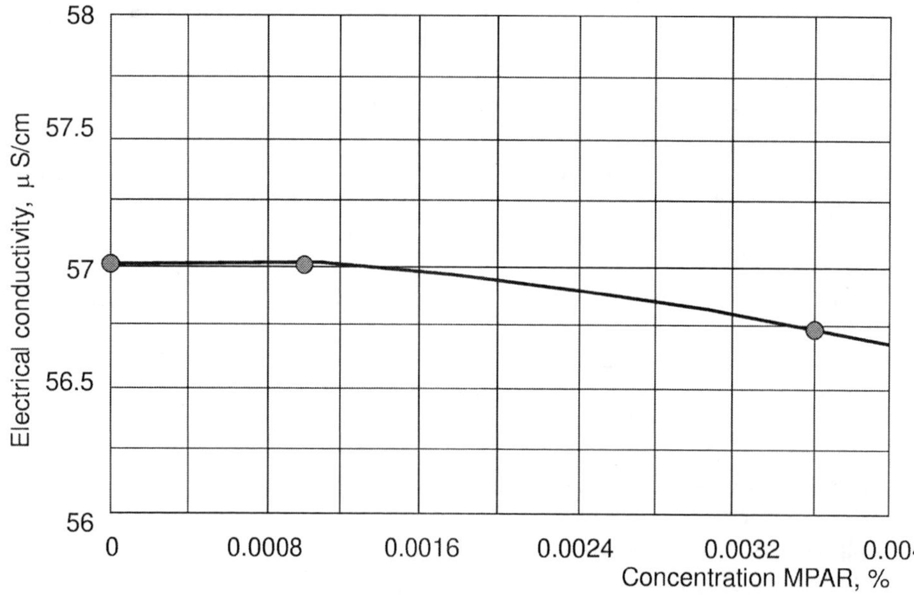


**Figure 1.** The dependence of the electrical conductivity on the concentration of MPAR

This indicates a change in the ionic strength of the water caused by the sorption of hydroxyl groups on the surface of the dimmers introduced into the liquid.

Thus, in the case of nanostructured water modification, an orientation interaction of its molecules with dimmers occurs, which translates the system into an activated state. In an aqueous medium containing dimmers, a new structure is formed, which depends on the interaction of the interfaces, significantly distorting its own structure of pure water, while the spatial hydrogen grid is not destroyed and its stability is not broken. Dimmers (nanomodifiers) act as stabilizers of the self-organization processes of the water system. The development of this process leads to the emergence of a secondary nanostructure - the fractal grid.

The obtained research results indicate the presence of a concentration interval of nanomodifiers - dimers ($10^{-3}...10^{-6}$ %), where the pH of the suspension decreases (Figure 2). The explained effect can be considered only in view of the change in the ionic production of water caused by the sorption on the surface of nanoparticles - dimmers of hydroxyl groups MPAR.

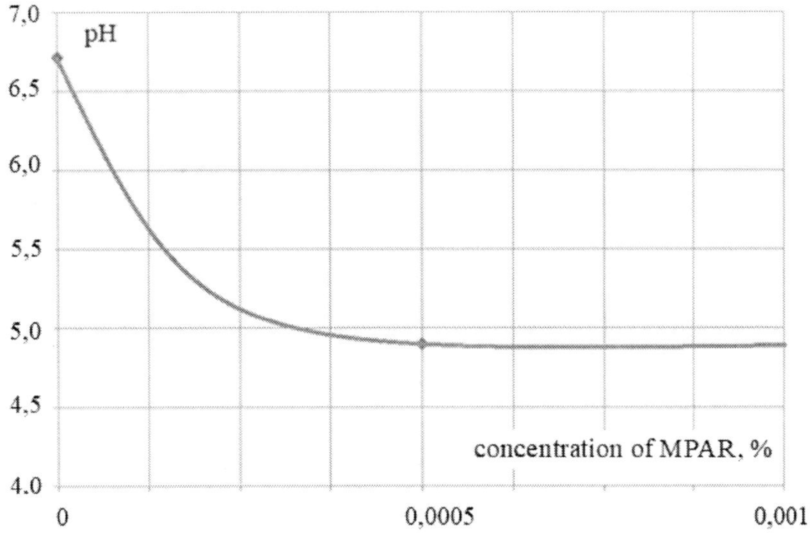


**Figure 2.** Changing the pH of the system depending on the concentration of MPAR

In the course of research, it was found that when using a superplasticizer in the amount of 0.17 % with a content of sodium oleate 0.00024% concrete strength was 150 % of the strength of concrete without additives, and in the absence of sodium oleate was 121 % (Figure 3).

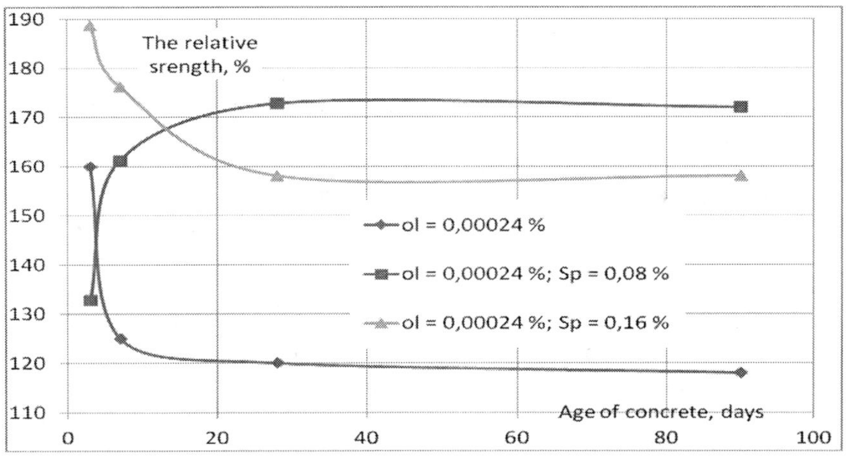


**Figure 3.** Influence of MPAR on the formation of concrete strength when using superplasticizer (Sp) and sodium oleate (ol)

When using a hyperplasticizer (Gp) in the amount of 0.17 % with a content of sodium oleate (ol) 0.0003% the strength of concrete was 240 % of the strength of concrete without additives. In the absence of sodium oleate, the strength of concrete containing 0.17 % of hyperplasticizer was 260 % of the strength of concrete without additives (Figure 4).

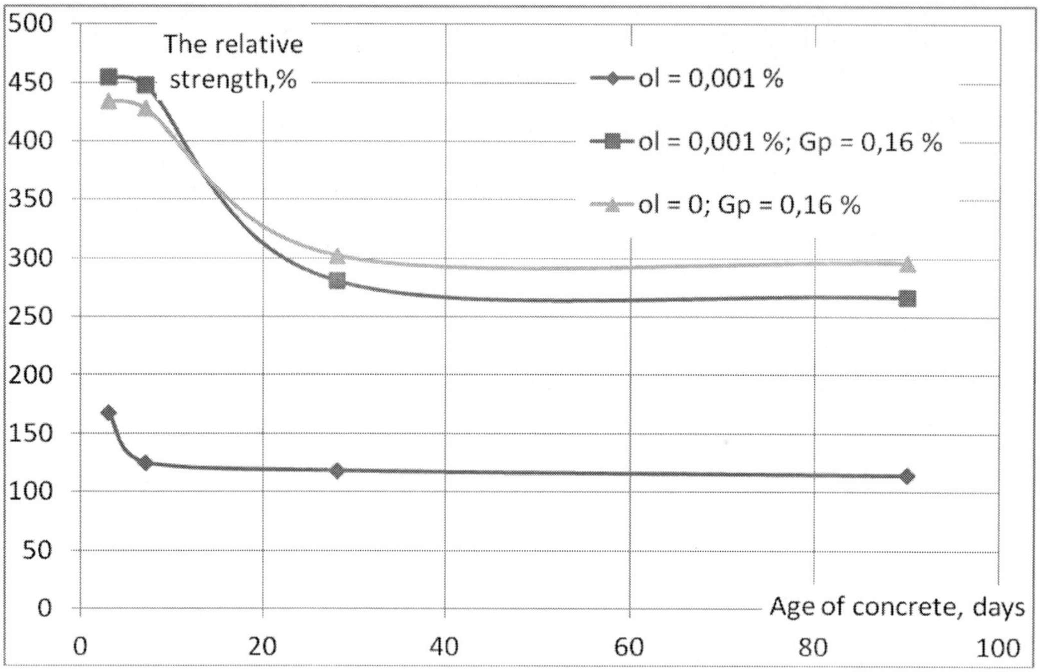


**Figure 4.** Influence of MPAR on the formation of concrete strength when using hyperplasticizer (Gp) and sodium oleate (ol)

When using filled oleate-based micelles in the amount of 0.0004 % the strength of sodium concrete was 198 % the strength of concrete without additives.

## 5. Conclusions

The results of the studies allow us to draw the following conclusions:

- the influence of surfactants on the formation of compressive strength of concrete exceeds the influence of solid nanomodifiers, in particular carbon nanotubes, silicon microxide and others;
- the use of aliphatic hydrophobic surfactants capable of forming micelles, most effectively as nanomodifiers of concrete;
- when using aliphatic hydrophobic surfactants, which are capable of forming micelles as separate modifiers of concrete, their costs are $10^{-4}$ ... $10^{-6}$ % By weight of cement. The increase in the strength of concrete at an early age reaches 90 %, and at the age of 28 days – 40 %. The mechanism of action of modifiers of this type and application is to structure the water due to hydrophobic hydration and the «low concentration effect»;
- when using aliphatic hydrophobic surfactants capable of forming micelles as filled micelles, their costs are $10^{-4}$-$10^{-6}$ wt.% the weight of cement (when filled with polymers). The strength of concrete at the age of 28 days reaches - 100 ... 200 %. The mechanism of action of modifiers of this type and application is consistent with the mechanism of micellar catalysis and the "low concentration effect";
- it is advisable to use aliphatic hydrophobic surfactants to enhance the strength of concrete at an early age. To increase the strength of concrete in older age, it is advisable to use micelles derived from aliphatic hydrophobic surfactants.

IOP Conf. Series: Materials Science and Engineering **907** (2020) 012038 doi:10.1088/1757-899X/907/1/012038

## 6. References


[1] Bazhenov Yu, Alimov L, Voronin V 2017 *Nanomodified cement concrete*. (Moscow: ASV) p 198 (in Russian)

[2] Tevyashev A, Shitikov E 2009 On the possibility of controlling the properties of cement concrete using nanomodifiers *Eastern-European Journal of Enterprise Technologies* **4/7** pp 35-40

[3] Tolmachev S, Belichenko E 2017 Prospects for the use of nanoparticles in concrete transport *Budivelni materialy ta vyroby* **1** pp 38–41 (in Russian)

[4] Suhanevich M V 2014 Features of composite construction of cement matrix for obtaining waterproofing coatings *Science Rise* **5/2** pp 99–107 (in Ukrainian)

[5] Jo B, Kim C, Tae G, Park J 2007 Characteristics of cement mortar with nano-$SiO_2$ particles *Construction and Building Materials* **21/6** pp 1351–1355

[6] Usherov-Marshak A, Zlatkovskyj O, Cyak M 2003 To the problem of compatibility in concrete technology *Proc. 42nd international seminar "Modelling and optimisation in materials science"* (Odessa: Odessa State Academy of Civil Engineering and Architecture) pp 6-7 (in Russian)

[7] Burlakova E, Konradov A 1999 Features of the action of ultra-low doses of biologically active substances and physical factors of low intensity *Rossiyskiy himicheskiy zhurnal* vol. XLIII **5** pp 3- 11 (in Russian)

[8] Palmyna N 2009 The mechanism of action of ultra-low doses *Himiya i zhizn* **2** pp 10-13 (in Russian)

[9] Myttel K 1980 *Micellization, solubilization and microemulsions* (Mosow: Moscow State University Press) p 598 (in Russian)

[10] Levashov A, Klyachkov N 2001 Micellar Enzymology: Methods and Technique *Proceedings of the Academy of Sciences* **10** pp 1638-1651 (in Russian)

[11] Shishkina A, Shishkin A 2016 Investigation of the influence of nanocatalysis on the formation of the strength of reaction powder concrete *Eastern-European Journal of Enterprise Technologies* **1/6** pp 55-60 (in Ukrainian)

[12] Shishkina A 2018 Research into effect of complex nanomodifiers on the strength of fine-grained concrete *Eastern-European Journal of Enterprise Technologies* **2/6** pp 29–33.

# Evaluation of the frost resistance of concrete in real operating conditions

**S D Lapovska[1,3], G Iu Krasnianskyi[2], V I Klapchenko[2] and I O Aznaurian[2]**

[1]Chair of Building Materials, Construction Technology Department, Kyiv National University of Construction and Architecture, Kyiv, Ukraine
[2]Chair of Physics, Engineering Systems and Ecology Department, Kyiv National University of Construction and Architecture, Kyiv, Ukraine
[3]State Enterprise 'Ukrainian research, project planning and design institute of building materials and products', Kyiv, Ukraine

aznaurian.io@knuba.edu.ua

**Abstract.** The article proposes a method for evaluating the frost resistance of concrete at actual operating temperatures, using measurement results at temperatures that are regulated by current standards. Frost resistance was evaluated by determining the amount of water freezing at different temperatures, based on the measured adsorption isobars and the obtained relationship between the freezing temperature of water in concrete pores and relative humidity. A comparison of the calculated values of the frost resistance of concrete with those got based on direct measurements showed the adequacy of the calculation model. To get information about the frost resistance of concrete during unilateral freezing, a conductometric method was used to determine the kinetics of moisture diffusion and ice formation. It is shown that the use of this method allows one to establish the propagation speed of the ice formation and water diffusion front and the corresponding freezing depth of concrete samples depending on their capillary-porous structure and initial storage conditions. In general, the studies conducted allowed us to get a more reliable picture of the behaviour of concrete under alternating temperature load than is provided for by current regulatory documents.

## 1. Introduction

The frost resistance of building materials is determined by the current regulatory documents, which are based on the standard method of fixing the number of comprehensive freezing and thawing cycles for specially manufactured specimens that have not lost operational properties during testing. However, this approach does not always meet the requirements for the production of building materials and has a number of significant limitations.

The main problem is the mismatch of laboratory research conditions with the conditions in which this material is located in real structures. First, the maximum negative temperatures at which the structures are operated usually differ from the temperature of -18 °C at which tests are carried out under the current standard. In addition, in most cases, concrete structures undergo unilateral freezing, and a correlation relationship between this process and the comprehensive freezing process has not yet been found.

Modern ideas about the mechanisms of concrete destruction under the influence of frost (e.g., [1, 2]) are based on Powers' work. Obviously, one of the main causes of concrete damage during freezing is the expansion of the water in the pores during its transition to ice. Powers suggested [3] that stresses leading to destruction can also cause hydraulic pressure when water moves from frozen areas. At the same time, there are other reasons that are not associated with an increase in the volume of water during crystallization, causing destructive deformation of concrete [4].

The super-cooled liquid in the gel pores has greater free energy than the ice in the capillaries. As a result, it is transferred to the capillaries with an increase in the volume of ice in them, which leads to additional internal stresses in the concrete. Moreover, due to the difference in salt concentration caused by freezing water in large pores, osmotic pressure arises, which also leads to concrete damage.

Since the root cause of the mechanisms of frost destruction of concrete is still the expansion of water during its transition to ice, it can be expected that the frost resistance of the material, expressed by the number of cycles, with a change in the maximum freezing temperature, should be inversely proportional to the volume of frozen water in the material at this maximum temperature. This is also consistent with the known semi-empirical correlations between frost resistance and ice content in concrete [5].

It is known that in concrete, which is subjected to unilateral freezing in real conditions of operation, because of the heterogeneity of moisture distribution, intensive mass transfer processes are taking place. These processes significantly affect the resistance of concrete to alternating temperature loads. At the same time, conductometric methods are highly sensitive to physicochemical processes occurring in concrete, in particular, such as changing the phase state of pore moisture, its chemical composition, concentration, and temperature [6].

Because of this, conductivity measurements can be used to obtain additional information on the frost resistance of concrete during unilateral freezing. The observed features of the conductivity behavior can be interpreted based on existing ideas about the kinetics of ice formation and moisture diffusion in capillary-porous materials [7, 8].

## 2. Experimental

Experimental verification of the proposed method for evaluating the frost resistance of concrete at different freezing temperatures was carried out on expanded clay concrete specimens. To prepare the concrete specimens, CEM I 42,5 N cement according to the Ukrainian standard DSTU B EN 197-1:2015, expanded clay according to DSTU B V. 2.7-17, quartz sand according to DSTU B V.2.7-32-95 were used. To determine frost resistance, concrete specimens were made according to DSTU B V. 2.7-214:2009 and DSTU B V. 2.7-18-95.

The dimensions of the specimens were 100×100×100 mm. Three sets of specimens of different density of six pieces each were made. In the set 1 – $\rho_1 = 1450$ kg·m$^{-3}$, in the set 2 – $\rho_2 = 1650$ kg·m$^{-3}$, in the set 3 – $\rho_3 = 1960$ kg·m$^{-3}$.

Examinations of frost resistance were conducted according to DSTU B V. 2.7-47-96. The frost resistance of concrete is equal to a certain number of freezing-thawing cycles of water-saturated specimens, at which concrete compressive strength is reduced by no more than 15 %, and specimen mass loss does not exceed 5 %.

Frost resistance determination method involves cyclic freezing of the specimens in the air with a temperature of -18 ± 2 °C for at least 4 hours and then defrosting under the water at room temperature for at least 4 hours. For research, frost resistance was also measured at temperatures from -5 to -40 °C, which differ from the standard.

The measurements of the equilibrium water-holding characteristics of the experimental expanded clay concrete specimens were carried out by the known method for the determination of adsorption isobars [9]. For this purpose, ground concrete specimens at a temperature $t_1$ were blown with air saturated with water vapour at a lower temperature $t_2$. The relative humidity $\varphi$ at which the specimens were located was calculated by the formula:

$$\varphi = \frac{p_S(t_2)}{p_S(t_1)}, \qquad (1)$$

where $p_s(t_1)$, $p_s(t_2)$ – saturated water vapor pressure at temperatures $t_1$ and $t_2$, respectively.

The equilibrium moisture contents of the specimens $U_\varphi$ were calculated by their masses measured by the weight method:

$$U_\varphi = \frac{m_\varphi - m_0}{m_0}, \tag{2}$$

where $m_\varphi$ – the mass of specimen in equilibrium with air having relative humidity $\varphi$, $m_0$ – dry mass.

The maximum moisture content $U_m$ of the specimens were calculated as the relative amount of moisture absorbed by the porous material upon its contact with water in relation to the dry mass $m_0$ of the material:

$$U_m = \frac{m_s - m_0}{m_0}, \tag{3}$$

where $m_s$ – mass of water-saturated specimen.

The kinetics of ice formation and moisture migration during unilateral freezing were investigated on hardened cement mortar specimens after 28 days of maturation. The dimensions of the specimens were 40×40×80 mm. Stainless steel electrodes with a diameter of 2.5 mm were placed into the upper and lower faces of the samples during formation at a distance of 1 cm from each other.

Electrical conductivity measurements were performed at an alternating current frequency of 10 kHz according to the method described in [6]. Specimens subjected to unilateral freezing were placed in a heat-insulating cassette. Before measuring, some samples were saturated with water at room temperature for 48 hours, and some were kept in the air. The passage of moisture was fixed by the salt mark method, for which a warm end of the sample was moistened with a 4 % solution of sodium chloride in water.

## 3. Results and discussion

In the paper, frost resistance of concrete at various freezing temperatures was evaluated by the experimental-analytical method, which is based on the assumption that frost resistance, expressed in the number of cycles (F), should be inversely proportional to the volume of frozen water at this temperature:

$$F(t_2) = F(t_1) \cdot \frac{V(t_1)}{V(t_2)}, \tag{4}$$

where $V(t_1)$ and $V(t_2)$ – volumes of frozen water at maximum freezing temperatures $t_1$ and $t_2$.

The amount of water frozen at different temperatures is uniquely determined by the thermodynamic characteristics of its bond in the material. The data needed to calculate $V(t)$ are most easily calculated from the water vapour adsorption isobar. Isobar takes into account the decrease in the freezing temperature of water both because of the curvature of the meniscus of water in the pores, and because of dissolved electrolytes. Therefore, there is no need to introduce any corrections to the change in the freezing temperature of the pore solution (at its low concentration). In addition, the adsorption isobar can be measured at any temperature, for example, at room temperature, since the chemical bond potential of water, which determines the decrease in the freezing temperature, is practically independent of temperature. The measured water vapour adsorption isobars are shown in figure 1.

The volume of frozen water increases with increasing negative temperature, although monotonously, but sharply nonlinear, since the adsorption isobar itself is usually nonlinear. Therefore, the change in frost resistance at different temperature ranges can be significantly different.

Since the isobar, as a rule, is set graphically (measured in the experiment), it is not possible to write the formula for calculating frost resistance at different temperatures in a general form. The calculation has to be carried out sequentially by determining the amount of frozen water and the corresponding frost resistance.

IOP Conf. Series: Materials Science and Engineering **907** (2020) 012039 doi:10.1088/1757-899X/907/1/012039

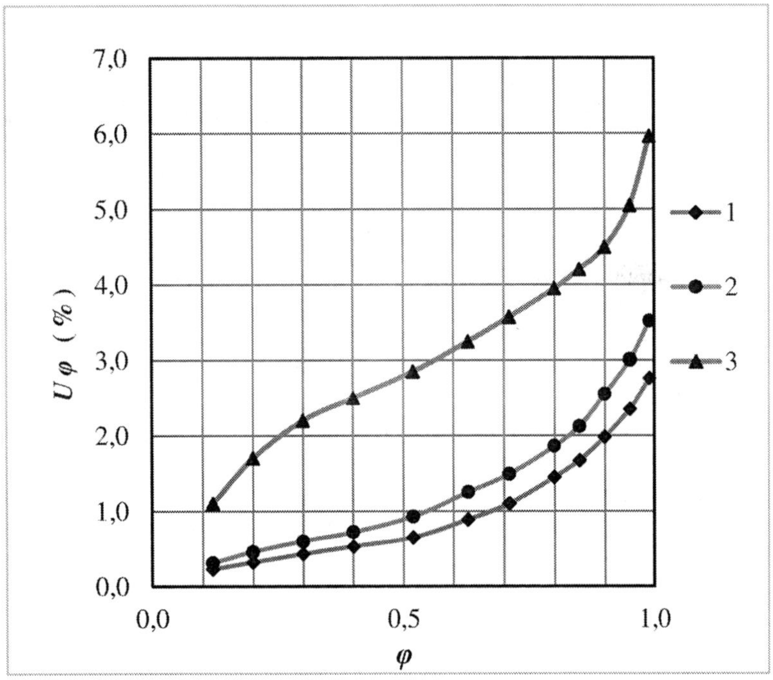


**Figure 1.** Adsorption isobars of expanded clay concrete specimens (the numbering of the curves corresponds to the table 1).

Evaluation of frost resistance was carried out according to the following scheme. Initially, the ratio of the volumes of water freezing in the specimen at given temperatures $t$ was determined:

$$\frac{V(t_1)}{V(t_2)} = \frac{U_m\text{-}U_{\varphi 1}}{U_m\text{-}U_{\varphi 2}}. \tag{5}$$

The expression for the relative humidity $\varphi$ corresponding to a given freezing temperature of water in the pores of the material can be derived from the Gibbs – Thomson equation. With a hemispherical interface between ice and a non-wetting liquid in a pore of radius $r$, the Gibbs – Thomson equation for the freezing temperature reduction can be written in the form [10,11]:

$$\Delta T = T_\infty - T = -T_\infty \cdot \frac{2\sigma_{il} \cdot \cos\theta}{\Delta H_m \cdot r \cdot \rho_i}, \tag{6}$$

where $T$ is the absolute melting temperature of the ice in a pore of radius $r$; $T_\infty$ is the bulk melting temperature of ice; $\sigma_{il}$ is the specific surface energy at the ice-liquid interface; $\Delta H_m$ is the specific melting enthalpy; $\rho_i$ is the density of the ice; $\theta$ is the contact angle (usually considered $\theta = 180°$).

Then, using the Ostwald – Freundlich equation:

$$\ln\varphi = \frac{2\sigma \cdot V_\mu}{RT \cdot r}, \tag{7}$$

where $\sigma$ is the liquid surface tension, $V_\mu$ is the molar volume, $R$ is the gas constant, for the value of relative humidity $\varphi$ at which the absolute freezing temperature is $T$, we get:

$$\varphi = \exp\left[\left(\frac{1}{T_\infty}-\frac{1}{T}\right)\cdot\frac{L\cdot\mu}{R}\right], \tag{8}$$

where $\mu$ is the molar mass of water.

Dependence (8) was obtained under the assumption that $\sigma = \sigma_{il}$ and when $\Delta H_m = L$ is taken into account, where $L$ is the specific melting heat of ice. A similar expression for the dependence of the phase transition temperature in the capillary on the relative vapour pressure was obtained in [12] from the Clausius – Clapeyron relation.

It is necessary to keep in mind that in the derivation of (8) it was assumed that the freezing temperature of the liquid is calculated in a water-saturated capillary-porous material with a rigid skeleton. We also note that in calculating the values of relative humidity corresponding to certain freezing temperatures and volumes of frozen water, the dependence of melting heat of ice $L$ on temperature was neglected. And although in the temperature range studied, $L$ can vary up to 20 % [13], this, as estimates show, does not introduce a significant error in the calculations.

After calculating the volumes of water freezing in the specimens at given temperatures, the values of frost resistance at different freezing temperatures were calculated using formula (4), choosing the value of $F$ at -18 °$C$ as the base point. The calculation was performed for each of the six specimens of each set. The experimental and calculated values of frost resistance of expanded clay concrete specimens at different temperatures are shown in the table and in figure 2.

**Table 1.** The experimental and calculated values of frost resistance of expanded clay concrete specimens at different freezing temperatures

| $t$ (°C) | $F$, cycles - average values | | | | | |
|---|---|---|---|---|---|---|
| | set 1 ($U_m$=3,02%) | | set 2 ($U_m$=4,12%) | | set 3 ($U_m$=6,55%) | |
| | exp. | calc. | exp. | calc. | exp. | calc. |
| −5 | | 49 | 78 | 77 | | 123 |
| −10 | 36 | 32 | 59 | 55 | 98 | 90 |
| −18 | 23 | - | 41 | - | 75 | - |
| −35 | | 17 | 34 | 32 | | 60 |
| −40 | | 16 | 25 | 30 | | 57 |

The frost resistance values found by direct measurements and calculation differ for each set no more than the experimental frost resistance values for specimens of one set. These data confirm the possibility of using the proposed method of evaluation of the frost resistance of concrete at different freezing temperatures and, accordingly, the fairness of the model calculation. Moreover, they show that, apparently, the same mechanism of frost destruction prevails in the studied temperature range from 0 to -40 °$C$.

Based on the research data, it is possible to evaluate the frost resistance of concrete and other building materials at any temperature using the measurement results at a temperature of -18 °$C$, which is regulated by current standards. Such a calculation can significantly save time and costs compared with those in a direct experiment.

For a more detailed study of the mechanisms of frost destruction of building materials, it is necessary to have at the same time information on the kinetics of ice formation and moisture migration in the process of their unilateral freezing. The following shows the possibility of determining the propagation velocity of the ice formation front and moisture diffusion in building materials using the conductometric method. The arrangement of electrode pairs in the tested specimens is shown in figure 3, and the results of measurements of the electrical conductivity σ are shown in figure 4 and figure 5.

As seen from figure 4, the electrical conductivity of the specimen, which was before the experiment under conditions of air storage, varies in different sections of the specimen according to different laws. The same dependences σ (τ) are observed for a water-saturated specimen with the only difference being that the changes in conductivity that occur in the corresponding time intervals are much less pronounced.

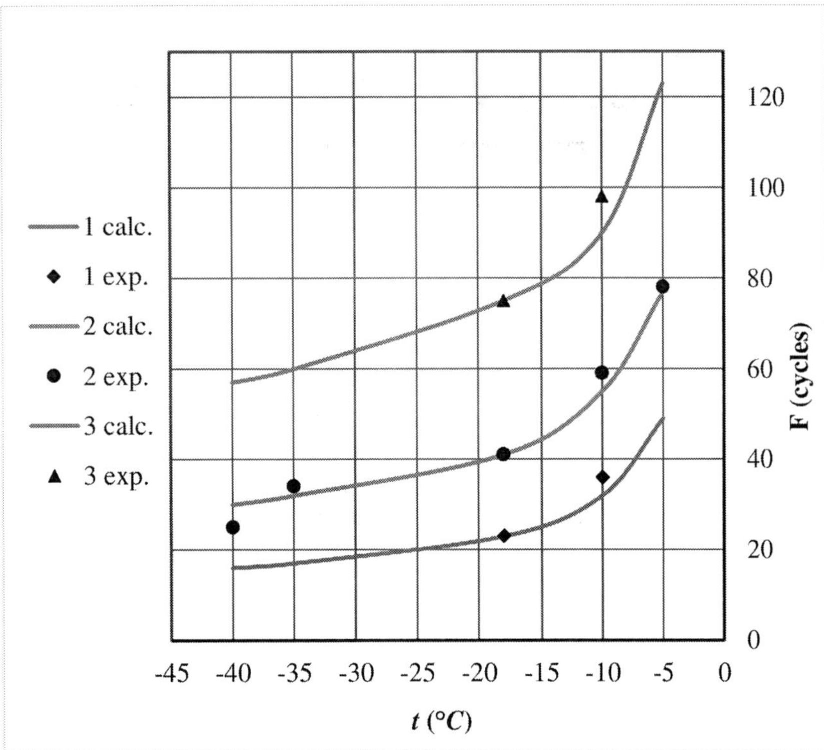


**Figure 2.** Dependence of frost resistance of expanded clay concrete specimens on freezing temperature (the numbering of the curves corresponds to the table).

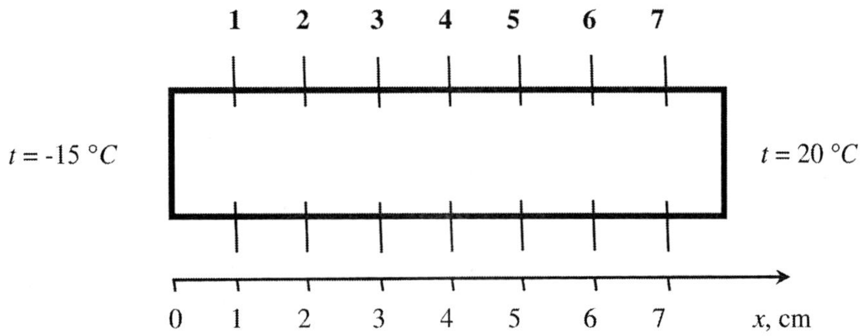


**Figure 3.** Experimental hardened cement mortar specimen (1...7 – electrode pair number).

At the initial time, the temperature is the same along the entire length of the specimen, and the moisture in its pores is in equilibrium. This corresponds to almost coinciding values of σ for all sections of the specimen. Then, the process of ice formation, starting from the frosty face of the specimen, is accompanied by the establishment of a temperature gradient and the appearance of thermal and concentration diffusion of moisture. In this case, the corresponding diffusion flows are directed to the side opposite to the direction of ice formation front propagation.

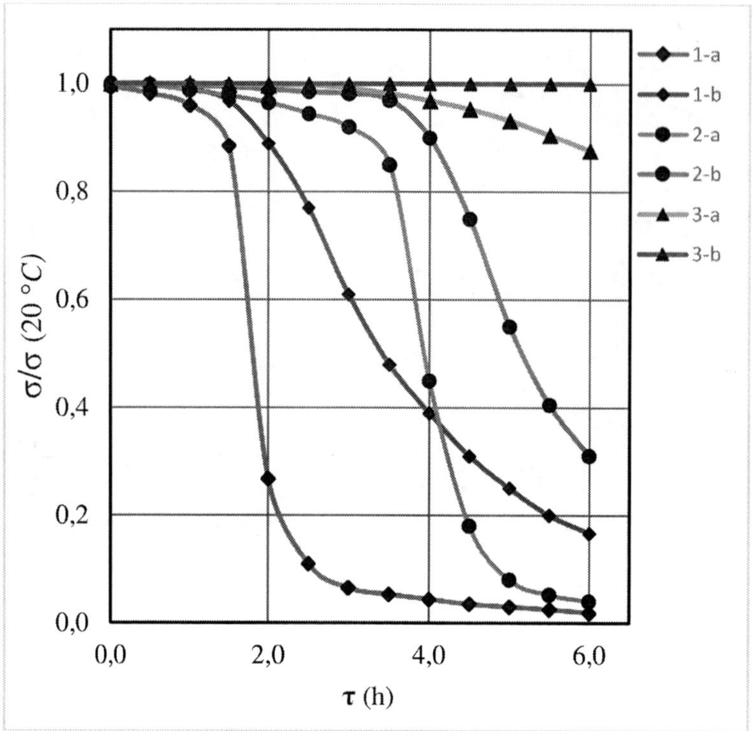


**Figure 4.** Dependence of electrical conductivity of mortar specimen on freezing time: a – water-saturated specimen; b – air-stored specimen (the numbering of the curves corresponds to electrode pair numbers).

It is known [14] that when the temperature changes the electric conductivity of cement mortar and concrete changes according to the exponential law. Thus, the reduction of temperature in some section of the specimen over time should lead to the fall of the σ. The conductivity should also be reduced because of a decrease in moisture content in the corresponding zone caused by thermal moisture transfer. A lower temperature and a larger temperature gradient near the cold face of the specimen cause redistribution of moisture over the volume of the specimen so that a concentration gradient is created that coincides in direction with the temperature gradient. At the same time, the indicated decrease in moisture content can be compensated by the displacement of unfrozen water to the considered region during ice formation. The observed slight decrease in conductivity at the initial stage of cooling shows the predominant contribution of the first two of the described mechanisms.

A sharp decrease in the conductivity at the locations of the electrode pairs 1 and 2 shows the achievement of these sections of the specimen by the ice formation front since the specific electrical conductivity of ice is much lower than the specific electrical conductivity of pore moisture. A smaller change in conductivity in the indicated time interval for a water-saturated specimen can be explained based on the following considerations.

When water-saturated concrete is frozen, ice formed in the surface zone creates increased pressure in the displaced liquid phase, because of which the ice formation temperature decreases. Thereby, at the same temperature, the ice content of the water-saturated specimen should be lower and its contribution to the change in conductivity, decreases.

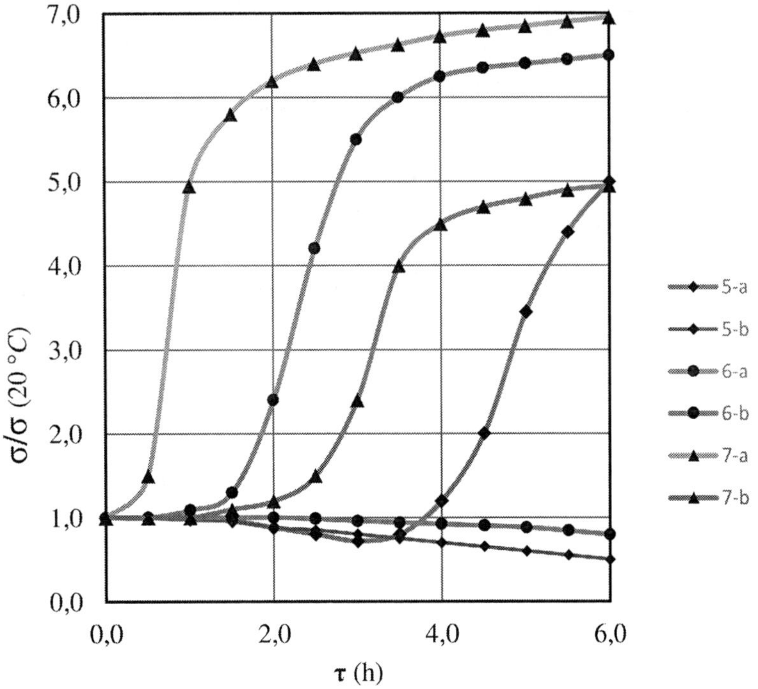


**Figure 5.** Dependence of electrical conductivity of mortar specimen on freezing time: a – water-saturated specimen; b – air-stored specimen (the numbering of the curves corresponds to electrode pair numbers).

An analysis of experimental data shows that during the measurements, the ice front does not reach the zone of the third electrode pair, although, based on the kinetics of ice formation, the duration of the experiment for this may be sufficient. This circumstance, apparently, is explained by the following. By the time the measurements are completed, the concentration of salt coming from the wetted face of the specimen increases in the indicated zone, which leads to a slowdown in the freezing process. This also allows us to conclude that the nature of changes in conductivity in the zones of the 5th, 6th, and 7th electrode pairs (figure 5) should reflect the kinetics of moisture transfer in the specimens during unilateral freezing.

Indeed, a sharp increase in electrical conductivity in sections 6 and 7 is apparently associated with a growth in moisture content and an increase in the concentration of the salt solution. By the time the salt mark arrives in section 6, the temperature in it has time to become lower than in section 7, which leads to lower values of conductivity hereafter. In this time interval, the influence of salt flow in section 5 has not yet been observed. And since the temperature here decreases, a slight decrease in the conductivity occurs (curve 5-a).

It follows from figure 5 that changes in the conductivity in the corresponding sections of a water-saturated sample occur much more slowly and their value is much smaller. To explain these features, it is necessary to take into account that the change in moisture content here can only be caused by thermal and moisture conductivity, whose role at high temperatures and low-temperature gradients in the considered region of the specimen should be small. In this case, the movement of the salt mark occurs because of the diffusion of salt from the wetted face of the specimen. Thus, an increase in the electrical conductivity in section 7 is apparently associated mainly with a growth in the salt concentration, and a decrease in the electrical conductivity in sections 5 and 6 should occur because of a decrease in temperature.

Comparing the measurement results, we can conclude that the average speed of the ice formation front during unilateral freezing of a mortar specimen of the investigated composition does not depend on the initial moisture content and 6 hours after the start of freezing is 0.4 cm/h. At the same time, the average speed of the salt mark in such a specimen for the same time decreases when the specimen is saturated with water, changing from 0.5 cm/h for an air storage specimen to 0.2 cm/h for a specimen subjected to water saturation in within 48 hours.

## 4. Conclusions

The proposed experimental-analytical method for evaluating the frost resistance of concrete at different freezing temperatures allows in a small amount of time to get more complete information about the behavior of the material in frosty weather under real operating conditions than is provided for by existing methods. An analysis of the dependence of frost resistance on temperature also makes it possible to identify the temperature regions where it changes most strongly, and to shift, if necessary, these regions to lower or higher temperatures due to the change of the composition and technology of concrete production.

Moreover, the developed combined method for the independent measurement of the kinetics of moisture diffusion and the ice formation allows one to determine the rates of these processes depending on the composition (capillary-porous structure) of the specimens and the initial storage conditions. Such information can give a more reliable picture of the behavior of concrete under alternating temperature load in conditions of different initial moisture content (including in hydraulic structures) than current regulatory documents allow.

## 5. References

[1] Ramachandran V S, Feldman R F and Beaudoin J J 1981 *Concrete Science: Treatise on Current Research* (London: Heyden)

[2] Neville A M and Brooks J J 2010 *Concrete Technology* (Harlow: Prentice Hall)

[3] Powers T C 1945 Basic considerations pertaining to freezing and thawing tests *Proc. Am. Concrete Inst.* **41** 245-272

[4] Powers T C and Helmuth R A 1953 Theory of volume changes in hardened Portland cement paste during freezing *Proc. Highw. Res. Board* **32** 285-297

[5] Dvorkin L, Dvorkin O and Ribakov Yu 2013 *Multi-Parametric Concrete Compositions Design* (New York: Nova Science Publishers Inc.)

[6] Krasnianskyi G, Aznaurian I and Kucherova G 2013 Methods of electrophysical research of concrete in the early stages of hardening *Mistobuduvannya ta terytorialne planuvannya* **50** 310- 315

[7] Feldman G M 1988 *Moisture Movement in Thawed and Freezing Soils* (Novosibirsk: Nauka)

[8] Gorchakov G I, Kapkin M M and Skramtaev B G 1965 *Improving the Frost Resistance of Concrete in the Industrial and Hydraulic Structures* (Moscow: Stroyizdat)

[9] Gregg S J and Sing K S W 1982 *Adsorption, Surface Area and Porosity* (London: Academic Press)

[10] Jackson C L and McKenna G B 1990 The melting behavior of organic materials confined in porous solids *J. Chem. Phys.* **93** 9002-9011

[11] Webber J B W 2010 Studies of nano-structured liquids in confined geometries and at surfaces *Progress in Nuclear Magnetic Resonance Spectroscopy* **56** 78-93

[12] Krasilnikov K G, Nikitina L V and Skoblinskaya N N 1980 *Physico-chemistry of Own Deformations of Cement Stone* (Moscow: Stroyizdat)

[13] Feistel R and Wagner W 2006 A new equation of state for $H_2O$ ice Ih *J. of Phys. Chem. Reference Data* **35** 1021-47

[14] Bernatsky A F, Celebrovsky Yu V and Chunchin V A 1980 *Electrical Properties of Concrete* ed Yu N Vershinina (Moscow: Energy)

IOP Conf. Series: Materials Science and Engineering **907** (2020) 012040 doi:10.1088/1757-899X/907/1/012040

# Optimization of the composition of hydrophobized cellular concrete according to its moisture-transporting and water-holding characteristics

**S D Lapovska[1,3], V I Klapchenko[2], G Iu Krasnianskyi[2], Yu G Gasan[1] and I O Kuznetsova[2]**

[1]Department of Building Materials, Construction Technology, Kyiv National University of Construction and Architecture, Kyiv, Ukraine
[2]Department of Physics, Engineering Systems and Ecology, Kyiv National University of Construction and Architecture, Kyiv, Ukraine
[3]State Enterprise "Ukrainian research, project planning and design institute of building materials and products", Kyiv, Ukraine

kuznetsova.io@knuba.edu.ua

**Abstract.** The article is devoted to the study of the possibility of obtaining information about frost resistance and strength of concrete based on the analysis of its moisture-transporting and equilibrium water-holding properties. The dependences of the moisture diffusion coefficient, equilibrium moisture content at various values of relative humidity and porosity of autoclaved cellular concrete specimens on the concentration of water-repellent additive were studied. A comparison of the results of these measurements with the data on frost resistance, got by the direct method according to the current standards, have been made. It is shown, that moisture-transporting and water-holding characteristics can be used to assign concrete compositions optimal in frost resistance and strength. The mechanisms of the influence of water repellent on the frost resistance at its various concentrations are considered. It has been established that the composition of cellular concrete containing 2% water-repellent additive is optimal.

## 1. Introduction

The problems of energy efficiency, resource-saving, lowering operating costs and improving the comfort of housing remain their primary importance in the practice of construction. Increasing requirements for the heat-shielding properties of the external building envelope makes it necessary to use highly efficient heat-insulating and structural heat-insulating materials such as autoclaved cellular concrete for their installation. Products made from autoclaved cellular concrete are distinguished by fairly good strength, high heat-insulating ability and allow to get significant energy savings necessary for heating facilities while maintaining a healthy indoor climate [1-4].

To increase the competitiveness of cellular concrete in modern conditions, the urgent task is to further improve the physical and technical properties of these materials and create energy-saving technologies for their manufacture. Modern building standards impose several mandatory requirements on the wall fencing materials, among which one of the most important is high strength and frost resistance. For such porous building materials as cellular concrete, the values of these indicators vary depending on the moisture content in them: the strength and frost resistance of the material in a water-saturated state decreases, and the thermal conductivity increases [5]. It is possible to reduce the indicators of sorption moisture content and water absorption of cellular concrete to a certain optimal value by optimizing its structure.

The durability of concrete under alternating freezing-thawing is the main factor determining the possibility of its use in cold climates. It is directly related to the moisture properties of concrete, which depend on the value and nature of its porosity, the state of the surface of the solid phase. The lower the total porosity, the higher the frost resistance. At the same time, to increase it, small spherical pores, serving as reservoirs, into which excess water formed during freezing can migrate must evenly be distributed in concrete [6, 7].

One of the ways to optimize the structure of building materials is their hydrophobization using various organic compounds that can give the solid phase of the material water repellent properties. At the same time, the role of water repellent additives in increasing the frost resistance of concrete has not been adequately studied and is usually associated with the difficulty of water suctioning into concrete and its migration, and with an increase in the number of closed pores that are not filled with water when concrete is saturated [8, 9].

The magnitude and nature of the porosity of the material, which determines its frost resistance, can be established based on a study of its equilibrium water-holding and moisture-transporting properties, which reflect the kinetics of moisture migration in the material.

In this paper, the results of a study of the effect of hydrophobization on the frost resistance of cellular concrete are presented and the possibility of assigning an optimal concrete composition from this point of view according to its water-holding and moisture-transporting characteristics is shown.

## 2. Experimental

### *2.1. Autoclaved cellular concrete specimens' preparation*

To prepare the concrete specimens, CEM I 42,5 N cement according to the Ukrainian standard DSTU B EN 197-1:2015, quicklime according to DSTU B V.2.7-90:2011 with active CaO + MgO content of 80%, quartz sand according to DSTU B V.2.7-32-95 with a quartz content of 85%, aluminum powder according to GOST 5494-95 and water according to GOST 23732 were used. For volume hydrophobization of cellular concrete, polydimethylsiloxane emulsion was used. To determine frost resistance and compressive strength, concrete specimens were made according to DSTU B V.2.7-45:2010 and DSTU B V.2.7-214:2009. The dimensions of the specimens were 100×100×100 mm.

To conduct studies of the water-holding and moisture-transporting properties of cellular concrete, an individual series of experimental specimens 50×50×50 mm in size were made. Three specimens of each series, together with specially selected steel plates for damping positive buoyancy, were used to measure the diffusion coefficient of water. The three remaining specimens of each series, after crushing in pressing equipment, were fractionated using sieves, leaving a fraction with an average size of 4 mm to study the equilibrium water-holding properties.

### *2.2. Determination of cellular concrete properties*

*2.2.1. Determination of compressive strength and frost resistance.* Examinations of compressive strength and frost resistance were conducted according to the above Ukrainian standards. According to DSTU B V.2.7-45:2010, the frost resistance of concrete is equal to a certain number of freezing–thawing cycles of water-saturated specimens, at which concrete compressive strength is reduced by no more than 15%, and specimen mass loss does not exceed 5%. Frost resistance determination method involves cyclic freezing of the specimens in the air with a temperature of −18 ± 2°C for at least 4 h and then defrosting under the water at room temperature for at least 4 h. Each step of freezing and defrosting was a single cycle of the research. For research, percentage mass and compressive strength losses were determined compared to the samples without the freezing–defrosting procedure after $25^{th}$ cycles, and then after each $5^{th}$ cycles.

*2.2.2. Determination of water-holding and moisture-transporting characteristics*. The measurements of the equilibrium water-holding characteristics of the experimental cellular concrete specimens were carried out by the known method for the determination of adsorption isobars [10]. For this purpose, ground concrete specimens (fraction 3...5 mm) at a temperature $t_1$, were blown with air saturated with water vapour at a lower temperature $t_2$. The relative humidity $\varphi$ at which the specimens were located was calculated by the formula:

$$\varphi = \frac{p_s(t_2)}{p_s(t_1)}, \tag{1}$$

in which $p_s(t_1)$, $p_s(t_2)$ – saturated water vapour pressure at temperatures $t_1$ and $t_2$, respectively. The equilibrium moisture contents of the specimens $U_\varphi$ were calculated by their masses measured by the weight method:

$$U_\varphi = \frac{m_\varphi - m_0}{m_0}, \tag{2}$$

where $m_\varphi$ – the mass of specimen in equilibrium with air having relative humidity $\varphi$, $m_0$ – dry mass. For this study, not a complete adsorption isobar was determined, but only the moisture content at three fixed values of relative humidity $\varphi$: 0.3, 0.6, and 0.95.

The moisture diffusion coefficient $D$ of the studied cellular concrete specimens was determined by the kinetics of capillary impregnation [11] of specimens completely immersed in water. In this method, the maximum moisture content $U_m$ of the samples is automatically determined as the relative amount of moisture absorbed by the porous material upon its contact with water in relation to the dry mass $m_0$ of the material:

$$U_m = \frac{m_s - m_0}{m_0}, \tag{3}$$

where $m_s$ – the mass of the water-saturated specimen. The maximum moisture content allows to determine the open porosity $P_O$ of dispersed materials – the ratio of the absorbed moisture volume $V_{H_2O}$ to the volume of the specimen $V_0$:

$$P_O = \frac{V_{H_2O}}{V_0} = \frac{(m_s - m_0)\cdot\rho_0}{\rho_{H_2O}\cdot m_0} = \frac{\rho_0}{\rho_{H_2O}}\cdot\frac{m_s - m_0}{m_0} = \frac{\rho_0}{\rho_{H_2O}}\cdot U_m, \tag{4}$$

where $\rho_0$, $\rho_{H_2O}$ – the density of the dry specimen and water, respectively.

The value of the closed porosity $P_c$ can be calculated from experimental data. Its value is equal to the difference between the total $P$ and the open porosity $P_O$: $P_c = P - P_O$. The total porosity $P$ is determined by the average density of the specimens $\rho_0$ and the average density of the solid phase $\rho_1$ of the mixture raw components:

$$P = 1 - \frac{\rho_0}{\rho_1}. \tag{5}$$

By the component composition of the mixture, for $\rho_1$ we find $\rho_1 \approx 2600$ kg·m$^{-3}$, which gives almost the same total porosity for all series of specimens $P = 80\%$.

## 3. Results and discussion

The obtained values of frost resistance $F$, compressive strength $R$ and average density $\rho_0$ of the studied cellular concrete specimens at various concentrations $c$ of water repellent are presented in table 1 and figure 1. Established water-holding and moisture-transporting characteristics are shown in table 2, figure 2 and figure 3.

**Table 1.** Frost resistance, compressive strength and average density of cellular concrete at various concentrations of water repellent.

| *c* (%) | *F* (cycles) | *R* (MPa) | $\rho_0$ (kg·m$^{-3}$) |
|---|---|---|---|
| 0 | 25 | 2.2 | 535 |
| 1 | 30 | 2.4 | 530 |
| 2 | 50 | 3.5 | 540 |
| 3 | 35 | 3.2 | 537 |
| 4 | 40 | 2.9 | 531 |
| 6 | 75 | 2.6 | 522 |

As follows from table 1 and figure 1, the frost resistance of cellular concrete specimens increases in the entire studied range of water repellent concentrations, except for the range of 2...3%, where it falls. Moreover, at $c > 2...3\%$, the strength of the samples decreases markedly.

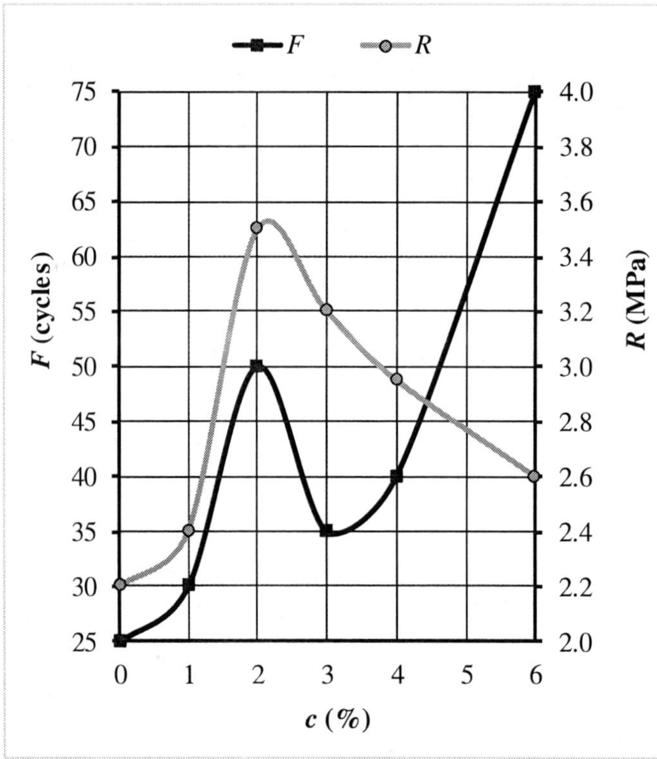


**Figure 1.** Dependences of frost resistance and compressive strength of cellular concrete on the concentration of water repellent.

Following the existing ideas (e.g., [12, 13]), the destruction of concrete under the influence of frost is caused not only by an increase in the volume of water during freezing. The hydraulic pressure of water when it is squeezed out from the freezing front, the crystallization pressure of ice during its aggregation, the arising osmotic pressure are the main destructive factors that determine concrete damage. The manifestation of each of these factors depends on the conditions of freezing, water saturation of concrete, its structure, the state of the surface of the solid phase and other parameters.

When a water repellent additive is introduced into the concrete, it is adsorbed in the form of monomolecular layers on the walls of large pores and capillaries, which are formed during the setting and expansion of cellular concrete mass. This leads to an increase in the number of closed pores that are not filled with water (closed porosity $P_c$), which can act as reserve ones and increase the frost resistance of concrete. An increase in the number of fully or partially closed pores with a change in the water repellent concentration between 0...2% is indicated by a decrease in the mass transfer coefficient and open porosity $P_O$ of the specimens in this range (figure 3). A slight increase in porosity in the 0...1% area is associated with air entrainment during introducing a water repellent additive.

**Table 2.** Water-holding and moisture-transporting characteristics of cellular concrete at various concentrations of water repellent.

| $c$ (%) | $U_\varphi$ (%) at $\varphi$ | | | $D \cdot 10^8$ | $U_m$ (%) | $P_O$ (%) | $P_c$ (%) |
|---|---|---|---|---|---|---|---|
| | 0.3 | 0.6 | 0.95 | (m²/sec) | | | |
| 0 | 0.22 | 2.6 | 4.5 | 48.0 | 81 | 43 | 37 |
| 1 | 0.23 | 1.9 | 3.7 | 7.4 | 92 | 48 | 32 |
| 2 | 0.22 | 1.4 | 1.9 | 1.0 | 66 | 36 | 44 |
| 3 | 0.24 | 2.1 | 2.7 | 24.0 | 102 | 54 | 26 |
| 4 | 0.23 | 2.5 | 4.2 | 0.9 | 84 | 44 | 36 |
| 6 | 0.22 | 2.3 | 5.1 | 0.1 | 66 | 34 | 46 |

From table 2 it can be seen that those series of specimens where the closed porosity is maximum have the highest frost resistance. This is a series with a concentration of water repellent additives of 2% and 6%.

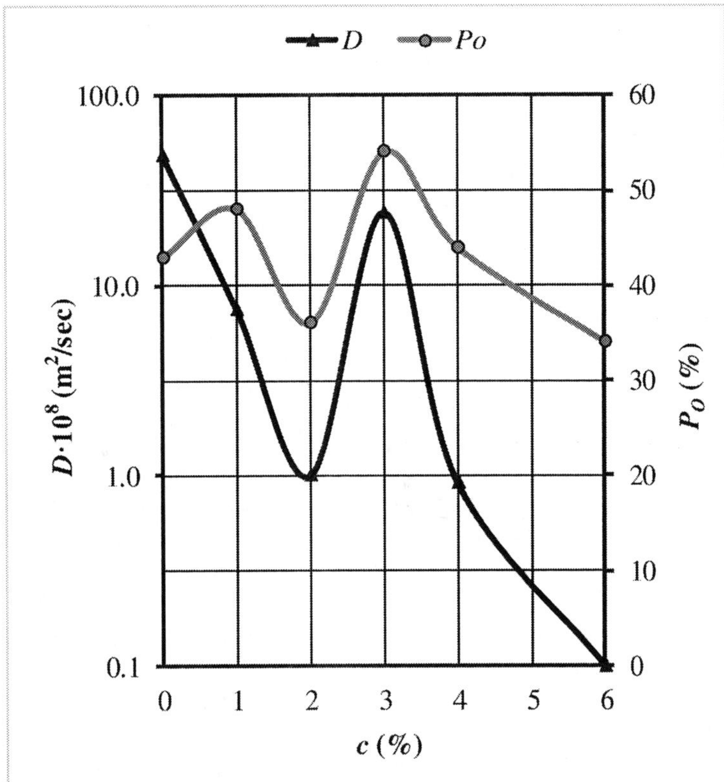


**Figure 2.** Dependences of the coefficient of moisture diffusion and open porosity of cellular concrete on the concentration of water repellent.

As was shown in [14], the use of polydimethylsiloxane emulsion additives in the composition of cellular concrete mixtures in an amount of up to 2% promotes the formation of closed pores with a smaller diameter and a smaller thickness of the inter-pore partitions as compared to the composition without additives. This should lead [15] to a decrease in hydraulic pressure and, therefore, also contribute to an increase in the frost resistance of concrete in this concentration range of the water repellent.

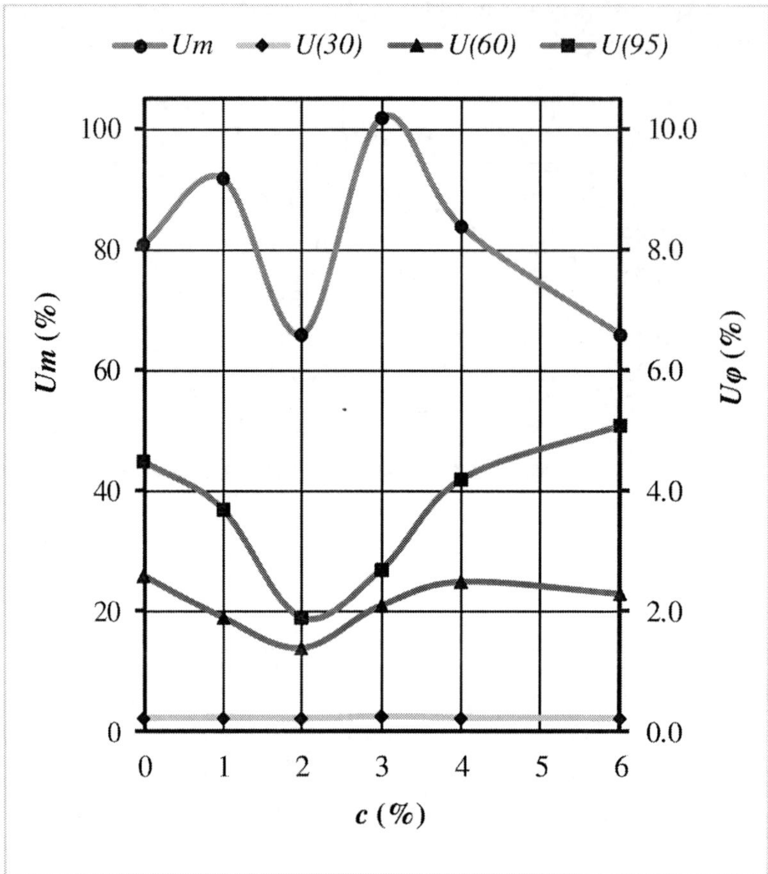


**Figure 3.** Dependences of the moisture content of cellular concrete at various values of relative humidity on the concentration of water repellent.

Another mechanism for increasing the frost resistance of cellular concrete during its hydrophobization can be a significant weakening of osmosis [6], which is responsible for the formation of ice lenses in the mouths of capillaries. Clusters of water molecules involved in osmosis can form at $t < 0°C$ only in the near-wall water layers, the structure of which is changed in the presence of powerful unsaturated fields of the solid phase surface. According to sorption measurements, the near-wall film thickness decreases with an increase in the concentration of water repellent from 0 to 2%, as indicated by a decrease in the equilibrium moisture content of the specimens (figure 2). The decrease in adsorption occurring at $\varphi = 0.6...0.95$ by the capillary condensation mechanism during hydrophobization is explained by a decrease in the affinity between water and the surface of the solid phase which leads to a decrease in the capillary potential. A decrease in ice formation in the mouths of capillaries leads to a weakening of such destructive factors as the hydrostatic pressure of unfrozen water and the crystallization pressure of the ice. As a result, frost resistance grows.

It should be noted that, as can be seen from figure 2, hydrophobization practically does not change the moisture content of concrete samples at $\varphi = 0.3$. According to [10], the value of $U_\varphi$ at $\varphi = 0.3$ characterizes the capacity of a monolayer of water molecules on the solid phase surface. Its immutability shows that hydrophobization does not affect the formation of the microstructure of autoclaved cellular concrete. The addition of a water repellent affects only the meso- and macro-structure of concrete, because of which the contact angle of wetting liquid with the solid phase surfaces changes. This effect should increase with an increasing concentration of water repellent.

The existence of near-wall water layers with a changed structure can also determine the mechanism of destruction, which consists in the appearance of additional tensile stresses during the formation of ice in the capillaries. This can lead to the breaking of hydrogen bonds between the near-ice and near-wall films. The hydrophobization of the surface, causing a decrease in the near-wall film and the specified tensile stresses, should contribute to an increase in frost resistance.

With an increasing concentration of water repellent between 2...3%, the frost resistance of cellular concrete decreases. To explain this, the authors proposed a mechanism based on the wedging of micropores and recesses on the surface of large pores with water repellent molecules.

The effect of disjoining pressure [16, 17] is known, which consists in increasing the potential energy of molecules in a thin film between two interphase surfaces. If the film thickness is less than or equal to two radiuses of action of intermolecular forces, then the excess energy of the molecules in it is not fully compensated. Molecules of the film draw molecules with lower energy from the environment, which leads to the appearance of disjoining pressure. In cellular concrete, wedging can be carried out by water repellent molecules in recesses on the surface of large pores. Water repellent molecules interact with water molecules, leading to the irreversibility of the wedging. The interaction of water repellent molecules and water in micropores can cause an increase in their size.

Thus, the wedging effect should determine an increase in the effective radius of the pores involved in moisture transfer, and an increase in the number of active adsorption centres on the surface of the solid phase of cellular concrete. This leads to an increase in equilibrium moisture content, moisture diffusion coefficient and open porosity of the specimens at a concentration of water repellent 2...3%. An increase in the moisture-transporting and water-holding characteristics of the material indicates a decrease in the relative number of closed pores and growth in the thickness of the near-wall layers of water. As a result, the factors discussed above that contribute to frost destruction of concrete are strengthened, and frost resistance decreases.

With a water repellent content of over 3%, the frost resistance of cellular concrete specimens is determined by the competing effects of increasing adsorption and decreasing porosity and mass transfer coefficient, which prevails. The decrease in moisture transfer characteristics is apparently because at such concentrations the hydrophobic effect becomes so significant that the penetration of water into the mouth of the pores is difficult.

A decrease in adsorption at a relative humidity of 0.6 and a water repellent concentration of over 5% can be explained as follows. With a decrease in vapour pressure during capillary condensation, following the Kelvin equation, only pores of small radius are filled with water. And at high concentrations of water repellent, such pores can be clogged by its molecules.

Starting with a concentration of water repellent of 3% frost resistance of cellular concrete increases. However, the amount of water repellent additive that can be incorporated into concrete is limited by a decrease in its strength because of increased air entrainment into the concrete mixture, and worsening hydration and hydrothermal synthesis conditions. Moreover, in the process of autoclave processing, recrystallization and enlargement of new formations occur [14], which causes a violation of the structure.

## 4. Conclusions

The dependences of the moisture diffusion coefficient, equilibrium moisture content at various values of relative air humidity and porosity of autoclaved cellular concrete specimens on the concentration of water-repellent additives were studied. A comparison of the results of these measurements with the data on frost resistance, got by the direct method according to the current standards, have been made. It is shown, that moisture-transporting and water-holding characteristics can be used to assign concrete compositions that are optimal in frost resistance and strength. The mechanisms of the influence of hydrophobization on the frost resistance of cellular concrete are considered.

It was shown that at additive concentrations of 0...2% the increase in frost resistance is because of a growth in the number of closed reserve pores of small diameter, a decrease in tensile stresses in the capillaries and ice formation in their mouths. The drop in frost resistance in the concentration range of 2...3% is associated with an increase in open porosity due to the wedging effect of the water repellent molecules in the recesses on the surface of large pores. At additive concentrations above 3%, frost resistance increases because of a significant hydrophobic effect that impedes the penetration of water into the pore mouths. At the same time, the strength of cellular concrete samples decreases due to increased air entrainment into the concrete mixture and deterioration of hydration and hydrothermal synthesis conditions. It was found that the concrete composition containing 2% of water repellent additive is optimal in frost resistance and strength.

## References

[1] van Boggelen W and van Boggelen J 2018 Sustainable building solutions with new generation autoclaved aerated concrete panel applications *Ce Papers* **2** 513-525 https://doi.org/10.1002/cepa.825
[2] Paleari M and Miliani A 2018 The sustainability of wall solutions: Life Cycle Assessment (LCA) of different solutions for external closures *Ce Papers* **2** 489-94 https://doi.org/10.1002/cepa.879
[3] Zapotoczna-Sytek G and Balkovic S 2013 *Autoklawizowany beton komórkowy. Technologia – Właściwości – Zastosowanie* (Warszawa: Naukowe PWN)
[4] Lapovska S D and Voloshyna T M 2004 The efficient building material for warm, single-layer walling *Build. Mater. and Prod.* **5** 11-13
[5] Lapovskaya S D 2012 To the question of the influence of operational humidity on the properties of autoclaved cellular concrete *Build. Mater. and Prod.* **3** 17-21
[6] Powers T C and Helmuth R A 1953 Theory of volume changes in hardened portland cement paste during freezing *Proc. Highw. Res. Board* **32** 285-297
[7] Udachkin I B, Lapovskaya S D, Yuskovich V I and Semididko A O 1986 Improving the quality of cellular concrete by controlling the process of porosity *Rheology of Concrete Mixtures and Its Tech. Tasks* (Riga) pp 190-191
[8] Bazhenov Yu M 1987 *Concrete Technology* (Moscow: Vysshaya Shkola)
[9] Soloviev V I 1990 *Concrete with Water-Repellent Additives* (Alma-Ata: Nauka)
[10] Gregg S J and Sing K S W 1982 *Adsorption, Surface Area and Porosity* (London: Academic Press)
[11] Kazanskii V M and Klapchenko V I 1981 Method for measuring the coefficient of moisture diffusion in dispersed bodies according to the kinetics of capillary impregnation *Indust. Heat Eng.* **3(№5)** 92-5
[12] Ramachandran V S, Feldman R F and Beaudoin J J 1981 *Concrete Science: Treatise on Current Research* (London: Heyden)
[13] Neville A M and Brooks J J 2010 *Concrete Technology* (Harlow: Prentice Hall)
[14] Lapovska S and Demchenko T 2018 Influence of siliconorganic hydrophobizing additives on the durability of autoclaved aerated concrete *Build. Mater. and Prod.* **97** (1-2) 50-53
[15] Powers T C 1945 A working hypothesis for further studied on frost resistance of concrete *Proc. Am. Concrete Inst.* **41** 245-272
[16] Derjaguin B V and Churaev N V 1976 The definition of disjoining pressure, its importance in the equilibrium and flow of thin films *Colloid J. USSR* **38** 402-410
[17] Adamson A and Gast A 1997 *Physical Chemistry of Surfaces* (John Wiley and Sons Inc.)

IOP Conf. Series: Materials Science and Engineering **907** (2020) 012041 doi:10.1088/1757-899X/907/1/012041

# A source of thermal energy for roasting phosphogypsum – incinerators flue gas

**V I Vinnichenko[1], O P Krot[2], O Y Krot[1]**

[1]Department of mechanization of construction processes, Kharkiv National University of Civil Engineering and Architecture, Sumskaya str., 40, Kharkiv, Ukraine
[2]Department of safety life and environmental engineering, Kharkiv National University of Civil Engineering and Architecture, Sumskaya str., 40, Kharkiv, Ukraine

vvinnichenko@ukr.net

**Abstract.** The existing technologies for phosphogypsum utilization during its handling into building materials have been analyzed. Functioning equipment for joint utilization of municipal solid waste and phosphogypsum has been presented. An additional product of the installation, in addition to electrical and thermal energy, is gypsum. The waste gases from the incinerator are a heat source for roasting phosphogypsum. Fuel for the handling of phosphogypsum into gypsum is not used. A methodology for determining the specific amount of gases from a waste incinerator required for roasting phosphogypsum has been described in the article. The quantitative indicators for the reduction of polluting gases during the joint utilization of two waste types compared with operating independently waste treatment have been presented.

## 1. Introduction

Phosphogypsum is one of the most multi-tonnage solid wastes of mineral fertiliser manufacturing. There are four types of phosphogypsum in Ukraine: phosphogypsum from apatite concentrate has a shelf life 10-30 years in the tailings; phosphogypsum from phosphorites on the dump with a shelf life of less than ten years; phosphogypsum from phosphorites that was recently formed and dump phosphogypsum from uranium-containing phosphorites obtained. Millions of tons of phosphogypsum accumulated around the world at the moment. The phosphogypsum chemical composition is mainly represented by calcium sulfate dihydrate. This composition determines its using instead of natural gypsum. Scientists around the world have conducted studies that confirm the possibility of obtaining building materials and products from phosphogypsum. But the volume of phosphogypsum is processed into building materials is less than ten percent from total production. The mass phosphogypsum utilization instead of natural gypsum stone is constrained by the lack of cost-effective technological processes for processing these wastes [1-2]. This is explained by the significant consumption of thermal energy for the treatment of phosphogypsum in comparison with natural gypsum [3]. An increase of combustible fuel consumption has often been accompanied by increasing the greenhouse gas emissions, such as carbon dioxide and water vapor.

Technological lines are equipped with a large number of machines, which increases the technology intensity and therefore capital and operating costs. Many different types of technologies for the recycling industrial waste phosphogypsum have been patented in the world [4, 5]. But all of them are economically and environmentally pointless because some proposed technologies have high energy intensity. One of the most pressing tasks of modern society is the search for ways to reduce emissions of gaseous pollutants into the atmosphere [6-9].

During the experimental studies, it was found that the rational temperature of the gas stream for roasting phosphogypsum is a temperature of 350°C [10]. A hypothesis has been got that the heat flow of the gases from the incinerator should be sent to the roasting phosphogypsum on this basis. A method has been proposed for using the heat of exhaust gases from waste incineration for roasting phosphogypsum as a result [10]. This excludes a need to use fuel. Since fuel is not used for the processing of phosphogypsum into building materials, there are no products of fuel combustion. This means that gaseous pollutants generated by fuel combustion did not pollute the atmosphere.

The purpose of this article is to determine the environmental parameters for roasting phosphogypsum using heat from the household waste incineration.

Operating principle of the installation is as follows. The finely ground phosphogypsum 1 (fig.1) through a hopper with a feeder and a conveyor system is fed to the roller crusher 2. After that, phosphogypsum is sent by a conveyor 3 in a drum-dryer 4. The dried phosphogypsum is sent to the elevator 5, then - con-veyor for dry phosphogypsum 7. Wet gases 6 are released into the atmosphere. Dry phosphogypsum 8 is fed to a mill-dryer 9. In the proposed version, a bowl-roll mill-dryer is used, developed at the Department of Mechanization of Construction Processes of Kharkiv National University of Civil Engineering and Architecture. Directly above the mill is a through separator. The mill-dryer 9 is equipped with an external combustion chamber 10.

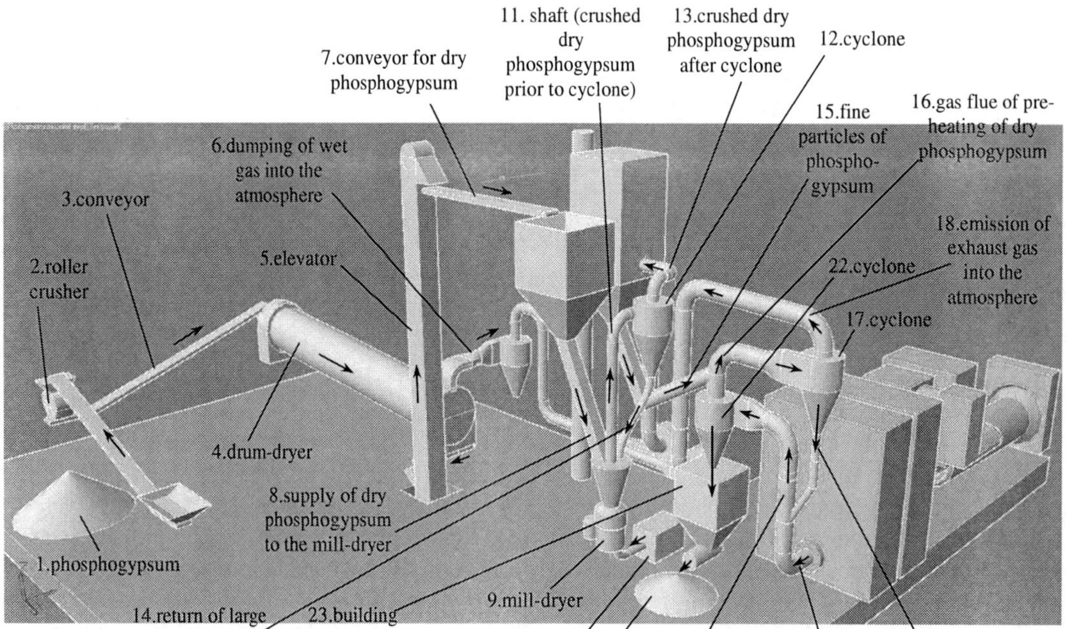


**Figure 1.** Complex of roasting phosphogypsum with the use of heat from waste incineration.

The fine particles of the material, passing the separator, flow up-ward into the shaft of the dryer 11, where effective heat exchange takes place in the suspended state. Shaft 11 is also used for further particle fractionation. From the shaft 11, the flow of the gas with small particles enters the centrifugal cyclone 12, in which the particles are separated from the gas that has received physical moisture from the material. The wet heat carrier through the flue gas line 13, the dust removal system and the smoke exhauster (not shown in the diagram) is discharged into the atmosphere. The dried material, precipitated, is collected in a bunker with a shutter at the bottom. By means of a screw feeder 15, material is fed into the lower part of the gas flue of pre-heating of the dry phosphogypsum 16. The material is picked up by a stream of hot gases that exit the cyclone 22 and is carried upward into the cy-clone 17. At this time the material is heated and dried. In the cy-clone 17, the particles are separated from the wet gases. Gases are discharged through the pipe 18 into the atmosphere, and the hot and absolutely dry particles of phosphogypsum from the cyclone 17 are drained into the dehydrator 20. Here is the transformation of phosphogypsum into hemi water gypsum in a suspended condition in the flow of hot gas. Water is released from phosphogypsum.

During the dehydration reaction, the water molecules are broken with the ions $Ca_2^+$ and $SO_4^{2-}$ and the water is removed from the crystal lattice in the form of superheated steam. As a result, β-modification of calcium sulfate hemihydrate is formed. The chemical reaction of this process (dehydration):

$$CaSO_4{\bullet}2H_2O \rightarrow \beta\text{-}CaSO_4{\bullet}0{,}5H_2O+1{,}5H_2O\uparrow$$

In cyclone 22, the flow is separated: particles of hemi water gypsum down and collected in hopper 23. Wet hot gases escape up-ward from cyclone 22 to tube 16. The main amount of heat that is used in technology is realized in dehydrator 20 (Figure 1). According to the proposed scheme, this heat is supplied from a municipal waste incinerator (21 on Fig. 1).

## 2. Quantity determination of exhaust gases from a waste incineration required for roasting phosphogypsum

At the first stage, we must calculate the amount of gas that is produced in the waste incineration (at a specific temperature) will be required to receivel kilogram of the semi-hydrate gypsum ($CaSO_4{\cdot}0{,}5H_2O$- calcium sulphate hemihydrate) from phosphogypsum (calcium sulphate dihydrate - $CaSO_4{\cdot}2H_2O$). For this we have composed the thermal balance.

*2.1. Thermal balance*: **Heat input** (kJ/kg of gypsum).
*With gases from the incinerator:*

$$q_1 = V \cdot C \cdot t \tag{1}$$

where, V - the volume of gases emanating from the incinerator, $m^3$ of gas/kg of gypsum; C - heat capacity of gases; t - gas temperature.

With calcium sulphate dihydrate - $CaSO_4{\cdot}2H_2O$:
– a dry raw materials consumption, taking into account the substance losses, kg/kg:

$$G = \frac{G^T 100}{100 - a}; \tag{2}$$

*a* - fraction of dust, which is lost with exhaust gases, %; $G^T$ - theoretical dry raw materials consumption, *kg/kg*:

$$G^T = \frac{100}{100 - LIM} \tag{3}$$

L*IM* - loss on roasting of raw materials, %;
– wet raw materials consumption, *kg/kg*:

$$G_W = \frac{G \cdot 100}{100 - W}; \tag{4}$$

*W* - initial moisture content of raw materials, %;
– yield of physical moisture, *kg/kg*:

IOP Conf. Series: Materials Science and Engineering **907** (2020) 012041 doi:10.1088/1757-899X/907/1/012041

$$G_W^f = G_W - G \tag{5}$$

$$q_2 = \left(GC_C + G_W^f C_{H_2O}\right)t_C; \tag{6}$$

Total heat input:

$$\Sigma q = q_1 + q_2 \tag{7}$$

Flow structure for calculating thermal balance roasting phosphogypsum with the use of heat from waste incineration is presented in the figure 2.

**Heat Consumption** (kJ/kg gypsum)

– with the evaporation of physical moisture:

$$Q_1 = G_W^f\left(I_S + C_S t_S\right) \tag{8}$$

$I_S$ - enthalpy of steam; $C_S$ - heat capacity of steam; $t_S$ - steam temperature.

– with $CaSO_4 \cdot 0{,}5H_2O$ - calcium sulphate hemihydrate:

$$Q_2 = C_g t_g \tag{9}$$

$C_g$ - heat capacity of gypsum; $t_g$ - gypsum temperature.

– with the formation of semi- hydrate gypsum:

$$Q_3 = \frac{\Delta H^\circ_{298}}{M_{CaSO_4 \cdot 0{,}5H_2O}} \tag{10}$$

$\Delta H^\circ_{298}$ - reaction enthalpy, J/mol;

$\Delta H^\circ_{298} = \Sigma\Delta H^\circ_{298}$ reaction products – $\Sigma\Delta H^\circ_{298}$ starting materials.

– with exhaust gases:

$$Q_4 = V \cdot C_w \cdot t_w + V_v \cdot C_v \cdot t_w \tag{11}$$

$C_w$ - the heat capacity of the exhaust gases from the roasting phosphogypsum installation; $t_w$ - temperature of the exhaust gases from the roasting phosphogypsum installation.

– yield of chemically bound moisture, kg/kg:

$$G_W^X = G_C \frac{M_{1{,}5H_2O}\left(CaSO_4 \cdot 2H_2O\right)}{M_{CaSO_4 \cdot 2H_2O} 100} \tag{12}$$

$M_{1{,}5H_2O}$, $M_{CaSO_4 \cdot 2H_2O}$ - respectively molecular weights $1{,}5H_2O$ and $CaSO_4 \cdot 2H_2O$.

$CaSO_4 \cdot 2H_2O$, the dihydrate gypsum content in raw materials, %.

– total water vapour yield:

$$G_v = G_W^f + G_W^X \tag{13}$$

$$V_v = \frac{G_v}{\rho_v} \tag{14}$$

$\rho_V$ - the density of water vapour, kg/m$^3$.

– heat loss to the environment

$$Q_5 = (0{,}05 \ldots 0{,}1)q_1 \tag{15}$$

Total heat consumption:

$$\Sigma Q = Q_1 + Q_2 + Q_3 + Q_4 + Q_5 \tag{16}$$

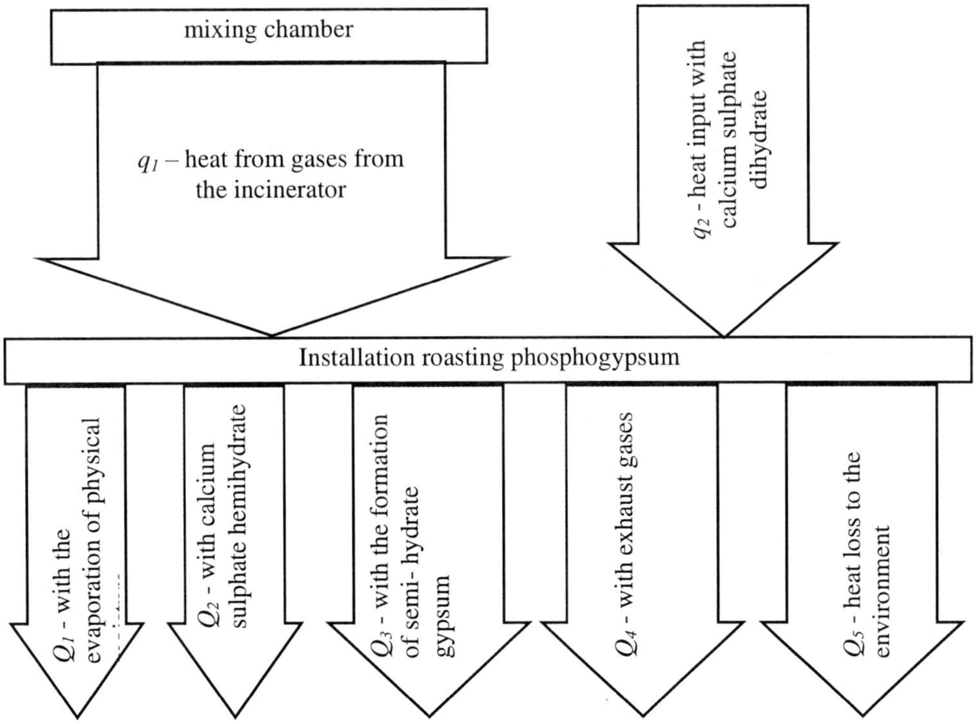


**Figure 2.** Flow structure for calculating thermal balance roasting phosphogypsum with the use of heat from waste incineration.

*2.2. Thermal Balance Equation*

$$\Sigma q = \Sigma Q \quad (17)$$

The volume of gases emanating from the incinerator, *$m^3$ of gas / kg of gypsum (V)* which must be submitted for roasting phosphogypsum is calculated from the thermal balance equation.

Our calculations showed that the 1.8 $m^3$ of gases per 1 kg of building gypsum is necessary for roasting phosphogypsum. When the installation productivity for the production of building gypsum from phosphogypsum is 10 t / h, the required hot gases volume from the incinerator is calculated:

$$10000 \cdot 1.8 = 18000 \text{ m}^3\text{/hour}$$

## 3. Results. Determination of environmental parameters

When roasting gypsum (using natural gas fuel), the total exhaust gas output [11, 12] during fuel combustion per 1 kg of $\beta$-semihydrate gypsum:

– carbon dioxide:

$$CO_2 = \vartheta^T_{CO_2} \cdot \rho_{CO_2} \cdot B_f \quad (18)$$

$$CO_2 = 1.036 \cdot 1.977 \cdot 0.040 = 0.082 \text{ , kg/kg}$$

$B_f$ - fuel consumption for roasting;

– water vapour

$$H_2O = \vartheta^T_{H_2O} \cdot \rho_{H_2O} \cdot B_f \quad (19)$$

$$H_2O = 2.01 \cdot 0.805 \cdot 0.04 = 0.065, \text{ kg/kg}$$

– oxygen

$$O_2 = \vartheta^T_{O_2} \cdot \rho_{O_2} \cdot B_f \quad (20)$$

$$O_2 = 1.96 \cdot 1.429 \cdot 0.04 = 0.11, \text{ kg/kg}$$

–nitrogen

$$N_2 = \vartheta^T_{N_2} \cdot \rho_{N_2} \cdot B_f \qquad (21)$$

$N_2 = 12.29 \cdot 1.251 \cdot 0.04 = 0.61$, kg/kg

Total gas output during waste incineration:

$G_T = 0.082 + 0.065 + 0.11 + 0.61 = 0.87$, kg/kg

0.87 kg of gas per 1 kg of gypsum – it is the amount of gas that would be emitted into the atmosphere if phosphogypsum was fired by the thermal energy of the fuel burned.

Since the gases of the incinerator are the source of thermal energy, the fuel is not used. This means that the products of combustion of this fuel do not pollute the atmosphere.

Thus, the roasting of *β*-semi-hydrate gypsum with the heat of the gases from the incinerator allows to reduce atmospheric emissions by approximately 870 kg of gases in the production of 1 ton of β-semi-hydrate gypsum.

If the line capacity is 10 tons of gypsum/hour, then the reduction in the level of air pollution by combustion products per year will be

$10 \cdot 24 \cdot 365 \cdot 870 = 76212000$, tons

In other words, the production of building gypsum from phosphogypsum without fuel costs, but using the thermal energy of gases from the solid waste incineration, will reduce emissions of polluting gases into the atmosphere by approximately 76 million tons per year.

## 4. Conclusions

The technological process of phosphogypsum dehydration consists of a several technological operations, which is aimed to changing the source material in order to give it the properties of a composite material. In the technology of gypsum binders that is intended for construction, the following main stages of the technological process are mainly used: crushing, drying, grinding or joint drying and grinding and dehydration. The recommended workflow depends on the physical properties of the raw materials, the type of equipment needed for dehydration and the quality requirements of the final product.

The method of using the waste gases of the incinerator as a source of thermal energy for roasting phosphogypsum is substantiated. It is shown that the combined disposal of solid waste and phosphogypsum will reduce greenhouse gases and other pollutants into the atmosphere.

## 5. References

[1] Ivanitskiy V V, Klykova L Y, Baykanov Z P, Pletnev V P 1983 Gypsum binders with increased strength and water resistance from phosphogypsum *Building materials* **9** 12–14

[2] Ivashchenko T H, Bondar O I, Novoselska L P and Vinnychenko V I 2017 *Phosphogypsum: environmentally friendly ways of disposal and use* (Kherson: Derzh. ekolohichna akademiya pislyadyplomnoyi osvity ta upravlinnya) p 220

[3] Gorakh S B, Kumthekar M B, Landage A B 2016 A Review of Effective Utilization of Waste Phosphogypsum as a Building Material *International Journal of Engineering Research* **5** pp 277–280

[4] Lutskiy D S, Litvinova T E, A S Ignatovich, Fialkovskiy I 2018 Complex Processing of Phosphogypsum - a Way of Recycling Dumps with Reception of Commodity Production of Wide Application *J. Ecol. Eng.* **19(2)** 221–225

[5] Grabas K, Pawełczyk A, Stręk W, Szełęg E, Stręk S 2019 Study on the Properties of Waste Apatite Phosphogypsum as a Raw Material of Prospective Applications *Waste Biomass Valor* **10** 3143–3155

[6] Integrated Pollution Prevention and Control: Reference Document on the Best Available. Techniques for Waste Incineration 2016 European commission p 602

[7] Zhang J, Smith K R, Ma Y, Ye S Jiang F 2000 Greenhouse gases and other airborne pollutants from household stoves in China: a database for emission factors *Atmospheric Environment* **34(26)** pp 4537–4549
[8] Stocker B D, Roth R, Joos F, Spahni Ro, Steinacher M, Zaehle S, Bouwman L, Xu-Ri and Prentice I C Multiple greenhouse-gas feedbacks from the land biosphere under future climate change scenarios 2013 *Nature Climate Change* **3** pp 666-672
[9] Review Practice Guidance for Review of National Communications and Biennial Reports of Developed 2018 *Country Parties Katowice climate change conference (COP 24)*
[10] Krot O P, Vinnichenko V I 2018 The use of heat from the incineration of municipal solid waste for processing of phosphogypsum *International Journal of Engineering and Technology (UAE)* Vol **7** No 4.3
[11] Vinnychenko V I, Ivashchenko T H, Ryazanov O M 2017 *Energy saving and ecology of building materials production* (Kherson: OLDI-PLYUS) p 211
[12] Duda W H 1975 *Cement data book* (Bauverlag GmbH, Wiesbaden and Berlin) p 444

IOP Conf. Series: Materials Science and Engineering **907** (2020) 012042 doi:10.1088/1757-899X/907/1/012042

# Special Features of Using Universal Technological Sets of Small-sized Equipment for Renovation of Buildings

**I A Emeljanova [1], D O Chayka [1], E S Lebedeva [2], D Yu Subota [1]**

[1]Department of mechanization of construction processes, Kharkov National University of Civil Engineering and Architecture, 40, Sumska str., Kharkov, 61000, Ukraine
[3]Department of Life Safety and Engineering Ecology, Kharkov National University of Civil Engineering and Architecture, 40, Sumska str., Kharkov, 61000, Ukraine

emeljanova-inga@ukr.net

**Abstract.** Nowadays, old utility lines are the most critical issue of society. Emergency condition of these utility lines results in their destruction and pollution of ecosystem and imposes threat to life of people residing within risk area. This article offers variants of mobile universal sets of small-sized equipment that allows restoration of destructed and emergency buildings through wet-mix shotcreting process. The universal nature of these sets of equipment allows performance of entire scope of renovation works from preparation of damaged surfaces to applying anti-corrosive coatings. Upon the use of such sets of equipment, it is possible to prepare and lay construction mixtures of different purpose directly at the place of works. Combination of process operation in time enhances the efficiency of equipment and reduces power consumption.

## 1. Introduction

Ukraine faces urgent concern of renovation and reconstruction of emergency construction facilities. In Ukraine, the problem of repair and reconstruction of objects with increased environmental and social risk is acute. The most urgent issue is old sewage systems of cities. Their worn state due to corrosion is evidenced by incident rate. On average, there are 300 accidents per 100 km of networks per year, resulting in pollution and destruction of urban infrastructure. The same holds for emergency bridges. Among 6 thousand bridges, maintained by Ukravtodor, one thousand in is emergence state and 2.5 thousand require total renovation. Problematic facilities of high social threat include buildings on the slopes and coastal zones etc. Currently, there are no uniform approaches and sets of equipment for quick response in emergency cases. This article describes sets of small-sized equipment for carrying out works at emergency facilities.

## 2. Review of the latest discoveries and publications

The issue of destruction of buildings of high human-induced risk is critical not only in Ukraine but also abroad. Works [1-3] are devoted to study of the behavior and methods of renovation of concrete sewerage conduits subject to wear due to biogenic sulphuric acid. Authors consider that shotcreting of surfaces with concrete mixtures and application of anti-corrosion polymeric coatings or the use of acid-resistant pipes (spools) with further injection of annular space are major methods of renovation. A considerable attention is paid to study of the design of hydraulic engineering works and methods of their strengthening [4]. The most commonly used method for casing of cast reinforced concrete under complicated conditions of construction is shotcreting technology which is widely used in the world due to its high efficiency [5]. Research papers [6, 7] are devoted to study of design of concrete mixers and improvement of the quality of concrete mixtures [6, 7]. Equipment for transportation of concrete mixtures is studied in research papers [8, 9]. It should be noted that there are many equipment for preparation of concrete mixtures, their transportation, casting, including is shotcreting. This equipment is manufactured by such leaders as Graco (USA), ASF Thomas (Germany), Watson-Marlow (UK), Welco (Japan), Brightwell (Canada), Tornado (Ukraine), StroyMechanics (Russia), Kyokuto (Japan), Monolithic Equipment (USA). Machines specified in review are considered as individual equipment, which may be included into composition of multi-functional process set. Studies devoted to creation of equipment sets that allow entire process cycle of renovation of emergency facilities is little known in foreign practice.

Specifics and basis for creation of sets of small-sized equipment for construction site are represented in research paper [10]. Authors consider the use of this equipment allows manufacture of multi-component modified mixtures at construction sites, which is very important in view of complexity of centralized delivery of special repair mortars from plants as this the case in conventional construction. The results of the analysis suggest that the creation of multifunctional mobile sets of small-sized equipment for carrying out renovation works is an urgent task, which is aimed at solving the problem of reconstruction of buildings of high human-induced risk.

## 3. Main part

The creation of high-tech small-sized equipment for renovation of emergency buildings is a very critical issue at this time. These facilities were built on average 40-50 years ago, operated in aggressive environment and are in a pre-emergency or emergency condition, which poses an increased risk to the environment and human life. The destruction of sewer pipelines is becoming more and more frequent, which results in penetration of impurities into groundwater, destruction the pavement and sinking of foundations. Collapses of bridge structures and other structures are less frequent. This requires a comprehensive approach, including the involvement of efficient equipment and qualified personnel.

Modern sets for small-sized equipment have modular design, which makes them flexible and suitable for using for wide range of purposes. The modular principle allows making special or universal sets of equipment from individual machine modules. Sets of small-sized equipment for certain types of work could be used according to the following schemes:

– for shotcreting works with the use of wet-mix shotcreting process: "metering unit → concrete mixer, operating in cascade mode → double-piston and flexible hose pump → nozzle with ring tip→ concreted surface".

– for off-form concreting by wet-mix shotcreting method with the use of fiber-concrete mixtures for manufacture of reinforced-concrete structures of complex geometric shape: "automatic fiber cutter → metering unit for complex concrete mixtures → three-shaft concrete mixer → piston-free hose concrete pump → shotcrete nozzle with ring tip → frame of concreted structure with embedded parts"

– for preparation of concrete mixtures and construction mixtures of different purposes: "metering unit for complex concrete mixtures → three-shaft or gravity concrete mixer".

Proposed sets of equipment will allow the following in case of their full mechanization and automation:

- creation of conditions for quick preparation to repair works in case of accidents at buildings due to mobility of equipment;

- disassembly of destructed elements of structures and preparation of surfaces that are subject to repair by means of additional tools on the basis of sets;

- preparation of construction mixtures by special recipes, including fast-curing mixtures, directly at the construction site, when they cannot be manufactured at the factory;

- minimization of dust formation from dry components upon wet-mix shotcreting at small scopes of work instead of dry, which creates a favorable working environment for builders and reduces the risk of their occupational diseases;

- application of protective anti-corrosion layers, in spite of basic concrete mixtures by means of nozzle with set of special nozzles;
- increase of efficiency of equipment up to 30% and reduction in power consumption up to 15-20% due to combination of all operation of process cycle in time;
- simplification of staff training and more qualified performance of works due to operation of one plant;
- creation of conditions for obtaining of high quality protective coatings, which have significant effect on durability and protection of support structures against corrosion.

Principal diagram of one of universal sets of small-sized equipment for renovation works at construction site is shown on figure 1.

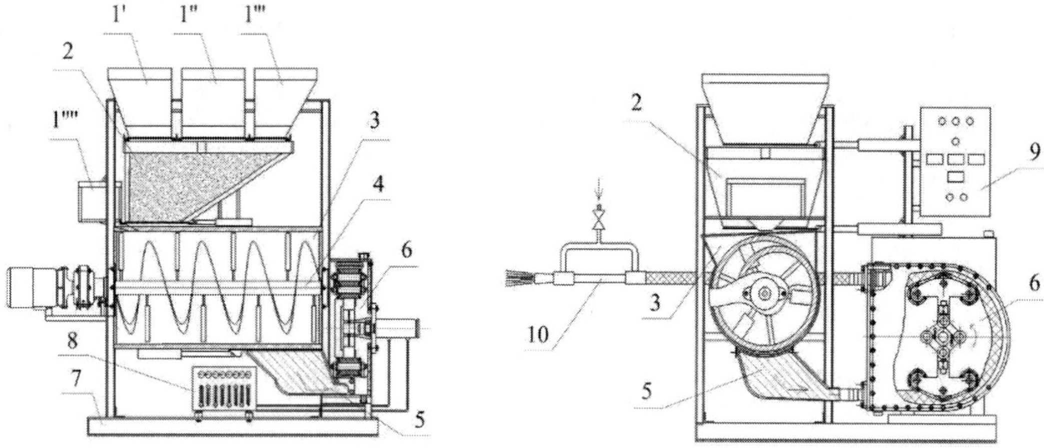


**Figure 1.** Modern universal sets of small-sized equipment for construction site.
1 is metering unit for multi-component concrete mixtures: 1' – gravel, 1" – sand, 1'" – cement, 1""– water; 2 is intermediate hopper; 3 is concrete mixer; 4 is ribbon shaft of mixer; 5 is charging chute; 6 is universal hose concrete pump; 7 is frame; 8 is hydraulic distribution unit ; 9 is automatic switchboard; 10 is shotcreting nozzle with ring tip

Machine module and component assemblies that ensure performance of entire scope of works at construction site upon construction of the buildings of different purpose are mounted on general frame. This set allows performance of all process operations, including batching of components for preparation of construction mixtures, their transportation and sputtering by wet shotcreting upon casting of monolithic reinforced concrete. If necessary, the set allows plastering works. Metering unit 1 is fitted with three hoppers with slide gate valves for: gravel (1'), sand (1"), concrete (1'"). Batched components from respective hoppers covered with slide gate valves enter into intermediate hopper 2, from which certain volume and weight of these components enters the housing of forced mixer 3. Concrete mixer is fitted with actuating element in the form of bonnet shaft 4 that ensures preparation of construction mixture with high degree of uniformity. Finished mixtures enters charging chute 5, from which it is transferred to shotcreting nozzle with ring tip through pipeline by means of universal hose concrete pump 6. Shotcreting process is performed by wet-mix method that ensures casting monolithic reinforced concrete provided that there is a reinforcing frame that is under construction. All these process operations are combined in time, which results in enhancement of efficiency of construction works by 15-20. This set could be used in cyclic and continuous concreting. Hydraulic device with electric control and possibility for fully automated PC-based operation process is one more advantage of this set.

Universal, pistonless, hose concrete pump or concrete forced mixer may be used as basic machine of set small-sized equipment. In this case, the performance of process set will be determined by performance of basic machine.

The process of process set is determined by performance of universal hose concrete pump according to the following relations [11]:

$$P_{concrete\ pump} = 3600 \cdot S_{hose} \cdot v_{avg} \cdot k_1 \cdot k_2 \cdot k_3 \qquad (1)$$

where $S_{hose}$ is cross section of hose in concrete pump, m$^2$; $v_{avg}$ is average flow of concrete mixture along flexible hose, m/sec; $k_1$ is coefficient that considers gradual increase in the force created by the clamping rollers of the rotor, which press hose in the working part of the pump; $k_2$ is coefficient that considers reliability of horse part of pump , considering tension and ultimate tensile strength; $k_3$ is coefficient that considers conditions for feeding mixture by concrete pump through flexible hose , considering its properties.

If basic machine is concrete mixer, the performance of set is determined as:

$$P_{concretemixer} = 3600(z_1 + z_2)V_{avg1} \cdot в \cdot \cos\alpha(R - r)K_{vol} \qquad (2)$$

where $z_1$ is a number of tuning of ribbon screw on mixer shaft; $z_2$ is number of blades of shaft; $V_{avg1}$ is an average rotation speed of mixer shaft, m/sec; $в=(в_1+в_2)$ is blade width and ribbon width, respectively; $R,r$ are maximum and minimum radiuses of ribbon shaft, respectively; $K_{vol}$ is volumetric efficiency of concrete mixer.

One of conditions for efficient use of above-mentioned set of equipment in wet-mix shotcreting process is maximum possible use of working space of basic mixer for preparation of necessary mixture, considering certain conditions for repair of building directly at construction site.

Working volume of mixer body is estimated by coefficient $K_{vol.}$

Volume coefficient $K_{vol}$ is determined, considering that the performance of concrete pump and concrete mixer modules should be equal in process set. This this case $P_{concrete\ pump} = P_{concrete\ mixer}$.

In this case, volume coefficient for offered process set is determined as follows:

- *For cyclic operation*

$$K_{vol.1} = \frac{S_{hose} \cdot v_{avg1} \cdot t_ц \cdot k_1 \cdot k_2 \cdot k_3}{V} \qquad (3)$$

where $S_{hose}$ is cross section of hose in concrete pump; $v_{avg1}$ is average flow of concrete mixture along flexible hose in concrete pump body; $k_1$, $k_2$, $k_3$ are coefficients that consider availability of pressure pulses upon discharging of mixture by concrete pump, conditions of mixture feeding, reliability of operation of concrete pump; V is volume of working part of mixer body (charging hopper); $t_c$ is time of operation cycle of concrete mixer;

- *For continuous operation*

$$K_{vol.2} = \frac{S_{avg} \cdot v_{avg1} \cdot t_ц \cdot k_1 \cdot k_2 \cdot k_3}{v_{avg2} \cdot (в_1 + в_2)(z_1 + z_2)\cos\alpha(R - r)} \qquad (4)$$

where $v_{avg\,2}$ is an average rotation speed of working element of mixer of ribbon shaft; $z_1$, $z_2$ is number of turning of shaft ribbon and blade, respectively; $(в_1+в_2)$ is a width of working part of ribbon shaft turn ($в_1$) and shaft blade ($в_2$); $R,r$ is maximum and minimum radiuses of ribbon shaft, respectively; $\alpha=\alpha_1=\alpha_2$ are installation angles of ribbon turns and blades on shaft against its horizontal axis.

IOP Conf. Series: Materials Science and Engineering **907** (2020) 012042 doi:10.1088/1757-899X/907/1/012042

Thus, volume coefficient of mixer depends on design parameters of modules of process set and their operation speeds, which should be considered upon development of process sets of small-sized equipment.

Total power consumption for transportation of concrete mixture through pipeline by means of hose concrete pump as basic machine are determined by the following dependency [12]:

$$P_{conc.pump} = k_s \cdot \left( \frac{\frac{3{,}45 \cdot \left( G_{conc.mix} \cdot k_{fric} + F_{press} \cdot \frac{k_{roll.fric.}}{r_{roll}} \right) \cdot n \cdot R_{avg}}{30 \cdot \eta_{avg}} + \frac{S_{hose} \cdot \Delta p \cdot v_{avg} \cdot k_{length}}{\eta_{\Gamma}}}{1000} \right), \quad (5)$$

where $k_s$ is safety factor; $G_{conc.mix}$ is a weight of concrete mixture affected by clamping rollers, H; $k_{fric}$ is a coefficient of friction due to constant effect between inner wall of flexible hose and concrete mixture during its movement; $F_{press}$ is a force of pressing rollers to outer surface of flexible hose in working space of pump, H; $k_{roll.fric.}$ is a coefficient of roller friction on hose surface, m; $r_{roll}$ is roller radius, m; n is rotor rotation rate, min$^{-1}$; $R_{avg}$ is an average distance between rotor axis and roller face, m; $\eta_{cp}$ is an efficiency of concrete pump; $S_{hose}$ is cross section on inner diameter of hose, located in body of concrete pump, m$^2$; $\Delta p$ is pressure differential at the ends of transport pipelines, Pa; $v_{avg}$ is an average speed of movement of concrete mixture through flexible hose, m/sec; $k_{length}$ is a coefficient that considers the length of transport pipeline ; $\eta_{\Gamma}$ is a coefficient of hydrodynamic loss in transport pipeline.

Upon the use of concrete mixer as basic machine, the performance of process set would be determined by the following dependency:

$$\mathrm{P}_{conc.mixer} = [\omega_{sh} \cdot \mathrm{M}_{sh} + F_{fric.} \cdot V_{abs} \cdot z_1 + \frac{c \cdot \rho_0 \cdot S_{avg}^{t} \cdot \omega_{sh}^{m} \cdot k_s \cdot z_{screw} \cdot (D^2 - D_{in.screw}^2)}{32 \cdot \pi^2 \cdot k_{\text{в}}} +$$

$$+ \frac{\pi \cdot c \cdot \rho_0 \cdot S_{avg} \cdot \omega_{\text{в}}^3 \cdot f_1 \cdot z_{screw} \cdot tg\alpha_{avg.screw} \cdot \sin \alpha_{avg.screw} (D_{in.screw}^5 - D_{out.screw}^5)}{80 \cdot k_{\text{в}}}] \frac{1}{1000 \cdot \eta_{shaft} \cdot \eta_{screw}}, \quad (6)$$

where $\omega_{sh}$ is angular rotation of blade shaft, c$^{-1}$; $\mathrm{M}_{\text{в}}$ is torque of blade shaft, N·m; $F_{fric}$ is friction force that arises upon movement of particle of concrete mixture along blade surface, H; $z_1$ is number of blades on shaft; $V_{abs}$ is absolute speed of movement of mixture particles along shaft blades, m/sec.; $c$ is coefficient of resistance to movement of blade upon mixing of concrete mixture in direction to its movement in circle; $S_{avg}$ is screw pitch along its average diameter, m; $z_{screw}$ is number of screw turns; $D_{in.screw}$, $D_{out.screw}$ is outer and inner diameter of screw, m; $\rho_0$ is average density of concrete mixture, kg/m$^3$; $f_1$ is a coefficient of friction of mixture upon movement of particles of concrete mixture along mixer body ; $\alpha_{avg.screw}$ is lifting angle of flight auger along average diameter, m; $k_{\text{в}}$ is return factor of mixture; $\eta_{shaft}, \eta_{screw}$ is efficiency of drive of blade shaft and ribbon screw.

Thus, process sets are assembled from machine modules, which, due to their mobility, allow renovations works with complete mechanization of separate operations and possibility for automated control of production process at small time consumption, depending on certain conditions of renovation works.

## 4. Conclusions

- Detailed analysis of emergency buildings within Ukraine was carried out.
- Using of modern construction technologies upon renovation of emergency buildings requires the development of respective equipment. For this purpose, new sets of equipment for renovation works based on modular principle were represented.
- Principal diagram of one of the variants for process set of small-sized equipment was proposed,

and its operation mode was described.

- Respective dependencies for calculation of basic parameters of operation of process sets were provided.
- Analytical dependencies for calculation of the basic parameters of technological sets operation are given.

**References**

[1] Wu L, Liu W V, Hu C 2018 Improving acid resistance of shotcrete for sewer tunnel rehabilitation in Edmonton *NASTT's No-Dig Show Conf.*

[2] Noeiaghaei T et al. 2017 Biogenic deterioration of concrete and its mitigation technologies *Construction and Building Materials* **149** 575-586.

[3] Goncharenko D F, Olejnik D J, Bondarenko D A 2014 The selection of design and technology solutions for inspection shafts construction at existing deep-laid water disposal networks *Water and Ecology* **57** (4) 59-68.

[4] Dams and other Hydraulic structures 2015 *Global asset protection services LLC, GAP.2.9.1* 1-5.

[5] Duc Thang Pham, Boris Bulgakov, Tang Van Lam 2016 Application of fine-grained shotcrete for the construction of underground tunnels *Vestnik MGSU: Research of building materials* **7** 81-90.

[6] Han D, Ferron R D 2016 Influence of high mixing intensity on rheology, hydration, and microstructure of fresh state cement paste *Cement and Concrete Research* **84** 95-106.

[7] Seán Moran 2017 Chapter 25 – mixers Process plant layout (second edition) 369-376.

[8] Riding K A, Vosahlik Ja, Feys D, Malone T, Lindquist W 2016 *Best practices for concrete pumping: A Report on Research* (K-TRAN: KSU-14-2. Desember. Kansas: Department of Transportation)

[9] Choi M, Ferraris C F, Martys N S, Van Bui K, Hamilton H R T, Lootens D 2015 *Research Needs to Advance Concrete Pumping Technology: NIST* Technical Note 1866

[10] Emeljanova I, Shevchenko V 2014 Technological sets of small-sized equipment for construction conditions *Proceedings of the International Scientific and Practical Internet Conference* (Ternopil: Krok) pp128-130

[11] Emeljanova I, Grigoryev A, Andrenko P, Chayka D 2018 Modeling of the Supplying Process of Building Mixture by Using of Peristaltic Concrete Pump *International Journal of Engineering Research in Africa* **38** 46-59.

[12] Kondrashchenko V I, Emelianova I A, Chaika D O 2019 Features of the Use of a Universal Hose Concrete Pump in the Construction Site. *IOP Conf. Series: Earth and Environmental Science* **272** 032107.

IOP Conf. Series: Materials Science and Engineering **907** (2020) 012043 doi:10.1088/1757-899X/907/1/012043

# The composite facing material for electromagnetic felds shielding

**V A Glyva[1], L O Levchenko[2], O V Panova[3], O M Tykhenko[4], M M Radomska[4]**

[1]Department of Civil and Industrial Safety, National Aviation University, Lubomyra Husara avenue, 1, Kyiv, 03058, Ukraine
[2]Department of Automation of projection of power processes and systems, National Technical University of Ukraine «Igor Sikorsky Kyiv Polytechnic Institute», Peremohy avenue, 37, Kyiv, 03056, Ukraine
[3]Department of Physics, Kyiv National University of Construction and Architecture, Povitroflotsky avenue, 31, Kyiv, 03037, Ukraine
[4]Department of Ecology, National Aviation University, Lubomyra Husara ave., 1, Kyiv, 03058, Ukraine

glyva.valentin@gmail.com

**Abstract.** The facing material was developed for the purpose of shielding man-made electromagnetic fields in production working and domestic conditions. The facing material was designed on the base of latex and fine iron ore concentrate as filler. The experimental data on shielding of electromagnetic fields of ultrahigh frequencies and magnetic field of industrial frequency were obtained for different material thicknesses and concentration of screening particles. The significant increase of the shielding coefficients was established and explained by the manifestation of the percolation effect - the increase of the material conductivity at the critical concentration of the conductive component. The calculation apparatus for determining the dependence on the shielding properties of the material on the concentration of shielding substance was proposed. The method of the screening substance critical concentration calculation is shown to enable designing materials of lower thickness, weight and self-cost.

## 1. Introduction

In the context of increasing anthropogenic impact on environment, much attention is paid to the protective properties of facing and decorative construction materials. Electromagnetic fields of anthropogenic origin are one of the most significant physical factors of adverse effects on humans. The most effective protection against their impact in buildings and structures is shielding. But the use of metal screens is not always possible. In many cases, it is technically non-feasible to mount the shield surface. In addition, these designs are complex, heavy and unaesthetic. In such conditions, especially for buildings that are already in operation, the most rational is the application of facing materials that prevent the penetration of electromagnetic fields of technogenic origin into residential, industrial buildings and separate premises. Such materials must have certain properties: sufficient efficiency in the critical frequency ranges of electromagnetic fields, applicability (ease of lining surfaces of different configurations, low mass and thickness), and reasonable cost. Traditional metallic shielding materials do not meet modern requirements, so the most promising materials for shielding electromagnetic fields of the wide frequency range are composite structures suitable for manufacturing electromagnetic screens with controlled protective properties.

## 2. Analysis of previous research

Much attention is currently paid to the development of composite materials for shielding electromagnetic fields. Most of them are designed for specific applications. In [1], the results of the development of the material with carbon nanotubes and graphene as shielding substances are presented. This material is intended for shielding electromagnetic fields only of ultra-high and high frequencies. In addition, it is expensive because of the high cost of the filler.

The same applies to the study [2], which shows the results of shielding the electromagnetic field with the frequency range of 8.2-12.4 GHz. The other paper describes the results of the development and properties study of the material for shielding electromagnetic fields in computers [3]. The use of composites based on carbon nanostructures is conditioned by their high efficiency in high-frequency spectrum [4]. Similar results were obtained in the study [5], in which iron-nickel ferrite particles were used as filler in the polymer matrix. This option is more acceptable because of greater availability of ferrite structures.

The main task in the protection against the influence of electromagnetic fields of ultra-high frequencies is to reduce the reflection coefficients. It is possible with the use of multilayer material [6]. But in manufacturing such surfaces it is necessary to maintain precisely the thickness of the layers, one of which must have the thickness equal to the quarter of the incident wave length. That is, such coatings are narrowband. Such facing could be applied for limited and specific tasks. Much attention is paid to the development of textile materials for shielding electromagnetic fields. In [7], a carbon fiber based material containing copper nanoparticles was described. Possessing high efficiency, such material, however, has great value. The study [8] describes the creation and testing of the material based on natural textile material soaked with magnetic fluid containing ferromagnetic nanoparticles. This material is highly efficiency for shielding magnetic and electric fields of industrial frequency (up to 8.6 - 8.9) and is suitable for manufacturing protective suits. Advantage of protective fabrics is small thickness, but for facing walls of buildings and rooms they are not suitable. In the analytical work [9], the efficiency and feasibility of using electromagnetic screens based on modified powder, nanostructured and film materials were analyzed. It is concluded that shielding surfaces should be produced for individual frequencies or ranges. This is unacceptable in the context of exposure to broadband electromagnetic fields. In [10] the results of development and investigation of the efficiency of metal-polymeric materials and shielding of electromagnetic fields of ultra-low and ultra-high frequencies are presented. The advantage of the design is the solidity of the material and small thickness (up to 10 mm). The screening substance used was iron ore, accumulated in the filters when grinding iron ore. There are many non-metallic and non-magnetic impurities in this material, so the shielding coefficients are not sufficient when the content of the shielding substance is low. But the direction of the work - the creation of materials for shielding electromagnetic fields of the most critical frequencies and frequency bands – is very promising. This requires the selection of the shielding substance of the desired electrophysical properties and dispersion and testing of protective properties of the composite with different content of shielding particles in the polymer matrix.

## 3. Problem statement

The aim of the study is to develop a composite metal-polymer material suitable for facing wall surfaces to shield electromagnetic fields of critical frequencies and frequencies ranges. To achieve this goal, the following main tasks are defined:

– to determine the material for manufacturing the polymeric material matrix and the shielding filler, as well as to study the filler dispersion;
– to develop manufacturing technology for material with different content of shielding filler;
– to study the shielding coefficients of the material for the magnetic field of industrial frequency and the electromagnetic field of ultra-high frequency;

IOP Conf. Series: Materials Science and Engineering **907** (2020) 012043 doi:10.1088/1757-899X/907/1/012043

- to analyze the obtained results and determine approaches to the design of electromagnetic screens with the required protective properties and promising areas of research to improve their efficiency.

## 4. Methods and materials

The shielding coefficients of the shielding materials (the ratio of the field strength (energy flux density) of the field source to the values of these parameters in the protected area) were determined by the method of direct measurements. Spectran 5035 Electromagnetic Spectrum Analyzer (Germany) was used to measure the magnetic field strengths at the industrial frequency. The maximum basic error of measurements did not exceed 1 dB. A geometrically closed cubic-shaped screen of 0.2 × 0.2 × 0.2 m in size was used to measure the material's shielding coefficient for the magnetic field of industrial frequency. To determine the ultra-high frequency electromagnetic field shielding coefficients (2.4-2.6 GHz), the 0.3 × 0.3 m flat screen was used, it closed the window in a solid metal sheet without any gaps, which would prevent the electromagnetic field from entering the screen. During the measurements the background values of the industrial frequency magnetic field did not exceed 0.13 μT (0.1 A/m). Background energy flux densities did not exceed 0.27 μW/cm$^2$. The measurements of the energy flux density were done with calibrated meter PZ-31 (Russian Federation). The maximum basic error of measurements did not exceed 2,7 dB.

## 5. Materials for manufacturing metal-polymeric material facing

To be applicable for facing large surfaces, the matrix material must be reasonably priced and easy-to-use. In addition, it must be mixed with the shielding filler and have sufficient strength and resistance to weathering. Such properties are typical for commonly used latex polymer. Latex-based material can be made both by adding a shielding substance during the latex production process [10] or into the ready-to-use liquid latex.

The iron ore concentrate obtained during its enrichment is selected as the screening substance. The average particle size of the concentrate is 150-200 μ m. The concentrate has the iron content (mainly in the form of magnetite) up to 82 %. The advantage of the iron ore is satisfactory electrical properties and low cost of the material. This enables its application in any concentration limited only by the mechanical properties of the final material.

## 6. Manufacturing technology of facing material for electromagnetic field shielding

At the initial stage, the iron ore concentrate was processed in a ball mill for one hour. Then, the particle-size analysis of the shielding substance was carried out by the sedimentation method, using torsion scales. It was necessary, because the shielding properties of metal polymers depend on the dispersion of metal and metal-containing particles [10]. For this purpose, the ratio between the relative amount of iron ore particles in the total mass of the filler ($-dQ/drQ_{max}$) and the particle sizes $r$ was determined (Fig. 1).

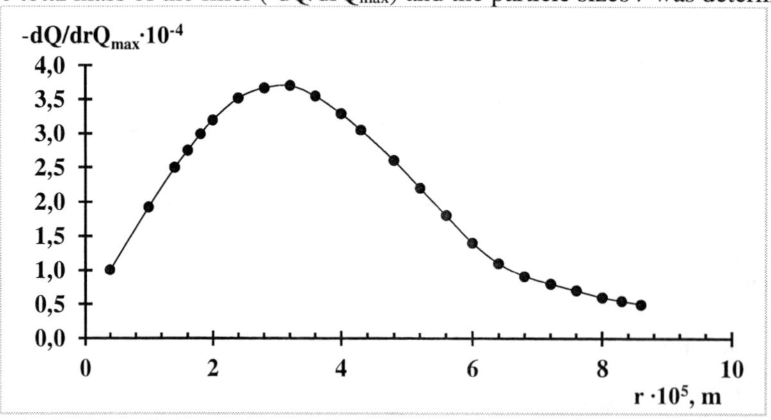


**Figure 1.** Distribution of the particles ($-dQ/drQ_{max}$) of iron ore concentrate by their size $r$

The data obtained indicate that the preferred fraction is particles of size 25-26 μm. This is important for the design of future shielding material with the desired shielding ratio of required or allowable thickness. It means that increasing the dispersion of the filler can reduce the thickness of the material. In the second step, the iron ore concentrate was added to the liquid latex in quantities of 5, 10, 15, 20 % (by weight). The resulting mixtures were thoroughly mixed in a special mixer and sonicated at the frequency of 23 kHz, with the amplitude of 45-50 μm for 10 minutes.

This treatment ensures a uniform distribution of the shielding particles in the latex and prevents them from sticking together. In order to obtain the final material, the metal-polymer mixture was squeezed on the rollers through the rectangular opening. The mixture was thermally treated between the rollers and the surface vulcanization was conducted there as well. The distance between the rollers was 5, 10, and 15 mm, which corresponds to the final thicknesses of the materials. The thickness of the vulcanized layer is 1.0-1.5 mm.

## 7. The study of the protective properties of facing materials for shielding electromagnetic fields

Testing of the protective properties of materials (shielding coefficients) was carried out in two stages. At the first stage, the shielding coefficients of the ultra-high frequency electromagnetic field and the contribution of electromagnetic waves reflection to the protective properties of materials were determined. The reflection measuring is necessary as the reflection of waves under real conditions can occur in undesirable directions and worsen the electromagnetic environment status both indoors and outdoors. Table 1 demonstrates the dependence of the shielding factor Ks on the content of the screening substance ρ, % (by weight) and the thickness of the screen. The energy flux density of the electromagnetic field source was 1300-1350 μW/cm$^2$.

**Table 1.** The dependence of the shielding factor of metal-polymer facing material on its thickness and the content of the shielding substance.

| ρ, % | Ks | | |
|---|---|---|---|
| | 5 mm | 10 mm | 15 mm |
| 5 | 3.2 | 7.4 | 12.3 |
| 10 | 6,8 | 14.5 | 33,4 |
| 15 | 22,0 | 48,2 | 92,2 |
| 20 | 34.5 | 77,2 | 125,5 |

It is well know, that the reflection coefficients do not depend on the thickness of the material, but are determined only by the electrophysical properties of its surface layer. Therefore, these values were measured for materials of the same thickness with various content of the screening substance. First of all, the simulations were performed by placing the antenna in the space between the source and the metal sheet with the opening, then - with the overlap of the opening by the screen and its absence (Table 2).

**Table 2.** The dependence of the electromagnetic waves reflection coefficients (Kr) of the facing material on the content of the shielding substance.

| ρ, % | 5 | 10 | 15 | 20 |
|---|---|---|---|---|
| Kr | 0.08 | 0.12 | 0.34 | 0.42 |

At the second stage, the shielding coefficients for the industrial frequency magnetic field were investigated (Table 3). The induction of the magnetic field source of the industrial frequency was 490-495 μT.

The results obtained indicate that the facing material developed is effective for shielding the ultra-low frequency magnetic field and the ultra-high frequency electromagnetic field. The shielding coefficients of the electrical component of the industrial frequency electromagnetic field were not

IOP Conf. Series: Materials Science and Engineering **907** (2020) 012043 doi:10.1088/1757-899X/907/1/012043

determined. This is because the shielding of the electric field is not a problem for any conductive material. In addition, it is believed that the magnetic field is more harmful to biological objects.

**Table 3.** The dependence of the shielding factor of the facing materials for the magnetic field of the industrial frequency on the content of the shielding substance.

| ρ, % | Ks | | | |
|---|---|---|---|---|
| | 5 mm | 10 mm | 15 mm | 20 mm |
| 5 | 1,2 | 2.4 | 3.3 | 5.5 |
| 10 | 3.4 | 5.8 | 7.2 | 11.7 |
| 15 | 13.6 | 22,4 | 32,2 | 44,2 |
| 20 | 34.5 | 20.5 | 49,6 | 98.4 |

## 8. Results and Discussions

The analysis of the testing results of the developed facing materials shows that they are efficient when used for shielding electromagnetic fields of ultra-low frequencies and electromagnetic fields of ultra-high frequencies. It is known that, even under production working conditions, magnetic fields of industrial frequency do not exceed the maximum permissible levels more than 2-3 times. Therefore, the proposed material has protective properties suitable for wide application. According to national and international standards the permissible values for ultra-high frequency electromagnetic fields are 10 μW/cm$^2$. That is, even thin screen with low content of the screening substance will ensure the regulatory conditions or people staying inside the premises. The results give possibility to optimize shielding of premises and buildings accounting the priority of the impact factor on the environment. The obtained results show that the shielding coefficients of magnetic and electromagnetic fields significantly increase at concentrations of shielding particles over 10 %. Simultaneously, the reflection coefficient of electromagnetic waves also grows. This can only be explained by the manifestation of the percolation effect (a sharp increase in the electrical conductivity of the material). With this concentration of conductive particles, the conductivity threshold is reached due to the contact between the particles and the formation of conductive chains.

This fact can be used to design materials with the required protective properties. The change in the real dielectric constant of the material is the evidence of reaching the threshold of electric current flow. On this basis, it is possible to calculate the volume fraction of the ferromagnetic filler in the total volume of the protective material. For the calculations of dielectric conductivity ε, the Odelevsky formula was applied, it determines the dependence of the real dielectric constant on the dimensions of the filler particles, their bulk concentration and the dielectric characteristics of the matrix components:

$$\varepsilon = \varepsilon_m \left[ 1 + \frac{k(\varepsilon_f - \varepsilon_m)}{(1 - k/v) \cdot F(\varepsilon_f - \varepsilon_m) + \varepsilon_m} \right]$$

where $\varepsilon_m$ is the dielectric constant of the matrix; $\varepsilon_f$ – dielectric constant of the filler; $k$ – volumetric concentration of the filler; $v$ – critical concentration of the filler above which the particles are in contact with each other; $F$ – depolarization index.

In turn, the depolarization factor can be obtained as

$$F = \frac{\ln(2l/d) - 1}{(l/d)^2},$$

where $l$ is the length of the particle; $d$ is the particle diameter.

From reference sources it is known that the relative dielectric constant of latex is 3.5-4.5. The main component of the mineral concentrate is magnetite, whose dielectric constant is 13.9-14.3. The geometric parameters of the screening particles of medium or mean dispersion are easily determined by the metallographic method. This method enables prediction of the protective properties depending on the parameters of the field that require shielding. The shielding coefficients of the developed facing materials make them usable not only for protection of people from electromagnetic influences. They are suitable for solving the problems of electromagnetic compatibility of electronic and electrical equipment and technical protection of information. It is important to determine the possibility of increasing the efficiency of the developed protective materials. This was carried out through the metallographic research. The surface of the material and its cross-section (the inner part of the section) were investigated.

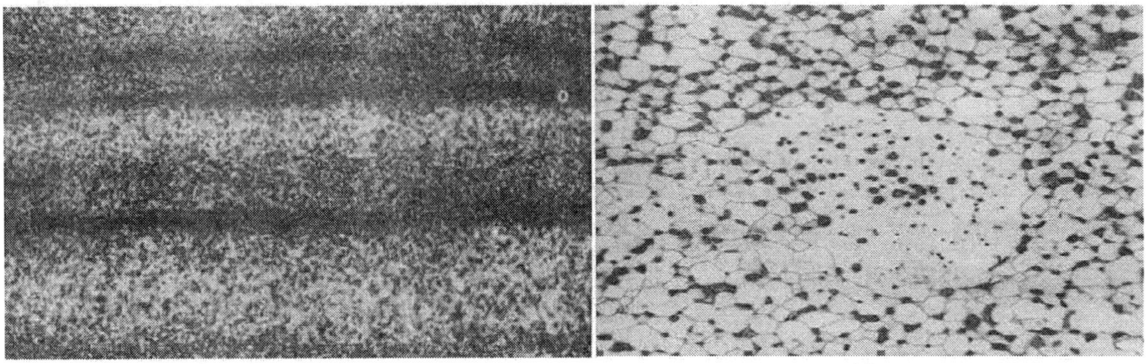

a) b)

**Figure 2.** Structure of metal-polymer material with the content of shielding substance of 15 %: a) material surface (× 150), b) material cross-section surface (× 250)

As it is seen in Fig. 2a) the distribution of the shielding particles on the surface of the material is uneven. It is most likely that this structure is formed because the rollers are not completely smooth. This has partially redistributed the fine iron ore concentrate. Fig. 2 b) indicates that the shielding particles are not evenly distributed in the inner part of the material and differ significantly in size. Preventing the redistribution of the shielding substance could be probably prevented by improving the quality of the rolling surface and their even heating. Increasing the dispersion and uniformity of the shielding filler distribution in the polymer body is possible by increasing the time and intensity of ultrasonic treatment of the original mixture. Thus, the ways to improve the developed materials are to increase the dispersion of the ferromagnetic filler and provide uniform distribution of the shielding particles in the body of the polymer material.

## 9. Conclusions

The developed latex-based material with fine ore concentrate filler material has small thickness, weight and cost with sufficient protective properties for domestic and industrial application.

The shielding coefficients of the electromagnetic fields of ultra-high frequencies (2.4-2.6 GHz) are 3.2-125.4 for the material 5-20 mm thick, containing 5-20 % shielding particles (by weight). The reflection coefficients are 0.28-0.42. Industrial frequency magnetic field shielding factors are 1.2-98.4.

The significant increase in the shielding coefficients at the concentrations of the shielding substance above 10% indicates the manifestation of percolation effect - the transition of the material conductivity threshold, which leads to the increase in the reflection coefficients of electromagnetic waves. The above presented calculation of the shielding substance critical concentration allows predicting the protective properties of the material, based on the electrophysical characteristics of the polymer matrix and the material of the shielding substance.

Improvement of the developed material is possible by increasing the dispersion of the shielding substance and the uniform distribution of the shielding particles in the polymer matrix. This will reduce the weight and thickness of the facing material and reduce its cost while maintaining its protective properties.

## References

[1] Gonza´lez M, Pozuelo J, and Baselga J 2018 Electromagnetic Shielding Materials in the GHz Range. *The Chemical Society of Japan and Wiley VCH Verlag GmbH and Co. KGaA Weinheim Wiley Online Library*; Scheme. Rec. Vol **18** pp 1-11.

[2] Kargar F, Barani Z, Balinskiy M G, Magana A S, Lewis J S and Balandin A A 2018 Graphene Composites with Dual Functionality: *Electromagnetic Shi elding and Thermal Management* URL: https://onlinelibrary.wiley.com/doi/full/10.1002/aelm.201800558

[3] Singh J 2015 Computer Generated Energy Effects on User s and Shielding Interference *International Journal of Innovative Research in Computer and Communication Engineering* Vol **3** Issue 10 pp 10022–10027

[4] Mondal S, Ganguly S, Das P, Khastgir D and Das N C 2017 Low percolation threshold and electromagnetic shielding effectiveness of ano structured carbon based ethylene methyl acrylate nanocomposites *Composites Part B: Engineering* **119** pp 41–56 https://doi.org/10.1016/j.compositesb.2017.03.022

[5] Yadav, Kuřitka, Vilčáková, Mac Hovský, Škoda, Urbánek et al 2019 Polypropylene Nanocomposite Filled with Spinel Ferrite $NiFe_2O_4$ Nanoparticles and In Situ Thermally Reduced Graphene Oxide for Electromagnetic Interference Shielding Application *Nanomaterials* **9** (4) p 621. https://doi.org/10.3390/nano9040621

[6] Belyaev A A, Bespalov E E, Lepeshkin V V 2015 Radar absorbing materials based on finishing materials for protection from microwave radiation of base stations of cellular communications *Proceedings of VIAM* No **6** pp 1-9 https://cyberleninka.ru/article/n/radiopogloschayuschie-materialy-na-osnove-otdelochnyh-stroitelnyh-materialov-dlya-zaschity-ot-svch-izlucheniya-bazovyh-stantsiy/viewer

[7] Jiao Y, Wan C, Zhang W, Bao W and Li J 2019 Carbon Fibers Encapsulated with Nano-Copper: A CoreShell Structured Composite for Antibacterial and Electromagnetic nterference Shielding Applications *Nanomaterials* **9** (3) p 460 https://doi.org/10.3390/nano9030460

[8] Glyva V, Barabash O, Kasatkina N, Katsman M, Levchenko L, Tykhenko O, Nikolaiev K, Panova O, Khalmuradov B and Khodakovskyy O 2020 Studying the shielding of an electromagnetic field by a textile material containing ferromagnetic nanostructures *Eastern-European Journal of Enterprise Technologies* Vol **1** No 10 (103) pp 26-31 https://doi.org/10.15587/1729-4061.2020.195232

[9] Lynkov L M and Byush V A 2019 New technologies for creating screens for electromagnetic radiation based on modified powder, nanostructured materials. *Reports of BSUIR* **2** pp 85-99

[10] Glyva V, Lyashok J, Matvieieva I, Frolov V, Levchenko L, Tykhenko O, Panova O, Khodakovskyy O, Khalmuradov B and Nikolaiev K 2018 Development and investigation of protective properties of the electromagnetic and oundproofing screen *Eastern-European Journal of Enterprise Technologies* Issue 6/5 (**96**) pp 54–61 https://doi.org/10.15587/1729-4061.2018.150778.

IOP Conf. Series: Materials Science and Engineering **907** (2020) 012044 doi:10.1088/1757-899X/907/1/012044

# Features of hardening and structure formation of concrete with high-dispersion mineral additives

**L N Butskaya, V A Hurkalenko, K V Latorets, O V Makarenko and L O Pershina**

Department of Physical and Chemical Mechanics and Technologies of Building Materials and Products, Kharkiv National University of Civil Engineering and Architecture, Sumska st. 40, 61002 Kharkiv, Ukraine

latorets.ev@gmail.com

**Abstract.** The article deals with the effect of micro-filler additives on the properties of high-strength concrete. The effectiveness of microcements – mixtures based on highly dispersed cement and silica fume – has been substantiated. Using the methods of thermokinetic analysis and thermoporometry, the processes of heat release and the formation of microporosity during the hardening of ordinary cement and microcements without additives and with the addition of silica fume are studied. The positive effect of silica fume on the hydration and formation of the structure of cement stone is proved. It is shown that the nature of the hydration interaction in the system of "microcement - silica fume" contributes to the compaction of the structure and reduce the development of shrinkage phenomena. It was revealed that the use of microcementives allows to increase the workability of concrete mixtures and to accelerate the rate of curing of high-strength concrete. According to the results of the studies, the high efficiency of the use of components of the type "microcement – silica fume" in the technology of high-strength and durable concrete has been proved.

## 1. Introduction

Modern concrete science has significantly expanded the capabilities of high-strength concrete technology through the use of organomineral complexes in the form of mineral and chemical additives at the stages of production of concrete mixtures and operation of concrete [1-3]. Such additives summarize the effect of microfillers with the pozzolanic reaction of $SiO_2$-containing products and the modifying effect of superplasticizers.

These include coal fly ash with a dispersion of up to 3000 $cm^2/g$ and with low pozzolanic activity [4- 6]. Amorphous silica fume (SF) with a particle size of almost two orders of magnitude higher than the size of cement particles is primarily a $SiO_2$-containing product. The high efficiency of such complexes is due to the combined influence of mineral and chemical additives. The presence of active $SiO_2$ determines the possibility of alkali-silica reactions (ASR) during the hardening of concrete and, as a result, damage to the structure of neoplasms during the formation of microcracks, the sizes of which reach 0.6 μm (figure 1) [7-8]. The probability of the course of such reactions is significantly increased when using chemical additives – hardening accelerators based on Na or K [8].

The elimination of these and other shortcomings is achieved by using the so-called microcements (MC) or microconcrete – mixtures based on finely dispersed cement and fine quartz sand [9-10].

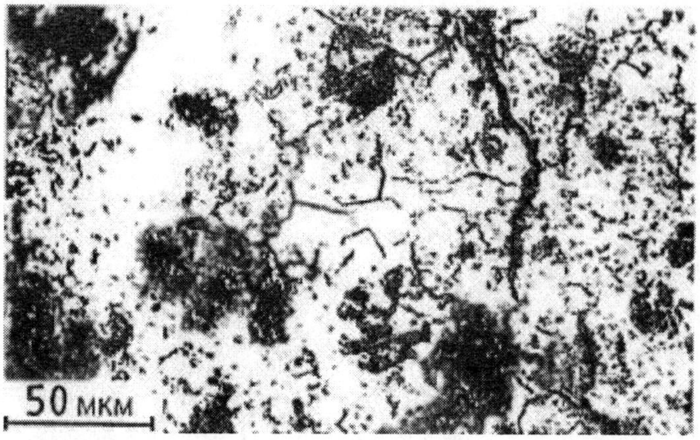


**Figure 1.** The occurrence of microcracks during alkali silica reactions

## 2. Materials and research methods

In this paper, we study the hardening and structure formation processes of MCs with microsilica as a filler. MCs based on Portland cement PC and slag Portland cement PCS were used (table 1).

Table 1 – Composition and properties of the MC components

| Characteristics | Cements | | | SF |
|---|---|---|---|---|
| | Ordinary cement C1 | PC | PCS | |
| Density, g/cm | 3.2 | 3.1 | 2.9 | 1.4 |
| Content, % <br> $SiO_2$ | 21.0 | 21.7 | 31.7 | 93.1 |
| $Al_2O_3$ | 4.6 | 4.1 | 11.4 | 1.1 |
| $Fe_2O_3$ | 6.1 | 1.6 | 0.6 | 0.7 |
| CaO | 56.8 | 65.8 | 44.3 | 0.4 |
| MgO | 4.7 | 0.8 | 7.0 | 0.7 |
| $K_2O$ | 0.54 | 0.83 | 0.37 | 1.04 |
| $Na_2O$ | 0.15 | 0.26 | 0.25 | 0.22 |
| $SO_3$ | 2.1 | 2.4 | 0.9 | 0.0 |
| The average particle diameter, microns | 13.2 | 5.0 | 7.9 | 0.15…0.2 |
| Specific surface area, $m^2/g$ | 1.0 | 2.3 | 1.5 | 20…22 |

The measurements were performed in the calorimetric center of building materials science on measuring and computing calorimetric complexes [11-12].

Thermokinetic analysis, the experimental basis of which is differential microcalorimetry, was used to study the early stages of hydration by the dependences of the rate of $dQ/d\tau = f(\tau)$ and the completeness of heat dissipation $Q = f(\tau)$ [11]. A Calve type microcalorimeter was used for the temperature range +5 ... + 100 $^0$C with such characteristics: sensitivity 0.05 cal/g. mm, temperature stabilization accuracy ± 10 $^0$C, time constant 20 s. Processing of the results includes obtaining the dependences of the speed and completeness of heat release on time with their subsequent kinetic study – an estimate of the beginning and duration of the induction period, the magnitude and time of reaching the maximum speed and completeness, and the forecast of heat release.

Thermoporometry is a method for assessing the size, volume, pore size distribution. The basis of the method is the dependence of the crystallization temperature of pore water on pore sizes. This dependence

IOP Conf. Series: Materials Science and Engineering **907** (2020) 012044 doi:10.1088/1757-899X/907/1/012044

is established from the thermodynamic equilibrium of the system "solid – water – ice", taking into account the state of the interphase surface and the well-known Laplace equation and it has the form [12]:

$$dR_p = \frac{A}{(\Delta T)^2} \cdot d(\Delta T), \quad (1)$$

where A is the methodological coefficient.

The volume dV of pores is found from the ratio:

$$dV = \frac{\Delta H}{W_a} \cdot d(\Delta T), \quad (2)$$

From (2) we obtain the values and distribution of pore volume depending on temperature. Pore size distribution is represented as $dV/dR = f(R)$. A feature of thermoporometry is the determination of parameters only of those pores that are filled with water and are in the range of 1.5 ... 100 nm, which is due to the fulfillment of the Kelvin equation.

The pore volume obtained by thermoporometry corresponds to the volume of free water in the material.

The amount of adsorbed water was determined by calculation, based on the polymolecular theory of adsorption and the volume and radius of the pores.

Chemically bound water corresponds to the difference between the initial content and the sum of free and adsorbed water.

We studied cement samples without (C1) and with the addition of silica fume (C2). For thermokinetic analysis, the sample weight was 1 g. The duration of the experiment was 1 day. In the method of thermoporometry, samples of 3, 6, 12, and 24 h of curing were tested. Their weight averaged 25 mg. W/T in all experiments was 0.5.

## 3. Discussion of the results

The kinetics of heat release of cement C1 characterizes it as quite active. Hydration of C2 proceeds less intensively, probably due to the reduced energetics of the formation of low-basic CSH (I) hydrosilicates during the interaction of $SiO_2$ and $Ca(OH)_2$ (figure 2, a). Already within 24 h, the heat release completeness is lower by 15 cal/g (figure 2, b). A distinctive feature of the hydration of cement C2 is the presence of a doublet of the main maximum heat. The bifurcation, apparently, is due to the interaction of $SiO_2$ and Ca $(OH)_2$ at the stage of formation of the cement stone structure at the age of 12 h. The value of the first heat release maximum for C2 samples is almost 2.5 times lower than for C1. This circumstance is probably associated with a high specific surface of silica fume. Water molecules are adsorbed on the surface of silica fume grains, which leads to a decrease in the number of molecules involved in cement hydration.

The study of formation of microporosity of cement stone was carried out using the method of thermoporometry. The research results are presented in figure 3.

A characteristic feature is the reduction in the size and volume of micropores with the addition of silica fume, which is due to the filling of a part of the pore space of cement stone with its particles.

An increase in the specific surface of cements also contributes to a decrease in the size of micropores and a decrease in their total volume (figure 3). The number of capillary pores decreases with a simultaneous increase in the number of gel micropores.

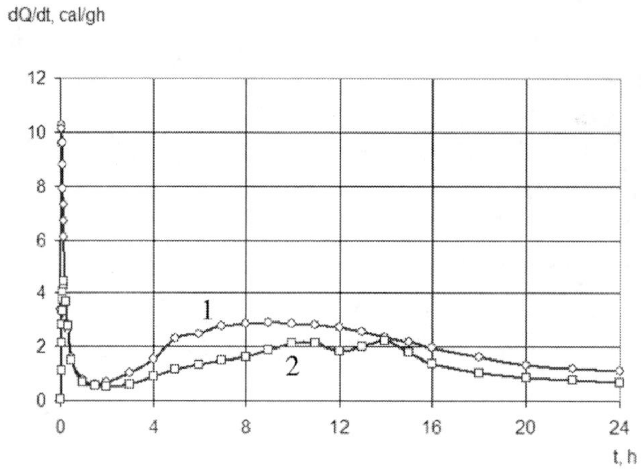


a)

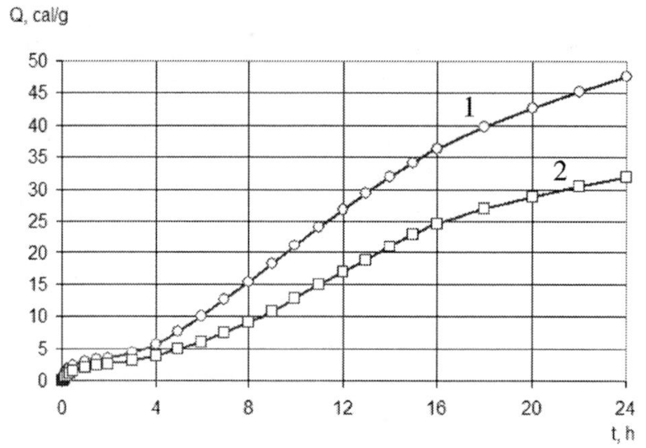


б)

Figure 2. Heat release during hydration of cement C1
a) heat release rate; b) heat of hydration
1 – control; 2 – with the addition of SF

The positive effect of silica fume on the formation of the structure of cement stone is, in our opinion, also due to the fact that even before the concrete mixture begins to harden, the bulk of the alkali of cement compounds will be bound by silica. The decrease in alkali content in the liquid phase caused by this process, in turn, inhibits the development of processes of internal corrosion of concrete [13], which contributes to an increase in durability.

The nature of the pore structure of samples of the obtained MC-SF systems was also investigated by the method of mercury porosimetry. It was noted that microcements contribute to a decrease in the number of capillary (diameter 0.01 ... 100 μm) and the growth of gel pores (diameter less than 0.01 μm), which is in good agreement with the data of thermoporometry. The conclusion about the formation of a denser and less defective structure of the hardened cement stone along with the high homogeneity of the neoplasms determines the high durability of concrete based on them. In [14], activation of the formation of hydrosilicates of the CSH (I) type in this case is noted, which affects the hardening kinetics and structural strength.

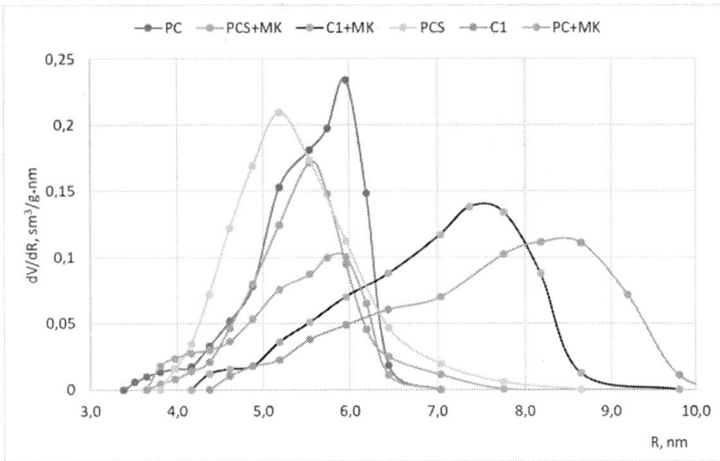

a)

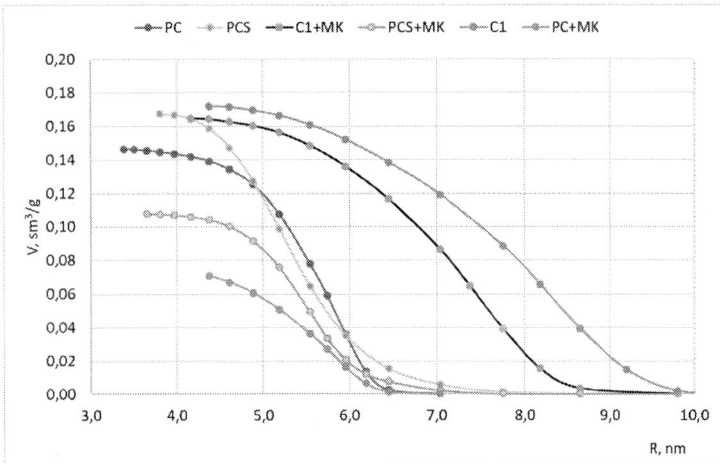

b)

Figure 3. Microporosity of cement stone
a) pore size distribution; b) pore volume

The nature of the hydration interaction in the system "microcement – silica fume" "MC-SF" contributes to the compaction of the structure (figure 3). This circumstance affects shrinkage phenomena. For the control composition at the PC, shrinkage at a relative humidity of 90-95% reaches 1.6 mm/m after 2 days. The introduction of silica fume causes a slower development of shrinkage. In 14 days, its value is 2.35 mm/m. Microcement RS in combination with silica fume determines the shrinkage value of 0.9 mm/m in only 2 days.

Concretes for comparative tests contained 430 kg/m$^3$ of "MT-SF" mixture, W/T = 0.37, the consumption of superplasticizer was 0.9% of the weight of cement.

When evaluating workability, it was found that the features of wetting and adsorption of the liquid phase by dispersed components caused a sharp decrease in this important technological characteristic with the introduction of SF and, on the contrary, its improvement in the case of MC, especially on a slag Portland cement basis.

The conclusion about the effect of the micro-fillers type on the strength of concrete is quite unambiguous - cements somewhat reduce it compared to silica fume. However, the rate of curing to a level of 80-90 MPa, i.e. for high-strength concrete, at microcements is much higher. At the age of 7 days, concrete at the MC-RS reaches 95% of brand strength. The increase in the content of micro-fillers ambiguously affects the kinetics of the strength, which, of course, is associated with the composition and properties of cement.

Thus, the studies have shown the high efficiency of using the type "MC-SF" components in the technology of high-strength and durable concrete.

## References

[1] Tang Van Lam, Ngo Suan Hung, Wu Kim Zien, Nguyen Chong Chyk, Bulgakov B I, Bazhenova O Yu, Galtseva N 2019 An Effect of Water-Binding Ratio and Complex Organic-Mineral Additives on Concrete Properties for Offshore Hydraulic Structures *Industrial and Civil Engineering* No **3** p 7–17 doi: 10.33622 / 0869-7019.2019.03.

[2] Mehta P K, Monteiro P J 2006 *Concrete – Structure, Properties and Materials* (Department of Civil and Environmental Engineering University of California at Berkeley) p 684

[3] Shah S P, Ahmad S H 1994 *High Performance Concrete: Properties and Applications* (Mc Graw Hill. Inc.) p 403

[4] Held M, Konig G 1992 Hochfester Beton bis B125 - Ein geeigneter Baustoff für hochbelastete Druckglieder *Beton-und Stahlbetonbau* Heft **2** S 41- 45

[5] Nicolay J, Stark J 1993 Dauerhaftigkeit von Weibbeton B85. Prüfungen an hochfesten Betonen *Beton* Heft **9** S 444-448

[6] Kern E. Technologie des hochfesten Betons // Beton. - Heft 3. - 1993, S.109-115.

[7] Stark J, Seyfarth K 2007 Secure prevention of ASR deterioration – concepts, practical application *Betonwerk + Fertigteil - Technik* **73** №2 pp 92-93

[8] Lagerblad B, Trägardh J 1994 Alkali-silica reaction in high strength concrete/Durability of high-performance concrete. *Proceedings of the RILEM International Workshop* (Vienna: February 14- 15) pp 143-153

[9] Rudert V, Strunge J, Wihler H-D 1994 Beton aus anderer Sicht - filigranes Mikrogefüge *Betonwerk + Fertigteil - Technik* Heft **9** S 86-93

[10] Hardl R 1995 *Veränderung des Betongefüges durch die Wirkung von Steinkohlenflugasche und ihr Einflüß auf die Betoneigenschaften* (Dissertation, RWTH Aachen, Institut für Bauforschung) S 107

[11] Usherov-Marshak A V, Sopov V P 1995 Calorimetry in the science of building materials *Journal of Therm. Analysis* Vol **43** № 2 pp 497-504

[12] Usherov-Marshak A V, Sopov V P, Kurdowski W 1999 Studies of capillary porosity of clinker phases during hydration *Journal of Therm. Analysis and Calorimetry* Vol **55** pp 1031-1037

[13] Rozental A N, Lyubarskaya G V 2012 Corrosion of Concrete during the Interaction of Alkalis with Aggregate Silica *Concrete and Reinforced Concrete* No **1** (6) S 50-60

[14] Sopov V, Pershina L, Butskaya L, Latorets E, Makarenko O 2017 The role of chemical admixtures in the formation of the structure of cement stone *Matec WEB of Conferences, 6$^{th}$ International Scientific Conference "Reliability and Durability of Railway Transport Engineering Structures and Buildings" (Transbud-2017)* Volume **116** 01018 pp 1-10 doi:10.1051/matecconf/201711601018.

# Contact formation mechanism between squeezed crystalline solids

**M A Volosyuk[1] and E M Protsenko[2]**

[1]Department of Physics, Kharkiv National Automobile and Highway University, Kharkiv, Ukraine
[2]Department of Graphics, Kharkiv National University of Civil Engineering and Architecture, Kharkiv, Ukraine

marina_volosyuk@ukr.net

**Abstract.** The mechanism was proposed and experimentally approved for the final stage of the high temperature contact formation after generation of closed pores in the contact plane. The mechanism is that the pores being healed by the dislocation-diffusion mechanism form assemblages of vacancy type prismatic dislocation loops which alternate with clusters of previously formed interstitial prismatic loops. Between the vacancy-type loops and interstitial ones the diffusion interaction is established resulting in recombination of defects with different signs, the contact formation, pore healing, and decreasing stress and dislocation density in the contact area.

## 1. Introduction

The problem of interaction of solids in contact and formation of the physical contact is multipronged and actual for modern technologies, namely, fritting, diffusion welding under pressure [1], nano-material preparation [2, 3], as well as under interaction of construction elements in various technology arrangements. The processes in the contact area are multifold and complex, these are determined both by the contacting solids' properties, and external conditions – temperature, loading, radiation influence, etc. [4-8]. Under conventional conditions, the rate of the contact formation is determined by temperature and loading. Irradiation of construction materials (for example, in nuclear power systems) promotes deformation processes [9, 10] stimulating formation of non-equilibrium point defect concentration, pore generation, and the pore structure evolution, irradiation induced swelling, dynamical recrystallization, dislocation structure evolution, etc. All these factors complicate substantially the processes in the contact area.

Under the contact formation, the solids adjoin by real surfaces which always have roughness of various geometry and scale [11]. Different contact types may be reduced to the extreme case that is the contact of a wedge (roughness) and a plane. That model is usually put into calculations and model experiments on the contact formation investigations.

The model "wedge-plane" allows applying a simple two-dimensional problem of the theory of elasticity for description of stress distribution in the contact plane where the local plastic deformation takes place [12]. It was experimentally shown in [13] that in the early stages, the contact formation process as a result of contortion and embedding the rough edges under certain conditions (high temperature, absence of irradiation, uniaxial compressive loading) occurs by the dislocation-diffusion mechanism, and its kinetics is controlled by thermal-fluctuation creeping away the dislocation assemblage.

In the present work on the base of the analysis of the contact geometry variations under uniaxial compression of crystalline solids, the substance transfer mechanism was proposed and experimentally approved for the final stage of the process where as a result of contortion and embedding the rough edges, the two-dimensional system of pores is formed.

## 2. Materials and methods

### *2.1. Materials and experiment*

The experiments were carried out on the samples made from galvanic purified and annealed copper. The samples are shown in figure 1. The sample sizes are $20 \times 10 \times 5$ mm.

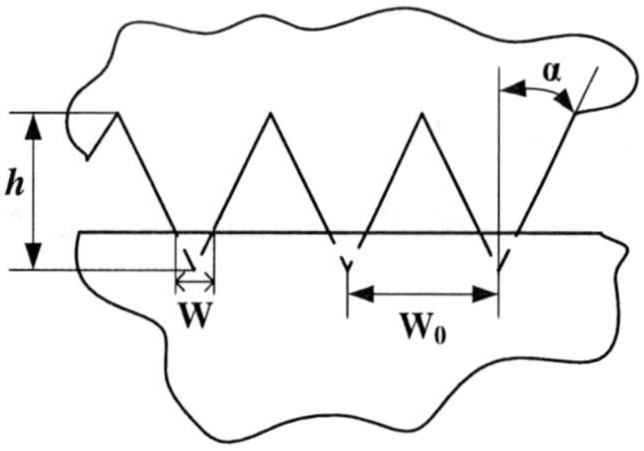


**Figure 1.** Samples for the study ($2\alpha$ is angle at the wedge vertex, $h$ is the wedge height, $W_0$ is distance between wedge vertices, $W$ is the contact width).

The rough edges (wedges) were cut on the sample surface and in cross-section looked like triangled jags of $h = 5 \cdot 10^{-5}$ m height distanced by $W_0 = 10^{-4}$ m from one another. Annealing under loading was carried out in vacuum $10^{-5}$ mm of mercury at temperature 800°C ($0.8T_m$). The width of a formed single contact, as in [12], was measured by the profilometer and averaged not less than by 200 measurements, also, progressively, during experiments by the contact electrical resistivity data.

### *2.2. Contact geometry evolution during formation and substance transfer mechanism at the final stage*

As the contact area expands, the stress distribution changes in the contact area. In the initial time of the contact formation, the loading is given as the force applied per unit length of the sharp wedge ($P$, N/m). After the wedge blunting and forming the contact with width $W$, the loading in the formed contact will be already distributed ($P/W$), and further, it will decrease with the contact growing. At that, both the efficiency of dislocation sources and the rate of contact growing caused by dislocation generation and thermo-activated motion are reduced. Direct observations of the dislocation structure evolution [14, 15] under the contact formation between squeezed $KCl$ single crystals in the "wedge-plane" geometry at 0.8 $T_m$ temperature show that after forming the quasi-stationary dislocation assemblage, the activity of dislocation sources and dislocation density in the contact decrease gradually. The activity of diffusion processes remains unchanged.

In contrast to the model applied in [12, 13] consisted in the system of flat sharp wedges ("comb") squeezed (crushed) into the plane half-space (figure 1), in the present work, after achieving the contact relative width $S = W/W_0 \geq 0.5$ and pore generation, we consider the model of uniaxially compressed plane system of residual voids forming a two-dimensional assemble (figure 2).

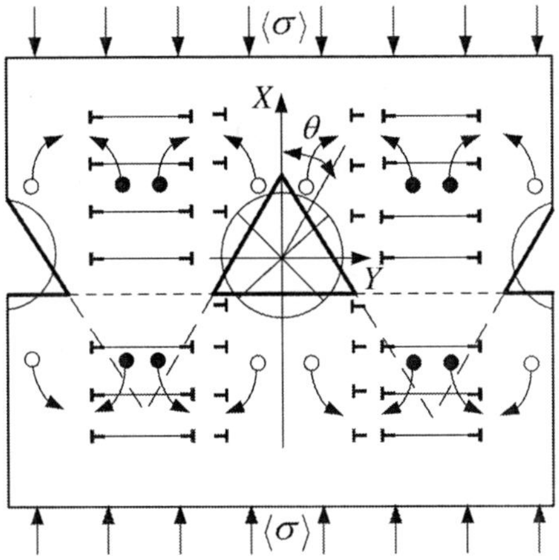


**Figure 2.** The scheme of point defect flows in the final stage of the contact formation (○ – vacancies, ● – interstitial atoms), $\langle\sigma\rangle = P/W_0$ is average stress in the contact, P is loading per unit of wedge length.

The problem with voids may be described (to a first approximation assuming the voids independent) using the simple solution of the plane problem for a single void in the uniaxially compressed wide band [16].

From the analysis of the mentioned solution [16] it follows that distribution of shear stress $\sigma_{xy}$ in Cartesian coordinates *XOY* with origin in the void center (figure 2) is that in points on the pore surface (for which the polar angle $\theta = 0°$, $180°$ и $\theta = \pm 90°$) $\sigma_{xy} = \sigma_{yx} = 0$. For the points with polar angle $\theta = \pm 45°$ and $\theta = \pm 135°$ we have $\sigma_{xy} \equiv (\sigma_{xy})_{max}$. As the distance from the void center $x$ increases, these stresses decrease: $\sigma_{xy}$ is proportional to $1/x$ (as in the wedge). The signs «+» and «–» concern to the stress on the right and on the left of the pore. With such stress distribution, under loading, shear dislocation loops [17-19] of different signs may be emitted from the pore surface to different sides of its center (figure 2); the edge components of these loops in total are equivalent to prismatic loops of vacancy type.

The physics of the mechanism proposed for the final stage of the contact formation consists in the following. Under healing by dislocation-diffusion mechanism, assemblages of vacancy-type loops are generated around the pore (figure 2). The vacancy loops and the void undergo as well diffusion dilution caused by their curvature, thus maintaining certain vacancy super-saturation in the neighborhood of the pore. Before the pore formation, the system "wedge-plane" acted efficiently. Each wedge generated interstitial loops. After the pore formation, in the contact plane of the crystal there occur alternating areas occupied by prismatic loops with different signs, i.e. the areas with different dilatation signs alternate. Between these areas, chemical potential gradients occur, resulting in counter-current flows of point defects: vacancies come from healed pores and vacancy loops, and interstitial atoms – from dissolving interstitial loops formed before and generated at present (figure 2). These flows lead to recombination of different sign defects and corresponding dislocation loops, increasing the contact area width (*S*), pore healing, stress relaxation, and decreasing dislocation density at the contact zone. In figure 2 the counter-current flows of opposite sign defects are marked by arrows.

The essential peculiarity of the mechanism under consideration is that unlike the mechanisms discussed in [12, 13] where the substance from the zone of contortion and rough edge embedding was incorporated into surrounding matrix resulting in stress increase and inhibiting the contact formation, in the process described here, due to the counter-current flows, this embedded substance is transferred finally into residual voids. At that, the contact area expands, and the stress created by embedded substance relaxes.

From the described mechanism it is obviously, that its kinetics will be determined by the degree of vacancy super-saturation generated in the crystal lattice by healed pores and vacancy dislocation loops, or other sources connected, for example, with irradiation.

The equation for dilution of an interstitial loop according to the theory [20] is

$$\frac{dR_l}{dt} = - \frac{2\pi}{b\ ln\frac{8R_l}{b}}\left[\frac{Gb\left(ln\frac{R_l}{b}+g\right)D_i\omega}{4\pi(1-\nu)kTR_l} + D_V \Delta C_V\right] \quad (1)$$

where $R_l$ is dislocation loop radius, $G$ is shear modulus, $b$ is Burgers vector, $g \cong 2 \div 3$ is value taking into account the energy of the dislocation nucleus, $D_V$ , $D_i$ are diffusion coefficients for vacancies and interstitials, respectively, at temperature $T$, $\omega$ is atomic volume, $\nu$ is Poisson coefficient, $k = 1.38 \cdot 10^{-23}$J/K is Boltzmann constant, $\Delta C_V$ is degree of vacancy super-saturation far from interstitial loops.

In [13] it was established that the observed kinetics of the contact formation is impossible to explain as being limited by dissolution of interstitial loops due to their curvature, that is connected with high activation energy of interstitial generation (4 ÷ 5 eV for copper).

In the model under consideration, where the plane system of healed pores is assumed, the interstitial loop dissolution takes place due to lattice vacancy supersaturation. Therefore, the first addend in (1) may be neglected, keeping only the second (osmotic) one $(D_V \Delta C_V)$ in square brackets, so we obtain the kinetics equation for growth of the contact relative width $S = W/W_0$, which is controlled by dissolution of interstitial loops. The scheme of such calculation was proposed in [12] and applied in [13]. The essence is in the following. After applying the loading, in the contact, interstitial dislocation loops are emitted and come away at the distance where loading stress is equal to Peierls stress $\sigma_p$, so the dislocation assemblage is formed.

In the model under consideration, dislocation loop generation is assumed athermic (inhomogeneous). As soon as the difference between external loading stress in the contact and the opposite stress from the dislocation assemblage exceeds the Peierls threshold, $\sigma_p$, a new dislocation loop is generated. Therefore, the generation rate is not included in the calculation. The process rate is assumed to be controlled by the dislocation loop dissolution rate.

The number of dislocations in the assemblage grows until, after a time, the opposite stress of the assemblage balances the stress from the external loading in the contact. Thus, the dislocation assemblage acquires a quasi-stationary state (with a number of dislocations in the assemblage), to which the contact relative width $S_{st}$ corresponds. The dislocation assemblage is in dynamical equilibrium. When dislocations go away from the assemblage dissolving by diffusion in the vacancy supersaturation field, new dislocation loops are generated in the contact resulting in its widening. Thus, two flows exist: the flow of interstitial dislocation loop dissolving, and the flow of interstitial dislocation loop generation resulting in the contact expansion. Equating the flows (taking into account the rates of interstitial loop dissolution from (1)) we obtain the equation of the contact growth kinetics in differential form. After its integration taking into account the initial condition (at $t = t_{st}$, $S = S_{st}$, where $S_{st}$is the contact area at the moment $t_{st}$, $t_{st}$is the formation time of a quasi-stationary dislocation assemble) we obtain the kinetic equation in integrated form:

$$S - S_{st} = BD_V \Delta C_V \langle\sigma\rangle(t - t_{st}),\ B = \frac{4\pi^2(1-\nu)}{[Gb-2\pi(1-\nu)\gamma]\cdot b\cdot ln(8R_l/b)}, \quad (2)$$

where $\gamma$ is surface energy density.

The equation (2) is linear by $\Delta C_V$,$\langle\sigma\rangle$, and $t$ .

## 3. Results and Discussion

The following dependences were recorded: $S(\langle t\rangle)$ at $\langle\sigma\rangle$ equal to 2.0·10$^7$ N/m$^2$, 4.0·10$^7$ N/m$^2$, 5.0·10$^7$ N/m$^2$, and 7.0·10$^7$ N/m$^2$; and $S(\langle\sigma\rangle)$ at fixed time exposures $t$ (10, 20, 30, 40, and 50 minutes). In figure 3 a, b, the obtained experimental dependences are shown in accordance with equation (2) as $(S - S_{st})(t)$ and $(S - S_{st})(\langle\sigma\rangle)$ (taking into account $t_{st} \approx 5$ minutes for our experiments). Taking into consideration inaccuracies in $S$ determination, the relative value of which is about $\pm(10 \div 15)\%$, we see that the dependences may be approximated by straight lines.

This allows, using the relation (2), estimating the value $D_V \Delta C_V$. Taking dispersion into account we obtain the range of values $D_V \Delta C_V = (10^{-21} \div 10^{-20})$ m²/s. If to take into consideration that copper self-diffusion coefficient at temperature 800°C is about $D = 4 \cdot 10^{-15}$ m²/s, and vacancy equilibrium concentration is about $2 \cdot 10^{-5}$ at this temperature, then taking into account $D = D_V C_V$ we obtain the estimation for $\Delta C_V$ in the range $10^{-10} \div 10^{-9}$.

On the other hand, it is possible to estimate the vacancy super-saturation degree $\Delta C_R$, which occurs in the system due to pore healing and diluting prismatic vacancy loops emitted by pores.

Let us apply the formula from [21]:

$$\Delta C_R = \frac{2\gamma}{R}\frac{\omega}{kT} C_0. \quad (3)$$

Substituting the values for copper ($\gamma \cong 0.1\ Gb$, $G = 4.15 \cdot 10^{10}$ N/m², $b = 3.6 \cdot 10^{-10}$ m, $\omega = 4.7 \cdot 10^{-29}$ m³, and $C_0 = 2 \cdot 10^{-5}$ for vacancy equilibrium concentration, and $T = 1073$ K experiment temperature) we obtain (taking $R \cong 3 \cdot 10^{-5}$ m): $\Delta C_R \cong 3 \cdot 10^{-9}$.

Comparing the estimations of $\Delta C_V$ and $\Delta C_R$ we see these are of the same order. That allows assuming the proposed physical model of the dislocation-diffusion mechanism for the contact final stage formation consisting in recombination of counter-coming flows of vacancies and interstitials is realistic. The mechanism becomes prevailing as the contact becomes formed and residual pores are generated, because during the process, not only the contact grows, but also stress relaxation between pores takes place in sub-boundary area, and dislocation density decreases due to recombination of dislocation loops of opposite signs.

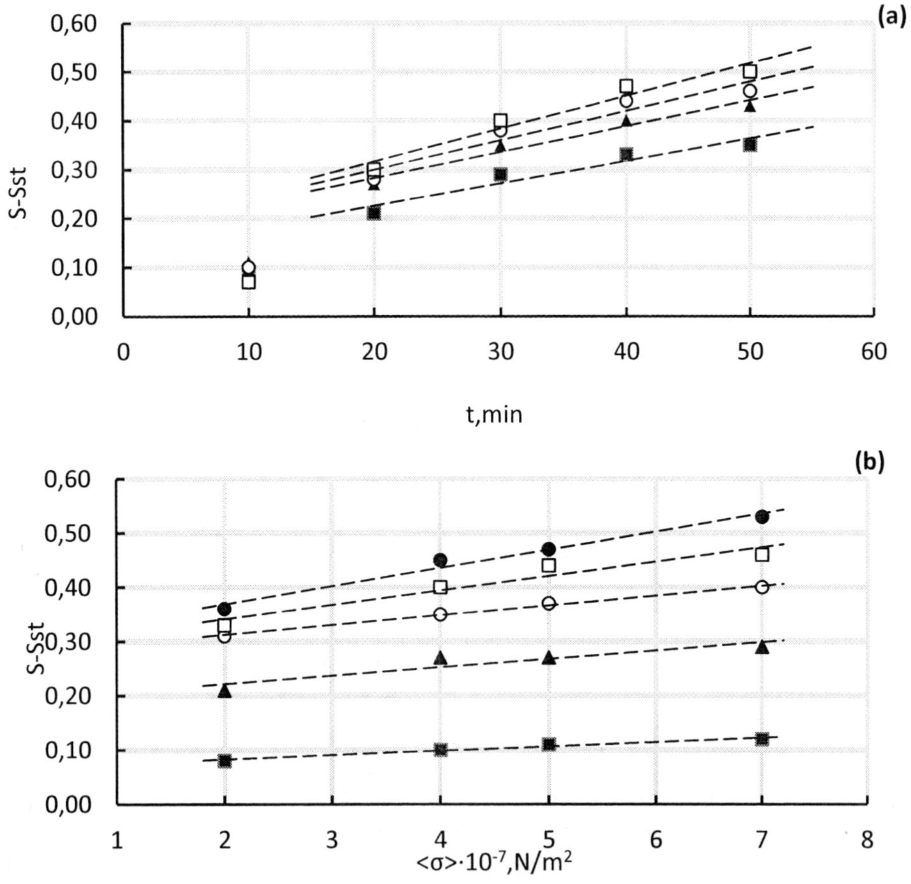


**Figure 3.** Dependences:
a) $(S - S_{st})(t)$ for $\langle\sigma\rangle$, ×10⁷ N/m²: ■ – 2.0, ▲ – 4.0, ○ – 5.0, □ – 7.0;
b) $(S - S_{st})(\langle\sigma\rangle)$ for t, minutes: ■ – 10, ▲ – 20, ○ – 30, □ – 40, ● – 50.

When the loop size becomes much less than the distance between pores, in the pore assemblage both coalescence and healing takes place which behavior will be determined by temperature and the loading value and character.

## 4. Conclusion

Analysis of variations of the contact geometry and spatial distribution of mechanical stress in the contact zone during the contact expansion was carried out. It was shown that activity of the dislocation-diffusion mechanism connected with dislocation thermo-fluctuation motion, at the contact relative value ($S = W/W_0$) more than 0.5 decreases gradually. Along with, the two-dimensional assemble of residual pores forms. The stress field around pores was analyzed, and it was shown that the pores can be healed by dislocation-diffusion mechanism at high temperature. The scheme of the dislocation-diffusion mechanism action was proposed for the contact formation due to recombination of diffusion counter-current flows of point defects: vacancies from pores and vacancy loops, and interstitial atoms – from the interstitial loop assemblages formed before.

## Acknowledgment

The authors are sincerely grateful to Dr. of Phys.-Math. Sci., Prof. V.G. Kononenko and to Dr. of Phys.-Math. Sci., Prof. Yu. I. Boyko for the discussion and constructive helpful consults.

## References

[1] Lyushinskiy A V 2006 *Diffusion welding of different materials* (Moscow: Akademia) p 208.
[2] Malygin G A 2007 Plasticity and strength of micro- and nano-crystalline materials (review). *Phys. Solid State* **49** 1013–1033
[3] Selyutina N S, Borodin E N and Petrov Y V 2018 Structural–Temporal Peculiarities of Dynamic Deformation of Nanostructured and Nanoscaled Metals. *Phys. Solid State* **60** 1813–1820
[4] Cui Y, Po G and Ghoniem N 2016 Temperature insensitivity of the flow stress in body-centered cubic micropillar crystals *Acta Mater.* **108** 128-137
[5] [5] Chembarisova R G, Dong Y and Alexandrov I V 2017 Mechanisms of high-speed defrmation of polycrystalline copper *Phys. Solid State* **59** 920–928
[6] Meyers M A 1994 *Dynamic behavior of materials* (John Wiley & Sons, N.Y.) p 668
[7] Golovin Y I, Tyurin A I, Aslanyan E G et al 2017 The physical and mechanical properties and local deformation micromechanisms in materials with different dependence of hardness on the depth of print *Phys. Solid State* **59** 1803–1811
[8] Betekhtin V I, Kadomtsev A G & Narykova M V 2020 Evolution of a Defect Structure during Creep Tests of Ultrafine-Grained Metals and Alloys Produced by Severe Plastic Deformation *Phys. Solid State* **62** 318–324
[9] Voyevodin V N , Neklyudov I M 2006 *Evolution of structure and phase state and radiation stability of construction materials* (Kiev: Naukova Dumka) p 376
[10] Parshin A M, Neklyudov I M, Kamyshanchenko N V et al 1998 *Physics of radiation phenomena and radiation materials science* (Moskov – Sankt - Peterburg - Belgorod: Belgorod State University publ.) p 378
[11] Khon Y A, Zapolsky H, Kaminskii P P et al 2019 Dynamic Spatial and Temporal Structures on the Surface of Solids under Load *Phys. Solid State* **61** 596–600
[12] Kononenko V G, Gostomelsky V S, Kononenko E V 1991 Diffusion-dislocation formation of the contact between squeezing rough surfaces [pressure welding] *Metallofizika* **12** 3–12
[13] Kononenko V G, Gostomelsky V S, Kononenko E V, Volosyuk M A 1991 Experimental investigations of diffusion-dislocation mechanism for formation of contact between squeezed rough surfaces *Metallofizika* **12** 13–21
[14] Boyko Yu I, Volosyuk M A, Kononenko V G 2013 Kinetics for dislocation structure formation in contact area of squeezed crystalline solids *Functional Materials* **20** 44–51

IOP Conf. Series: Materials Science and Engineering **907** (2020) 012045 doi:10.1088/1757-899X/907/1/012045

[15] Boyko Yu I, Volosyuk M A, Kononenko V G 2013 Mechanism of contact formation between squeezed crystalline solids *Problems of Atomic Science and Technology* **5** 163-167
[16] Timoshenko S P, Gudier J 1975 *Theory of elasticity* (Moscow: Nauka) p 576
[17] Warner D H, Curtin W A 2009 Origins and implications of temperature-dependent activation energy barriers for dislocation nucleation in face-centered cubic metals *Acta Materialia* **57** 4267– 4277
[18] Tschopp M A, McDowell D L 2007 Tension-compression asymmetry in homogeneous dislocation nucleation in single crystal copper *Appl. Phys. Lett.* **90** 121916–121919
[19] Zhu T, Li J, Van Vliet K J, Ogata S, Yip S and Suresh S 2004 Predictive modeling of nanoindentation-induced homogeneous dislocation nucleation in copper *Journal of the Mechanics and Physics of Solids* **52** 691–724
[20] Kosevich A M 1981 *Physical mechanics of real crystals* (Kiev: Naukova Dumka) p 290
[21] Betekhtin V I, Kadomtsev A G 2005 Evolution of microscopic cracks and pores in solids under loading *Phys. Solid State* **47** 825–831

IOP Conf. Series: Materials Science and Engineering **907** (2020) 012046 doi:10.1088/1757-899X/907/1/012046

# Utilization of chemical waste and by-products as one of technological progress directions

A I Gabitov[1], V A Ryazanova[1], L Z Rolnik[2], A S Salov[3], V A Timofeev[4]

[1]Department of Engineering Structures, Ufa State Petroleum Technical University, Mendeleev St., 195, Ufa, 450080, Russian Federation
[2]Department of General and Analytical Chemistry , Ufa State Petroleum Technical University, Mendeleev St., 195, Ufa, 450080, Russian Federation
[3]Department of Roads and Construction Technology, Ufa State Petroleum Technical University, Mendeleev St., 195, Ufa, 450080, Russian Federation
[4]Department of Applied Sciences, Ufa State Petroleum Technical University, Mendeleev St., 195, Ufa, 450080, Russian Federation

azat7@ufanet.ru

**Abstract.** Getting cement-free lime-containing binder by firing thereof and mixed grinding of sand together with solid sludge of still waste liquid being bulk waste of soda ash production is analyzed herein. The experience of making masonry autoclaved concrete blocks and construction of low-rise residential houses therefrom is presented. Lime-containing sludge of still waste liquid, i.e. waste of soda ash production, is shown to be prospective raw material in making construction materials as a substitute for lime.

Now basically four ways are applied for making industrial soda: the ammonia-soda process (the Solvay process), from the natural raw stock containing soda, from nephelines and by sodium hydroxide carbonization [1]. Over 50% of caustic soda in the world including Russia is made by ammonia-soda process, where the basic raw stock is sodium chloride and limestone. When making 1 ton of caustic soda up to 250 kg of solid waste (on a dry basis) is obtained and to be stored in sludge reservoirs requiring more areas therefor and taking environmental measures thereto. Solid waste contains useful components to be applied in the national economy, such as $CaCO_3$, $Ca(OH)_2$, $CaSO_4$, $CaCl_2$, $MgCO_3$, NaCl, etc. [2]. R&D was conducted in various times in the process development for preparing and utilization of solid soda waste as the raw stock for making lime, belite binders, feed additives, ameliorants and backfill materials [3]. However, actually the problem of utilization of solid soda waste of the soda ash production is not solved yet as it deals with some difficulties first stipulated by high chloride and alkali in sludge [2, 4].

One of the technological progress directions in making mineral binding materials is utilization of bulk waste of various industries, such as soda production waste in the Republic of Bashkortostan. Pilot production of lime-belite binders by “Promstroiproekt” Institute (Ufa) design was launched for making R&D in soda ash production site in 1981. Drained sludge reservoir by “Soyuzvodokanalproekt” Institute (Moscow) design was arranged for making sludge under the original data by the Institute of Kharkiv Research and Manufacturing Association “Karbonat” (Kharkiv) [2].

Production of ash soda in JSC “Bashkir Soda Company” (“BSC”) (Sterlitamak, Bashkortostan Republic) by the currently accepted ammonia-soda process is accompanied by vast waste as sludge of still waste liquid. 8-10$m^3$ of sludge containing 200-250 kg in solid phase is per 1 ton of production. This waste is delivered to special sludge basins where solid particles are sedimented (deposited).

Supernatant still waste liquid containing calcium chloride (up to 100g/l), sodium chloride (up to 40g/l), ammonia (up to 80g/l), calcium sulfate and calcium hydroxide (within solubility thereof) goes to water reservoirs or pumped into oil beds. Annual discharge of waste is about 400,000 $m^3$.

Sludge reservoirs of JSC "BSC" contain now tens of millions of tons (on a dry basis) of soda production waste being stored in vast agriculturally used areas. Main sludge reservoir area is 136 ha with 22 m in depth.

Solid sludge of still waste liquid is the sludge with humidity of 25-60% containing fine particles of calcium carbonates mainly. It also contains magnesium carbonate, calcium and magnesium hydroxide, admixtures of calcium sulfate and calcium and sodium chloride, ammonia and clay compounds. Chloride content depends on the material humidity being increased with depth of finding thereof.

A new product with the active lime similar properties obtained by burning solid sludge of still waste liquid at t = 850-950°C was found under research made by BashNIIstroi Institute together with the "Engineering structures" department specialists at the Ufa State Petroleum University in 1970-1980 s. It is characterized by activity (by CaO+MgO) of up to 60%, slaking temperature up to 86°C and high rate of slaking (2-9min). Cement-free lime-containing binder was obtained therefrom at mixed grinding with quartz sand.

According to the laboratory analyses the autoclaved aerated concrete, dense M50-M200 concrete, M200 silica bricks with the technical properties similar to that of standard materials may be made therefrom [5, 6]. This binder may also be used as the mineral plasticizer for making mortars and concretes. Technical specifications for solid sludge of still waste liquid and cement-free lime-containing binder, as well as recommendations for making dense concrete, mortar and autoclaved cellular concrete based on lime-containing binders were prepared; operating procedure for making small blocks made of autoclaved aerated concrete based on lime-containing binders was prepared; several inventor's certificates to cement-free binders of autoclaved and non-autoclaved aerated concrete and 4 foreign patents (in GDR, FRG, England and France) were obtained.

Burnt solid sludge of still waste liquid was experimentally found to be able to activate binding properties of granulated blast-furnace slag or ash left after the brown coal firing. Thoroughly selected mixes provide for getting cement-free non-autoclaved binders where there are absolutely no standard materials like lime and cement, and therewith the sand either partially or fully is substituted by ash or slag. Such binders may be used for making mortars or low-grade concrete (up to M200) [1].

Furthermore, the laboratory experiments demonstrated the possibilities to use solid sludge of still waste liquid as carbonate component to get non-autoclaved binders by the technology similar to Portland cement production of low-temperature synthesis at 1100-1200°C, i.e. by 350-250°C low than under the traditional technology (Inventor's certificate No.1076410 USSR) [7, 8]. According to X-ray examination results the resulting fired product contains calcium clorosilicates and chloroaluminates, i.e. compounds of minerals typical to alinite cement. Pilot-plant testing made by Tashkent "Promstroiproekt" and "Soda" Association together the possibility to get cement with the activity equal to 42.0MPa was proved (in accordance with GOST 810.4-76). The binder is specified by the accelerated gain in strength and expansion effect [9].

The laboratory results obtained in BashNIIstroi institute have been consistently checked in a production environment. In 1980 the pilot production of lime-containing binders from the fired soda production waste and quartz sand and items therefrom was launched in "Soda" Association (Sterlitamak) under the BashNIIstroi institute technology by design of institutes "Bashpromstroiproekt" (Ufa) and "NIOKHIM" (Kharkiv) [10, 11]. Initially these binders were used as a substitute to lime in making silica bricks in Sterlitamak Construction Materials Plant, as well as to improve plasticity and partial substitute for cement when making mortars in some companies of Sterlitamak and Ufa (e.g. "Sterlitamakstroi" trust and BNZS). In 1984 the second phase was launched, i.e. manufacture of small autoclaved aerated concrete blocks characterized by density of 600-700kg/$m^3$ and strength of 2.5…5.0 MPa. According to the project the small side blocks were made by casting technology with further cutting of the resulting expanded body to 150×300×500 mm blocks and autoclave treatment at saturated steam pressure of 0.8 MPa and isothermal processing for 8 hours.

After adoption of the technology the properties of items made of cement-free aerated-concrete did not differ from that of blocks manufactured with application of traditional binders like Portland cement and lime, and absolutely complied with GOST 21520-89 requirements.

Autoclaved aerated concrete blocks based on lime-containing binders were used for erecting residential and household buildings. Nine one-storey houses were built of such blocks in two villages of Ishimbai region as an experiment. Due to no special compositions the facing was made by usual plastering, the blocks therewith were saturated with water actually for the whole thickness. The houses were occupied in autumn of 1985.

Despite this when the houses were inspected after 6 months of upkeeping (in spring of 1986), the temperature inside the houses was found to be +22°C on the average, with outside temperature being 4-5°C and relative humidity of 62-70%. Aerated concrete moisture in the plastered sections of the wall was varying from 18 to 0.9% depending on sampling depth taken from the exterior wall, with the average value of sorption moisture in aerated concrete being 8.8%, i.e. slightly exceeding the nominal range of 8% as per Attachment D to SP 23-101-2004. Reduction of aerated concrete moisture in autumn-winter and spring seasons from water-saturated state to the value similar to equilibrium points to satisfactory moisture-yielding ability of the material in the course of house usage. During examination no freezing of walls except for angles between walls and concrete slabs where heat insulation of thermal bridge was not good enough were found. According to residents the exterior walls and houses on the whole were dry and housing conditions were comfortable [12].

The houses proved to be in good condition after 17 years of operation according to the inspection made in 2002.

As the production line for making cement-free binders was launched as the pilot one, it was designed as low-capacity. In 1986 the line was stopped for reconstruction to increase the rotary kiln capacity. Due to failed reconstruction the production was finally stopped because of lack of funding. But the short period of the pilot line activity demonstrated technical capabilities in making aerated concrete blocks of cement-free lime-containing binders from soda ash production waste, as well as use of cement-free binders for making mortars and silica bricks. The output production therewith was of similar quality with that of the items manufactured with application of lime and cement.

The problem of soda production waste was reconsidered by SUE BashNIIstroi institute in 2008. One of the complicated problem in the technology of producing the fired solid sludge of still waste liquid is extraction and delivery of water-saturated sludge before firing thereof, as under natural conditions the water evaporates from the sludge very slowly (for many years), with the only upper layers being dried out therewith. Now there are foreign-made press filters enabling to dewater the sludge and get the solid sludge of still waste liquid ready for firing. Should such press filters be installed in JSC "Soda", the technology may be significantly improved getting almost absolutely homogeneous finished product (binder). Under filtration of solid sludge content of chlorine salts is also reducing enabling expansion of areas for application of the obtained cement-free binder and items thereunder. Within the scope of this paper BashNIIstroi Institute in 2008 received a patent for "The method of getting autoclave-cured binder" from the filtered sludge [13].

Nowadays the researches of this direction are actively conducted by the Research and Educational Center of Innovation Technologies at the Architecture-Construction Institute of the Ufa State Petroleum Technical University. This Institute specialist provide for R&D support and control of quality of the materials and the work done in most sites of Ufa and the Republic of Bashkortostan. Innovation lab-tested concrete compositions with use of both waste and chemical and petrochemical by-products with high performance properties and being modified by a number of multifunctional chemical admixtures and modifiers developed and patented by the University fellows are being introduced [14]. Currently alternative nano-based binders and aggregates, as well as compositions of efficient superplasticizers are being developed together with vast expansion of automated processes in making compositions. Certificates of state registration of programs to optimize and assess the quality of the applied materials are obtained [15, 16, 17].

It should be mentioned that the raw stock for making experiments and pilot production in1980s was the sludge taken from sludge reservoirs where fresh sludge was not discharged to for many years [18]. This waste has great compositional inhomogeneity due to both small and rough particle segregation by area and in depth because the sludge was stored in the reservoir for a long period of time. The manufacturing technology is complicated by the raw stock inhomogeneity thereby reducing the finished product quality. Application of the fresh just filtered sludge and delivered directly from the manufacturing with the homogeneous chemical composition thereby increasing the quality of the binder produced is the optimum one.

Total demand in binders based on the fired solid sludge of still waste liquid of Bashkortostan silicate plants is 80,000t per year under the production output of about 130m bricks per year. Demand in binders based on the fired solid sludge of still waste liquid of Bashkortostan aerated concrete manufactures (JSC “Glavbashstroi” plant in Ufa, a plant in Agidel, JSC “Silicate Plant” in Sterlitamak) is about 50,000t under the proposed output of 250,000m$^3$ per year.

Consequently, lime-containing sludge of still waste liquid i.e. soda ash waste, was proved (by experiments, factory testing and successful running of pilot production) to be potential raw stock in making construction materials (aerated concrete, silica bricks, and mortars) as a substitute for lime [19]. Apart from saving the natural raw feed (to extract lime from limestone, in particular) the application of manufacturing waste would enable to solve environmental problems related to removal of sludge reservoirs from hundreds hectares of farmland occupied thereby.

**References:**

[1] Nasyrov R R, Daminev R R 2008 A procedure for main soda ash waste utilization *Bashkir Chemical Journal* Vol **15** No 3 p 95-100

[2] Sabitov K B, Voronin A V 2016 The outlook for utilization of JSC “BSC” soda ash waste as raw carbonate component in the manufacture of cement *Bulletin of the Academy of Science of the Republic of Bashkortostan* Vol .**21** No 3 (83) p 86-95

[3] *Waste to be applied in the construction* 1988 (Ufa: Bashkir Publishers) p 119

[4] Taylor Kh 1996 *Chemistry of cement* (Moscow: World) p 560

[5] Bedov A, Salov A, Gabitov A 2018 CAD methods of structural solutions for reinforced concrete frame *XXI International Scientific Conference on Advanced in Civil Engineering "Construction - The Formation of Living Environment" (FORM 2018)* (25–27 April 2018, Moscow, Russian Federation) Volume **365** p 1-8

[6] Bedov A I, Gabitov A I, Gaisin A M, Salov A S, Chernova A R 2018 CAD technologies under thermal properties analysis of wall cladding of framed buildings. *IOP Conference Series: Materials Science and Engineering: VII International Symposium Actual Problems of Computational Simulation in Civil Engineering* (1–8 July 2018, Novosibirsk, Russian Federation) Volume **465** p 1-8

[7] Babkov V V, Kuznetsov D V, Sakhibgareev R R, Chuykin A Ye, Khalimov R K, Gaisin A M 2006 Problems of durability of autoclave aerated concrete *Bashkir Chemical Journal* Vol. **13** No 2 p 97-99

[8] Krot A, Ryazanova V, Gabitov A., Salov A, Rolnik L 2018 Resource-saving technologies for advanced concrete in the Republic of Bashkortostan *MATEC Web of Conferences 7. "7th International Scientific Conference "Reliability and Durability of Railway Transport Engineering Structures and Buildings" (Transbud 2018)* Vol **230** Article number 03009

[9] Vinichenko V, Gabitov A I, Salov S A, Gaisin A M, Kuznetsov D V 2018 The heat loss calculating methods of external walls in the buildings reconstruction *MATEC Web of Conferences Volume 230 (2018) 7th International Scientific Conference “Reliability and Durability of Railway Transport Engineering Structures and Buildings” (Transbud-2018)* (Kharkiv, Ukraine, November 14-16) p 1-6

[10] Bedov A I, Babkov V V, Gabitov A I, Sakhibgareev R R, Salov A S 2014 Investigation of modified concrete with chemical additives *Concrete and reinforced concrete – future outlook:*

*scientific papers of the 3rd All-Russian (II International) conference on concrete and reinforced concrete* (May 12-16, 2014: Moscow: MGSU) Vol. **6** p 14-25

[11] Borisov I N, Vinnichenko V I, Ryazanov A N 2013 Energy efficient construction materials of dolomite and refuse coal *Energy efficient technology complexes and equipment for making construction materials. Interuniversity collection of papers* (BGTU named after Shukov. Belgorod) Issue **XX** p 114-117

[12] Oratovskaya A A, Galeeva L Sh, Ravilova L R 2008 A patent for invention No 2396227 *Procedure to get autoclaved binder*

[13] Bedov A, Gabitov A, Gaysin A, Salov A 2019 Engineering solutions for heart efficient exterior walls in climatic condition of the Republic of Bashkortostan *E3S Web of Conferences* **97** 02039

[14] Vagapov R F, Sinitsyn D A, Oratovskaya A A, Tanenbaum G V 2014 Use of industrial enterprises wastes while construction materials production by the example of Bashkortostan republic *News of SWSU* No 3 (**54**) p 76-82

[15] Mazitova A K, Maskova A R, Aminova G K, Builova Ye A, Nedopyokin D V, Stepanova L B *PVC plasticizer* A patent for invention RU 2681631 11.03.2019 Application No 2018111355 of 29.03.2018

[16] Pudovkin A N, Gabitov A I, Salov A S, Chernova A R *Program for calculating the grain size and grain composition of sand*: Certificate of state registration of the computer program No 2018617134 / Right holder FGBUU from UGNTU; claimed. 24.04.2018; reg. 19.06.2018

[17] Pudovkin A N, Ryazanova V A, Nedoseko I V, Salov A S *Program for calculating clay content in lumps*: Certificate of state registration of the computer program No 2018617277 / Right holder FGBUU from UGNTU; claimed 26.04.2018; reg. 21.06.2018

[18] Salov A S, Nedoseko I V, Gaitova A R, Yunusova S S *Program for calculating gypsum and slag-gypsum raw mixes to make small prior dewatered side phosphogypsum-based blocks*: Certificate of state registration of the computer program No 2015614300 / Right holder FGBUU from UGNTU; claimed 24.02.2015; reg. 14.04.2015

[19] Alp I, Deveci H, Süngün H 2008 Utilization of flotation wastes of copper slag as raw material in cement production *Journal of Hazardous Materials* Vol. **159** No 2-3 p 390-395

[20] Lyazat A, Nabi D 2014 Process parameters of production of non-autoclaved aerated concrete on the basis of complex use of ash and gypsum-containing wastes *Mediterranean Journal of Social Sciences* Vol. **5** No 23 p 2565-2571

IOP Conf. Series: Materials Science and Engineering **907** (2020) 012047 doi:10.1088/1757-899X/907/1/012047

# Innovative Thermal Insulation Materials with Dual Properties

**K Plakhotnikov[1], O Starkova[1], O Dedenova[2], D Bondarenko[1], O Demina[3]**

[1]Department of Computer Science and Information Technologies, Kharkiv National University of Civil Engineering and Architecture, Sumska str. 40, 61002 Kharkiv, Ukraine
[2]Department of Building Materials and Products, Kharkiv National University of Civil Engineering and Architecture, Sumska str. 40, 61002 Kharkiv, Ukraine
[3]Department of Physical Chemical Mechanics and Technologies of Building Materials and Products, Kharkiv National University of Civil Engineering and Architecture, Sumska str. 40, 61002 Kharkiv, Ukraine

kirillplahotnikov84@gmail.com

**Abstract**. The paper presents the results of development and research of an innovative thin-layer thermal insulation material that protects the building from heat loss due to thermal radiation. The material is a dry mortar of cement binder and filler of hollow aluminosilicate microspheres. A thin-layer coating made from this optimally combines the functions of effective heat protection and external finishing warm plastering that is advantageous when reconstructing and restoring historical objects with a lavish decoration. The conducted studies have shown that aluminosilicate microspheres in the composite thermal insulation coating are also involved in the reflection of infrared radiation. The use of Class 2 admixtures made it possible to synthesize additional crystalline hydrates with the aluminate phase of clinker brick, which contributed to the strengthening of the cement composite at the micro- and submicrolevels. This was confirmed by electron microscopy. For the developed composition of the mineral thin-layer thermal insulation coating, the cost of the mixture is almost 3 times lower than for the known composition of the thermal insulation coating on the polymer binder. In addition, unlike polymer analogs, there are no restrictions in terms of temperature and service life. The thermal insulation coating may be applied both exteriorly and interiorly with no condensation being formed between the wall and the coating. Applying the developed thin-layer thermal insulation coating on a brick wall measuring 10 m by 10 m in the climatic conditions of Kharkiv allows reducing heat consumption by 0.188% per 1 $m^2$ per year.

Currently, the demand for thermal insulation materials shows a tendency for steady growth. Their varieties are growing rapidly. The domestic market offers a wide range of thermal insulation materials, including products made of mineral and glass fibers, products made of foamed polymers, cellular concretes and more. While thick-layer thermal insulation materials have been known for a long time now, liquid thermal insulation materials in thin layers require further research. This group of construction materials is relatively new; there is almost no equipment or methods for its study.

Scientists Yu. M. Bazhenov, V. T. Yerofieiev, Ye. V. Korolov, Yu. V. Pukharenko, V. N. Sokiv and S. V. Fedosov are engaged in the development of high-performance thermal insulation materials by creating porous structures. The theoretical foundations of the structure formation in filled polymer composite materials are explored in the works of A. N. Bobryshev, V. T. Yerofieiev, Yu. H. Ivashchenko, V. F. Stroganov, B. Muller. The works by D.V. Oreshkin and Ye. V. Korolev on the use of hollow microspheres in high-strength lightweight concrete, grouting and masonry mortars confirm their effectiveness as a light filler. Of great interest are studies of methods for determining the thermophysical characteristics of insulating materials and structures, considered in the works by M. O. Bakanov and V. H. Kotlov.

The transfer of thermal energy occurs spontaneously from one isothermal surface to another towards the equalization of temperatures from high to low temperatures. Heat transfer can be performed by convection, thermal radiation or thermal conductivity. Since heat transfer in solids is due to the movement of phonons (a phonon is a quantum of vibrational motion of the crystal atoms) or electron gas along the crystal lattice, metals with a large number of free electrons and a perfect crystal lattice exhibit the maximum thermal conductivity. Thermal insulation materials that are amorphous or submicrocrystalline structures have a small number of free electrons, so thermal energy is transferred as a result of phonon interactions. The vibrations in noncrystalline solids are anharmonic in nature owing to the attenuation of longitudinal elastic waves in the course of their propagation.

Most thermal insulation materials are multi-component systems; therefore, heat transfer in these depends on the thermal conductivity of the solid phase, on the convection and on the thermal conductivity of the gas in the pores and the radiation in these. The relative contribution of these components varies depending on the porosity level. New energy saving standards have forced to increase the requirements for thermal protection of buildings, which has increased interest in the development and production of new non-conventional materials. Of particular interest at present are thermal insulation materials made in the form of dry mortars from inorganic binder and mineral or organic filler, the so-called warm plasters (WP). These materials may simultaneously function as both thermal insulation and decoration. Their thermal insulation performance is only 1.5 times lower than that of recognized insulation materials such as expanded polystyrene and mineral wool, and 10 times higher than that of brick (the thermal conductivity of WP is 0.06 W/(m·°C), of foam is 0.039 W/(m·°C), of brick 0.65 W/(m·°C)). This means that 5 cm of plaster on the wall will replace 3 to 4 cm of foam or 51 cm of brick. Thermal insulating plasters are easy to use. These are dry mortars that are simply diluted with water before use. Insulation using WP is performed quickly and does not require particularly high qualification of the contractor. Cement, lime and gypsum are used as binders in WP. WP are adhesive to all materials; there is no need to use meshes when using WP. They are quite strong, and their durability is comparable to that of the base. However, WP are not always advantageous to be used as the only layer of insulation. To ensure the specified resistance to heat transfer of a brick wall with a thickness of 51 cm in climatic zone I in Ukraine, a 9 to 12 cm thick layer of WP should be laid. Under such conditions, WP are used as additional insulation. In climatic zone IV in Odessa and Kherson regions, a 2.5 to 5 cm thick layer of WP is enough. If the main structure of the wall is made of more energy-efficient material such as expanded clay concrete or shell rock, WP only will be enough in any climatic zone. As additional thermal insulation, WP are easy to apply on existing buildings, in particular when reconstructing and restoring historical objects with a lavish decoration. Plaster cannot be replaced when applying insulation on curved surfaces of facades, arches, architectural details, window slopes and sloping surfaces. If necessary, WP can be used as internal insulation, provided a vapor barrier be installed.

The performed literature data analysis has confirmed the unique character of plasters and thermal insulation coatings that work based on the principle of reflective thermal insulation. It is recommended for both outdoor and indoor application. One may insulate everything that needs to be plastered and painted. These materials function both as thermal insulation and a leveling and finishing plaster. No special construction skills are required for installation. Therefore, the development of a thin-layer thermal insulation coating with high thermal and physical and mechanical properties based on the mineral binder and fillers from hollow microspheres with a reflective power is a priority at present.

The research objective is set to justify and develop the composition of a thin-layer thermal insulation coating on cement binder and fillers of aluminosilicate and glass hollow microspheres with high physical and mechanical and thermal insulation properties, particularly by means of reflecting infrared radiation by modifying the binder.

Thin-layer cement-based compounds quickly lose moisture from the surface owing to the evaporation and absorption of moisture by porous substrates. As a result, the hydration processes are insufficient and such material after curing has low physical and mechanical properties.

The main objective for improving the properties of the thin-layer coating is to increase the number of crystalline hydrates in the composition of cement stone, as well as to adjust the curing time and rheological properties, which may be done by using chemical admixtures. This effect can be produced, for example, by plasticizers, water-retaining admixtures, and salt solutions.

Chemical admixtures on the basis of cement components were classified in terms of the mechanism of their action by V.B. Ratinov and F.M. Ivanov [1]. Class 2 admixtures interact with aluminum- and iron-containing phases of cement with the formation of poorly soluble hydrated double salts. Their specific feature is that they all participate in the addition reactions that occur in solutions and crystallize as hydrated double salts from supersaturated solutions.

This effect is exerted by sulfate-, chlorine- and nitrate-containing admixtures, which lead to the formation of crystalline hydrates such as ettringite or other hydrated calcium sulfoaluminates, hydrated calcium oxychlorides, hydrated calcium aluminates and hydrated calcium nitroaluminates, which have, as noted, an equilibrium surface potential [2, 3].

In the microstructure of the cement paste there are cement particles with a predominant fraction by weight of 50 μm, which are the structural elements of the microstructure of the cement dough and cement stone.

Cement particles are represented by four main minerals: alite $C_3S$, belite $C_2S$, tricalcium aluminate $C_3A$ and tetracalcium aluminoferite $C_4AF$.

A clear idea of the structural elements of cement stone can be obtained from electron microscopic images (EMC).

According to the paper [4], on the surface of $C_3S$ there are flat hexagonal formations of Portlandite $Ca(OH)_2$, which grow from the bumps of growth in the form of cone-shaped crystals measuring 200-300 nm (2000-3000 Å). In the volume between the cement particles there are larger prismatic crystals of portlandite. Ettringite $3CaO{\cdot}Al_2O_3{\cdot}3CaSO_4{\cdot}31H_2O$ in free space has the form of individual long needle-shaped or prismatic crystals up to 10 μm and more, the thickness of which is up to 1-2 μm. In cement stone, the size of ettringite crystals is an order of magnitude smaller; the size of its blocks is up to 1 μm (1000 nm) at a thickness of 100-200 nm (1000-2000 Å). It is believed that these differences in the size of ettringite in the free water space and in the cement paste are due to the thickness of the aqueous layer between the cement particles in the cement paste during its mixing. The thickness of this layer averages 1-3 μm depending on the water-cement ratio.

The blocks of hydrated calcium monosulfoaluminate (CMS) $3C_3O{\cdot}Al_2O_3{\cdot}CaSO_4{\cdot}19H_2O$ have the shape of thick hexagonal plates; the size of the blocks is less than 1 (one) μm. The blocks of hexagonal hydrated calcium aluminates (HCA) $4C_3O{\cdot}Al_2O_3{\cdot}13H_2O$ are similar in shape [5].

Hydrated calcium aluminates $3C_3O{\cdot}Al_2O_3{\cdot}4H_2O$ are cubic in shape [5]; their size does not exceed 1 (one) μm.

In free space, the singularity of the growth of ettringite in the form of thin needle-shaped crystals is also manifested on the surface and in the layers of cement grains [4]. This suggests that ettringite can developed on the surface of glass microspheres by forming a kind of lattice, which serves as a structure that binds the individual microspheres during the first hours of cement hardening. The conducted studies using electron microscopy confirmed the suggested scientific hypothesis (figure.1).

IOP Conf. Series: Materials Science and Engineering **907** (2020) 012047 doi:10.1088/1757-899X/907/1/012047

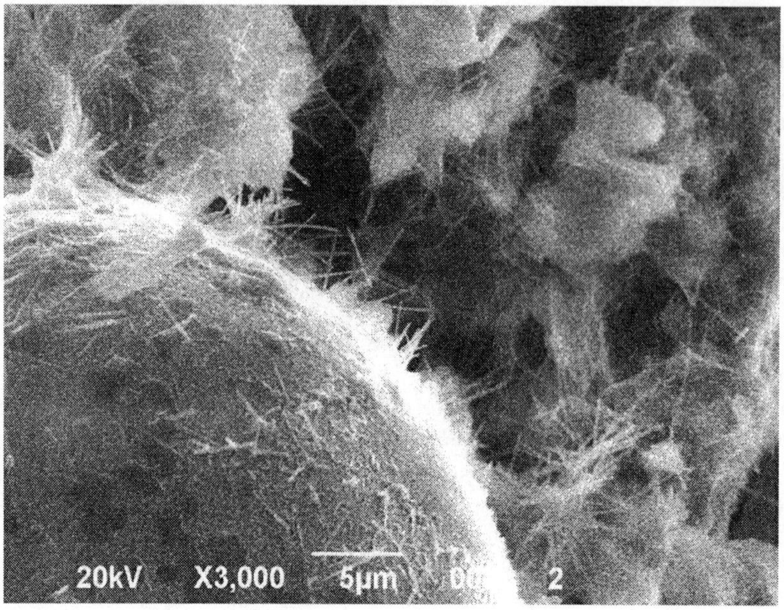


**Figure 1**. The growth of ettringite needles on the surface of glass microspheres

Hydrated calcium silicates in the form of thin felted structures will develop on aluminosilicate microspheres and serve as a basis for the synthesis of the hydroaluminate phase (figure. 2), which is confirmed by studies using electron microscopy.

**Figure 2**. Fibrous structure of hydrosilicate gel on the surface of aluminosilicate microspheres

As a result of experimental research and application of the mathematical planning method of the experiment, a rational composition and rational ratio of aluminosilicate and glass hollow microspheres, a ratio of Portland cement and hollow microspheres, carboxymethyl cellulose (Walocel with a molecular weight of 15,000) and a complex chemical compound were selected, which included a plasticizer, calcium nitrate and calcium chloride in the range of 0.05% to 0.09% and 1% to 2.8% of cement, respectively.

Chemical admixtures in the form of water-soluble compounds, just as Class 2 admixtures such as calcium chlorides and calcium nitrates, are capable under certain conditions to react with tricalcium aluminate and tetracalcium aluminoferrite with the formation of poorly soluble hydrated double salts. This is due to the fact that the joint presence in the concrete of two or more of these admixtures induces a "competition" between these for the right to participate in the reaction with calcium aluminates $3CaO{\times}Al_2O_3$ and calcium aluminoferrites $4CaO{\times}Al_2O_3{\times}Fe_2O_3$. The admixture, which with these components of cement forms the most poorly soluble salts promotes the greatest supersaturation and, as a result, crystallizes with the maximum speed, wins the competition. Hydrated calcium sulfoaluminate has similar properties in cement; it crystallizes upon interaction between aluminum in the composition of cement and calcium hydroxide and sulfates (gypsum in Portland cement) (figure.1).

As long as these reactions go on, other admixtures introduced into the concrete with mixing water such as $CaCl_2$ and $Ca(NO_3)_2$ wait, i.e. remain in the electrolyte of the cement paste in the form of the corresponding ions, which accelerates the hardening of silicate components (e.g. felted formations of the CSH phase) (Ref. figure.2) by changing the solubility of calcium silicates. These admixtures act as Class 1 admixtures, which do not contain any ions of the same name with cement and accelerate hardening by changing the ionic force of the solution. Since cement must contain gypsum, any of the above admixtures does not react with the aluminum-containing components of the cement until the gypsum is almost completely bound to the poorly soluble hydrated calcium sulfoaluminate. Further interaction with calcium nitrate and calcium chloride is possible, if after the formation of hydrated calcium sulfoaluminate in the solution of cement paste there remain tricalcium aluminate and tetracalcium aluminoferrite with the synthesis of calcium hydrochloro- and hydronitroaluminates, as evidenced by the calculated negative Gibbs free energies, -307.1 and -280.2 respectively.

A Seek thermal imager was used to determine the reflection efficiency of infrared radiation of the thermal insulation composition based on modified Portland cement and fillers made of aluminosilicate and glass microspheres. The Seeks wavelength spectral range operates in the wavelength range of 3.5 to 5.5 µm or 8 to 13.5 µm. To set up the experiment, a plastic container was used, on one part of the outer surface of which a composite thermal insulation material was applied. The other part of the outer surface of the container remained uncovered. The container was filled with water heated to a certain temperature. The temperature on the test surface was determined using a thermal imager. The temperature of the water poured into the container was 78°C; the outside temperature on the surface of the plastic container was 73°C; the temperature on the surface of the thermal insulation layer with a thickness of 1 mm was 51°C, and on the layer with a thickness of 2 mm was 28°C (figure. 3).

According to the obtained measurement results, it can be concluded that thin-layer coatings on the base of Portland cement filled with aluminosilicate and glass microspheres quite effectively reflect heat fluxes even at a thickness of 1-2 mm. The thermal conductivity coefficient of the thin-layer coating was calculated, which, depending on the content of components, ranged from 0.033 to 0.067 W/(m·°C).

A composite material was obtained in the form of a dry mortar with physical and mechanical properties and performance shown in Table 1.

To determine heat savings owing to the use of a thin-layer thermal insulation coating from the developed dry mortar, calculations were performed on the heat transfer resistance of the enclosing structure with reference to the thermal insulation (TI) coating containing glass microspheres according to the proposed method [6]. The building, for which the calculation was performed, operates in Kharkiv; the enclosing structure is shown by the form of a wall of pressed cement and sand brick with an area of 100 $m^2$ and a thickness of 0.5 m. The outside part of the wall is covered with a thermal insulation layer; the inside part is plastered with lime and sand mortar. When performing calculations, it was taken into consideration that the heat flux $q$ from the outside wall to the surrounding space has

IOP Conf. Series: Materials Science and Engineering **907** (2020) 012047 doi:10.1088/1757-899X/907/1/012047

two components: the convective component $q_k$ with heat transfer from the wall surface to the outside air and the radiation component $q_r$ which is transmitted remotely by infrared radiation.

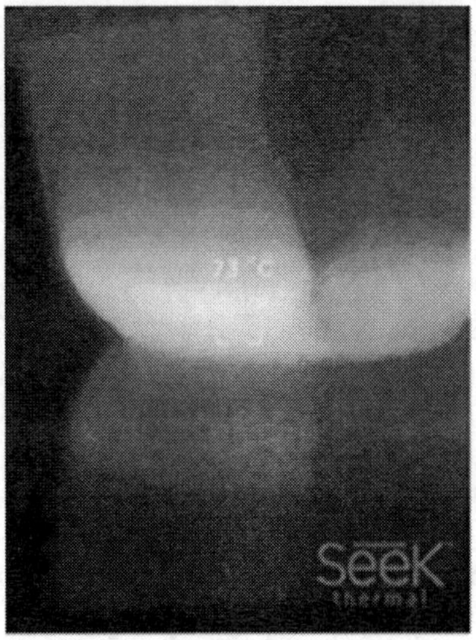

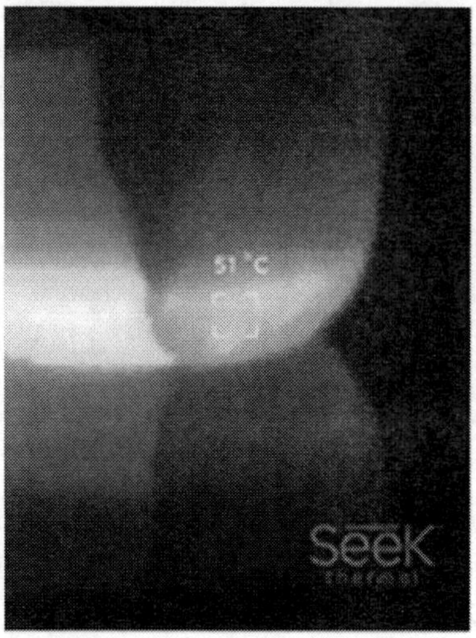

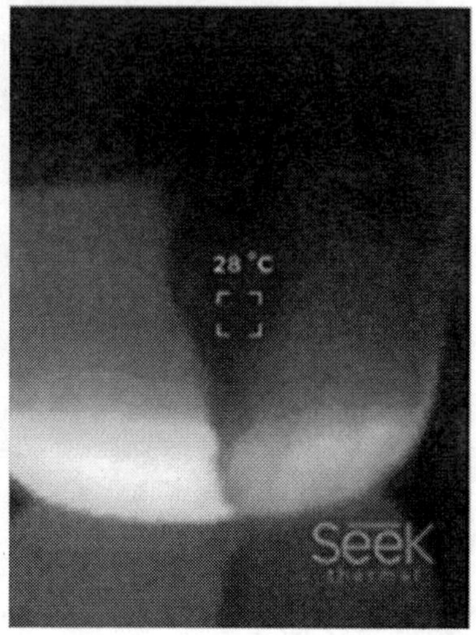


a b c

**Figure 3**. Measuring the surface temperature using a thermal imager: a) on the surface of the plastic container (73°C); b) on the surface of the coating 1 mm thick (51°C); c) on a surface of a thermal insulation coating 2 mm thick (28°C)

**Table 1.** Physical and mechanical properties and performance of the hardened composition of the thin-layer thermal insulation coating

| Parameter | Test results | Requirements according to ДСТУ-П Б В.2.7-126: 2011 |
|---|---|---|
| Average density, g/cm$^3$ | 0.42 | - |
| Tensile strength at bending, MPa | 1.65 | 0.5 |
| Compressive strength, MPa | 1.75 | 1.0 |
| Crack resistance | No cracks | No cracks in the layer with a thickness equal to the recommended maximum applied thickness in one go |
| Adhesion to the base, MPa | 1,33 | 0.2 |
| Frost resistance F, cycles, at least | 25 | - |
| Vapor permeability, not less than, mg/m·h·Pa | 0.07 | 0.05 |

Since the thermal insulation coating consisting of hollow glass microspheres is an optical system, it leads to a change in the emissivity of the wall to reduce the radiation component of heat loss and reduce the amount of heat flux from the wall surface. As the coating changes the structure and magnitudes of heat loss from the wall surface, the heat transfer resistance of the wall changes. We define the change in heat transfer for the enclosing structure with and without coating at the indoor air temperature $T_i$. The emissivity of the thermal insulation coating has a lower emissivity $\varepsilon_t$ than the emissivity of the outer surface of the uncoated wall $\varepsilon_0$. From the classical definition of heat transfer resistance [7], we obtain the heat transfer resistance of an uncoated wall and the heat transfer resistance of a wall with a thermal insulation coating containing hollow microspheres.

The intrinsic thermal resistance of the thermal insulation (TI) coating with a thickness of 1 mm and thermal conductivity $\lambda = 0.067$ W/(m·°C) will be $Rt = 0.0150$ m$^2$·°C/W. The effective emissivity of brickwork is taken as $\varepsilon$ = 0.95. The effective emissivity of the coating consisting of 90% of hollow glass and aluminosilicate microspheres by volume is taken as $\varepsilon$ = 0.85. The thermal resistance of the wall without TI coating is $Rt$ = 0.8118 $m^2 \cdot °C$ /W, and with TI coating is $R_t^{ti}$ = 0.913 $m^2 \cdot °C$ /W.

The average annual heat consumption for the outside wall at the admitted heat transfer resistance is $Q$ = 770860.6 W/year, and the average annual heat consumption for the entire surface of the wall with TI coating is $Q$ = 625770 W/year. Heat savings by applying a TI coating containing glass hollow microspheres on a wall with an area of 100 $m^2$ in the climatic conditions of Kharkiv allows reducing heat consumption by 18.82% per year or by 0.188% per 1 $m^2$ per year.

To sum up, it should be noted as follows.

The developed thermal insulation material optimally combines the functions of effective heat protection and external finishing warm plastering. It is easy to apply on existing buildings, in particular when reconstructing and restoring historical objects with a lavish decoration as additional insulation. Plaster cannot be replaced when applying insulation on curved surfaces of facades, arches, architectural details, window slopes and sloping surfaces. The most effective application of this thermal insulation material is planned as a finishing plaster for indoor thermal insulation, in particular where intense infrared radiation (heating radiators) is present and for the installation of "warm floors". For the developed composition of the mineral thin-layer thermal insulation coating, the cost of the mixture is almost 3 times lower than for the known composition of the thermal insulation coating on the polymer binder. The economic efficiency from the introduction of the developed composition for the arrangement of a thin-layer thermal insulation coating is about 70 UAH/$m^2$.

The developed thin-layer thermal insulation coating does not require positive temperatures when applied (which are not below +50°C) or heated buildings for storage, has unlimited service life, unlike polymeric analogs (not exceeding 10 years). The thermal insulation compound may be applied both exteriorly and interiorly with no condensation being formed between the wall and the coating. Applying a thin-layer thermal insulation coating on a brick wall measuring 10 m by 10 m in the climatic conditions of Kharkiv allows reducing heat consumption by 0.188% per 1 $m^2$ per year.

**References:**

[1] Ratinov V B, Ivanov F M 1969 *Chemistry in construction* (Moscow: Stroyizdat) p 199
[2] Larionova Z M, Nikitina L V, Garashin V R 1977 *Phase composition and strength of cement stone and concrete* (Moscow: Stroyizdat) p 263
[3] Plakhotnikov K, Bondarenko D, Dedenyova E, Saliia M, Kostuk T 2018 Heat-insulating materials in modern building industry *MATEC Web of Conferences.* doi: https://doi.org/10.1051/matecconf/201823002024.
[4] Shangina N N 1998 *Prediction of the physical and mechanical characteristics of concrete taking into account the donor-acceptor surface properties of fillers and aggregates*: abstract, Dissertation for the degree of Doctor of technical sciences (St.Petersburg) p 45
[5] Shpynova L G et al. 1971 *Physical and chemical fundamentals of the formation of the structure of cement stone* (Lviv: Vyshcha shkola) p 157
[6] *Methodology for calculating the heat transfer resistance of the building enclosing structure with respect to Moutrical coating* Aailable at: http://geum.ru/next/art-200091.php.
[7] Lykov A V 1978 *Heat and mass transfer* (Handbook) (Moscow: Enerhiya) p 480

IOP Conf. Series: Materials Science and Engineering **907** (2020) 012048 doi:10.1088/1757-899X/907/1/012048

# Comparative evaluation of quality of hard and special paving grade bitumen's

**Y Pyrig, A Galkin, S Oksak**

Department of Technologies of Road-Building Materials and Chemistry, Kharkiv National Highway University, Kharkiv, Ukraine

pirig2000@gmail.com

**Abstract**. The application of hard paving grade bitumen or polymer modified bitumen is one of the solutions to prevent a creep of road pavement. The implementation of special hard bitumen and multigrade bitumen starts at 80-th of the last century in a purpose to construct high module asphalt pavement. The results of comparative evaluation of conventional and fundamental (cohesion, adhesion, viscosity) quality indicators for hard paving grade bitumen and multigrade bitumen are given in the article. Low value of temperature susceptibility and a wide plasticity interval are obtained as characteristics of multigrade bitumen. The same is inherent to the air-blown bitumen, with a "gel" type structure. Low ductility and a significant change in softening point temperature after hardening can be considered as disadvantages of multigrade bitumen.

## 1. Introduction

The early failure of flexible road pavements is stipulated by steady increasing in amount of traffic load and axial loads on a pavement all over the world. Early failure reduces the terms between overhauls as a common durability of a road pavement. Permanent deformation (as a rutting, waves etc.) initiated by a decline in quality of a road pavement as by a residual deformation of a base layers of a road construction is a wide spread reason of low durability of road pavement.

The implementation of low penetration bitumen binders (hard paving grade bitumen, polymer modified bitumen and bitumen with warm mix additives) is one of ways to reach a high quality of road pavements and reduce permanent deformations. Mastic asphalts with natural bitumen are the first road pavements in a history. The obtained pavements were characterized by high strength and durability because the low penetration (0 ... 40 × 0.1 mm) and extremely high softening point (≥ 90 ... 100 °C) of the binder in their composition. [1]. Road bitumen with penetration range 40...70 × 0.1 mm was implemented in a road building industry in a result of increasing in an oil refining capacity and chemistry production of bitumen at the beginning of the 20-th century [1–2]. The further development of road industry and designing of new types of asphalt concrete led to the common use of bitumen with penetration range from 100...200 × 0.1 mm to 200...300 × 0.1 mm.

The returning of hard bitumen in road mixtures design got in trend in 50-th 70-th last century. This was a consequence of a few main factors [3]. At first, using of dense Middle Eastern oil (as a result of oil crisis) led to decreasing in penetration of bitumen as a product of oil refinery. At second, the increasing in traffic load caused significant residual deformations of existed road pavements that made a choice of bitumen with high penetration doubtful. At third, the development of oil refining technology allowed producing of high-quality road bitumen with lower penetration. Intention to improve the durability of the road pavements raised more sever demanding for the bitumen temperatures of softening point and breaking point to extend the plasticity interval. It was found that the most common ways to increase softening point temperature and to decrease breaking point temperature are a polymer modification and an improvement in technological processes of bitumen production [3].

In 1984 Royal Dutch Shell PLC introduced the bitumen "multigrade" with lower temperature susceptibility (higher Penetration Index value) and wider plasticity range correspondingly. This bitumen was branded as "Shell Multiphalte" [4]. Since 1985 a global testing of the pavements with multigrade bitumen has been started in different countries [5]. The results of testing of a few years old pavements in field proved an absence of the residual deformations, such as rutting, and a good durability of the pavements. The gathered experience helped to implement this type of bitumen in road industry for other countries [6].

The low temperature susceptibility of Shell Multiphalte bitumen is proved by comparative analysis with conventional road bitumen and polymer modified bitumen, presented in a research report [6]. However for the multigrade bitumen significant changes of the standard properties after hardening by RTFOT are registered. In compare with conventional road bitumen the average value of softening point temperature for the multigrade bitumen is higher on 3.5 °C, average residual penetration for the multigrade bitumen is lover on 5.5 %. The viscosity changes with temperature are similar for both Shell Multiphalte and polymer modified bitumen, which causes the higher temperatures for the multigrade bitumen at mixing with aggregates, and higher temperatures at the compaction (round 19-27 °C more than for the paving grade bitumen) [5, 6]. Due this multigrade bitumen are taking an intermediate position between conventional bitumen and polymer modified bitumen [7].

At the end of 90-th an extended research of multigrade bitumen has been carried out in Poland [8, 9]. The obtained data prove the efficiency of the multigrade bitumen as a component of the High Stiffness Module Asphalt concrete.

Nowadays the multigrade bitumen properties are normalized with European EN 13924-2:2014 [10] and Australian AS 2008-2013 [11] standards. Multigrade bitumen applies as a component of the thin layers of asphalt concrete and in a composition of High Stiffness Module Asphalt concrete.

The process of harmonization of the European standards for test methods and quality of the pavement and special bitumen has been started in Ukraine since 2016. Considering this the finding differences and peculiarities of these binders technology in contrast with conventional air blown road bitumen is a high priority research topic.

## 2. Materials and methodology

Special bitumen multigrade Unibit 35/50, low penetration bitumen 20/30 and 35/50 supplied by Polish "LOTOS Asfalt" are chosen for the research. For comparative analysis bitumen (Bo) is obtained in a laboratory with a petroleum binder of low viscosity (penetration at 25 °C is 214 × 0.1 mm, softening point temperature is 37.2 °C). To obtain bitumen Bo petroleum binder of low viscosity was blown by air flow with a rate 5 l/min at temperature 200 ± 5 °C in compressor oxidizer until its penetration got close to the value of multigrade bitumen, chosen for the research.

The common quality indexes are obtained for this bitumen (penetration, softening point temperature, Fraass breaking point temperature, ductility and properties changing after hardening). In addition, a number of specific tests are performed: cohesion test, adhesion test, temperature susceptibility and viscosity tests, which can provide more comprehensive evaluation of advantages and disadvantages of multigrade bitumen.

Cohesion test is performed in a planar shear condition [12] with finding of the maximal stress at deformation rate of 1 $s^{-1}$ at 25 °C. For this purpose, the binder sample in a layer of 200 um. is placed between two parallel polymer tapes, making a bound. Stretching of this bound with a constant speed initiates sample deformation with a shear rate that can be set in a range 0.2 $s^{-1}$ – 4.2 $s^{-1}$. Maximal stress is determined as a maximal value of the resisting force at deformation, divided on a binder sample (spot) area.

Adhesion is estimated by the amount of binder that stays on a surface after exposing in water medium at specific conditions [13]. For this purpose glass surface 70 × 25 mm is covered with 0.35 g of bitumen (at this condition the thickness of the bitumen layer is round 200 um). The samples (glass plate with a bitumen layer) are heating at 85 °C higher than a softening point temperature during 30 minutes in a horizontal position (bitumen surface on a top). After cooling samples are placing in a water bath with distilled water at a temperature 75 °C during 25 min. Adhesion is calculated as a percent of the surface covered with bitumen after the test.

Temperature susceptibility of bitumen is evaluated by the Penetration Index, which is calculated with the conventional equation (1) and equation modified by a softening point temperature replacing. In a conventional equation the softening point temperature is defined by Ring and Ball test. In a modified equation the temperature at which penetration reaches 800 × 0.1 mm value ($T_{800}$) is set as a softening point temperature.

$$IP = \frac{(20 \times T) + (500 \times \lg P_{25}) - 1952}{T - (50 \times \lg P_{25}) + 120} \quad (1)$$

In the equation $P_{25}$ is a penetration at 25 °C, T is a softening point temperature by Ring and Ball test ($T_{R\&B}$), or a temperature at 800 × 0.1 mm penetration ($T_{800}$). $T_{800}$ temperature is found by graph of penetration changing within a test temperature set 5 °C, 15 °C, 25 °C и 35 °C [14].

Evaluation of the colloidal structure type of bitumen is performed with "Coefficient of Standard Properties" ($C_{STD}$) by equation (2).

$$C_{STD} = \frac{(T_{R\&B} - T_{Fraass}) \cdot L}{25 \cdot D} \quad (2)$$

In the equation $T_{R\&B}$ is a softening point temperature by Ring and Ball test, $T_{Fraass}$ is a breaking point temperature by Fraass, L is a length of a "neck" part of the dumbbell form for the ductility test (3 cm), 25 is a ductility test temperature, D is a ductility. When the value of $C_{STD}$ is higher than 0.13 bitumen is classified as a "gel" type. Value of $C_{STD}$ is lower than 0.08 points on a "sol" type of bitumen. Intermediate values of $C_{STD}$ refer to a "sol-gel" bitumen type.

Binder viscosity obtained with a rotary rheometer in a wide range of temperatures (from 90 °C to 170 °C) and shear rates (from 0.005 $s^{-1}$ to 300 $s^{-1}$).

The indicators of quality of binders, chosen for the analysis, displayed in Table 1.

## 3. Results and Discussion

The experimentally obtained data points on both convincing advantages and disadvantages of multigrade bitumen. As an advantage can be considered lower temperature susceptibility, that is a result of high softening point temperature, and low Fraass breaking point temperature. Due this the plasticity interval of Unibit 35/50 is significantly wider than the one of hard bitumen 20/30 and 35/50 (on 15.9 °C and 18.3 °C respectively).

The change of the penetration with a temperature was obtained to estimate the temperature susceptibility of the chosen bitumens (Figure 1). The shown data prove the low temperature susceptibility of the multigrade bitumen thus temperature dependency of its penetration flatter than others. Such type of dependency characterizes a "gel" structural type of bitumen (in this research work it is bitumen Bo) and a polymer modified bitumen. Wherein the softening point temperature and a temperature at 800 × 0.1 mm penetration for the multigrade bitumen are close, that is inherent to the "sol" type bitumen.

By the Penetration Index value, obtained with softening point temperature and with a temperature at 800 × 0.1 mm penetration, multigrade bitumen can be characterized as a "sol-gel" type, but with a Coefficient of Standard Properties, that with softening and breaking point temperatures taking in a consideration ductility at 25 °C [15], multigrade bitumen is a typical sample of the "gel" structure binders ($C_{STD}$ = 1.91). This can explain the close values of the softening point temperature and Fraass breaking point temperature for both multigrade bitumen and Bo bitumen. And the last one can be classified as a "gel" type bitumen only, with $C_{STD} = 0.83$.

**Table 1.** Bitumen properties.

| Index | | Value | | | |
|---|---|---|---|---|---|
| | | 20/30 | 35/50 | Unibit 35/50 | Bo |
| Penetration at 25 °C, × 0.1 mm | | 26 | 44 | 35 | 39 |
| Penetration at 0 °C, × 0.1 mm | | 10 | 15 | 21 | 18 |
| Softening point temperature by Ring and Ball test ($T_{R\&B}$), °C | | 62.1 | 53.2 | 63.5 | 63.7 |
| Fraass breaking temperature ($T_{Fraass}$), °C | | -6.5 | -13.0 | -21.0 | -19.0 |
| Ductility at 25 °C, cm | | 25 | > 150 | 5.3 | 12 |
| Penetration Index, calculated with a softening point temperature ($PI_{TR\&B}$) | | -0.02 | -0.72 | 0.85 | 1.13 |
| Plasticity Interval ($T_{R\&B} - T_{Fraass}$), °C | | 68.6 | 66.2 | 84.5 | 82.7 |
| Coefficient of Standard Properties ($C_{STD}$) | | 0.33 | < 0.05 | 1.91 | 0.83 |
| Temperature at 800 × 0.1 mm penetration ($T_{800}$), °C | | 67.0 | 57.0 | 63.0 | 69.5 |
| Temperature difference $\Delta T = T_{R\&B} - T_{800}$, °C | | -4.9 | -3.8 | 0.5 | -5.8 |
| Penetration Index, calculated with a $T_{800}$ temperature ($PI_{T800}$) | | 0.82 | 0.11 | 0.76 | 2.13 |
| Hardening at 163 °C during 5 hours | Retained penetration, % | 88.5 | 81.8 | 94.0 | 87.0 |
| | Increase in softening point, °C | 4.3 | 4.8 | 7.3 | 4.7 |
| | Change of mass, % | 0.36 | 0.36 | 0.44 | -2.4 |
| Adhesion to the glass surface at 75 °C, % | | 88.8 | 62.0 | 73.0 | 74.7 |
| Cohesion at 25 °C and at a shear rate 1 $s^{-1}$, MPa | | - | - | 0.248 | 0.271 |

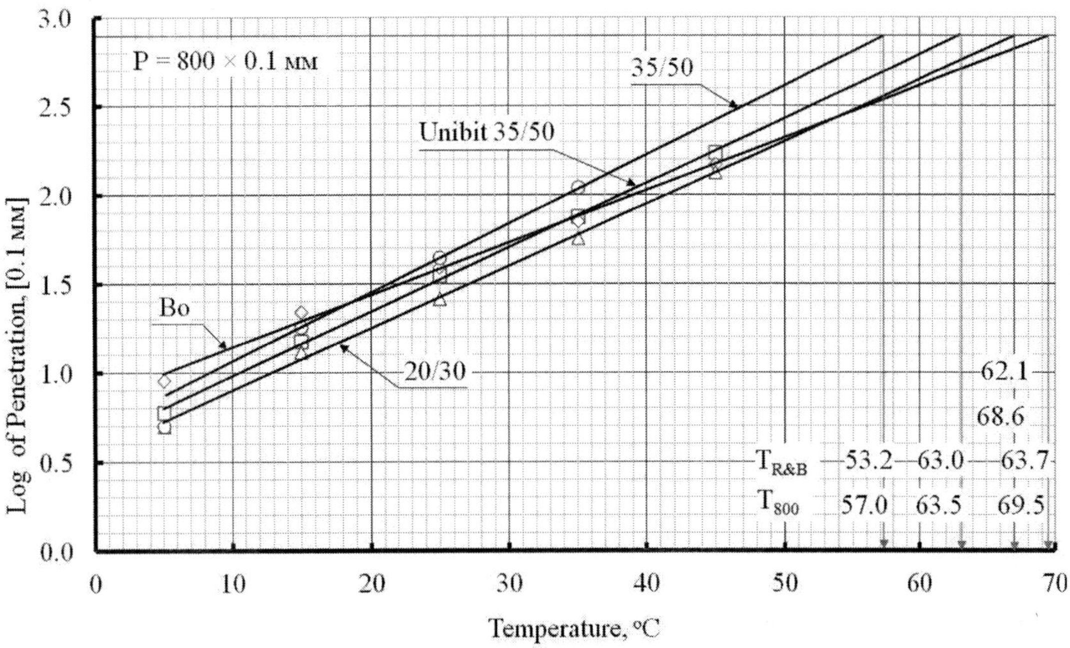


**Figure 1.** Temperature penetration dependencies of bitumens.

One more evidence that colloidal structure of multigrade bitumen even closer to a "gel" type than the one of specially prepared in laboratory bitumen Bo is speed-cohesion dependencies of these binders (Figure 2). With comparison of the properties of Unibit 35/50 and Bo bitumens that have the same penetration differ in cohesion. For multigrade bitumen cohesion is lower in 1.09 times than for

Bo bitumen. This is in an agreement with data, obtained by Kolbanovskaya in [16], who noticed that cohesion values decreasing with changing of the bitumen structure from "sol" to "gel" type.

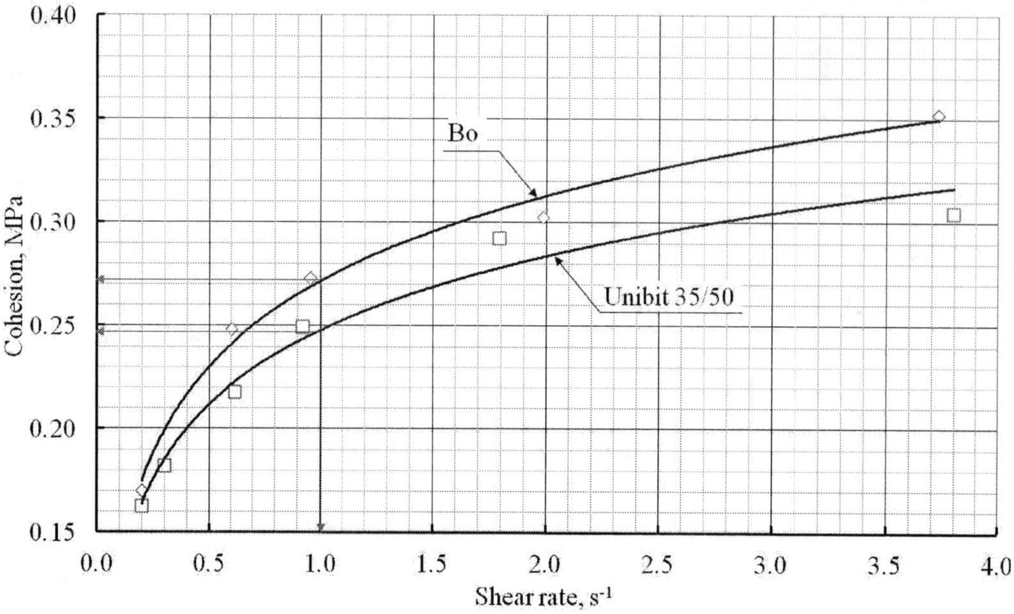


**Figure 2.** Influence of the shear rate on a cohesion of bitumen at 25 °C.

In comparison with hard distillated bitumen, Unibit 35/50 is characterized by a higher quality at low temperature. The penetration at 0 °C of multigrade bitumen is in 2.1 and 1.4 times higher than the one of hard bitumen 20/30 and 35/50 correspondingly and Fraass breaking point temperature is lower (on 14.5 and 8.0 °C correspondingly). This may lead to conclusion that asphalt concrete with Unibit 35/50 bitumen may be more flexible at 0 °C and may have higher crack resistance at winter temperatures. Low temperature performance of the multigrade bitumen is similar to the one of the high polymer modified bitumen [15].

Contra verse the low ductility (5.3 cm) at 25 °C can be considered as an obvious disadvantage of the multigrade bitumen. This ductility is much lower than the one for the hard bitumen 20/30 (25 cm) and air-blown bitumen Bo (12 cm). Taking into account the fact that ductility is stipulated by a chemical composition of the bitumen (concentration of resins and aromatic hydrocarbons) and indicates the cracking resistance of the asphalt concrete, it can be concluded that asphalt concrete with multigrade bitumen may have low cracking resistance.

Ductility is not included as a quality indicator in the European standards on paving grade bitumens (EN 12591 и EN 13924-2). This is grounded on a high ductility (more than 100 – 150 cm) [17] that inherent to the European distilled bitumens, but can't be spread on the multigrade bitumen. The last one differs in a technology of producing and can't reach the same ductility, so the normalizing of ductility values can be reasonable in this case.

The change of the bitumen quality indicators after hardening is the important technological parameter that characterizes the changes in bitumen in a process of asphalt concrete production. The data, obtained for the multigrade bitumen after hardening with TFOT method (5 hours storage at 163 °C) have a dual character. On the one hand the penetration of the multigrade bitumen stays on the same level. Such a high value of the retained penetration (94 %) is inherent to the sol type bitumen. For the multigrade bitumen it may caused by a low initial penetration of the binder. On the other hand the softening point temperature for the multigrade bitumen after hardening rises up significantly (on a 7.3 °C). Sharply change in a softening point temperature indicates a gel type of the bitumen.

IOP Conf. Series: Materials Science and Engineering **907** (2020) 012048 doi:10.1088/1757-899X/907/1/012048

The instability of the multigrade bitumen can be an explanation of obtained data. Instability can be confirmed by an increasing in mass after hardening (on 0.44 %) and by an Oliensis spot test. The last method used to evaluate the homogeneity of the binder and consists in a visualization of the structure of the spot that formed by a drop of the binder dissolved in gas on a filter paper (Figure 3). With this test method the multigrade bitumen in spite of the air-blown bitumen shows few halos that point on heterogeneity of its composition.

20/30 Unibit 35/50 Bo

**Figure 3.** The evaluation of the homogeneity of a binder with an Oliensis spot test.

Certainly with the increasing in a softening point temperature the technological temperatures of the mixing and compaction of asphalt concrete with multigrade bitumen must be increased. The evaluation of the technological temperatures of the multigrade bitumen is performed with the obtained temperature dependencies of the viscosity (Figure 4).

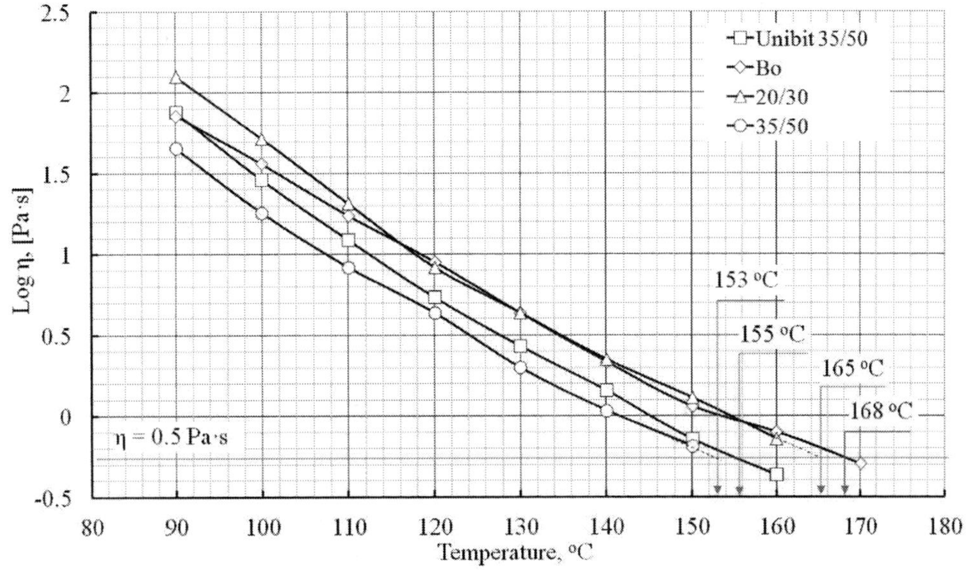


**Figure 4.** Temperature dependencies of the viscosity of binders.

The 0.5 Pa·s viscosity is set as a criteria of the heating temperature for bitumen. Trough the all considered bitumen the highest heating temperatures are required for the air-blown bitumen Bo (168 °C) and 20/30 bitumen (165 °C). The technology temperature of the multigrade bitumen (155 °C) is close to the one of the hard 35/50 bitumen (153 °C). Because this required technological temperatures of multigrade bitumen can't be taken as its disadvantage.

The suppliers of multigrade bitumen positioning it as a binder with lower temperature susceptibility that allows producing of asphalt concrete with higher cracking resistance at low temperature, higher shear resistance at high summer temperature and high resistance to hardening. These properties are reaching by a specific composition of the primary products and schemes of its producing.

The estimation of the multigrade bitumen as a component of the asphalt concrete is provided with mastic asphalts. According to the properties shown in Table 2 the mastic asphalt with multigrade bitumen is characterized by the lowest depth of the indenter immersion, which can guarantee the high resistance to the residual deformations. The similar depth of the indenter immersion for the mastic asphalt is gained with using polymer modified bitumen as a binder. With this the compressive strength at 20 °C of the mastic asphalt with the multigrade bitumen is the lowest.

**Table 2.** Mastic asphalt properties

| Bitumen | Bitumen content, % | Residual porosity, % | Depth of the indenter (2 cm$^2$) immersion after 30 min of loading at 40 °C, mm | Increasing in the depth of the indenter (2 cm$^2$) immersion after 30 min of loading at 40 °C, mm | Compressive strength at 20 °C, MPa |
|---|---|---|---|---|---|
| 20/30 | 10 | 1.89 | 3.06 | 0.52 | 4.52 |
| 35/50 | 10 | 1.73 | 7.38 | 1.98 | 3.16 |
| Unibit 35/50 | 10 | 1.93 | 2.73 | 0.38 | 2.30 |
| Bo | 10 | 1.47 | 3.36 | 0.56 | 4.20 |

Such a contradiction in obtained results can be caused by a low temperature susceptibility of the multigrade bitumen. The test temperature of the indenter immersion is 40 °C, while the compressive strength is measuring at 20 °C. Despite a low strength at 20 °C of the asphalt concrete with multigrade bitumen, its strength at higher and lower temperatures can be higher than the asphalt concrete with conventional bitumen. The strength at 0 °C and 50 °C is obtained for two mastic asphalts with Unibit 35/50 and hard bitumen 35/50 to prove this assumption. The temperature dependencies of the strength for this asphalt concretes are shown in Figure 5.

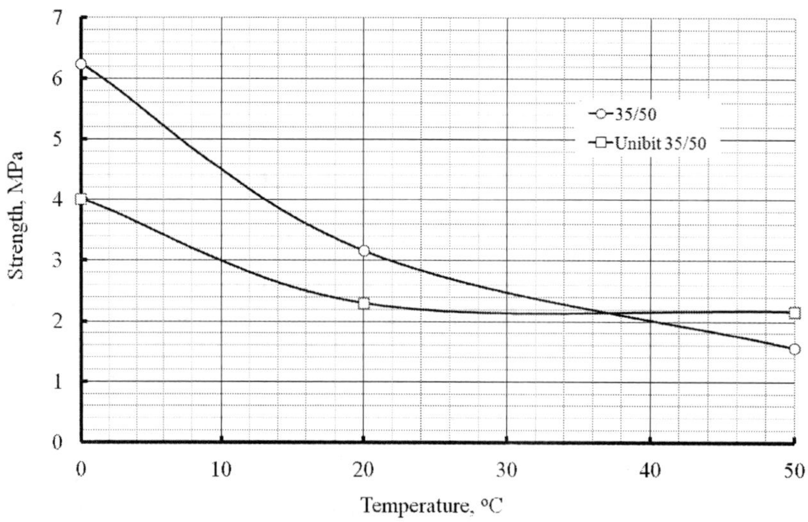


**Figure 5.** Temperature dependencies of the strength of asphalt concretes.

The asphalt concrete with multigrade bitumen demonstrates higher performance in a range of high temperatures in terms of compressive strength then the asphalt concrete with distilled bitumen. Despite the difference in penetration of these bitumens (35 for Unibit 35/50 and 44 for hard 35/50) the fact of switching their strength dependencies with a temperature proves the advantages of the multigrade bitumen at high environmental temperatures, other words at conditions when you can obtain significant residual deformations of the asphalt concrete pavement.

The low temperature characteristics in this research work can't be validated. The lower strength at

0 °C of the asphalt concrete with multigrade bitumen points on possibly higher cracking resistance, but this requires further research.

## 4. Conclusion

Multigrade bitumen is characterized by a wider Plasticity Interval in relation to air-blown and distillated paving grade bitumens according the obtained data on evaluation of their conventional and fundamental (adhesion, cohesion, viscosity) properties. The increasing in plasticity range is reached by both softening point temperature increasing and Fraass breaking point decreasing. The advantages of the gel type bitumen (higher softening point temperature, wider Plasticity Interval) as the advantages of the sol type bitumen (higher retained penetration, high cohesion) are both inherent to the multigrade bitumen. As a disadvantage of the multigrade bitumen can be considered its low ductility at 25 °C, significant change in softening point temperature after hardening and a low homogeneity.

The lowest depth of the indenter immersion for the asphalt concrete with multigrade bitumen as a high compressive strength at 50 °C proves its high performance at high environmental temperatures, which can results in low residual deformations or their absence on the pavement with multigrade bitumen.

## 5. References

[1] *Petroleum, Asphalt and Natural Gas* 1918 (Kansas City Testing Laboratory) 248
[2] Shiperovich V L 1929 *Oil bitumen and their application in the road business* 40
[3] Korte J-F 2002 Tverdye bitumy i vysokomodul'nye asfal'tobetony na ih osnove *Nauka i tehnika v dorozhnoj otrasli. Science and Engineering for Highways* **3** 30–31
[4] Read J, Whiteoak D, Hunter R N 2003 *The shell bitumen handbook* 460
[5] Koole R C, Valkering C P, Lancon D J 1992 Development of a multigrade bitumen to alleviate permanent deformation *Local Government Engineers Association of Queensland Journal* **10**(1) 8
[6] Nicholls J C 1994 Assessment of multiphalte, the Shell multigrade bitumen. *TRL Project Report* (PR 61) 23
[7] Zolotarev V A, Bratchun V I 2003 Modificirovannye bitumnye vjazhushhie, special'nye bitumy s dobavkami v dorozhnom stroitel'stve. *Vsemirnaja dorozhnaja associacija. Tehnicheskij komitet «Nezhestkie dorogi»* (S8) 229
[8] Sybilski D, Mularzuk R, Palys M, Kaczycka J 2000 The First Application of Multigrade Bitumen in Poland *In Proceedings of the papers submitted for review at 2nd Eurasphalt and Eurobitume congress* (Held 20-22 September 2000, Barcelona, Spain) Book **2** Session 2 599-605
[9] Sybilski D 2011 Prace badawcze laboratoryjne i w pejnej skali nad zastosowaniem betonu asfaltowego o wysokim module sztywnosci w nawierzchni drogowej *Drogi i mosty* **1-2** 81-119
[10] EN 13924-2:2014 Bitumen and bituminous binders – Specification framework for special paving grade bitumen. Multigrade paving grade bitumens *Brussels, European committee for standardization* 34
[11] AS 2008 – 2013 Australian Standard. Bitumen for pavements *Sydney, NSW 2001, Australia* 8
[12] Zolotarev V A, Pyrig Ja I, Galkin A V 2013 Opredelenie kogezii bitumnyh vjazhushhih na kogeziometre KHD-1 *Vestnik DonNASA* **1** 71–77
[13] Pyrig, J I, Galkin A V 2019 Sravnitel'naja ocenka adgezii bitumnyh vjazhushhih, opredelennoj po metodam, normiruemym nacional'nymi standartami Ukrainy. *Bulletin of Kharkov National Automobile and Highway University* **86**
[14] Pyrig J I 2015 O pokazatele temperaturnoj chuvstvitel'nosti bitumov. *Bulletin of Kharkov National Automobile and Highway University* **69** 128 – 133
[15] Zolotarev V A 2009 Bitumy, modificirovannye polimerami i asfal'topolimerbetony. *Road Building equipment* 16–23.
[16] Kolbanovskaja A S, Mihajlov V V 1973 Dorozhnye bitumy 264
[17] Hoiberg A J 1965 Bituminous materials: Asphalts, tars and pitches 604

IOP Conf. Series: Materials Science and Engineering **907** (2020) 012049 doi:10.1088/1757-899X/907/1/012049

# Manufacture of construction materials by energy-saving technology through the example of the Bashkir region

**A I Gabitov[1], V A Ryazanova[1], A S Salov[2], A M Gaisin[1], A A Timofeev[3]**

[1]Department of Engineering Structures, Ufa State Petroleum Technical University, Mendeleev St., 195, Ufa, 450080, Russian Federation
[2]Department of Roads and Construction Technology, Ufa State Petroleum Technical University, Mendeleev St., 195, Ufa, 450080, Russian Federation
[3]Department of Applied Sciences, Ufa State Petroleum Technical University, Mendeleev St., 195, Ufa, 450080, Russian Federation

salov@list.ru

**Abstract.** Information on the history and formation of the production and application of efficient concrete walling items and road facilities in the construction practice is presented herein by studying historical data. Ways to improve quality of small walling items and road facilities to comply with current energy-saving requirements are analyzed from the historical point of view. Vibropressed technology became widespread over 40 years ago in North America and Europe, being characterized by high capacity and automation of equipment with abilities to manufacture wide spectrum of products by size, shape, finish and color distinguished by low material and energy intensity of production, efficiency of application thereof in energy-saving wall constructions. Special features in making vibropressed concrete items by Besser equipment in the Bashkortostan construction companies are presented.

Concrete masonry was originated in the USA where large and heavy blocks were made by formation of lime and moist sand mixture with further steam-drying [1].

The first factory-made concrete blocks ("artificial Ranger's blocks") appeared in the commercial advertising in 1832 in England. Solid blocks were made from powdered lime, small aggregates and boiling water for rapid solidification. They were used in 1830s when erecting a number of buildings in London and Brighton, Pall Mall houses and the Royal College of Surgeons, in particular that are still staying [2].

The "artificial blocks" were more actively used in the practice of West European countries where there was a long-standing tradition in pise building erecting. The first examples of laying the concrete (not the green clay) to the form were known since 1830 but the new material became known in France and England after 1850 only [1, 2].

The solid blocks were not widely used because of weight thereof. In 1866 the technology for hollow block-making with lesser weight was improved. Several patents were granted in the USA and UK within 1866-1876. However, there was no mass production of concrete blocks till 1900.

It was exactly in this year when Harmon S. Palmer got a patent in the USA to his machine made of cast iron with the removing channel and regulated walls thereby originating the modern industry of making concrete blocks [3]. Manufacturing of economic hollow concrete blocks was certainly preceded by invention of the improved technology of grinding and burning of Portland cement. The cement was made from lime and clay mix burnt at high temperature with further fine grinding thereof [3, 4].

Production of both Portland cement and concrete blocks was closely connected and both these branches were developed rapidly in the first decade of XX century.

Initially the concrete blocks dimensions were 610x305x205mm, while the capacity of the first machines amounted to 200 blocks per 8 hours [4].

From the first quarter of XX century in the world construction practice the concrete blocks were manufactured and applied in masonry constructions in the USA, Canada, Great Britain, France, Germany, Italy and many other countries. The world construction boom in 80s demonstrated increasing application of concrete blocks that due to low energy intensity of production successfully compete with ceramic items significantly reducing sales markets of the latter.

The American ORCO Block Co., Inc. (Stanton, California) is one of the biggest manufacturers of concrete items in the USA, producing over 500 various concrete items, such as wall concrete blocks, facing blocks, paving stones and hardscape products.

Range of hollow wall items made the company includes 190mm, 140mm and 240mm blocks of 190mm, 140mm and 90mm in height. Facing blocks are made with various surface finishing of face side and of different colors.

The similar range of concrete items is made in the USA by RCP Block & Brick, Western Block Company, Blocklite, Quality Block Company and others [5].

The demand for concrete blocks is noted recently in Great Britain. Many manufacturers of concrete blocks such as Evered Concrete Products, ARC Powell Duffrym Brick increased the production output and expanded sales markets.

Wide range of hollow wall concrete blocks is produced in Germany, by such companies as Schlosser und Co. GmbH. and G.P.O. Stoelok Aussenhandel GmbH.

Many French construction materials manufacturers produce wall blocks of dense concretes. In particular, Damiani Freres is making large blocks of various type and size to be applied in load-bearing and non-load-bearing walls. AF-A blocks with two perforated vertical cavities are 400×200×200mm in nominal size and over 20kg in weight, while AF-A blocks are 400×200×130mm in size and over 12kg in weight [6].

Wall blocks and small blocks of light concrete with expanded aggregates and of cell concrete became widely applied both in our country and abroad. They are produced by a large number of foreign companies as Transswerke Maurin and Sikabrick (Germany), Carrieres Biallais and Siporex (France), Leca Beton and Planblock (Italy) and others [7].

Some countries produce blocks of light concrete with hollows filled by heat-insulating materials. According to experts from Italy and Germany such blocks are differed by improved heat-insulating properties (by 43%) as compared with blocks without hollows filling. Together with light concrete blocks with hollows filled with efficient heat-insulating materials, the three-layer blocks with the heat-insulating layer in the middle are manufactured. Such blocks are produced by Association pour development du bloc and ADB (France), Bau-Innovation (Austria).

Until 1990 the share of wall concrete blocks amounted to about 2.5% of all wall materials in the USSR territory (about 2 billion of conventional bricks per year) [6, 7]. In the middle of 1990s the production and application of small wall concrete items by various technologies (casting, vibroforming and vibropressing) has started in Russia and in Bashkortostan, in particular.

Development of road construction is closely connected with continuous improvement of structural solutions for paving flags, being analyzed by A.A. Timofeev in his monograph [8].

Small concrete blocks have started to apply in road construction from the first half of XX century when together with the construction of roads from cast-in-situ concrete pavement the West European countries began to use prefabricated pavements of relatively small size. In some countries the pavements were made of small precast slabs [9].

The experience showed that application of concrete road blocks was reducing the concrete consumption as compared with cast-in-situ concrete pavement providing reliability thereof and enabling making prefabricated blocks.

A new pavement system from concrete elements was accepted in Poland in 1934, with several trial pavement sections being launched. The pavement elements (50×20×20 cm) were made of three-layer concrete: cement-sand ratio of the upper layer was 1:2, the middle – 1:3, and the lower – 1:5. Granite or crushed limestone was used as a coarse aggregate.

The blocks were made by using the metal shaped formwork. They were placed on 2-5cm sand blanket.

In the USSR the application of pavement slabs dated back to 1946 when reinforced 50×50×6 and 50×70×7cm slabs were applied in Chelyabinsk and Nizhny Tagil as walkways, landing platforms for trams and bridge approaches [9, 10].

In 1956 Dormostproekt (Moscow) engineered prefabricated pavements with the following dimensions 30×30×6, 50×50×8,75×75×6 and 150×75×6 cm, with the latter two thereof being engineered as the reinforced ones [10].

In 1962 Mosinzhproekt proposed vibration-rolled concrete 320×160×15cm slabs with application of rebars at 3.9kg/m$^2$ [9, 10].

In 50-60s large size plates were basically used as prefabricated foot-path paving in Moscow, while 20x20, 30x30, 33x33 and 40x40cm plates were applied in cities of Lithuania, Ukraine, Krasnodar and Yerevan [11].

The experience in making walkways in the former Lithuanian SSR was of great interest. In 1964 Gosstroi of the Lithuanian SSR adopted a resolution on total application of cement-concrete slabs in making walkways in cities and settlements. Application of such slabs enabled to work all the year round, made repairing easy and improved urban health conditions. As a rule, 50×50×6, 33×33×5 and 25×25×5 cm slabs were made of two-layer concrete.

Other kinds of pavement were also applied, with prefabricated foot-path paving as basalt 30x30x5cm slabs being widely used in Bulgaria both in big cities (Sofia, Plovdiv, Varna) and in the countryside [9,11].

Walkways made of stone or concrete tiles are very popular in Great Britain. Stone tiles are made from hard sandstones with up to 0.6-0.9m in size and 50-70mm in width.

Prefabricated concrete pavement became widely used in the Netherlands. The tiles manufactured are of 30×30 and 25×25cm in size, with the former being widely applied. The thickness varies from 3 to 10cm. Tiles with 4.5-5cm thickness are more often used for walkways with usual pedestrian flow, while tile thickness is increased with the intensive flow, accordingly [12].

Apart from usual grey tiles the Netherlands has arranged manufacture of tiles, with the upper part thereof being painted yellow, red, blue and white. Special rounded tiles with notches for water and air pass are applied for paving sections near planted borders of walkways. They are of the same size (except for notches) as usual tiles, so they may be laid in any place and be replaced if necessary.

Walkway tiles applied in the Netherlands are specially characterized by being made of two-layer concrete. The solid upper layer thereof is made of concrete with either gravel or basalt aggregates. The lower layer is made of cement mortar or concrete with similar solid aggregates.

In 1993-1995 the authors [10, 12] considered possibilities to use fiber-reinforced concrete as a special layer on the surface of paving slabs thereby providing increased wear-resistance and strength thereof thus enabling to use them in areas with severe climatic conditions.

Precast concrete slabs are widely used in Germany for pavements for several decades. They are of 30×30, 35×35, 40×40 and 50×50 cm in plan dimension with 4.5-7 cm in thickness depending on loading thereon and durability requirements [12]. They are also used in parkway areas. Walkway pavements of 80x80x10cm and width of up to 5.6m are provided in residential districts. As a rule they are made with no rebars inside. Also in crossings and intersections the colored slabs with rough surface are applied apart from usual ones.

Manufacture of small concrete prismatic slabs to be used as paving for walkways and cycle lanes began in 1976 in the Netherlands: these slabs look like concrete paving 21.1×10.5 mm and 21.1×6.9 mm blocks, with 7-12mm being the thickness thereof. Fully mechanized and automated manufacture of these slabs by special concrete molding machines is arranged in this country.

Pavement made of this concrete paving blocks was widely used due to low cost thereof. Nowadays maintenance and repair, as well as repairing of underground facilities and good appearance are considered to be of great advantage.

Cement concrete slabs with plan dimensions of 40×60 and 61×61 cm and 10.2 cm in width, as well as 45×45 and 61×92 cm of the same width are being used for the last years in the USA for paving areas, and local drives [11, 12]. These slabs are made of regular Portland cement with sand, gravel and small pieces of metal (chip, filings, debris, etc.) used as aggregates. More than 2mln $m^2$ of such slabs for walkways and local drives was annually made in the middle of 1990s in the USA. Pavements made of such slabs are environmentally friendly and good for health [13].

The experience in making precast footway slabs both in the USSR and abroad was considered as the basis for making All-Union State standard in our country first in 1972 with amendments thereto in 1981 (GOST 17608-81) and 1991.

Several Russian plants began manufacturing of reinforced footway slabs made of concrete of wet consistency with filtration of excess water therefrom. Thereby due to high performance properties and durability thereof they may applied for making precast footway and roads in the North and Siberia [12, 13].

For a period from 1990 to 2005 over 15 inventor's certificates and patents were obtained to the procedures and technological equipment for making small concrete wall items and road facilities in Russia [14].

The proposed engineering methods are based on principles of vibration or pressing of concrete mix and are differed by efficiency, degree of pressure and strength properties of the items made.

One of the world leading manufacturers of vibropressed equipment is the American Besser Company. Wall items and road and walkway facilities, landscape and architectural products and concrete pipes are made by this equipment.

Since 1993 the construction companies of the Republic of Bashkortostan including OJSC "Bashneftezavodstroi" and OJSC "Krupnopanelnoe Domostroenie" have bought and launched several Besser line to make vibropressed concrete items with a total capacity about 300mln conventional bricks per year [12, 14].

Since 1996 up to now proceeding from large scope of both theoretical and experimental researches of masonries made of hollow vibropressed concrete Besser blocks made by Ufa companies, the specialists from the "Construction Materials" department at Ufa State Petroleum Technical University headed by Professor V.V.Babkov have proposed the procedures to calculate the bearing capacity of hollow walls and that of the similar walls with both reinforced and non-reinforced frame installed inside by applying concrete coating to the block cavities [FSBEI of Higher Education USPTU, Architectural and Construction Institute: materials for design and construction. Manual of engineering solutions for the construction of residential and public buildings applying concrete vibropressed "Besser" wall blocks, Ufa, 2018; 15, 16].

Therefore, the technologies for making small concrete wall items and road facilities may be stated to become widely and rapidly applied since the middle of the XX century.

Principal advantages of concrete items are high durability and compressive strength thereof, low energy requirements of production, and applicability of local construction materials for making thereof.

**References:**

1. Cowan H J 1982 *Structural and environmental design in the XIX-XX centuries* (Moscow: Stroiizdat) p 359
2. Davey N 1961 *A History of Building Materials* (London: Phoenix House) p 260
3. Cowan H J 1982 *The master builders* (Moscow: Stroiizdat) p 240
4. Jope E M 1961 *Studies of Building History* (London: Odhams) p 287
5. Allsopp B 1965 *A History of Classical Architecture* (London: Pitman) p 215

6. Veiner T M, Ruzhansky S Ya 1993 Manufacture of wall blocks (by USA companies experience) *Construction materials* No **5** p 29-31
7. Babkov V V, Kolesnik G S, Gabitov A I et al. 2001 *Rock and reinforced masonry structures under vibropressed concrete items* (Ufa: State publishing house of scientific and technical literature “Reaktiv”) p 182
8. Chistova T P 1993 *Concrete wall stones and blocks* (Review. Series: Construction materials, Moscow: VNIINTPI) **2** p 77
9. Gabitov A I, Bedov A I 2006 *Engineering, Renovation and Strengthening of Masonry and Reinforced Masonry Structures* (Moscow) p 568
10. Bedov A I, Gabitov A I, Salov A S, Gaisin A M, Khabibullina L I 2017 The usage of computer modeling to optimizing the cross-sections of reinforced concrete frame elements *Building and reconstruction* No 6 (**74**) p 3-12
11. Bedov A, Salov A, Gabitov A 2018 CAD methods of structural solutions for reinforced concrete frame *XXI International Scientific Conference on Advanced in Civil Engineering "Construction - The Formation of Living Environment" (FORM 2018)* (25–27 April 2018, Moscow, Russian Federation) Vol. **365** p 1-8
12. Krot A, Ryazanova V, Gabitov A, Salov A, Rolnik L 2018 Resource-saving technologies for advanced concrete in the Republic of Bashkortostan *MATEC Web of Conferences 7. "7th International Scientific Conference "Reliability and Durability of Railway Transport Engineering Structures and Buildings" (Transbud 2018)* Vol. **230** Article number 03009
13. Bedov A I, Gabitov A I, Salov A S, Biktasheva A R 2020 Increase of energy performance of residential buildings with enclosing structures made of masonries with application of ceramic blocks *Journal of Physics: Conference Series* **1425** (1) 012042
14. Bedov A, Gabitov A, Gaysin A, Salov A 2019 Engineering solutions for heart efficient exterior walls in climatic condition of the Republic of Bashkortostan *E3S Web of Conferences* **97** 02039
15. *Rock and reinforced masonry structures based on vibropressed concrete items*: 2000 (Design standards): TSN 51-303-00. RB: Effective 3.07.2000 by the Ministry of Housing and Construction of RB: replaced Recommendations for design strength of masonries of vibropressed concrete items. (Publ.house. Ufa: BashNIIstroi) p 27
16. Bedov A I, Gabitov A I 2005 Rock/reinforced masonry structures. Design, strengthening and restoration *Manual for construction students, and design and constructional engineers* (Ufa)

IOP Conf. Series: Materials Science and Engineering **907** (2020) 012050 doi:10.1088/1757-899X/907/1/012050

# The research of tubular mixer with improved design

**S Epoyan[1], G Sukhorukov[1], V Volkov[2] and O Haiduchok[1]**

[1]Department of Water Supply, Sewage and Hydraulics, Kharkiv National University of Civil Engineering and Architecture, Kharkiv, Ukraine
[2]Kharkivvodokanal Municipal Enterprise, Kharkiv, Ukraine

vkg.knuca@ukr.net

**Abstract.** Modern technological schemes in water supply for clarifying surface water include using reagents as coagulants, flocculants, etc. To mix them with the source water on plants usually use different types of mixing devices. Tubular mixer is one of such devices. Our research work involves the study of a tubular mixer with improved design, which can mix several reagents with source water. The paper presents theoretical and experimental studies of the tubular mixer with improved design for mixing several reagents with natural water. It is theoretically and experimentally substantiated the feasibility of using this tubular mixer for mixing several reagents with natural water over a short mixing time. The improved design of tubular mixer can mix several reagents with natural water in a short mixing time, intensify the mixing process and improve quality of water after treatment.

## 1. Introduction

The most important sectors that determine the level of urban development are: water supply, water and wastewater treatments. Nowadays, a big problem is to supply clean water with high drinking quality to consumers. Water clarification becomes crucial in the integrated system of water supply in cities and villages. Modern technological schemes in water supply for clarifying surface water include using reagents before sedimentation and filtration [1-5]. To mix them with the source water on plants are usually used different types of mixing devices. [6,7]. Tubular mixer is one of such devices. The efficiency of these facilities influences on the efficiency of water clarification. A very important condition for improving the efficiency and depth of coagulation is the rapid and intensity of mixing process in the mixer. Therefore, mixing the source water with the reagents should be complete and rapid. Tubular mixers are widely used in water treatment [8-12], but they have got disadvantages. They are used to mix only one reagent with water and can’t adjust the intensity of mixing process. Therefore, increasing the efficiency of mixing process is a very important task.

We proposed a tubular mixer with improved design [13]. This tubular mixer consists of four sections: two sections for supply and distribution reagents, another two sections have got mixing elements. They are made from porous polymer concrete and have got different thickness, material and size of aggregate.

The aim of the research is to study the tubular mixer with improved design for mixing reagents with water.

## 2. Material and Methods

A method was proposed to intensify the mixing process for water with reagents. Design changes and the possibility of entering several types of reagents in one mixer will improve the efficiency of water clarification at water treatment plants [13].

The phenomenon of water velocity in porous surroundings in hydraulics is called water filtration. Filtration can be laminar and turbulent.

For laminar filtration, the main dependency is the Darcy equation [14]:

$$V = k \cdot i, \tag{1}$$

where, V is filtration velocity, cm/s; k is filtration constant, cm/s; i is piezometric level.

We use the dependence (1) if:

$$Vd < 0{,}01 \div 0{,}07, \tag{2}$$

where, d is average size (diameter) of particle, sm. Filtration takes place through this particles.

If the condition (2) is not implemented, then turbulent filtration takes place.

Filtration velocity is the part of flow which rate to the the intersection area of the porous surroundings where filtration takes place.

The equation (2) shows that the critical velocity for laminar filtration can't be more than:

$$V = \frac{0{,}01 \div 0{,}07}{d} \tag{3}$$

At an average size of gravel material of porous surroundings of 2.5 and 3.0 cm, the critical velocity for laminar filtration will accordingly be: 0.004 - 0.028 cm/s and 0.0033 - 0.0233 cm/s.

In this case, when the porous surroundings are placed in the tubular mixer and water velocity in it is more than 0.028 cm/s, then turbulent filtration will occur.

For turbulent filtration conditions, its velocity (m/s) is determined by the equation:

$$V_t = K_t \cdot I^{0{,}5}, \tag{4}$$

where, $K_t$ is coefficient of turbulent filtration m/s; I is hydraulic level.

Based on the calculation of pressure filter mounds we have decided to use the equation of turbulent filtration flow with uniform motion [14]:

$$Q = K_t \cdot \omega\sqrt{I}\,, \tag{5}$$

where Q is water flow rate through the filter embankment, $m^3/s$; $\omega$ is the living area of the intersection of the filter embankment, $m^2$.

Equations (4) and (5) show that:

$$Q = V_T \cdot \omega \tag{6}$$

If we know water flow rate (Q) and living area of the intersection ($\omega$) of the filter embankment, that it is possible to determine the rate of turbulent filtration:

$$V_T = \frac{Q}{\omega} \tag{7}$$

And after that we can use the equation (4) to determine the hydraulic level:

$$I = \left(\frac{V_T}{K_T}\right)^2, \tag{8}$$

where, $K_T$ is determined by the follow equation (cm/s):

$$K_T = 18\rho\sqrt{d}\,, \tag{9}$$

where, $\rho$ is coefficient of porosity. For gravel this coefficient is $\rho = 0{,}4$; d is diameter of filter material, cm.

Thus, the head loss in the porous polymer concrete partition can be determined by the equation:

$$\Delta h = \left(\frac{V_T{}^2}{K_T^2}\right) \cdot \delta\ , \tag{10}$$

where $\delta$ is thickness of porous polymer concrete (cm).

All mixing devices should satisfy the efficiency and intensity of mixing process, but there is another economic issue. In this case, this is the head loss in the porous area, which influence on pressure of pumps.

The intensity of mixing process is the energy $N_P$ that must be spent on mixing an unit of volume ($N_P/V$) or mass ($N_P/\rho Q$) of liquid per unit time to obtain the desired quality of the mixture.

The energy $N_P$, which can be used for mixing process in a porous area, we can find by the follow equation:

$$N_\Pi = \Delta P \cdot V \ , \tag{11}$$

where $\Delta P$ is pressure drop, Pa; V is volumetric flow rate, $m^3/s$.

Therefore, the mixing intensity in the porous area depends entirely on the pressure drop:

$$I = \frac{N_\Pi}{V} = \frac{\Delta P \cdot V}{V} = \Delta P \tag{12}$$

In this way, the greater loss of head in a porous area is, the greater the intensity of mixing process will be.

## 3. Experimental and numerical database

For experimental research, we have created a model of a tubular mixer with a constant diameter of 100 mm in the laboratory of the Department of WaterSupply, Sewerage and Hydraulics of Kharkiv National University of Civil Engineering and Architecture. The scheme of the experimental setup is shown in Figure 1.

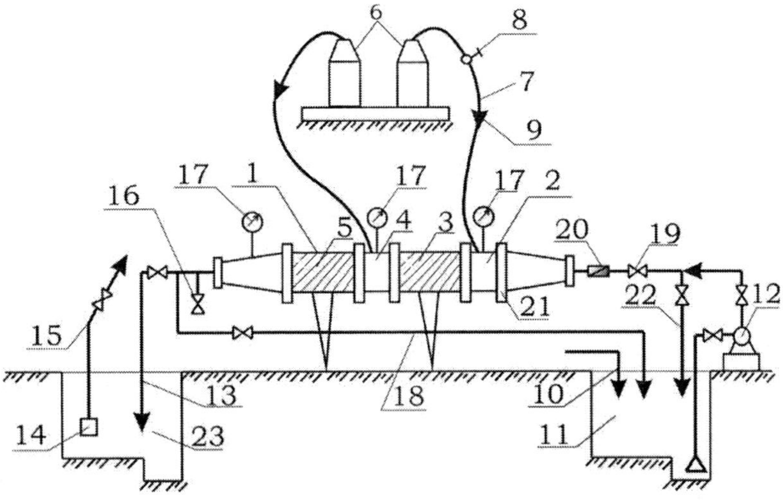


**Figure 1.** Scheme of the experimental setup.

This scheme includes: a tubular mixer (1) which has four sections: first section (2) is for distribution part of sulfuric acid aluminum; the second section (3) is the first mixing element, which made from porous polymer concrete with a gravel aggregate with average size ($d_{aver}$ = 25 mm) and a length of 80 mm; third section (4) is for distribution another part of sulfuric acid; fourth section (5) is the second mixing element, which made from porous polymer concrete with gravel aggregate ($d_{aver}$ = 30 mm) and a length of 120 mm and Mariotte vessel (6). The hoses (7) regulate the flow of coagulant (8). Tap water through the pipe (10) supplies to the tank (11) and from there by the pump (12) is transported into the tubular mixer (1). Through the pipe (13) volume of water moves into the collecting tank (23).From there by the submerged pump 14 through the pipeline 15 is discharged into the sewer. The tubular mixer (1) has a sampler (16), pressure gauges (17) and a return water supply line (18). The mixer is equipped with a shut-off valve (19) and a water counter (20). All sections of the tubular mixer is connected with the flanges (21). The pipe (22) is used to mix the water in the tank (11).

The experiments were carried out at a water temperature of from 7 to 17 ° C (at different times of the year).The flow speed in the mixing elements was from 0.6 to 1.2 m/s. Suspended solids concentration in source water was from 50 to 140 mg/dm$^3$. The silt was from the Siverskyi Donets River and we used them as an opacifier. Mixing elements were made from gravel on epoxy resin brand ED - 20. This material is characterized by high strength, good chemical resistance to the aggressive effect for water with reagents, and, finally, permitted by the Ministry of Health of Ukraine for using in water supply systems.

The first stage of the research was carried out to determine the pressure loss in the mixing elements. They were determined by the difference between the values of manometers (17), which were installed before and after the mixing element (Fig. 1). Water from the tank (11) was transferred by the pump (12) into the mixer (1), and then through the pipes (18) again returned to the tank (11).

The second stage of this research was to determine the efficiency of the tubular mixer. In the tank (11) was added the sludge and the pump (12) mixed it with water. After mixing, we determined the dose of coagulant (sulfuric acid aluminum) using the device "Drop" for trial coagulation.

Experimental studies were performed according to the method, which described in the work [13].

The experimental dependences of the pressure drop on the flow velocity in the tubular mixer are shown in Table 1. A Table 2 shows the experimental and theoretical databases of the pressure drop in the mixing elements. The theoretical values of the pressure drop in the mixing elements were determined by equations (7), (9) and (10).

The Table 1 shows that with increasing velocity of water the pressure drop increases and reaches considerable values. The water temperature slightly affects the viscosity of the water. Unfortunately, it was not possible to fix it with the help of gauges. To reduce the magnitude of the pressure drop in the mixing elements, materials with high porosity should be used as a filler for porous polymer concrete.

**Table 1.** Experimental values of pressure drop in tubular mixer and in mixing elements

| | The water velocity in the mixer, m/s | Tempe rature, °C | Manometer readings, kPa | | | Pressure drop kPa | | Pressure drop kPa in the mixer |
|---|---|---|---|---|---|---|---|---|
| | | | 1 | 2 | 3 | *1st mixing element* | *2nd mixing element* | |
| 1 | 0.6 | 7 | 69.6 | 46.6 | 12.1 | 23 | 34.5 | 57.5 |
| 2 | 0.6 | 9 | 69.7 | 46.4 | 12.1 | 23.1 | 34.5 | 57.6 |
| 3 | 0.6 | 14 | 69.4 | 46.3 | 12.1 | 23.1 | 34.2 | 57.3 |
| 4 | 0.6 | 17 | 69.4 | 46.5 | 12.1 | 22.9 | 34.4 | 57.3 |
| 5 | 0.8 | 7 | 105.0 | 63.9 | 12.9 | 41.1 | 51 | 92.1 |
| 6 | 0.8 | 9 | 104.6 | 64 | 12.9 | 40.6 | 51.1 | 91.7 |
| 7 | 0.8 | 14 | 104.7 | 63.8 | 12.9 | 40.9 | 50.9 | 91.8 |
| 8 | 0.8 | 17 | 104.8 | 63.9 | 12.9 | 40.9 | 51 | 91.9 |
| 9 | 1.0 | 7 | 157.1 | 93.5 | 14.2 | 63.6 | 79.3 | 142.9 |
| 10 | 1.0 | 9 | 157.2 | 93.7 | 14.2 | 63.5 | 79.5 | 143.0 |
| 11 | 1.0 | 14 | 157.0 | 93.5 | 14.2 | 63.5 | 79.3 | 142.8 |
| 12 | 1.0 | 17 | 156.9 | 93.5 | 14.2 | 63.4 | 79.3 | 142.7 |
| 13 | 1.2 | 7 | 222.1 | 130.0 | 15.8 | 92.1 | 114.2 | 206.3 |
| 14 | 1.2 | 9 | 222.0 | 130.0 | 15.8 | 92.0 | 114.2 | 206.2 |
| 15 | 1.2 | 14 | 222.1 | 130.1 | 15.8 | 92.0 | 114.3 | 206.3 |
| 16 | 1.2 | 17 | 222.0 | 130.1 | 15.8 | 91.9 | 114.3 | 206.2 |

The Table 2 shows that the experimental and theoretical values of the pressure drop in the mixing elements has difference (from 4.4 to 5.6%), but remain constantly above the theoretical. This can be explained by the fact that pore channels, in the connected porous structure, have a smaller diameter than the bulk material.

**Table 2.** Experimental and theoretical values of the pressure drop in the mixing elements

| | The water velocity in the mixer, m/s | Experimental determination of pressure drop kPa | | Theoretical value of the pressure drop, kPa | | Deviation,% | |
|---|---|---|---|---|---|---|---|
| | | *1st mixing element* | *2nd mixing element* | *1st mixing element* | *2nd mixing element* | *1st mixing element* | *2nd mixing element* |
| 1 | 0.6 | 23 | 34.5 | 21.8 | 32.7 | 5.2 | 5.2 |
| 2 | 0.6 | 23.1 | 34.5 | 21.8 | 32.7 | 5.6 | 5.2 |
| 3 | 0.6 | 23.1 | 34.2 | 21.8 | 32.7 | 5.6 | 4.4 |
| 4 | 0.6 | 22.9 | 34.4 | 21.8 | 32.7 | 4.8 | 4.9 |
| 5 | 0.8 | 41.1 | 51.0 | 38.8 | 48.4 | 5.6 | 5.1 |
| 6 | 0.8 | 40.6 | 51.1 | 38.8 | 48.4 | 4.4 | 5.3 |
| 7 | 0.8 | 40.9 | 50.9 | 38.8 | 48.4 | 5.1 | 4.9 |
| 8 | 0.8 | 40.9 | 51.0 | 38.8 | 48.4 | 5.1 | 5.1 |
| 9 | 1.0 | 63.6 | 79.3 | 60.5 | 75.7 | 4.9 | 4.5 |
| 10 | 1.0 | 63.5 | 79.5 | 60.5 | 75.7 | 4.7 | 4.8 |
| 11 | 1.0 | 63.5 | 79.3 | 60.5 | 75.7 | 4.7 | 4.5 |
| 12 | 1.0 | 63.4 | 79.3 | 60.5 | 75.7 | 4.6 | 4.5 |
| 13 | 1.2 | 92.1 | 114.2 | 87.2 | 108.9 | 5.3 | 4.6 |
| 14 | 1.2 | 92.0 | 114.2 | 87.2 | 108.9 | 5.2 | 4.6 |
| 15 | 1.2 | 92.0 | 114.3 | 87.2 | 108.9 | 5.2 | 4.7 |
| 16 | 1.2 | 91.9 | 114.3 | 87.2 | 108.9 | 5.1 | 4.7 |

Based on the Table 1 we have made graphs of the dependence of the pressure drop in the mixing elements and in the tubular mixers on the flow velocity in the device, which is presented in Figure 2.

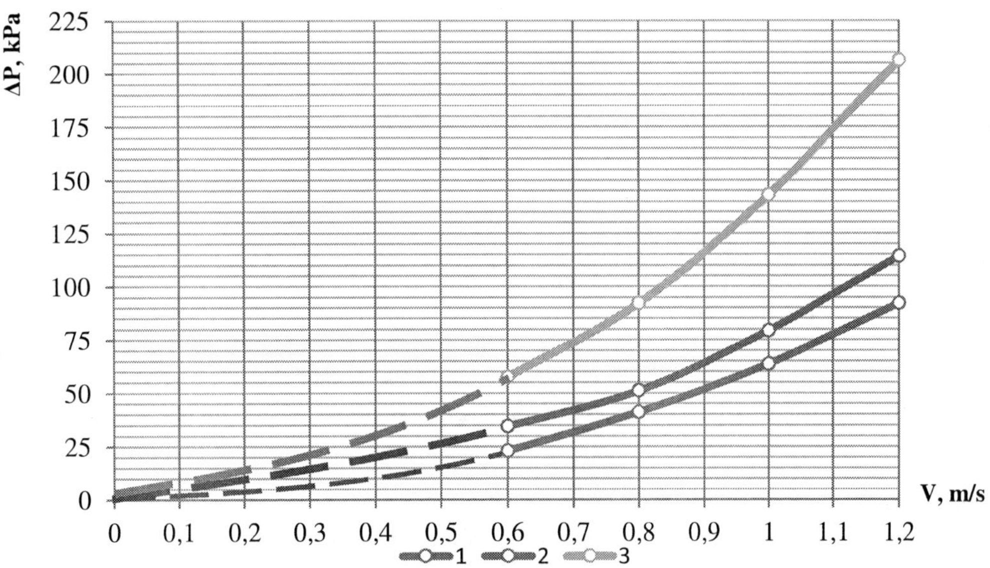


1 is the 1st mixing element; 2 is the 2nd mixing element and 3 is tubular mixer

**Figure 2.** Dependences of the differential pressure ΔP in the mixing elements and in the tubular mixer on the flow velocity V in the mixer

From the Figure 2 we have concluded that results connect well with the theory. The pressure drop is proportional to the square of the fluid flow velocity.

In the Table 3 results of experiments that were conducted to determine the efficiency of the tubular mixer is shown.

**Table 3.** Improving the efficiency of the tubular mixer

| | Turbidity of source water, (NTU) | Temperature | The velocity of the water in a mixer (m/s) | Turbidity of clarified water (NOC) | | Coefficient of deposition efficiency, K |
|---|---|---|---|---|---|---|
| | | | | $C_y$ | $C_T$ | |
| 1 | 171.0 | 17 | 0.9 | 13.6 | 11.8 | 1.15 |
| 2 | 86.0 | 17 | 0.9 | 13.7 | 11.8 | 1.15 |
| 3 | 223.6 | 15 | 0.9 | 14.6 | 12.8 | 1.14 |
| 4 | 240.8 | 12 | 0.9 | 16.8 | 14.7 | 1.14 |
| 5 | 103.2 | 10 | 0.9 | 22.5 | 19.9 | 1.13 |
| 6 | 137.6 | 7 | 0.9 | 24.0 | 21.4 | 1.12 |
| 7 | 171.0 | 17 | 1.0 | 13.0 | 11.1 | 1.17 |
| 8 | 86.0 | 17 | 1.0 | 13.1 | 11.2 | 1.17 |
| 9 | 223.6 | 15 | 1.0 | 14.3 | 12.3 | 1.16 |
| 10 | 240.8 | 12 | 1.0 | 16.2 | 14.0 | 1.16 |
| 11 | 103.2 | 10 | 1.0 | 21.8 | 18.8 | 1.16 |
| 12 | 137.6 | 7 | 1.0 | 23.3 | 20.3 | 1.15 |
| 13 | 171.0 | 17 | 1.1 | 13.5 | 11.8 | 1.14 |
| 14 | 86.0 | 17 | 1.1 | 13.5 | 11.8 | 1.14 |
| 15 | 223.6 | 15 | 1.1 | 14.8 | 13.0 | 1.14 |
| 16 | 240.8 | 12 | 1.1 | 16.5 | 14.6 | 1.13 |
| 17 | 103.2 | 10 | 1.1 | 24.4 | 21.8 | 1.12 |
| 18 | 137.6 | 7 | 1.1 | 22.8 | 20.4 | 1.12 |

The Table 3 shows that the efficiency of the tubular mixer is almost constant in the range of velocity of 0.9 ÷ 1.1 m/s and temperature of 7-17 °C, but a slightly better effect is observed at a velocity of 1 m/s.

## 4. Discussion

These studies are continuing of previous studies on this topic. They considered increasing the efficiency of mixing process of water and reagents. It can effect on clarifying water and get water with high-drinking quality after water treatment plant [16, 17].

## 5. Conclusion

A tubular mixer with improved design in which it is possible to enter several types of reagents for mixing with water and control the intensity of mixing is proposed and investigated. The expediency of using the tubular mixer with improved design for water velocity 1 m/s has been experimentally confirmed. The expediency of using materials for aggregate bound porous structures of mixing elements of a tubular mixer with improved design with materials of high density is shown. This structure can significantly reduce the pressure drop in the mixer.

## References

[1] Epoyan S M 2010 *Water supply and treatment surface water*: A textbook (Kharkiv: Factor) p 192.
[2] Tuhai A M 2009 *Water supply*: A textbook (Kyiv: Znannia) p 735.
[3] Bratby J 2016 *Coagulation and Flocculation in Water and Wastewater Treatment* (London - IWA Publishing) p 538

[4] Huck P and Sozanski M 2015 *Designing and Optimizing Drinking Water Treatment Processes – A Guid to Conducting Investigations* (London - IWA Publishing) p 200
[5] Orlov V, Martynov S and Kunitskiy S 2016 Energy saving in water treatment technologies with polystyrene foam filters *Journal of Water and Land Development* **31** 119-122
[6] Kulskyi L *Natural Water Treatment Technology* (Kyiv: Vyshcha shkola) p 352
[7] Apeltsyna E 1978 *TsBNTI MZhKH RSFSR Ser. Water supply and sewage* **3** p 41
[8] Bogdanov V, Hristoforov E and Klotsung V 1989 *Efficient low volume mixers* (Leningrad: Himiya) p 224.
[9] Frog B and Pervov A 2014 *Water treatment*: Textbook for students (Moscow: Association of building universities) p 512.
[10] Mordanov S, Nikylin V, Petsura S and Syirometnikov S 2011 *Vodoochistka* **10** 35-39
[11] Laptev A, Farahov T and Dudarovskaya O 2016 Comparative assessment of energy efficiency of static packed flow mixers *Vestnik Kazanskogo gosudarstvennogo energeticheskogo universiteta* **1** 68-83
[12] Flyayshli M, Gryutter T, Fisher T and Koller V 2004 Static mixer with profiled layers *Patent №2221631 RF MPK B01F 5/00* (Russia) p 2
[13] Epoyan S, Volkov V, Sukhorukov G and Haiduchok O 2020 Improved tubular mixer and its research method *Scientific Bulletin of Civil Engineering* **99**
[14] Bolshakov V 1984 Handbook of hydraulics (Kyiv: Vyshcha shkola) p 343
[15] Farahov T and Laptev A 2011 *Vestnik Kazanskogo gosudarstvennogo energeticheskogo universiteta* **4** 20-24
[16] Epoyan S, Yarkin V, Sukhorukov G and Babenko S 2018 Corridor-type baffled mixing basin with cross porous barriers *Science and Transport Progress. Bulletin of Dnipropetrovsk National University of Railway Transport* **1** 24-32
[17] Epoyan S, Yarkin V, Sukhorukov G and Sizova N 2019 Improving sedimentation efficiency through the use of an Improved design mixer in the preparation of drinking water *Scientific Bulletin of Civil Engineering* **98** 181-187.

# Fillers for modification of polyester powder coating

**V I Gots, O V Lastivka, O O Tomin, S A Tymoshenko**

Department of Technology of building constructions and wares, Kyiv National University of Construction and Architecture, Povitroflotsky avenue 31, Kyiv, Ukraine

oles.lastivka@gmail.com

**Abstract.** In the article effectiveness of the use of fillers to improve the physical and mechanical properties of the coating varied depending on the average size of the filler particles and crystalline shape was studied. According to the indicators of impact strength to back impact and bending strength of the coating, the most effective is the use of microsilica with an average size of 2.9 μm. As the average filler size increases, the gloss and impact strength of the coating reduce. To increase the bending strength of the coating, it is advisable to use fillers in the form of wollastonite, which is characterized by a fibrous crystal shape, which improves the physical and mechanical properties of the coating and prevents cracking, acting as a reinforcing agent. In addition, the analysis of the obtained study results shows that the use of the studied Ukrainian-made fillers in the powder paint composition promotes obtaining a covering with adjustable physical and mechanical characteristics.

## 1. Introduction

With the strengthening of environmental requirements and the need to adapt Ukrainian legislation on quality and safety of construction materials to EU requirements, the priority direction of development is to minimize the negative impact of harmful substances in their composition on consumer health and the environment. This issue is especially acute for goods of the paint and varnish industry using volatile organic compounds [1-3]. In recent years, the world has paid considerable attention to the environmental aspects of production and use of paint and varnish materials, which has led to a steady increase in production and use of powder paints [4]. It should be noted that in Ukraine the powder paints are also becoming more common due to their environmental friendliness by virtue of the absence of harmful solvents, ease of use and storage, as well as obtaining a durable coating [5]. The factor of their efficiency is the durability of coatings. In most cases the coatings of powder paints show higher protective properties and have a longer service life than coatings of solutions and dispersions of polymers [6]. This is due to the fact that powder paints do not contain mediators of film formation (solvents, water) and surfactants, and are prepared mainly on thermosetting oligomers and polymers of crystalline structure with high chemical resistance. Therefore, further numerous experiments to improve the prescriptions will contribute to the rapid development of technology of manufacturing powder coatings and their application to construction products, which will reduce economic costs and increase the durability of the material.

## 2. Analysis of the latest research works and publications

The prescribed composition of the powder coating consists of five main components: polymer resin, hardener, pigments, functional additives and filler. In general, the polymer resin and hardener play a major role in providing the required mechanical properties and durability of the powder coating. The introduction of pigments in paint and varnish materials is the main method of regulating the decorative properties of coatings - color and opacity (coverability). The additives are used to regulate the technological properties of powder paint and performance characteristics of the obtained coating. The fillers are used for two main purposes: the first is to reduce the cost of the material by replacing the most expensive component in the form of a polymer resin; the second is to modify the functional properties of the material, such as hardness, gloss, bending and impact strength, modulus of elasticity; permeability; corrosion, wear resistance; fire resistance.

The term "filler" is very broad and covers a wide range of materials. Most of fillers used in powder paints are inorganic and are usually extracted from rocks or ores, and with subsequent processing are converted into powders [7].

The most common fillers used in prescribed compositions of powder paints are barium sulfate ($BaSO_4$), calcium carbonate ($CaCO_3$) and talc ($Mg_3Si_4O_{10}(OH)_2$). As a rule, increasing their content in the powder paint helps to reduce the viscosity and maximum elongation of the system with increasing hardness of the coating [8]. In addition to their commonality, each filler also has its own unique effect. For example, the advantages of barium sulfate include its high corrosion and weather resistance [9]. Calcite contributes to the uniformity of the dispersion of the powder paint and the ability to obtain a matting effect of the coating. Talc has a high latent ability, matting effect and helps to obtain a satin type of coating.

However, the nomenclature of fillers by type and chemical structure is quite diverse, which is of great interest in the study of their effect on the properties of powder coatings [10]. This is especially true of Ukrainian-made fillers, as the main components for the production of powder paint are imported from abroad. Therefore, determining the effectiveness of the use of various Ukrainian-made fillers in systems of powder paints and coatings provides valuable information about their choice.

## 3. Definition of the goal and objective of the research

The objective of the work is to study the physical and mechanical properties of decorative and protective powder coatings using different types of fillers.

*Materials and methods of studies. Rationale for the choice of raw materials for powder coatings.* Prescribed composition of powder paint (PP) consisted of the following components: film-forming component, filler, pigments and functional additives.

Crylcoat 2441-3T carboxyl-containing polyester resin produces by Alnex was used as a film-forming component. Its characteristics are given in the Table 1. For carboxyl-containing polyester resin, the use of a structure-forming hardener is also mandatory. This paper uses TGIC produced by Huangshan.

**Table 1.** Characteristics of the film-forming component.

| Type of the resin | Indexes | | | | | |
|---|---|---|---|---|---|---|
| | Appearan-ce | Gloss 200/600, % | Viscosity (cP at 200 °C) | Color Gardnes (10 % in DMF) | Acid Value (mg KoH/g) | Temperature cure and time |
| Allnex 2441-3T | Opaque granules | 67 | 4000-5200 | 10 | 30-35 | 200°C 10 min |

K-2190 titanium dioxide produced by Kronos was used as the white pigment. Resiflow PV88 agent produced by Estron chemical in the amount of 1 % by weight of PP was used as a rheological additive. Benzoin produced by Estron chemical in the amount of 0.6 % by weight of PP was used as a degasser.

Fillers of different chemical nature were used in the studies: 1 - salt in the form of *barium sulfate* produced by Changsha Lianda, *lime* produced by the Private Joint-Stock Company "Sumiagroprombud"; 2 - oxide in the form of *ZnO* produced by EverZinc; 3 - silicates in the form of *metakaolin* of Glukhovets deposit, *microsilica no. 1* produced by the Limited Liability Company

"Teravion", *microsilica no. 2* produced by the Limited Liability Company "Mining Engineering Company", *wollastonite* produced by Minco LTD, 4 - man-made product in the form of *fly-ash* produced by Ladyzhyn Thermal Power Station. Characteristics of fillers are given in the Table 2.

**Table 2.** Characteristics of the fillers.

| Fillers | Appearance | pH value | Oil absorption, G/100g | Humidity, % | The average particle size, μm | Crystalline form of the filler |
|---|---|---|---|---|---|---|
| Barium sulfate | white powder | 7,86 | 16,1 | 0,074 | 1,87 | cubical |
| Fly ash | grey powder | 6,7 | 19,4 | 0,042 | 16,43 | spherical |
| Microsilica *No* 1 | grey powder | 6,9 | 27,3 | 0,15 | 15,2 | spherical |
| Microsilica *No* 2 | grey powder | 6,8 | 25,1 | 0,12 | 2,9 | spherical |
| Wollastonite | white powder | 8,2 | 38,4 | 0,1 | 14,02 | needlelike |
| Lime | white powder | 8,6 | 17,2 | 0,21 | 5,47 | cubical |
| Metakaolin | white powder | 6,1 | 35,4 | 0,03 | 10,03 | lamellar |
| ZnO | white powder | 8,4 | 31,2 | 0,062 | 2,32 | cubical |

*Study methods.* The study of the properties of decorative and protective powder coatings using different types of fillers was carried out in the following sequence:

1. Using different types of fillers, the powder paint (color RAL 9016) was applied on plates (sized 150x60 mm) made of steel St3. The powder paint was applied by electrostatic method according to ISO 1514: 2016 using a spray gun Start 50.
2. Polymerization of the powder coating on the samples-plates was carried out in a polymerization furnace at a temperature of 200 °C within the period of 10 min.
3. Study of the coating gloss was performed according to ISO 2813: 2014.
4. Study of the coating adhesion was performed by the cross cut method according to ISO 2409-2013.
5. Determination of the impact strength of the coating to back impact was performed according to ISO 6272-2.
6. Determination of the bending strength of the coating was performed according to ISO 1519: 2011.

## 4. Basic part of the researches

Experimental studies of the effect of fillers of different chemical nature on gloss, adhesion, impact and bending strength of the coating were conducted in the laboratory of the Limited Liability Company "Lacover". The compositions of the studied paint and varnish powder coatings are given in the Table 3.

As a result of study, it is established that the change of chemical nature, average size of particles and crystalline form of filler in the composition of powder paint affects the physical and mechanical properties of the coating.

It is shown that the use of ***barium sulfate*** in the powder paint (control composition) contributes to a high gloss of the coating, which is 91 % (Fig. 1) while providing impact strength to back impact at the level of 30 cm / kg (Fig. 2), adhesion of the Gt0 class (Fig. 3) and bending strength, which corresponds to the diameter of the bending shaft of 10 mm (Fig. 4).

However, the use of filler in the form of ***lime*** in the powder paint composition reduces the gloss and impact strength to back impact of the coating (Fig. 1, Fig. 2) to 82 % and 20 cm/kg, respectively, compared with the control composition (filler in the form of barium sulfate) - gloss 91%, impact strength to back impact 30 cm/kg. This can be explained by the change in diffusion reflection of light by increasing the average size of the filler particles in the form of chalk compared to barium sulfate (Table 2) and the deterioration of the powder paint dispersion uniformity, which determines the reduction of gloss and impact strength. It should be noted that the adhesion and flexural strength of the coating does not change compared to the control composition, which is Gt0 and 10 mm, respectively.

**Table 3.** Compositions of the powder coating.

| No | Resin | Fillers | Compositions of the powder coating, % | | | | |
|---|---|---|---|---|---|---|---|
| | | | Resin | Hardening agent | Filler | $TiO_2$ | Additive |
| 1 | Crylcoat 2441-3T | Barium sulfate | 55,8 | 4,2 | 18,4 | 20,0 | 1,6 |
| 2 | | Fly ash | 55,8 | 4,2 | 18,4 | 20,0 | 1,6 |
| 3 | | Microsilica *No* 1 | 55,8 | 4,2 | 18,4 | 20,0 | 1,6 |
| 4 | | Microsilica *No* 2 | 55,8 | 4,2 | 18,4 | 20,0 | 1,6 |
| 5 | | Wollastonite | 55,8 | 4,2 | 18,4 | 20,0 | 1,6 |
| 6 | | Lime | 55,8 | 4,2 | 18,4 | 20,0 | 1,6 |
| 7 | | Metakaolin | 55,8 | 4,2 | 18,4 | 20,0 | 1,6 |
| 8 | | ZnO | 55,8 | 4,2 | 18,4 | 20,0 | 1,6 |

The use of oxide in the form of ***ZnO*** to improve the mechanical properties of the powder coating is ineffective. Thus, the introduction of ZnO into the powder paint composition leads to a decrease in impact strength to 10 cm/kg (Fig. 2) and flexural strength, which corresponds to a diameter of the bending shaft of 12 mm (Fig. 4). The adhesion of the coating using ZnO corresponds to the Gt0 class (Fig. 3). It should be noted that the use of ZnO in the powder paint also reduces the gloss to 77 %.

In contrast to ZnO, the use of silicates in the form of ***microsilica no. 1*** to improve the mechanical properties of the powder coating is effective. Thus, with the introduction of microsilica no. 1 there is an increase in impact strength to 50 cm/kg (Fig. 2) and bending strength (8 mm) of the coating compared to the control composition. This is due to the increase in the compaction coefficient of the system by changing the crystalline form of the filler from cubic (ZnO) to spherical (microsilica), which determines the increase in the mechanical properties of the coating. The adhesion of the coating using microsilica corresponds to the Gt0 class (Fig. 3). The gloss of the coating is 84 %.

However, increasing the average size of the filler particles (Table 2) from 2.9 μm (microsilica no. 1) to 15.2 μm (***microsilica no. 2***) leads to a decrease in the mechanical properties of the coating. Thus, the use of microsilica no. 2 with an average particles size of 17.2 μm helps to reduce impact strength to back impact (10 kg / cm) and bending strength (14 mm) compared to the control composition. It should be noted that increasing the size of the filler particles leads to a decrease in the gloss of the coating to 64%.

The use of filler in the form of ***fly ash*** reduces the gloss to 51 % compared to the control composition - gloss 92 % (Fig. 1). This is explained by the change in the diffusion reflection of light by increasing the average size of the filler particles in the form of fly ash compared to barium sulfate (Table 2), which determines the decrease in the gloss of the coating. The impact strength to back impact and the bending strength of the coating using fly ash is similar to the strength of the control composition, which is 30 kg/cm and 10 mm, respectively (Fig. 2, Fig. 4). The adhesion of the coating using fly ash corresponds to the Gt0 class (Fig. 3).

The introduction of the filler in the form of ***wollastonite*** also reduces the gloss from 94 % (control composition) to 62 % by increasing the average size of the filler particles from 1.87 (barium sulfate) to 14.0 (wollastonite). There is an increase in the bending strength of the coating, which corresponds to the diameter of the bending shaft 8 mm (Fig. 4). This is due to the change in the crystalline form of the filler from cubic (barium sulfate) to fibrous, which improves the physical and mechanical properties of the coating and prevents the formation of cracks, acting as a reinforcing agent. The adhesion of the coating using wollastonite corresponds to the Gt0 class (Fig. 3).

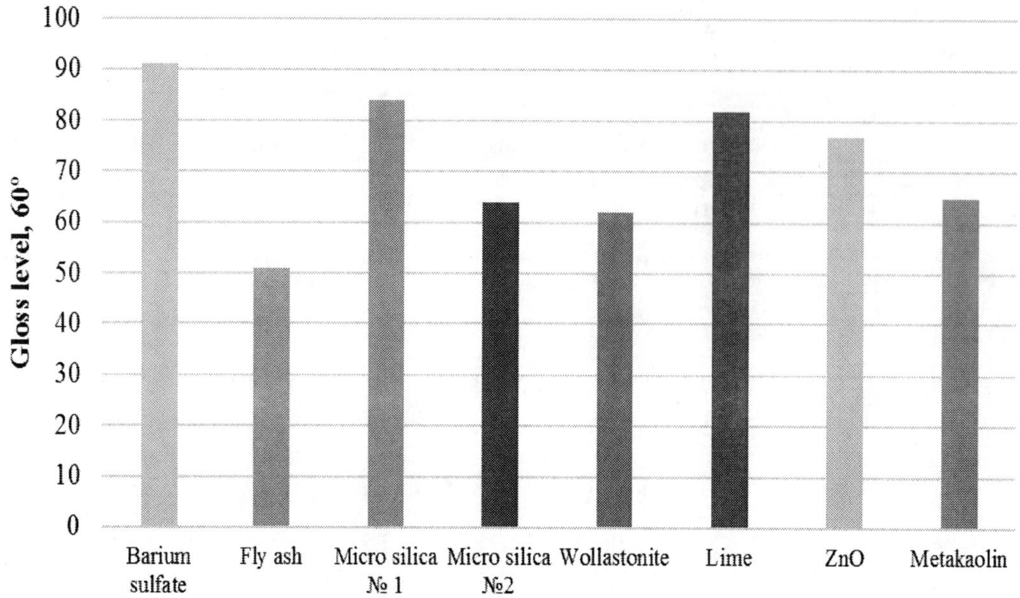


**Figure 1.** Gloss level of the powder coating depending of the type of filler (Table 2).

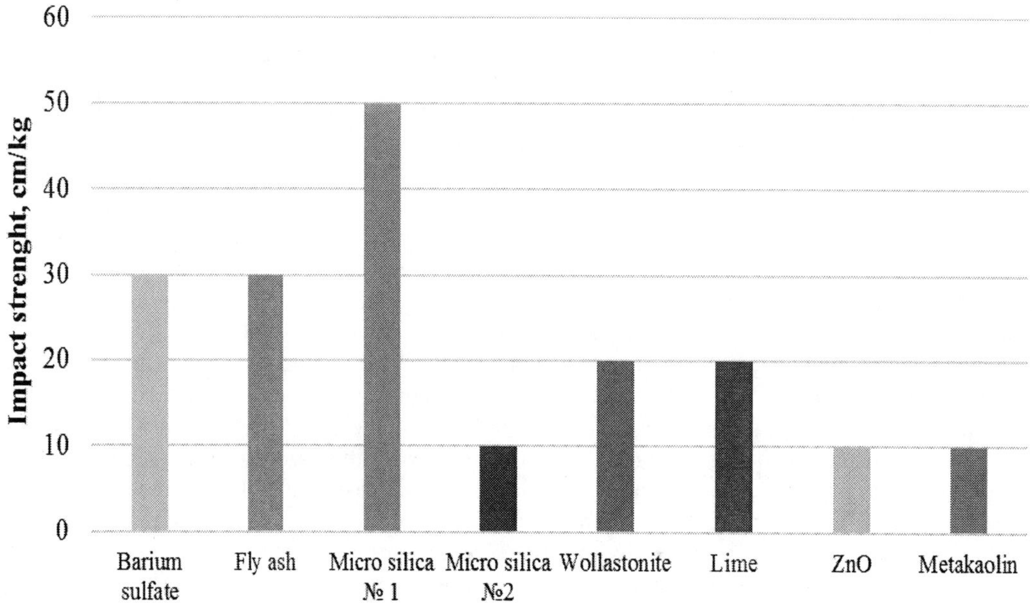


**Figure 2.** Impact strength of the powder coating depending of the type of filler (Table 2).

The use of ***metakaolin*** to improve the mechanical properties of the powder coating is ineffective. Thus, the introduction of metakaolin into the powder paint composition reduces the impact strength to back impact (10 cm/kg) and bending strength, which corresponds to the diameter of the bending shaft of 12 mm (Fig. 2, Fig. 4). The adhesion of the coating using metakaolin corresponds to the Gt0 class (Fig. 3). However, its introduction helps to reduce the gloss of the coating to 65 % (Fig. 1) by increasing the average size of the filler particles (Table 2).

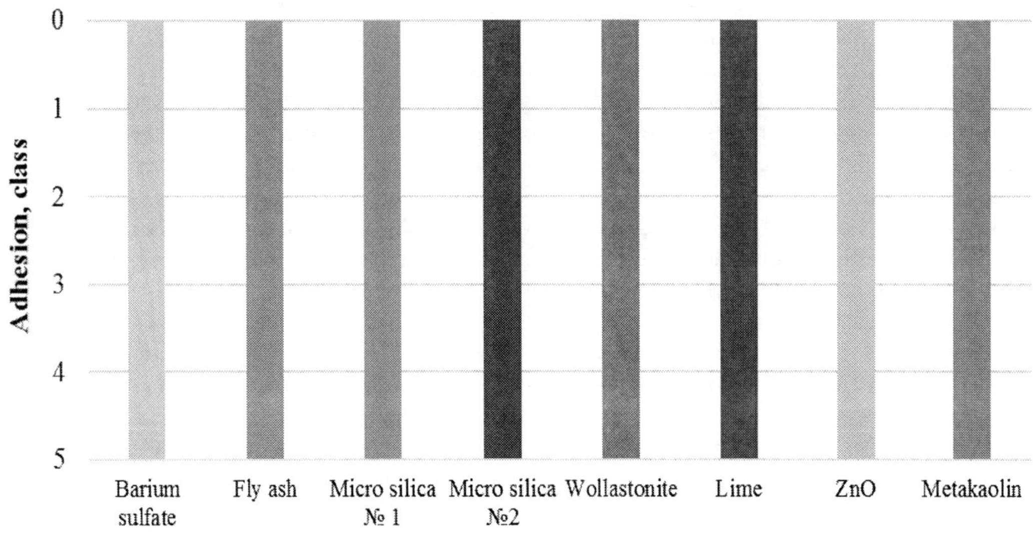


**Figure 3.** Adhesion of the powder coating depending of the type of filler (Table 2).

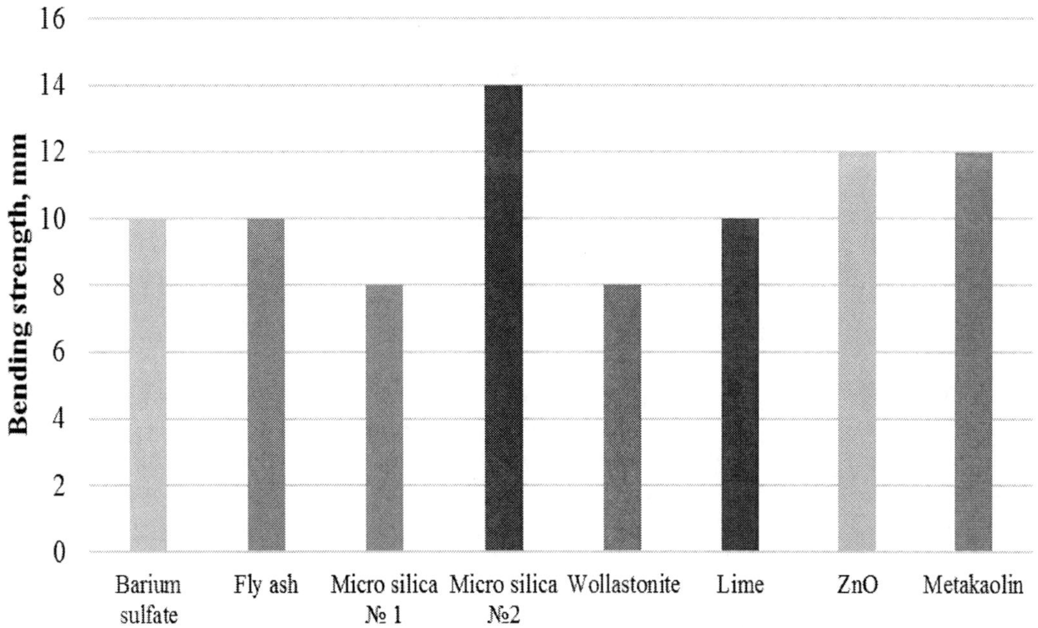


**Figure 4.** Bending strength of the powder coating depending of the type of filler (Table 2).

Thus, the replacement of fillers in the powder paint composition affects the change in gloss, impact strength to back impact and bending strength of the coating depending on the average size and crystalline shape of the filler. Increasing the average size of the filler particles reduces the gloss of the coating by changing the diffuse reflection of light on the surface of the coating. Changing the crystalline shape of the filler from cubic to spherical helps to obtain a denser structure of the system with increasing mechanical properties of the coating. When using a filler of fibrous structure, there is an increase in the bending strength of the coating.

IOP Conf. Series: Materials Science and Engineering **907** (2020) 012051 doi:10.1088/1757-899X/907/1/012051

## 5.Conclusions

The study has found that the effectiveness of the use of fillers to improve the physical and mechanical properties of the coating varied depending on the average size of the filler particles and crystalline shape. According to the indicators of impact strength to back impact and bending strength of the coating, the most effective is the use of microsilica with an average size of 2.9 μm. As the average filler size increases, the gloss and impact strength of the coating reduce. To increase the bending strength of the coating, it is advisable to use fillers in the form of wollastonite, which is characterized by a fibrous crystal shape, which improves the physical and mechanical properties of the coating and prevents cracking, acting as a reinforcing agent. In addition, the analysis of the obtained study results shows that the use of the studied Ukrainian-made fillers in the powder paint composition promotes obtaining a covering with adjustable physical and mechanical characteristics.

## References

[1] Al-Sherrawi M 2018 Corrosion as a source of destruction in construction *International Journal of Civil Engineering and Technology* 306-314
[2] Krivenko P, Pushkareva K, Baranovsky V 2004 *Building Materials Science*: *Textbook* 455
[3] Golterova T, Obukhova N 2018 Problem problems of economics of design and construction *Scientific herald of construction* 284-288
[4] Svidersky V, Karavaev T 2010 State, structure and prospects of development of the market of paint and varnish production in Ukraine *Paints and varnishes and their application* 8-16
[5] Puig M, Gimeno J 2018 Anticorrosive Properties Enhancement in Powder Coating Duplex Systems by Means of ZMP Anticorrosive Pigment *Assessment by Electrochemical Techniques* 1993-1999
[6] Gots V, Lastivka O, Tomin O, Kovalchuk O 2019 Influence of film-forming components on the corrosion resistance of powder coating *Tech Publications Ltd, Switzerland* 143-152
[7] Xanthos M, Carraher C, Craver C 2000 Polymer processing, in Applied Polymer Chemistry *Elsevier, Oxford* 355–371
[8] Mirabedini S, Kiamanesh A 2013 The Effect of Micro and Nano-Sized Particles on Mechanical and Adhesion Properties of a Clear Polyester Powder Coating *Progress in Organic Coatings* 1625-1632
[9] Lafabrier A, Fahs A, Louarn G 2014 Experimental Evidence of the Interface. Interphase Formation between Powder Coating and Composite Material *Progress in Organic Coatings* 1137-1144
[10] Tadmor Z, Gogos C 2006 Principles of Polymer Processing *John Wiley & Sons, Inc*. 14-17.

IOP Conf. Series: Materials Science and Engineering **907** (2020) 012052 doi:10.1088/1757-899X/907/1/012052

# Ecological efficiency photocatalytic concrete

**V P Sopov[1], N S Shyshko[1], V I Kondrashchenko[2], Zhang Yihe[3]**

[1]Department of Physical and Chemical Mechanics and Technology of Building Materials and Products, Kharkiv National University of Civil Engineering and Architecture, Sumska st., 40, Kharkiv, 61002 Ukraine
[2]Department of Building Materials and Technologies, Russian University of Transport, Minaevsky Lane, 2, Moscow, 127055 Russian Federation
[3]School of Materials Science and technology, China University of Geosciences, 29 Xueyuan Rd, Beijing, 100192 China

vsopov@ukr.net

**Abstract.** The process of urbanization is accompanied by the formation of a high density of street highways, buildings, people, transport and industrial enterprises, high energy consumption, an increase in the amount of waste and the release of a large amount of pollutants into the atmosphere and the aquatic environment. Air pollution is one of the most serious environmental problems in cities and poses a threat to vegetation, animals, materials and human health. One of the main environmental pollutants is nitrogen oxides ($NO_x$). Finding ways to remove $NO_x$ from the surrounding air is a necessity today, as it will help to improve the ecology of large cities, as well as reduce $O_3$ concentration at ground level. Titanium Dioxide ($TiO_2$) is a natural semiconductor material that is widely used in many industries. In construction, titanium dioxide is used to create building materials with a self-cleaning surface. In recent years, a fairly powerful photocatalytic effect of surfaces with titanium dioxide has been discovered. Self-cleaning combined with the photocatalytic effect of $TiO_2$ makes it an ideal additive in the production of building materials for urban environments. Depending on the application, $TiO_2$ can be used as a coating, additive in concrete, gypsum or paint. $TiO_2$ can be used for surfaces of paving elements or building facades, retaining walls, tunnels. In this paper, we study the effect of the addition of titanium dioxide on the formation of the physicochemical properties of concrete.

## 1. Introduction

The first use of concrete with self-cleaning properties dates back to 1996, when Italcementi participated in the construction of the church Dives in Misericordia in Rome (completed in 2003). This project involved the construction of a complex structure of three huge white sails, which are assembled from precast concrete. In such a building, it was required to use concrete, unique in its properties. Concrete of high strength and durability, had to retain white color indefinitely due to the self-cleaning properties of the surface.

It is known that the main reason for staining cement materials, which are a porous body, is the accumulation of colored organic compounds in the surface layers of materials. Cement composites with $TiO_2$ nanoparticles retain their esthetic characteristics, especially color, for a long time even under the influence of an aggressive urban environment.

After the construction of the church in Rome, photocatalytic concrete was used in other European architectural projects in France, Belgium and Italy. Concrete was also used as a material that controls the ecology and air quality of megacities [1].

Already, Ukraine has faced the problem of controlling air pollution from vehicle emissions, especially in growing urban areas.

The main air pollutant in the city is motor vehicles, and the main danger to humans is traffic fumes. They contain more than 200 chemical compounds.

Harmful ultrafine particles from car exhaust can lead to asthma, allergies, heart disease, lung cancer. According to the World Health Organization, annually about 3 million deaths are due to air pollution, largely due to car exhaust.

New developments in concrete offer the inclusion of photocatalytic $TiO_2$ nanoparticles in concrete, which can reduce environmental air pollution (for example, by decreasing the content of nitrogen oxides when exposed to ultraviolet radiation) and promote self-cleaning of the concrete surface.

As a result of exposure to sunlight, titanium dioxide nanoparticles act as a photocatalyst, which turns water vapor and atmospheric oxygen into atomic oxygen. The active oxygen that is released as a result of this reaction is enough to kill bacteria, oxidize and decompose organic pollutants, and deodorize rooms. Houses made of cement composites with titanium dioxide nanoparticles retain their color for a long time, despite the influence of an aggressive urban environment.

The $TiO_2$ photocatalyst activates ultraviolet (UV) radiation to oxidize air pollutants such as nitrogen oxides ($NO_x$) and volatile organic compounds (VOCs) [2]. The application of the photocatalytic effect of $TiO_2$ on a concrete coating to remove pollutants from the air is a promising alternative for removing pollutants from the street. This article compared various ways of using $TiO_2$ in concrete and evaluated the effectiveness of protective coating materials when removing pollutants from the atmosphere, preserving self-cleaning properties.

## 2. Architectural shells of glass fiber reinforced concrete

The experience of construction in our country and abroad shows that fiber-reinforced concrete is one of the promising building materials of the 21st century [3]. The study of dispersed reinforced concrete allows the production of lightweight building structures with increased strength.

The use of fiberglass in concrete as a fiber has shown that a strong and relatively light composite with a high fracture toughness and a fairly wide range of applications is obtained. [4].

Glass fiber reinforced concrete combines the positive properties of concrete regarding high compressive strength and, at the same time, thanks to the inclusion of fibreglass, it has increased crack resistance, water resistance, and operational stability.

Fiber concrete, in comparison with concrete of a similar brand, can have a tensile strength at bending higher by 4-5 times, axial tension higher by 3-4 times, impact strength more than 15-20 times, and this is important for the operation of a significant part structural elements.

The main advantages of fiberglass concrete products are their increased decorativeness, architectural expressiveness and plasticity, which can be used to improve the appearance of buildings. All this contributed to an increase in interest in fiberglass concrete and the expansion of its effective use in construction.

The development of the organization of the technology for the manufacture of fiber-reinforced concrete is based largely on basic research. Scientists from Japan, the USA, Norway, Germany, France, Great Britain, and China made a great contribution to the development of various methods for producing fiber-reinforced concrete.

Currently, interest in the study of fiber-reinforced concrete manufacturing technologies among scientists from around the world is due to the need to obtain building materials based on local raw materials with enhanced performance characteristics that provide high energy efficiency, environmental safety and low cost.

This primarily relates to the facades of buildings and constructions that form the outer shell of the building and serves different purposes. The facade protects the interior from environmental conditions such as wind, rain and sun. Another key aspect of a modern facade cladding is to creation of a unique appearance for the building. A facade system usually consists of the facade cladding, a two-dimensional element that is supported by a primary and/or secondary structure [5]. Various materials for the cladding have been used in architecture ranging from more traditional envelope materials such as glass, metal, stone, timber, concrete, etc. to materials that have been developed during the last decades such as glass fiber reinforced plastic (GRP), glass fiber reinforced concrete (GRC).

The ability to produce thin-walled (1-2 cm thick) glass fiber reinforced concrete shells of almost any shape and configuration provides wide prospects for their application for cladding facades of residential and industrial buildings, as supporting structures or decor elements.

The mechanical strength of glass fiber reinforced concrete is 150% higher than the strength of ordinary, its frost resistance is 50% higher, and the likelihood of cracking is several times lower. It is worth noting that the weight of a structure made of such concrete is reduced several times. Three types of glass fiber reinforced concrete are used in construction: lightweight; medium strength; high strength.

Lightweight glass fiber reinforced concrete is used in the construction of small private houses, utility rooms, to create partitions between rooms and other non-load-bearing structures. Nanostructured lightweight concrete is used in bridge construction.

Glass fiber reinforced concrete of medium strength is used in the construction of buildings, bridges and roadbed.

Glass fiber reinforced concrete of increased strength is used to create load-bearing structures of structures, elevator shafts and where high strength is required. A variety of requirements are imposed on concrete as the main building material in various projects. But almost always the designer is interested in access to structural concrete with a minimum specific gravity, while maintaining or even developing the bearing capacity of parts made of such concrete.

To improve the properties of architectural shells made of glass fiber reinforced concrete and provide them with environmental characteristics, it is proposed to use the addition of titanium dioxide.

Titanium dioxide $TiO_2$ is a natural compound that is used in a variety of applications, from consumer products to architectural coatings. It is also the best semiconductor material for photocatalytic applications [6]. $TiO_2$ occurs in nature as rutile, anatase and brookite. Rutile is the most common and thermodynamically stable form and is mainly used as a colorful pigment. Brookite is rare and has few common commercial uses [7-8]. Anatase is chemically stable at room temperature, non-toxic and economically beneficial as a photocatalyst compared to other semiconducting materials such as zinc oxide (ZnO) [9-10]. The most common, commercially available form of $TiO_2$ for photocatalysis is titanium dioxide with an average primary particle size of 21 nm and an anatase-rutile ratio of 80/20 [11- 12]. Many researchers studying photocatalytic materials have used this commercial product for direct testing or for comparison with other semiconducting materials.

## 3. Photocatalysis in concrete

The photocatalytic properties of titanium dioxide in the anatase phase have been used to create various materials since the mid-1990s. [13-14]. At first, titanium dioxide was used as a white pigment in the paint and varnish industry, it allows not only to obtain coatings of various colors, but also significantly improve their properties. Then, self-cleaning phenomena, the so-called "lotus effect" [15], were detected on surfaces treated with titanium dioxide, due to their high hydrophilicity (Fig. 1).

Currently, the possibilities of using titanium dioxide for the decomposition of organic and inorganic substances on the surface of materials, the so-called process of photooxidation [16] (Fig. 2).

IOP Conf. Series: Materials Science and Engineering **907** (2020) 012052 doi:10.1088/1757-899X/907/1/012052

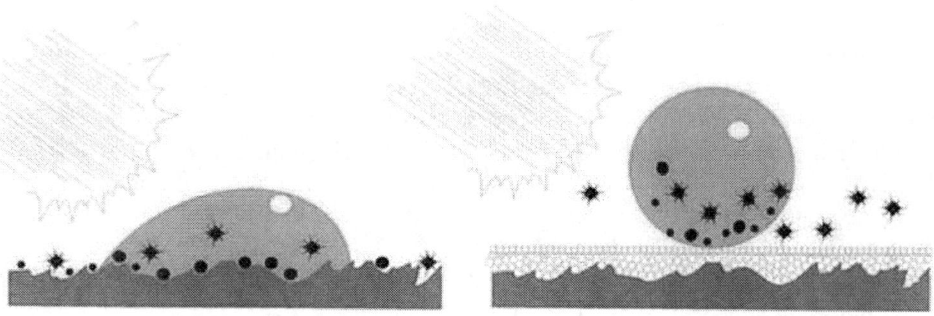

a) b)

**Figure 1.** Surface self-cleaning mechanism with $TiO_2$ photocatalyst
a) surface without $TiO_2$; b) surface with $TiO_2$

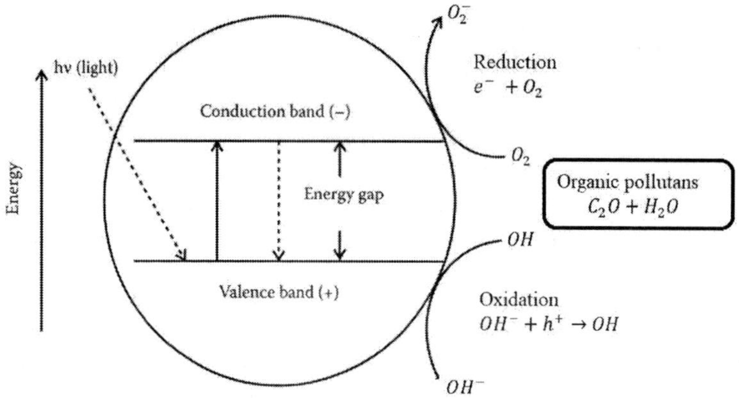


**Figure 2.** Mechanism of photocatalysis

Photocatalysis is a process that occurs on the surface of a semiconductor that is exposed to light. When a photon whose energy is equal to or greater than the band gap (3.2 eB для $TiO_2$) of a semiconductor is absorbed, an electron ($e^-$) moves from the valence band to the conduction band. The result is the presence of a "hole" in the valence band ($h^+$). $H^-$ i $e^-$ are strong oxidizing agents and reducing agents, respectively. An electron-hole pairs can react with electron donors or acceptors adsorbed to the surface of a semiconductor. If the reaction does not occur, the pair of electron-hole pairs recombine and the energy dissipates as heat. Oxygen and water adsorbed on a semiconductor surface, catalyzed with the formation of reactive species, superoxide anion $O^2$ and hydroxyl radical ($OH$):

$$h^+ + H_2O \rightarrow OH^- + H^-$$
$$e^- + O_2 \rightarrow O_2^-$$

Hydroxyl radicals and superoxide anions are strong oxidizing and reducing agents. They can react with contaminant molecules adsorbed onto a photocatalytic surface, such as $NO_x$ or hydrocarbons, to form nitric acid, carboxylic acid or carbon dioxide.

Thus, the use of $TiO_2$ in concrete is aimed at achieving two main effects - self-cleaning from dirt and purification of the atmosphere due to the oxidation of nitrogen oxides ($NO_x$) as a result of photocatalysis.

Photocatalytic cement-based materials were used in architectural projects in Germany, Belgium, Italy, Spain, France, and Japan [16-18]. For example, in France, "green technologies" using $TiO_2$ were used in the construction of the "Porte de Vanves" metro station in Paris. In Japan, photocatalytic surfaces are used at airports to control harmful substances in the air. For the first time in Italy, photocatalytic coatings were used in the "Via Porpora" tunnel in Milan.

## 4. Materials and methods

For research, raw materials were used, which are usually used in the manufacture of fiberglass concrete products:

1. Portland cement CEM I 52,5;
2. Sand $M_к$=1;
3. Alkali-resistant fiberglass with fiber length 12 mm;
4. Chemical additive Sika ViscoCrete-5-600;
5. Titanium dioxide anatase-rutile ratio 80/20.

In laboratory conditions, studies have been carried out of the photocatalytic removal of air pollutants. Most of these studies have focused on the ability of concrete containing $TiO_2$ to remove one type of pollutant. As a result, the test materials were usually exposed to air containing only one interesting pollutant, rather than a mixture of pollutants similar to that found in urban air. In previous laboratory studies, realistic environmental conditions or air mixtures were rarely used, making it difficult to estimate expected air emissions. However, the of previous laboratory test results can be used to better understand the mechanisms leading to the oxidation of $NO_x$ and other pollutants, and what parameters affect the photocatalytic process.

There is a discrepancy in the literature [2, 8-12] regarding the mechanisms that occur during heterogeneous NOx photocatalysis. An understanding of the mechanisms involved in these reactions is necessary in order to correctly describe and evaluate the kinetic parameters that control the reaction process.

It is generally accepted in the literature that $OH^-$ and $O_2$ are reactive species created on the surface of a photocatalytic material. Two mechanisms are proposed for the destruction of NO, each of which includes one of these reactive species as an initiator of the removal process.

Several researchers [17-18] have proposed a mechanism involving hydroxyl radicals, and this is the most common mechanism in the literature. The proposed mechanism is such

$$NO_{ads} + OH^- \rightarrow HNO_2$$
$$HNO_2 + OH^- \rightarrow NO_{2,ads} + H_2O$$
$$NO_{2,ads} + OH^- \rightarrow HNO_3$$

Several researchers have studied the effect of irradiation on the photocatalytic activity of $TiO_2$. The photocatalytic activity of $TiO_2$ is affected by the intensity and nature of light. Despite the large differences in the experimental conditions used to test photocatalytic materials and the variations of the most tested materials, it is difficult to quantify the results of various studies. However, some qualitative results have been found in various studies [19-20]. The photocatalytic activity of materials containing $TiO_2$ increases linearly with light intensity at low light intensity until the light intensity reaches about one equivalent of the sun (1 mW $cm^{-2}$).

The activity of air purification with titanium dioxide used in building materials was determined by the oxidation of NO and $NO_2$ to $NO_3$. Particular attention is paid to these pollutants, since they are the main components of traffic fumes and play an important role in the formation of smog in large cities.

The experimenta set-up (Fig. 3) consists of a plastic container in which the concrete sample with a $TiO_2$ coating is located. The container is covered with glass that transmits ultraviolet radiation. Air from an internal combustion engine that contains nitrogen oxides is blown into a container on a concrete surface. Ambient temperature – 23±2 °C and relative humidity – 50%. The light intensity is 10 W/$m^3$ with a wavelength in the range 300...460 nm. Concentration is measured at the exit of the container. Measurements are taken at 7:00. At the beginning, the initial concentration of NO and NO2 in the air of the container is measured. To activate photocatalytic reactions, sunlight or an artificial source of ultraviolet radiation is used.

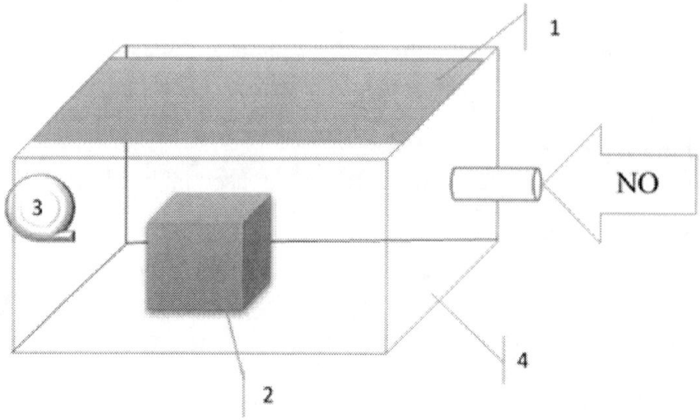


**Figure 3.** Experimental set-up for studying the photocatalytic effect
1 – quartz glass; 2 – concrete sample; 3 – gas analyzer; 4 – container.

## 5. Results and discussion

After reaching the maximum concentration of $NO_x$ = 5 g/m$^3$ in the container, the engine turned off. The reading of the gas analyzer was carried out with an interval of 1 hour. The measurement results are shown in Fig. 4.

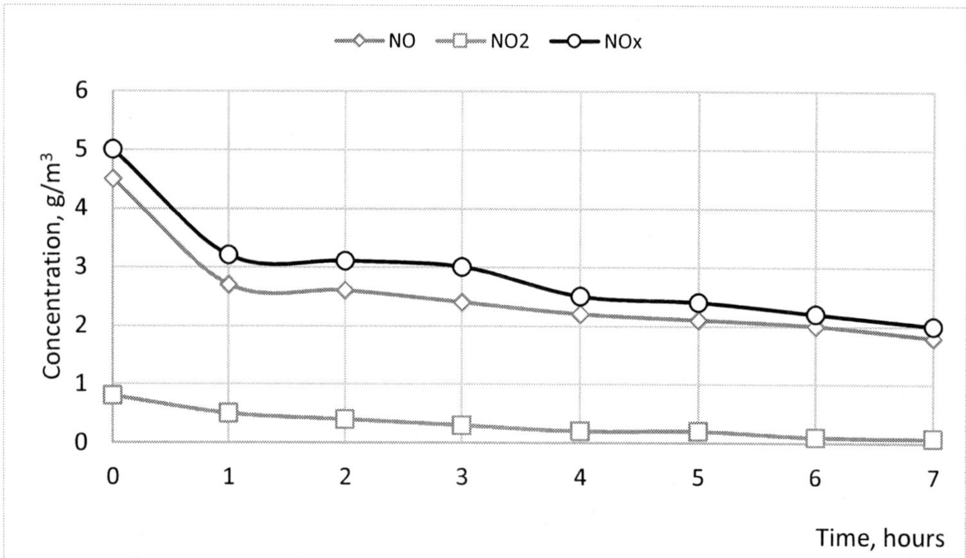


**Figure 4.** The effect of titanium dioxide on the change in the concentration of pollutants

The nature of the decrease in the concentration of $NO_x$ depends on the size of the concrete surface, the initial concentration of $NO_x$, light intensity and ambient temperature. A 6-fold increase in the photocatalytic surface contributes to a decrease in NOx concentration to 82%.

In fig. 5 shows the effect of the thickness of the $TiO_2$ layer, which is deposited on concrete samples, on the $NO_x$ reduction efficiency. The test was performed at 30% humidity, ultraviolet radiation intensity of 2,4 mW/cm$^2$ and low pollutant feed rate of 3 min$^{-1}$.

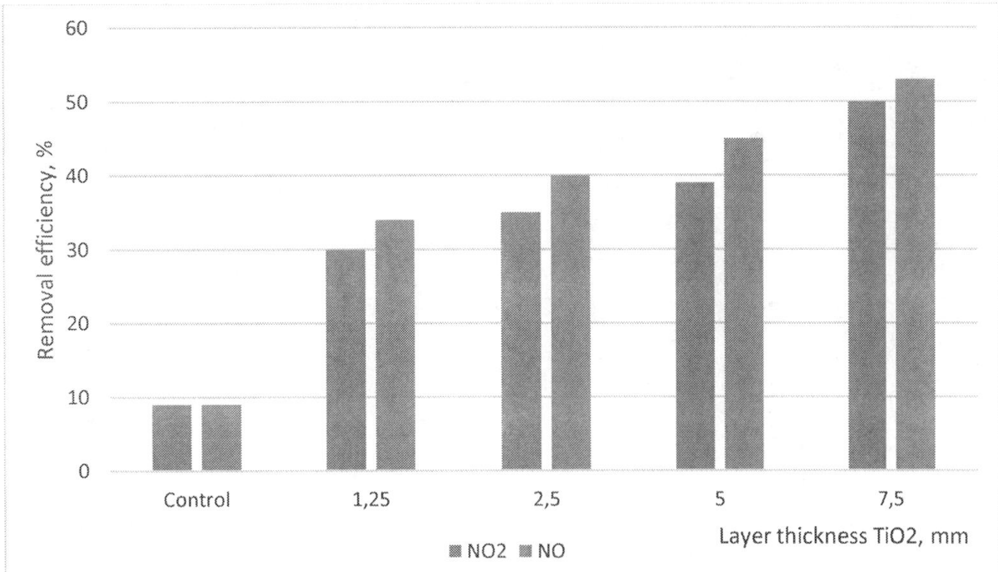


Figure 5. Effect of $TiO_2$ layer thickness on $NO_x$ reduction efficiency

As can be seen in fig. 5, an increase in the layer thickness had a positive effect on the $NO_x$ reduction efficiency, since an increase in the thickness led to an increase in the $NO_x$ removal efficiency. The results show an average NOx reduction efficiency of 34% for a 12.5 mm $TiO_2$ layer thickness compared to a 62% reduction for a sample with a 75 mm layer thicknessThis indicates the possibility of applying a $TiO_2$ layer on the surface of road surfaces, since UV light can penetrate to a considerable depth, enhancing the photocatalytic efficiency of the sample.

Operational properties were evaluated by standard physical and mechanical studies. The expiremental results showed that the addition of $TiO_2$ to the concrete composition increases the compressive strength of concrete by 5-7%, reduces water absorption by 1%.

In addition, the introduction of $TiO_2$ nanoparticles leads to an increase in the strength of fine-grained concrete: bending after 7 days by 1.8 times, after 28 days by 1.2 times; compression 7 days 1.7 times, and after 28 days 1.5 times.

**Table 1** Compressive strength of concrete after

| Samples | Days | Compressive strength, MPa |
|---|---|---|
| Control sample | 7 | 67,5 |
| $TiO_2$ sample | 7 | 72 |
| Control sample | 28 | 74 |
| $TiO_2$ sample | 28 | 77,5 |

The study of the structure of concrete by optical microscopy is shown in Fig. 6.

As shown by microscopic studies, the introduction of $TiO_2$ contributes to the compaction of the concrete structure. The presence of a surface coating from a $TiO_2$ layer in the form of a white coating is observed (Fig. 6, b).

A study by thermoprometry [19] showed that with the introduction of nanodispersed titanium dioxide additives, the porosity of fine-grained concrete is redistributed in size (Fig. 7) - a decrease in the size and volume of capillary pores and an increase in the fraction of gel pores.

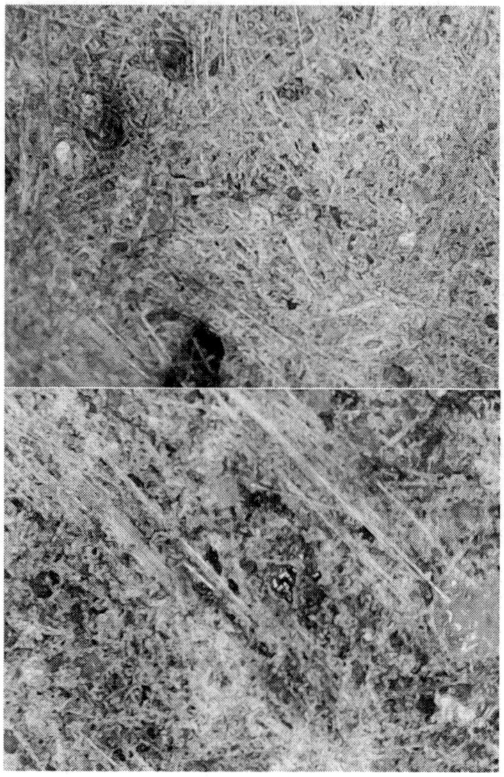
a)

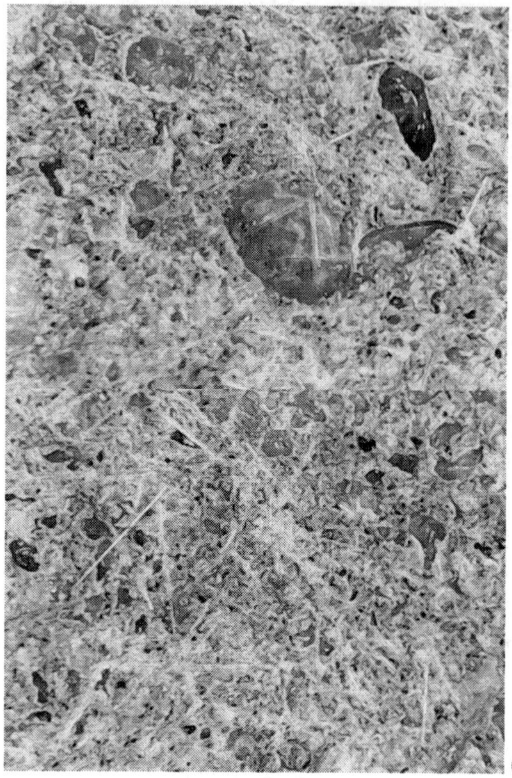
б)

**Figure 6.** Microscopic examinations: a) control sample; b) $TiO_2$ sample

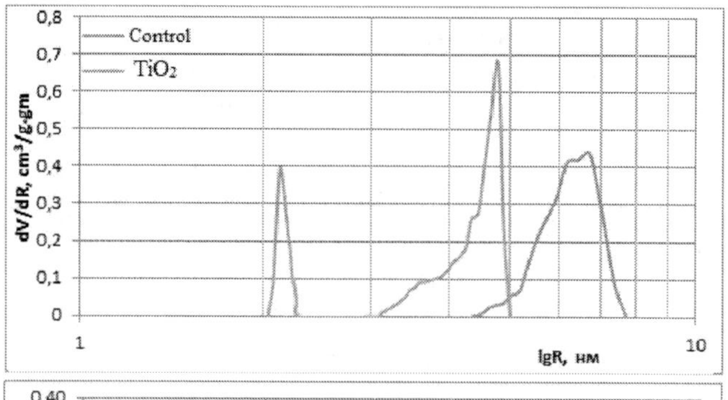

a)

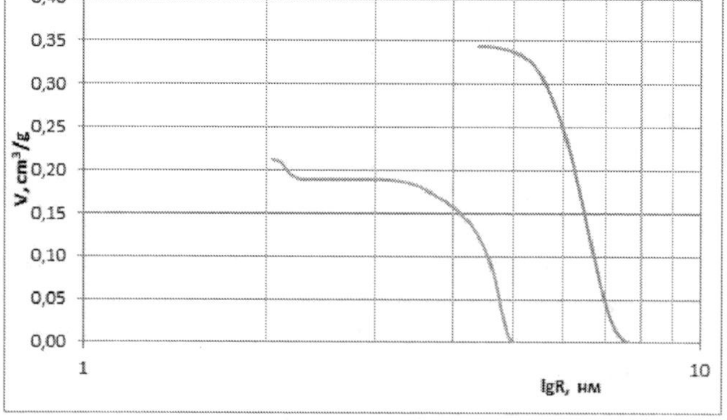

b)

**Figure 7.** Microporosity of cement stone: a) pore size distribution; b) pore volume

A sharp decrease in micropore volume is due to a change in the nature of adsorption processes at the solid phase – pore liquid interface and the filling of large pores with particles of titanium dioxide.

Thus, after mixing, $TiO_2$ very well compacts the structure of cement stone in the same way as silica fume or other fine fillers. In addition, $TiO_2$ helps to increase erosion resistance and reduce the amount of pollutants that penetrate the pores.

## 6. Conclusions

The use of titanium dioxide will significantly improve the environmental situation of large cities in terms of cleaning surfaces from dirt and dust and air purification from organic compounds and $NO_x$. Titanium dioxide helps to improve the physical and mechanical properties of glass fiber rinforced concrete, densification of its structure. Gives specific properties to its surface, which promotes self-cleaning from dust and dirt. TiO2 is an effective material for removing NOx pollutants from the air stream when used in the composition of sustainable Fiberglass Concrete with a coefficient of performance from 34% to 62%.

## 7. Reference

[1] Vojlokov I A 2009 Fiber concrete - history, regulatory framework, problems and solutions *ALIT Inform international analytical review* № **2**

[2] Hamidi F and Aslani F 2019 $TiO_2$-based Photocatalytic Cementitious Composites: Materials, Properties, Influential Parameters, and Assessment Techniques *Nanomaterials* **9** 1444 doi:10.3390/nano9101444

[3] DBC 56-97 «Design and the main provisions of the technology for the production of fiber-reinforced concrete structures».

[4] Anaczkij F I *Glass fiber reinforced concrete is a modern material with a wide range of properties and areas of effective application in construction* URL: http://www.master-grad.ru.

[5] Kharchevnikov V I 1987 Fundamentals of structure formation of fiberglass polymer concrete *Bulletin of universities. Construction and architecture* № **11** p 62-66

[6] de Lorenzis L, Miller B, Nanni A 2001 Bond of FRP laminates to concrete *ACI Material Journal* Vol **98** p 256-264

[7] Haselbach L 2008 *The Engineering Guide to LEED-New Construction* (McGraw-Hill, New York)

[8] Taoda H 2004 The newest trends in development of photocatalytic materials *Engineering Materials* (Kogyo-Zairyo) **52** (11) pp 32-37

[9] Agrios A G, Pichat P 2005 State of the art and perspectives on materials and applications of photocatalysis over $TiO_2$ *Journal Appl Electrochem* **35** pp 655–663

[10] Chen J, Poon C-S 2009 Photocatalytic construction and building materials: From fundamentals to applications *Building and Environment* **44** pp 1899-1906

[11] Dylla H, Hassan M M, Schmitt M, Rupnow T, and Mohammad L 2011 Laboratory Investigation of the Effect of Mixed Nitrogen Dioxide and Nitrogen Oxide Gases on Titanium Dioxide Photocatalytic Eficiency in Concrete Pavements *Journal of Materials in Civil Engineering* Vol. **23** No 7 pp 1087–1093

[12] Ohtani B, Ogawa Y, and Nishimoto S 1997 Photocatalytic Activity of Amorphous-Anatase Mixture of Titanium (IV) Oxide Particles Suspended in Aqueous Solutions *Journal of Physical Chemistry* Vol. **101** pp 3746–3752

[13] Montes F, Valavala S, and Haselbach L M 2005 A New Test Method for Porosity Measurements of Portland Cement Pervious Concrete *Journal of ASTM International* Vol. **2** No 1 pp 1–13

[14] Ballari M M, Yu Q L, Brouwers H J H 2011 Experimental study of the NO and $NO_2$ degradation by photocatalytically active concrete *Catal Today* **161** pp 175–180

[15] Bengtsson N, Castellote M 2010 Photocatalytic Activity for NO Degradation by Construction Materials: Parametric Study and Multivariable Correlations *J Adv Oxid Technol* **13** pp 341-349

[16] Cros C J, Terpeluk A L, Crain N E, Juenger M C G, Corsi R L 2015 Influence of environmental factors on removal of oxides of nitrogen by a photocatalytic coating *Journal of the Air & Waste Management Association* **65** pp 937-947
[17] Dehn F, Bahnemann D, Bilger B 2004 Development of photocatalytically active coatings for concrete substrates *RILEM International Symposium on Environment-Conscious Materials and Systems for Sustainable Development* pp 347-352
[18] Jayapalan A R, Lee B Y, Fredrich S M, Kurtis K E 2010 Influence of additions of anatase $TiO_2$ nanoparticles on early-age properties of cement-based materials *Transportation Research Record: Journal of the Transportation Research* **2141** pp 41–46
[19] Lee B Y, Thomas J J, Treager M, Kurtis K E 2010 Influence of $TiO_2$ nanoparticles on early $C_3S$ hydratation *Journal of the American Ceramic Society* **93** pp 3399-3405
[20] Usherov-Marshak AV, Sopov V P 1994 Thermoporometry of cement stone *Colloid Journal* Vol. **56** No 4 pp 600-603

# Tensile strength of the pipe material determination according to the ultimate tensile strength

**B Younis[1], L Saienko[2], I Kazimahomedov[3] and S Salem[3]**

[1]Department structural Mechanics, Kharkiv National University of Civil Engineering and Architecture, Sumska Street 40, 61000, Kharkiv, Ukraine
[2]Department of Mechanization of Construction Processes, Kharkiv National University of Civil Engineering and Architecture, Sumska Street 40, 61000, Kharkiv, Ukraine
[3]Department of building Materials and Products, Kharkiv National University of Civil Engineering and Architecture, Sumska Street 40, 61000, Kharkiv, Ukraine

sesonoor@yahoo.com

**Abstract.** A new calculation method for the strength of concrete pipes, developed, helps to determine the stresses in the walls of concrete pipes under the combined action of an external (compressing pipe) load and internal pressure from the fluid transported under pressure. An analysis of the nature and magnitude of the stresses is necessary to choose the material of the pipe walls, as well as to establish a possible combination of ultimate external. New formulas make it possible to determine the stresses in the pipe walls under the combined action of an external load and internal pressure. Based on the developed formulas, it was found that the main dangerous stresses in the pipes are tensile stresses. Using the obtained equations allows to determine the range of application of pipes made of modified concrete and to select the necessary concrete strength based on operational loads. There are two methods for determining the bearing capacity of pipeline: three-edge bearing method, in which the maximum breaking load (ultimate), and hydro-static loading method to determine the maximum internal pressure. The design of the test bench for pipes was developed in order to determine the maximum internal pressure on the pipe walls.

## 1. Introduction

The modern requirements to the quality of construction products are not inferior to the demands for the newest architectural elements [1-4]. In addition to the high requirements for physico-mechanical characteristics and durability of concrete and reinforced concrete pipes, the requirements for the quality and casting techniques are set.

In today's economic environment, designing for long term, sustainable project performance is imperative for the Engineer. Unlike some alternate pipe materials, concrete pipe has a proven track record of performance. Concrete pipe will not rust, burn, tear, buckle or deflect, and is immune to most environmental elements. Corps of Engineers has recommended precast concrete pipe for a design life of 70-100 years, and there are numerous examples of installations that have exceeded these parameters. Precast concrete drainage products have a reputation for strength and durability. They will not burn, corrode prematurely, deflect or move off grade to reduce hydraulic performance, or collapse under loads designed into the pipe structure. Comprised of the world's most commonly used building materials, precast concrete infrastructure is quickly integrated into ecosystems. This is clearly demonstrated by the use of three-sided precast boxes used to accommodate the natural channels of streams at road crossings, and precast concrete pipe for storm sewers and outfalls in valleys and shorelines.

Today, being recognized as a green material or product is growing in importance to many specifiers. Concrete pipe is suitable for LEED projects and it fits sustainable development.

Unlike plastic pipe, concrete is produced with benign, natural materials. Manufacturing of concrete consumes less energy than plastic fabrication. It's also recyclable and has little if any environmental impact. And, when you use local resources, concrete can also provide lower fuel cost for delivery [5, 6].

Reinforced concrete pipes have been in wide-spread use for many decades in the world. They have often been used as open channels (non-pressurized pipes), particularly for the conveyance of sewage and storm water. They have generally shown reliable long-term performance. In the late 1890's, the manufacturing of precast concrete pipes became a recognizable industry in Ontario. The Ontario Concrete Pipe Association (OCPA) was founded in 1957. OCPA has joined several industrial and governmental agencies to establish standards for the manufacture of high quality concrete pipe products. The Canadian Concrete Pipe Association (CCPA) was founded in 1992 (OCPA, 2010). Reinforced concrete (RC) pipes are widely used as open channels (non-pressurized pipes) for sewage and storm water conveyance. RC pipes have generally achieved a reliable long-term performance. Depending on multiple parameters (e.g. pipe diameter, pipe wall thickness, required strength, etc.) the pipe may have up to three welded reinforcement cages in order to resist anticipated loads. Each cage is an assembled unit of steel reinforcement consisting of circumferential and longitudinal bars or wires. The fabrication process of steel cage reinforcement is time- and labour-consuming. Thus, eliminating the steel cage reinforcement will yield an overall reduction in the production cost of precast concrete pipes [5, 7].

Purpose of the work – to determine the stresses in the walls of concrete pipes under the combined action of an external (compressing pipe) load (see formula №5), and internal pressure from the fluid transported under pressure (see formula №6). An analysis of the nature and magnitude of the stresses is necessary to select the material of the pipe walls, as well as to establish a possible combination of the maximum external loads (the pipeline depth – the magnitude of the working fluid pressure). These formulas are not identical in terms of initial data: in the Lyame formula, the dependence is taken according to the inner and outer diameters.

In the Klein formula [3, 8], the average diameter of the pipe is adopted. Wall thickness is not taken into account. It turned out to be impossible to select the wall thickness under the combined action of internal pressure and external soil load, so we modified the formulas. A new indicator has been introduced: the ratio of diameter to wall thickness, which, when switching from one diameter to another for pipes, retains its average values with some deviations. The new formulas developed allow to determine the stresses in the walls of pipes under the combined action of external load and internal pressure. The formulas showed that the main stresses in the pipe walls are tensile stresses. It follows from this that the guidelines given in the standards for concrete strength in the form of a concrete grade for compressive strength do not correspond to the actual work of the material in the product. Based on the regulatory requirements for reinforced concrete pipes (breaking load) according to the developed formulas, the necessary material strength was calculated ($R_{bt}$), which, depending on the magnitude of the working pressure, should be in the range of 50-70kg/cm$^2$. As for ordinary concrete tensile strength $R_{bt}$=1…4 MPa, the problem arises of increasing this indicator by modifying concrete.

## 2. Experimental

There are two methods for determining the bearing capacity of pipeline:

1. Three-edge bearing method, in which the maximum breaking load (ultimate) $P$ is established for 1 running meter of a pipe.

The structural load test, also known as the three-edge bearing test, is a destructive test that has been commonly adopted to evaluate the crack load and ultimate load of RCP. The test has over 100 years of history indicating its success in RCP assessment. This is a primary acceptance test among all study areas, though the examination process and interpretation of the test results are slightly different. In common, the testing procedure is to apply load evenly on a vertical plane along the pipe and the pipe axis. The load definition and terminology vary among study areas and are translated for consistency in Figure 1 illustrates load progression during the three-edge bearing test and the critical load definitions. The load in most procedures is applied to the pipe either in continuous or discrete manner. The inspector examines the crack development by measuring the crack width and length using a crack gauge. Subsequently, the load is carried to the design crack load and in some study areas, the pipe is loaded to failure [9].

**Figure 1.** Three-edge bearing test.

2. Hydro-static loading method to determine the maximum internal pressure $q$. For testing concrete and reinforced concrete pipes with hydrostatic pressure, a setup is used, the diagram of which is shown in Figure 2.

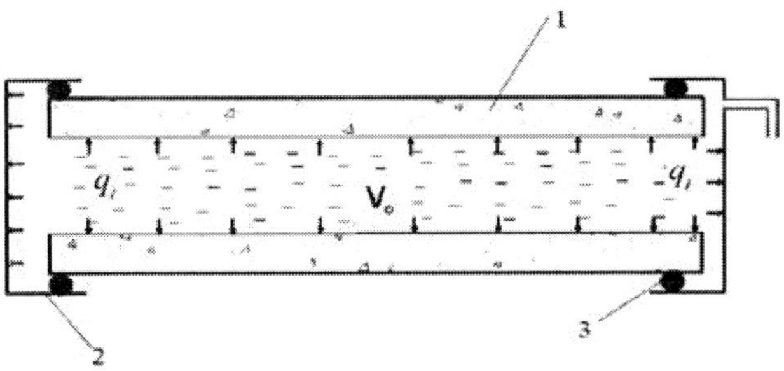


**Figure 2.** Scheme of a standard installation for hydrostatic testing of pipes:
1 – concrete pipe; 2 – end plug 3 – rubber sealing ring; $V_o$ – test fluid volume.

The test is carried out as follows: two pipes (1) horizontally mounted on the bench are filled with water. At the ends, plugs (2) with rubber O-rings (3) are installed. When creating hydrostatic pressure inside, water exerts a lot of pressure on the end caps. The pressure of the rubber rings (their compression) should be such that the pressure of the rubber on the concrete is always (during the tests) greater than the maximum test pressure, i.e. at the place of installation of the seals, their pressure on concrete is greater than the test pressure $q_m$, which distorts the test results.

To increase the reliability of indicators and greater convenience of testing, a new method and device for testing was developed.

Figure 3 shows the studied pipes subjected to hydrostatic pressure test.

The proposed test procedure has the following advantages over those already known:

– less water used in the pipes,

– automatic maintenance of the pressure of the elastic sealant on concrete, proportional to the value of the test pressure due to the fact that the pressure in the sealing annular chamber is 20% higher than inside the pipe (test pressure), which eliminates the overload of concrete.

This is especially important for unreinforced concrete pipes.

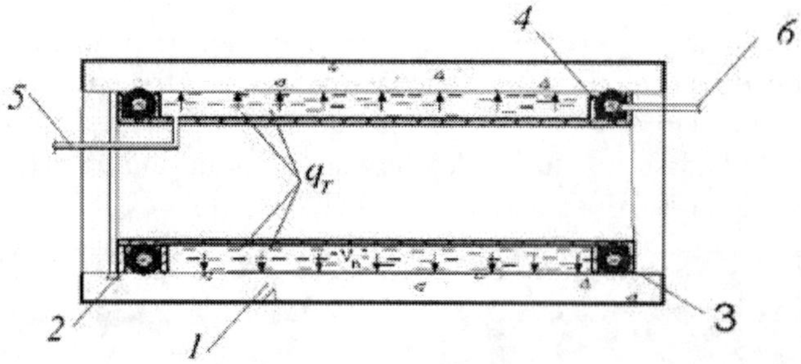


**Figure 3.** The developed scheme of hydrostatic testing of pipes:
1 – concrete pipe; 2 – ring sealing chamber; 3 – working cavity; 4 – thrust ring; 5 – input test fluid; 6 – introduction of fluid into the annular chamber; $V_n$ – volume of the test fluid.

Improved methods and devices are a new solution compared to the known ones, where concrete mixture impregnation can be used.

The design of the test bench for pipes was developed in order to determine the maximum internal pressure on the pipe walls (Figure 4).

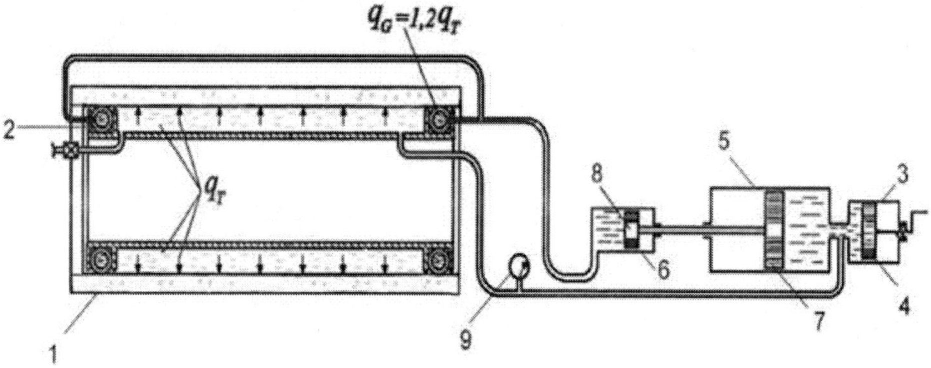


**Figure 4.** Installation diagram for hydrostatic testing of pipes with automated backpressure in the sealing ring:
1 – tested pipe; 2 – annular chambers; 3 – working cylinder; 4 – working piston; 5 and 6 – pressure conversion cylinders; 7 - piston under operating pressure $q_T$; 8 – piston transmitting back pressure $q_G$ in the annular hose; 9 – pressure gauge.

The stand has a steel cylinder placed inside the test pipe (1), sealing elements in the form of elastic annular chambers (2) installed at both ends in the space between the cylinder and the pipe (1), a device in the form of twin cylinders (5), (6) with pistons (7) and (8). A test pressure is applied to the piston 7, $q_T$ which under the piston 8 is converted into a back pressure $q_G$. The pressure in the system is controlled by a manometer (9).

Ealing pipes with a known method of testing internal pressure (Figure 5). So that water does not flow out through the sealing ring, the pressure of the ring $P$ should be 1.25 greater than the test pressure $q$; $P \geq 1.25\ q$.

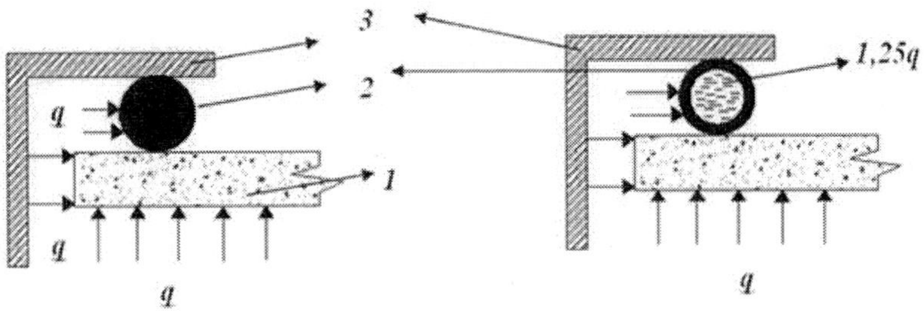


**Figure 5.** Pipe sealing:
1 – test pipe, 2 – sealing ring, 3 – steel sheath
*P* – compression force of the pipe ring, *q* – test pressure

Since the value of the limiting pressure ($q_{max}$) is not known, during testing it is necessary to significantly compress the ring ($P \geq q$), due to which the pressure of the ring *P* creates accordingly an excessive force in concrete and, therefore, distorts the test results. In Japan, to reduce the pressure of the sealant, an additional device with a separate pump is used, which creates electronics back pressure, presses the ring with force, which accordingly maintains the level of $P=1.25q$.

In the developed stand, a device with two pistons of different diameters is used for this. Due to the difference in diameters, a different pressure is created under each piston, and by selecting their diameters, the pressure inside the ring is $P= 1.25q$.

Calculation of operating parameters of the converter device (Figure 6) The task of research is to obtain pressure in the sealing chamber (1) $q_G$, exceeding the test pressure $q_T$ by "*n*" times, i.e. $q_G = nq_T$.

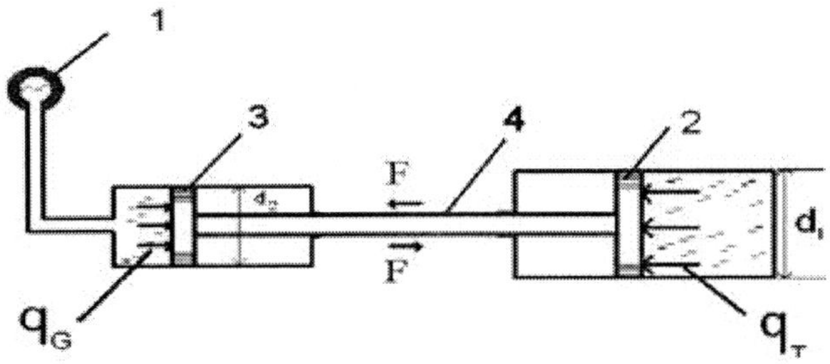


**Figure 6.** Pressure conversion diagram in a sealing hose:
$Q_T$ – test pressure; $q_G$ – pressure in the sealing hose (back pressure)
1 – annular chamber; 2 – piston $d_1$ under operating pressure $q_T$, 3 – piston with a diameter $d_2$, transmitting pressure $q_G$ to hose 1; 4 – rod.

In the piston connecting rod 2 ($d_1$) and 3 ($d_2$) a force *F* arises, which is equal to "0" under equilibrium conditions, when the total pressure on the area *S* of the pistons on each side is balanced ie $\sum q_G = \sum q_T$, then in equilibrium:

$$F = q_m \cdot S_1 = q_G \cdot S_2, \tag{1}$$

$$S = \frac{\pi d^2}{4}. \tag{2}$$

The increase in pressure in the ring is achieved by reducing the diameter of the piston ($d_2$) compared to the piston ($d_1$):

$$q_G = \frac{q_T \cdot S_1}{S_2}, \quad (3)$$

then

$$q_G = \frac{q_T \cdot d_1^2}{d_2}, \quad (4)$$

The pilot plant was made with the following dimensions of the pressure transducer: $d_1$= 60 mm and $d_2$ = 54,5 mm, which gives the ratio $q_G$= $1{,}2q_m$.

Normal stress determination ($\sigma_p$ and $\sigma_q$) based on values *P* and *q* material strength using the obtained values *P* and *q*. Use the formulas:

$$\sigma_P = \frac{1.1 \cdot P \cdot r_{av}}{b \cdot c^2}, \quad (5)$$

where *p* – reduced loading, $r_{av}$ – average pipe radius, *c* – wall thickness.
Under the action of internal pressure *q* (kg\см$^2$) we have:

$$\sigma_q = q \frac{r_H^2 + r_B^2}{r_H^2 - r_B^2}, \quad (6)$$

where *q* – internal pressure, kg/см$^2$, $r_H$ and $r_в$ – respectively the outer and inner radius of the pipe.

Due to the fact that formulas (5) and (6) show various geometric parameters that cannot be used in the design of pipes, the ratio $n = \frac{d}{c}$ allows us to bring formulas (5) and (6) to the form:

$$\sigma_p = A\frac{p}{d}, \text{ where } A = 0.0055n \cdot (n+1), \quad (7)$$

$$\sigma_p = q \cdot B, \text{ where } B = \left[\frac{n^2}{2n+2} + 1\right] \quad (8)$$

The values of *A* and *B* are given in tabular form [3, 11]. Showed that for concrete and reinforced concrete pipe ratio $\frac{d}{c}$ lies in a narrow range in which the table values *A* and *B* can be replaced by analytical expressions:

$$A = 0.007618 \cdot n^{1.9} \quad (9)$$

$$B = 0.78 \cdot n^{0.85} \quad (10)$$

In order to simplify the dependencies (9) and (10), additional studies were carried out that showed the possibility of using the coefficients *A* and *B* in a narrow range of actually used pipes (n = 6 ... 9). Which makes it possible to use the calculator in engineering calculations.
In this range for external load *(P)*:

$$A = 0.0064 \cdot n^2 \quad (11)$$

$$\sigma_p = 0.0064 \cdot n^2 \cdot \frac{p}{d} \quad (12)$$

For internal pressure:

$$B = 0.491 \cdot n + 0.621 \quad (13)$$

$$\sigma_q = q \cdot (0.491 \cdot n + 0.621) \quad (14)$$

## 3. Results

Table 1 shows the comparative results of calculations by the exact formulas (7), (8) and the proposed ones (11), (13). The calculated deviations do not exceed 2.4%.

**Table 1.** The values of the coefficients A and B in the interval n = 6 ... 9.

| $n=\frac{d}{c}$ | $A=0.055n\cdot(n+1)$ | $A=0.0064\cdot n^2$ | Deviation by A: ΔA | Deviation by A: Relative, % | $B=\left[\frac{n^2}{2n+2}+1\right]$ | $B=0.491\cdot n+0.621$ | Deviation by B: ΔB | Deviation by B: Relative, % |
|---|---|---|---|---|---|---|---|---|
| 6 | 0.230 | 0.230 | 0 | 0 | 3.571 | 3.567 | 0.006 | 0.10 |
| 7 | 0.308 | 0.313 | 0.005 | 1.6 | 4.062 | 4.058 | 0.011 | 0.09 |
| 8 | 0.396 | 0.409 | 0.010 | 2.5 | 4.555 | 4.459 | 0.012 | 0.13 |
| 9 | 0.495 | 0.518 | 0.020 | 2.4 | 5.050 | 5.04 | 0 | 0.19 |

Example 1: Determine the voltage limit $\sigma_p$ for a concrete pipe with an inner diameter of $d$ = 1000 mm, wall thickness = 89.5 mm, ultimate test load $p$ = 5970 kg/running metre. Pipes are made on the SIOME hyro press machine. Substitute the source data:

$$n=\frac{d}{c}=\frac{1000}{89.5}=11.17 \tag{15}$$

Using the formula (12), we have: $\sigma_p=0.0064\cdot 11.17^2\cdot\frac{5970}{1000}=4.7\ MPa$.

Example 2: Determine the voltage limit $\sigma_q$ for a concrete pipe with an inner diameter of $d$ = 1000 mm, wall thickness = 89.5 mm, if the test hydro-static pressure = 6.9 kg/cm$^2$. Pipes are made on the SIOME hyro press machine. Substitute the source data:

$$n=\frac{d}{c}=\frac{1000}{89.5}=11.17 \tag{16}$$

According to the formula (14): $\sigma_q=6.9\cdot(0.491\cdot 11.17+0.621)=42.15\ kg/cm^2=4.21\ MPa$.

Example 3: Determine the maximum pressure that a modified concrete pipe can withstand $R_{bt}=72$ kg/cm$^2$, $d$=1000 mm, wall thickness = 89.5 mm. Design tensile strength: $R_p=0.6\cdot R_{bt}=0.6\cdot 72=43.2\ kg/cm^2$

According to the formula (14):

$$43.2=q\cdot(0.491\cdot 11.17+0.621),$$
$$q=\frac{43.2}{(0.491\cdot 11.17+0.621)}=7.07\ kg/cm^2 \tag{17}$$

The result shown in Example 3 showed that the use of modified concrete makes it possible to produce pipes for pressure pipelines.

Conclusion – conversion of known dependencies, allowing to determine the normal annular stresses from the external load P and internal pressure $q$, allowed to obtain new transformed dependencies for calculating stresses in pipes under the combined action of internal pressure and external (reduced) load. Using the obtained equations allows you to determine the range of the application of pipes of modified concrete and to select the necessary indicator of its strength.

IOP Conf. Series: Materials Science and Engineering **907** (2020) 012053 doi:10.1088/1757-899X/907/1/012053

## 4. References

[1] Runova R F, Gotz V I and Sanytsky M A 2008 *Design materials of the new generation and technology in the future*

[2] Jamrozy Z 2000 *Concrete and its technologies* (Scientific publishing house Pwn., Warsaw)

[3] Shumakov I and Basheer Y 2014 *International Journal of Engineering Science and Innovative Technology* **3** 6 562-567

[4] Nikiforov S N 1955 *Theory of Elasticity and Plasticity* (Stroyizdat, Moscow)

[5] Schladweiler J C 2018 *The History of Sanitary Sewer*

[6] Asid N J and Wardoyo W 2018 Head loss detection for irrigation pipe system in Poncokusumo Agropolitant, Malang *In AIP Conference Proceedings* **1977** (1) 050001

[7] Mohamed N (2015) *Experimental and Numerical Study on Full-Scale Precast Steel Fibre-Reinforced Concrete Pipes*

[8] Klein G K 1969 *Calculation of underground pipelines* (Moscow: Stroyizdat)

[9] Younis B, Kazimahomedov I, Salem S, Kostuk T and Dedenyova E 2018 Casting of concrete and reinforced concrete pipes by vibro-vacuum technique *In MATEC Web of Conferences* **230** 03022

[10] Wong L S and Nehdi M L 2018 Critical Analysis of International Precast Concrete Pipe Standards *Infrastructures* **3** (3) 18

[11] Younis B and Saienko L 2019 Dependence of Lamé transformation for concrete pipeline design calculation *In IOP Conference Series: Materials Science and Engineering* **708** (1) 012116

IOP Conf. Series: Materials Science and Engineering **907** (2020) 012054 doi:10.1088/1757-899X/907/1/012054

# Compatibility analysis of superplastisizers with different types of cements

**S M Tolmachov[1], O A Belichenko[1], D S Tolmachov[2]**

[1]Department of the Technology of Road Building Material and Chemistry, Kharkov National Automobile und Highway University, ul. Yaroslava Mudrogo, 25, 61002 Kharkov, Ukraine
[2]LLC "Modern infrastructure technologies", ul. Gvardijtsiv-Shy`ronintsiv, 33, office 84, 61170 Kharkov, Ukraine

Tolmachov.serg@gmail.com

**Abstract.** The article considers the urgent issue of improving the quality and durability of road cement concrete. It is shown that the decrease of the quality of concrete mixtures and concrete is often due to incompatibility between cement and superplasticizer. This can happen if the same superplasticizer is used, but different cements of the same plant. This also happens if superplasticizers of the same type, but of different manufacturers, with the same cement are used. Therefore, it is important to quickly assess the compatibility of cement and superplasticizer at the stage of concrete mix design. The article proposes to evaluate the compatibility by changing the properties of cement paste without superplasticizer and with it. It is shown that compatibility can be established within a few hours. Based on the experiments, it was established which superplasticizers should be used for cements of a certain mineral composition.

## 1. Introduction

Many scientists have studied the problem of ensuring the durability of heavy concrete. Among the various ways to increase durability, several main ones can be distinguished.

1. Improving the interfacial zone "hardened cement paste - aggregate". To do this, use clean aggregates, without dust, and also process aggregates with water or aqueous solutions of salts [1 – 3].

2. Increasing the density of cement mortars and concrete. For this, four or more fractions of high strength aggregates are used, as well as microfillers and nanoparticles [4 – 8].

3. Increasing the strength of mortars and concrete. Superplasticizers and cements with high activity are used [4, 9 – 12].

4. Creating a backup pore system using air-entraining additives [13 – 15].

There are other ways to increase the durability of concrete, however, in all cases, the main components of concrete, which are all impacts, and which ensure the durability of concrete, are cement and superplasticizer. In construction, especially road construction, the methods described above are not often used, because their application does not always lead to a positive result. Studies over the past two to three decades have shown that the reason of this is the poor compatibility of cements and superplasticizers. With the right choice of cement and superplasticizer, the durability of concrete will be ensured. This choice must be made at the stage of concrete mix design. Therefore, the urgent task of this study is to develop a simple method for assessing the compatibility of cements and superplasticizers. The main objective of this study is a quick assessment of the compatibility of the additive and cement.

## 2. Status of the issue

In the construction of paved concrete, the replacement of cement or superplasticizer in concrete mixture often leads to serious problems. This is because the construction of roads takes place in different temperature and humidity conditions. For example, in spring and autumn, relative humidity is quite high, and the air temperature is in the range + 5...+ 25 °C. In this case, the use of finely ground cements with superplasticizers based on polycarboxylates or melaminesulfonated formaldehydes gives good results. In the summer, when the air temperature reaches + 30...35 °C, it is more efficient to use cements of low dispersion and superplasticizers based on lignosulfonates. The need to ensure the high speed of modern road construction leads to the fact that cements of the same type but different plants are used in production. Also during the construction process, the organization that performs the work can replace one superplasticizer to another. Such a replacement of cement from one plant to cement from another plant, or replacement of a superplasticizer from one manufacturer to a similar but different manufacturer, can lead to negative results. For example, the effectiveness of a superplasticizer can be greatly reduced, which will affect the mobility of concrete mixtures or the strength of concrete. Concrete mixtures may also deteriorate over time. This will lead to a decrease of workability of mixtures, decompaction of concrete and will affect the operational characteristics of the finished concrete pavement. All these changes are caused by one reason, which in science and practice is called "the compatibility of cement and superplasticizer."

The concept of "compatibility" is closely related to the concept of "effectiveness of the additive." Some researchers believe that these two concepts are identical, but this is not so. In the normative documents of Ukraine it is indicated that "the effectiveness of the additive" is the degree of change in the value of the main indicator (or indicators) of the quality of cement composites when the additive is introduced. For example, this may be a decrease of the water-cement ratio or an increase of the slump of the concrete mix cone test.

With the incompatibility of cement and additives, its effectiveness is significantly reduced. Therefore, a decrease in the effectiveness of the additive is a consequence, and the "compatibility of cement and additives" is the reason for this decrease in effectiveness.

There are several definitions of "compatibility." For example, A.V. Usherov-Marshak defines «compatibility» as the ability of an additive to provide the specified properties of concrete mixtures and concrete and to preserve these properties for a certain time. This applies to all composite materials based on cement and superplasticizer.

The main property is the ability of a particular superplasticizer when interacting with a specific cement to provide a certain plasticizing effect, which depends on the chemical basis of the superplasticizer. Most often, three types of superplasticizing additives are used in road concrete technology: modified (sugar-free) lignosulfonates (LS) and their mixtures with other additives; melamine formaldehyde sulfonate (MFS) or naphthalene formaldehyde sulfonate (NFS); polycarboxylates (PC). The most powerful plasticizing effect is PC. MFS or NFS are less effective, and LS are the weakest plasticizers.

A fairly well-known and effective method for assessing the compatibility of cement and superplasticizer is the calorimetric method proposed by A.V. Usherov-Marshak [16, 17]. A special technique was developed and calorimeters were used. A relative disadvantage of the method is the need to have such equipment and specialists who can work on it. In addition, the measurement process itself continues for at least one day. Different sizes of cement particles, changes in test temperature, and fluctuations in the mineral composition of cement can also influence the quality of research.

Therefore, we propose to evaluate the compatibility of cement and additives more simply and quickly - by changing the standard properties of cement paste. There are devices for this in every laboratory. Another goal of this study is an attempt to establish patterns of compatibility of superplasticizers with cements of various mineral composition.

## 3. Materials and methods of research

For research, we selected cements from the same plant, which have the same clinker, slightly different dispersion, but different slag contents. These are cements: PC II/A-S-400 (CEM II/A-S 32.5 R), PC II/A-S-500 (CEM II/A-S 42.5 N), PC I-500 R-N (CEM I 42.5 R). Cement CEM II/A-S 32.5 R has a specific surface area of 380 $m^2/kg$. The slag content in the cement is 18 %. The specific surface area of the cement CEM II/A-S 42.5 N is 410 $m^2/kg$, and the slag content is 12 %. Cement CEM I 42,5 R has a specific surface area of 390 $m^2/kg$, there is no slag in cement.

As superplasticizers that relate to LS, the following additives were used: BV18 from BASF and Sika 2607A from Sika. The following additives were used as superplasticizers such as MFS and NFS: FK 88 from MC Bauchemie, FM 21 from BASF. The effect of superplasticizers like PC was investigated. These are additives Glenium 27 from BASF, Relaxol Super PC from Budindustria (Ukraine), Fk 59 from MC Bauchemie, Sika LTP S9 from Sika.

The change in the normal consistency and setting time of the cement paste was investigated. To do this, use the standard test method using a Vicka device.

## 4. Experimental research

Tests of normal consistency (NC), as well as the beginning of setting (BS) and the end of setting (AS) of the cement paste were conducted.

**Table 1.** Properties of cement CEM II/A-S 32.5 R

| № | Type and amount of additive, % by weight of cement | Normal consistency, % | Setting time | | Note |
|---|---|---|---|---|---|
| | | | beginning of setting | end of setting | |
| 1 | Without additives | 29,0 | 2 h 40 min | 3 h 40 min | water separation |
| 2 | Muraplast FK 88 – 0,5 % | 26,5 | 3 h 45 min | 5 h 30 min | no water separation |
| 3 | FK 88 – 1,0 % | 25,5 | 5 h 10 min | 6 h 50 min | no water separation |
| 4 | BASF FM 21 – 0,5 % | 26,5 | 3 h 05 min | 4 h 25 min | strong water separation |
| 5 | FM 21 – 1,0 % | 26,25 | 4 h 20 min | 6 h 00 min | strong water separation |
| 6 | BASF BV18 – 0,5 % | 27,5 | 4 h 30 min | 6 h 30 min | no water separation |
| 7 | BASF BV18 – 1,0 % | 26,5 | 4 h 00 min | 5 h 45 min | no water separation |
| 8 | Relaxol Super PC – 0,5 % | 26,0 | 1 h 30 min | 3 h 00 min | no water separation |
| 9 | Relaxol Super PC – 1,0 % | 24,5 | 1 h 45 min | 4 h 00 min | no water separation |
| 10 | BASF Glenium 27 – 1,0% | 24,5 | 4 h 25 min | 6 h 10 min | poor water separation |

If we consider the performance of superplasticizers with cement CEM II/A-S 32.5 R, we can see that the use of FK 88 (type MFS) leads to a decrease in NC by 2.5...3.5 % (Table 1). At the same time, the beginning of cement setting is extended by 1 h 05 m. With an increase of the amount of this additive to 1.0 %, the beginning of setting extended by 2 h 30 m. The end of cement setting is extended by 1 h 50 m...3 h 10 m respectively. In this case, water cement is not observed on the cement test. This indicates that this additive is very well combined with this cement. However, another additive of the same type FM 21, despite a decrease in NC by 2.5 %, slightly lengthens BS and AS. To significantly extend the setting time, it is necessary to increase the consumption of additives. At the same time, the cement paste with this additive shows water separation, which is very dangerous, especially for road mortars and concrete. Water separation can lead to flaking and layer-by-layer destruction of concrete during operation. Therefore, this additive is poorly combined with this cement.

Additive BV18, type LS, has a slightly lower water-lowering effect than superplasticizers type MFS, but it leads to a significant extension of the setting time of the cement paste and does not cause water separation. Therefore, this additive is well combined with CEM II/A-S 32.5 R. PC additives are also combined in different ways with this cement. Despite a strong decrease in water consumption (3...4.5 %) and the absence of water separation from the cement paste, the addition of Super PC reduces the setting time. This is especially noticeable when the consumption of this additive is 0.5 % by weight of cement. This is dangerous for concrete mixtures and can lead to a quick loss of workability and poor compaction of concrete. Therefore, we can say that the Super PC additive is not compatibility with the CEM II/A-S 32.5 R. Its use requires strict implementation of the concrete mixture transportation modes. Glenium 27 is better combined with this cement. The use of this additive in an amount of 1.0 % contributes to a significant reduction in water consumption and lengthening the setting time of cement paste. However, at the same time, water separation from the cement paste is observed, which is a negative result of the action of the additive and indicates insufficient combination with this cement.

**Table 2.** Properties of cement CEM II/A-S 42,5 N

| № | Type and amount of additive, % by weight of cement | Normal consistency, % | Setting time | |
|---|---|---|---|---|
| | | | beginning of setting | end of setting |
| 1 | Without additives | 31,75 | 3 h 00 min | 4 h 00 min |
| 2 | FK 88 – 0,5 % | 29,0 | 3 h 15 min | 4 h 45 min |
| 3 | FK 88 – 1,0 % | 28,0 | 4 h 05 min | 6 h 15 min |
| 4 | FM 21 – 0,5 % | 29,25 | 3 h 20 min | 4 h 15 min |
| 5 | FM 21 – 1,0 % | 28,25 | 4 h 00 min | 5 h 15 min |
| 6 | BV18 – 0,5 % | 29,0 | 4 h 10 min | 5 h 20 min |
| 7 | BV18 – 1,0 % | 28,75 | 4 h 40 min | 5 h 05 min |
| 8 | Sm 21 – 0,5 % | 29,5 | 3 h 50 min | 5 h 15 min |
| 9 | Sm 21 – 1,0 % | 28,0 | 4 h 45 min | 6 h 00 min |
| 10 | Relaxol Super PC – 1,0 % | 26,75 | 0 h 40 min | 1 h 20 min |
| 11 | BASF Glenium 27 – 1,0 % | 26,75 | 2 h 20 min | 4 h 05 min |

A study of the compatibility of additives with cement CEM II/A-S 42,5 N showed that water separation from cement paste with additives of different types was absent in all compositions. At the same time, it is obvious that for higher-quality and more dispersed cement, which is a CEM II/A-S 42,5 N, all superplasticizers of types LS and MFS or NFS are well compatible (Table 2). This is evidenced by a rather large decrease in NC cement test (2.75...3.75) with these additives compared to the cement paste without additives. Such a decrease in water content corresponds to the effectiveness of the additives of these types.

Additives type PC, despite a strong decrease in NC, which is due to their high efficiency, are poorly combined with this cement. The use of these additives, despite their high content in cement paste, not only does not extend the setting time of the cement paste, but even reduces them. Super PC additive is especially poorly compatible with the CEM II/A-S 42,5 N. When it is introduced into the cement paste, its setting time is sharply reduced. Despite this, given the high efficiency of the polycarboxylates and the ability to significantly reduce water consumption, as well as the lack of water separation from cement paste, these additives remain attractive for use in monolithic concrete. However, this requires additional research.

Studies of the compatibility of cement CEM I-42.5 R with additives showed that water separation from the cement paste did not occur. An analysis of the results shows that the additive FM 21, which is of the MFS type, does not combine well with this cement (Table 3). This can be seen from the weak water-lowering effect (reduction of NC is not more than 2 %), which is small for additives of this type). In addition, the introduction of this additive into the cement paste slightly reduces the setting time. The remaining additives combine well with this type of cement. Moreover, the effect of their action is already apparent when the consumption of additives is 0.5 % by weight of cement, which is less than for other types of cement.

IOP Conf. Series: Materials Science and Engineering **907** (2020) 012054 doi:10.1088/1757-899X/907/1/012054

**Table 3.** Properties of cement CEM I 42.5 R

| № | Type and amount of additive, % by weight of cement | Normal consistency, % | Setting time | |
|---|---|---|---|---|
| | | | beginning of setting | end of setting |
| 1 | Without additives | 30,0 | 2 h 30 min | 3 h 50 min |
| 2 | Fk 59 – 0,5 % | 27,0 | 4 h 00 min | 6 h 40 min |
| 3 | Relaxol Super PC – 0,5 % | 27,0 | 3 h 40 min | 5 h 10 min |
| 4 | FM 21 – 0,5 % | 28,0 | 2 h 30 min | 3 h 15 min |
| 5 | Sika LTP S9 – 0,5 % | 27,0 | 4 h 00 min | more 6 h |
| 6 | Sika 2607A – 0,5 % | 28,0 | 4 h 30 min | more 6 h |

2. Assessment of the water-retaining capacity of cement paste with additives (water separation).

**Figure 1.** Strong water separation on the surface of cement paste and on glass with the addition of FM 21

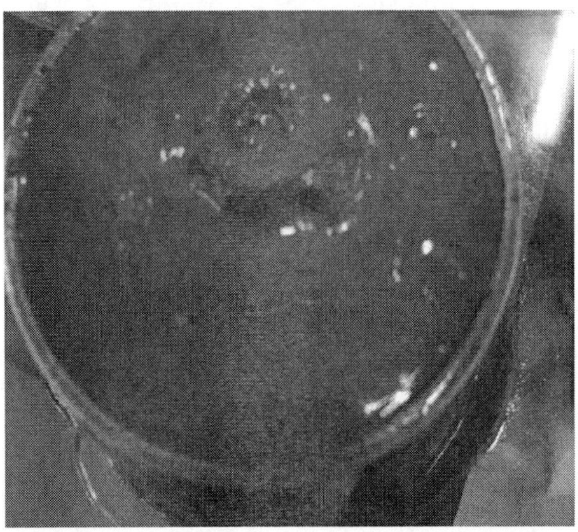

**Figure 2.** Water separation on glass with additive Glenium 27

**Figure 3.** No water separation with the addition of FK 88

The initial assessment of the compatibility of cement and superplasticizer additives is conveniently carried out visually. This allows you to most quickly evaluate compatibility. If immediately after compaction of the cement paste in the ring of the Vica device, or after a certain time (from 10...15 minutes to 1 hour), a layer of water appears on the surface of the cement paste in the ring or under the ring (water separation occurs), then the cement is poorly or weakly combined with additive. The water-retaining capacity of this cement paste is low.

For comparison, we take different additives: Fm 21, Fk 88 and Glenium 27 and one type of cement CEM II/A-S 32.5 R. In the case of using the Fm 21 additive, strong water separation can be observed both on the surface of the cement paste sample and on the glass of the device under cement paste (Fig. 1). This indicates a poor water-retaining capacity of the cement paste and the poor compatibility of this additive and this type of cement. In the second case, when using the additive Glenium 27, water separation can be observed only on the glass of the device. It is practically absent on the surface of the cement paste sample (Fig. 2). This indicates a weak compatibility of the additive Glenium 27 and cement. The Fk 88 additive has good compatibility with this cement, as evidenced by the lack of water separation on the glass of the device (Fig. 3).

## Conclusions

The analysis of the compatibility of different superplasticizers with cements of different types allows us to draw the following conclusions.

1. Superplasticizers such as MFS and NFS as well as LS type are well combined with cement type CEM II/A-S 32.5 R. PC superplasticizers combine poorly with this cement.
2. Superficialisers such as MFS or NFS, as well as LS type, are well combined with the cement type CEM II/A-S 42,5 N. Polycarboxylate-based superplasticizers are poorly combined with this cement.
3. With cement CEM I-42.5 R, superplasticizers such as PC and LS are well combined. Superplasticizers like MFS and NFS are poorly combined with this cement.
4. With cements containing slag, it is most effective to use superplasticizers such as MFS or NFS, as well as type LS.
5. With cements in which mineral additives, such as slag, are absent, it is most effective to use superplasticizers like PC and LS.

## References

[1] Ickovich S M 1983 *Aggregates for concrete* (Minsk: Higher school)

[2] Vinogradov B N 1979 *Effect of aggregates on concrete properties* (Moscow: Stroyizdat)

[3] Dobshits L M, Portnov I G, Solomatov V I 1999 *Frost resistance of concrete transport structures* (Moscow: Stroyizdat)

[4] Sanytsky M, Marushchak U, Olevych Y, Novytskyi Y 2020 Nano-modified Ultra-rapid Hardenin Portland Cement Compositions for High Strength Concretes *Lecture Notes in Civil Engineering* **47** 392-399.

[5] Bazhenov Yu M, Demyanova V S, Kalashnikov V I 2006 *Modified high-quality concrete* (Moscow: Publishing House Association of Construction Universities)

[6] Netesa N I, Kiryash V G 2001 Efficiency of concrete mixtures with rational grain composition *Bulletin of the Dnieper State Academy of Civil Engineering and Architecture* **5** 41-46

[7] Dvorkin L J, Lushkinova N V, Runova R F, Troyan V V 2007 *Metakaolin in mortars and concretes* (Kiev: KNUBiA publishing house)

[8] Tolmachov S N, Belichenko O A 2014 *The use of carbon colloidal nanoparticles in fine-grained cement concrete* (Harkov: HNADU).

[9] Sinajko N P, Likhopud A P, Sopov V P, Tolmachev S N, 2004 The mechanism of action of the additives against a frost *Montazhnye i Spetsial'nie Raboty v Stroitel'stve* **4** 12-16.

[10] Stark I, Whitch B 2004 *The durability of concrete* (Kiev)

[11] Usherov-Marshak A V 2005 *Chemical and mineral additives in concrete* (Kharkov: Color)

[12] Batrakov V G 1998 *Modified Concretes. Theorium and practice* (Moskow)

[13] Shejnin A M, Ekkel S V 2001 Development and implementation of road concrete with high strength and frost resistance *Science and technology in the road industry*, **3** 15-18.

[14] Ekkel S V 2016 Some suggestions to supplement current standards for road concrete *Concrete Technology* **7-8** 50-59

[15] Tolmachov S M, Brazhnik G V, Belichenko O A, Tolmachov D S 2019 The effect of the mobility of the concrete mixture on the air content and frost resistance of concrete.*IOP Conf. Series: Materials Science and Engineering* **708** 012109 doi:10.1088/1757-899X/708/1/012109

[16] Usherov-Marshak A V, Ciak M 2009 Compatibility - the topic of concrete science and the resource of concrete technology *Construction Materials* **10** 12-15.

[17] Usherov-Marshak A V, Kabus A V 2012 *Days of modern concrete* (Zaporozhye: Publishing House "Budindustriya, LTD") 12-18.

IOP Conf. Series: Materials Science and Engineering **907** (2020) 012055 doi:10.1088/1757-899X/907/1/012055

# Complex multifunctional additive for anchoring grout based on alkali-activated portland cement

**P V Krivenko[1], O M Petropavlovskyi[1], I I Rudenko[1], O P Konstantynovskyi[1] and A V Kovalchuk[2]**

[1]Scientific Research Institute for Binders and Materials, Kyiv National University of Construction and Architecture, 03037 Povitroflotskyi Avenue, 31, Kyiv, Ukraine
[2]"Fomalhaut-Polimin" Ltd., Ukraine, 03680 Pshenichnaya str., 2-A, Kyiv, Ukraine

alexandrkp@gmail.com

**Abstract.** Complex multifunctional additive (further, CA) which consists of aluminum powder, surfactant, salt-electrolyte ($NaNO_3$) and calcium sulfate hemihydrate ($CaSO_4 \cdot 0.5H_2O$) is proposed for ensuring necessary properties of anchoring grouts based on alkali-activated portland cement (further, AAPC) presented by the system «ordinary portland cement clinker - sodium metasilicate». Specified consistency of fresh AAPC mortar along with it strength after hardening are provided by water reducing due to application of modified polyethylene glycol characterized by stability of molecular structure in AAPC hydration medium. CA influence on shrinkage mitigation in AAPC mortar is explained both water-reducing effect and gas release with volume increasing during formation of dispersive-coagulation structure. Expansion of crystallization-condensation structure is ensured due to stress of calcium hydrosulfoaluminate $3CaO \cdot Al_2O_3 \cdot 3CaSO_4 \cdot 32H_2O$ and calcium hydronitroaluminate $3CaO \cdot Al_2O_3 \cdot Ca(NO_3)_2 \cdot 10H_2O$ which form due to presence of $CaSO_4 \cdot 0{,}5H_2O$ and $NaNO_3$ in CA. Time matching of gas release with structure formation insures dense microstructure due to filling of pores by hydrated phases that causes lower excess stress during crystallization. CA ensures necessary performances of AAPC anchoring grout: consistency 190 mm; workability retention time 15 min; tensile strength in bending / compressive strength 6.1 / 25.7 MPa and 12.9 / 68.5 MPa in 1 d and 28 d agreeably; adhesion 0.9 MPa; linear extension up to +0.37 mm/m.

## 1. Introduction

According to requirements [1] and experience [2] dry mixes for anchoring application should meet requirements which are difficult to combine: consistency of fresh mortar, workability retention time, strength, adhesion and expansion. Alkali-activated cements (further, AAC's) are perspective also for such materials in view of modern tendencies concerning efficient consumption of raw materials and energy. At that, AAC mortars and concretes are characterized by increased strength [4], sulphate-resistance [5], freeze-thaw resistance [6] and weather resistance [7, 8] in comparison with analogues based on ordinary portland cement clinker (further, OPC clinker) cements.

AAPC [9] differs among the known types of AAC's [10]. The feasibility of an alkaline component in AAPC is based primarily on the idea of accelerated structure formation, especially at the early stages of hardening, and thus ensuring high early strength of artificial stone. Application of soluble sodium silicates provides the highest activity of AAPC due to formed low-calcium hydrosilicates and alkaline as well as alkaline earth hydroaluminosilicates during hydration [11]. Hence, AAPC is relevant for anchoring grouts.

Nowadays performances of fresh and hardened mortars can be provided primarily by modifiers. It was shown the effectiveness of their different nature (organic and inorganic) as well as of their complexes concerning enhancement of strength [12, 13], water impermeability [14], corrosion resistance [15], freeze-thaw resistance [16, 17], crack resistance [18] and water resistance [19].

Control of mortar properties, including proper deformations, is possible due to surfactants as organic admixtures of plasticizer group. Positive effect of surfactants on mortar is caused firstly by reduction of water content (water-reducing effect) [20]. Most of surfactants are ineffective for AAC's and therefore the principles for their choice were proposed [21, 22]. Thus, the highest effect in this case can be provided by sodium lignosulfonate [23], polyethers [21], polyols and as well as by other acyclic low and high molecular compounds [24]. Additional effects of such compounds are known. For example, decreasing of AAC's shrinkage in the presence of polypropylene glycol (polyether) is caused by redistribution of porous space and higher volume of pores with diameter of 0.1–1 μm, the capillary pressure of which is much smaller than that of capillaries with smaller size [25].

Hydrosulfoaluminate expansion is traditional mean to regulate proper deformations while using of inorganic additives. Increasing volume of hardened mortar is ensured due to formation of hydrosulfoaluminate, i.e. ettringite ($3CaO{\cdot}Al_2O_3{\cdot}3CaSO_4{\cdot}32H_2O$) [2, 26], while sulfate-containing substances are applied. For example, the effectiveness of calcium sulfate hemihydrate ($CaSO_4{\cdot}0.5H_2O$) was shown for AAC as well unlike to calcium sulfate dehydrate ($CaSO_4{\cdot}2H_2O$) [27, 28].

Regulation of mortars proper deformations can be also realized due to fiber reinforcement [29, 30].

Effective mean for mitigation of proper deformations in AAC mortars is complex multifunctional additives (CA's) which contain both organic and inorganic substances. The actuality of specified method when using salts-electrolytes, i.e. $Na_2SO_4$ and $NaNO_3$, was shown [31, 32, 33]. The effectiveness of these CA's was explained by their co-acting for reduction of water content, acceleration of crystallization, alteration of porous structure as well as morphology of hydrated phases.

Well-known mean for mitigation of proper deformations in mortars is expansion due to release of gas (hydrogen) when using aluminum powder [26]. Volume increasing in such way is widely used in technology of aerated concretes and mortars based both on traditional cements [34, 35] and AAC's [36, 37].

The above results allow to predict higher effectiveness of CA's in the system «surfactant - salt-electrolyte», which was proposed for anchor AAPC grouts [32], due to involving of aluminum powder and calcium sulfate hemihydrate ($CaSO_4{\cdot}0.5H_2O$). Supposed in this case CA provides effect on proper deformations in AAPC mortar during different stages structure formation.

Aluminum powder should ensure expansion of AAPC mortar during formation of dispersive-coagulation structure. Unlike to aerated material technology gas release should happen in dispersive-coagulation structure of AAPC mortar which is characterized by a certain plastic strength [38]. At that, gas release will be accompanied by expansion with stressing of initial structure.

Interaction of $CaSO_4{\cdot}0{,}5H_2O$ with aluminate phases ensures expansion of AAPC mortar during crystallization-condensation structure formation as a result of ettringite formation [26]. However, calcium sulfate hemihydrate decelerates plastic strength due to formation of ettringitte cover on surface of aluminate phases during dispersive-coagulation stage [27, 28].

It is well-known, surfactants also decelerate plastic strength of AAC pastes [21]. Moreover, depending on molecular structure and content, surfactants can ensure both acceleration and deceleration of gas release [39, 40].

It can be predicted that the retarding effect of $CaSO_4{\cdot}0{,}5H_2O$ and surfactant on plastic strength of AAPC mortar can be minimized by presence of salt-electrolyte in CA. It was shown, that sodium nitrate ($NaNO_3$) is more efficient in comparison with other salts-electrolytes [31, 32].

It was assumed that combination of aluminum powder, surfactant, salt-electrolyte ($NaNO_3$) and $CaSO_4{\cdot}0{,}5H_2O$ in CA composition can ensure time matching of gas release with structure formation of AAPC mortar. Such mean allows control of fresh and hardened mortars performances due to reduction of water content, formation of effective porous structure and morphology of hydrated phases.

So the aim of this research was to investigate the co-application of aluminium powder, $CaSO_4{\cdot}0{,}5H_2O$, surfactant and $NaNO_3$ in CA for ensuring necessary properties of AAPC anchoring grout.

## 2. Raw materials and testing techniques

The known AAPC composition was used as base of mortar [41], %: OPC clinker – 97.6, sodium metasilicate – 2.4 (0.70 % by $Na_2O$). OPC clinker (content, %: CaO – 66,15; $SiO_2$ – 22,61; $Al_2O_3$ – 5,29; $Fe_2O_3$ – 3,93; MgO – 0,84; $Na_2O$ – 0,15; $K_2O$ – 0,98; $SO_3$ – 0,50), specific surface= 450 $m^2$/kg (by Blaine), was used as AAPC aluminosilicate component. Five-water sodium metasilicate $Na_2SiO_3 \cdot 5H_2O$ (dry state) was used as alkaline component.

The reference composition of AAPC mortar (AAPC – 40 %, mix of aggregates and filler – 60 %) was chosen according to the proper functionality [42].

The modifiers for AAPC mortar were presented by:

- surfactants of different types having water-reducing effect: 1 - sodium lignosulphonate (further, LST) according to CAS № 8061-51-6 (pH ≥ 8.5); 2 - polycarboxylate («JK-04PP», Jiankai); 3 - polyacrylate («Vinavil fluxe», Mapei); 4 - admixture based on modified polyether polyethylene glycol («Melflux PP100F», Mapei); 5 - sodium gluconate according to CAS № 527-07-1; 6 - sorbitol ($C_6H_{14}O_6$) according to CAS № 50-70-4; 7 - admixture based on hydroxyethylated fatty alcohol («OC-25», JSC «Barva»);
- aluminum powder according to CAS № 7429-90-5 (content, %: active aluminum – 85.00 - 93.00, Fe – 0.50, Si – 0.40, Cu – 0.05, Mn – 0.05, moisture – 0.20, fatty admixtures – 3.80), water covering≥ 7000 $cm^2$/g;
- calcium sulfate hemihydrate ($CaSO_4 \cdot 0.5H_2O$) according to the national standard of Ukraine DSTU B V.2.7-104: 2000;
- sodium nitrate ($NaNO_3$) according to CAS № 7631-99-4.

The aggregates were presented by: quartz sands fr.< 0.63 mm and fr.= 0.63…1.25 mm; washed granite screening fr.= 1.25…2.5 mm.

Milled sand fr.< 0.16 mm (sieve residue, %: < 0.05 mm – 80.0, 0.05 mm - 9.5, 0.063 mm - 7.0, 0.10 mm - 2.5, 0.16 mm - 1.0) was used as filler.

Optimal ratio between aggregates and filler was calculated by approaching of total grain-size curve to ideal one for enhancement of AAPC mortar (strength, proper deformations). Optimal proportion between aggregates and filler guarantees the lowest interparticle porosity along with minimal summary surface of grains [43, 44], %: milled sand – 25, sand fr.< 0.63 mm – 20, sand fr.= 0.63 - 1.25 mm – 30, granite screening – 25.

Water-reducing effect of surfactants was evaluated by W/C ratio and compressive strength of AAPC mortar characterized by constant consistency measured according to the national standard of Ukraine DSTU B V.2.7-126:2011.

Proper deformations in AAPC mortar were fixed from start of mixing with water were determined due to dial test indicator in rectangular profile assembled form with dimensions of 40×40×170 mm. Form filled by fresh mortar was closed by cover, which could to move along inner surface, and was placed between tips of dial test indicator. The step in measuring was 1 h.

The AAPC mortar strength and proper deformations in it from initial technological moisture down to equilibrium moisture were determined on specimens 40x40x160 mm. After 1 d hardening in forms with an insulated surface, the samples were removed from forms and stored for 7 d under the normal conditions (t= 20 ± 2 °C, R.H. = 95 ± 5 %). Then the samples were stored over saturated solution of potassium carbonate ($NH_4NO_3$) at t= 20 ± 2 °C till the control age. The length of samples after 1 d was taken as the initial one (zero).

The microstructure of AAPC after hardening during 14 d was carried out by electronic microscope with microanalyzer.

## 3. Results and discussions

**The effectiveness of surfactants in AAPC mortar was researched. The content of surfactants was 0.5 % by mass of OPC clinker (further, %). The compressive strength of the reference AAPC mortar (W/C = 0.47) corresponds to values of 17.1 and 56.4 MPa after 1 d and 28 d agreeably while ensuring fresh mortar consistency of (190±10) mm. Admixture based on modified polyether is characterized by the highest water-reducing effect. Compressive strength of**

**modified AAPC mortar increases from 17.1 up to 30.5 MPa after 1 d and from 56.4 up to 74.3 MPa after 28 d** regarding **to the reference AAPC mortar while decreasing of W/C ratio from 0.47 down to 0.38 (figure 1).**

Surfactant based on polycarboxylate is characterized by lesser water-reducing effect than the modified polyether. Thus, compressive strength of AAPC mortar, modified by the this polyester, increases up to 27.3 and 70.5 MPa after 1 d and 28 d agreeably while decreasing of W/C ratio down to 0.38. Compressive strength of AAPC mortar, modified by polyacrylate, corresponds to values of 25.7 and 67.1 MPa agreeably with W/C ratio of 0.4.

LST ensures increasing of AAPC mortar compressive strength up to 22.4 MPa and 64.3 MPa after 1 d and 28 d agreeably while decreasing of W/C ratio from 0.47 down to 0.43. Compressive strength of AAPC mortar, modified by sodium gluconate, corresponds to values of 20.5 and 61.7 MPa agreeably with W/C ratio of 0.44.

**Polyol and hydroxyethylated fatty alcohol are characterized by the lowest effectiveness. Thus, sorbitol ensures increasing of AAPC mortar compressive strength up to 18.3 MPa 59.5 MPa after 1 and 28 d agreeably while decreasing W/C ratio down to 0.46. Compressive strength of AAPC mortar, modified by hydroxyethylated fatty alcohol, corresponds to values of 17.5 MPa after 1 d and 57.3 MPa after 28 d along with similar decreasing of W/C ratio.**

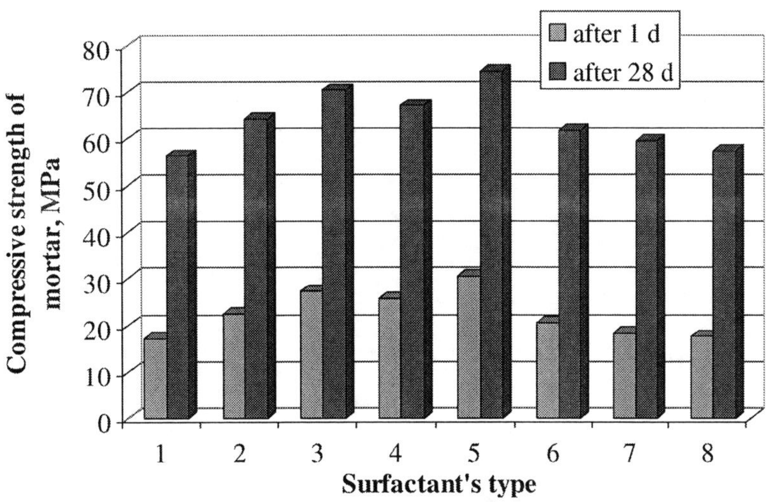


**Figure 1.** Effect of surfactants on AAPC mortar strength: 1- without surfactant; 2 - LST; 3 - polycarboxylate; 4 - polyacrylate; 5 - modified polyether; 6 - sodium gluconate; 7 - polyol; 8 - hydroxyethylated fatty alcohol

Hence water-reducing effect of surfactants depending on type reduces in the order: modified polyether > polyester > LST > sodium gluconate > polyol > hydroxyethylated fatty alcohol. This data corresponds to previous results [28, 29].

The effect of CA composition «surfactant - aluminum powder - $CaSO_4 \cdot 0,5H_2O$ - $NaNO_3$» on proper deformations in AAPC mortar was researched. The content of aluminum powder was 0.01 % to ensure expansion while minimal negative effect on AAPC hardened mortar [38]. Contents of $CaSO_4 \cdot 0,5H_2O$ (9.0 %) [41] and $NaNO_3$ (2.5 %) [31, 32] were accepted basing on previous investigations. It must be noted that CA ensures insignificant decreasing of AAPC mortar compressive strength regarding to one modified only by «Melflux PP100F», i.e. from 30.5 down to 27.6 MPa after 1 d and from 74.3 down to 71.5 MPa after 28 d while W/C= 0.38.

The reference AAPC mortar is characterized by early shrinkage of dispersive-coagulation structure (-1.73 mm/m after 7 h) due to water evaporation (figure 2). At that, CA ensures expansion of AAPC mortar which begins after 1 h since mixing with water and continues during next hour mainly. As result the expansion of AAPC mortar corresponds to value of 0.435 mm/m after 7 h.

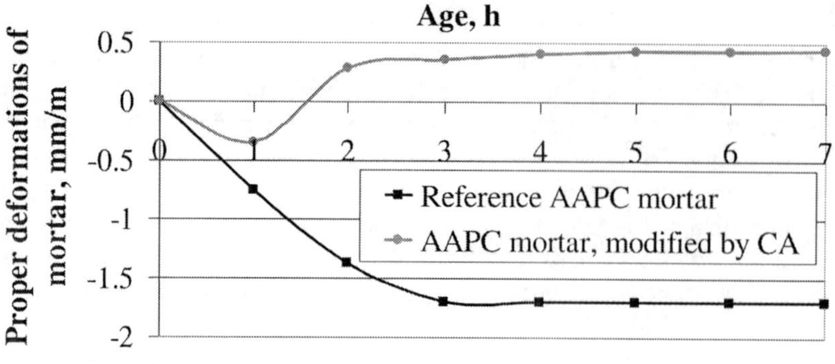


**Figure 2**. Early age proper deformations of AAPC mortar

The further investigations of the reference AAPC mortar showed increasing shrinkage up to value of -2.075 mm/m after 80 d (figure 3 a). Instead of that, the proper deformations of AAPC mortar, modified by CA, despite insignificant initial decreasing from +0.430 down to +0.381 mm/m after 6 d, increases up to +0.454 mm/m after 14 d (figure 3 b). Thus, the modified AAPC mortar characterizes by insignificant decreasing of expansion deformation down to +0.385 mm/m after 80 d.

a)

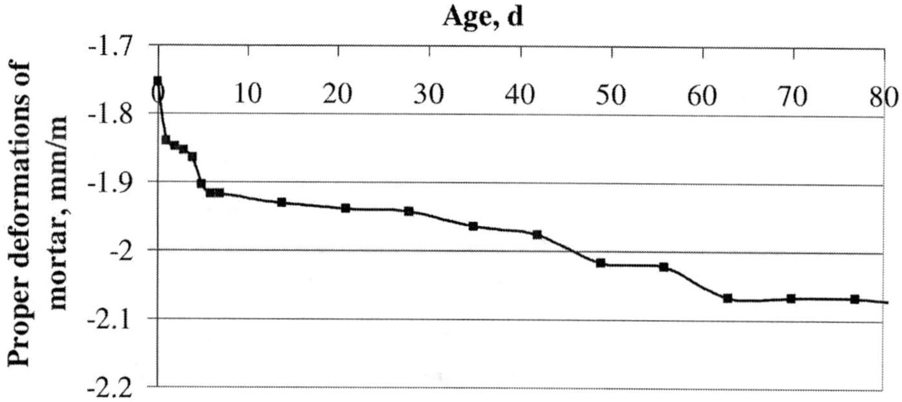


b)

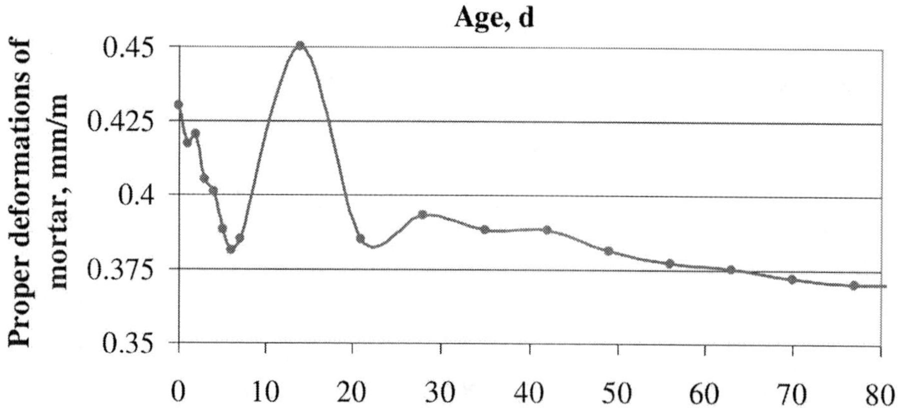


**Figure 3**. Proper deformations of the reference (a) and modified AAPC mortar (b)

So it can be established that the proper deformations of AAPC mortar can be regulated by mean of CA due to less content of water as well as effect on structure formation. Expansion of AAPC mortar dispersive-coagulation structure takes place due to gas release. But the proper deformations of AAPC mortar during further stages of structure formation is caused by effect of CA on crystallization processes in artificial stone.

The features of hydrated phases in AAPC, modified by CA, were investigated by electronic microscopy (figure 4).

Sub-microcrystal (gel-like) low-calcium hydrosilicates (content in probe, %: CaO – 41.19, $SiO_2$ – 53.42) are identified in the reference AAPC (figure 4 a), which cause increasing of AAPC mortar shrinkage.

Crystalline structures of ettringite (content in probe, %: CaO – 28.37, $Al_2O_3$ – 9.18, $SO_3$ – 19.40) (figure 4 b) [19] and calcium hydronitroaluminate $3CaO{\cdot}Al_2O_3{\cdot}Ca(NO_3)_2{\cdot}10H_2O$ (content in probe, %: CaO – 38.26, $Al_2O_3$ – 17.87, $N_2O_5$ – 21.87) (figure 4 c) [45] were identified in pores of modified AAPC after 14 d of hardening. The specified hydrated products ensure expansion of AAPC mortar.

**Figure 4.** SEM images and probe analysis of AAPC: reference (without CA) (a); modified by CA with formation of ettringite $3CaO{\cdot}Al_2O_3{\cdot}3CaSO_4{\cdot}32H_2O$ (b) and calcium hydronitroaluminate $3CaO{\cdot}Al_2O_3{\cdot}Ca(NO_3)_2{\cdot}10H_2O$ (c)

The properties of AAPC anchoring grout with specified CA, that is produced by dry mix technology, are in compliance with mandatory basic requirements [1]: consistency 190 mm; workability retention time 15 min; tensile strength in bending / compressive strength of mortar 6.1 / 25.7 MPa and 12.9 / 68.5 MPa after 1 d and 28 d agreeably; adhesion 0.9 MPa; linear extension up to +0.37 mm/m.

## 4. Conclusions

1. The AAPC mortar with adjustable proper deformations can be obtained due to application of CA containing aluminum powder, calcium sulfate hemihydrate, surfactant and sodium nitrate as ingredients. Specified CA also provides regulation of strength of AAPC mortar while ensuring specified consistency.

2. AAPC mortar expansion while strength increasing is explained by reduction of water content, formation of effective porous structure, time matching of gas release with structure formation with

additional structure stressing due to formation of ettringite and calcium hydronitroaluminate in hydrated AAPC.

3. The characteristics of anchoring AAPC grout, containing specified CA, are in compliance with mandatory requirements including on the order greater expansion (+0.37 mm/m) regarding to analogue, modified by CA system «surfactant - salt-electrolyte». The obtained results indicate the need for optimization of CA composition and its content.

## References

[1] DSTU B V.2.7-126:2011: Mixes building dry modified. National standard of Ukraine

[2] Ma C, Tan Y, Li E, Dai Y and Yang M 2015 High-performance grouting mortar based on mineral admixtures *Advances in Materials Science and Engineering* **260** 1–11 https://doi.org/10.1155/2015/425456

[3] Fang Z and He S 2012 High-Performance Grouts for Rock Anchor *Key Engineering Materials Trans Tech Publications* **517** 946-951 https://doi.org/10.4028/www.scientific.net/KEM.517.946

[4] Chistyakov V, Grankovskii I and Gots V 1986 Structure formation upon hardening of slag-alkali binder *Journal of applied chemistry of the USSR* **59** (3 pt 1) 542–546

[5] Kryvenko P, Guzii S, Kovalchuk O and Kyrychok V 2016 Sulfate resistance of alkali activated cements *Materials Science Forum* **865** 95–106 https://doi.org/10.4028/www.scientific.net/MSF.865.95

[6] Krivenko P., Petropavlovskyi O., Kovalchuk O. 2018 A comparative study on the influence of metakaolin and kaolin additives on properties and structure of the alkali-activated slag cement and concrete *Eastern-European Journal of Enterprise Technologies* **1(6-91)** 33–39 https://doi.org/10.15587/1729-4061.2018.119624

[7] Pushkarova E, Gots V and Gonchar O 2006 Stability of hydrosulfoaluminosiljcate compounds and durability of an artificial stone based on them *Brittle Matrix Composites* **8** 399–408

[8] Pushkarova E, Gots V and Gonchar O 2007 Stability of hydrosulfoaluminosiljcate compounds and durability of an artificial stone based on them (Book Chapter) *Brittle Matrix Composites* **8** 399–408 https://doi.org/10.1533/9780857093080.399

[9] DSTU B V.2.7-181:2009: Alkaline cements. Technical conditions. National standart of Ukraine

[10] Krivenko P, Petropavlovskyi O, Kovalchuk O, Lapovska S and Pasko A 2018 Design of the composition of alkali activated portland cement using mineral additives of technogenic origin *Eastern-European Journal of Enterprise Technologies* 4 (6–94) https://doi.org/10.15587/1729-4061.2018.140324

[11] Gelevera A and Munzer K 1994 Alkaline portland and slag portland cements *First Int. conf. on alkaline cements and concretes* (Kiev: Vipol Stock company) 173–180

[12] Kropyvnytska T, Rucinska T, Ivashchyshyn H and Kotiv R 2019 Development of Eco-Efficient Composite Cements with High Early Strength *International Conference Current Issues of Civil and Environmental Engineering Lviv-Košice–Rzeszów* (CEE 2019: Proceedings of CEE 2019) 211–218 https://doi.org/10.1007/978-3-030-27011-7_27

[13] Krivenko P, Sanytsky M and Kropyvnytska T 2019 The effect of nanosilica on the early strength of alkali-activated portland composite cements *Solid State Phenomena* **296** 21–26 https://doi.org/10.4028/www.scientific.net/SSP.296.21

[14] Plugin A, Pluhin O, Borziak O and Kaliuzhna O 2020 The Mechanism of a Penetrative Action for Portland Cement-Based Waterproofing Compositions *Lecture Notes in Civil Engineering* **47** 34–41 https://doi.org/10.1007/978-3-030-27011-7_5

[15] Borziak O, Plugin A, Chepurna S, Zavalniy O and Dudin O 2019 The effect of added finely dispersed calcite on the corrosion resistance of cement compositions *IOP Conference Series: Materials Science and Engineering* **708** **(1)** 012080 https://doi.org/10.1088/1757-899X/708/1/012080

[16] Moskalenko O and Runova R 2016 Ice Formation as an Indicator of Frost-Resistance on the Concrete Containing Slag Cement in Conditions of Freezing and Thawing *Materials Science Forum* **865** 145–150 https://doi.org/10.4028/www.scientific.net/MSF.865.145

[17] Borziak O, Chepurna S, Zidkova T, Zhyhlo A and Ismagilov A 2018 Use of a highly dispersed chalk additive for the production of concrete for transport structures *MATEC Web of Conferences* **230** 03003 https://doi.org/10.1051/matecconf/201823003003
[18] Chepurna S, Borziak O and Zubenko S 2019 Concretes, modified by the addition of high-diffused chalk, for small architectural forms *Materials Science Forum* **968** 82–88 https://doi.org/10.4028/www.scientific.net/MSF.968.82
[19] Sanytsky M, Kropyvnytska T Fischer H, Kondratieva N 2019 Performance of low carbon modified composite gypsum binders with increased water resistance *Chemistry & Chemical Technology* **13 (4)** 495–502 https://doi.org/10.23939/chcht13.04.495
[20] Batrakov V G 1998 *Modified concretes. Theory and practice* 2-d ed (Moscow: Tehnoproekt) p 768
[21] Kryvenko P, Runova R, Rudenko I, Skoryk V and Omelchuk V 2017 Analysis of plasticizer effectiveness during alkaline cement structure formation *Eastern-European journal of Enterprise Technologies* **4(6-88)** 35–41 https://doi.org/10.15587/1729-4061.2017.106803
[22] Runova R, Gots V, Rudenko I, Konstantynovskyi O and Lastivka O 2018 The efficiency of plasticizing surfactants in alkali-activated cement mortars and concretes *MATEC Web of Conferences* **230** 03016 https://doi.org/10.1051/matecconf/201823003016
[23] Krivenko P, Petropavlovskii O, Vozniuk G and Pushkar V 2010 Constructive properties of the concretes made with alkali-activated cements of new generation *First Intern. Conf. on Advances of Chemically-activated Materials (CAM' 2010 - China)* 139–146
[24] Krivenko P V, Runova R F, Sanitskiy M A and Rudenko I I 2015 *Alcali-activated cements* (Kiev: OOO Osnova) p 448
[25] Palacios M and Puertas F 2007 Effect of shrinkage-reducing admixtures on the properties of alkali-activated slag mortars and pastes *Cement and Concrete Research* **37(5)** 691–702 https://doi.org/10.1016/j.cemconres.2006.11.021
[26] Shtark Y and Vicht B 2008 *Cement and lime* [ed P Krivenko] p 480
[27] Gots V, Runova R, Rudenko I, Konstantynovskyi O and Lastivka O 2015 Effect of calcium sulfate modifications on properties of alkaline cements *19. Ibausil Weimar* 1-1087–1-1094
[28] Lastivka O 2015 *Modificated concretes based on alkaline slag portland cement in monolith building* PhD thesis (Eng)
[29] Mastali M, Kinnunen P, Dalvand A, Mohammadi Firouz R and Illikainen M 2018 Drying shrinkage in alkali-activated binders – A critical review *Construction and Building Materials* **190** 533–550 https://doi.org/10.1016/j.conbuildmat.2018.09.125
[30] Turba Y, Solodkyy S and Markiv T 2020 Strength and Fracture Toughness of Cement Concrete, Dispersedly Reinforced by Combination of Polypropylene Fibers of Two Types *Proc. International conference* **47** 488–494 https://doi.org/10.1007/978-3-030-27011-7_62
[31] Krivenko P, Gots V, Petropavlovskyi O, Rudenko I, Konstantynovskyi O and Kovalchuk A 2019 Development of solutions concerning regulation of proper deformations in alkali-activated cements, Eastern-European journal of Enterprise Technologies **5 (6-101)** 24–32 https://doi.org/10.15587/1729-4061.2019.181150
[32] Krivenko P V, Petropavlovskyi O M, Rudenko I I, Konstantynovskyi O P and Kovalchuk A A 2019 Alkali activated portland cement with adjustable proper deformations for anchoring application *IOP Conference Series: Materials Science and Engineering* **708** 012090 https://doi.org/10.1088/1757-899X/708/1/012090
[33] Krivenko P, Petropavlovskyi O, Rudenko I and Konstantynovskyi O 2019 The influence of complex additive on strength and proper deformations of alkali-activated slag cements *Materials Science Forum* **968**, 13–19 https://doi.org/10.4028/www.scientific.net/MSF.968.13
[34] Sethy K, Nayak G and Nanda S 2018 Aerated Concrete: A Revolutionary Construction Material *International Journal of Engineering Technology Science and Research* **5 (4)** 988–993
[35] Haris A 2016 The influence aluminium waste addition on density and compressive strength mortar *ARPN Journal of Engineering and Applied Sciences* **11 (13)** 8057–8062

[36] Font A, Borrachero M, Soriano L, Monzó J, Mellado A and Payá J 2018 New eco-cellular concretes: sustainable and energy-efficient materials *Green chemistry* **20** 4684–4694 https://doi.org/10.1039/c8gc02066c
[37] Kumar M, Saxena M and Singh N 2019 Influence of some additives on the properties of fly ash based geopolymer cement mortars *SN Applied Sciences* https://doi.org/10.1007/s42452-019-0506-4
[38] Wang F, Liu Z and Hu S 2010 Early age volume change of cement asphalt mortar in the presence of aluminum powder *Materials and Structures* **43(4)** 493–498
[39] Remnev V 2011 Principle of plasticizers effect on water-solid ration and gas release in slate-ash binder while aerated concrete producing *Almanac of modern science and education* **1 (44)** 79-83
[40] Baranovkaya E, Mechai A, Dovzhik N and Kolpaschikov V 2018 Research effect of organic admixture based on polycarboxylate lignine on rheological properties of fresh aerated concrete *proceeding BGTU* **2 (2)** 77–82
[41] Rudenko I I, Konstantynovskyi O P, Kovalchuk A V, Nikolainko M V, Obremsky D 2018 Efficiency of redispersible polymer powders in mortars for anchoring application based on alkali activated Portland cements *Key Engineering Materials* **761** 27–30 https://doi.org/10.4028/www.scientific.net/KEM.761.27
[42] Dubrovina G, Goncharov Yu, Gubskaya A and Nazarov D 2015 Cement dry mix: innovation in the producing and application: [production-practical textbook] (Minsk: Kovcheg) p 247
[43] Kropyvnytska T, Sanytsky M, Rucinska T and Rykhlitska O 2019 Development of nanomodified rapid hardening clinker-efficient concretes based on composite Portland cements Eastern-European *Journal of Enterprise Technologies* **6(6-102)** 38–48 https://doi.org/10.15587/1729-4061.2019.185111
[44] Markiv T, Sobol K, Petrovska N and Hunyak O 2020 The Effect of Porous Pozzolanic Polydisperse Mineral Components on Properties of Concrete *Lecture Notes in Civil Engineering* **47** 275–282 https://doi.org/10.1007/978-3-030-27011-7_35
[45] Plugin A and Runova R 2018 Bonding calcium chloride and calcium nitrate into stable hydration portland cement products: Stability conditions of calcium hydrochloraluminates and calcium hydronitroaluminates *International Journal of Engineering Research in Africa* **36** 69-73 https://doi.org/10.4028/www.scientific.net/JERA.36.69

**Acknowledgments**
Authors would like to acknowledge the contribution of the COST Action CA15202 «Self-Healing concrete: the path to sustainable construction» of the European Union's framework program HORIZON2020, http://www.cost.eu/COST_Actions/ca/CA15202. The authors also express their gratitude to the Ministry of Education and Science of Ukraine for financial support of this research that is carried out within the budgetary financing of topic with registration No 1020U001010 and implementation period 2020 - 2022.

IOP Conf. Series: Materials Science and Engineering **907** (2020) 012056 doi:10.1088/1757-899X/907/1/012056

# Research of technologies for restoration of the concrete protective layer of reinforced concrete constructions during the reconstruction of the buildings and structures

**O M Galinsky[1], O S Molodid[2], N V Sharikina[2], R O Plokhuta[2]**

[1]Department of Construction Organization and Management, Kyiv National University of Civil Engineering and Architecture, Kyiv, 31 Avenue Povitroflotsky, 03037, Ukraine
[2]Department of Construction Technologies, Kyiv National University of Civil Engineering and Architecture, Kyiv, 31 Avenue Povitroflotsky, 03037, Ukraine

agalin@ukr.net

**Abstract:** The use of reinforced concrete in modern construction is conditioned by its mechanical and physical properties. However, over time, under the influence of all sorts of factors, structures collapse and require repair work to restore quality indicators and geometric configuration. Typically, the destruction of the structure begins with the upper layers of concrete and extends in deep. Therefore, this article presents the results of experimental studies on the restoration of the surface of concrete specimens and determines the dependence of the adhesive strength of recovered concrete with the repair mix depending of the method of preparation of the surface of the base structure. Experimental studies were planned and carried out in five series, in which was changed the way the base was prepared: applying two-component glue "Erorip" (TM "MAPEI") to the base; application of two-component "Edmock" glue (Composite LLC); with wetting the concrete surface and restoring it with a modified concrete mixture; application of "Consolid 1" and "Consolid 2" compositions ("Composite" LLC) with an interval of 24 hours; moistening the surface of the samples with water (control samples). In addition, the strength of adhesion of recovered concrete with the repair mix depending of the spatial position of the repair site was determined. For this purpose, the repair mixture was put into the formwork on the upper, lateral or lower surface of the samples. After 28 days after laying the repair mixture, the average values of adhesion strength of the recovered layers with the base were determined. It was found that in the process of restoration of the upper surface of the samples all the methods of preparation of the base gave a good adhesion of layers of concrete, during the restoration of the lateral surface the best results were obtained in series No. 1 and No. 3, while the restoration of the lower surface is effective in case preparation of the basis as for the series No 3. The results of the obtained experimental research indicates a further need to investigate the method of restoring the lateral and lower surfaces of concrete and reinforced concrete structures.

## 1. Introduction

Reinforced concrete constructions are the most common ones in buildings and structures. Reinforced concrete constructions have gained such popularity due to the ability to give them any kind and shape. Under the influence of all kinds of factors, including climatic, mechanical, non-observance of clear technological regulations for the execution of works, etc., and also due to errors at the stage of design, concrete constructions collapse, lose their durability and do not meet the requirements of regulatory documents, and therefore do not fully comply its functional purpose. During the reconstruction of buildings and structures, they usually perform a complex of construction -assembly works aimed at restoring or changing the geometric dimensions of constructions and restoring their operational properties in the case of damage. The destruction of constructions, as a rule, begins with the protective layer of concrete and extends in deep, resulting in corrosion of the reinforcement and subsequent destruction of the reinforced concrete construction [8, 11].

## 2. Analysis of recent research and publications

In the process of restoring the protective layer of reinforced concrete constructions, an important indicator is the strength of adhesion of the repair mortar to the existing construction. Analyzing the scientific and technical literature, it was found that the adhesion strength depends on the method of preparation of the concrete surface [3, 4, 11] and the spatial position of the repair area relative to the horizon line [4].

In most cases, the repair of cracks on reinforced concrete constructions is carried out prior to the restoration of the protective layer of concrete. Such works can be carried out according to the investigated technology [5, 6]. According to research data, it is advisable to use the developed surface impregnation technology for gluing cracks with widths from 0.05 to 0.4 mm, classical injection technology at openings from 0 to 0.5 mm, 05 to 2.0 mm – combined technology. When using impregnation technology, a film of polymer composition is formed on the surface of the construction, which can subsequently serve as an adhesive layer to restore the protective layer of the construction. Typically, the restoration of the surface of reinforced concrete constructions is performed by putting a concrete repair mixture into the formwork. This method is also used to ruggedize reinforced concrete constructions, by increasing their cross-section [9, 10, 12].

## 3. Setting objectives

Research of technology of restoration of a protective concrete layer of reinforced concrete constructions, based on establishing of durability of adhesion of the restored concrete with the recovering mix depending on a position of a surface of a construction and a method of its preparation.

## 4. Presentment of the main material

Prior to the beginning of the experimental studies, a methodology was developed that involved the restoration of concrete specimens on three types of surfaces, which were determined by the different spatial position of the repair site (top, side, and bottom) in different ways of preparing the base. In this case, it was assumed that the concrete mixture was putted in formwork structures installed at a certain distance from the existing construction.

For the experimental studies there were used samples made of B30 concrete with the geometric dimensions: 270 x 120 x 120 mm. The surface of the specimens to be renewed was cleaned with an angle grinder, followed with dedusting by compressed air.

After completion of the preparatory work, the series of concrete samples was renewed. The work was performed at a temperature of 20 ± 2 °C and a relative humidity of 50 ± 5%.

The concrete mix used to restore the structures had the following component composition (based on 1 $m^3$ of concrete mix): cement (PC II / A-W-500) – 395 kg; river sand (1.5 mm) – 870 kg; crushed stone (2-5 mm) – 1075 kg; water – 170 kg; plasticizer – 7.5 kg.

According to the methodology of experimental studies the restoration of samples was assumed in five series, with different ways preparation of the basis.

The first series of preparation of the basis consisted of applying by brush to the cleaned dry surface of two-component glue "Erorip" produced by TM "MAPEI", to which after 40 min was placed a repair mixture 2 cm thick.

The second series is similar to the first one, using as a adhesive layer two-component glue "Edmock" manufactured by "Composite" LLC, to which a concrete mixture 2 cm thick was placed after 40 min.

The third series of preparation of the basis consisted of moistening the prepared surface with water and arranging the samples in the formwork, followed by the placement of the concrete-modified mixture.The repair concrete mix had the following component composition (based on 1 $m^3$ of concrete mix): cement (PC II / A-W-500) – 450 kg; river sand (1.5 mm) – 750 kg; screening granite (2-5 mm) – 250 kg; crushed stone (5-10 mm) – 800 kg; water – 200 kg; Dynamon SR 3 (1.3%) – 5.9 kg; Mapecure SRA (1%) – 4.5 kg. Modifying additives used in the mixture, manufactured by TM "MAPEI".

In the fourth series, one layer of the "Consolid 1" composition manufactured by "Composite" LLC, prepared according to the manufacturer's recipe, was applied to the cleaned dry concrete surface. After 24 h, another layer of "Consolid 2" composition was applied to the surface. Next, the samples were placed in the formwork structure and after 2 h in the space between the formwork and the prepared surface was placed a concrete mixture 2 cm thick.

The fifth series of preparation of the base – the control one. The surfaces of the specimens, after cleaning and dedusting, were moistened within formwork with water and placed the concrete mix.

For restoration of the upper surface of the specimens, the formwork structure was installed on the lateral faces on four sides `Figure 1a` and the surfaces of the specimens was prepared for the subsequent placement of a concrete mixture 2 cm thick. During the laying, the mixture was pinned and then compacted, by impacting on the formwork structure by the low-speed vibrating machine. The compaction time was 10-15 s. After 24 h, the specimens were removal from the formwork and their renewed surface moistened twice a day with water for 5 days.

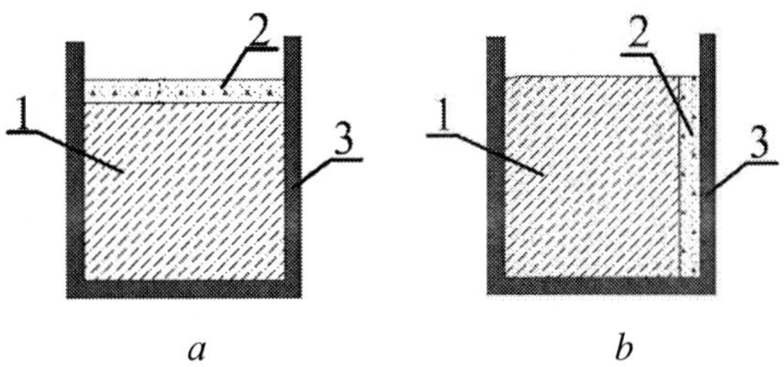


**Figure 1.** Layout of the specimen relative to the formwork during recovery: *a* – of the upper surface; *b* – of the lateral surface, where: *1* – concrete specimen; *2* – repair mixture; *3* – formwork

For restoration of the lateral surface of concrete specimens, the formwork structure was installed in such a way that a space of 2 cm was formed on one of the lateral faces between the prepared specimen and the formwork `Figure 1b`. The concrete mixture, while filling the formwork, was pinned, compacted for 10–15 s, by impacting by low-speed vibrating machine to the lateral surface of the formwork structure. After 48 h, the specimens were removed from the formwork and their renewed surfaces moistened twice a day with water for 5 days.

To restore the lower surfaces of concrete specimens was used technology [7, 9] `Figure 2`.

The formwork design consisted of a bottom and four side faces. In this case, two faces was made of transparent plexiglass and served to control the filling of the formwork with a concrete mixture, and the others were made of plastic. In one of the plastic faces, a hole with a diameter of 35 mm was formed in the middle, into which a metal funnel was glued `Figure 3`. To prevent leakage of repair mortar from formwork, its joints were sealed with hot-melt glue. At the bottom of the formwork, metal supports 2 cm high were placed in the corners to simulate the blank space between the concrete specimen and the formwork. The bodies of the samples had grooves at the place of installation of the funnel for free passage of the mortar into the formwork.

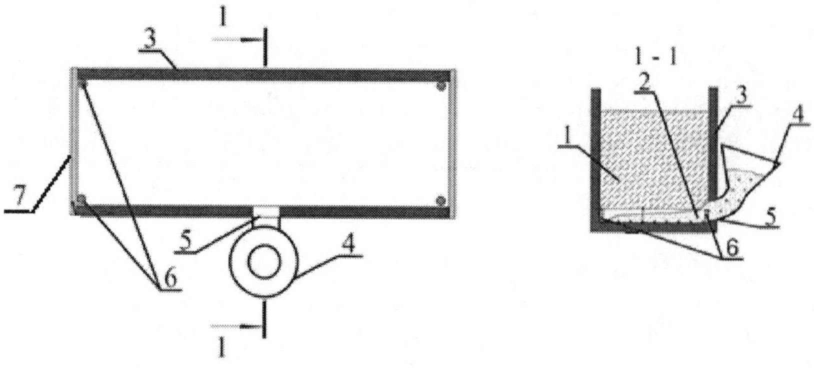


*a* *b*

**Figure 2**. Scheme of restoration of the lower surface of the sample:
*a* – top view; *b* – cross-section 1- 1, where: *1* – concrete sample; *2* – repair mortar; *3* – formwork: *4* – funnel; *5* – hole in the formwork; *6* – metal supports; *7* – transparent plexiglass

*a* *b*

**Figure 3.** Formwork for restoration of the lower surfaces of concrete specimens:
*a* – general view; *b* – side view

Previous studies have found that it is impossible to fill the formwork cavity qualitatively while restoring the lower surface of the specimens with the proposed concrete mix, which is due to its rigidity. That is why in order to increase the mobility of the repair mortar, it was decided to increase the amount of water in the composition by 1.5 times. The prepared repair mortar was fed into the funnel by small portions and carried out with a permanent pinning. The flow of the mixture was controlled through the transparent plexiglass. After complete filling of the formwork with the mortar, it was compacted with a vibrating machine at low speed revolutions for 10–15 s. During the vibration, the mortar remaining in the funnel was filling the formwork additionally. After that the funnel was removed and the hole was closed with a stopper. The recovered samples were removed from the formwork after 2 days. Subsequently, the samples were moistened with water twice a day for five days.

After 28 days after the restoration of the surfaces of the samples the adhesion strength of the restored layers with the base was determined. For this purpose, the restored surface was cleaned and dedusted. Metallic plates with catch members were glued to this surface with epoxy glue.

Determination of adhesive strength was made using a special device `Figure 4` by a known method [1].

**Figure 4.** The device for determining the adhesion strength of the layers of the construction

The results of the experimental tests are given in table. 1.

**Table 1.** Values of the strength of the adhesion of layers of the construction

| № series of the study | Adhesion strength of the protective layer and base of the recovered samples | | | | | |
|---|---|---|---|---|---|---|
| | From above | | Laterally | | From the bottom | |
| | MPa | Kind of destruction | MPa | Kind of destruction | MPa | Kind of destruction |
| 1 | 1,42 | Cohesive type (in the layer of the new concrete) | 1,15 | Cohesive type (in the layer of the new concrete) | 0,37 | Adhesive type (in the junction of layers) |
| 2 | 0,82 | Adhesive type (in the junction of layers) | 0,7 | Adhesive type (in the junction of layers) | 0,57 | Adhesive type (in the junction of layers) |
| 3 | 0,98 | Cohesive type (in the layer of the new concrete) | 1,1 | Cohesive type (in the layer of the new concrete) | 0,82 | Adhesive - cohesive type |
| 4 | 0,91 | Adhesive type (in the junction of layers) | 0,76 | Adhesive type (in the junction of layers) | 0 | It came off during the sawing |
| 5 | 0,7 | Adhesive type (in the junction of layers) | 0,4 | Adhesive type (in the junction of layers) | 0 | It came off during the sawing |

According to the results of the study given in table 1, a diagram of adhesion strength of recovered concrete and repair mortar was constructed `Figure 5`.

IOP Conf. Series: Materials Science and Engineering **907** (2020) 012056 doi:10.1088/1757-899X/907/1/012056

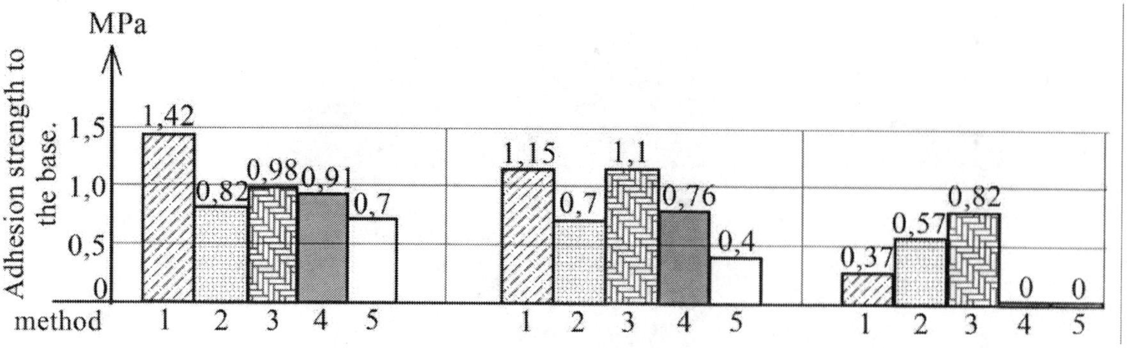


*a* *b* *c*

**Figure 5.** Diagram of adhesion strength of reclaimed concrete to the base, depending on the method of surface preparation and the spatial position of the repair site: *a* – from above; *b* – laterally; *c* – from the bottom

After analysis of the results of the research, it was found that during the restoration of the upper surfaces of the samples, the best adhesion strength between the layers was observed in the series of studies No. 1 and 3 with the cohesive type of destruction; that is, tears were formed in the layer of a new concrete `Figure 6a`. In the series of studies No. 2, 4 and 5, tears were formed at the junction of two layers – it is the adhesive type of destruction `Figure 6b`.

*a* *b*

**Figure 6.** Type of destruction of specimens: *a* – cohesive destruction; *b* – adhesive destruction

In the case of the restoration of the lateral surfaces of specimens series 1 and 3, the destruction occurred in the body of recovered concrete. There were the destructions of the specimens on the border of layers in other series of studies.

No bonding of the layers was observed at all on the samples of series 4 and 5, which were recovered from below. During the preparation of the specimens for adhesion tests, the recovered layer was separated. In research series № 3, the highest adhesive strength was observed. In series 2 and 3, the destructions occurred at the junction of two layers.

## 5. Conclusions

As a result of experimental studies, it is established that the highest adhesion strength of concrete layers is observed when using the glue "Erorip" and in a series of studies on the restoration by the modified concrete mixture.

The adhesive strength of concrete layers when using two-component "Edmock" glue is in the range of 0.57–0.82 MPa. In all cases, the destruction occurred at the junction of "old" and "new" concrete.

The results obtained in a series of studies No. 4 (surface preparation with "Consolid" 1 and "Consolid" 2 compositions) revealed that in the case of the restoration of the upper surfaces of the specimens, the average adhesion strength was 0.91 MPa, of the laterals – 0.76 MPa, lower – no grip. In all cases, the destruction occurred at the junction of the layers.

A series of studies of control samples (the method of surface preparation was to moisten the surface of the base) was able to establish that in the process of restoration of the upper surfaces of the samples, the average value of adhesion was 0.7 MPa, in laterals – 0.4 MPa, the lower – no grip. In this case, the destruction in all tests occurred at the junction of the layers.

The lowest adhesion strength in all series of specimens was observed during the restoration of the lower surfaces of concrete specimens. This can be explained by the fact that in the process of compaction of the mortar on the upper surface of the last that adjacent to the concrete sample, cement milk is formed, which prevents the connection of two layers of concrete.

## 6. References

[1] DSTU B EN 1015-12: 2012 *Test methods for mortar. Part 12. Determination of adhesive strength of plaster solutions with bases (EN 1015-12: 2000, IDT)* [Valid from 10/1/2013] (Kyiv: Ministry of Regional Development of Ukraine, 2013) p 13

[2] DSTU B B.3.1-2: 2016 *Repair and reinforcement of load-bearing and enclosing building structures and foundations of buildings and structures.* [In place of DBN B.3.1-1-2002. Valid 01.04.2017] (Kyiv: UkrNDNC, 2017) p 72

[3] Duhanin P V 2001 *Improvement of technology of repair of reinforced concrete structures of urban sewage treatment facilities* (dis. Cand. tech. Sciences: 05.23.08. Rostov state. building. Univ. Rostov-on-Don) p 137

[4] Kononenko A N 2008 *Technology of repair and restoration of damaged surfaces of reinforced concrete structures* (dis. Cand. tech. of sciences. 05.23.08. Kharkiv) p 135

[5] Molodid O S, Plokhuta R O 2018 *Experimental studies of the technology of bonding cracks of reinforced concrete structures with impregnation* (International scientific and technical journal: Modern technologies, materials and constructions in construction, Vinnytsia) Vol **25** No 2 pp 45–50.

[6] Molodid O S, Plokhuta R O 2019 *Technology of repair of cracks in reinforced concrete structures depending on the width of their opening.* (Coll. of sciences. Ave.: Ways to increase the efficiency of construction in the conditions of formation of market relations, Kiev) No **39** part 2 pp 154–162

[7] Molodid O S, Sharikina N V 2018 *Experimental studies of the technology of restoration of the lower surface of reinforced concrete structures using formwork.* (Coll. of sciences. works: Ways of improving the efficiency of construction in the conditions of formation of market relations, Kiev). No **35** pp 173

[8] ODM 218.3.100-2017 *Recommendations for the use of materials for the repair of concrete and reinforced concrete structures of transport structures* [Introduced for the first time] (LLC STROYMaterials Research Laboratory 2018) p 191

[9] Savitsky A N, Pshinko A N and Savitsky N V 2006 *Technology of repair of reinforced concrete structures by highly mobile repair mixtures* (Sat. scientific tr.: Construction. Material science. Mechanical Engineering. Series: Innovative technologies for the life cycle of residential, industrial and transport facilities. Dnepropetrovsk: PGAAS) Issue **37** pp 431–437.

[10] Savievsky V V 2018 *Reconstruction of buildings and structures* (Kiev: Lira-K Publishing House) p 320

[11] EN 1504 *Product and system for the protection and repair of concrete structures - Defines the general principles for the use of products and systems for the repair and protection of concrete* Was approved by CEN on 2 June 2005 p 214

[12] Aguilar J A 1995 *Case studies of the rehabilitation of existing reinforced concrete buildings in Mexico City* (thesis at the University of Texas at Austin) p 259

IOP Conf. Series: Materials Science and Engineering **907** (2020) 012057 doi:10.1088/1757-899X/907/1/012057

# Modeling of thermal conductivity of reed products

**Yu V Tsapko[1], A Yu Tsapko[2] and O P Bondarenko[3]**

[1]Institute for Binders and Materials, Kyiv National University of Construction and Architecture, Povitroflotskyj avenue, 31, 03037, Kyiv, Ukraine and Department of Technology and Design of Wood Products, National University of Life and Environmental Sciences of Ukraine, Heroiv Oborony str., 15, 03041, Kiyv, Ukraine
[2]Department of Technology and Design of Wood Products Department, National University of Life and Environmental Sciences of Ukraine, Heroiv Oborony str., 15, 03041, Kiyv, Ukraine
[3]Department of Building Materials, Kyiv National University of Construction and Architecture, Povitroflotskyj ave., 31, 03037, Kyiv, Ukraine

juriyts@ukr.net

**Abstract.** The present work researches processes of heat transfer by samples of mats manufactured from reed. Due to the unique properties of the reed, such as medium density, low thermal conductivity, relatively high weather resistance, high chemical resistance, possibility to produce parts on-site, cost-effectiveness and others, reed products are widely used in building construction despite the high rate of development of new technologies. The mechanisms of heat insulation in the result of energy transfer through the material are established, which makes it possible to influence this process. It is proven, that the mechanisms are conditioned upon the increased porosity of the material. Thus, decreasing volume weight results in decreasing thermal conductivity, and vice versa. The simulation of the heat transfer process with the flameproof coating is carried out, the dependencies of the thermophysical coefficients on the temperature are determined. Based on the obtained dependencies, the coefficient of thermal conductivity for the products made of dry pine wood is calculated and makes 0.056 W / (m · K). Features of slowing down the process of heat transfer to material made of wood wool and glue binding agent with the formation of pores were studied. This is explained by the fact that there is no movement of air in large pores, accompanied by heat transfer. Thus, there are grounds to argue for the possibility of directed regulation of the processes of the formation of thermal insulation products using the reeds characterized by voids in the stems.

## 1. Introduction

Reed belongs to the group of organic heat-insulating materials and has quite many unique properties, such as low volume weight, low thermal conductivity, relatively high weather resistance, high strength, and elasticity.

Reed products are widely used in low-height housing construction. Reed slabs are used mainly for the construction of walls, partitions, floors, and roofs, that are protected by the heat-insulating and sound-proof material. Due to the tubular structure of the stem, the reeds have low weight and poor thermal conductivity. According to the heat-insulating properties, walls made from the 10 cm thick reeds, plastered on both sides, corresponding to the brick walls with 2 bricks width. Due to the presence of flint deposits in the cells of reed, the reed is exposed to decomposition less than wood. Slabs made from the pressed reed do not burn open flames under the influence of fire but only smolder. Low volume weight of the reed allows the manufacturing of slabs, fascines, and mats used as a wall building material, as well as for filling of the inter-floor ceilings, as overlapping and roofing material. Production of construction products from reed can be organized easily and with low expenses in places where reed grows. The reed nodes are thickened joint rings, which are located along the stem at a distance of 15-20 cm from each other. There are partitions inside the rings that do not leak air and water. Such a structure of the reeds causes a low thermal conductivity and sufficient strength of products made from reeds.

Thus, the use of reed products, especially in construction, requires the determination of certain properties, in particular, the thermophysical characteristics necessary for the design and manufacture of heat-insulating products. This explains the necessity of research in this area.

## 2. Analysis of recent research and publications

In recent years, among the works devoted to the use of vegetable raw materials in construction, the works directed to the development of panel slabs for the production of heat-insulating constructions and wooden wall panels are known [1, 2]. The technologies are based on the pressing of a mixture of vegetable fibers with mineral additives, using various natural materials (asbestos, mica, basalt) mixed with hydrophobic components [3, 4]. In the works [5], the impact of the amount of vegetable fiber (flax fiber – cotton fiber) on the density and elasticity of the materials obtained by aeration sedimentation was studied. Also, the effect of the binding agent on the properties of flexible heat-insulating materials was researched, but the question concerning the thermal conductivity that decreases the quality of the gained results remains unanswered.

In the works [6], the thermal conductivity of insulating wooden and fiber slabs was evaluated at different temperatures and relative humidity. It was established that accurate datasets on the thermal behavior of insulation materials are crucial in numerical modeling approaches that will improve the correct construction of enclosure structures. However, it is not specified how the thermal conductivity was determined. In the works [7], data on production technology, thermophysical properties of material made from hemp and gypsum binding agent are presented and the possibility of its use as a heat-insulating material is shown. But the issue of the combined effect of the components on heat-insulation remains uncovered. Materials, listed in the works [8], are manufactured on the base of basalt fiber and are characterized by high heat-insulating properties. But the technology of its production, the method of determination of this value and the volumetric characteristics are not described.

In the direction of these studies [9], a mathematical model was proposed describing the dynamics of the heat transfer and conservation properties on a fibrous insulation coating, taking into account the "internal" features of the insulator (granularity and porosity of fibrous insulation made from a mixture of natural and synthetic fibers). However, this model does not take into account how the change in pore shape affects the heat transfer to the structure itself.

The heat-insulating materials manufactured from the mixed carpet waste with the solution of a colemanite ore, one of the boron minerals and the solution with added colemanite waste [10] were also studied. It is shown that the established optimal ratios enable the correction of the content of the components to ensure the heat-insulation process. In addition to gaining knowledge about permanent panels, the understanding of non-uniform plates has been improved due to the identification of two major components [11]. This knowledge of the internal microstructure of a slab made of the bark of wood allowed to create a numerical model for thermal conductivity based on methods of finite differences.

Thus, the use of reeds in building construction requires the determination of certain properties, especially the thermophysical characteristics necessary for the manufacture of heat insulation products, which is the subject of this works.

## 3. The purpose of this work

The purpose of this work is to determine the thermophysical characteristics of reed construction products for the determination of conditions of thermal conductivity suppression of building construction.

## 4. Materials and methods of research

For the study of thermal conductivity, the reed samples of average sizes were used: 10 mm in diameter and 310 mm in height, bound in 150 × 150 × 25 mm mats (Figure 1).

**Figure 1.** Model sample of a heat-insulating mat from reeds.

To obtain the values of the thermal conductivity of vegetable raw materials, special equipment was developed and manufactured; a flat electric heater was used to simulate a low-calorie heat source (Figure 2).

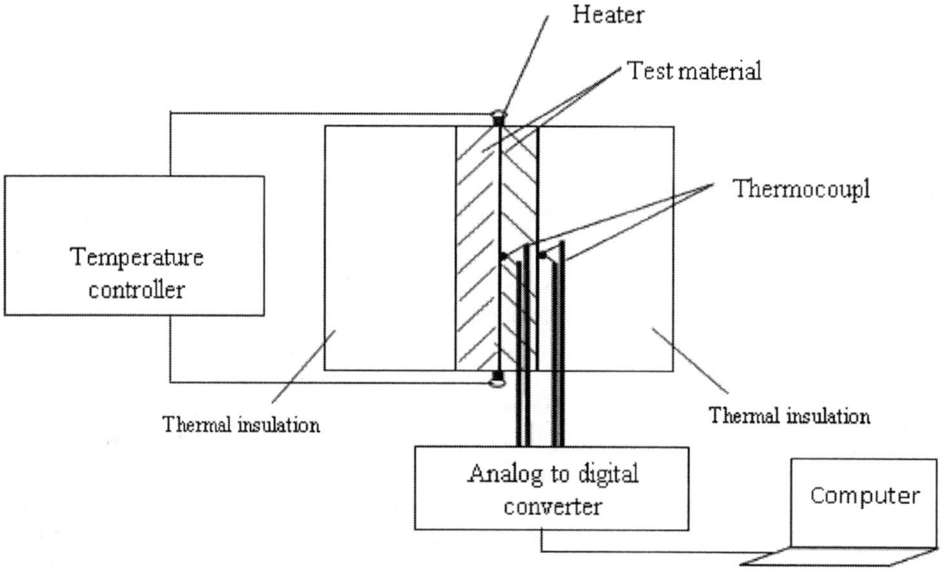


**Figure 2.** Device for research of thermal conductivity of the mat made from the reeds.

The heater was manufactured as follows: a nichrome wire with a resistance of 83 ohms was applied to a 100 × 100 mm electrical insulation plate, a tension of 24.5 volts was supplied to it. The heater was placed in a heat-insulating plate to minimize the heat losses around the perimeter.

A heater with a thermocouple was placed between the test plates made of the reed, and a control thermocouple was placed on the reverse wall of the reed sample. The sample was fixed so that the end of the thermocouple was pressed against the inner surface of the sample. The electric heating was switched on, the temperature of the heater and the one on the back surface of the sample were measured.

When the temperature reached 70°C, the heater was switched off, and the temperature continued to be measured to the value of $0{,}5T_{max}$ on the reverse surface of the sample. The measured values determined the heat-insulation properties of the sample mat made from the reeds.

IOP Conf. Series: Materials Science and Engineering **907** (2020) 012057 doi:10.1088/1757-899X/907/1/012057

The criterion for determining the thermal conductivity of the reed under thermal action is the value equal to $0.5T_{max}$ on the reverse surface of the mat sample.

## 5. Modeling of the heat transfer parameters for reed construction products.

Studies on the modeling process of thermal conductivity of the mat made from the reeds under thermal action were performed using fundamental principles of the mathematical physics [12].

A method for solving the thermal conductivity problem for a plate is proposed to determine the thermophysical characteristics of the reed mat. A semi-infinite body at temperature T0 is given. One of the surfaces is heated by a constant heat flow Q = Const. The temperature changes in one direction (Figure 2). Find the temperature distribution in this direction at any given time.

The differential equation describing this process looks as follows:

$$\frac{\partial^2 T(x,\tau)}{\partial x^2} - \frac{1}{\varphi^2}\frac{\partial T(x,\tau)}{\partial \tau} = 0, \ (\tau>0;\ 0<x<\infty), \tag{1}$$

with the following initial and boundary conditions:

$$T(x,0) = T_0, \tag{2}$$

$$\lambda \frac{\partial T(x,\tau)}{\partial x} = q = \text{const}, \tag{3}$$

$$T(\infty,\tau) = 0, \tag{4}$$

$$\frac{\partial T(\infty,\tau)}{\partial x} = 0 \tag{5}$$

where is the $T_0$ – initial temperature of the reed, °C; $T(x, \tau)$ – temperature pattern of the mat made from the reed at points $x$ at time $\tau$, °C; $\varphi = \sqrt{a}$; $a$ – coefficient of temperature conductivity of the mat made from the reed, m²/s; $\tau$ – residence time of the sample in a high-temperature environment, s; $q$ – heat flow, W/m²; $\lambda$ – coefficient of thermal conductivity of the reed mat, W/(m·°C).

The solution of the equation (1) with the initial and boundary conditions (2) – (5) is given in the works [16] as follows:

$$T(x,\tau) - T_0 = \frac{2\cdot q\cdot \varphi\cdot\sqrt{\tau}}{\lambda}\, ierfc\,\frac{x}{2\varphi\cdot\sqrt{\tau}}, \tag{6}$$

where is the

$$ierfc\ x = \int_x^\infty erfc\,\xi d\xi = \frac{1}{\sqrt{\pi}}e^{-x^2} - x erfc\ x \text{ – error integral.} \tag{7}$$

If the temperature is measured in the plane of the heater ($x$=0), the equation below follows from (6):

$$T(0,\tau) - T_0 = \frac{2\cdot q\cdot\varphi\cdot\sqrt{\tau}}{\lambda\cdot\sqrt{\pi}}, \tag{8}$$

as the right part of the equation (7) by $x=0$ makes $\pi^{-0,5}$.

We denote the relation as an equation:

$$\frac{\lambda}{\varphi}=b, \tag{9}$$

where is the $b$ – coefficient of thermal activity, that characterizes the product's thermal capacity, $W\cdot s^{1/2}/(м^2\cdot K)$.

We introduce the solution (9) into (8) and obtain

$$T(0,\tau)-T_0=\frac{2\cdot q\cdot\sqrt{\tau}}{b\cdot\sqrt{\pi}}. \tag{10}$$

The function (10), in the $\tau^{0,5}=f(T-T_0)$ coordinate system, is a straight line passing through the origin of the coordinates with the following tangent of the angle of slope:

$$tg\,\alpha=\frac{b\cdot\sqrt{\pi}}{2q}. \tag{11}$$

The equation for calculation of the coefficient of activity follows from the function (11):

$$b=\frac{2q\cdot\sqrt{\tau}}{\sqrt{\pi}\cdot(T(0,\tau)-T_0)}. \tag{12}$$

As the function (12) shows, the maximum value of the coefficient of the thermal activity is possible at the value $x=0$, which means at the largest value of the heating device temperature.

To ensure the same heat quantity for every experiment, it is necessary to provide constant resistance of the electric heater, tension, supplied to this heater, and heat duration. The heat flow density of the heater is equal to:

$$q=\frac{U^2}{R\cdot 2S}=\frac{P}{2S}, \tag{13}$$

where is the $R$, $U$, $S$ – are values of resistance, tension and contact area of the heater.

The coefficient of thermal conductivity is determined by the delay time; for different moments of time $\tau_1$ and $\tau_2$ with the condition $\tau_2>\tau_1$ it can be written as the equality between equation (6) and (8):

$$\frac{2\cdot q\cdot\varphi\cdot\sqrt{\tau_1}}{\lambda\cdot\sqrt{\pi}}=\frac{2\cdot q\cdot\varphi\cdot\sqrt{\tau_2}}{\lambda}ierfc\frac{x}{2\varphi\cdot\sqrt{\tau_2}}. \tag{14}$$

After transformations we get the following function:

$$\frac{\sqrt{\tau_1}}{\sqrt{\tau_2}\cdot\sqrt{\pi}}=ierfc\frac{x}{2\varphi\cdot\sqrt{\tau_2}}. \tag{15}$$

The values of the left side of the equation are determined using the formula (15), which includes the experimentally measured values. Using the table [13], the corresponding value of the number can be found, which is an argument of the function *ierfc*, the value of which allows determining the coefficient of thermal conductivity, using the right side of the equation (15):

$$a = \frac{1}{4 \cdot \tau_2} \left( \frac{x}{B} \right)^2, \tag{16}$$

as $\varphi = \sqrt{a}$ .

The coefficient of the mat thermal conductivity can be calculated using the following equation:

$$\lambda = b \cdot \sqrt{a}. \tag{17}$$

Accordingly, the specific heat capacity of the mat made from the reed is found from the ratio:

$$c = \frac{\lambda}{a \cdot \rho}, \tag{18}$$

where is the ρ – density of the reed product, kg/m$^3$.

In any case, a complex determination of the thermophysical characteristics of a reed product based on the solution (6) involves knowledge of the nature of the change in temperature over time at any two points of the testing sample.

These dependencies (16) – (18) are adequate to those obtained for the determination of thermophysical characteristics in [14], and the results of determination of thermophysical characteristics coincide within the margin of error of studies.

Thus, the obtained dependencies allow calculating the thermophysical characteristics of the wood during thermal action.

## 6. Research results

To establish the thermophysical characteristics of the mat made from reeds, the study of the thermal conductivity under the action of the heating device (Figure 3) was completed. The studies on the maximum temperature (0.5*T*, °C) and the duration of the induction time of temperature transfer through the mat layer made of reeds were performed using the above-mentioned method and equipment, and the respective results are shown in Figure 4.

**Figure 3.** The process of determining the thermal conductivity of the mat sample made from reeds under the action of a heater.

As Figure 4 shows, the action of the heater led to the intense heat transfer and a slight increase in temperature on the reverse surface of the sample lasting for about 1800 s. As a result of the tests, it was found that the thermal conductivity of this sample is characterized by the reeds itself.

It is established that the mechanism of thermal insulation by the transfer of energy through the material is the braking, caused by air barriers, which makes the influence on this process possible.

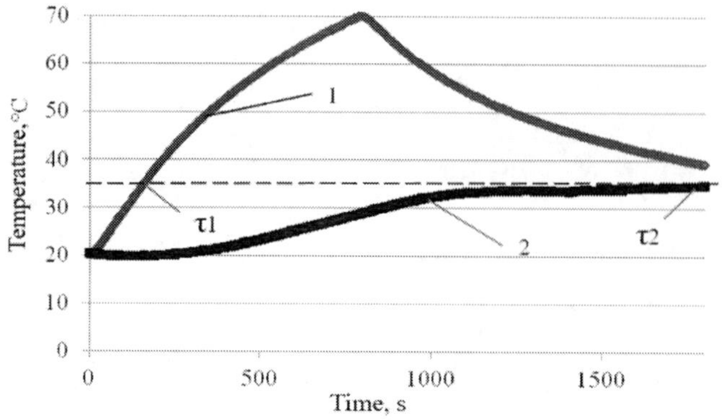


**Figure 4.** Results of tests of thermal conductivity of the mat made of reeds:
1 – heating curve, 2 – temperature value on the reverse surface of the mat.
The points $\tau_1$ – correspond to the average value of the temperature of the heating curve and $\tau_2$ – correspond to the average value of the temperature on the reverse surface.

The thermophysical characteristics of the reed products were calculated using the above-described method (Table 1) and the results of the measured temperature.

**Table 1**. Thermophysical characteristics of the reed products

| Name of the material | Thickness, mm | Mass, gr | Estimated characteristics of the reed products | | | | |
|---|---|---|---|---|---|---|---|
| | | | Density, ρ, kg/m³ | Thermal activity, W·s$^{1/2}$/(m$^2$·K) | Temperature conductivity, m²/s | Thermal conductivity λ, W/(m·K) | Heat capacity, kJ/(kg·K) |
| Mat made from reeds 180 × 155 mm | 25 | 96 | 137.6 | 200.1 | $0.08 \cdot 10^{-6}$ | 0.056 | 5.08 |

Studies have shown that a sample of an insulation product made from reeds reveals thermophysical properties, namely, the coefficient of thermal conductivity (Table 1), which approximates to one of the thermal insulation materials (λ = 0.07 W / (m·K) [14-16].

## 7. Conclusions and prospects for further studies

Thermal conductivity is the process of transferring heat energy from heated parts of a room to fewer warm ones, and energy exchange will occur until the temperature is balanced. The coefficients of thermal conductivity of some building materials depend on many factors: the nature of the material, its structure, the degree of porosity, the nature of the pores, the humidity and the average temperature at which the heat transfer occurs. This corresponds to the data reported in the works [5, 6], whose authors also relate the effectiveness of thermal insulation materials from organic raw materials.

In contrast to the results of the authors' studies [5, 9], the obtained data on the influence of the structure on the heat transfer process and changes in the insulating properties allow us to state the following:

– the main regulator of the process is the density and porosity of the material since the high density and low porosity leads to a rapid equilibration of temperatures, and with increased humidity and wetting of the walls of the building, their open throat factor will be higher;

– a significant influence on the process of thermal conductivity in the application of wood material is in the direction of orientation of the natural material.

The research results of the thermal conductivity process in the material made from the reeds (Figure 1) indicate an ambiguous nature. Such ambiguity cannot be solved within the framework of the present research, as additional experiments would be required to obtain more reliable data. In particular, this implies the availability of data sufficient for the qualitative completion of the heat transfer process and determination of the starting point of time for the fall of heat resistance. Such determination will allow us to investigate the transformation of the surface of the material produced based on reeds, moving in the direction of high temperature with increasing time of transfer, as well as to identify those variables that significantly affect the beginning of the process transformation.

This work is a continuation of the research presented in [1], that explores the mechanism of transfer and insulation of high temperature by the organic natural material.

This work was prepared as part of the research work "Development of flame retardant materials for natural flammable roof structures (straw and reeds)" of the Ministry of Education and Science of Ukraine.

**References**

[1] Tsapko Yu, Zavialov D, Bondarenko O, Pinchevs'ka, O, Marchenko N, Guzii S 2019 Design of fire-resistant heat- and soundproofing wood wool panels *Eastern-European Journal of Enterprise Technologies* vol 3/10 (**99**) pp 24-31 doi: 10.15587/1729-4061.2019.166375

[2] Tsapko Yu, Zavialov D, Bondarenko O, Marchenco N, Mazurchuk S, Horbachova O 2019 Determination of thermal and physical characteristics of dead pine wood thermal insulation products *Eastern-European Journal of Enterprise Technologies* vol 4/10 (**100**) pp 37-43 doi: 10.15587/1729-4061.2019.175346

[3] Czajkowski Ł, Olek W, Weres J, Guzenda R 2016 Thermal properties of wood-based panels: thermal conductivity identification with inverse modeling *European Journal of Wood and Wood Products* vol **74**(4) pp 577-584

[4] Mathis D, Blanchet P, Landry V, Lagière Ph 2019 Thermal characterization of bio-based phase changing materials in decorative wood-based panels for thermal energy storage *Green Energy & Environment* vol **4**(1) pp 56-65

[5] Babashov V, Bespalov A, Istomin A, Varrik N 2017 Heat and Sound Insulation Material Prepared Using Plant Raw Material *Refractories and Industrial Ceramics* vol **58**(2) 208-213.

[6] Troppová E, Švehlík M, Tippner J, Wimmer R 2015 Influence of temperature and moisture content on the thermal conductivity of wood-based fibreboards *Materials and Structures* vol **48** (12) pp 4077–4083

[7] Brencis R, Pleiksnis S, Skujans J, Adamovics A, Gross U 2017 Lightweight composite building materials with hemp (Cannabis sativa L.) additives *Chemical Engineering Transactions* vol **57** pp 1375-1380

[8] Li Z., Ma, J, Ma, H, Xu X 2018 Properties and Applications of Basalt Fiber and Its Composites *IOP Conference Series: Earth and Environmental Science* vol **186** (2) 012052.

[9] Grickus A, Guseynov S 2015 On one mathematical model for dynamics of propagation and retention of heat over new fibre insulation coating *Vide. Tehnologija. Resursi – Environment, Technology, Resources* vol **3** pp 82-86

[10] Erdoğan Y 2016 Production of an insulation material from carpet and boron wastes *Bulletin of the Mineral Research and Exploration* vol **152** pp 197-202.

[11] Barbu M-C, Plank B, Richter K, Petutschnigg A 2016 Evaluation of relationships between particle orientation and thermal conductivity in bark insulation board by means of CT and discrete modeling *Case Studies in Nondestructive Testing and Evaluation* vol **6** pp 21-29.

[12] Samarsky A, Vabishchevich V 2003 *Computational heat transfer* (Moscow: URSS editorial) p 784

[13] Janke E, Emde F, Lesh F 1964 *Special functions (formulas, graphs, tables)* (Moscow: Nauka) p 344

[14] Bobrov Yu, Ovcharenko E, Shoykhet B, Petukhova E 2003 *Thermal insulation materials and structures: Textbook for secondary vocational schools* (Moscow: INFRA-M) p 268.

[15] Tsapko Yu, Tsapko A and Bondarenko O 2019 Effect of a flame-retardant coating on the burning parameters of wood samples *Eastern-European Journal of Enterprise Technologies* vol 2/10 (**98**) pp 49-54 doi: 10.15587/1729-4061.2019.163591
[16] Tsapko Yu, Bondarenko O, Tsapko A 2019 Research of the efficiency of the fire fighting roof composition for cane *Materials Science Forum* vol **968** pp 61-67 doi: 10.4028/www.scientific.net/MSF.968.61

# Reduction of formaldehyde in ambient air through application of special acrylic co-polymer in paint-and-lacquer coatings

**R F Runova[1], V V Pipa[1], Yu L Nosovsky[1], M A Laboda[2] and I B Monfort[3]**

[1]Department of Technology of building constructions and wares, Kyiv National University of Construction and Architecture (KNUCA), Kyiv, Ukraine
[2]Department of Raw Materials for Coatings, Dry Building Mixes and Adhesives, Group of Companies LKB LLC, Kharkiv, Ukraine
[3]Technical Group Architectural Coatings, EMEAI Dow Coatings Materials, Dow Europe GmbH, Horgen, Switzerland

pipa.volodymyr@gmail.com

**Abstract.** Application of modern coatings in interior decoration is becoming increasingly popular. With their help, new decorative effects are created and implemented, which contribute to the desired aesthetic pleasure. The development of the chemical industry allows to produce a wide range of decorative coatings, paints and water-based enamels, as well as other finishing building materials that are widely used by designers and craftsmen in the design of premises. As a result, manufacturing companies face a clear challenge in creating safe materials for human health. It is known that the use of preservatives, plasticizers and opalescent additives in coatings are a source of harmful volatile organic compounds (VOCs), in particular phenol-formaldehyde resins, which provoke dangerous human diseases, and formaldehyde itself ($H_2$–C=O) is a carcinogen and can cause lungs, nasopharyngeal cancer and cause acute myeloid leukemia. Formulation models of coatings based on PRIMAL™ SF-208 ER acrylic copolymer dispersion have been developed, which, among the required operational and quality characteristics, have the ability to reduce the content of formaldehyde. There is defined the regularity of the increase in the amount of phenol-formaldehyde absorbed by the paint from the ambient air from the increase in the quantitative content of the dispersion in the paint.

## 1. Introduction

There are more and more challenges that threaten human life and health today than ever before. We are not insured against some dangerous influences, but it is quite possible to avoid danger due to the introduction of modern scientific knowledge in production technologies.

It is known that about 90% of the time city dwellers spend in residential and office premises, the level of air pollution in which is several times higher than in the open air. Not least because the main source of contamination are building materials and paints, which may contain phenol-formaldehyde resins [1] $[C_6H_3(OH)\text{-}CH_2\text{-}]n$ – macromolecular compounds formed by the interaction of phenol $C_6H_5OH$ with formaldehyde ($H_2$–C=O) in the presence of acids (HCl, etc.) or alkalis (NaOH, $NH_4OH$) as catalysts [2]. Both formaldehyde and phenol used in the production of phenol-formaldehyde resin [3] are toxic and flammable substances. Phenol-formaldehyde resin has a negative effect on the human body: it causes eczema and dermatitis. Chronic formaldehyde poisoning can be manifested by loss of appetite, weight loss, weakness, persistent headaches, palpitations, insomnia. But most importantly, formaldehyde $H_2$-C=O according to the IARC (International Agency for Research on Cancer) classification belongs to group I of known carcinogens which can cause lung, nasopharyngeal cancer and acute myeloid leukemia [4].

## 2.Analysis of Recent Research and Publications

According to the Directive 2004/42/EC of the Council of European Parliament dated April 1, 2004, phenol-phomaldehydes are also volatile organic compounds (VOCs) [5] and subject to certain restrictions.

It is known that in the production of water-soluble paints, which include primers, putties, decorative coatings, paints, enamels and decorative plasters for facade insulation systems (ETICS), as the main binder are polymer dispersions, which can be platoon -polymers by main types of origin: acrylic, styrene acrylic, styrene butadiene, vinyl acetate, the most common of which is styrene acrylic copolymer [6].

One of the most important indicators of the quality of such co-polymer dispersions is the minimum temperature of film formation (MFFT). To form a film from the dispersion that is part of the paint, it is necessary that the drying temperature was higher than the MFFT [7]. If the drying temperature is lower, it will lead to irreversible coating defects. As a rule, dispersions traditionally have MFFT + 8 °C to 22 °C. To reduce the MFFT, paint manufacturers traditionally use external plasticizers, or organic solvents, which are also called film-forming additives, or opalescent agents [7], which after the end of their action to create a film evaporate, forming a non-stick surface. Despite the fact that they are inert to solutes, plasticizers are able to penetrate the human body in the form of evaporation through the skin, mouth and lungs. Another source of formaldehyde is container preservatives, which traditionally contain either free formaldehyde or its donors.

Socially responsible companies, understanding the future and importance of safe products, are already working to improve materials to reduce the content of harmful substances. For example, DAW SE (Germany) has developed and launched three lines of E.L.F. plus (Emissions- und Lösemittel-Frei®, which means without solvents and harmful emissions plus without preservatives) – interior paints, primers and plasters [8].

As for the US market, American scientists, for example, have developed an interesting technology for the production of so-called ceramic coatings. Ceramic wall paint is a fairly recent alternative to traditional water-based paints. These paints are based on acrylate binders and are not only high strength, but also low VOC content. Ceramic paint acquires its characteristics due to the tiny ceramic "beads" used as pigment fillers. These "beads" have a tiny size and a perfect cylindrical shape and due to this pigment filler, there is formed a perfectly smooth coating with very high resistance to wet abrasion [9].

For example, the Michigan company O'Leary Paint, along with some other manufacturers, offers a line of premium enamels "Ceramic Coat Interior Premium", which uses the above-mentioned technology of ceramic paints [10].

In addition, recent studies on the development of modified phenol-formaldehyde resins indicate a decrease in the content of free formaldehyde [11].

It should be noted that some Ukrainian companies are already working to reduce phenol-formaldehyde in coatings. So, KOMPOZIT™ worked to create such a paint "PRIME AIR" [12].

## 3. Purpose and Tasks

Taking into account the successful experience of global companies in VOCs reduction in coatings that limit the content of plasticizers, coalescents and container preservatives, it is proposed to create working models of formulations for semi-gloss and matte coatings. When developing such formulations, avoid the use of coalescents and container preservatives. The coating must meet quality requirements and be able to absorb formaldehyde present in the ambient air.

## 4. Raw Materials and Methods

PRIMAL™ SF-208 ER acrylic co-polymer dispersion from DOW Chemical was used as a binder [13]. This dispersion is designed to create paints in response to growing concerns about indoor air quality (IAQ-Indoor Air Quality) and the health effects associated with exposure to formaldehyde. The dispersion does not contain alkylphenol ethoxylates (APEO-free) [14] and when developing coatings using it does not require the introduction of coalescents, as the MFFT is 4°C. The properties of coatings made of this material allow you to independently form a film, which during operation provides irreversible binding of formaldehyde to the indoor air, thus making it safe. The main properties of the dispersion are given in (table 1).

**Table1.** Basic properties of PRIMAL™ SF-208 ER dispersion

| Property | Typical Values |
|---|---|
| Appearance | Opaque, white to off-white liquid |
| Solids, by weight, % | 48 |
| pH | 8.8 |
| Viscosity (Brookfield, LV 2#, 60 rpm, 25°C), cps | <500 |
| Minimum Film Forming Temperature, ±2C$^0$ | 4 |
| Density (g/ml), wet | 1.06 |
| Storage precautions | Protect from freezing |

The principle of action of the dispersion film is the absorption of functional monomers of carbon and hydrogen from formaldehyde and their return to the air in the form of water vapor.

In this work, field tests were performed according to the methods of DOW Chemicals, in order to determine the properties of the coating to absorb formaldehyde.

Static and dynamic testing methods were used to determine the formaldehyde content. The essence of the static method was to use a room with the same properties, which remain unchanged (table 2). The formaldehyde content was checked using a Formaldegidmeter htV. The equipment of this type allows to perform measurements in a wide range and with sufficient accuracy quickly enough, in 60 ÷ 180 sec.

**Table 2.** Characteristics of the room for determination of the content of formaldehyde by static method according to the method of DOW Chemicals

| Room characteristics | Room features |
|---|---|
| Room type | Manager office |
| Room description and its components | The walls made of ceramic block and mineral plaster. Lightweight concrete and laminate floor. Ceiling made of reinforced concrete and mineral putty. Aluminum profile window. |
| Possible source of formaldehyde | Doors and furniture made of MDF plates. Polyurethane plinth, vinyl glue |
| Room volume | 31.4m$^3$ |
| Total area of surfaces (walls, floor, ceiling) | 50.9 m$^2$ |
| Total area of window and door | 3.2 m$^2$ |
| Painted area | 10.3 m$^2$ |

Samples of semi-gloss and matte coatings, in the amount of 7 kg each, were prepared with the help of laboratory dissolver DISPERMAT® CN20 with the same dosing and dispersing regimen. The composition of the formulations is shown in (table 3).

## 5. Results and Discussion

Measurements were performed by simulating two types of conditions. In the first case, the window and door were closed, and in the second case, the window was periodically opened for ventilation. The measurement results are shown in Figures 1 and 2.

**Table 3.** Paint formulations for formaldehyde abatement testing [a]

| Recipe code | VP12/19 | VP15/19 | VP18/19 | VP21/19 |
|---|---|---|---|---|
| Component, kg/to | Semigloss paint | | Matt paint | |
| Water | 183.79 | 194.65 | 220.89 | 216.33 |
| BYK-024 | 1.96 | 1.98 | 1.64 | 1.63 |
| Orotan 731AER | 12.79 | 12.87 | 21.21 | 21.19 |
| Acrysol RM-5000 | 33.47 | 23.99 | 33.86 | 31.25 |
| Tioxide TR 92 | 178.97 | 178.92 | 281.66 | 280.55 |
| Minex S6 | 153.27 | 152.56 | 262.88 | 259.24 |
| Primal TM SF 208 ER | - | 426.80 | - | 157.17 |
| Usual acrylic dispersion | 414.75 | - | 161.06 | - |
| Acrysol TM RM-8W | 20.00 | 7.23 | 16.00 | 31.84 |
| RocimaTM 564 | 1.00 | 1.00 | 0.80 | 0.80 |
| Total, kg | 1000 | 1000 | 1000 | 1000 |
| Density, gr/m$^3$ | 1.34 | 1.34 | 1.61 | 1.61 |
| PVC, % | 37 | 37 | 72 | 72 |
| VS, % | 38 | 38 | 38 | 38 |

[a] Reproduction of formulation is possible under conditions of observance of a certain sequence of dosing of components, a mode of turns of a mill and time of dispersion.

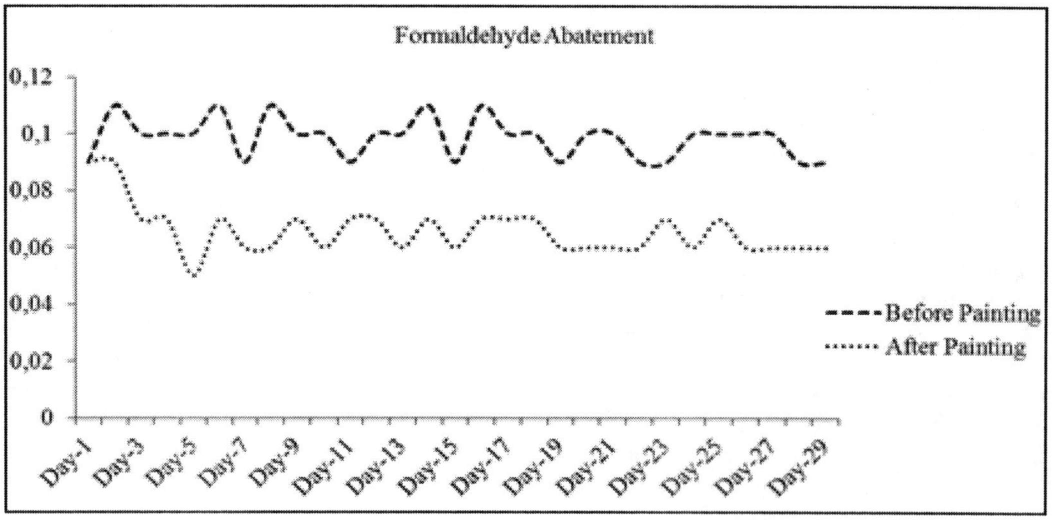


**Figure 1.** Static method in closed room

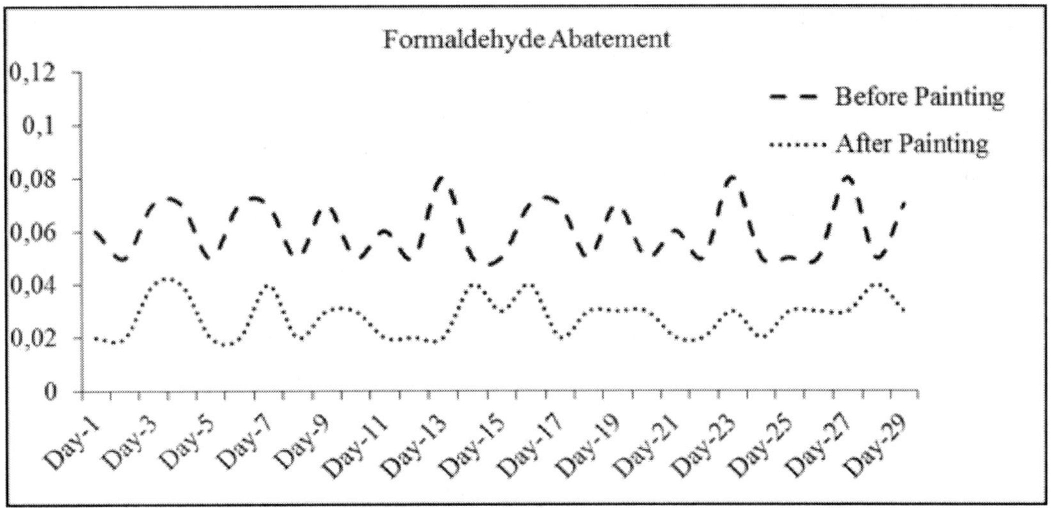


**Figure 2.** Static method in a ventilated room

As can be seen from the diagram (Figure 1), the concentration of formaldehyde has decreased by more than 20% since the third day. The dynamics of the measurement within 29 days from the moment of coating shows that formaldehyde can be absorbed in the amount of more than 32% compared to conventional paint.

The results of formaldehyde measurements in a ventilated room (Figure 2) are somewhat different and indicate a decrease in concentration by 50%.

Dynamic tests were performed with the help of photoelectric photometry method according to the requirements of DIN EN 482:2015 [15], EN 689:2018 [16] and NIOSH Formaldehyde Method No.2016 [17]. For this purpose, models of the rooms were made, which were painted with the above-mentioned samples of paints. The FP-31 gas analyzer ("Tritech Instruments") was connected to them by probes. The research results are shown in (table 4).

**Table 4.** The results of the study for determination of formaldehyde by the dynamic method

| Property of paints | PVC =37%, VS=38% Semigloss paints | | PVC =72%, VS=38% Matt paints | |
|---|---|---|---|---|
| Type of binder | Usual acrylic dispersion | SF208ER | Usual acrylic dispersion | SF208ER |
| Recipe cod | VP12/19 | VP15/19 | VP18/19 | VP21/19 |
| | H2–C=O concentration the lower, the better | | | |
| 2 hours | 68 μg/m³ | 17 μg/m³ | 83 μg/m³ | 14μg/m³ |
| 48 hours | 88 μg/m³ | 35 μg/m³ | 85 μg/m³ | 33 μg/m³ |
| | H2–C=O reduction the higher, the better | | | |
| 2 hours | 19% | 82% | 9% | 87% |
| 48 hours | 2% | 61 % | 6% | 69% |

According to the results, it can be concluded that semi-gloss coatings, which were made on the basis of PRIMAL™ SF-208 ER dispersion give a significantly lower concentration of formaldehyde – 4 times in two hours and 2.5 times in 48 hours. Regarding the reduction percentage, the results indicate an intensive process of formaldehyde absorption by the coating film and its neutralization. A similar effect occurs when testing matte paints, but in contrast to semi-gloss, where the content of acrylic co-polymer is much higher, this manifestation is not the same, although quite significant in comparison with traditional paints.

## 6. Summary

Basic formulation models of semi-gloss and matte coatings were created without the use of coalescents and container preservatives. The measurement results suggest that the obtained coatings are able to absorb formaldehyde from the surrounding air and neutralize it. The test results coincide with the experimental data from DOW Chemical. Their analysis provides grounds for further implementation of the Primal™ SF-208 ER dispersion in the field of paint coatings and further study and research of the possibilities of reduction of formaldehyde in paints and coatings for both internal and external use. It is established that when increasing the amount of co-polymer in the formulation to 43%, with the pigment volume concentration (PVC) in the range of 35-40%, the coating has the ability to remove more formaldehyde than in case of matte paint, where the polymer concentration is 16%, and PVC is 72%. Thus, semi-gloss paint in the amount of 1 gram can remove up to 1.34 mg ($H_2$–C=O) from 1 m$^3$ of air compared to 1 gram of matte paint, which can remove only 0.53 mg ($H_2$–C=O).

## References

[1] Yelle D J and Ralph J 2016 Characterizing phenolformaldehyde adhesive cure chemistry within the wood cell wall *J. Int., of Adhesion and Adhesives* (Amsterdam: Elsevier Ltd) pp 26-36
[2] Fink J K 2013 Reactive Polymers Fundamentals and Applications *Guide to Industrial Polymers, 2nd* (Amsterdam: Elsevier Ltd) p 576
[3] Kumar R N and Pizzi A 2019 *Adhesives for Wood and Lignocellulosic Materials* (Beverly, MA: Scrivener Publishing LLC) p 491
[4] Kaden D A, Mandin C, Nielsen G D and Wolkoff P 2010 Selected pollutants *WHO guidelines for indoor air quality* (Geneva: World Health Organization) p 484
[5] Winkelaar A 2009 *Coatings Basics* (Hannover: Vincentz) p 152
[6] Runova R F, Gots V I, Gelevera O G, Konstantynovskyi O P, Nosovsky Yu L and Pipa V V 2017 *Basics of production of walls and finishing materials* (Kyiv: Osnova LLC) p 524
[7] Brock T, Groteklaes M and Mischke P 2000 *European Coatings Handbook* (Hannover: Vincentz) p 548
[8] Klump-Bickert B 2018 Grün mischt mit – Nachhaltigkeit in der DAW Gruppe *Klimaneutralität Hessen 5 Jahre weiter* (Wiesbaden: Springer Vieweg) pp 136-140
[9] The Spruce URL: https://www.thespruce.com
[10] O'Leary Paint URL: https://www.olearypaint.com
[11] Chubinsky A *et al* 2019 *IOP Conf. Ser.: Earth Environ. Sci.* **316** 012023
[12] Kompozit URL: https://kompozit.ua
[13] Dow Chemical Corporate URL: https://www.dow.com/en-us/document
[14] Kovarova J, Blahova J, Divisova L and Svobodova Z 2013 Alkylphenol ephoxylates and alkylphenols - update information on occurrence, fate and toxicity in aquatic environment *Pol. J. Vet. Sci.* **16** (4) pp 763-772
[15] CEN - EN 482:2012 Workplace exposure - General requirements for the performance of procedures for the measurement of chemical agents (Brussels:CEN)
[16] CEN - EN 689:2018 Workplace exposure - Measurement of exposure by inhalation to chemical agents - Strategy for testing compliance with occupational exposure limit values (Brussels: CEN)
[17] Campisano R, Hall K, Griggs J, Willison S, Reimer S, Mash H, Magnuson M, Boczek L and Rhodes E 2017 Selected Analytical Methods for Environmental Remediation and Recovery (Washington, DC: EPA) pp 82-83

## Acknowledgments

The authors express special gratitude to the team of DOW Chemical specialists who kindly provided test methods, exam results and shared their presentations. And also, especial gratitude to O P Konstantynovskyi, associate professor of KNUCA, for valuable recommendations concerning the staging of the experiment.

IOP Conf. Series: Materials Science and Engineering **907** (2020) 012059 doi:10.1088/1757-899X/907/1/012059

# Express-method of determination hidroactiviti of Portland cement

**A P Nikolaev[1], E V Kondrachenko[1] and V I Kondrachenko[2]**

[1] Department of Technology of Building Production and Building Materials, O. M. Beketov National University of Urban Economy, M. Bazhanova St., 17, Kharkov, 61002, Ukraine;
[2]Department Building Materials and Technologies, Russian University of Transport, Obraztsova St., House. 9, Building 9, Moscow, GSP-4, 127994, Russia;

apnikolaev1946@gmail.com,

**Abstract.** Basic problems are touched for existing methods of Portland cement binding properties control. Some of them consist in the fact that direct methods for Portland cement clinker and Portland cement physical-mechanical activity determination are too time-consuming, which does not enable timely correction of their properties directly in the course of production process. Others are connected with insufficient definitiveness of cement behavior with water depending on the value of its phase-forming minerals condition regulated parameters in their totality. We introduced the term "hydroactivity of Portland cement" for quantitative characterization of its part capable under certain conditions to transfer to liquid phase after cement mixing with water. An instant control method for produced Portland cement was developed based on direct connection between its hydroactivity and binding properties. Terms were substantiated for binder material hydroactivity determination under which the obtained results characterize more exactly the initial condition of material under analysis. A stage-wise description is given of application of the proposed shortcut method of hydroactivity determination for Portland cement clinkers, various classes of Portland cement and certain model systems. The proposed control method for properties of Portland cement clinker and Portland cement enables production of cement with assured activity under optimal relation between production expenses and obtained product quality.

## Introduction

The essence of Portland cement (PC) quality control lies in determination of its parameters matching to valid requirements [1]. Methodology of this control includes two basic lines. One of them is connected with determination of PC behavior under interaction with water, such as under hardening [2]. Another one implies control of condition parameters of Portland cement clinker (PCC) and PC and subsequent prediction of their behavior, issuing from the results of such control [3, 4]. The first approach is too time-consuming, which makes difficult timely control action on clinker furnace and milling unit product quality control. Besides, the very part of PC interaction activity with water and obtained results of its physical-mechanical strength are blurred by properties of contact area formed on quartz filler surface.

The advantage of another approach lies in its promptitude. It was just this circumstance that enabled PCC burning process adjustment depending on appearance of unabsorbed calcium oxide ($CaO_{free}$) and crystallization pattern of its minerals determined by petrography directly at furnace head. Nevertheless,

IOP Conf. Series: Materials Science and Engineering **907** (2020) 012059 doi:10.1088/1757-899X/907/1/012059

the authors suppose that currently applied methods of PC minerals condition parameters estimate do not ensure in the totality of their results the ability to unambiguously predict cement behavior under its interaction with water.

**Analysis of recent research**

It should be noted that within last several decades methods of PC physical-mechanical properties control never actually changed as regards their essence [2]. At the same time methods of PCC and PC condition determination have been substantially improved. Thus, for instance, for determination of PCC and other cement materials phase composition X-ray phase analysis was widely applied using Rietveld method [5]. This method permits quite prompt determination by progressive approximation of not only quantitative phase composition of tested PCC, but also of crystallographic peculiarities of its phase-forming minerals structure. Besides, for evaluation of PC grinding fineness in addition to its specific surface determined by air permeability [6], laser analyzers are widely used which also enable to obtain cement particle size distribution function as well [7]. Nevertheless, as far as known to authors, Rietveld method fails to discover small catalytic-active impurities in tested samples, whereas laser analyzer and Blaine method are unable to determine condition of PC particles surface. Meanwhile, these factors are substantially important in formation of PC binding properties [8].

It follows from the above that a search of new efficient PC running quality control methods is prospective in analysis of PC behavior under accelerated interaction with water, as the latter is closely connected with all parameters defining condition of both Portland cement clinker and Portland cement.

**Purpose and object of study**

Development of a new method of PCC and PC behavior control under their interaction with water, which enables to obtain real-tine information on current condition of these materials quality in the course of their manufacture.

**Theoretical substantiation and development of method**

The authors suppose that for online control of PCC and PC quality it is most feasible to make use of their ability of interaction with water, by determining it at an accelerated rate. The selected rate must ensure participation of a significant part of tested sample in formation of the results of said ability. Such an approach will enable real-tine determination of all apparent factors connected with condition and behavior of phase-forming cement minerals and concomitant additives in tested materials.

The proposed method may be based on known truth that 28rh hardening day ultimate compression strength of small-beam cement samples is in direct proportion to the number of hydrate compounds forming cement rock [8]. These compounds are formed in interaction of calcium silicates, aluminates and alumoferrites with water [9]. Their source in cement is Portland cement clinker and mineral additives which become able of demonstrating their latent activity due to alkaline properties of PC clinker parts [10]. Hydrate new formations are basis of hardening cement grout which unites cement fillers into a conglomerate strengthening in time.

To enable disclosure of PC binding properties, transfer of products of its minerals hydration, hydrolysis and dissolution into mixing water is necessary, but not sufficient, as certain terms are also required for solid formation from cement grout liquid phase. In case these terms do not meet their required values, PC binding resource may not be efficiently utilized. An example is PC hardening under too high water /cement ratio (W/C).

Within the frames of the above provisions, the results of small-beam cement sample strength determination from various cements may be interpreted as collation of the value of their active part which transferred to mixing water and was successfully realized as hardening binding material. The dynamics of this phenomenon may be divided into two simultaneous processes, one of which is connected to PC mineral-forming materials being absorbed by mixing water, whereas the other one consists in cement grout solid phase formation on their basis. The first process depends on PC condition parameters, such as its phase composition, structure defects of phase-forming minerals, especially at phase borders, and appearance of catalytic-action impurities in them. The second process mostly depends on parameters of the cement grout being formed and on environment conditions. Both the first and the second processes may act as limiting factors as regards final results of PC hardening. This is due to the fact that dynamics of cement minerals matter transfer into liquid phase may also depend on release rate of new formations from it.

Our analysis shows that under the existing terms of PC physical-mechanical activity determination the obtained results depend on mutual behavior of tested material and products of its reaction with water. We may obtain necessary terms under which analysis results would better characterize initial material by itself, by way of PC active kneading at high W/C values. Measuring the residual amount of PC solid phase after such kneading would enable to define its part which transferred to liquid phase under experimental conditions, i.e., showed activity. Tne result obtained in this way would not actually depend on the behavior of PC reaction products with water. To designate this PC property we offer the term "*Portland cement hydroactivity*". This term reflects the amount of PC matter which is able of transferring into water after its mixing. This amount may be expressed in per cent of tested material initial mass. This parameter, as distinct to its applicable counterparts, firstly characterizes the properties of PC matter by itself. For instance, the applied value of *PC physical-mechanical or binding activity* characterizes behavior of not only initial material, but also of forming cement grout and contact between cement rock and surface of added filler [8]. Other applied characteristics, such as *PC hydraulic and hydration activity*, are also connected with cement grout behavior in water and with condition of its liquid phase [8].

The ability of real-time determination of a parameter whose value firstly characterized properties of PCC and PC by themselves enables online control of condition of these binders directly in the course of their manufacture and consumption. The application of the control method may be presented as PCC and PC matter activity testing with water. In this case the totality of all properties inherent in tested materials takes part in formation of tested activity value. See below a description of the proposed determination sequence of cement matter hydroactive part.

**Equipment and materials**

In our work we applied: analytical balance, constant rpm impeller mixer, centrifuge, drying cabinet, muffle furnace below 1000 °C, photoelectric colorimeter, known concentration calcium chloride solution, double-distilled water, PCC, CEM 42.5 and CEM 32.5.

**Results and discussion**

*Results of method execution* Necessary amount (100 mg) of tested material sample was taken by quartering and brought to impeller mixer vessel with 100 ml of double-distilled water. Slurry concentration was 1 g/l, kneading time 10 minutes. After kneading the slurry was immediately brought into centrifuge working area to separate liquid and solid phases. After 3 minutes centrifuging liquid part was carefully drained, whereas vial with wet sediment brought into drying cabinet at 105 °C and dried up to constant mass. As drying took about 10 minutes and during this time a certain part of sample bound some water, the ignition loss of obtained dry residue at 1000°C was also estimated. After that we calculated how much initial binder transferred to water under experimental conditions. The obtained results for samples of PCC, CEM 42.5, CEM 32.5 are shown in Table 1.

The authors have earlier developed a determination method for amount of calcium capable of transferring into water under similar terms [11]. If we use this method, having previously determined total calcium content in initial samples, we may estimate the degree of PC matter absorption at interaction with

IOP Conf. Series: Materials Science and Engineering **907** (2020) 012059 doi:10.1088/1757-899X/907/1/012059

water issuing from the amount of calcium that transferred into water relative to its initial content, with due consideration of the fact that it is uniformly distributed in fine-grained meal

In the present work calcium content in cement slurry liquid phase was measured after centrifuging of the same samples than before as described in [11]. Calcium content in initial samples was determined beforehand. Results of second method application are also shown in Table 1.

**Table 1.** Results of binder hydroactivity determination

| nn | Cement types | Initial mass (mg) | Mass residue (mg) | Hydroactivity (%) | Initial CaO mass (mg) | CaO mass in liquid phase (mg) | Hydroactivity (%) |
|---|---|---|---|---|---|---|---|
| 1 | PCC | 100 | 69.4 | 30.6 | 66.7 | 20.2 | 30.3 |
| 2 | CEM 42.5 | 100 | 71.2 | 28.8 | 60.6 | 17.4 | 28.7 |
| 3 | CEM 32.5 | 100 | 75.9 | 24.1 | 56.1 | 13.9 | 24.8 |

As we see, the results of this measurement do not actually differ from previous one.

To determine the character of connection between PC hydroactivity and its physical-mechanical properties we used model systems obtained by mixing CEM 32.5 and CEM 42.5, gradually substituting less active cement for more active. After that we measured hydroactivity of obtained model cements and determined the hardness of cement-sand samples prepared on their basis. Thew samples hardened within 12 hours after their formation at 40 oC with cement/sand ratio 1:1 and W/C = 0.45. The revealed dependence between hydroactivity of model cements and hardness of samples made on their basis is shown in Fig.1.

Control of physical-mechanical properties of tested PCs by the value of cement rock hardness was already applied in earlier works by authors [12]. This method is simple and easily accessible.

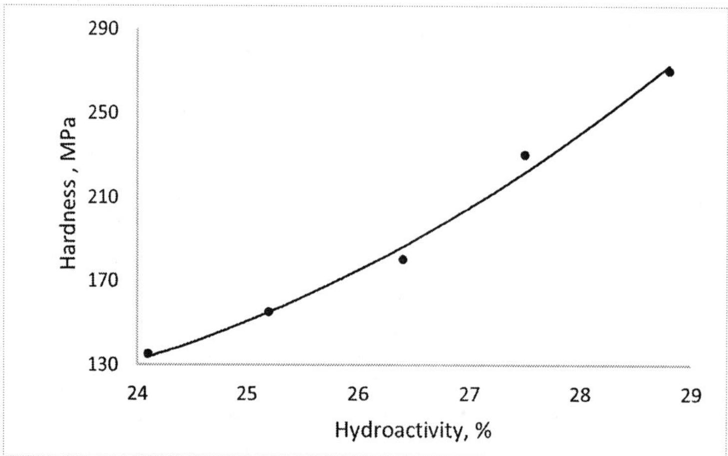


Fig. 1 Connection between hydroactivity of used model cement systems and hardness of cement-sand grouts made on their basis

As seen from the plot, the obtained dependence is direct and sufficiently monotonous.

The materials of our research prove that PC absorption with water may be controlled by weight method using drying and calcination of solid residue, as well as by results of calcium double determination in initial sample and in liquid phase after kneading. Calcium content in PC is netter determined by plant online quantometer, whereas in liquid phase preferred methods are ionometry; flame photometry; chemical methods or photoelectric colorimetry with special indicators. Single sample analysis never takes more than one hour.

**Conclusion**

The newly introduced parameter of hydroactivity first of all characterizes the condition of PCC and PC matter by itself.

The proposed control method for this parameter enables obtaining, on a par with PCC and PC characteristics, test of cement plant equipment operation efficiency. This capability permits output of cement with warranted properties at optimal relation between production expenses and quality of final product.

PC hydroactivity determination method may be also applied for accelerated quality control of cement incoming to plants manufacturing concrete and reinforced concrete articles.

**References**

[1] Nikolaev A P and Kondrashchenko E V 2020 *The topicality of internal control of Portland cement binding properties* (Vistnyk Odeskoii Derzhavnoii Akademii Budivbytstva i Arkhitektury) vol **78** pp 108-117
[2] DIN EN 196-1 2005 *Methods of testing cement Part 1* (Determination of strength» German version)
[3] Malova E Y and Manuilo E V et al. 2011 *Mineralogical composition determination of clinkers by a combined analysis method* (Vestnik Altaiskogo Gosudarstvennogo Tekhnichesko go Iniverrsiteta imebi Polzunova) vol **1** pp 79-83
[4] Torre A and Cabeza A et.al. 2001 Full phase analyses of Portland cement clincer by penetrating synchrotron powder diffraction *Analytical Chemistry* vol **73** 2 pp 151–156
[5] Ed R A and Young N Y 1993 *The Rietveld Method* (Oxford. University press) p 298
[6] DIN EN 196-6 2010 *Methods of testing cement-Part 6* (Determination of fineness». German version)
[7] Gerasimov A V 2014 Particle size analysis *Tsement i ego Primenenie* vol **3** pp 73-75
[8] Volzhenskiy A V and Burov A V et al. 1979 *Mineral Binder Compounds* (Moscow: Stroyizdat Publishers)
[9] Kurdowski W 2010 *Chemistry of cement and concrete* (Warsawa: Scientific Publishing PWN)
[10] Taylor Harold F W 1990 *Cement* Chemistry (London: Academic Press)
[11] Nilolaev A P and Kondrashchenko E V 2017 *Portland Cement Binding Properties: Control and Analysis* (Kharkov. A.N. Beketov Kharkov National University of Municipal Services Press)
[12] Nikolaev A P and Kondrashchenko E V 2019 On the issue of water activation for Portland cement *IOP Conference Series: Materials Science and Engineering, Reliability and Durability of Railway Transport Engineering Structures and Buildings* p 708

IOP Conf. Series: Materials Science and Engineering **907** (2020) 012060 doi:10.1088/1757-899X/907/1/012060

# Investigation of adhesive-strength characteristics of fire-retardant epoxy polymers modified with metal-containing additives

**O Hryhorenko[1], Ye Zolkina[1], N Saienko[2], Y Popov[2], R Bikov[2]**

[1]Department fire and technogenic safety of facilities and technologies, National University of Civil Defense of Ukraine, Chernishevska Street 94, 61023, Kharkiv, Ukraine,
[2]Department of General Chemistry, Kharkiv National University of Civil Engineering and Architecture, Sumska Street 40, 61000, Kharkiv, Ukraine

hryhorenko81@gmail.com

**Abstract.** Composite materials based on epoxy oligomers due to their unique properties are widely used in various fields of construction and mechanical engineering, including fire retardant coatings. The introduction of flame retardants and dispersed inert fillers as a rule, affects the structure of epoxy polymers. When addressing issues related to the creation of flame-retardant, composite materials, there is insufficient data on the impact of dispersed mineral fillers on their performance properties. Attention is mainly paid to the indicators of flammability, toxicity of combustion products and thermophysical properties of the coating. Therefore, the aim of this work is to conduct experimental studies of changes in the adhesive and strength characteristics of fire-retardant epoxy polymer from the content of metal-containing additives. As a result of experimental studies, it was found of that changing the component composition of the composition can vary widely the adhesive strength characteristics of fire-retardant epoxy polymers. The obtained data must be taken into account when developing epoxyamine coatings for fire protection of building structures and technological communications.

## 1. Introduction

Composite materials based on epoxy oligomers due to their unique properties are widely used in various fields of construction and mechanical engineering. The main disadvantage of epoxy composite materials is their flammability. This problem is solved by the introduction of various fillers and flame retardants. The tendency of epoxy resins to carbonization allows to obtain on their basis fire-retardant coatings of intumescent type. However, the filling of the composition leads to a change in their properties, including operational.

It is known that the performance characteristics of epoxy polymers are largely determined by the technological properties of the polymeric binder, structuring processes, adhesive interactions at the interface between polymer-solid, intermolecular interactions and other factors [1-8].

The introduction of fillers, as a rule, significantly affects the structuring processes, chemical and water resistance, adhesive strength, rheological properties and structure of epoxy polymers. However, in addressing issues related to the creation of non-combustible, composite materials, there is insufficient data on the impact of dispersed mineral fillers on their performance properties. Attention is mainly paid to the indicators of flammability, swelling rate, smoke formation, toxicity of combustion products and thermophysical properties of the coating [1, 9-13].

Previous studies [10] allowed us to obtain results that reflect the influence of the nature and content of metal-containing additives on the swelling rate and the value of the coke residue of fire-retardant epoxyamine coatings of low flammability.

The dependence of the swelling epoxy polymer content on the additive content was investigated experimentally and the resistance of the polymer to the action of a celite rod heated to 950 ° C was studied.

The change of thermomechanical properties [14] of epoxy polymers modified with metal-containing additives has been studied. It is shown that the rate of gas evolution in the temperature range after the transition of the system to the viscous state until the phase transition from the plastic state to curing, determines the multiplicity of swelling of the intumescent composition.

In [15-19] the results of research of thermal and thermo-oxidative destruction of epoxy polymers are given. However, no less important are the studies of their mechanical properties [20-23].

## 2. Unresolved issues

These studies give an idea of the mechanisms of formation of a swollen oblique layer on the surface of the epoxy polymer, as well as the effect of metal-containing additives on the processes of thermal and thermo-oxidative destruction, thermomechanical properties of the polymer and others. However, no less important characteristics of fire-retardant coatings are also their performance properties.

The adhesive strength of a polymeric binder with a substrate is a decisive factor in obtaining compositions intended for the protection of building structures. Given that fire-retardant epoxyamine compositions are complex highly filled systems, it is of interest to study the effect of metal-containing additives on their adhesive strength, impact viscosity and destructive bending stress.

The purpose of this work is to conduct experimental studies of changes in the adhesive strength characteristics of fire-retardant epoxy polymer from the content of metal-containing additives.

## 3. Main part

An epoxy oligomer-based composition was used as the base flame retardant epoxy polymer ED-20, and monocyanethylenediethylenetriamine was used as a hardener [1, 10, 11] Monoammonium phosphate (MAP) ($NH_4H_2PO_4$) and activated basalt husk (ABS) $SiO_2$ – 30.2; FeO + $Fe_2O_3$ – 16.1; $TiO_2$ – 1.6; $Al_2O_3$ – 14.0; CaO – 9.6; MgO – 4.1) were used to reduce flammability.

Copper oxide (II), zinc oxide (II), vanadium oxide (V) and bentonite [10, 13] were used as metal-containing additives (clay-based material with a percentage by weight: $SiO_2$ – 72.5; $TiO_2$ – 0.27; $Al_2O_3$ – 14.45; $Fe_2O_3$ – 1.23; CuO – 1.5; MgO – 2.8; $K_2O$ – 0.29; $Na_2O$ – 1.55). Additives were introduced into the composition of epoxy polymers in an amount of from 5 to 20 mass parts (m.p.).

The adhesive strength was determined experimentally by the method of uniform separation from each other of two metal washers with a base diameter of 25 mm, made of steel grade St3. The essence of the method is to determine the magnitude of the destructive load $P_{dest}$, when stretching two washers, the bases of which are pre-glued with the investigated epoxy polymers, forces perpendicular to the bonding plane. The nature of the destruction can be both adhesive or cohesive, and mixed [6, 24]. The studies were conducted on a series of 10 samples. The largest and smallest value of the destructive load $P_{dest}$ from the series were not taken into account in further calculations. The remaining 8 values of the destructive load $P_{dest}$ used to determine the average value of the destructive load $P_{dest}$.

Tensile strength $\sigma_{t.st.}$ was determined based on the average value of the destructive load $P_{dest}$ and the cross-sectional area of the washer base $F$ [20]:

$$\sigma_{t.st.} = \frac{P_{destr.a.}}{F}. \quad (1)$$

Figure 1 shows the results of experimental studies of the effect of metal-containing additives on the tensile strength $\sigma_{t.st.}$ fire-retardant epoxy polymers for steel grade St3.

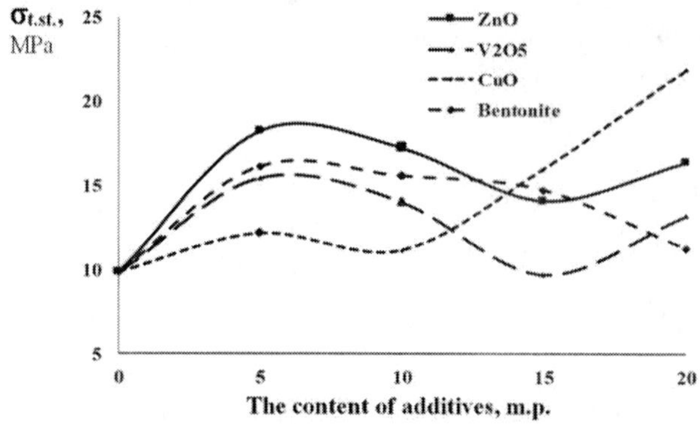


**Figure 1.** The dependence of the tensile strength $\sigma_{t.st.}$ fire-retardant epoxy polymers for steel St3 from the content of metal-containing additives: ZnO, $V_2O_5$, CuO and bentonite.

From the presented data it is seen that all metal-containing additives increase the adhesive strength of the fire-resistant epoxy polymer to varying degrees. The introduction of a small number of additives up to 5 m.p. leads to a sharp increase in adhesive strength: ZnO 1.8 times, $V_2O_5$ and bentonite – 1.6 times, CuO – 1.2 times. Further increase in the content of additives from 5 to 15 m.p. leads to a decrease in adhesive strength (except CuO). Filling the polymer matrix with metal-containing additives (except bentonite) from 15 to 20 m.p. also leads to an increase in adhesive strength by 1.3 to 2.1 times compared with fire-retardant epoxy polymer without additives. Introduction 20 m.p. bentonite has almost no effect on the adhesive strength of the flame retardant epoxy polymer.

The effect of metal-containing additives on the change in impact strength without notch according to ISO 179 and destructive bending stress was also studied $\sigma_{dbs}$ according to ISO 1209 of the investigated fire-retardant epoxy polymer from the content of additives.

The essence of the method of testing the impact strength without incision is the effect of pendulum dill on the test sample, which is placed on two supports. The impact energy spent on the destruction of the sample is recorded as the difference between the initial stock of potential energy of the pendulum and the energy remaining in the pendulum after the destruction of the test sample [21, 24].

Studies of the impact strength were performed on samples without incision type 1/179-1 with dimensions of 80×10×4 mm, ten samples were used in each series of tests. The distance between the supports on which the sample is 62 mm. The pendulum hit was carried out on a sample placed "flat".

The results of studies of the effect of metal-containing additives on the change in impact strength are presented in Figure 2.

From the presented data (Figure 2) it is seen that metal-containing additives affect the impact strength of the fire-resistant epoxy polymer is not the same. The addition of bentonite in small amounts (up to 10 m.p by weight) leads to a 1.5-fold increase in impact strength, but a further increase in the additive content gradually reduces this index. Introduction to the composition of the fire-retardant epoxy polymer additives of oxides of metals of transition valence $V_2O_5$, CuO in amounts up to 5 m.p. leads to a decrease in the impact viscosity of the samples. The largest decrease is observed for epoxy polymer with the addition of $V_2O_5$ – 1.5 times, for CuO – 1.1 times. For ZnO there is a slight increase of 1.05 times. A further increase in the content of ZnO and CuO additives leads to a sharp increase in impact strength by 1.8 and 1.6 times, respectively. Increasing the content of $V_2O_5$ to 10 m.p. does not significantly change the impact viscosity compared to filled with 5 wt.h. epoxy polymer. Increasing the content of additives to 15 m.p. leads to a decrease in impact viscosity in all cases except epoxy polymer with the addition of $V_2O_5$. The most significant decrease is observed for epoxy polymer with the addition of ZnO. For epoxy polymer with $V_2O_5$ there is an increase in impact strength by 1.2 times compared to fire-retardant epoxy polymer without additives. Further increase the content of additives to 20 m.p. leads to a gradual decrease in impact viscosity in all cases.

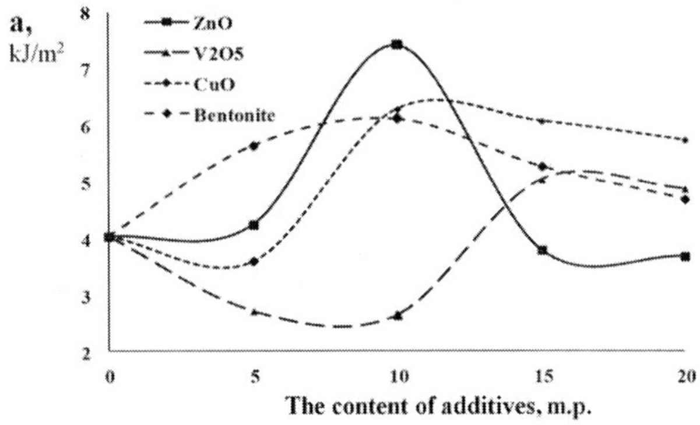


**Figure 2.** Dependence of the impact strength of fire-retardant epoxy polymers on the content of metal-containing additives: ZnO, $V_2O_5$, CuO and bentonite.

The essence of the test method to determine the destructive stress during bending $\sigma_{dbs}$ is the effect on the sample, which is scattered on the two supports, short-term load applied in the middle between the supports [21, 23].

Investigation of the index of destructive stress during bending $\sigma_{dbs}$ were performed on samples with dimensions of 80×10×4 mm according to the method described in [21, 23].

The results of studies of the effect of metal-containing additives on the change in the destructive stress during bending $\sigma_{dbs}$ presented in Figure 3.

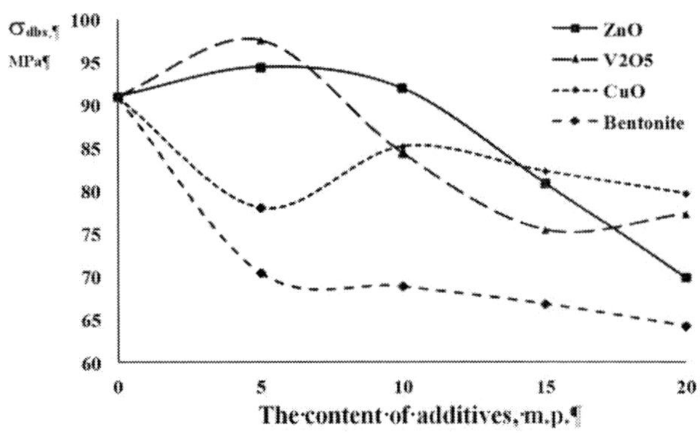


**Figure 3.** Dependence of destructive stress during bending $\sigma_{dbs}$ fire-retardant epoxy polymers from the content of metal-containing additives: ZnO, $V_2O_5$, CuO and bentonite.

As can be seen from the graphs in Figure 3 increase in the content of metal-containing additives with 5 до 20 m.p. leads to a gradual decrease in the destructive stress during bending $\sigma_{dbs}$ fire-retardant epoxy polymers 1.23, 1.15, 1.12 and 1.29 times for additives ZnO, $V_2O_5$, CuO and bentonite, respectively. Introduction of small amounts to 5 m.p additives ZnO and $V_2O_5$ has almost no effect on the destructive stress, the growth is slightly – 1.04 and 1.07 times, respectively. Add to 5 m.p. CuO and bentonite, on the other hand, reduce this Figure by 1.14 and 1.23 times. A further increase in the content of additives leads to a deterioration of the destructive stress during bending $\sigma_{dbs}$.

From the obtained results it is seen that by changing the component composition of the composition it is possible to vary the adhesion-strength characteristics of fire-retardant epoxy polymers.

## 4 Conclusion

The paper presents the results of studies of the effect of metal-containing additives – ZnO, $V_2O_5$, CuO and bentonite – on the adhesive strength, impact viscosity and destructive stress in bending of fire-retardant epoxy polymer based on oligomer ED-20.

As a result of experimental studies, it was found that metal-containing additives on the adhesive and strength characteristics of fire-retardant epoxy polymers do not affect equally. The adhesive strength of the studied samples, mainly, increases, achieving the best performance when filling to 5 m.p., except for samples with the addition of CuO, for which the best performance is observed when filling to 20 m.p. The results of studies of the effect of metal-containing additives on the change in impact strength indicate the possibility of adjusting this characteristic in a wide range. The best values were achieved during the introduction 10 m.p. ZnO, receiving an increase in impact strength of 1.8 times. The introduction of flame retardant epoxy polymers of metal-containing additives leads to a gradual decrease in the destructive stress during bending.

From the conducted experimental researches it is visible that changing component structure of a composition it is possible to vary in wide limits adhesive-strength characteristics of fire-retardant epoxy polymers. The obtained data must be taken into account in the development of epoxyamine coatings for fire protection of building structures and technological communications.

## References

[1] Saienko N, Bykov R and Muratov U 2016 *Problems of Fire Safety* **40** 174-178 http://nbuv.gov.ua/UJRN/Ppb_2016_40_31
[2] Su L, Zeng X, He H, Tao Q and Komarneni S 2017 *J. App. Clay Sci.* **148** 103-108
[3] Danchenko Yu M, Obizhenko T M, Umanska T I and Barabash E S 2018 *Scientific Bulletin of Civil Engineering* **94** (4) 160-170 https://vestnik-construction.com.ua/images/pdf/4_94_2018/28.pdf
[4] Andronov V A, Danchenko I M and Bukhman O M 2012 *Problems of Fire Safety* **31** 10-18 http://repositsc.nuczu.edu.ua/bitstream/123456789/3690/1/29.pdf
[5] Danchenko Yu, Kachomanova and Barabash E *Chemistry and Chemical Technology* **12** (2) 188-195 http://ena.lp.edu.ua/handle/ntb/45161
[6] Fedoseev M S, Derzhavinskaya L F and Tsvetkov R V 2014 *Perspective materials* **4** 30-36 https://elibrary.ru/download/elibrary_21350149_49885537.pdf
[7] Starokadomsky D L and Pahlov E M 2015 *Composites and nanostructures* **7** (7) 41-51 http://www.issp.ac.ru/journal/composites/2015/Starokadomskii.pdf
[8] Danchenko Yu M 2017 *Polymer materials and technologies* **3** (2) 56-63
[9] Kovalov A, Otrosh Y, Vedula S, Danilin O and Kovalevska T 2019 *Nauk.Visn. Nats. Hirn. Univ.* **3** 46-533
[10] Hryhorenko O and Zolkina Y 2018 *Problems of Fire Safety* **43** 31-37 https://nuczu.edu.ua/sciencearchive/ProblemsOfFireSafety/vol44/Hryhorenko.pdf
[11] Spirina-Smilka E, Jakovleva R, Saienko N, Dovbish A and Ribka Je 2011 *Problems of Fire Safety* **30** 247-252
[12] Berezovskiy A 2014 *Visnyk Cherkaskogo derzhavnogo tehnologichnogo universitetu* **4** 66-71 http://nbuv.gov.ua/UJRN/Vchdtu_2014_4_14
[13] Eremina T Yu, Rob M V and Dmitrieva Yu N 2012 *Fire and Explosion Safety* **21** (7) 52-56
[14] Hryhorenko O, Mikhailuk A and Zolkina Y 2018 *Fire Safety* **44** 15-19 http://repositsc.nuczu.edu.ua/handle/123456789/8761

[15] Berezovsky A, Maladyka I, Popov Yu, Sayenko N (2012) *Fire safety* **20** 27-31 https://journal.ldubgd.edu.ua/index.php/PB/article/view/683
[16] Buketov A, Maruschak P, Sapronov O, Brailo M, Leshchenko O, Bencheikh L and Menou A 2016 *Mol. Cryst. and Liq. Cryst.* **628** 167-179
[17] Mostovoi A, Bunenkov P and Panova L 2016 *Inorg. Mater. Appl. Res.* **7** 768-772
[18] Mostovoi A, Nurtazina A, Kadykova Y and Bekeshev A 2019 *Inorg. Mater. Appl. Res.* **10** 1135-1139 https://doi.org/10.1134/S2075113319050228
[19] Shen J, Huang W, Wu L, Hu Y and Ye M 2007 *Compos. Part A: Appl. Sci. and Manuf.* **38** 1331-1336
[20] Skrypnyk O and Zolotov S 2011 *Mun. Econ. of Cities* **101** 41-46 https://eprints.kname.edu.ua/23338/1/41-46%20%D0%A1%D0%BA%D1%80%D0%B8%D0%BF%D0%BD%D0%B8%D0%BA%20%D0%95%D0%A1.pdf
[21] Buketov A and Skyrdenko V 2015 *Phys.-Chem. Mech. of Mater.* **51** 43-49 http://dspace.nbuv.gov.ua/handle/123456789/134330
[22] Huang P, Zheng S, Huang J, Guo Q and Zhu W 1997 *Polym.* **38** 5565-5571
[23] Jakovleva R, Popov Yu, Saienko N, Shevtsova K 2007 *Advances in Chemistry and Chemical Technology* **21** 6 (74) 23-28 https://cyberleninka.ru/article/n/ognebiostoykie-epoksidnye-kompozitsii-dlya-zaschity-drevesiny
[24] Danchenko Y, Andronov V, Barabash E, Obigenko T, Rybka E, Meleshchenko R and Romin A 2017 Research of the intramolecular interactions and structure in epoxyamine composites with dispersed oxides *Eastern-European Journal of Enterprise Technologies* **6** (12) 4 http://journals.uran.ua/eejet/article/view/120998

IOP Conf. Series: Materials Science and Engineering **907** (2020) 012061 doi:10.1088/1757-899X/907/1/012061

# Theoretical substantiation and calculations of water flow to ranney water intakes and drainages under protection from submergence of the urban territories and buildings by ground water

**S V Telyma[1], E O Oliynyk[1], O S Voloshkina[2] and I M Vildman[3]**

[1]Department of Applied Hydrodynamics, Institute Hydromechanics of NAS of Ukraine, M. Kapnist str., 8/4, Kyiv, 03057 Ukraine
[2]Department of Ecology, Kyiv National University of Building and Arkhitecture, Povitroflotsky avenue, 31, Kyiv, 03680 Ukraine
[3]VSP "Institute of innovation education", Kyiv National University of Building and Arhitecture, Povitroflotsky avenue, 31, Kyiv, 03680 Ukraine

sertelyma@gmail.com

**Abstract.** The methodic of evaluation of the innerdrain hydraulic of the water flow along the length of the drains of the ranney drainage (water uptake) on summary inflow to them is developed. The methodic is based on the mathematical models which allow more fully to take into account the complex real flow character and the hydraulic conditions in the zone of the influence ranney water intakes and drainages.The different cases of solution of joint problem of flow in aquifer and in the drains-rays are considered and presented as the system of the analytic dependences. Obtained results showed that the inflow to drain-ray is distributed uneven along its length and the main inflow is formed in the first half of the drain-ray. Therefore, in some cases the influence of innerdrain hydraulic should be taken into account when calculating the parameters of drainage and evaluation of their operation

## Introduction

When studying the work of underground water intakes and drainages an important problem in the calculation of the inflow of groundwater to them concludes into account the movement of water in these structures. However it should be noted that in most calculations of filtration to horizontal water intakes and drainages the inflow along the length of the drains as well as other hydraulic parameters were assumed to be unchanged and in some cases at small inflows and others the influence of water flow inside the drains will be negligible and can be neglected. These assumptions were used at development of the calculation methods based on the theory flow resistances [1,2]. The efficiency of numerical methods using was presented under solution of the problem of protection from submergence of ground waters urban territory in Kharkov [3]. Comparative analysis of modeling data and regime observations of drawdowns on the protection areas showed a good agreement betweem ones. But as shown by the results of experiments in more complex filtration schemes of inflow to underground horizontal structures reliable results of calculations of filtration inflow to the drainage can be obtained only by taking into account the hydraulics of water flow inside the drainage. Therefore the methods of filtration calculation of ranney water intakes (drainages) based on the method of filtration resistances and on a position of uniform distribution of the intensity of inflow *q(y)* along drains and in some other cases require additional improving and scientific justification.

## Materials and methods

As known the main mathematical model of the ajoint flow in aquifer and inside of the horizontal drainages consists of the well-known system of Saint-Venan equations which describes in hydraulic formulation the variable motion of the channel flow [4,5]:

$$\frac{\partial Q}{\partial t}+\frac{\partial\left(Q^2/A\right)}{\partial x}+gA\left(\frac{\partial Z}{\partial x}+S_f+S_e\right)+L+W_fB=0, \tag{1}$$

$$\frac{\partial Q}{\partial x}+\frac{\partial\left(A+A_0\right)}{\partial t}=q, \tag{2}$$

$$S_f=\frac{n^2|Q|Q}{2{,}2A^2R^{4/3}}=\frac{|Q|Q}{K_C^2},\qquad S_e=\frac{K_C}{2g}\frac{\partial\left(\frac{Q}{A}\right)^2}{\partial x},\qquad W_f=C_W|V_r|V_r, \tag{3}$$

where $S_f$ – friction, which is determined from the known Manning equation, $S_e$ – local flow gradient, $Q(x,t)$ – flow through the cross section, $A(h)$, $A_0$ – respectively the active and inactive cross-sectional areas, $K_C$ – the coefficient of water conductivity of the channel, $K_e$ – expansion coefficient (minus sign) and compression one (plus sign), $B$ – the width of the channel along the water surface, $W_f$ – correction to the wind, $q$ – outflow from the channel, $z(x,t)$ – depth of flow, $x$ – the distance along the channel, $R\approx\frac{A}{B}$ – hydraulic radius, $L$ – lateral possible inflow, $n$ – roughness.

The value of $q$ in equation (2) is calculated from the solution of the planned filtration problem which is described by the known equation of transient filtration[6,7]:

$$\mu\frac{\partial H}{\partial t}=\frac{\partial}{\partial x}\left(T\frac{\partial H}{\partial x}\right)+\frac{\partial}{\partial y}\left(T\frac{\partial H}{\partial y}\right)+\varepsilon(x,y,t), \tag{4}$$

where $H(x,y,t)$ – the level of groundwater, $\mu(x,t)$ – storage coefficient, $T$ – coefficient of water transmissivity, $\varepsilon$ – flow rate per unit surface.

An important problem in the joint interaction of the given equations is in the changing of the model of their conjugation. In [6] the following conjugation criterias are proposed:

a) in the case of hydraulic communication we have:

$$\frac{\partial H}{\partial\vec{u}}=\lambda\left(H-\left(z+z_g\right)\right), \tag{5}$$

where $\vec{u}$ – normal to the boundary of conjugation, $z_g$ – marking of the bottom (river, canal), $\lambda(x,t)$ – a parameter that takes into account the hydraulic friction which in general will be variable along the length of the drain. Some recommendations relative given parameter are proposed in [4,7].

b) in the absence of hydraulic connection:

$$Q_n\to=Bk_n\frac{z-z_{кр}}{z+z_k},\qquad z>z_{кр}, \tag{6}$$

$$Q_n\to=0,\qquad z<z_{кр}, \tag{7}$$

where $B(z)$ – the width of the river, $k_n$ – vertical filtration coefficient, $z_k$ – the thickness of the impermable layer, $z_{кр}$ – critical depth below which water infiltration stops.

As a rule the system of equations (1), (2), and (4) is realized numerically under the following initial and boundary conditions:

$$z(x,0)=z^0(x),\qquad Q(x,0)=Q^0(x),$$

$$Q(0,t) = Q_0(t), \qquad Q(L,0) = Q_L(0)$$

taking into account the conjugation models.

The joined interaction of the filtration flow to underground water intakes (drainages) and the movement of fluid inside them were investigated when solving various problems of reclamation, protection of areas from flooding, water supply, etc [5]. However mainly these studies were conducted empirically but the impotance of solving this problem lies in the fact that the existing features of the formation of filtration of the uneven inflow to the drains-rays of the finite length can significantly affect on the parameters of the flow inside these drains which must be taken into account in the calculations [7, 8].

With the existing theoretical justification for the calculated dependences on the basis of the known method of filtration resistances it is assumed that the inflow intensity is uniformly distributed along the length of drains *q(y)* and at the same time in some cases the equality of the heads along the length of the drainage is not always true[7,8].

In the first approximation if we accept *q(y) = const* which as shown by field studies in some cases is possible (with a small number of drains – rays, high-quality filter, etc.) the hydraulic losses due the friction and discharge of the drainage area will be next [7,8]:

$$\Delta h_y = \frac{Q_0^2}{2gw^2}\left[2\left(\frac{y}{l}\right)^2 + \frac{\lambda y}{3d_\partial}\left(\frac{y}{l}\right)^2\right], \tag{8}$$

$$Q(y) = \frac{Q_0}{l}y, \tag{9}$$

and along the full length of the drain $y = l$ we have

$$\Delta h_e = \frac{Q_0^2}{2gw^2}\left(2 + \frac{\lambda l}{3d_\partial}\right), \tag{10}$$

$$Q(l) = Q_0 = ql. \tag{11}$$

Analysis of the numerical results of field studies shows that in many practical cases the distribution of the inflow along the length of the drain can be taken by linear law that is in this approximation we have:

$$q(y) = \frac{2Q_0}{l^2}(l - y), \tag{12}$$

where $Q_0$ – flow of the entire drain length *l*.

The water flow (flow rate) in section *y* is obtained after integrating the equation (12):

$$Q(y) = 2Q_0\frac{y}{l} - Q_0\left(\frac{y}{l}\right)^2. \tag{13}$$

Checking the boundary conditions of equation (13) shows that at the beginning of the drain at $y = 0$ we have $Q = 0$, and at the end of the drain at $y = l$ we have $Q = Q_0$ that is the equation is correctly integrated. As expected the analysis of equation (13) shows that the most intensive inflow will be from 0 to 0.5 in the first half of the drain which receives 75% $Q_0$ about and only 25% $Q_0$ in the second half of the drain.

Finally omitting the intermediate laying-outs the head pressure losses on the drain area with a length *y* at the linear distribution of the inflow along the drain length are:

$$\Delta h_y = \frac{Q_0^2}{2gw}\left\{8\left[\frac{1}{4}\left(\frac{y}{l}\right)^4 - \left(\frac{y}{l}\right)^3 + \left(\frac{y}{l}\right)^2\right] + \right. \tag{14}$$
$$\left. + \frac{\lambda l}{d_\partial}\left[\frac{1}{5}\left(\frac{y}{l}\right)^5 - \left(\frac{y}{l}\right)^4 + \frac{4}{3}\left(\frac{y}{l}\right)^3\right]\right\},$$

and along the full length $y = l$ we have

$$\Delta h_l = \frac{Q_0^2}{2gw}\left(2 + \frac{8}{15}\frac{\lambda l}{d_g}\right), \tag{15}$$

where $d_\partial = 2r_\partial$, $w$ – the cross-sectional area of the drain.

To determine the flow rate $Q_0$ of each drain-ray the dependencies are used in which the head (level) in the drains is assumed to be constant along theirs length.

For more accurate calculations of this problem we consider the results based on the solution of the above mathematical models which describe and take into account the joint interaction of water flow in the drain and filtration inflow to it. After linearization and simple transformations of the equations of hydraulics and filtration taking into account of the linear dependence of the evaporation intensity to the depths are reduced to a stationary form:

$$\pi C d^{5/2}\frac{d^2H}{ds^2} + 32kh_c\sqrt{\left|\frac{dH}{ds}\right|}\left.\frac{dh}{dx}\right|_{x=0}, \tag{16}$$

$$kh_c\frac{d^2h}{dx^2} - \varepsilon_0\left(\frac{h-h_k}{M-h_k}\right) = 0, \tag{17}$$

where $M$ – the ordinate of the surface of the earth, $h_c$ – the average thickness of the filtration flow, $h_k$ – critical level below which no evaporation from the groundwater surface occurs, $\varepsilon_0$ – intensity of evaporation on the surface of the earth, $\left|\frac{dH}{ds}\right|_C$ – the average value of the head gradient along the length of the drain.

The solution of linearized equation (17) with linearized boundary conditions allows to determine the parameters of the filtration flow in the area of the drain and has the form:

$$h = h_k + \frac{(H-h_k)ch\sqrt{\alpha}(x-L/2)}{ch(\sqrt{\alpha}\,L/2) + 2\sqrt{\alpha}\Phi sh(\sqrt{\alpha}\,L/2)}, \quad \alpha = \frac{\varepsilon_0}{kh_c(M-h_K)}. \tag{18}$$

From expression (18) we find $\left.\frac{dh}{dx}\right|_{x=0}$

$$\left.\frac{dh}{dx}\right|_{x=0} = \frac{\sqrt{\alpha}(H-h_k)sh(\sqrt{\alpha}\,L/2)}{ch(\sqrt{\alpha}\,L/2 + 2\Phi\sqrt{\alpha}sh(\sqrt{\alpha}\,L/2))}. \tag{19}$$

Substituting expression (19) into (16) we obtain a linear equation with respect to the function *H(y)* whose solution under given boundary conditions is obtained in the form:

$$H(y) = h_k + \frac{(H-h_k)ch\sqrt{\beta}(y-l_1)}{ch(\sqrt{\beta}l_1)}, \quad \beta = \frac{32\sqrt{\alpha}Kh_c\sqrt{\frac{dH}{dS}}sh(\sqrt{\alpha}\,L/2)}{\pi C d^{5/2}\left[ch(\sqrt{\alpha}\,L/2) + 2\Phi\sqrt{\alpha}sh(\sqrt{\alpha}\,L/2)\right]}. \tag{20}$$

Thus by formulas (20) and (18) one can find the values of the piezometric head along the length of the pipe-drain *H(y)* and the value of the level *h(x)* in different sections orthogonal to its axis.

The comparative analysis with using the examples of calculations showed that the results of the analytical calculations according to the formulas closely coincide with the results of numerical modeling of the proposed general equations at $t \to \infty$.

Thus when solving an external (filtration) problem it is necessary to take on the drain *(x = 0)* the value of the head which changes along the coordinate *y*. In this case the groundwater head *h* value and the drain flow rate *q* will also change and depend on the coordinate *y*. That to solve the external problem consider the profile (one-dimensional) model of filtration to the imperfect drain in homogeneous soil, namely:

$$T\frac{\partial H}{\partial x^2} = \mu\frac{\partial H}{\partial t}, \qquad T = km_c. \tag{21}$$

Equation (21) is solved under initial and boundary conditions which on an imperfect drainage are taken according to the expression for the head $H_g^0$ which changes inside the drain along its length *y*:

$$x = 0, \qquad H - 2\Phi_\partial\frac{\partial H}{\partial x} = H_\partial^0 = H_{0\partial} - x(y)q_0^2; \tag{22}$$

and the depth (level) of water is assumed to be constant $H_k$ that is at $x = L$ we have a boundary condition $H = H_k$.

The solution of equation (21) under the given boundary conditions is made for the stationary model of filtration in aquifer $\left(\frac{\partial H}{\partial t} = 0\right)$.

The following equation is obtained to determine the level *H* at section *L*:

$$H = H_\partial^0(y) + \frac{\overline{C}_1\overline{x} + 1}{\overline{C}_1 + 1}\left(H_k - H_\partial^0(y)\right) \tag{23}$$

To determine the flow (magnitude of inflow to the drain in each section) we obtain the dependence:

$$q(y) = \frac{T\left(H_k - H_\partial^0(y)\right)}{L + 2\Phi_\partial} \tag{24}$$

where $T = km_c$; $m_c = m_{\partial c} + \frac{h_k - m_\partial}{2}$, $h_k$ – the depth of water to the ground water level, $m_{\partial c}$ – the average depth of the drain position with respect to the impermeable layer, $\Phi_d$- filtration resistance.

Thus the total drain-ray discharge of length *l* will be

$$Q = q_c l. \tag{25}$$

where $q_c$ – the value of the specific inflow to the drain-ray determined by the formula (24) by averaging several values $q(y)$ along the length of the drain *l*.

Recall that in the conditions of lateral inflow from different sources of the flow the flow rate of each i-horizontal ray (drain) of the ranney water intake or drainage is determined by the following universal dependence [1,7,8]:

$$Q_{\partial i} = q_i l_i = \frac{2\pi T S_{\partial i} l_i}{F_i + \Phi_i}. \tag{26}$$

Then the total inflow of the water intake (drainage) consisting of $N(N \geq 1)$ rays will be equal to

$$Q = \sum_{n=1}^{N} Q_{\partial i}. \tag{27}$$

Here $F_i$ and $\Phi_i$ are known external and internal resistancess of the i-th drain of length $l$ which do not depend on the regime of water flow in the drains- rays.

The calculation of the drawdown $S_{\partial i}$ in the i-th ray (drain) in the case of taking into account the interdrain hydraulics (parameters of water flow in the drain) in formulas (26) and (27) should be taken as follows:

$$S_{gi}(y) = H_k - H^0_{\partial i}(y) \tag{28}$$

$$H^0_{\partial i}(y) = H_{0\partial i} - \gamma_i \cdot q^2_{0i},$$

Where $H_{0\partial i}$ and $q_{0i}$ – the mean value of the level in the drain and the specific flow rate determined by the disregard of the interdrain hydraulics, i.e at $\gamma_i q^2_{oi} = 0$.

As in the previous case that to take into account the flow inside the drain we consider a joint solution of the equation in the outer region (filtration problem) and the continuity flow of water in the inner region (hydraulic problem). Then for steady filtration we have the following system of equations, namely: an equation describing the non-stationary fluid flow in a horizontal drain-ray (21) and an equation for discharge:

$$q = \frac{k\left(H_k - H^0_\partial(y)\right)}{\overline{\Phi}_\partial}. \tag{29}$$

In solving this problem we assume that the filtration resistance $\overline{\Phi}_\partial$ does not change along the length of the drain-ray, the slope of the drain is taken equal to 0, however if necessary it is easy to take into account in the problem; the drain works by the full cross-section, the inflow to it is distributed evently along the drain, the resistence coefficient $\lambda = const$ and is equal to its average value along the length of the drain.

Thus the radial velocity of water inflow into the drain will be determined as:

$$V_{r_\partial} = \frac{q}{2\pi r_\partial} = \frac{k\left(H_k - H^0_\partial(y)\right)}{2\pi r_\partial \overline{\Phi}_\partial}. \tag{30}$$

The given system of equations is solved under following boundary conditions $y = 0,\ V = 0;$ $y = l,\ H^0_\partial = H_{\partial 0}$.

As a result of this solution the following dependencies are proposed to determine the distribution of average velocity and head along the length of the drain – ray $l$ :

$$\overline{V}_{r_\partial} = A\left\{\left[\exp(a+b)\frac{y}{2}\right] - \exp\left[(b-a)\frac{y}{2}\right]\right\}, \tag{31}$$

$$H^0_\partial(y) = H_k - \left(\overline{\Phi}_\partial \frac{\omega A}{2k}\right)\left\{(a+b)\exp\left[(a+b)\frac{y}{2}\right] + (a-b)\exp\left[(b-a)\frac{y}{2}\right]\right\}; \tag{32}$$

$$A = \frac{2k\left(H_k - H_{\partial 0}\right)}{\overline{\Phi}_\partial \omega\left\{(a+b)\exp\left[(a+b)\frac{l}{2}\right] + (a-b)\exp\left[(b-a)\frac{l}{2}\right]\right\}} \tag{33}$$

$$b = \frac{\overline{V}_{r_\partial}}{\overline{\Phi}_\partial \omega}, \qquad \xi = \frac{\lambda}{2r_\partial}, \qquad a = \sqrt{b^2 + \xi b}.$$

If the drain is working in the submerged regime then the average velocity can be determined by the expression:

IOP Conf. Series: Materials Science and Engineering **907** (2020) 012061 doi:10.1088/1757-899X/907/1/012061

$$\overline{V}_{r_g} = \sqrt{4r_{\partial}g\left(H_k - H_{\partial 0}\right) \Big/ \lambda l \left(1 - \frac{1}{ch\left(ab/2\right)}\right)}. \tag{34}$$

Since equation (33) is transcendental the velocity $\overline{V}_{r_\partial}$ must be determined by selection that is to find such value at which equality (33) is satisfied.

The results of the calculation of the piezometric head $H$ along the length of the drainage pipe according to formulas (20) and (18) were compared with the corresponding data obtained numerically. Good agreement of the calculative results was obtained [7, 8].

## Conclusions

The analysis of the existing models and methods of calculation of underground water intakes and drainages shows a significant possible influence on the formation of filtration inflow the hydraulics of the flow inside these structures. It is shown that the special influence of interdrain hydraulics on the formation of filtration inflow can be observed in perforated drain-rays and therefore the influence of interdrain hydraulics should be taken into account when calculating the parameters of drainage and evaluation of their operation [9].

Based on the analysis of existing and the proposed models which take into account the joint influence of water movement in the outer filtration area and in the inner region inside the drain the methodic of calculating the parameters of drains-rays in particular their discharges was developed taking into account the possible loss of pressure due to additional hydraulic resistance arising when water is moving in perforated drains-rays. The results of investigations presented in given work are the part of the general methodic of calculations of the water inflow to ranney water intakes and drainages which includes the calculations on the base of method of flow resistances, numerical modeling of the plane and space problema of water inflow and evaluation of the influence of innerdrain hydraulics on the inflow character to them. Developed methodic allows sucssesfully to solve the complex filtration problems in the areas of influence of ranney water intakes and drainages and may be widely used in projects and building of these structures.

## References

[1] Olijnik O Ja Tugay A M and Olijnik J O 2013 Calculation of radial underflow intakes based on filtration resistance *Collection of scientific papers "Problems of water supply, sewerage and hydraulics"* (Kyiv, KNUBA) **22** p 49-56

[2] Telyma S V 2002 Analysis of methods of calculation of ranney drainages *Visnyk CHDTU* (Cherkasy) **1** p 90-96

[3] Telyma S V 2003 Research of efficiency of application of ranney drainages. *Scientific and technical collection"Municipal economy of the towns"* (Kharkiv, HDAKH) p121-126

[4] Shternlikht D V 1991 *Hydraulics Book 2* (Moscow, Energoatom) p 367

[5] Khublaryan M G 1974 On the joint solution of the problem inflow to he drain and the fluid flow inside it. *Collection of scientific papers "Improving the methods of hydrogeological and soil-reclamation studies of irrigated lands"* (Moscow, VNIIGiM) **2** p 81-95

[6] Kremez V S 1983 Joint solution the problem of groundwater filtration and water flow in a drain pipe *Hydraulica and hydrotekhnica* (Kyiv) vol **37** p 29-33

[7] Oliynyk O Ja and Polyakov V L 1987 *Drainage of Overmelloving lands* (Kyiv, Naukova dumka) p 280

[8] Oliynyk O Ja 1981 *Geohydrodynamics of drainage* (Kiev, Naukova dumka) p 284

[9] *Drainage with fiber filters to protect areas from flooding* 2000 (Kiev, IGM NASU) p 332

IOP Conf. Series: Materials Science and Engineering **907** (2020) 012062 doi:10.1088/1757-899X/907/1/012062

# A study of the alkali-silica reaction in recycled glass concrete

**V P Sopov[1], O I Korkh[1], M Yu Izbash[2]**

[1]Department of Physical Chemical Mechanics and Technologies of Building Materials and Products, Kharkiv National University of Civil Engineering and Architecture, Sumska str. 40, Kharkiv, 61000 Ukraine
[2]Department of Architectural Structures, Kharkiv National University of Civil Engineering and Architecture, Sumska str. 40, Kharkiv, 61000 Ukraine

vsopov@ukr.net

**Abstract.** Recently, it has been paid more and more attention to the concept of sustainable development. One of these technologies is the production of "green concrete". It can be created by replacing natural concrete aggregate with secondary one. Therefore, white glass cullet was used as a replacement for 30% of coarse and fine aggregates in this study to design "green concrete". It was designed four concrete mixes for the investigation. Measurement of the fresh concrete mixes' performance showed that the use of glass aggregates reduces its workability. But it can be increased by using chemical admixtures. The highest strength - 28.94% more than the control sample, was obtained by a sample that contained coarse of glass aggregate and had low water-cement ratio. Sample №3 had lower strength only for 1.62% than sample №1 among mixtures with normal W/c ratio. Исследование фазовых превращений показало

## 1. Introduction

The concept of sustainable development combines three main areas: economic, environmental and social, which are aimed at the unity of mankind with the environment [1]. One of the important aspects of sustainable development is the development of solid waste processing technologies. The construction industry has the greatest potential for this because of the high resource intensity of production. Therefore, the replacement of natural components with secondary resources in concrete technology is an urgent task. Such technologies are defined as "green" because they use secondary resources and the production process does not lead to environmental destruction [2].

One of the potential secondary resources for concrete production is cullet. It can be used not only as coarse or fine aggregate [3-5], but also as active mineral admixture [6, 7]. But there is a serious problem for the widespread use of glass in concrete is caused by the reaction of alkali in cement with silica from aggregates.

Today, there is a general definition for chemical reactions "silica acid - alkali" that is formulated the American Institute of Concrete as ASR (alkali–silica reaction) – development reactions of expanding between aggregate containing active silica and alkalis obtained from cement hydration products or from external sources (groundwater, alkaline solutions contained or used in the construction). Si – O – Si bonds in silica are destroying during the interaction of alkalis with the concrete aggregate under the action of $OH^-$ ions. As a result, an alkali silica gel with a loose structure is formed, which tends to absorb water from other parts of the concrete. This causes local internal expansion and tensile stresses within the concrete [8, 9]. Some forms of silica can react so slowly that the gel has enough time to expand into the pore network without causing an accumulation of internal stresses. Therefore, it is difficult to predict the potential reactivity of the aggregate. Moreover, the reactivity of various forms of silica in concrete aggregate depends on its physical properties, that is, on the degree of ordering of the silicate structure. Amorphous glass is quite reactive because of active ion-exchange surface. Therefore, glass usage in concrete mixes can be a potential threat to its durability [10, 11].

Microcracks can form in glass aggregates when they are crushed or when preparing a concrete mixture. This leads to an increase in the active specific surface of the aggregate. It was previously noted that an increase in the size of glass particles increases the risk of interaction between silica and alkalis due to an increase in the active surface [12, 13]. But the authors of [14, 15] found that ASR occurs only on the inner surface of cracks of glass particles (Figure 1).

a) b)

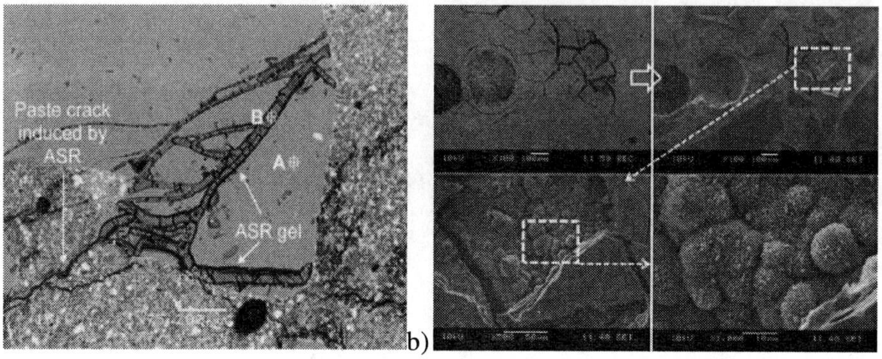


**Figure 1.** Gel formation during ASR passage in cracks on a glass surface a) [14] b) [15]

There is also an opinion [16] that the color of glass can affect its crack resistance. So, for example, transparent glass is more susceptible to microcracks. It was determined that brown glass particles did not have cracks even after crushing and could be considered as a non-reactive aggregate.

## 2 Experimental programs

### *2.1. Materials and mix proportions*

In this research, Portland cement CEM I manufactured by Heildeberg cement Ukraine that conformed to the requirements set out within EN 197-1 was used as a binder. There were two types of aggregate: natural and recycled. As a coarse aggregate was used granite crushed stone particle size 5–10 mm originated from the Kremenchug quarry and white cullet fraction 5–10 mm. As for fine aggregate, river sand obtained from Bezlyudovka quarry and the cullet fraction 0,16–2,5 mm was also used. The cullet was crushed and sieved. Production of concrete for this test was conducted at the concrete laboratory at Kharkiv National University of Civil Engineering and Architecture. The preparation of the concrete mixture was carried out in accordance with DSTU 2.7-214: 2009 (Ukrainian standard 2009). Three samples (1-3) with a water/cement ratio (W/C = 0.5) and one sample (4) with low W/C were made. Concrete mixture No. 1 was considered as a control sample. In sample 2, 30% of fine aggregate was replaced with a fine fraction of transparent cullet, in sample 4, 30% of coarse aggregate was replaced with the corresponding fraction of transparent cullet. Concrete mixture 3 contained 30% of coarse and fine glass particles. The compositions of concrete are shown in Table 1.

### *2.2. Specimen preparation and measurement*

Production of concrete samples was carried out by mixing for 5 minutes the corresponding mixtures, molding cubes with dimensions of 10 × 10 × 10 cm, followed by hardening in a chamber at t=18±2 °C and air humidity of 80–90%. The workability tests of the concrete mixture were carried out on the slump of a standard cone in accordance with EN 12350-2. The physical-mechanical properties of the samples were determined according to DSTU B V.2.7-214: 2009. The compressive strength was determined at the age of 1, 3, and 28 days. To determine hydrated neoplasms, samples 2 and 4 were investigated using differential thermal analysis DTA in the temperature range of 20 ... 1000 °C. DTA studies were carried out on the solution part of the samples after 1 month of hardening, taken from the surface and center of the sample.

**Table 1.** Composition of concrete mixtures (kg/m$^3$)

| Component | Number of compositions | | | |
|---|---|---|---|---|
| | 1 | 2 | 3 | 4 |
| Aggregate | 770 | 770 | 539 | 539 |
| Sand | 490 | 343 | 343 | 490 |
| Cement | 280 | 280 | 280 | 280 |
| Crushed glass (coarse) | - | - | 231 | 231 |
| Glass sand | - | 147 | 147 | - |
| Water | 140 | 140 | 140 | 90 |
| W/C ratio | 0,5 | 0,5 | 0,5 | 0,32 |

## 3. Results and discussion

During the study, the basic properties of concrete mixtures with glass aggregates were determined. Evaluation of the workability of the concrete mixture was carried out immediately after its preparation. A draft of 70 mm was obtained for sample No. 1. In concrete mixtures with coarse and fine glass aggregate, a decrease in cone precipitation in sample No. 2 to 40 mm, No. 3 to 20 mm was observed. This could be caused by an increase in the interparticle friction force due to the angular shape of the cullet particles. The compressive strength of concrete with glass aggregates was slightly lower than that of the control sample, but this difference is not significant (Figure 2).

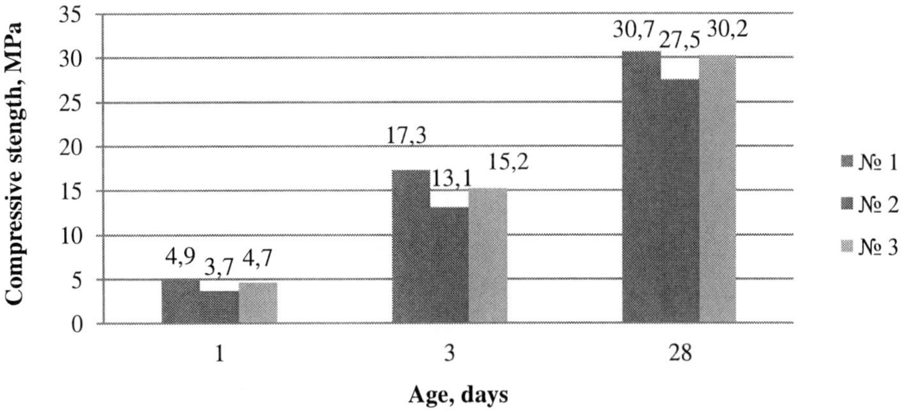


**Figure 2**. Compressive strength of concretes of composition 1-3

The most effective from the point of view of compressive strength was composition No. 3. The difference with the control sample was only 1.62%. A comparative assessment of the increase in the compressive strength of concrete of composition No. 4 and the control composition is shown in Figure 3. The compressive strength of concrete with glass aggregate is 28.9% higher than that of the control composition. This may be due to a lower water-cement ratio and, as a consequence, a decrease in the development of the ASR reaction.

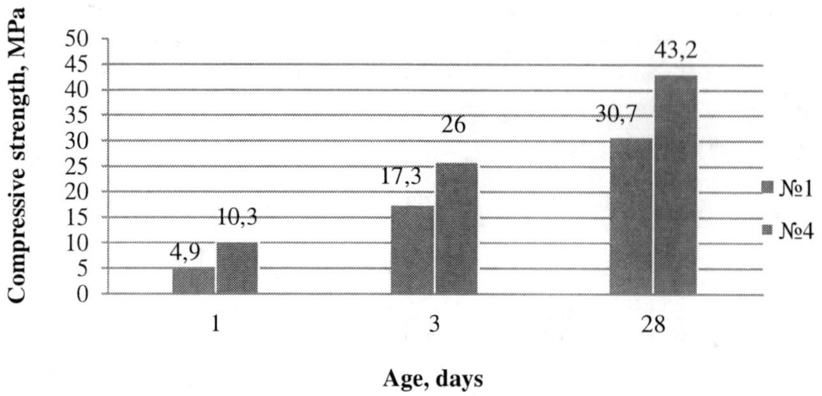


**Figure 3**. Compressive strength of concretes of composition 1, 4

The results of thermal analysis of samples No. 2 and No. 4 are shown in Fig. 4 and in tables 2-3. The endoeffect area in the range of 60 ... 215°C, characteristic for the hydrosilicates of the tobermorite group and ettringite, in sample No. 4 was SdTA = -206.75 square grams with weight loss m = 48.89 mg. In sample No. 2, the endoeffect area in the same region was SdTA = -84.56 square grams with weight loss m = 29.42 mg.

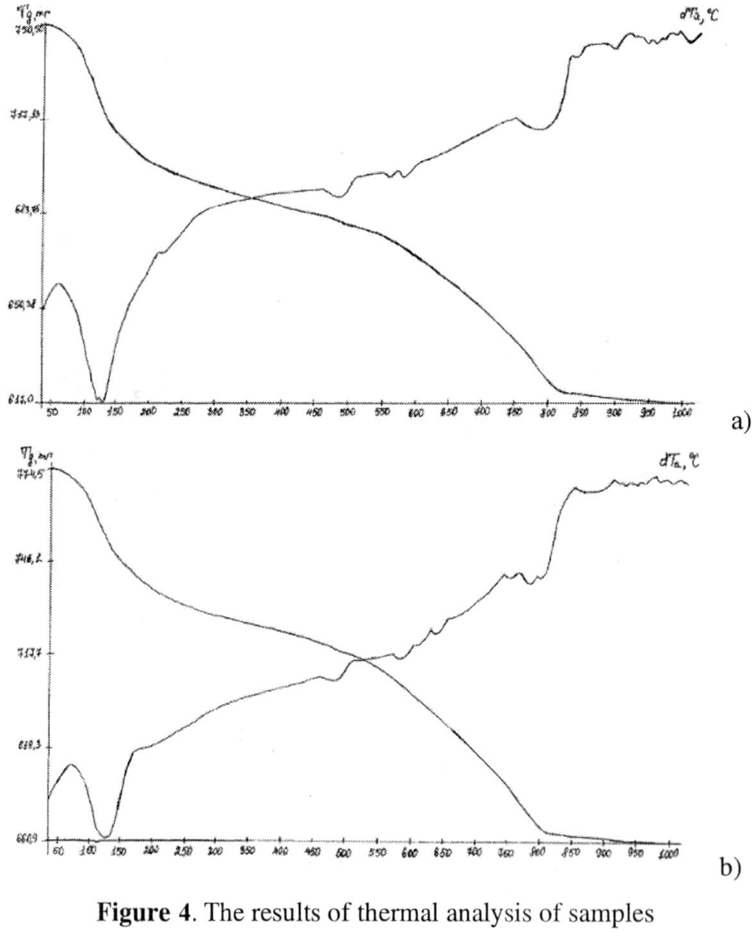


**Figure 4**. The results of thermal analysis of samples
a) No. 4; b) No. 2

**Table 2.** Weight loss.

| Sample № | Weight, mg | Mass losses which are corresponded to endo-effects, mg | | | |
|---|---|---|---|---|---|
| | | 60…215 °C | 460…510 °C | 570…600°C | 760…850 °C |
| 4 | 750,5 | 48,89 | 3,97 | 7,04 | 12,60 |
| 2 | 774,5 | 29,42 | 3,02 | 4,41 | 11,37 |

Thus, the amount of tobermorite gel in sample No. 4 exceeds its amount in sample No. 2 by 2.44 times, which may indicate a more complete passage of hydrolysis processes in cement. At the same time, the width of the endoeffect in sample No. 2 is much smaller than in sample No. 4, and its temperature range is shorter. This may indicate that the amount of ettringite in sample No. 4 is less than in sample No. 2.

The endoeffect area in the range 460 ... 510°C (crystalline form of portlandite) in sample No. 4 is SdTA = -9.54 square grams with mass loss m = 3.97 mg. In sample No. 2, SdTA = -5.8 square grams with weight loss m = 3.02 mg. Obviously, the amount of portlandite in sample No. 2 was 3.12 mg, which is 27% less than in sample No. 4. This may indicate a more complete binding of calcium hydroxide in sample No. 2, or a more complete passage of hydrolysis reactions in sample No. 4.

The endoeffect in the range of 570 ... 600 °C in both samples is probably due to the inversion conversion of β-quartz unstable at these temperatures to a more stable α-quartz, stable to temperatures of 870 ... 900 °C. In sample No. 4, the area of this effect is SdTA = -4.80 square grams with a mass loss of m = 7.04 mg.

In sample No. 2, the area of this effect is much smaller and amounts to SdTA = -1.79 square grams with weight loss m = 4.41 mg. This may indicate that in sample No. 2, quartz to a much lesser extent entered into a pozzolanic reaction with cement lime with the formation of calcium hydrosilicates.

In the range of 760 ... 850 °C, in both samples endo effects are evident, which are typical for medium and low basic hydrosilicates. The total area of these effects in sample No. 4 is SdTA=-11.84 square grams. with mass loss m = 12.13 + 0.47 = 12.60 mg. The total area of these effects in sample No. 2 is SdTA = -9.81 square grams. with weight loss m = 11.37 mg. This indicates that the amount of low-basic calcium hydrosilicates in sample No. 4 is greater than in sample No. 2.

It can be seen from the table 2 that in sample №4, the number of hydrosilicates of varying degrees of basicity significantly exceeds their number in sample №2. The amount of portlandite in sample №4 also exceeds its content in sample №2. This may indicate that the degree of cement hydration in sample №4 is higher than in sample №2, and the high content of hydrosilicates confirms that $Ca(OH)_2$ in sample №4 more intensely enters the pozzolanic reaction than in sample №2, which leads to the formation of hydrosilicates.

## 4. Conclusions

The results obtained in this experimental study demonstrate that glass aggregates can be used as a substitute for natural ones, but no more than 30% of their content. It is also recommended to use chemical additives to reduce water/cement ratio, which may influence on the development of destructive reactions. A longer experiment is supposed to be carried out to determine the development of the alkali-silica reaction and how it may effect on the strength of concrete in the long term.

Partial replacement of coarse and fine natural aggregates with cullet aggregates reduces the overall compressive strength. Partial replacement of only coarse aggregate while reducing the water-cement ratio leads to an increase in compressive strength of concrete and a more complete course of hydration processes.

It should be noted that the conclusions herein may be limited to the scope of the work.

## References

[1] Brodhag C, Taliere S 2006 Sustainable development strategies: Tools for policy coherence. *Natural Resources Forum* 136-145

[2] Obla K 2009 *The Ind. Conc. J.* (India) **24** 26-28
[3] Malik M, Bashir M, Ahmad S, Tariq T and Chowdhary U 2013 *J. of Eng.* **3** 8-13
[4] HweeTan K and Du H 2013 *Cem. and Conc. Comp.* **35** 109-17
[5] Bentchikou M, Bouriah S, Hamdane N, Debieb F and Boukendakdji O 2017 *Ener. Proc.* **139** 499–504
[6] Khan S U, Nuruddin M F, Ayub T and Shafiq N 2014 Effects of Different Mineral Admixtures on the Properties of Fresh Concrete *The Scientific World Journal* **4** doi: 10.1155/2014/986567
[7] Gautam S, Kumar A and Afaque M 2016 *Int. Res. J. of Eng. and Tech* **03** 500-503
[8] Giannini E R and Folliard K J 2013 A Rapid Test to Determine Alkali-Silica Reactivity of Aggregates Using Autoclaved Concrete Prisms (Portland Cement Association, Skokie, Illinois, USA) p 21 doi: 10.13140/RG.2.2.16188.74885
[9] Stark J and Wicht B 2013 *Dauerhaftigkeit von Beton* (Springer-Verlag, Berlin, Heidelberg) p 473 doi: 10.1007/978-3-642-35278-2
[10] Meyer C, Egosi N, and Andela C 2001 Concrete with waste glass as aggregate *Proc. of the Int. Sym. Concr. Tech. Unit of ASCE and University of Dundee* p 179-180
[11] Saand A *et al* 2017 Utilization of waste glass as partial replacement of fine aggregate in concrete *Engineering Science and Technology International Research Journal* **1** (1) p 28-32
[12] Park S, Lee B and Kim J 2004 Studies on mechanical properties of concrete containing waste glass aggregate *Cement and Concrete Research* 34 (**12**) 2181–2189
[13] Mavroulidou M, Candengue D and Ramachandran S 2011 Use of glass cullet aggregate in concrete with binders counteracting Alkali-Silica reaction *Proceedings of the 12th international conference on environmental science and technology* (Rhodes, Greece 8-10 September 2011) pp A-1202–A-1209
[14] Maraghechi H, Shafaatian S, Fischer G, and Rajabipour F 2012 How does fly ash mitigate alkali–silica reaction (ASR) in accelerated mortar bar test (ASTM C1567)? *Cem. and Conc. Comp.* **34** 41-47
[15] Giaccio G, Torrijos M C, Milanesi C *et al.* 2019 Alkali–silica reaction in plain and fibre concretes in field conditions *Mater Struct* **52** 31 https://doi.org/10.1617/s11527-019-1332-2
[16] Gorospe K, Booya E, Ghaednia H and Das S 2019 Effect of various glass aggregates on the shrinkage and expansion of cement mortar *Constr. Build. Mater.* **210** 301–311 https://doi.org/10.1016/j.conbuildmat.2019.03.192

**Acknowledgments**
The authors are grateful to prof. Usherov-Marshak A.V. for informative consultations on the subject of work, prof. Tolmachev S.N. for assistance in conducting experimental research on thermal analysis.

# Development of the tram network in the big city (on the example of Lviv)

**B S Posatskyi, T M Mazur, E I Korol**

Department of urban planning, Institute of Architecture and Design, National University of Lviv Polytechnic, 12 S. Bandera street, Lviv, Ukraine

nicmaztam@gmail.com

**Abstract.** The article presents an analysis of the development of the Lviv tram network from the end of the XIX $^{th}$ century till nowadays and for the near future. The stages of construction of tram depots and development of tram lines in the context of the city territory servicing, their changes, and connection with the configuration of the city street network are shown. City planning approaches to further development of the tram network infrastructure are presented.

## 1. Introduction

In the developed countries of the world, modern transport policy and practice of urban transport planning are based on the concept of development of the intermodal transport systems. The idea of intermodally involves the balanced and coordinated use of cars and various modes of public passenger transport. Public transport not only offers a wide range of ways to travel in the cities but also integrates well with developed pedestrian areas and gives the urban environment a humanistic look and scale [1].

Tram is one of the oldest types of public passenger transport in the city. From the late 1950s, the increasing level of motorization and the emergence of new types of passenger transport (trolleybus, bus, shuttle taxis) are gradually pushing the tram out of the city streets. However, due to its high transport capacity, the introduction of modern technologies in the production of tramcars, the improvement of their environment-friendly characteristics and, consequently, a more favorable impact on the urban environment, the tram transport has regained its popularity and undergoes a real renaissance over the last decades. Trams and light rail systems are rapidly returning to major European cities, where tram tracks have been mostly dismantled for the past 60 years. Today, rail transport remains the only way to preserve large cities with buildings of varying density, especially their central historic districts [2,3].

## 2. The tram network in Lviv: history and modernity

The tram network in Lviv has a long history of formation and development. The first trial rides on a horse-drawn tram in Lviv took place in 1879 [4]. The regular electric tram line on the main Railway Station – Stryiskyi Park, started in 1894. Lviv became the first in Galicia and the fourth in Austria-Hungary (after Vienna, Budapest, and Prague) to establish a regular tram route. The construction of tram tracks and the construction of a depot were the successive stages of the development of tram infrastructure in the city.

To serve the needs of the electric tram and the Regional General Exhibition in Stryiskyi Park at the beginning of the Vuletska Street (now - Academician Sakharov Street) the tram depot and the first city power station on the bank of the Mlynivka River was built, which was then flowing in an open channel.

The location for the power plant was chosen based on the needs for using water to cool the steam turbines and equidistant from the terminal stops of the first line of the electric tram (Main Railway Station – Stryiskyi Park). The 300 kW power plant, together with the first tram depot, was built by Siemens-Halske, a firm which was engaged in the construction of the tram lines in Lviv in 1894. This technical solution was very advanced for that time, as it provided a reliable supply of electricity for the tram needs [5]. The project of a complex of buildings of the tram depots and a power plant at the intersection of Pelchynska and Vuletska Streets (now Vitovskyi and Academician Sakharov Streets) was designed "in a brick style" by the famous Lviv architect - A. Kamenobrodskyi [6]. The first tram line was extended to the eastern border of the city already in the following 1895, the so-called Lychakiv Slingshot with a branch to the Lychakiv Cemetery; back then the length of the two lines was 8.3 km. In 1906 was built a new line on the Konovalets Street and thus appeared the first three radial directions of the Lviv tram lines.

**A.**

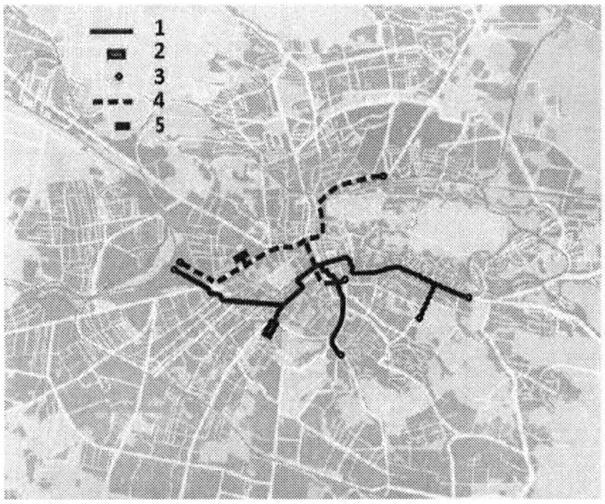


**B.**

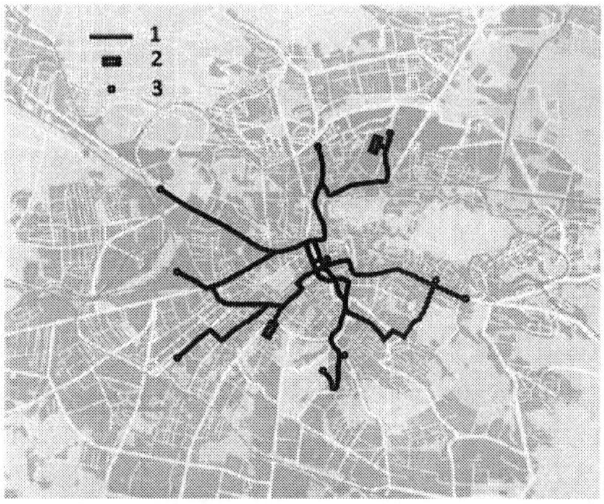


**Figure.1** Lviv. Electric tram network (according to D. Kovalev): **A -** 1895**; B** - 1909
1 - Electric tramline; 2 - depot of the electric tram; 3 - terminal tram stop: 4 - horse tramline; 5 - horse tram station

**A.**

**B.**

**Figure 2** Lviv. The tram depot at the intersection of D. Vitovskyi and Academician Sakharov Streets: **A** - 1906; **B** – 1927

For the new mean of transport - electric tram, the old depot was expanded, and new depots have been built. The new tramway hangar, built-in 1906 at the intersection of D. Vitovskyi and Academician Sakharov Streets, was larger than the first building (1894) and had a big production shop covered by metal trusses with a glass roof lamp. A particular example of industrial architecture in the form of

European Art Nouveau (secession) was the tram depot at Gabrielievka, built in 1908 on the Novoyi Rizni Street (now Promyslova Street). The architect Y. Bogutskyi embodied the structural basis of the building of new at that time material of reinforced concrete in the original flowing forms of the Secession style, popular during that time in Lviv. Exceptionally spectacular was the main facade of the depot house with decorative metal gates, completed with a large parabolic attic on the two pylons [6].

The new depot was intended to meet the needs of the tramway traffic of the well-developed tram network. As of 1909, radial directions were built on the Shevchenko Street (to the so-called Janivska Slingshot), Zamarstynivska, B. Khmelnytskyi, Promyslova. Track on Gorodotska Street connected S. Bandera and T. Shevchenko Streets, and the upper part of Lychakivska Street got a connection with Ivan Franko and Zelena Streets. The new tram tracks have been built on both sides of Svoboda Avenue and became ring junctions between Horodotska Street and the diameter West-East (Main Station - Lychakivska Street).

In 1926 - 1927 the need for the reconstruction of an old tram depot at the corner of D. Vitovskyi and Academician Sakharov Streets occured again - the building of one more industrial pavilion, administrative building, and the surrounding fence. The project was executed by A. Kamenobrodskyi in classical forms, thus achieving the stylistic similarity between the buildings of the late XIXth - early XXth centuries [6].

From the time of construction during the 1920s the tram depot at the corner of Vitovskyi and Academician Sakharov Streets was transformed into car repair shops. They were engaged in the repair of rolling stock by the end of the XXth century. In 1911 the tram track from the so-called Lychakiv Slingshot was brought to the Lychakiv Railway Station, thus creating the second transport hub in Lviv, which existed before the Second World War in 1939. Taking into account the tram connection to the Pidzamche railway station, it can be stated that the tram connected the three main Lviv railway stations (Main station, Pidzamche, Lychakiv) with the suburban transport network. Using modern terminology, already at the beginning of the XXth century three transport hubs have been formed in Lviv [7].

Further development of the tram network in Lviv took place under the newly established Polish state in 1920 - 1939. In 1925, a tram line was built from Kropyvnytskyi Square along Horodotska Street to the intersection with Kulparkivska Street on the western outskirts of the city. The tram track was also laid from the Torgova Square along the Shpytalna Street and P. Kulish Street with the turn-on Pid Dubom Street along the railway track to Zamarstynivska Street. A significant addition to the tram network in the city center occurred in 1928 when a two-lane line was built along Kyivska and Vitovskyi Streets between S. Bandera and Ivan Franko Streets. This line formed the southern part of the tram ring around the historic center, which is still heavily used today [7].

The growth of tram connections led to the further development of infrastructure, which is why a new tram depot (today №1) was built near the freight railway station on Horodotska Street, 183 in the 1930s [8]. The production part of the depot consisted of four parallel hangars, each with four tracks, which together made 16 tracks. The construction of the large building was made in a monolithic reinforced concrete frame, the cross spans of hangars covered parabolic reinforced concrete forms with roof lights to provide natural light. In addition to the premises for the maintenance of trams (a dispensary, which conducted daily inspections of trams and workshops for their overhaul), the depot also had two administrative buildings. The traction electrical substation №3, built in the early 1950s, was also located there [8].

Thus, as of 1939, the length of tram lines in Lviv was 32,812 km, with a total length of 60,872 km. During the years 1935-1939, two-lane tram lines with a length of 5,690 km (17.6%) were reconstructed by eliminating one-lane lines on the streets of 29 Lystopada (now E. Konovalets), V. Poninskyi (now Ivan Franko) and Kozelnytska [9]. In the late 1940s tram tracks have been located on both sides of Svoboda Avenue and their conjunction made it possible to turn the tram.

The tramway network began to change after the Second World War during the Soviet period of the city's development. In the early 1950s, streets began to be prepared for the introduction of trolleybuses and, accordingly, the tramway on Svoboda Avenue was closed. Instead, in 1951, was built a new two-lane line from the Torgova Square (behind the Opera House) through the squares of Yaroslav Osmoysl and Danylo Halytskyi, and then along Pidvalna Street to the junction with Ruska Street.

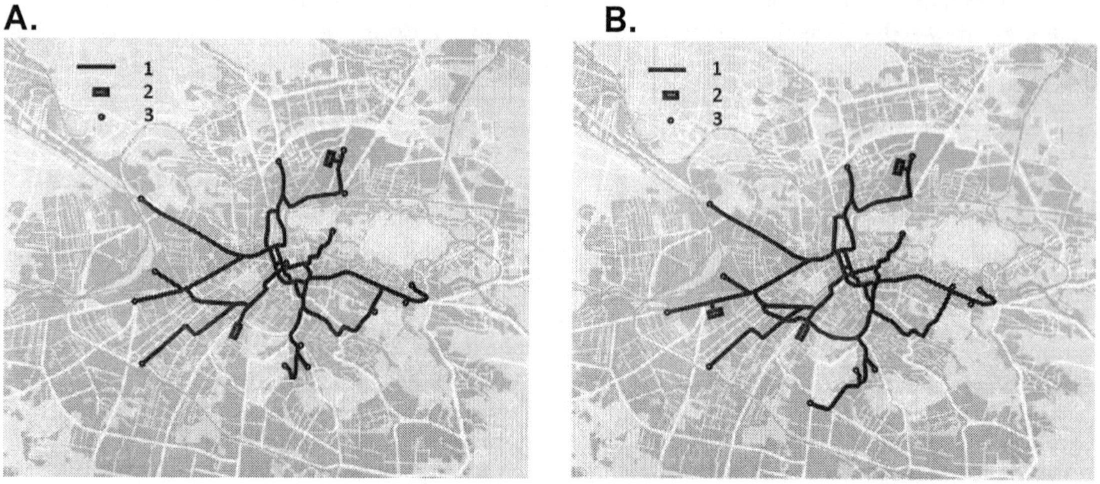


**Figure. 3.** Lviv. Tramlines network (by D. Kovalev): **A** - 1925; **B** - 1936
1 - electric tramline; 2 - depot of the electric tram; 3 - terminal tram stop

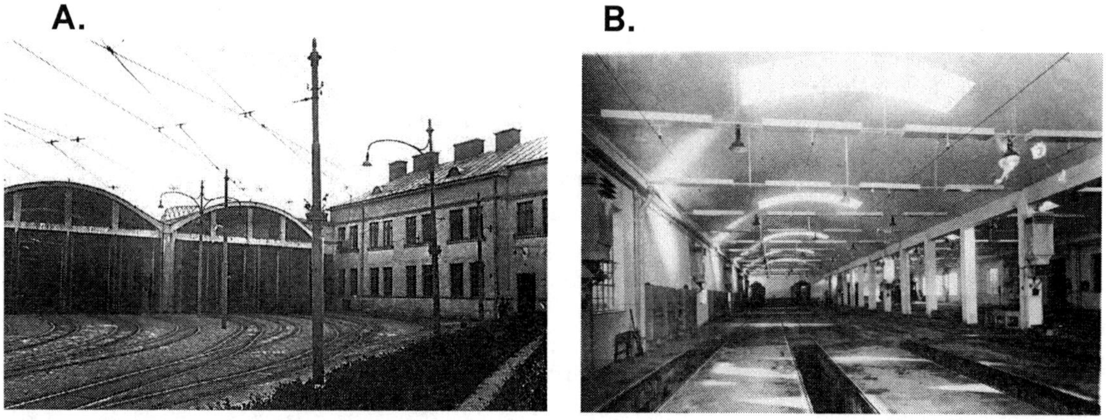


**Figure. 4**. Lviv. Tram Depot No. 1 on Horodotska Street: **A** - 1931; **B** – 2010

The new line, laid in the northeastern part of the boulevards of the late XIXth century, covered the medieval historic center of Lviv with a ring and connected four main radial directions: Horodotska, Bogdan Khmelnytskyi, Lychakivska, and Ivan Franko streets [7]. During the 1970s, tram lines in the center of Lviv continued to decline; the tracks that led to Vysokyi Zamok Park and the upper part of Stryiskyi Park by Ivan Franko Street have been dismantled. Today, from the perspective of the tourist traffic development, it can be argued that back then the city lost an attractive tourist object - a tram, which was going up the steep streets of M. Lysenko and M. Kryvonis to the plateau of the High Castle (ukr.: "Vysokyi Zamok"). Rlatively recent track dismantling took place in 2003 on the Halytska square and Knyaz Roman Street due to the reconstruction of the square for the installation of the monument to Danylo Halytskyi [7].

At the same time, in the 1970s have been developed projects that foresaw the development of the tram network in Lviv. Based on an elaborated scheme for the development of all types of public urban transport (developed by the Dipromisto Institute in 1975), the technical and economic foundation for the construction of an underground tram was developed. It was planned to create three tram diameters: from the south to the east (Knyagyni Olhy Street - Lychakivska Street), from the southeast to the north (residential district of Sykhiv - Zamarstynivska Street), from the southeast to the west (residential area of Sykhiv - residential areas Levandivka and Ryasne). Within the central part of Lviv, these diameters were outlined to be laid in tunnels, and at their intersections was planned the construction of underground stations in two levels with escalator crossings and underground lobbies [10].

The concept of an underground tram in Lviv began to be realized with the construction of a landline on a separate lane along the radial streets of Academician Sakharov and Knyahynya Olga to the ring street - Naukova. The portal of the underground entrance was to be located on the territory of the old depot at the intersection of the streets of D. Vitovskyi and Academician Sakharov. The two-lane line was commissioned in 1987 and significantly improved the city's southern sector transport connection with its historic center. This track was the first to go beyond the built-up during the pre-war period area of the city (until 1939). It is important to note that the population of the city from 1939 to 1989 increased from 320,000 to 786,000 inhabitants [11].

In the eastern part of Lviv in 1984 a two-lane line was built from the I. Mechnikov Street to the Palace of students on A. Vakhanianyn Street, also providing a convenient connection with the forest park "Pohulyanka" [10]. As of 2007, the total length of the tram lines in the city was already 76.43 km, including the passenger lines 67.3 km [12].

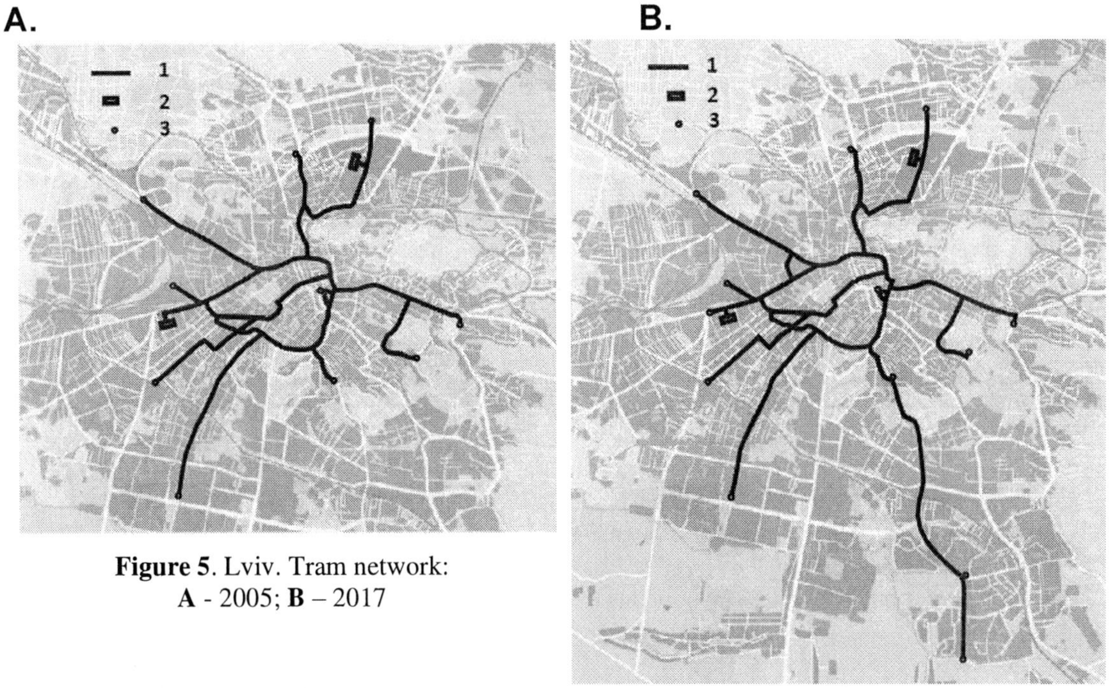


**Figure 5**. Lviv. Tram network:
**A** - 2005; **B** – 2017

Along with the development of the Lviv tramway network infrastructure, appeared the question of the reconstruction of the oldest depot at the intersection of D. Vitovskyi and Academician Sakharov Streets. In 1994, a decision was made to open a depot in one of the hangars of the museum, but even nowadays it is still not fully operational. The question of changing the function of the old tram depot has been repeatedly raised in professional circles and by the Lviv City Council [13]. In 2017, another decision was made to reconstruct it. For this purpose, the depot is leased for 50 years to set up a community center for the development of innovative technologies, recreation, and cultural events [14].

As mentioned above, as early as the middle of the 1970s, it was planned to build a tramline from the city center to the then-designed Sykhiv housing district of 120,000 inhabitants. It became possible to return to the implementation of this project only at the beginning of the XXIst century after the construction of the bridge over the tracks of the Persenkivka Railway Station, thus ensuring the unhindered transport connection between Sykhiv and central part of the city.

IOP Conf. Series: Materials Science and Engineering **907** (2020) 012063 doi:10.1088/1757-899X/907/1/012063

The Sykhiv tram track was commissioned in two stages. The first began on the V. Stus Street in the historical center of Lviv and ended in 2016 at the intersection of Chervona Kalyna Avenue with the Sykhivska Street. In the first year of operation, this route was used by almost 5 million passengers. In the second phase in 2017, the line was completed till the final planned stop on the V. Vernadsky Street. As of 2019, nearly 20 million people have benefited from this route, as two lines have been introduced: the first (№8) - Soborna Square - V. Vernadsky Str.; second (№4) - Main railway station - V. Vernadsky Str. [15]. Thus, during 2010 - 2018 the volume of passenger traffic for trams in Lviv increased from 49808 thousand people to 58808 thousand people (similarly for trolleybuses - from 25876 thousand people to 31220 thousand people) [16].

It should be mentioned that the development of the Lviv tramway network at the end of the XXth century and the beginning of the XXIst century still occurs by increasing the length of the radial directions in the absence of their annular connections in the new city districts, which covered the historic area of the city with the ring. Therefore, the proposal for the continuation of the Sykhiv tram line from V. Vernadsky to Naukova Streets with the connection to the bus station on Stryiska Street is being currently discussed. This line will connect the Sykhiv housing district with the residential area along Naukova Street between Stryiska and Knyahynya Olga Streets [17]. And the construction of tram tracks on V. Vernadsky Street will allow taking tram routes to the Arena Lviv stadium, which is now insufficiently connected by public transport.

## 3. Conclusions

At the present stage, the object of strategic urban planning in the field of land use, construction, and development of transport systems should not be a city within its administrative boundaries, but an agglomeration (metropolitan area), the boundaries of which are determined by the epure of regular passenger traffic.

The Lviv City Council plans a comprehensive approach to the development of the Lviv tramway network through its development in the residential areas of the 1960s and 1980s, where the number of inhabitants continues to grow due to the building densification and increase in the height of new residential buildings. It is planned to lay the tram tracks for the new routes, such as tracks on A. Lincoln Street (from Zamarstynivska St. to Promyslova Street), from Promyslova Street to B. Khmelnytskyi Street, on I. Mykolaychuk Street (from Promyslova Street to P. Orlyk Street), on Pasichna Street (from Lychakivska St. to Medova Pechera St.), on Lychakivska Street (to the town of Vynnyky). "Lvivelectrotrans" also plans to continue the route of tram №8, putting the track on Knyahynya Olga Street (from the Aquapark to Truskavetska Street, and Stryiska Street), and on V. Vernadsky Street (from Stryjska Str. to Chervona Kalyna Avenue) [18].

This approach will encompass both the planned changes in the administrative structure of Lviv for 2020 after the organization of the Lviv United Territorial Community and, accordingly, the prospects for a significant increase in the population and territory of the city.

## 4. References:


[1] Vuchic V R 2017 *Transportation for Livable Cities* (New York: Routledge) p 388

[2] *In praise of the tram: Britain's lost network and the future of transport Cities / The Guardian* 2018 URL: https://www.theguardian.com/cities/2018/aug/24/in-praise-of-the-tram-britains-lost-network-and-the-future-of-transport

[3] Kohl S 2015 *150 years of tram in Berlin* URL: https://t2m.org/newsletter/view-from-the-street/150-years-of-trams-in-berlin/

[4] Kotlobulatova I P 2009 *Dates and events in the history of Lviv* (Lviv: Avers) p 208

[5] Kryzhanivskyi A 2015 *History of electrification of Lviv region* (Lviv: Halytska vydavnycha spilka) p 360

[6] Biriulov Yu O 2008 *Arkhitektura Lvova: Chas i styli. XIII – XXI* st. (Lviv: Tsentr Yevropy) p 720

[7] Posatskyi B S, Mazur T M, Korol Ye I 2018 Tram network in the historical center of a large city (on the example of Lviv) *Mistobuduvannia ta terytorialne planuvannia* **67** (K.: KNUBA) pp 391– 397
[8] Liahushkin A, Yankivskyi D 2018 *Lviv tram depot. History and contemporary's* URL: https://photo-lviv.in.ua/lvivski-tramvajni-depo-istoriya-i-suchasnist/
[9] *Lwów 1934 – 1939. Development paths of the urban economy* 1939 (Lwów) p 80
[10] Trehubova T O, Mykh R M 1989 *Lviv. Architectural and historical essay* (Kiev: Budivelnyk) p 272
[11] Lozynskyi R 2005 *Ethnic warehouse of the population of Lviv* (Lviv: Vydav. Tsentr LNU imeni Ivana Franka) p 357
[12] Nazaruk M M 2008 *Lviv in the twentieth century: socio-ecological analysis* (Lviv: Ukr. akad. drukarstva, vyd. tsentr LNU im. Ivana Franka) p 348
[13] Mazur T M, Korol Ye I 2018 The problem of preservation and modern use of industrial culture objects (on the example of Lviv) *National scientific space: perspective, innovations, technologies: Proceedings of the V All-Ukrainian correspondence scientific-practical conference* (Kharkiv) pp.15-26
[14] Rievunova I 2017 Why invest $ 7 million in an old depot. What will be the innovative space of LEM Station in the Lviv tram depot URL: https://zaxid.net/navishhovkladati 7_mln_u_stare_depo_n1436739
[15] *Three years ago, a tram was launched in Lviv to Sykhiv* 2019 URL: http://tvoemisto.tv/news/try_roky_tomu_u_lvovi_zapustyly_tramvay_na_syhiv_video_104524.html
[16] *Transport, information society. The structure of the freight turnover* URL: http://www.lv.ukrstat.gov.ua/ukr/themes/09/theme_09.php?code=9
[17] *The petition to extend the tram from Sykhiv to Naukova received the required number of votes* 2018 URL: http://tvoemisto.tv/news/petytsiya_pro_prodovzhennya_tramvaya_z_syhova_do_naukovoi_nabrala_neobhidnu_kilkist_golosiv_97298.html
[18] Rodak K 2019 New tram and trolleybus lines are planned to be built in Lviv URL: https://zaxid.net/u_lvovi_planuyut_proklasti_novi_tramvayni_i_troleybusni_liniyi_n1474625

IOP Conf. Series: Materials Science and Engineering **907** (2020) 012064 doi:10.1088/1757-899X/907/1/012064

# Improvement of the safety of multi-floor housing

**T Tkachenko[1], V Mileikovskyi[2], V Dziubenko[1], and O. Tkachenko[3]**

[1]Department of Labour and Environmental Protection, Kyiv National University of Construction and Architecture, Ukraine
[2]Department of Heat and Gas Supply and Ventilation, Kyiv National University of Construction and Architecture, Ukraine
[3]Department of Architectural Design of Civil Buildings and Structures, Kyiv National University of Construction and Architecture, Ukraine

tkachenkoknuba@gmail.com

**Abstract.** To improve the safety in common areas of multi-storey buildings (lift halls, common corridors), it is proposed to renovate interiors of the areas. Such areas usually have no permanent ventilation. Thus, they can accumulate pathogenic microorganisms and viruses. The effective solution for reducing the danger is phytoncides plants. For successful phytodesign, an assortment of phytoncides plants of 11 species has been developed: *Aspidistra elatior, Aglaonema "Silver queen", Aglaonema "Maria", Chlorophytum comosum, Chlorophytum capense, Dracena marginata, Monstera deliciosa, Philodendron scandens, Sansevieria triaeffieria trichelifera, Zamiaculcas zamiifolia.* The offered assortment completely corresponds to climatic features of premises. For the normal growth and development of plants in the absence of natural light, three options for additional effective illumination are proposed.

## 1. Introduction

The main components of the sustainable development [1-2] concept are social, economic and environmental. The social component, among other things, is about improving people's health. The residential area of the modern city has individual and multi-apartment buildings. In the first kind of apartments, all rooms can be easily fitted for comfortable living of residents. They can set comfortable climate, illumination etc. and take care of their safety. It is possible to use energy efficient equipment for microclimate [3-4] and energy supply [5-6]. However, there are problems even in such kind of buildings, especially if builders and assembly workers made mistakes or dwellers of neighbouring houses violate building and sanitary standards. Examples are the supply of radon [7-11] and moisture [12-13], the growth of fungi and so on. All these factors contribute to the development of the sick building syndrome [14-19], especially if the proper air exchange is not provided on the premises to save energy.

In apartment houses, it is more difficult to create a safe environment due to dependence on collective systems, possible unauthorized interventions in the systems and violations of sanitary-hygienic requirements by neighbours. Public areas are especially dangerous: lobbies, lift halls, inter-corridors, stairwells etc. Together with outerwear, pets and during coughing or sneezing, microorganisms and viruses are constantly brought there. When used improperly (smoking, storage of garbage, old things, bicycles, sleighs, baby carriages etc.), harmful substances and microorganisms enter the area. It worsens people's health.

IOP Conf. Series: Materials Science and Engineering **907** (2020) 012064 doi:10.1088/1757-899X/907/1/012064

In modern high-rise building lift halls and inter-corridors are performed without natural lighting and ventilation, which leads to the absence of insulation and close to zero air exchange. This poses a direct threat to the uncontrolled growth of pathogens and the spread of viral infections, especially airborne (*Adenoviridae, Coronaviridae, Orthomyxoviridae*, *Picornaviridae*, *Paramyxoviridae* etc.). The growth of fungi leads to the release of strong carcinogens – mycotoxins. Chemical methods of disinfection incl. chlorine and spirits are harmful to humans causing intoxication and irritations. Lift halls are usually separated from other rooms. They are not a part of fire escape routes.

One of the effective methods of dealing with these threats is biological – phytodesign of premises. As early as the 1930s, Boris Tokin discovered phytoncides – volatile biologically active substances released by plants that kill or slow down the growth of bacteria, fungi, and protozoa [20]. Professor D. Verderevsky hypothesized about the effect of phytoncides on viruses. His follower N. Moldovan proved the influence of phytoncides against individual viruses. After that, the antiviral effect was confirmed by many scientists [21].

NASA scientists have proven that different types of houseplants can absorb harmful emissions. However, different species respond to the volatile organic matter in different ways. Some prefer to absorb formaldehyde from the atmosphere, while others absorb acids or toluene. NASA experts have calculated the total coefficient of effectiveness of air purification by plants. It was obtained taking into account the degree of danger of the absorbed gases, the width of their spectrum, and the rate of their absorption. This coefficient is expressed in units from 0 to 10 [22]. Based on the analysis, we propose using this biological method of phytodesign for public areas of apartment buildings.

**The purpose of the work** is the recommendations of a renovation of interiors for public areas of apartment buildings.

**Tasks of work**:

1. Selection of the optimal range of phytoncidal tropical and subtropical plants based on many years of author's research taking into account the climatic conditions of the experimental interiors;
2. Propositions for additional effective illumination to adapt the limiting abiotic factor of illumination in the public areas without natural illumination.

## 2. Object and research methods

This work is based on long-term studies of the ecology and volatile properties of houseplants [23-28]. Preferably, these are tropical and subtropical plants, which are divided into three groups:

1. By morphological and physiological characteristics;
2. By climatic factors;
3. By useful properties (Figure 1).

Combining knowledge of climatic features of the interior and plant classification is the key to successful and efficient phytodesign. The efficiency and success of phytodesign are determined by the life expectancy of plants in the interior while preserving decorative qualities and beneficial properties. This is especially important for the creation of disinfection interiors. The ability of phytoncide emission, harmful substances absorption from the air, as well as dust catching and maintenance of comfortable air humidity depends on the morphological and physiological properties of plants.

Plant phytoncidal properties were determined by the method of passive sedimentation using sanitary indicative bacteria [29]. Determination of the influence of phytoncides on moulds was determined by the method of L. A. Grape [30]. The common areas in apartment buildings were considered:

- lift lounges, 2×4 m on the floor plan;
- apartment areas, 2×7 m on the floor plan.

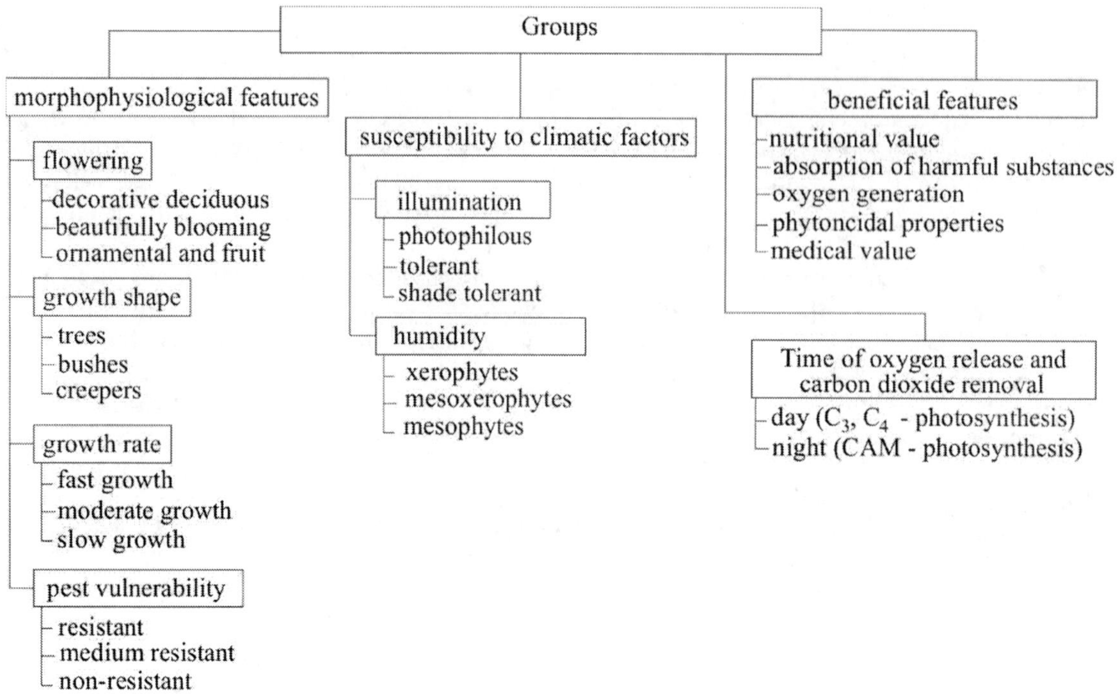


**Figure 1.** Classification of plants by useful properties.

There is no daylight, which promotes the development of pathogenic microflora. Therefore, it was decided to create or renovate interiors of these premises achieving the goals of healthy environment and effective growth of plants with minimum maintenance and energy consumption. For interiors of this type it is necessary to select plants that meet the following requirements:

- decorative-deciduous (have a significant amount of leaf mass, which can precipitate dust, release phytoncides and moisten the room air);
- trees and shrubs with a compact or ampelous shaped crown, which does not interfere with the movement of people and the movement of household and sports items: bicycles, baby carriages, sledges, scooters, etc.;
- moderately growing plants whose crown requires minimal care and pruning;
- resistant to diseases and pests;
- non-poisonous;
- shade durable (able to grow in the interior without daylight);
- xerophytic (resistant to dry air and substrate);
- phytoncydal;
- ability to absorb harmful substances from the air;
- unpretentious in care.

## 3. Research results

The author's research has found that for effective air purification of a room of 18 $m^2$, 5 - 7 plants with a height from 60 cm to 1.0 - 1.5 m are needed [24]. We have proposed the range of plants (Table 1), which consists of 11 species. For a room of 8 $m^2$ 2 - 3 plants are enough. For the convenience of moving people, we recommend using ampelous plants in wall pots: *Philodendron scandens* and *Chlorophytum*. For visual improvement, five staggered pots are enough (Figure 2).

For a room of 14 $m^2$ four plants are enough to place along the free wall. Here we recommend choosing small trees and tall bushes, for example, *Dracena marginata*, *Aspidistra elatior*, *Philodendron scandens* (on support), *Zamiaculcas zamiifolia* (Figure 3 and Figure 4). It is possible to create any composition of plants from the range offered to fit the liking of residents. The main thing is to place the plants so that they do not interfere with the movement of residents with any things.

**Table 1.** An assortment of phytoncidal plants for the creation of disinfection interiors for multi-apartment buildings

| Name of the plant | Form of growth | Height, m | Coefficient of the effectiveness of air purification [22] |
|---|---|---|---|
| | *Asparagaceae* family | | |
| *Aspidistra elatior* | bush | 0.7 | data is missing |
| *Dracena marginata* | small tree | 1.5-2.0 | 7.8 |
| *Chlorophytum comosum* | bush | 0.35 | 7.8 |
| *Chlorophytum capense* | bush | 0.35 | 7.8 |
| *Sansevieria trifasciata Laurentii* | bush | 0.60 | 6.8 |
| | *Araceae* family | | |
| *Aglaonema "Silver queen"* | bush | 0.50 | 6.8 |
| *Aglaonema "Maria"* | bush | 0.60 | 6.8 |
| *Monstera deliciosa* | liana | ≥ 0.6-2.0 | data is missing |
| *Philodendron scandens* | liana | ≥ 0,6-2,0 | 7,0 |
| *Zamiaculcas zamiifolia* | bush | 0,5-1,0 | data is missing |
| | *Araliaceae* family | | |
| *Schefflera "Nora"* | small tree | 2,0 | 8,0 |

Our offered assortment consists of shade-resistant plants. Nevertheless, no plant can live without light. In the absence of natural illumination, artificial lighting must be used, which satisfies the requirements of effective photosynthesis of plants, safety and human comfort.

For this purpose, the spectrum of chlorophyll absorption and human eye perception was analyzed, as well as the spectrum of UV radiation that is safe and dangerous for humans. Effective and safe illumination for plants can be provided by LED phytolamps. However, they do not meet the conditions of human comfort due to the lack of green light, which is most effectively felt by human eyes.

The photosynthesis surfaces of plants reflect the green radiation. This forms the green colour of leaves. Because of this, when the phytolamps are illuminated, the decorative qualities of the plants are reduced and leaves appear grey or black. There are three effective illumination options available from currently available sources at the market:

- a combination of general lighting with phytolamps and green LED lamps;
- combination of general illumination with white lamps and ultraviolet LED lamps;
- general white illumination and local illumination of the plant area by ultraviolet LED lamps or phytolamps.

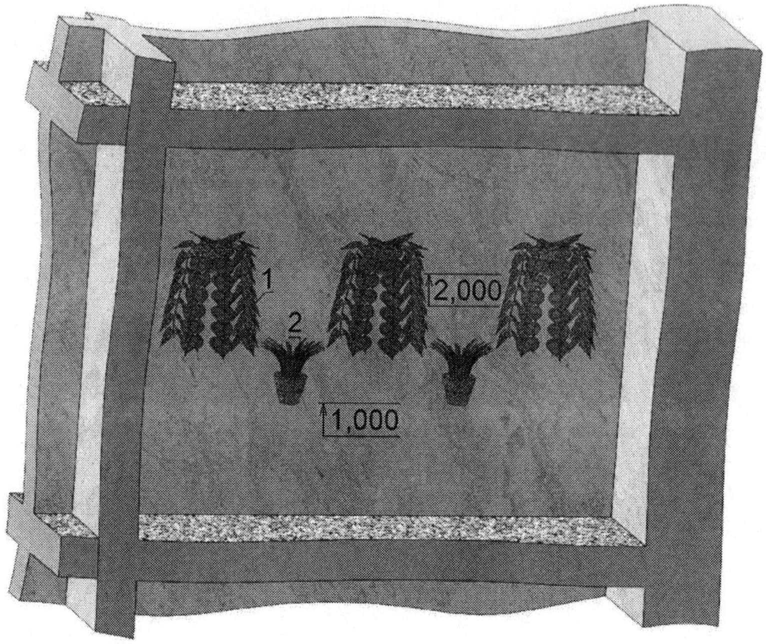


**Figure 2.** Example of phytodesign by five plants in wall pots:
1 - *Philodendron scandens*; 2 - *Chlorophytum.*

**Figure 3.** Example of phytodesign by four plants in wall pots:
1 – *Dracena marginata*; 2 – *Zamiaculcas zamiifolia*; 3 – *Aspidistra elatior*; 4 - *Philodendron scandens.*

IOP Conf. Series: Materials Science and Engineering **907** (2020) 012064 doi:10.1088/1757-899X/907/1/012064

**Figure 4.** Second example of phytodesign by four plants in wall pots:
1 – *Dracena marginata*; 2 – *Zamiaculcas zamiifolia*; 3 – *Aspidistra elatior*; 4 - *Philodendron scandens*.

To fit the natural biorhythm of the plants, phytolamps and ultraviolet lamps should be switched off at deep night.

## 4. Conclusions
For the first time, we have offered an assortment of phytoncidal shade-resistant plants for sanitary interiors in common areas of apartment buildings, consisting of 11 species: *Aspidistra elatior*, *Aglaonema "Silver queen"*, *Aglaonema "Maria"*, *Chlorophytum comosum*, *Chlorophytum capense*, *Dracena marginata*, *Monstera deliciosa*, *Philodendron scandens*, *Sansevieria trifasciata Laurentii*, *Schefflera "Nora"*, *Zamiaculcas zamiifolia*. Indoor plant placement options are suggested for the typical types of common areas. To preserve the decorative and physiological features of the plants, there are three options of effective artificial illuminations. Thanks to the beneficial properties of plants, the proper phytodesign can enhance the safety of multi-storey housing in a short time with minimal cost.

## 5. References
[1] Smith M J 2020 Journal of maps **16** i-iii
[2] Sadovenko A, Maslovska L, Sereda V, Tymochko T 2011 *Sustainable Development of Society* (Kyiv)
[3] Voznyak O, Korbut V, Davydenko B and Sukholova I 2019 Proc. of CEE 2019. *Advances in Resource-saving Technologies and Materials in Civil and Environmental Engineering* (Cham: Springer) pp 526-533
[4] Korbut V,Vozniak O, Sukholova I, Myroniuk K 2017 *SSP Journal of Civil Engineering* **12** pp 15-22
[5] Smirnov S V and Moiseenko V V 1995 *Applied Solar Energy* **31** pp 66-71
[6] Smirnov S V and Mojseenko V V 1993 *Geliotekhnika* **29** pp 40-43

[7] Lebed O O, Voloshkina O S, Myslinchuk V O, Shchuryk V O and Lysytsya A V 2019 *Ukrainian Journal of Ecology* **9** pp 552–560
[8] Sirocic A P, Stanko D, Sakac N, Dogancic D and Trojko T 2020 *Applied Sciences* **10** 2341
[9] Usikalu M R, Onumejor C A, Achuka J A, Akinpelu A, Omeje M and Adagunodo T A 2020 *Cogent Engineering* **7** 1759396
[10] Yarmoshenko I V, Malinovsky G P, Onishchenko A D and Vasilyev A V 2020 *Radiacionnaa Gigiena* **12** pp 56-65
[11] Narasimhamurthy K N, Ashok G V, Nagaiah N, Shiva Prasad N G and Prema A N 2020 *Radiation Protection and Environment* **43** pp 21-5
[12] Vornanen-Winqvist C, Jarvi K, Andersson M A, Duchaine C, Letourneau V, Kedves O, Kredics L, Mikkola R, Kurnitski J and Salonen H 2020 *Environment International* **141** 105781
[13] Lee J H, Yeo M S 2020 *Sustainability* **12** p 4033
[14] Tkachenko T and Mileikovskyi V 2019 *Procedia Environmental Science, Engineering and Management* **6** pp 405-411
[15] Quoc C H, Huong G V and Duc H N 2020 *International Journal of Environmental Research and Public Health* **17** 3635
[16] Tran V V, Park D and Lee Y-C 2020 *International Journal of Environmental Research and Public Health* **17** 2927
[17] Alomirah H F and Moda H M 2020 *International Journal of Environmental Research and Public Health* **17** 1972
[18] Nakayama Y, Nakaoka H, Suzuki N, Tsumura K, Hanazato M, Todaka E and Mori C 2019 *Environmental Health and Preventive Medicine* **24** 1-10
[19] Gladyszewska-Fiedoruk K 2019 *Environmental and Climate Technologies* **23** 1-8
[20] Tokin B P 1980 *Healing Plant Poisons. The Story of Phytoncides* (Leningrad: Leningrad University)
[21] Tulchinskaya V P and Yurgelaitis N G 1981 *Plants Against Microbes* (Kyiv: Urozhai)
[22] Van der Neer Y 2009 *All about indoor plants that clean the air* (Saint Petersburg: Crystal Book)
[23] Tkachenko T N 1998 *Proc. of the 3rd Int. Scientific Conf. "Industrial Botany. Status and Prospects for Development"* (Donetsk: Multipress) pp 21–23
[24] Tkachenko T N Gornitskaia I P and Varenko Yu S 1999 *Vestnik of Hygiene and Epidemiology* **3** 121-125
[25] Tkachenko T M 2006 *Zbirnyk naukovykh prac Donetskoho Derzhavnoho Universytetu Upravlinnia. Seriia "Derzh. Upravlinnia"* **69** 194–202
[26] Tkachenko T M and Marova S F 2007 *Innovative technologies and mechanisms of public administration at the regional level. Proceedings of the Scientific and Practical Conference, November 28 and December 21, 2006* (Kharkiv: Mahistr) pp 53–56
[27] Tkachenko T M, Marova S F and Kuskov A E 2007 *Ist All-Ukrainian Scientific and Practical Conf.: "The Genesis of Creativity in the Life of a Person", April 8–9, 2007* (Horlivka) pp 50–56
[28] Tkachenko T N 2008 *Proceeding of the DonNACEA*. **70** 107–110
[29] Isakov Yu F 1985 On the Unification of Microbiological (Bacteriological) Methods of Research Used in Clinical-Diagnostic and Educational-Preventive Institutions. *Order of the Ministry of Health of USSR 22 April 1985 No 535*
[30] Vinogradova L A 1985 *Methods for indicating the biocenosis of pathogenic and potentially pathogenic microorganisms in environmental objects* (Moscow: MNIIG) pp 42–53

IOP Conf. Series: Materials Science and Engineering **907** (2020) 012065 doi:10.1088/1757-899X/907/1/012065

# Typological and Functional Planning Features of Sports and Tourist Hotels

**D S Borodai, A S Borodai, S P Borodai, Ya O Borodai**

Department of Architecture and Engineering Survey, Sumy National Agrarian University, Ukraine

dimaboroday@gmail.com

**Abstract.** The study determines the relevance of breaking down sports and tourist hotels into categories, and provides the main factors that justify the feasibility of a typological classification. The paper presents the analysis of scientific works and publications on this issue. The authors have identified the following five hotel types: on-site hotels, resort hotels, hotels targeted at sports, ski hotels and hotels located on sports tourism routes. Moreover, they have determined the main features and differences of each hotel type. The paper focuses on accommodation conditions, urban planning, and architectural and space-planning characteristics of certain types of hotels. The authors give the examples of existing sports and tourist hotels with their unique sports focus, determine the bed capacity, functional and planning structure of all types of sports and tourist hotels, as well as the main functional assemblies, their filling, proportional spacing in the building, the relationship between premises. They have developed the functional and building stacking diagrams of each type of sports and tourist hotels.

## 1. Introduction

Hotels are an integral part of modern tourism infrastructure. The growth of a social role of tourism and the dynamics of its development across the globe have made it advisable to break down hotels into categories based on their purpose and functional profile. Sports and tourist hotels occupy a special place in the common typological classification of tourist hotels. The hotels of this type cater to the needs of people, whose main goals are rest, leisure and recreational activities, as well as health improvement.

The main factors influencing the development of infrastructure of sports and tourist hotels are: dynamic development, improvement resort and recreational areas, reshaping of views of a modern man regarding a healthy lifestyle, growth in the popularity of sport and its massive scale, increase in the investment potential and attractiveness of many sport kinds, transformation of sports into a highly profitable business. The above factors have become a prerequisite not only for the development of infrastructure of sports and tourist hotels, but also for their differentiation by purpose. This differentiation corresponds to the social needs of tourists in this direction.

Aspects related to the typological classification and functional planning features of sports and tourist hotels are the most pressing challenge and burning issue nowadays due to the constantly changing conditions of the provision of tourist services and requirements of the modern tourist market. The solution to this issue and implementation of the results in actual designing will satisfy the social needs of various groups of tourists and athletes in this direction, as well as address the issue of feasibility of operation of hotels of different types, including the standpoint of economic potential.

## 2. Review of recent research and publications

The works of such authors as Judy A. Ciguo, Cathy A. Enz, Josefa D. Martín-Santana, Eva Reinares-Lara, Pedro Reinares Lara, and Annabel J. Wharton, M. Mustapić, A. Vlahov, Yu. V. Lysenko, L. H. Lukianova, V. I. Tsybukh focus on the issues related to the architecture and typological classification of hotels and their differentiation by purpose.

In their paper "A Proposal for a New System of Classification for Hotels Based on Information Taken from a Multi-sponsor Loyalty Program" D. Martín-Santana, Eva Reinares-Lara, Pedro Reinares Lara propose an innovative classification of hotels based on guests' assessments of the attributes that influence the choice of a particular hotel, sociological research on the compliance of requirements of tourists with the services provided by hotels in Spain [1]. The article by M. Mustapić and A. Vlahov "Importance of Design in Architecture for Positioning Hotels on Tourism Market" substantiates the importance of design in the field of tourism, especially in terms of hotel facilities, where it has become one of the key factors influencing business performance [2]. The work of L. H. Lukianova and V. I. Tsybukh "Recreational Complexes" provides for the typology of recreational environment, the targeted arrangement of space of recreational complexes, taking into account various types and forms of recreational activities, and types of recreational complexes and their space-planning design, in accordance with the pre-assigned functional program – medical treatment, recreation, tourism, including tourist hotels [3].

In terms of classification and typology, special attention should be given to the studies of Yu. V. Lysenkaia "Analysis of the Typology of Hotel Complex Operators (the Case of Marriott International, Inc.)", "Study of the Hotel Classification System" and "Some Issues on the Hotel Classification System Formation", which provide for an in-depth analysis of the common system of classification of hotel facilities and propose a new system of hotel classification that meets the design requirements [4].

The above studies cover a wide range of problems in the field of architecture and typological classification of hotels, but the issues of differentiation, and architectural and space-planning design of sports and tourist hotels remain understudied.

## 3. Research results

The functional and planning arrangement of hotels depends directly on their purpose and functional profile. The active dynamics of the hotel and tourist business development, expansion of a range of hotel services, improvement in the level of comfort and service have resulted in advisability of breaking down hotels by categories based on their purpose The purpose and functional profile of a hotel determine its place in the urban planning structure, capacity, comfort level, functional relationship between rooms, their composition, and the structure of hotel room capacity.

During the study of the functional planning and typological features of sports and tourist hotels, it is essential as a matter of priority to determine their place in the common typological classification of hotel facilities. There are two main groups of hotels: hotels intended for designated purpose (or target hotels) and transit hotels. Target hotels include business-type hotels (business hotels), hotels for recreation, leisure and rest ("vacation-interval" hotels), resort hotels (holiday centers) and other places of accommodation. Hotels for recreation, leisure and rest include the following types: tourist hotels, resort hotels, recreation centers, holiday centers. Tourist hotels are divided into sports and tourist, excursion and tourist and targeted tourist hotels (hotels, campsites, floating hotels, boatels). Transit hotels, in turn, are divided into hotels located on air, rail, water and road routes [5].

Sports and tourist hotels are the hotels designed to accommodate and serve athletes, sports teams and delegations, as well as tourists who prefer active recreation and leisure sports [6]. It is expedient to locate the hotels of this type as part of resorts, recreational and sports complexes and centers, near tourist facilities, in suburban recreational areas. The location of sports and tourist hotels depends on the sports infrastructure development in a particular region. An important factor is also the favorability of natural and climatic conditions for the development of a particular sport.

The main criteria that characterize the typological features of hotels are: room capacity, floor number, comfort level, functional purpose and profile, design solutions, and operation modes.

Such features as floor number, design solutions, and operation modes are common to all types of hotels, including those for sports and tourism purposes.

In order to make the typological classification of sports and tourist hotels, it is important to determine the criteria for room capacity and differentiation by functional features.

Room capacity of a hotel is characterized by the number of beds or permanent sleeping accommodation. The room capacity of sports and tourist hotels is calculated in accordance with the following parameters: the total number of population of a particular region, existing indicator of the number of hotel beds per 1,000 inhabitants, standardized capacity ratio, average coefficient of the quality of natural and landscape resources of a region, room capacity of sports facility, or indicator of intensity of sports and tourism route, number of facilities of this type scheduled in the region. To determine the room capacity of sports and tourist hotels, it is advisable to use the following formula:

$$M = \frac{(P+pv)(N-n)k}{1000e},$$

where $M$ – room capacity of a sports and tourist hotel; $P$ – total number of population of a particular region; $p$ – room capacity of a sports facility; one-time daily intensity of a sports and tourist route; $v$ – coefficient of average occupancy of a sports facility by tourists and athletes, who have arrived in a particular region; $N$ – standardized ratio of the number of hotel beds per 1,000 residents; $n$ – existing indicator of the number of hotel beds per 1,000 inhabitants of a particular region; $k$ – coefficient of the quality of natural and landscape resources of a particular region for certain sports functions; $e$ – number of sports and tourist hotels in a particular region [7].

The diverse specifics of the operation of sports and tourist hotels have made it expedient to break down them into categories based on functional features and profile. Thus, it is advisable to define five types of sports and tourist hotels: on-site sports and tourist hotels; sports and tourist hotels of a resort type (sports and tourist resort hotels); ski sports and tourist hotels; hotels with certain sports orientation; hotels located on sports tourism routes [6].

**On-site sports and tourist hotels** have general sports orientation, without certain sports specifics. Hotels of this type should be located near multi-purpose sports centers or large-scale sports facilities designed for a large number of spectators (stadiums, racetracks, athletics arenas, etc.). The optimal room capacity of such hotels is 250-1500 beds, depending on the local socio-economic, demographic factors and sports infrastructure development of a particular region. The functional and planning structure ensures the uniform allocation of all functional units and assemblies in the building: unit of reception and auxiliary facilities (8-10% of the total area), public catering unit with a restaurant and café (10% of the total area), unit of entertainment, and cultural and community service facilities (5-10% of the total area), unit of conference and meeting rooms (3-5 % of the total area), unit of sports and fitness areas includes a gym, pool, sauna, fitness room, medical service (10-15% of the total area), the administrative and housekeeping unit (5-7% of the total area). The room capacity of this hotel type provides for single and double standard rooms, suites, junior suites and apartments. The examples of this hotel type are: NH Barcelona Stadium Hotel (Barcelona, Spain), Hellas Hotel (the town of Yuzhnyi, Ukraine), Naftovyk Hotel (the town of Akhtyrka, Ukraine).

**Sports and tourist hotels of a resort type (sports and tourist resort hotels):** the optimal capacity of such hotels is 300-1000 beds. This type of hotels caters to a relatively long stay and rest in one place, and sometimes provides the possibility of preventive treatment and health improvement. It is expedient to locate such hotels in resort areas with favorable natural and climatic conditions to provide guests with the opportunity to participate in certain kinds of sports: on the sea coasts with a moderately warm climate in winter, in mountainous areas, near healing and mineral springs. The peculiarity of this hotel type is the accommodation of sports (in particular, football) teams during the winter off-season and the arrangement of training camps for athletes. Therefore, these hotels are independent of seasonal fluctuations: during the holiday season they render services to tourists, and during the off-season period

– to professional athletes. Thus, sports and tourist hotels of a resort type have some advantages in comparison with standard resort hotels from the standpoint of economic potential.

The area of sports and tourist hotels of a resort type is of an impressive size (up to 10 or 15 hectares). Therefore, these hotels have a wide range of sports fields and grounds. In percentage terms, the area of sports infrastructure of this hotel type is up to 50% of the total area and contains: outdoor swimming pools, standard-sized football pitches, golf courses, tennis courts, mini-football, volleyball, basketball, squash, water parks, etc. A characteristic feature of the first-class and top-class hotels located on the sea coasts is the availability of their own beach.

The functional and planning structure of sports and tourist resort hotels assumes a well-developed sports infrastructure both outside and inside. It is advisable for hotels of this type to arrange a great number of rooms for treatment and health improvement: SPA centers, massage rooms, catering facilities offering diet food. During the holiday season, the functional and planning structure of these hotels provides tourists with a wide range of entertainment and cultural facilities (halls for recreation, leisure activities and rest, billiard rooms, night clubs, children's play rooms, conference and meeting rooms, multi-purpose halls). The functional units are allocated (in the total hotel area) as follows: reception and auxiliary premises – 5%, catering – 10%, sports and recreation – 10-15%, entertainment – 5-10%, entertainment, and cultural and community service – 5-10%; administrative and housekeeping –5%, accommodation - 45-50%.

The functional and planning arrangement of sports and tourist resort hotel units have the following features:

- availability of two or more restaurants and cafes (self-service buffets, waiter service, national cuisine) with a capacity of 90-100% of the total room capacity;
- availability of lobby area with a bar in the unit of reception and auxiliary premises;
- availability of several conference or meeting rooms of different capacity (10-500 persons), tour and travel desk or office (18-20 $m^2$), currency exchange office (8-10 $m^2$) or bank branch or outlet (20-25 $m^2$);
- availability of indoor swimming pools (at least 17×8 m in size), fitness centers and spa centers with Finnish saunas, Russian saunas, massage and relaxation rooms;
- possibility of arranging separate cottages with rooms with modern amenities in the hotel area.

Sports and tourist resort hotels have standard double and single rooms. They may provide family-type rooms or additional third beds in the room. The number of suites and apartments is minimal, and depends on the hotel category. There are a lot of sports and tourist hotels of a resort type in countries with the well-developed resort infrastructure and a mild favorable climate in winter (Turkey, Israel, Cyprus, Egypt). The examples of such hotels are: Adora Golf Resort Hotel (Belek, Turkey), Concorde El Salam Sport Hotel (Sharm El-Sheikh, Egypt), Sport 4 FC Hotel (Eilat, Israel).

Ski sports and tourist hotels designed to accommodate and serve tourists and athletes-skiers. In recent years, skiing both at the professional and amateur level has gained a high popularity across the globe. This direction is currently being developed, and occupies a leading position in the field of winter tourism. The above factors have led to the creation of a network of ski sports and tourist hotels. In summer, these hotels are used by tourists for mountain hiking and sightseeing activities. Ski sports and tourist hotels include ski tourist complexes, and are located in close proximity to areas with the developed ski infrastructure: near lifts and ski trails. Their capacity is 100-600 beds. As a whole, the room capacity of this hotel type depends on the development of ski complex infrastructure, its scale, lift number, as well as the demographic and socio-economic factors and capabilities of the region. The functional structure of this hotel type includes public catering facilities (restaurant or cafe), sports and health facilities (saunas, gym), and a range of cultural and consumer services. Ski sports and tourist hotels provide ski and snowboard rental and other sports equipment necessary for skiing. The functional units are allocated in this way: unit of reception and auxiliary rooms – 6%, unit of public catering – 8-10%; unit of cultural and consumer services – 5%; unit of sports, recreation and entertainment facilities – 6-8%; administrative unit – 3-5%; unit of housekeeping and technical premises – 5%; accommodation area – 65-70%. Ski sports and tourist hotels rarely have conference or meeting rooms, since this type of hotel mainly caters to tourists who prefer skiing, and is not targeted at professional sports delegations and associations. It is advisable to provide stores to sell and maintain ski equipment, ski passes, premises of a ski school, a medical center (emergency room). The entertainment unit often includes the recreation and leisure facilities that reflect the local flavor (cafes and restaurants of national cuisine, specific baths, vats, etc.). The room capacity of this hotel type includes 1, 2, 3-bed standard rooms, as well as junior suites and suites. It is reasonable to arrange reserve rooms-dormitories for 6-10 beds in ski sports and tourist hotels to accommodate youth groups. But it is mandatory to arrange separate premises or storerooms for storing and drying ski equipment. The network of ski sports and tourist hotels is most widespread in European countries with mountain terrain, where skiing and mountain tourism is traditionally developed (Italy, Austria, Switzerland, France, Andori). In recent years, the infrastructure of ski sports and the construction of ski sports and tourist hotels have been developing dynamically in the countries of Eastern Europe (Bulgaria, Romania, Slovakia, Poland, Ukraine). The examples of this type of hotel are: Ruka Hovi Hotel (Ruka, Finland), Ice Angels Hotel (Borovets, Bulgaria), Bukovel Hotel Complex (Polianitsa, Ukraine).

Hotels with certain or specific sports orientation are located in the area of sport centers that focus on certain types of sports – tennis, golf, biathlon, football, etc. These hotels are also part of specialized sports complexes or a sports base, or camp. Hotels with specific sports orientation render services to professional athletes and athletes who prefer specific kinds of sports. The capacity of this hotel type is 50-300 beds. Hotels located in the area of sports and training camps accommodate athletes and representatives of sports teams. Specifics of various sport kinds affect the functional and planning structure and features of the sports infrastructure development in the hotel. Therefore, special attention is given to the arrangement of sports and recreation facilities. In the structure of functional zoning, sports grounds and facilities occupy up to 70 % of the total area, since they are the main focus of the sports complex and the corresponding sports and tourist hotel. Accordingly, if the hotel is targeted at football players, its sports infrastructure includes training football pitches, playgrounds for mini-football, and it is possible to build a stadium for arranging practice games. If the hotel's focus is ski-biathlon, there is some potential for this hotel to develop the infrastructure of ski and lick-roller trails. The units of the functional and planning structure of this hotel type are allocated as follows: public catering unit (restaurant or cafe, canteen that provides special food to athletes with a capacity of 60-75% of the total area with a full service cycle); unit of sports and recreation facilities (specialized sports hall of up to 200 $m^2$, specialized gym with the total area of 64 $m^2$, special sports equipment rental desk, coaching rooms, rooms for methodical classes, massage rooms, sauna, medical center, rest rooms, etc.); unit of cultural and consumer services (conference hall with a capacity of up to 200 seats, press center, shops selling special sports equipment, hairdressing salon); accommodation unit (80% of the total area – double standard rooms). A special feature of this hotel type is a catering unit that meets special dietary needs of athletes and gives a sports team chef the opportunity to cook in the kitchen. It is advisable to provide a small percentage of single rooms and suites for representatives of the management and coaching staff. The examples of sports and tourist hotels with specific sports orientation are: Superior Golf Hotel (Kharkiv, Ukraine), Biathlon Hotel (Ufa, Russia).

Hotels located on sports tourism routes are in the midway points of routes – the point of departure (initial), the point of destination and intermediate point. The most popular routes are: cycling, hiking, water, and mountain. Tourist clubs and organizations arrange such routes taking into consideration the specifics of local natural and climatic conditions (the availability of water bodies, mountains, interesting tourist attractions and landscape formations). It is advisable to locate hotels in the points of departure and destination, and to create a network of tourist bases, campsites, tent sites in the intermediate points. Since the main goals of certain tourist categories (cyclists engaged in water sport) are to have a rest, relax and experience nature, and therefore, their stay at hotels is short-term, and there is no need to create a network of hotels with modern amenities and a wide range of services. It is expedient to design budget or economy class hotels located on sports tourism routes which are of a budget class and economy class (1, 2-star), with a capacity of 30-100 beds (depending on intensity of a certain tourism route). The functional and planning arrangement of hotels located on sports tourism routes represents: entrance unit

IOP Conf. Series: Materials Science and Engineering **907** (2020) 012065 doi:10.1088/1757-899X/907/1/012065

with a lobby and reception desk (total area of up to 100 m$^2$ with maximum capacity); unit of administrative premises with offices of the director, reception officer and accounting space (area of up to 80 m$^2$ with maximum capacity); catering unit with a cafe or self-service buffet and incomplete production cycle (capacity – 50-60% of total area); unit of sports and recreation facilities with the sports equipment rental and repair office (total area of up to 25-35 m$^2$), rooms for instructors (total area of up to 8-12 m$^2$), medical center (t total area of up to 12 m$^2$); production and economic premises; accommodation area. It is possible to arrange a recreation room and sauna. Thus, the main features of the functional and planning structure of this hotel type are catering facilities with a limited range of services, as well as sports equipment rental and repair offices.

Hotels of this type are situated on small land plots (up to 0.5-0.8 ha), with the allocation of entrance, recreation, parking and sports areas. Depending on the specifics of sports tourism routes, the hotel area may provide berths for boats, kayaks, parking for bicycles, etc. The room capacity of this hotel type includes 2-4-bed rooms of 2 and 3 categories. The examples of hotels located on sports and tourism routes are Silmaci (Latvia), Bili Horvaty (Yaremcha, Ukraine).

The typological classification, and functional and planning features of sports and tourist hotels are shown in Table 1.

**Table 1.** Typological classification of sports and tourist hotels

| No | Sports and tourist hotel type | Optimal category | Optimal capacity | Functional structure | Examples |
|---|---|---|---|---|---|
| 1 | On-site | 3-4 stars | 250-1500 beds | 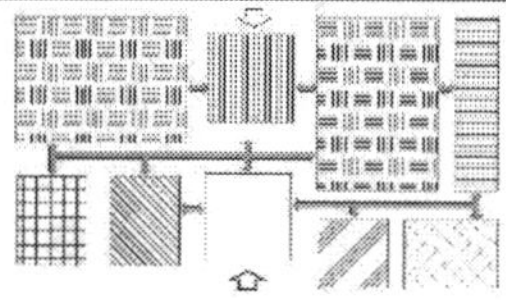 | 1. NH Barcelona Stadium Hotel (Spain) 2. Hellas Hotel (Ukraine) |
| 2 | Of a resort type | 4-5 stars | 300-1000 beds | 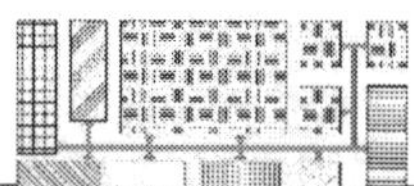 | 1. Adora Golf Resort (Turkey) 2. Sport 4 FC (Turkey) |
| 3 | Ski sports and tourist | 2-4 stars | 100-600 beds | 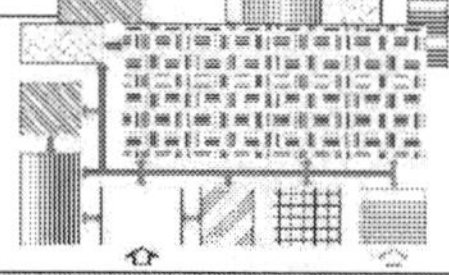 | 1.Ruka Hovi Hotel (Finland) 2. Ice Angels Hotel (Bulgaria) |
| 4 | With a certain sports orientation | 2-3 stars | 50-300 beds | 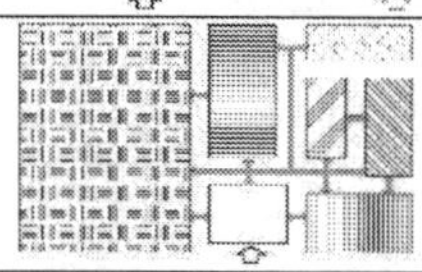 | 1. Superior Golf Hotel (Ukraine) 2. Biathlon Hotel (Russia) |
| 5 | Located on the routes of sports tourism | 1-2 stars | 30-100 beds | 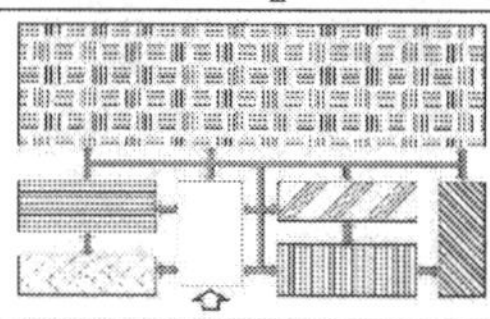 | 1. Silmaci (Latvia) 2. Bili Horvaty Hotel (Ukraine) |

Unit of reception and auxiliary rooms; Sports and recreation unit; Accommodation unit; Administrative unit; Unit of cultural and consumer services; Unit of household and technical premises; Public catering unit; Unit of entertainment facilities

## 4. Conclusions

Thus, this study has resulted in the identification of the socio-economic, urban planning and functional factors that determine the need to break down sports and tourist hotels into categories. The authors have defined five types of hotels, which correspond to different operation modes. These modes cater to the needs and interests of tourists and athletes according to the functional purpose: sports events, inter-season training camps for professional sports teams, active sports tourism, ski tourism, etc. The systematization of the typological classification of sports and tourist hotels makes it possible to define the main criteria for the hotel architectural and planning arrangement: capacity, comfort level, functional structure, room stock. As a result, this classification justifies the economic feasibility of operation of sports and tourist hotels.

## References

[1] Martín-Santana JD, Reinares-Lara E, Reinares-Lara P 2017 A proposal for a new system of classification for hotels based on information taken from a multi-sponsor loyalty program. *Cuadernos de Gestión* **17** (1) 15-36
[2] Mustapić M and Vlahov A. 2015 Importance of design in architecture for positioning hotels on tourism market. *Acta turistica*, **27** (2)
[3] Lukianova L.H., Tsybukh V.I. 2004 Recreation Complexes. K.: *Higher School*
[4] Lusenka Yu. V. 2007 *Methodology for Forming and Evaluating the Functional and Planning Organization of Hotel Complexes at the Initial Stages of Design* (Thesis for a Doctor's Degree in Architecture: (18.00. 01). KNUBA)
[5] Rohlev H.Y. 2003 Fundamentals of Hotel Management. K.: VIRA-R, AltPress. 2003.
[6] Borodai D. S. 2012 Functional and Typological Features of Sports and Tourist hotels. *Modern Problems of Architecture and Urban Planning*, **30** 249-54
[7] Borodai D.S. 2014 Method of Calculating the Capacity of Tourist Hotels. *Modern Problems of Architecture and Urban Planning*. Nov. **27** 261.

IOP Conf. Series: Materials Science and Engineering **907** (2020) 012066 doi:10.1088/1757-899X/907/1/012066

# Principles of reconstruction of multifunctional stadiums on the example of the Kiev stadium of CSC Armed Forces of Ukraine

**A N Chub[1], I I Kreizer[2]**

[1]Architecturally building workshop Limited Liability Company "Institute Kharkivproekt", Nauki avenue, 38, of. 300, Kharkiv, Ukraine
[2]Department of the Fundamentals of Architecture, Kharkiv National University of Civil Engineering and Architecture, str. Sumska, 40, Kharkiv, Ukraine

irinakreizer57@gmail.com

**Abstract.** The most common to date are multifunctional sports complexes. Their multifunctionality, in addition to competing in various sports, provides the opportunity for a variety of grand events, including corporate brand companies, concerts of the most popular music groups and performers, as well as meetings and conferences of religious denominations. When designing and reconstruction stadiums, it is important to anticipate activities and lay out ideas that may be implemented in the future. Although multifunctionality is ensured by the ability to organize and conduct various public events, the stadium must retain a number of features that will help it to fulfill its essential function of generating appropriate revenue. For modern sports facilities it is necessary to perform two main functions - the city-forming and the function of the development of the multifunctional urban environment. Today, work is underway on the reconstruction of the Kiev stadium of the CSC AF of Ukraine, located on Povitroflotsky Prospekt. The reconstructed site is located in the city's central planning zone, in the northern part of Solomenskiy administrative district on Povitroflotsky prospect, 6. The capacity of the facility is 20,000 seats. The updated CSK AFU Stadium will be a football and athletic arena. The project meets modern European and world standards. This stadium, located in the center of the capital, together with a spare (warm-up) and separate thrower stadium, as well as a special athletics arena, can become the main base for the preparation of the national athletics team of Ukraine, the athletes of Kiev, and the place of qualitative holding championships in athletics of Ukraine.

## 1. Introduction

The topic of new construction and reconstruction of existing multifunctional stadiums is still relevant. Trends in the development of the morphology of modern stadiums, the innovativeness of architectural and design solutions for sports facilities are being studied all over the world, the requirements for spectators' safety are being improved all the time, and the infrastructure of stadium visitors' service is being improved. Conducting world-class, continental or regional competitions puts forward their respective requirements for the organizers of these events and should, of course, be taken into account in the concepts of new stadium or reconstruction projects.

IOP Conf. Series: Materials Science and Engineering **907** (2020) 012066 doi:10.1088/1757-899X/907/1/012066

## 2. Review of current research

A large number of publications by both Ukrainian and foreign scientists are devoted to the problems of studying and researching the reconstruction of existing stadiums. The most relevant are the researches of scientists from Germany, such as: S. Klos, K. Güppert, T. Schober, M. King and others [1]-[ 4].

Much attention in the research of Ukrainian scientists is paid to the state of the national regulatory framework. Compared to foreign reference guidelines for the design of modern stadiums [5]-[7], Ukrainian building standards are far behind [8]-[9]. The research of reconstruction of existing Ukrainian stadiums was carried out by: V V Kutsevich, D F Goncharenko, G G Zubko, O V Starkov and others [10-13]. It should be noted that the latest modernization of the national regulatory framework was in the form of annex №1 from October 1, 2010 to the State Building Regulations "Sports and Physical Culture and Health Facilities". This made it possible to partially harmonize Ukrainian design requirements with the European stadiums, two of which were built and two were reconstructed to participate in the EURO 2012 Football Championship.

It should be noted that the aforementioned annex to the State Sports and Sports Center "Sports and fitness facilities" only partially satisfies the need to modernize the national regulatory framework for the design of sports facilities and only for football stadiums. Stadiums that specialize in other sports, such as athletics, require further changes to the state building codes mentioned.

## 3. Statement of the problem

Today there is a constant monitoring of the world experience of competitions. Accordingly, the regulatory framework of UEFA, FIFA, IAAF is changing. Ukrainian architects involved in the modernization of existing multifunctional stadiums need to address the issue of compliance of design solutions with new changes. It is necessary to take into account the condition of these stadiums; to predict their future existence; take into account the possibilities of holding international competitions in the future; as well as their existence after these competitions.

## 4. Principles of work

Based on the experience gained from participating in the preparation for EURO 2012, namely, the design and reconstruction of Metallist Stadium in Kharkiv (figure 1), one of the authors of the article developed and implemented the principles of reconstruction of multifunctional stadiums [14].

**Figure 1.** Reconstruction project for Metallist Stadium in Kharkiv.

In the process of working on the project for the reconstruction of Metallist Stadium, four principles were formulated:

1. The principle of developing a new quality of interaction of the existing stadium, which is reconstructed with the urban structure of the city.
2. The principle of preserving the historical and cultural value of the stadium or its place in the city structure.
3. The principle of compliance of the morphology and planning structure of an existing multifunctional stadium with the applicable standards and standards by which the stadium is to be certified.
4. The principle of forecasting the continuous sustainable development of the stadium in the future.

The developed principles were formulated on the basis of the methodology of decision making regarding the reconstruction of the existing stadium (Figure 2).

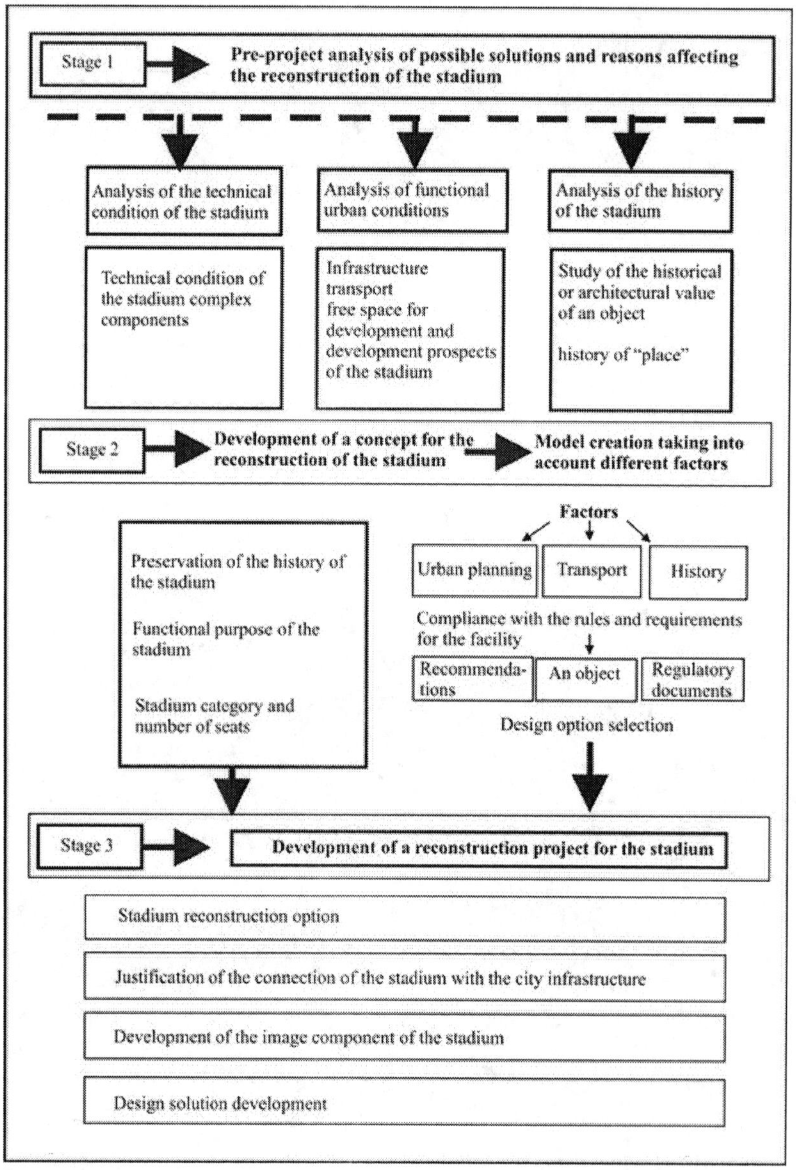


**Figure 2.** Methodology of the formation of a solution for the reconstruction of an existing stadium

## 5. Discussion of results

The gained experience and developed principles are implemented in the following works on reconstruction of multifunctional stadiums, namely, in the framework of cooperation between the Federation of Athletics of Ukraine and the Ministry of Defense of Ukraine on the construction of a new object of athletics infrastructure in the framework of work on the reconstruction of the Kiev stadium of CSC AFU prospectus. The reconstruction of the stadium of the Central sports club of the Armed Forces of Ukraine, Kyiv is relevant and urgent [15]. This object has a history of existence. The stadium, formerly called CSCA, was built in 1964-1965 (architects: MM Shmorgun, YK Belous, II Ponomaryov). The facility had a capacity of 12,000 seats. The arena hosted matches of the USSR Army Second League Championship football team. In the years of independence, the stadium began to accept members of the major league championships of Ukraine, in addition, the youth national teams of Ukraine played here. In 2004, the stadium was declared an emergency, and in 2009, CSCA football club stopped performing at the stadium.

The reconstructed site is located in the central planning zone of the city of Kyiv, in the northern part of the Solomenskiy administrative district on Povitroflotsky avenue (Figure 3). The site includes a number of existing sports facilities, namely: a stadium with a football field, treadmills, four triples and lighting masts (all in unsatisfactory condition), swimming pool, tennis courts, training football fields, a complex of sports halls for playing sports, outdoor fields for athletics, auxiliary technical buildings, administrative and sports unit.

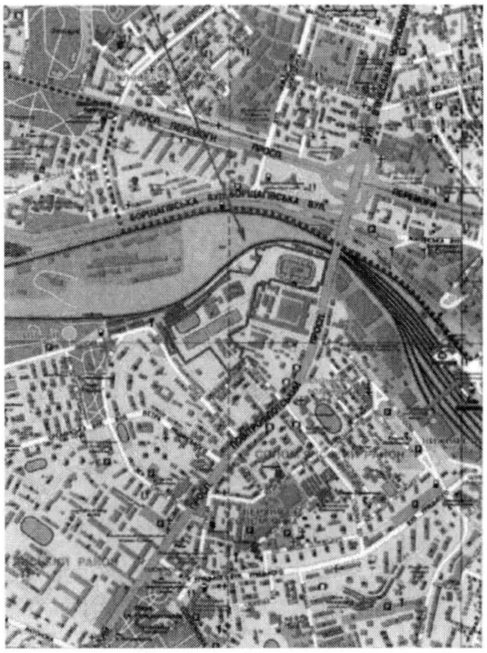

**Figure 3.** Layout scheme of a site in the city of Kiev.

According to the Planning Restrictions Scheme of the Master Plan of Kyiv until 2020, the site is located in an area where there are no enterprises with harmful emissions. The site does not fall within the restricted area. Landslides and groundwater flooding are not possible in this area. According to the scheme of greening and recreational areas of the Master Plan of Kyiv by 2020, part of the site (stadium) is assigned to sports parks. After reconstruction the complex will be provided for the possibility of holding competitions at continental, national and regional levels.

The project provides the following facilities and services:

- central sports arena with stands for 20,000 spectator seats with a complex of all necessary premises, namely: spectator service premises, locker rooms, VIP spectators complex, restaurants, conference rooms, competition service infrastructure, fitness center, media center, administrative sports club complex, two-level parking under the western and eastern stands, technical facilities required; which includes;

- space intended specifically for athletics competitions consists of the following flat structures;
- a standard 400-meter 8-meter track and field track meet the IAAF requirements for running straight (100m and 110m hurdles) and in a circle equipped with an obstacle course inside the oval.

A football field of 105.0 × 68.0 m inscribed in the running oval, which meets UEFA requirements: The 100 m and 110 m barrier lanes are located on two longitudinal sides.

Sectors inside the oval for track and field athletics for javelin throwing competitions (both ways), discus throwing, hammering, kernel bouncing and pole vaulting. Along the straight 100-meter paths from the north and south are platforms with divergent paths and sand pits for long jumps (each in two directions). The architectural solution received positive results from aerodynamic studies on the construction of the roof above the stands and relatively wind conditions in the field of athletics competitions, which are particularly sensitive to the direction and strength of the wind. The project takes into account the conditions for athletes with disabilities.

From the south, the sports arena is covered by the construction of a 130-meter special athletics arena.

The updated CSC AFU Stadium will be a football and athletic arena, and athletics will be its main sport. The project meets modern European and world standards. Only with such stadiums can we be optimistic about the future and become strong players in the European and World Athletics market in relation to competitions of the highest international level. This stadium, located in the center of the capital, together with a training jogging complex, a training complex for throwing a spear, a disk, a shot put and others, can become the main base of preparation of the national athletics team of Ukraine, athletes of Kiev, and a place of qualitative holding of championships of continental and national levels.

Separately, it is important to note that the location of the stadium with regard to the connection with the two airports of the city is very important for the international level competitions - with the Zhuliany Airport, and the airport of Boryspil - by the railway shuttle.

It is very important to increase the level of general funding for the training of high-class army athletes. It is necessary to update and reconstruct the unique bases of Olympic training, which are subordinated to the Defense Ministry. In addition to general and special training of servicemen of the Armed Forces of Ukraine in the program of purely military training, it is also necessary to rehabilitate employees and members of their families.

The stadium must be of the IAAF category, which will allow them to host level competitions for the European Championship and the European Cup inclusive. Also, let's not forget about the ambitious plans of the army to hold Invites Games in Ukraine - Igor Neskorynykh (Figure 4-5).

**Figure 4.** Concept of reconstruction of the stadium of the Central sports club of the Armed Forces of Ukraine.

**Figure 5.** Concept of reconstruction of the stadium of the Central sports club of the Armed Forces of Ukraine

The authors of the article express their gratitude to the staff of the Architectural Construction Workshop N2 of the LTD «Kharkivproekt Institute» for their help and provision of design and archival materials.

**References**

[1] Klos S 2012 Die Hüter der weiβen Elefanten *Bautechnik* **10** pp 719–24
[2] Güppert K 2012 Under large roofs – Interdisciplinary design of stadiums *Bautechnik* **10** pp 694- 700
[3] Schöne L and Arndt J 2012 Stadium in blue-construction of «Grand Stade du Havre» *Bautechnik* **10** pp 686-693
[4] King M and Mungall G 2012 Aquatics Centre, London 2012. Olimpic and Paralympics Games *Bautechnik* **10** pp 701-711
[5] UEFA EURO 2012tm Stadium Guide Revision 2 01/02/2010
[6] IAAF Track and Field Facilities Manual 2008 Edition
[7] *Guide to Safety at Sports Grounds* 2008 Fifth edition (United Kingdom)
[8] DBN B.2.2-13-2003 *Sports and fitness facilities* (Kyiv: State Committee of Ukraine for Construction and Architecture) as amended **1** from 10/1/2010
[9] DBN A.2.2-3: 2014 *State Building Standards of Ukraine. Composition and content of project documentation for construction*
[10] Kutsevich VV 2014 Architectural typology of public buildings and structures. Modern tendencies of development *Modern problems of architecture and urban planning Scientific and Technical Collection* (Kiev) **35** pp 376-84
[11] Goncharenko D, Evel S, Zubko E and Starkova O 2013 *Construction and reconstruction of stadiums* (Kharkov-Color) 352 p
[12] Chub A N 2017 Studying the Problem of Construction and Reconstruction of Stadiums in Specialized Literature *Scientific Works of NUAS* **III (66)** pp 97-102
[13] Chub A N and Kreizer I I 2019 *Plateforme scientifique europeenne* doi: 10.36074/20.09.2019 v2
[14] Chub A 2018 *Principles of reconstruction of multifunctional sports arena on the example of Metalist stadium* (Kharkiv: Kharkiv National University of Civil Engineering and Architecture) p 20
[15] *CSC ZSU Stadium: from the construction of the 1960s to a modern athletics complex* URL: http://uaf.org.ua/index.php?option=com_content&view=article&id=14850:stadion-tssk-zsu-vid-sporudi-1960-kh-rokiv-do-suchasnogo-legkoatletichnogo-kompleksu&catid=40:uanews

IOP Conf. Series: Materials Science and Engineering **907** (2020) 012067 doi:10.1088/1757-899X/907/1/012067

# Utilization of sorted secondary PET waste - raw materials in the context of sustainable development of the modern city

**M S Malovanyy, S D Synelnikov, O A Nagurskiy, K M Soloviy and I S Tymchuk**

Viacheslav Chornovil Institute of Sustainable Development, Lviv Polytechnic National University, S. Bandery Str., 12, Lviv, 79013, Ukraine

E-mail: myroslav.mal@gmail.com

**Abstract.** It is established that the lack of technologies for the complete utilization of secondary PET – raw material sorted from municipal solid waste - raw materials is a threat to the sustainable development of the modern city. One of the promising ways of PET disposal is to use it to encapsulate mineral fertilizers. Under the laboratory conditions, the technology of PET modification and capsule application to mineral fertilizer particles under fluidized bed conditions has been successfully tested. Conducted pilot tests of the release of nutrition elements from mineral fertilizers encapsulated with modified PET, confirmed the uniformity and quality of the coating. Pilot studies have confirmed the effectiveness of the new form of obtained fertilizers.

## 1. Introduction

Sustainable development of a modern city is impossible without successful functioning of the disposal and utilization systems of the population`s life activity products (sewage drains [1, 2], municipal solid waste [3, 4]). Recently it has become especially important for Ukraine to ensure that the municipal solid waste management system (MSW) is functioning properly. Underestimation of this aspect has often caused the destabilization of the environment, the emergence of environmental and social threats for basically all cities of Ukraine, and for some of them (Lviv, Kyiv, Ternopil) the problem has become especially acute.

The successful operation of a MSW management system is impossible without sorting separate types of waste at first, combined with the same technology of their utilization in separate containers. However, the higher MSW treatment culture, the more waste categories are sorted. But in addition to sorting, it is important to ensure that effective sorted waste utilization technologies are implemented. Traditionally known are technologies for recycling glass, metal, paper, textiles. Technologies for using the organic part of waste for biogas, composting or incineration are widely used. Problematic and not always solved (despite the seemingly simple implementation) is the problem of recycling plastic bottles made from polyethylene terephthalate (PET bottles).

Units for the primary processing of sorted PET bottles with obtaining granules or so-called «chips» are widely implemented. But this raw material does not always find a place to be used precisely because of the huge array of PET bottles that have immensely come into our lives. Even in a technically highly developed country like Japan, the recycling rate of plastic bottles is only 40%, which is not enough, especially given the huge mass of them.

Each year, the world produces around 485 billion PET products, of which 46% is for water packaging. In 2017, approximately 635 million tons of plastic waste have accumulated in the world, 79% of this waste is stored in landfills or in the natural environment, about half of which is from PET plastic products [5]. If such trends of production and waste management continue in the future, there will be about 12 billion tons of PET plastic in landfills and the environment by 2050. Annually, 25 million tons of PET waste are generated in the European Union, 26 million tons in the US, 8,8 million tons in China, and 9 million tons in Japan [6].

The urgent task is to find new technologies for the reuse of PET plastics, which together with existing technologies would help to minimize the problem. In 2018, the United Nations, in conjunction with the Ellen McArthur Foundation, announced a global commitment to the New Plastic Economy, with more than 290 signatories. This obligation encourages countries to use a cyclical economy model that closes the plastic production cycle and encourages innovation in reuse or recycling. One of these innovations, which opens large volumes of recycling of PET, is the use of it to encapsulate mineral fertilizers.

Mineral fertilizers applied into soils cannot be completely absorbed by plants, some of them get into the environment, polluting the atmosphere, groundwater and surface water. A new promising form of fertilizer is the prolonged-action fertilizer that is achieved by encapsulating traditional granular fertilizers with a water-permeable capsule. The application of the fertilizer shell (capsules) on the surface of the fertilizer`s granules reduces the process of transition of the nutrients into the soil environment, which helps to increase the coefficient of their absorption by plants. Accordingly, the required dose of mineral fertilizers, the frequency of their application, the loss of fertilizers not assimilated by plants to the environment (which leads to its contamination) etc. are reduced. Secondary PET raw material can be used as a capsule-forming material, provided that it is soluble in organic solvents, which plays a crucial role in the process of creating a film-forming composition and coating the mineral fertilizer granules. We investigated the possibility of modifying PET by implementing an alcoholysis reaction using diethylene glycol as a reagent. The result is a solubility of the modified PET in ethyl acetate, sufficient for the implementation of the process of capsule formation in the fluidized bed apparatus.

## 2. Physical model of encapsulation process

The encapsulation of granular fertilizers is the application on their surface of the polymer shell by the method of spraying a liquid film-forming agent onto a layer of particles. To ensure good coverage, the particles in the layer must be constantly stirred. In technologies of encapsulation of solid dispersed materials, as a rule, devices of three types are used, which are capable of providing the required mode of motion for particles: plate, drum and fluidized state [7]. During encapsulation there is a complex heat-mass transfer between the solid phase (particle surface), liquid (film-forming agent) and air. Figure 1 shows a diagram of such interaction on the example of a single particle.

The flow of heated air *I* contacts the solid part *III*, on the surface of which is a solution of the film-forming agent. The result is a process of heat transfer from the gas medium to the solid phase surface, which is characterized by the coefficient of heat transfer $\alpha$. Supplied thermal energy is spent on the evaporation of the solvent from the surface of the particle with an intensity characterized by the coefficient of mass transfer $\beta$.

For the experimental studies we used a cylindrical apparatus of periodic action. Such devices are effective in encapsulating materials that are prone to sticking. Moving the entire cross section of the apparatus at a constant speed of air promotes a more active behavior of the particles, preventing them from sticking [7]. The film-forming solution consisted of the following components: ethyl acetate 87% (w), modified PET 10% (w), hydrolysis lignin 3% (w). Taking into account the solvent content in the film-forming agent using the mathematical model [7], the values of mass transfer coefficients were determined, the values of which for the granular mineral fertilizers studied are given in Table 1.

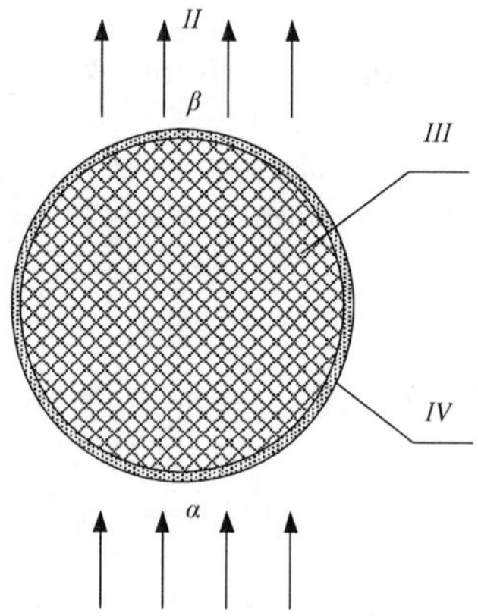


**Figure 1.** Scheme of heat-mass exchange in a system of solid body - liquid - gas during encapsulation: *I* - air flow that supplies heat, *II* - air flow that removes solvent vapors, *III* - solid particle, *IV* - film-forming solution

**Table 1.** The values of the coefficients of mass transfer $\beta$ during the encapsulation of granular mineral fertilizers

| Material | Consumption of film-forming agent, $10^4$ kg/s | Amount of evaporated solvent, $10^4$ kg/s | Mass delivery coefficient, m/s | |
|---|---|---|---|---|
| Ammonium nitrate | 1,0 | 0,87 | 0,246 | 0,251 |
| | 3,0 | 2,61 | 0,252 | |
| | 5,0 | 4,35 | 0,256 | |
| Nitroamophos | 1,0 | 0,87 | 0,195 | 0,198 |
| | 3,0 | 2,61 | 0,197 | |
| | 5,0 | 4,35 | 0,201 | |

Based on the obtained mass delivery coefficients, the maximum consumption of $P_{max}$ film-forming agent ($10^4$ kg/s · kg of fertilizers) was calculated: ammonium nitrate - 20,512, nitroamophos - 22,857. According to the obtained technological parameters, the encapsulation of granular fertilizers (ammonium nitrate and nitroamophos) was carried out in a cylindrical apparatus of periodic action. The coverage was 10% and 20% by weight of fertilizers, which corresponds to the following average film thicknesses on the particle surface ($10^5$ m): ammonium nitrate – 5,74 and 11,48; nitroamophos – 5,23 and 10,46. The properties of the obtained fertilizers were tested experimentally.

## 3. Testing of encapsulated fertilizers properties

The solubility of encapsulated fertilizers was investigated experimentally by a conductometric method according to the requirements of European standard EN 13266: 2001 [8]. The results of the studies in graphical form are shown in Fig. 2.

The results obtained (Fig. 2) indicate that the kinetic dissolution curves have a predicted character, the process runs smoothly without abrupt downturns or rises. This is proof of a uniform, high-quality coating that enables the production of long-acting mineral fertilizers with the required release time.

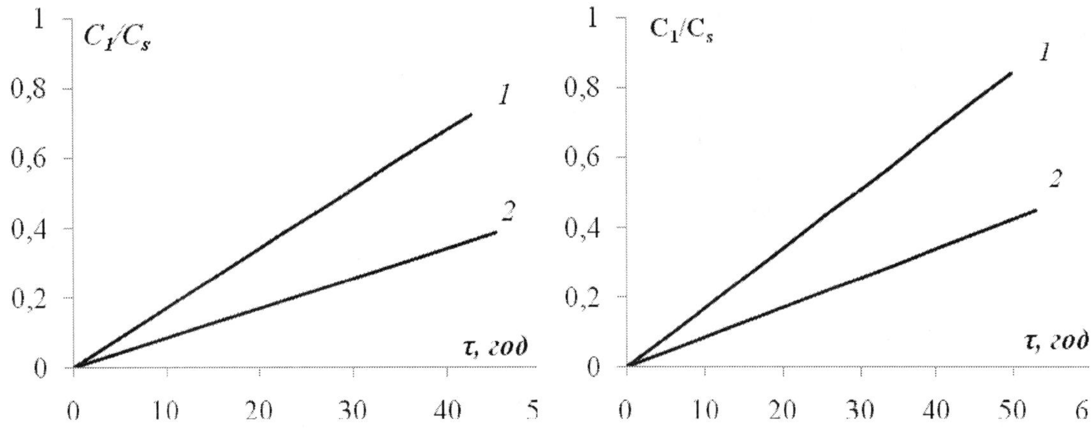


a) ammonium nitrate b) nitroamophos

**Figure 2.** Dissolution kinetics of capsule coated fertilizers based on modified PET of different thicknesses (% wt): 1 - 10, 2 - 20. $C_l$ - concentration of released component over a certain period of time, kg/m$^3$; $C_s$ - maximum concentration of released component, kg/m$^3$.

## 4. Pilot studies of encapsulated fertilizers.

Small-scale studies of the effectiveness of the use of mineral fertilizers encapsulated with modified PET were carried out using such crops as potatoes, soybeans, and maize [9]. The analysis of the results of field studies shows that in case of application of encapsulated mineral fertilizers in agroecosystems of crops such as potatoes, soybeans and maize, the rate of fertilizers nutrient elements absorption by plants is increased by an average of 4,5%. Atmospheric and hydrosphere losses in the case of encapsulated fertilizers are reduced compared to granular fertilizers from 47% to 74%. We found no dependence of nitrate content in potato tubers on the type of fertilizer applied. This indicator did not change significantly, it was much lower than the limit.

The use of prolonged-action encapsulated mineral fertilizers has shown significantly higher environmental and agronomic efficiency than granular fertilizers.

- in the case of potato cultivation, the environmental efficiency of the use of encapsulated PET mineral fertilizers is increased by 74% compared to granulated fertilizers. The application of 1 ton of active ingredient of fertilizers in a capsule form increases the yield of potatoes to 20,8 tons, whereas for granular fertilizers it is 11,5 tons.
- in the case of soybean cultivation, the environmental efficiency of the use of encapsulated PET mineral fertilizers is increased by 47,5% compared to granular fertilizers. The application of 1 ton of the active ingredient of fertilizers in a capsule form increases the yield of soybeans to 4,4 tons, while for granular fertilizers it is 1 tons.
- in the case of maize cultivation, the environmental efficiency of the use of encapsulated PET mineral fertilizers is increased by 47,5% compared to granulated fertilizers. The application of 1 ton of active ingredient of fertilizers in a capsule form increases the yield of soybeans to 7,5 tons whereas for granular fertilizers it is 4,5 tons.

The use of encapsulated fertilizers reduces the need for the active substance by 20%. The introduction of encapsulated fertilizers would reduce the production of these fertilizers by 20% and, accordingly, reduce the use of natural resources, reduce the scale of production and, accordingly, the scale of environmental pollution from these industries.

## 5. Conclusions.

It is established that the imperfect system of separate collection of solid waste is a significant threat to the sustainable development of the modern city, as well as the lack of technologies for full utilization of sorted waste. To a large extent, this applies to PET recyclables, with the lack of full recycling technologies experienced by all countries in the world. It has been established that modified secondary PET raw material can be a promising material for mineral fertilizer encapsulation. In this case, it is possible to utilize the hazardous component of the MSW and to obtain long-acting encapsulated fertilizers that reduce environmental pollution and produce a significant agro-environmental effect.

Under the laboratory conditions, the technology of PET modification and capsule application to mineral fertilizer particles under fluidized bed conditions has been successfully tested. By calculated technological parameters we obtained encapsulated film based on modified PET fertilizer with predicted characteristics. Conducted pilot tests of the release of nutrition elements from mineral fertilizers encapsulated with modified PET, confirmed the uniformity and quality of the coating, which makes it possible to obtain long-acting mineral fertilizers with the required time of release of the nutrition elements.

The analysis of the results of pilot studies of mineral fertilizers encapsulated with modified PET shows that in the case of the use of such mineral fertilizers in the agroecosystems of crops of potatoes, soybeans and maize, the absorption rate of fertilizers by plants increases by an average of 4,5%. Atmospheric and hydrosphere losses in the case of encapsulated fertilizers are reduced compared to granular fertilizers from 47% to 74%.

## References

[1] Iurchenko V, Lebedeva E and Brigada E 2016 *Procedia Engineering* **134** 181
[2] Jurchenko V and Brigada E 2014 *Water and Ecology* **1** 51
[3] Popovych V, Stepova K and Prydatko O. 2018 *MATEC Web of Conferences* **247** 00025
[4] Malovanyy M, Zhuk V, Sliusar V and Sereda A 2018 *Eastern-European Journal of Enterprise Technologies* **1(10)** 23
[5] Geyer R, Jambeck J and Law K 2017 *Science Advances* **3(7)** 1
[6] Jambeck J, Geyer R, Wilcox C, Siegler T, Perryman M, Andrady A and Law K 2015 *Science* **347(6223)** 768
[7] Nahurskyi O A 2012 *Patterns of encapsulation of substances in a state of fluidization and their diffusion release* (Lviv: Vydavnytstvo Natsionalnoho universytetu «Lvivska politekhnika»)
[8] EN 13266:2017 *Slow-release fertilizers Determination of the release of the nutrients* Method for coated fertilizers
[9] Synelnikov S, Soloviy K, Malovanyy M, Tymchuk I and Nahurskyy O 2019 *Environmental problems* **4(4)** 222

# "Green" principles of sustainable development of road and transport infrastructure of the cities of Ukraine

**N V Vnukova, G M Zhelnovach and O V Kozlovskyi**

Department of Ecology, Kharkiv National Automobile and Highway University, Kharkiv, Ukraine

vnukovanv@ukr.net

**Abstract.** The paper is devoted to the development of criteria for the implementation of the principles of "green" and sustainable development of road and transport infrastructure of the cities of Ukraine on the basis of world experience, as well as own research. It is established that the world experience offers a number of systems for assessing environmental friendliness and sustainability of the development of transport infrastructure, which cannot be applied unambiguously and without adaptation in the cities of Ukraine. Therefore, environmentally significant criteria for assessing compliance with the requirements of "green" standards for the development of the city's transport infrastructure in the context of assessing its life cycle stages, which in the future will help the environmentally balanced development of the city's transport system were developed and proposed in the paper. As an example of the practical application of the criteria for "green" construction and assessment of environmental sustainability for the development of the road network, a mechanism was developed and proposed to assess the environmental efficiency of the development of the urban bicycle transport network based on estimates of reduced fuel consumption, pollutant emissions and noise levels by replacing motor work with cycling.

## 1. Introduction

Among the global challenges that humanity now faces, one of the most urgent is to ensure the sustainable development of society in the global context in accordance with the Sustainable Development Goals [1]. They are a kind of call for action aimed at improving well-being and protection of the planet. At the same time, states recognize that measures to eliminate poverty should be taken in parallel with efforts to increase economic growth and solve whole host of issues in the field of education, health, social protection and employment, as well as combating climate change and environmental protection. The analysis of global environmental issues showed that for the world community the problem of urbanization is very acute [2]. According to forecasts, by 2030, more than 60 % of the world's population will live in cities [3], which maximizes the need to develop effective measures for the implementation of Goal 11 "Sustainable cities and towns". This problem is relevant for the most cities of Ukraine. The state has not developed a targeted integrated strategy for sustainable development of cities and settlements, being the integral part of the national document "Strategy for Sustainable Development of Ukraine until 2030" [4]. The analysis performed showed that the key criterion for ensuring sustainable, multi-vector and balanced development of urbanized areas is the sustainable development and operation of their street-road network in accordance with the criteria for "green" construction [5–6].

The purpose of this paper is to develop approaches to the implementation of the principle of sustainable development of cities through the optimization of their road transport infrastructure by introducing standards for the "green" construction of the road network and assessing the environmental efficiency of the development of the bicycle transport network.

## 2. "Green" standards for sustainable development of the urban road network

In recent years, green construction has become increasingly relevant, which is being implemented in the context of sustainable urban development.

The main objective of the "green" construction is the greening of design and construction technologies, which can significantly reduce operating costs, contribute to the implementation of innovative technologies and encourages the search for innovative solutions.

To unify the requirements for "green" construction, "green" standards were developed containing specific requirements (criteria) for the construction of buildings, structures, transport infrastructure and rating system for their implementation.

The most common certification systems for green or ecological building in the world are the British Environmental Assessment Method (BREEAM) –UK, Demarche Haute Qualité Environnementale (HQE) – France, Leadership in Energy and Environmental Design (LEED) – USA, EcoProfile – Denmark, German Sustainable Building Council (DGNB) – Germany, Comprehensive Assessment System for Built Environment Efficiency (CASBEE) – Japan, Green Building Initiative (GBI) – Canada [7].

In the area of ensuring sustainable environmentally-oriented development of the transport infrastructure of cities of the EU countries, "green" criteria and standards are applied according to the series of guides "Transport and Environmental Reporting Mechanism" (TERM), in particular: TERM 01 Total energy consumption in transport depending on its type ; TERM 02 GHG emissions in the transport sector, depending on type; TERM 03 Air pollutant emissions in the transport sector depending; TERM 04 Exceeding air quality requirements; TERM05 Exposure to traffic noise and irritation from it; TERM 06 Fragmentation of ecosystems and habitats by transport infrastructure; TERM09 Number of fatal accidents and serious injuries (caused to soil, air and the marine environment); TERM 11a Generation of waste from end-of-life vehicles; TERM 11b Oil waste and used tires from vehicles [8].

In the USA, almost the same approaches are used to assess the environmental friendliness and sustainability of the development of transport infrastructure in the context of "green" construction. Section 502 of the U.S. Federal Clean Water Act defines green infrastructure as "... a set of measures that use plant or soil systems, permeable coatings or other permeable surfaces or substrates, the collection and reuse of rainwater, or landscape design for storage, infiltration or evaporation of storm water and reduce flows to sewers or surface waters [9].

The Sustainable Highway Construction project , implemented in the context of the "Joint Research Program" of the Transport Research Council, defines modern practice in the field of sustainability of road construction and provides a guide for practitioners to help inform, implement and evaluate effective sustainable highway construction [10].

The US Federal Highway Administration (FHWA), the US Environmental Protection Agency (EPA), other federal transportation and environmental agencies, industry, trade associations, academics and contractors to promote ecological, sustainable and green road construction are implementing partnership in the ranks of the Green Highways (GHP) program is an alliance [11].

The Asphalt Research Consortium (ACR) research program, created in collaboration with FHWA with private institutions and several universities, is exploring potential ways to make asphalt pavements more environmentally friendly and sustainable. In the future, it is expected to lead to increased road safety and reduced environmental impact at the stages of the construction life cycle [12].

It has been established that in foreign countries principles and approaches for the practical implementation of the "green" and sustainable development of the road transport infrastructure of cities are actively developing, although unified approaches and criteria for implementing such activities have not been developed.

Based on our own studies, it was found that the application of the principles of "green" development of transport infrastructure in Ukrainian cities is most effective when taking into account the life cycle of its constituent parts in the context of taking into account environmentally significant criteria for its construction and operation. The developed system of criteria according to the groups of the environmental management system and quality management, material and energy conservation, quality and comfort of the living environment is given in table 1. [13–16].

**Table 1.** Environmentally significant criteria for assessing compliance with the requirements of the "green" standards of the transport infrastructure of the city in the context of assessing its life cycle.

| **Criterion** | **Life Cycle Stage** | | |
|---|---|---|---|
| | **D** | **C** | **O** |
| *Environmental and Quality Management Systems* | | | |
| Implementation of the environmental management system (ISO 14001: 2004 Environmental management systems. Requirements with guidance for use) | + | + | + |
| Implementation of the quality management system (ISO 9001: 2015 Quality management systems. Requirements) | + | + | + |
| Informing the public about the implementation of the object construction project | + | + | – |
| *Rational nature management and landscape arrangement* | | | |
| Selecting a site (highway) for the construction of the facility | + | + | – |
| Reclamation, sanation and rehabilitation of the territory, landscape arrangement and conservation and / or restoration of soil and vegetation cover | + | + | + |
| Integrated storm water management systems | + | + | + |
| Rational water use: reducing the consumption of drinking water for domestic use | | + | + |
| Prevention of surface and groundwater pollution during construction and operation | + | + | + |
| Waste management and recycling | + | + | + |
| Development and implementation of solutions to ensure animal migration routes | + | + | |
| Development and implementation of the solution to preserve and / or provide landscape attractiveness | + | + | + |
| Use of the best available technologies from the list (register) of innovative technologies | + | + | + |
| *Material and energy saving* | | | |
| Energy Efficiency Solutions Developed and Implemented | + | + | + |
| Use of materials mined, processed and produced in the region where the facility is located | + | + | + |
| Use of recycled materials generated in the process of construction, reconstruction, overhaul, repair and maintenance of roads | + | + | + |
| Assurance of the excavation work balance | + | + | – |
| Use of the best available technologies from the list (register) of innovative technologies of State Highway Service of Ukraine | + | + | + |
| Availability of energy efficiency certificates for the applied technologies and equipment | – | + | + |
| Use of warm mix asphalt technology | + | + | + |
| Quality and comfort of the living environment | | | |
| Creation of forest shelter belts along roads of resistant species of trees and shrubs | + | + | + |
| Use of materials with a low level of emission of harmful substances and the non-use of asbestos-containing materials | + | + | + |
| Ensuring minimization and monitoring of the level of air pollution in residential areas | + | + | + |
| Ensuring minimization of the level of pollution of water bodies | + | + | + |
| Ensuring minimization of soil pollution in residential areas | + | + | + |
| Provision of regulatory noise levels in residential areas | + | + | + |
| Assurance of the reduction of vibration level | + | + | + |
| Use of the best available technologies from the list (register) of innovative technologies | + | + | + |

**Note.** D – design stage, C – construction stage, O – operation stage.

The application of the proposed approach to ensure the sustainable development and operation of the road network of Ukrainian cities according to the "green" standards and criteria in the context of the stage assessment of its life cycle taking into account regulatory and legislative requirements in the long term can lead to the integrated reduction in the technogenic load generated by the road network.

## 3. The mechanism for assessing the environmental efficiency of the development of the bicycle transport network of cities

In the framework of the integrated sustainable development of the transport infrastructure of urban areas, the important factor is the development of cycling. It helps to redistribute passenger and traffic flows, reduce operating costs, reduce exhaust emissions and improve the acoustic climate, as well as improve overall fitness and improve public health [17].

The key criterion for the possibility of mass development of bicycle traffic in cities is the formation of the optimal bicycle transport network that ensures safe, comfortable and encouraged use of bicycle transport as an alternative to car trips.

The main tasks of the development of the bicycle transport network include the following:

– improving the convenience of movement over distances of up to 10-15 km;
– increasing the accessibility of territories;
– the solution for transport, environmental and social problems;
– reduction of health care costs;
– improving the quality of the environment by reducing the number of trips by car at distances of up to 10-15 km.

The generally accepted approaches used in the design of bicycle infrastructure include the following activities:

1) Identification of opportunities to use the territory of the settlement, urban district to ensure the movement of cyclists, including:

– improving the layout through the reorganization and reconstruction of existing transport infrastructure facilities to increase their capacity (including reducing or increasing lanes, reconstructing intersections, creating separate streets, intersections at different levels);

– the search for redistribution of cycling and pedestrian traffic using territories located off-road (including green areas, exclusion lanes along railway lines).

2) Improving the efficiency of trips by differentiating cycling by distance, speed, time; combining and dividing the movement of cyclists; development of intermodality; reorganization of traffic.

3) Introduction of new transport solutions and types of transport services for the population.

4) Analysis of existing conditions and prospects for the development and deployment of bicycle transport infrastructure, assessment of the regulatory framework necessary for the functioning and development of bicycle transport infrastructure, and assessment of the volume of financing of transport infrastructure, taking into account the development of bicycle transport [18].

The legislation of Ukraine establishes a number of criteria and requirements for the development of the urban bicycle transport network [19–20], the main of which are given below.

Bicycle paths and bicycle lanes should be arranged on the territories of residential and industrial areas, communal and warehouse areas, on main roads and streets of continuous and regulated traffic, streets and local roads, settlement and rural streets (roads), which provide access for cyclists to residential, public buildings, industrial enterprises, public visiting facilities, open parking lots and garages, etc.

Bicycle lanes are designed only for one-way traffic, and bicycle lanes should be arranged primarily one-way on both sides of the street. If there is a building on one side of the street, a two-way bicycle path on the built-up side should be arranged. If there is cycling on one-way streets, it should be provided in both directions, including using the oncoming bike lane.

The values of indicators and the main calculated parameters of bicycle paths for settlements should be determined in accordance with table 2.

**Table 2**. Parameters of bicycle paths in Ukraine.

| Estimated speed, km/h | Minimum radius of the curve in plan, m | | Minimum radius of the convex curve, m | Minimum radius of the concave curve, m | Maximum longitudinal slope, ‰ | Braking distance on the wet surface, m |
|---|---|---|---|---|---|---|
| | asphalt/ concrete | crushed stone /ground coatings | | | | |
| 20 | 10 | 15 | 40 | 25 | | 15 |
| 30 | 20 | 35 | 80 | 50 | 40 | 25 |
| 40 | 30 | 70 | 150 | 100 | | 40 |

**Note** The speed of 40 km/h is used when calculating the movement on the slopes.

Under the conditions of historically developed urban development, the longitudinal slope of bicycle paths can be increased to 60 ‰, and in hilly and mountainous areas - up to 100 ‰. In areas with the longitudinal slope of more than 60 ‰, it is necessary to avoid double-sided bicycle lanes and provide for the extension of one-sided bicycle lanes and tracks by 0.5 m. On sections of slopes with a longitudinal slope of more than 60 ‰ it is recommended to use bicycle lanes on sections of slopes - bicycle or bicycle-pedestrian paths.

The selection of the form of organization of cycling, depending on the categories of streets and roads, is carried out in accordance with table 3.

**Table 3.** Forms of organizing cycling.

| Street category | | Forms of organizing cycling: Bike path | Bike lane | Bike and pedestrian path | Mixed traffic on the roadway |
|---|---|---|---|---|---|
| Trunk roads | | × | | | |
| Main streets of municipal significance | Continuous motion | × | | × | |
| | Adjustable movement | × | × | × | |
| Main streets of district significance | Adjustable movement | × | × | | |
| Streets and local roads | Residential streets | × | × | | × |
| | Departmental streets and roads | × | × | | × |
| | Pedestrian streets | × | | × | |
| | Driveways | | | | × |

**Note.** Joint bicycle and pedestrian path is arranged according to the total traffic intensity of not more than 75 units/h.

Within the framework of landscape-recreational territories (parks and forest parks), as well as along water bodies and railways, bicycle and bicycle-pedestrian paths are provided mainly with two-way traffic. At the same time, measures should be taken to separate the flows of cyclists and pedestrians in accordance with the requirements (Table 4).

**Table 4.** Minimum width of bicycle lanes and paths.

| Forms of organizing cycling | Minimum width, m: With new construction | Under reconstruction |
|---|---|---|
| Bike lane | 1.85 | 1.5 |
| One-Way Bike Lane | 1.85 | 1.5 |
| Bike lane with two-way traffic on both sides of the street | 2.5 | 2.0 |
| Bike lane with two-way traffic on one side of the street | 3.0 | 2.5 |
| Joint bike and pedestrian path | 3.0 | 2.5 |

IOP Conf. Series: Materials Science and Engineering **907** (2020) 012068 doi:10.1088/1757-899X/907/1/012068

The main criterion for the development of the bicycle transport network, which is a network of bicycle transport routes, is its effectiveness, based on the principles of:

1) Safety – the cycling infrastructure should provide a minimum risk of injury or traumas, as well as a sense of safety among cyclists.

2) Consistency – the bicycle transport infrastructure should be a unified system linking the main places of the start of trips and destinations, be continuous, uniform in terms of movement, have information signs, and allow you to choose options for the route of movement.

3) Straightness and uniformity of movement – the cycling route should have a minimum number of sections with a change in direction. Cyclists should not linger at intersections with traffic flows. Cyclists must be able to move at maximum speed

4) Attractiveness – the bicycle transport infrastructure should provide lighting, aesthetics, integration with the surrounding space, access to service and trade facilities.

5) Comfort – the bicycle transport infrastructure must ensure the quality of coverage, minimal deviations, the exclusion of complex maneuvers, minimizing the need to dismount, minimal interference from vehicles and pedestrians [21].

The analysis of the above criteria for the development of the bicycle transport network carried out in the work indicates the impossibility of their systematic accounting for the integrated implementation of such activities, including in the context of ensuring the environmental safety of urban areas. Therefore, to assess the potential effectiveness of the development of the bicycle transport network, it is advisable to use the method of convolution into the integral criterion (indicator) of single indicators characterizing the impact of bicycle transportation on the functioning of the urban transport system, the environment and public health. The integral criterion P is determined by the formula:

$$\mathrm{P} = \frac{S_1 \cdot \alpha_1 + S_2 \cdot \alpha_2 + \cdots S_n \cdot \alpha_n}{\alpha_1 + \alpha_2 + \cdots \alpha_n} \rightarrow max, \tag{1}$$

where $\alpha_1$, $\alpha_2$, ... $\alpha_n$ – weighting factors of the *i* performance indicator; $S_1$, $S_2$, ... $S_n$ – the relative values of the performance indicators of the certain scenario of the development of military-technical cooperation compared with the base, score [22–23].

As basic indicators for assessing the expected environmental performance from the implementation of the set of measures for the development of cycling it is advisable to consider:

– reduction of pollutant emissions (pollutants) by motor vehicles when it is replaced by cycling ($S_1$, t/year);

– reduction of fuel consumption by vehicles when it is replaced by cycling ($S_2$, t/year);

– reduction of traffic noise in the residential area due to the development of cycling ($S_3$, dBA) [22, 24].

As one of the input parameters for carrying out such comprehensive assessment, it is recommended to consider the rate of decrease in motor transport work performed by all types of urban transport, primarily personal automobile transport, on cycling according to the formula:

$$\Delta Z = \frac{\delta \cdot K_v \cdot W(T)}{K_c} = N \cdot \Delta L, \tag{2}$$

where δ is the proportion of cyclists who previously used a personal car; $K_v$ – regional coefficient of passenger capacity of cycling vehicles, pass/bicycle; $K_c$ is the coefficient of filling the passenger compartment, pedestrian/car; *N* is total number of personal cars, auto.; $\Delta L$ is change in the average annual mileage of 1 car associated with the replacement of part of road trips by bicycle rides, km; $W(T)$ is the volume of cycling work performed at the time $T$, bicycles / km, calculated by the formula:

$$W(T) = k_u \cdot \sum_{i=1}^{n} L_i \cdot \int_{t=0}^{T} S_i(t) \cdot dt, \tag{3}$$

where $L$ is length of the *i* section of bicycle transport network, km; $S_i(t)$ is dependence of traffic on *i* section of bicycle transport network from time to time slot under consideration, bicycle/min.; $T$ is the time interval for which the transport work is calculated, hour; $k_u$ is the uncertainty coefficient, taking into account the proportion of movements of cyclists outside the considered bicycle transport network; $n$ is the number of sections of the bicycle transport network [17].

For the calculation of $S_1$ must use a technique [25], which allows to calculate the pollutant emissions during starting and warming up the motor vehicle, as well as running exhaust emissions which magnitude inversely to bicycle transport work done ΔZ. Expressing the magnitude of the reduction in total emissions of pollutants from vehicles through ΔZ, we obtain the following formula:

$$\Delta m_i = \sum_i \cdot \sum_j \cdot \sum_k \cdot \sum_l \cdot \sum_n \frac{\delta \cdot W(T) \cdot K_{bike}}{K_c \cdot 1000} \cdot (m_{1ijk} \cdot l_{kl} \cdot m_{2ijn} \cdot t_{cr} \cdot \Pi_j \cdot \alpha_j \cdot N_{ikl}), \quad (4)$$

where $m_{1ijk}$ is the mileage emission of the *i* pollutant of a vehicle of the *j* design type when driving along streets and roads of *k* category, g/km; $l_{kl}$ is length of the *l* section of streets and roads of the *k* category, km; $N_{jkl}$ is traffic intensity of the vehicle of *j* design type on the *l* section of streets and roads of the *k* category during the day, thousand cars/day; $m_{2ijn}$ is emission of *i* pollutant when starting and warming up the engine of the motor vehicle of *j* design type for *n* period of the year, g/min; $t_{cr}$ is the warm-up time of the engine, min; $p_j$ is the number of cold starts per day of the vehicle of *j* design type; $\alpha_j$ is the exit coefficient of the vehicle of the *j* design type [25] .

The reduction in fuel consumption ($S_2$) consumed by the vehicle is calculated according to the methodology [26].

The value of the efficiency indicator $S_3$ is advisable to determine the value of the noise characteristics of traffic flows according to the formula [27]:

$$L_{7,5} = 9{,}5 \cdot \lg(N_0) + 12{,}64 \cdot \lg(v) + 7{,}98 \cdot \lg(1 + P_r) + 11{,}39, \quad (5)$$

where $N_0$ is the estimated maximum traffic intensity, car/h; $R_g$ is the share of freight vehicles and buses, %; *v* is speed, km/h.

From formula 5 it follows that the value of noise characteristics of traffic flows will be affected by the intensity, speed and composition of the traffic flow, which depend on the adopted architectural and planning decisions and the conditions for organizing traffic.

Typical measures for urban areas are the following measures to limit the use of cars in the city:

– redevelopment of the typical city street into the bicycle-pedestrian zone – there is no motor noise in this case, and the noise is taken equal to the background;

– limiting the speed of vehicles to 30 km/h with the organization of combined automobile-bicycle traffic, while the value of noise characteristics of traffic flows can be calculated by the formula:

$$L_{7,5} = 9{,}5 \cdot \lg(N_0 - N_1) + 12{,}64 \cdot \lg(v - v_1) + 7{,}98 \cdot \lg(1 + P_r) + 11{,}39, \quad (6)$$

where $N_1$ is the estimated maximum traffic intensity after the creation of the bicycle transport network, car/h; $v_1$ is speed after restrictive measures, km/h;

– the structure of the bike path (or bike lane) instead of the rightmost lane of the driving way on the streets with a different number of lanes of the driving way, while the value of noise characteristics of traffic flows can be calculated by the formula:

$$L_{7,5} = 9{,}5 \cdot \lg(N_0 - N_1) + 12{,}64 \cdot \lg(v) + 7{,}98 \cdot \lg(1 + P_r) + 11{,}39 - \Delta L_{dis}, \quad (7)$$

where $\Delta L_{dis}$ is the correction taking into account the attenuation of traffic noise with distance (the magnitude of the cylindrical divergence), dBA, which is of the greatest importance when implementing the measure and is calculated by the formula:

$$\Delta L_{dis} = 10 \cdot lg\left(\frac{R_0 + L_B}{R_0}\right), \quad (8)$$

where $L_v$ is the width of the created bike lane, m; $R_0 = 7.5$ m.

The value of the weighting coefficients for these criteria, determined by the method of expert estimates, are given in table 5 [17].

**Table 5.** Coefficients of weighting indicators.

| Weight indicator | Name | Significance factors, options | | | | | |
|---|---|---|---|---|---|---|---|
| | | 1 | 2 | 3 | 4 | 5 | 6 |
| Pollutant emission reduction | $\alpha_1$ | 0.05 | 0.025 | 0.075 | 0.025 | 0.5 | 0.05 |
| Fuel economy | $\alpha_2$ | 0.1 | 0.15 | 0.5 | 0.025 | 0.125 | 0.2 |
| Noise level reduction | $\alpha_3$ | 0.05 | 0.025 | 0.025 | 0.5 | 0.075 | 0.05 |

Thus, all three of the above-mentioned measures contribute to improving the environmental situation in large cities, potentially formed by the development of the bicycle transport network and the noticeable motor transport work on the bicycle transport. The proposed integral indicator of the ecological efficiency of the development of the municipal bicycle network helps to combine the above criteria for objective management decisions to ensure sustainable development of urban areas.

## 4. Conclusions

Based on the goal of the article, which is to develop approaches to implementing the principle of sustainable development of cities through the optimization of their road transport infrastructure by introducing standards for the green construction of the road network and assessing the environmental efficiency of the development of the bicycle transport network, a general approach was proposed for such works, based on the stage assessment of the stages of the life cycle of objects taking into account regulatory and legislative requirements. In prospect, when implementing this approach, the integral decrease in the technogenic load formed by the street-road network is expected. To detail the practical approach to the implementation of the principle of sustainable or "green" construction, the paper proposed the criterion for the environmental efficiency of the development of the municipal bicycle transport network, based on the replacement of motor transport work with bicycle transport and the subsequent reduction in fuel consumption, pollutant emissions and acoustic load. The proposed indicator can be recommended for use for the objective adoption of managerial decisions to ensure the sustainable development of urban areas.

## 5. References

[1] Kroll C, Warchold A & Pradhan P 2019 Sustainable Development Goals (SDGs): Are we successful in turning trade-offs into synergies? *Palgrave Commune* **5** p 140.

[2] Lu Y, Nakicenovic N, Visbeck M, Stevance A-S 2015 Policy: five priorities for the UN sustainable development goals – comment *Nature* **520** pp 432-433.

[3] Leeson G W 2018 The Growth, Ageing and Urbanisation of our World *Population Ageing* **11** pp107-115

[4] Decree of the President of Ukraine On the Sustainable Development Goals of Ukraine until 2030 2019 *Official Journal of the President of Ukraine* **21** pp 17.

[5] Hull A, O'Holleran C 2014 Bicycle infrastructure: can good design encourage cycling? *Urban, Planning and Transport Research* **2(1)** pp 369-406.

[6] Rajiv K, Roa G & Teiborlang R 2018 Green Road Approach in Indian Road Construction for the Sustainable Development *Urban Ecology, Water Quality and Climate Change* pp 107-119.

[7] Nikiforova V A, Vidishheva E A, Nikiforova A A, Vidishheva D D (2016) Features of the application of modern environmental technologies in construction activities *Sistemy. Metody. Tekhnologii* **4 (32)** pp 209-218.

[8] The contribution of transport to air quality TERM 2012: Transport indicators tracking progress towards environmental targets in Europe 2012 *European Environment Agency* p 92.

[9] Copeland C 2016 Clean Water Act: A Summary of the Law *Library of Congress. Congressional Research Service* p 10.

[10] Pafidan R, Rooshd R M & Sahamir S R 2015 Sustainable design and construction elements in green highway *Jurnal Teknologi* **77** pp 13-18.

[11] Federal Highway Administration Research and Technology *Federal Highway Administration* URL: http://www.tfhrc.gov.
[12] Asphalt Research Consortium *University of Nevada Reno* URL: http:// www.arc.unr.edu.
[13] ISO 14001:2004 Environmental management systems. Requirements with guidance for use 2004 *ISO/TC 207/SC 1 Environmental management systems* p 23.
[14] ISO 9001:2015 Quality management systems – Requirements 2015 ISO/TC 176/SC 2 Quality systems p 29.
[15] Komurlu R, Gurgun A P & Arditi D 2015 Evaluation of LEED Requirements for Site Properties in Developing Country-Specific Certification *Procedia Engineering* **118** pp 1169-1176.
[16] Zhelnovach G M 2014 Environmentally significant parameters of a technogenic linear object (motor-road) *Tekhnika ta Enerhetyka* **196** pp 396-400.
[17] Trofimenko Yu V, Galy`shev A B 2016 Technique of quality estimation efficiencies of cycle transport networking the large city *Izvestiya Samarskogo nauchnogo czentra Rossijskoj akademii nauk* **4(5)** pp 948–958.
[18] Galy`shev A B, Trofimenko Yu V 2015 Solving the environmental and socio-economic problems of large cities through the development of cycling *Mezhdunarodny`j nauchny`j studencheskij vestnik* **3-3** pp 318-319.
[19] Law of Ukraine On the improvement of settlements 2005 *Official Bulletin of Ukraine* **40** p 12.
[20] State Standard of Ukraine 8906:2019 Planning and design of bicycle infrastructure. General requirements 2020 *Derzhavne pidpryiemstvo «Derzhavnyi dorozhnii naukovo-doslidnyi instytut imeni M.P.Shulhina»* № **311**.
[21] Dekoster J, Schollaert U 1999 *Cycling: the way ahead for towns and cities* (Luxembourg: Office for Official Publications of the European Communities) p 63.
[22] Vnukova N V, Zhelnovach G M 2011 Selection of autotransport systems environmentally significant parameters to assess environmental hazard of roadside area *Ekolohichna bezpeka* **12** pp119-123.
[23] Hryhoruk P M, Tkachenko I S 2012 Methods of building an integral indicator *Ekonomiko-matematychne modeliuvannia* **4** pp 34-38.
[24] Merkisz-Guranowska A, Merkisz Jy, Kozak M & Jacyna M 2013 Development of sustainable road transport system *Urban Transport XIX* pp 507-517.
[25] Donchenko V V 2008 *Settlement instruction (methodology) for the inventory of emissions of pollutants from vehicles on the territory of major cities* (Moscow – Avtopolis-plyus) p 84.
[26] Turovsky` V F 2008 *Fuel and lubricant consumption rates in automobile transport: guidelines* (Moscow – Avtopolis-plyus) p 79.
[27] Shubin I L, Czukernikov I Ye 2015 *Basics of designing noise screens* (Moscow – Publishing house Bastet) p 208.

IOP Conf. Series: Materials Science and Engineering **907** (2020) 012069 doi:10.1088/1757-899X/907/1/012069

# Introduction of innovative water treatment technologies - the key to sustainable development of the modern city

**K.I. Petrushka, M S Malovanyy, I M Petrushka, V D Pohrebennyk and V I Mokryy**

Viacheslav Chornovil Institute of Sustainable Development, Lviv Polytechnic National University, S. Bandery Str., 12, Lviv, 79013, Ukraine

myroslav.mal@gmail.com

**Abstract.** The mechanism of conductivity of natural ion exchangers and resin KU-2 in ammonium and chlorine ions. On the basis of experimental data obtained dependence for pure solution and ionit-solution systems allowed to set the concentration of isoconductive solution (in which the concentration of electrical conductivity of the solution is the same as in the ionic). The method of measuring of resistance of a layer of grains of granulated ionite after removal of equilibrium solution by centrifugation is suggested. On the basis of experimental studies the types membrane backfill and determined the optimal modes of implementation of electrodialysis in the cell with membranous backfilling. The conducted researches give the possibility to implement the principle model approach to describe the conductive properties of ion-exchange materials and its experimental verification. The mathematical model of the conductivity of ion-exchange resins and columns, which combines the theoretical approaches of three wired and microheterogeneous model. To check the adequacy of the developed model was investigated the electrical conductivity of synthetic and natural ion exchangers materials, as well as columns of these materials in solutions of different salts. This has allowed us to develop a model to describe the electrical conductivity of ion-exchange materials and to measure the proportion of current which flows through the gel, solution and a mixed channel conductivity in an ion- exchange system containing two-chargers ions. The algorithm of calculation of physical-chemical parameters of water purification process by electrodialysis with ionite membranes with use of ionization in the form of granules, as a intermembrane backfilling. The use of the obtained results will predict parameters of process of ion-exchange purification of model solutions of wastewater and multicomponent industrial solutions.

## 1.Introduction

Water treatment and water supply are important elements of human activity in industry and in everyday life, including the rational functioning of the urban area. Cleaning and recycling are the most effective ways of treating wastewater. Achieving a high degree of water purification allows it to be reused in the production cycle. When recycling water it is possible to reuse dissolved substances. At present, membrane methods of water purification using electrodialysis are increasingly used in industry and everyday life.

With modern anthropogenic loads on the hydrosphere, increasing the quality requirements for drinking and industrial water, such traditional coagulation and settling water purification technologies are not effective enough. The use of membrane technology in industry opens wide prospects for the creation of fundamentally new energy-efficient, environmentally friendly technological schemes.

Today, water treatment systems use membrane technologies that allow to obtain water of almost any composition with a high degree of purification (removal of impurities dissolved in water can reach the value of 99.8%).

The current environmental situation contributes to the increased use of membrane systems. This is primarily due to the increased requirements for the quality of drinking water - the content of organochlorine compounds, pathogenic bacteria, fluorides, nitrates, etc. Modern membranes demonstrate unquestionable efficiency and versatility in the purification of water from various types of pollution.

The main advantages of membrane technology in comparison with traditional, physical and chemical filtration methods are:

- high degree of water purification;
- constant quality of the filtrate at the output, regardless of the change of the input composition;
- no chemical reagents required;
- technology allows you to get the maximum amount of clean water from the inlet stream.

Thus, two environmental problems are solved simultaneously: reducing the use of natural resources and reducing the amount of discharge.

Wastewater treatment using membranes makes it possible to apply waste-free water technology, low energy costs; all types of membranes are made of polymeric and corrosion-resistant materials and are therefore durable. Membrane-based installations take up minimal space, are fully automated and do not require constant maintenance.

Electrodialysis is the process of separation of salts ions in the membrane apparatus which is carried out under the influence of constant electric current. Electrodialysis is used to demineralize drinking water. The main equipment is electrodialysis, consisting of cation and anionite membranes. Electrodialysis with the resin membranes is one of the main methods in membrane technology. It is the combined method in which the processes of electrolysis and dialysis are combined [1].

The advantage of electrodialysis before other methods is the absence of phase conversion of water, which occurs during distillation, freezing or use of gas hydrated method.

The method of electrodialysis has the following drawbacks: formation of sediments of calcium carbonate, magnesium and gypsum hydroxide in the case of polarization; "Poisoning" of the conomonmetabolic membranes iron, Mangan, and anionoexchange – organic substances contained in the treated water; In the case of the installation on the current with the maximum density, lower than the optimum values, the cost of the process significantly increases; Due to the absence of large single-power devices, specific capital and operating costs are increased, high selectivity of the membranes is not enough, availability of concentrated solutions and, again, the need for Water purification [2].

It is used to produce ultra clean water. An important point in the development of electrodialysis with ionic membranes was its use for demineralization of brackish water.

Most of the processes of electrodialysis are carried out in solutions, which are cleared from stiffening ions, or the concentration of which in these solutions is low.

Sewage and mine water in case of their insufficient purification getting into surface reservoirs, groundwater, and aquifers pose a threat of pollution of these environments, as Hydrosphere is one of the components of the biosphere-respectively and threat to the environment in general [3].

The process of water purification by electrodialysis is based on the division of ionized substances under the influence of Electromotive force, which is created in solution on both sides of membranes. It is due to the migration of ions through the membrane under the effect of the attached potential difference. In order to exclude the diffusion process, it is necessary that ionopermeable membranes are owned by the selectivity, i.e. the ability to pass ions with a charge of one sign. In other words, the positively charged membranes (anionactive) should pass only the anions, but negatively charged (cation-active) – only cations (Fig. 1).

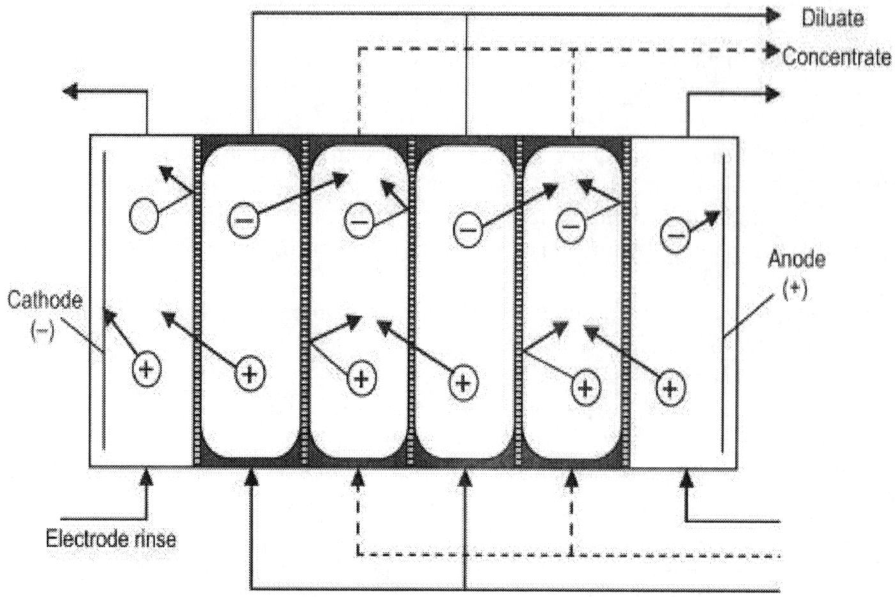


**Figure 1**. Schematic diagram of the process of electrodialysis.

Electrodialysis differs from conventional electrolysis in that semipermeable partitions (membranes) are installed between the electrodes, the pore sizes of which allow the penetration of solute ions through them, but impede the passage of larger particles, as shown in Fig. 2. Then, due to the ordering of the movement of ions, anions of solutes are accumulated in the anode compartment and cations in the cation, and thus the necessary desalination of water is achieved. From the formed concentrated solutions regeneration of valuable substances is carried out [4,5].

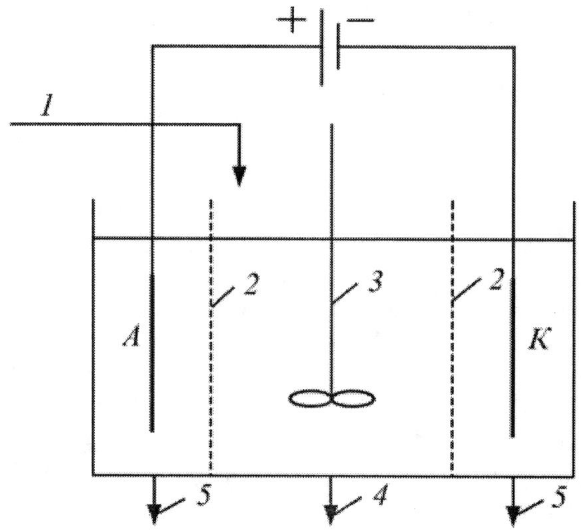


**Figure 2**. Electrodialysis: 1 - sewage (liquid) supply, 2 - semipermeable partitions (membranes); 3 - stirrer; 4 - release of purified liquid; 5 - releases of concentrated solutions.

The efficiency of electrodialyzers increases with the arrangement of partitions of ionic materials, which are films made on the basis of polymeric materials with the addition of powders of ion-exchange resins.

IOP Conf. Series: Materials Science and Engineering **907** (2020) 012069 doi:10.1088/1757-899X/907/1/012069

**2.Details experimental**

We studied the system of ion exchange resin KU 2 - solution of NaCl and system of ion exchange resin KU 2 - solution $NH_4Cl$. The results to be received during the experiments will allow us to develop an algorithm calculating physical - chemical parameters of water treatment by electrodialysis with ionite membranes using resins in the form of granules as intermembrane backfilling.

The resin was placed in the measuring cell (U - form tube) and was set in the balance with test solution. The dependencies of the pure solution reverse resistance and the resin-solution system on the concentration of solution (1 / R) were established experimentally and were presented on the same graph. The intersection point of the obtained dependencies for a clean solution and the resin-solution system allows determining the concentration of iso-conducting solution (the concentration at which the electrical conductivity of the solution is the same as in the resin).

After that the resin was transferred to a centrifuge cell (ion exchange cell with a porous bottom and two platinum electrodes, soldered against each other in the cell walls). The equilibrium solution was removed by centrifugation.

The centrifugation of the cell lasted for 15 minutes at 373 - fold increase in gravity. Using the bridge alternating current (1000 Hz) we measured the cell resistance (Rx), on the base of which the electrical conductivity of the resin was calculated. The obtained results made it possible to study the concentration dependence of the electrical conductivity of the resin in the test solutions (Fig.3).

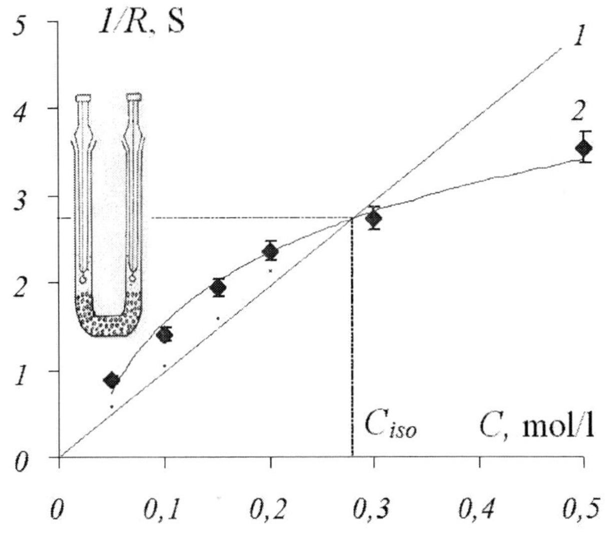


**Figure 3.** Determining of isoconductivity concentration of KU-2 resin through U- shaped cells: 1 - solution conductivity; 2 - resin conductivity with a balanced solution.

The dependencies obtained for pure solution and for the ionite-solution system allowed to determine the concentration of the isoconducting solution. The intersection point of these dependencies for a clean solution and for the resin-solution system allowed to define the concentration of isoconducting solution in Fig.4.

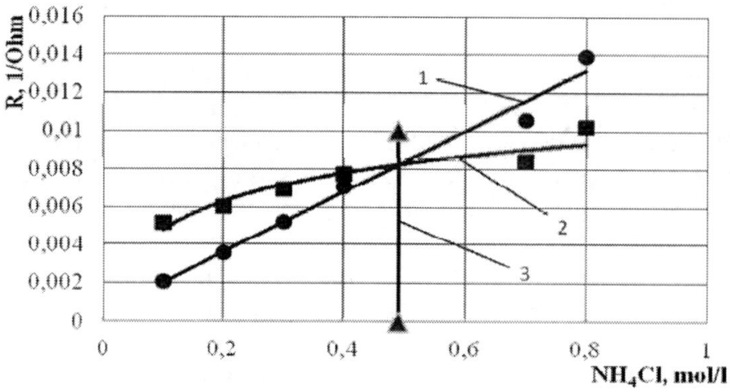


**Figure 4**. Determination of the isoelectric conductivity point of KU-2 resin in a U-shaped tube: 1 - $NH_4Cl$ solution; 2 – ionite, 3- line of isoelectric conductivity.

The obtained data can be used to calculate the electrodialysis process of solutions with $NH_4Cl$, using intermembrane filling by the ion exchanger KU-2.

## 3.Results and discussion

The necessary condition for successful application of the gaskets from insulating materials in electrodialysis is the preliminary research of conductivity of materials that will be used for the intermembrane backfilling. Determination of electrical conductivity comes to measurement of resistance because it is inversely proportional to the resistance [6].

Electrometric method of determining the electrical conductivity allows to accelerate and to increase the accuracy of its determination in comparison to other methods [7].

On the border of phase separation electrode-liquid the polarization of contact voltage occurs, which is caused by a potential jump and has a polarity opposite to the polarity of the applied voltage. At the same time molecules seem to form a double layer, whose properties differ from the properties of molecules, polarized at a greater distance from the phase separation surface.

Therefore the criterion of the possibility for application of such an approach in a particular experiment can be one that takes into account the impact of the potential jump on the measurement results [8].

In completely dry state ion exchange resins have a high electrical resistance. When the resin swells, there appears the ability to conduct electrical current.

The known methods of measurement of specific electrical conductivity of granular resin make impossible to carry out the measurements in a wide range of concentrations of the equilibrium solution. That is why we proposed the approach to measure the resistance of the grains layer of granular resin after the removal of the equilibrium solution by centrifugation.

The purpose of setting optimal types of backfilling and optimum implementation modes of electrodialysis with intermembrane backfilling we conducted studies that were aimed at developing a model approach to describe the electroconductive properties of ion exchange materials and its experimental testing.

We developed a mathematical model of the conductivity of ion exchange resin and columns, which combines the theoretical approaches of three leading and microheterogeneous models. To check the adequacy of the developed model, the electrical conductivity of synthetic and natural cation exchange materials, as well as columns with these materials in solutions of different salts, was investigated.

This allowed us to develop a model for describing the electrical conductivity of ion-exchange materials and to estimate the fractions of current flowing through the gel, solution, and mixed conduction channel in an ion-exchange system containing two-charge ions.

## 4.Conclusions

On the basis of experimental data the mechanism of conductivity of natural cation-exchangers and resin of KU-2 in the form of ions of ammonium and chlorine is set. The obtained dependence for pure solution and the ionite system-solution enable to determine the concentration dependence of the specific electrical conductivity of resin in the studied solutions.

Defined and graphically confirmed the value of the concentration of isoconductive solution for which the conductivity of the solution is the same as in the ionic. This value is supported by the intersection of the obtained dependencies for Clean solution and the ionite system – solution.

Experimental studies have identified the types of intermembrane backfill and established optimal modes of implementation of electrodialysis in the cell with intermembrane backfill. The conducted researches have implemented the principle of model approach for the description of electrically conductive properties of ion-exchange materials and its experimental verification.

The method of measuring of resistance layer of grains of granulated ionite after removal of equilibrium solution by centrifugation has been developed and proposed.

The obtained results gave an opportunity to study the concentration dependence of the specific electrical conductivity of resin in the studied solutions. A mathematical model is intended for forecasting the diffusion of one and two-rechargers in ion-exchange column.

A model for describing the conductivity of ion-exchange materials and measuring the fraction of current flowing through the gel, solution, and mixed conduction channel in an ion-exchange system containing two-charge ions.

The developed algorithm of calculation of physical-chemical parameters of water purification process by electrodialysis with ionic membranes using ionization in the form of granules as a membrane filling is developed.

Using the obtained results will predict parameters of process of ion-exchange purification of model solutions of wastewater and multicomponent industrial solutions.

## References

[1] Wang Q, Yang P and Cong W 2011 Cation-exchange membrane fouling and cleaning in bipolar membrane electrodialysis of industrial glutamate production wastewater *Separation and Purification Technology* **79**(1) pp 103–113.

[2] Daniilidis A, Vermaas D, Herber R and Nijmeijer K 2014 Experimentally obtainable energy from mixing river water, seawater or brines with reverse electrodialysis *Renew. Energy* **64** pp 123-131 doi: http://dx.doi.org/10.1016/j.renene.2013.11.001

[3] Tedesco M, Brauns E, Cipollina A, Micale G, Modica P, Russo G and Helsen J 2015 Reverse, Electrodialysis with saline waters and concentrated brines: a laboratory investigation towards technology scale-up *J. Memb. Sci.* **492** pp 9-20

[4] Gomelya M, Hrabitchenko V, Trokhymenko A and Shabliy T 2016 Research into ion exchange softening of highly mineralized waters *Easten-Europen journal of Enterprise Technologies* **4/10** (82) pp 4–9

[5] Tufa R A, Curcio E, Van Baak W, Veerman J, Grasman S and Fontananova E et al. 2014 Potential of brackish water and brine for energy generation by salinity gradient powerreverse electrodialysis (SGP-RE) RSC Adv. **4** pp 42617–42623 doi: 10.15587/1729-4061.2016.75338

[6] Dah-Yu Kao 2016 Study on the leachable behavior of cation exchange resins *Journal of Nuclear Science and Technology* **53** (6) pp 921–927

[7] Morflyuk V F, Churkin V V 2013 Objective determination of the electrical conductivity of the humidifying solution *Texnologiya i texnika drukarstva: zbirnyk naukovyx pracz* (VPI NTUU "KPI") **1**(39) pp 106–111

[8] Inamuddin, Khan S A, Siddiqui W A and Khan A A 2007 Synthesis, characterization and ion-exchange properties of a new and novel «organic-inorganic» hybrid cation-exchanger: Nylon-6,6, Zr(IV) phosphate *Talanta* **71** pp 841-847

# Aspects of preserving historic landscapes of palace and park ensembles of the Romantic period

**E T Cherkasova, A A Zavorina**

Department of Reconstruction, Restoration architectural objects, Kharkiv National University of Civil Engineering and Architecture. 61002, Kharkov, Ukraine

Cherkasova168@gmail.com

**Annotation**. Using the example of historical palace and park ensembles in the north-eastern region of Ukraine, the article clarifies the methodology for identifying complex monuments. The relevance of the research topic is due to: the need to reevaluate complex monuments on the basis of clarifying the substantive aspects of the concept of authenticity. In the work, on the example of the most studied palace and park ensembles of the Kharkov region, the characteristic features of the formation of the mansion culture are described and the most striking examples of ensembles that determine the historical and cultural value and regional features of the development of architecture and urban planning of this typological group of objects are presented. The criteria for assessing the specific components of palace and park ensembles, as well as the degree of their physical preservation established at the stage of identifying the monument, affect the level of historical and cultural value and the integrity of the architectural - planning and spatial composition of the ensemble At the stage of registration of the monument, the database is filled with detailed information that ensures the completeness of the external and internal factors influencing the design object, for optimal adaptation of the monument to modern use.

## 1. Introduction

The palace and park ensembles of the Romantic period are complex objects of cultural and natural heritage. A change in architectural ideology, the impact of natural and historical factors in the development of the historical environment of cities and populated areas led to multiple transformations of the physical structure and functional use of architectural ensembles, a change in the spatial composition of landscape gardening, and in some cases to significant losses in the material component of the architecture of ensembles and vegetation of mansion parks. Statistical information about the suburban ensembles of the Romantic period preserved in the north-eastern region of Ukraine indicates a significant number of lost objects. According to the literature sources of the beginning of the twentieth century the number of mansions examined and studied, the former Kharkov province counted about 70 objects. In the second half of the twentieth century, according to the results of an on-site examination, in preparation for the publication of the Code of Monuments of History and Culture of Ukraine, about 40 objects with different degrees of physical safety and authenticity were recorded. Among those, at the beginning of the 19th century, only 17 complexes were classified as protected objects of local and national value. The relevance of the study is based on the necessity for a detailed fixation and reassessment of the architectural and urban planning heritage, the identification of ensembles in the modern structure of the residential landscape, the identification of complex objects of cultural heritage to compile an information database on cultural heritage sites, and the clarification of a number of provisions for identifying classification features and areas of modern use.

## 2. Materials substantiating historical and cultural values

The history of creation, the features of the formation and development trends of palace and park ensembles in the history of architecture of Ukraine are considered at the stages of the formation of secular educational ideology and culture of the New Time. The period of the creation of palace and park ensembles has historically been associated with the stages of the formation of classicism in architecture (1770s - 1 third of the 19th century) and the implementation of administrative-territorial reform in a vast region of the north-eastern regions of Ukraine. The term "mansion culture" characterizes an important phenomenon for the era of classicism, which for a long time determined the sustainable development of the state, the formation of its economic, social and cultural space. Manor estates were widespread throughout the territory of the former Russian Empire, and appeared as centers of educational culture, points of dissemination science, art, literature and poetry.

### *2.1. The degree of knowledge of the palace and park ensembles of the period of classicism*

Manor culture, as a phenomenon associated with the formation of the architecture of the palace and park ensembles in large and medium-sized estates of the Kharkov nobility, developed chronologically in the 1770-1780s. The creation of palace and park ensembles was a common phenomenon and the main type of suburban manor construction until the middle of the 19th century.

Suburban mansions were built according to the well-known scheme, based on the including religious buildings, residential buildings, service and household courtyards in a single complex, surrounded by rural nature and the artificially created landscape park. The significance of the landscape park in the compositional structure of the suburban mansion consisted of expressing the artistic ideas that the enlightened literature preached in its own language. The beginning of the study of the architecture of nobles' mansions dates back to the end of the 19th century.

In the famous monograph of the architect-artist G.K. Lukomsky "The elder mansions of the Kharkov province" was studied in detail and presented in the photographs, the historical picture of the formation of the ancestorial mansions of the Kharkov nobility. Among periodicals of the late 19th - early 20th centuries the most widely popular publications were "The Capital and the Estate", "Among Collectors", "The Old Years", which contained numerous articles of a historical and literary content about rich noble mansions (G. K. Lukomsky, N. N. Wrangel, Y. I. Shamurin). Architect and artist G.K. Lukomsky, a well-known researcher of the cultural life of the province, has repeatedly noted the high level of development of the estate culture in Slobozhanshchyna. This was confirmed by references to works of art and architecture of a high artistic level, owned by famous representatives of the capital's school and their students.

Of great importance in the development of suburban manor construction was the activity of A.A. Palitsin (1749-1816), a well-known translator and writer, head of the educational group "Popov Academy" in Slobozhanshchyna. The main directions of literary activity of A.A. Palitsin were translations of works of Western European literature [1]. During 1802-1816 he published seven works of educational literature in Kharkov, St. Petersburg and Moscow, which hold the main place in understanding the poetics of the romantic art culture. Poetic symbols of literature at the beginning of the 19th century received reflection in associative images of the natural environment of country estates when creating landscape-style parks, in artistic images of paintings that adorned the interiors of palace buildings.

The poem "Gardens or the Art of Decorating Rural Views" by Jacques Delil's first appeared in print in the Russian Empire in 1803 translated by A.A. Palitsin, which in 1814 was reprinted by the printing house of Kharkov University. Translated by A.F. Voeikova, this poem was distributed in Russia only in 1816 [2]. The popularity of this publication in Russia by the famous writer, academician D.S. Likhachev explains that the new principles of gardening described in the poem did not require complex philosophical rethinking. Having familiarized with the poem by J. Delil, one could immediately begin to create a landscape park. The most important thing for him was the perceived spectacular essence, reflecting the aesthetic principles of creating a natural environment [3]. This poem can be considered a practical guide to creating landscapes of manor parks [4].

**Figure 1**. The palace of the estate ensemble in the Old Merchik, 1776-778.
arch. V. Bazhenov, P. Yaroslavsky. North and south facades. Photo of 1910. Current state.

A.A. Palitsin owns the idea of creating a landscape park in the palace ensemble of Old Merchik - the estate of the leader of the nobility of the Kharkov province G.R. Szydlowski (Figure 1). The two-story palace on a high basement in the style of early classicism was built by the governorate architect P.A. Yaroslavsky, whose project, in the opinion of I.E. Grabar, performed by the famous architect V.I. Bazhenov. [5]. A. Palitsin built outbuildings in the Old Merchik, located by the kurdoner along the axis of the palace, a temple; in the landscape park - "stone arena, altanka, reservoir, rotunda in the form of a round temple with arches" [6]. The park was created simultaneously with the construction of the palace. This is mentioned by V.I. Yaroslavsky in his notes, naming the date of the main work, - not earlier than 1776-1778. Old Merchik was a place frequented by lovers of art and nature. In the estate lived and worked as a French teacher, a famous writer Jean Vernet; during the last years of his life, a well-known landscape painter, founder of a watercolor portrait, academician of the Imperial Academy of Arts P.F. Sokolov (1787-1848) often visited Old Merchik [7].

Researchers of the estate ensemble in the Old Merchik of the post-war period (1946-1947) first presented measurements of the complex of manor buildings stored in the funds of the Gosstroy of the Ukrainian SSR. In 1968, A. Ryabushin measurements of the manor house in Stary Merchik were first published [8].

For the period 1969 - 1979 all objects accessible for study were identified, and the territorial parts of the park were studied in detail, preserving the layouts, small architectural forms and landscape groups of vegetation [9]. The compositional space of the estate in Old Merchik is formed in the vast visual basin of the Mokry Merchik river valley. The palace occupies the highest place in the area of the main streets built in one-story residential buildings. The composition of the landscape includes all objects that are natural and natural objects of a landscape park, a water mirror of ponds, and rural buildings. Production areas that appeared in the structure of the estate in the 19th century, were situated outside the manors' boundaries aside from the main visual disclosures of the palace buildings. The individuality and originality of the park areas were determined by natural features, the skill of the gardener and the tastes of the owner. In small-scale estates, special importance was attached to the use of the local assortment of vegetation of trees and shrubs, and to the techniques of arranging landscape groups.

Among the landscape parks of the palace estates in Slobozhanshchyna, the Old Merchik estate park was most famous. In contrast to the large palace and park ensembles, the manor - park complexes of Kharkiv oblast occupied relatively small territories. The area of the park of the Old Merchik estate was 69 hectares, in the Osnovyantsy park - 46 hectares, in the Sharovka estate - 70 hectares, Natalievka - 55 hectares, in Kiyanitsa - 55 hectares.

The complex of manor buildings dominated in the space of the estate and was the center of the composition, from which the storyline of the perception of the park's landscapes unfolded on the ground. In the system of organizing the territory of the estate, one can see the three-layer scheme traditional for classicism: a regular stall, pond or river, the natural environment of the estate. The regularity zone of the estate complex was continued in the composition of the stalls in front of the southern facade of the palace, which became the transitional zone to the landscape park. The planning system of park roads arranged along terraced slopes fixed the location of the "moving point of view" on the system of park landscapes, views and panoramas of different levels of perception being revealed to the viewer. The water surface of artificial ponds stretched along the slopes was a necessary spatial zone, which provided the depth and multifaceted nature of the perception of landscapes. The conditions for orienting the open space determine the favorable conditions for the black-and-white composition and color of individual landscape groups throughout the daylight hours, especially expressive in the morning and evening hours. By the middle of the 19th century changes in the composition of the space of the central part of the estate can be traced back from the old central-axial planning with the location of the palace on the edge of the terrace and the layout of the park on two levels.

**Figure 2.** Panorama of the central part of the estate in Sharovka, 19th - early 20th century. Photo 1915

*2.2. Historical transformations of the architectural imagery and structure of the park and garden landscape*

At the second stage of the development of the estate culture in the period from the 1880s to the beginning of the 20th century, in the practice of suburban construction, a new typology of suburban residential estates is formed. The traditions of the suburban manor building of the classic period received a peculiar development in the complexes of romantic country cottages and eclectic mansions, which were created using the architecture of historical styles: neo-Renaissance, false Gothic, stylization in the spirit of national architecture and wooden architecture. The layout of the estates of the second half of the 19th - early 20th centuries is based on the zoning system, with the leading importance of the ornamental garden, to which subordinate geographically separate working villages, household yards, complexes of service buildings, technical structures and workshops, combined into an integral architectural composition (ensembles Sharovka, "Natalievka" Figure 2).

The new era has changed the attitude to the aesthetics of landscape gardening, enriching the gardens with a variety of natural forms of vegetation introduced in botanical gardens throughout the 19th century. Collecting of art works, usual for the romantic era, in landscape gardening art was expressed in the creation of a variety of landscapes based on exotic and local species of vegetation, in the enrichment of the typology and forms of garden buildings that form the architectural accents of the landscape gardening landscape. The architectural and spatial composition was based on the maximum consideration of the natural features of the area, the organic relationship of architectural structures and the natural environment. Mandatory was the orientation of the palace buildings on the natural landscape, free planning techniques, a variety of historical forms and architectural solutions of ceremonial and utilitarian buildings, small architectural forms.

A special stylistic and symbolic meaning was borne by green spaces, which were distinguished by the decorative diversity of vegetation of local and exotic species, their layout, taking into account the biological and decorative characteristics of plants, which changed significantly compared to the landscape parks of classicism. The new aesthetics of the park and garden landscape was aimed at revealing the decorative possibilities of plants from different regions and countries, following the example of using techniques and forms of historical styles in architecture. In the Kharkov region this became possible thanks to the efforts of I.N. Karazin and I.I. Karazin, their work on the introduction and acclimatization of new plant forms. Planting material grown in the nursery of Osnovyantsy Park in the Kharkov region was distributed to other regions of Ukraine. Decorative dendrology, as a scientific direction in landscape gardening art, gave impetus to the development of architecturally eclectic landscape gardens in suburban palace and park ensembles.

## 3. Identification of complex monuments of architectural and urban planning heritage

In the methodology for the preservation of cultural heritage objects in the process of studying, evaluating, adapting and identifying the direction of modern use, the basic stage is the identification of a historical object. At the stage of preliminary field studies and subsequent comprehensive studies with the use of reliable sources of information, the value characteristics of the studied objects are established. The main in assessing the monument as an object of urban development, are the historical, cultural and urban value of the territory, which determine the protection subject of the complex monument [10]. The heterogeneity of the assessment criteria for the species components of the complex monument (objects of history, architecture, park and garden art, landscape), the degree of their physical preservation, to varying degrees, affect the integrated assessment of the consistency of the architectural and spatial composition of the ensemble.

At the stage of urban planning analysis, the definition of authenticity is replaced by a substantiation of the identity of documentary information about the monument based on a comparison with the results of field survey and visual assessment of the qualities of the architectural and spatial composition. In the theoretical literature and normative recommendations on the preservation and restoration of the architectural and urban planning heritage that have appeared in recent years, a number of changes have been outlined in the interpretation of the significance of the authenticity of the urban heritage.

When analyzing regulatory documents of an international level, changes are observed in the criteria for assessing heritage in connection with the rethinking of the concept of protecting a monument as a single object, awareness of the principles of heritage conservation, based on the criteria of diversity and value of heritage objects, the need to maintain interconnection with the traditional environment of the monument.

A number of sources question the position on the universality of the category of authenticity recommended by the Venice Charter of Restoration (1964); the provision on the conformity of authenticity indicators for the restoration of individual monuments, including for objects that have retained their original appearance, features of historical morphology, traditional functional use [11]. Due to the expansion of the range of protected objects of cultural heritage and the urgency of the problem of preserving the historical urban environment, the historical context and traditions of urban culture, the idea of authenticity is shifting towards revealing the identity of information confirming the authenticity of the values of individual and territorial objects of cultural heritage.

In article 9 of the "Nara Document on Authenticity", adopted in 1994, the meaning of authenticity is considered in the context of awareness of the values of heritage, and, as stated in the document, "the ability to recognize them largely depends on the sources of information about these values, their veracity and reliability. Knowledge, understanding and interpretation of information sources, in relation to the initial and subsequent characteristics of cultural heritage, are a necessary basis for determining all aspects of authenticity. " Sources of information contain information: "about the form and design, materials and substance, the use of the monument and its functions, traditions and technologies, the location and surroundings of the monument, its spirit and expressiveness, as well as other internal and external factors" [12]. Thus, the criterion of authenticity, as an indicator of the authenticity of information, confirms the assessment levels of the monument according to certain criteria: historical, archaeological, artistic, scientific, utilitarian (typological), and also substantiates the level of the integral historical and cultural value of the object of cultural heritage.

The semantic emphasis characterizing the authenticity of information sources was enshrined in the international document published in 1996, "Principles for the Registration of Monuments, Groups of Buildings and Landmarks" [13], which aims to ensure the collection of information on the physical configuration, condition and use of monuments, groups of buildings and places of interest at different times. The content of the document indicates that the registration procedure is an essential part of the conservation process. Registers of monuments, groups of buildings and places of interest - may include tangible and intangible certificates, which are an integral part of the documentation that helps to understand the heritage and its inherent values. Detailed information placed in the registers of monuments is aimed at ensuring the completeness of information that cannot be obtained from a direct study of the monument, a group of buildings or a landmark itself. Registries should include partial or complete information about the state of the monument, its external and internal parameters, including groups of monuments located on the site; the external design, structures and their condition, maintenance services and equipment, topographic and natural features of the site; the construction history of the conservation of the structure, its maintenance and repair; assessment of risk and possible damage as a result of human activities, natural phenomena, environmental pollution and land use in adjacent areas [13].

The form of the proposed register of documentation on the object of world cultural heritage requires detailing for its exercising at the national level, in accordance with applicable legal documents. The accumulation of the necessary information and the creation of a database of heritage objects are aimed at optimizing scientific and design work and improving all areas of monuments and conservation activities. [15]. The country palace and park ensembles, as complex monuments of history, architecture, urban planning, landscape gardening art and landscape, in the architectural heritage of Ukraine are adjacent to the typology of territorial objects of cultural heritage and are considered as monuments of urban planning. In the methodology for identifying complex monuments, the most crucial stage in their study is a component-wise analysis of the components of the ensemble, aimed at maintaining the integrity of the compositional structure and spatial organization of the entire ensemble.

## 4. Conclusions

In the architecture of suburban palace and park ensembles of romanticism, two stages of historical development can be traced, indicating a change in the style of architecture from classicism to historicism, the continuous development of the structure and spatial organization of ensembles, a change in the natural forms of vegetation and the aesthetics of landscape gardening. The artistic level of manor architecture and garden art in Slobozhanshchyna became possible thanks to the joint activities of Kharkov architects, artists, writers, botanists and their students under the influence of Western European cultural ties and traditions of the capital's architectural schools.

Regional features of the architecture of suburban palace and park ensembles manifested themselves in the planning decision - when choosing the location of the estates; in the typology of manor buildings; in the stylistic features of the architecture of classicism, pseudo-Gothic and "wooden classicism"; in the composition of the garden and park landscape, focused on the use of vegetation of local origin. Elements of historical ensembles that are most subjected to temporary transformations, to varying degrees, have preserved the stylistic and compositional integrity of the architectural space and artificial natural environment. Based on the attraction of genuine documents and materials of historical research, the authors substantiate the level of historical and cultural value of the territory of the most studied palace and park ensembles: Old Merchik (1770-1780s) with a landscape park of the late 18th - early 19th centuries; Sharovka (19th- early 20th centuries); "Natalievka" in the village Vladimirovka (beginning of the 20th century).

An important aspect of the complex urban planning studies, preceding the development of scientific and design documentation for the conservation and restoration of the monument, is to find a balance between the architectural environment and the natural environment, the inclusion of historical landscapes of palace and park ensembles in the natural framework and recreational areas of regional resettlement systems. The optimal direction of the modern use of these ensembles is museumification, which determines the conditions for a complex of works on the conservation, restoration and adaptation of objects for tourist use.

## 5. References

[1] Sumtsov N F 1888 *The Cultural Corner of the Kharkov Province (Popov Academy)* (Kharkov: Kharkov collection) p100-111
[2] Lotman Yu A 1987 *"Gardens" Delil in per. A.F. Voeikova and their place in Russian literature* (Leningrad: Science) p 191-209, 226.
[3] Likhachev D S 1987 *Jacques Delille - Gardening Teacher* (Print AN SSR) p 210-213
[4] Delille J 1816 *Gardens or the Art of Decorating Rural Views* (SPb) p 10
[5] Grabar I E, Kazhdan T P, Gunkin G I 1951 *Unknown and Alleged Buildings of Bazhenov M* (Print AN SSR) 389 p
[6] Yaroslavsky V I 1887 *Notes* (Kiev antiquity T XIX) p 133
[7] 1896-1918, 1909 *Russian Biographical Dictionary in 25 volumes* (SPb-M) p 66-67
[8] Ryabushin A V 1969 A Masterpiece of Classicism in the Kharkov *Region Construction and Architecture* **8** p 34-39
[9] Cherkasova E T 2010 *Architectural Culture of the Region* (Kharkov: Fort Publishing) 124 p
[10] Pribega L 2009 *Preservation and Restoration of the Architectural and Urban Planning Heritage. Methodological Aspect* (Kiev: Mistetstvo) 304 p
[11] Flamenitska O A, Moiseynko Z V 2016 Authenticity and Accessibility: a Concept and a Problem of Architectural Restoration *Architekturniy visnik*: KNUBA **9** p 83-92
[12] *Nara Document on Authenticity* http://www.icomos.org/charters/nara-e.pdf
[13] 2008 *Principles of Concluding Cases of Documentary Fixation of Monuments, Architectural Ensembles and Places of Interest: International Principles of Protection of Immovable Cultural Heritage* (Kiev: Feniks) p 142-145
[14] 2004 *Cultural Landscape As An Object Of Heritage* Ed. Vedenina Yu A, Kuleshovoy M E (SPb: Institute of Heritage) 620 p
[15] Sadovsky I V 1913 *Guta Estate. The Research Experience of the Natural - Historical and Technical and Economic Conditions of the South Russian Economy* (Kharkiv) p 493
[16] Lukomsky G K 1917 *Ancient estates of the Kharkov province* (Kleinmichel)

# The Modern Shopping Centers of Kharkiv in Visitors' Perception

**S A Smolenska and A S Borysenko**

Department of Architectural Design, Kharkiv National University of Civil Engineering and Architecture, Sumskaya str. 40, Kharkiv, 61002 Ukraine

argeciy@gmail.com

**Abstract.** Article is dedicated to revealing the qualitative characteristics of environment of five modern shopping centers of Kharkiv. Analysis is based on user reviews published on Google Maps. Review texts underwent semantic analysis with division of results under 4 qualitative categories, namely "commercial attractiveness", "comfortability", "energization of activity", "aesthetic articulation of the place". Results confirm the existence of three patterns of preferences' distribution that correspond to three types of shopping mall by structure (by B. Maitland), these being introverted, integrated and specialized or thematic type. The figures were then compared with results from similar analysis of foreign centers. Three shopping centers of Kharkiv received positive reviews in numbers comparable with foreign ones, all of the prevalent introverted type, signalling that competition between shopping complexes in Kharkiv is not high enough for their owners to seek new spatial concepts for retail objects. Meanwhile, developed retail markets of USA and Western Europe indicate the diversification of type and spatial characteristics of shopping centers. An adoption of shopping center organization methods from western practice and its belated spread in Ukraine lead to lag in new types' introduction. Particularly an idea of thematic narrative concept is a promising but underused resource in Kharkiv's situation.

## 1. Introduction

The article interprets shopping center as a building complex of commercial and public use, united by common pedestrian space (further referred as a "connective recreational space") into a singular complex system of large urban scale and influence that plays a significant part in citizens' preferences and way of life.

Current stage of shopping centers' development in USA and Western Europe is characterized by decreased visitors' interest to formerly popular objects. It is caused by a number of problems: the development of electronic commerce, retail real estate market's oversaturation by similar building types, the changes in consumers' preferences etc. This leads to researchers and developers seeking new design concepts, including an idea of stimulating the social and recreational use of shopping centers.

Last two decades saw an active construction of new shopping centers influenced by foreign examples in cities of Ukraine. While following the western way of development, it would be useful to evade the same mistakes and issues that are now being dealt with by shopping centers' owners and designers in countries more advanced in that regard. With this goal in mind, it is deemed necessary to take into account the existing experience and progressive trends in studying this building type.

IOP Conf. Series: Materials Science and Engineering **907** (2020) 012071 doi:10.1088/1757-899X/907/1/012071

A number of modern trends in shopping center design is linked to the practices of placemaking, dealing with people's relationship to an environment they live in. Therefore, a study of visitors' perception of five popular shopping centers of Kharkiv constructed in last 15 years is considered relevant. An analysis is based on text reviews published on Google Maps public comment section. Said analysis is considered an acceptable use in accordance to service's rules [1].

The following study is based on a general theory and typology of shopping centers developed by B. Maitland and D. Gosling [2]. Said typology defines such types as an introverted, integrated and specialized or thematic type. Modern issues in shopping centers' development are highlighted by V. Reimers [3], J. Stillerman and R. Salcedo [4], J. Beyard et al [5]. Examples of social studies utilized for defining specifics of shopping centers' perception by visitors are found in works of N. Beddington [6] and C. Teller [7]. Different sets of qualities influencing shopping center's attractiveness to visitors were proposed by C. Teller and T. Reutterer [8], S. Bagdare and R. Jain [9], R. Gomes and F. Paula [10]. It is important to note that currently there are no studies of qualitative attractiveness evaluation of Kharkiv's shopping centers from visitors' point of view, and the data on the issue is limited.

## 2. Materials and methods

### *2.1. Methods*

The study utilizes semantic analysis of visitors' reviews on five modern shopping centers of Kharkiv – 200 reviews per center, 1000 reviews in total. Said reviews are published on Google Maps service, sorted by service as relevant and containing an analyzable textual part. Additional analysis is conducted on 1000 reviews on five foreign shopping centers that are representative of different shopping center types and situated in English-speaking countries (200 per center). Both sets of data are then compared.

Semantic analysis diagnoses either positive or negative character of visitors' impressions of each of five objects and its corresponding qualities as summed under qualitative categories. Subsequently it allows to visualize the patterns of Kharkiv's visitors' preferences' distribution in regard to different qualities of shopping centers in general.

### *2.2. Limitations*

The sample of respondents is limited by Google Maps service users. Their impressions are published post-factum, therefore relate to the shopping center patronage (connected to the decision of repeat visit) more than to the shopping center retention value (denoting time spent on site). The results obtained are limited by the object of analysis, namely existing types of Kharkiv's shopping centers and their qualitative characteristics. Therefore the study is supplemented by an analysis of five representative examples of foreign shopping centers for comparison and verification purposes. Additionally, an on-site visitors' survey on the centers' premises is planned in the future.

### *2.3. Choice of qualitative categories*

A number of classification variations exists for qualitative characteristics connected to shopping centers' attractiveness to visitors. Authors propose a new set of qualities based on placemaking practices and three general aspects of environment characteristics' perception by human, according to D. Canter [11] (physical attributes, activities, conceptions). Therefore it is proposed to designate supposed aspects of shopping centers' attractiveness to visitors under following qualitative categories: "comfortability" (corresponding to "physical attributes of environment"), "energization of activity" (linked to "activities") and "aesthetic articulation of the place" (connected to "conceptions" level). The suggested qualitative categories' division is based on the possibilities of architectural expression and support of the qualities included. A basic "commercial attractiveness" is a fourth qualitative category that is connected to primary reason for a visit, an availability and diversity of commercial offer. It is mostly out of focus of the current study aside from its general role in center's attractiveness to visitors, as it is an aspect well covered by other studies and less dependent on architectural and structural specifics of its expression.

The "commercial attractiveness" category is largely defined by two indicators – the quality of tenant mix and the quality of catering options. These indicators may have differing relative values signifying different priorities in shopping center's market positioning and logic. The phrases "good selection of shops", "can get everything in one place", "excellent restaurants" or "lacking food and shopping options" are demonstrative for this category.

The "comfortability" category is comprised of three main qualities – an accessibility of site, convenience of connective recreational space, qualities of parking. Site accessibility describes both pedestrian accessibility of shopping center and its connections to public transportation infrastructure. Convenience of connective recreational space includes the convenience of internal planning and navigation, security, availability of places for a short rest, additional services and a comfortable sanitation infrastructure. The qualities of parking describe impressions of comfort, size, ease of navigation and technical facilities of center's automobile infrastructure. The "comfortability" category is often characterized in reviews by phrases like "a warm place in a cold weather", "clean and well lit", "makes shopping easier" or "place is unsafe".

The "energization of activity" category includes such characteristics as a social activation, entertainment options and connection with urban spaces. A social activation describes the shopping center's qualities connected to willing or necessary interaction with other visitors, a liveliness, crowded or deserted character of shopping center's spaces. Connection with urban spaces assumes an existence of shopping center not as an insular entity, but as an open system, an organic part of urban environment that allows an exchange of visitors with adjacent places. The examples of phrases from reviews that fall under this category are "nice place to hang out with friends", "like coming here to people watch", "bustling energy of the place" or "too crowded".

An "aesthetic articulation of the place" category is defined by such indicators as an atmosphere of internal spaces, architecture and prominent details of shopping center. Reviews speaking about an atmosphere of being inside the complex relate general impressions from aesthetic qualities defining place as a distinct entity on conceptual level. This sense of shopping center as an articulated meaningful place lived-in by people above its basic function is also emergent in reviews on architectural qualities of building as a whole and aesthetic impressions of distinct remarkable details of its environment. This category is characterized by such expressions as "beautiful place", "striking architecture", "nice building", "great sights", "neat design" or "soulless", "dull, looks like everything else".

### *2.4. Materials*

Analysis is conducted on reviews describing visitors' impressions of five popular shopping centers of Kharkiv. These are "French Boulevard" (opened in 2013, situated in an urban fabric far from the city center), "Ave Plaza" (opened in 2011 in the city center), "Dafi" (opened in 2008 in residential district), "Karavan" (opened in 2007 in residential district) and "Magelan" (opened in 2011 on city outskirts).

Additionally, reviews on five prominent foreign shopping centers of different types are analyzed for comparison purpose. Said centers are "Faneuil Hall Marketplace" (opened in 1976 in Boston, USA, a festival market/mall in a historic place), "Eaton Centre" (opened in 1977 in Toronto, Canada, an urban covered mall tightly integrated into the system of downtown urban spaces), "Mall of America" (opened in 1992 in Bloomington, USA, a third mall by size in USA, with distinct role of entertainment functions), "Cabot Circus" (opened in 1992 in Bristol, Great Britain, an integrated urban mall with mixed covered and open spaces) and "Coal Drops Yard" (opened in 2018 in London, Great Britain, an open urban shopping center close to a thematic type, based on an industrial zone renewal).

## 3. Results

An analysis of reviews allows to divide them into positive and negative ones, which are subsequently distributed by qualitative categories per center (Figure 1).

Three among five of Kharkiv's shopping centers received positive reviews measuring around 90 %, namely "French Boulevard", "Dafi" and "Karavan". "Ave Plaza" and "Magelan" received only 70,33 % and 68,36 % of positive reviews correspondingly. Moreover, "Magelan" has the least amount of reviews in total – 3683 reviews by 30.01.2020 in comparison to "Ave Plaza"'s 6761 reviews, while both opened in 2011. Meanwhile, "French Boulevard" that opened in 2013 has 18307 reviews by the same date. This marked difference denotes "Magelan"'s lesser popularity among inhabitants of Kharkiv.

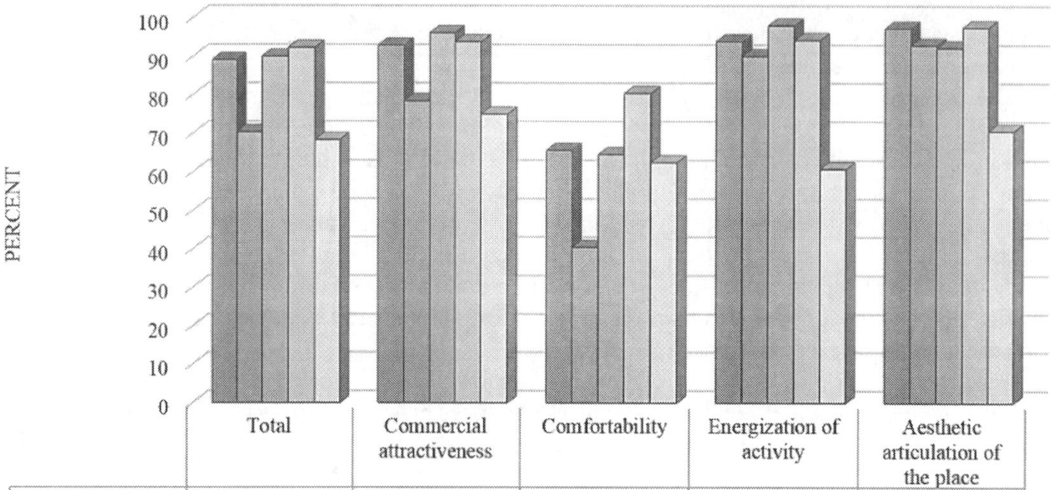


| | Total | Commercial attractiveness | Comfortability | Energization of activity | Aesthetic articulation of the place |
|---|---|---|---|---|---|
| French Boulevard | 89,15 | 92,92 | 65,66 | 93,96 | 97,37 |
| Ave Plaza | 70,33 | 78,41 | 40,43 | 90,13 | 92,98 |
| Dafi | 90,02 | 96,12 | 64,58 | 98,08 | 92,31 |
| Karavan | 92,29 | 93,84 | 80,43 | 94,32 | 97,62 |
| Magelan | 68,36 | 75 | 62,5 | 60,84 | 70,69 |

**Figure 1.** Ratios of positive reviews on five shopping centers of Kharkiv

Three shopping centers have mostly positive reviews in "commercial attractiveness" category (more than 90 %), while for "Ave Plaza" they number around 78,41 % and "Magelan" has only 75 % of positive reviews, indicating a certain level of visitors' dissatisfaction by existing commercial offer of these two centers. The "comfortability" category received only 59,69 % positive reviews on average. The quality deficiencies are most apparent in this category when present. The dissatisfaction by overcrowding of centers (or, in fewer cases, their deserted state) increases the negative impression of "energization of activity" category, while the availability of additional functions as well as different possibilities for leisure activities are almost universally positively received. The drawback is the lack of an interesting specific for leisure that can cause negative reviews. The "aesthetic articulation of the place" category has more positive reviews in general. The negative reaction from visitors is usually caused by the standard look and non-welcoming atmosphere of internal space or, in fewer cases, an architecture of the building itself when it is considered not appropriate to surrounding urban environment.

The next stage of study is focused on determining a percentage ratio of reviews that signifies visitors' attention to different qualitative categories, divided per center (Figure 2).

Taking all data on Kharkiv's shopping centers into account, the generally low level of attention to "aesthetic articulation of the place" category becomes apparent. It is relatively high in reviews on "Ave Plaza" (12,67 %) and "French Boulevard" (13,19 %). The latter shopping center also has the most amount of reviews in "energization of activity" category (31,6 %), followed by "Karavan" and "Dafi" (29,06 % and 27,6 % correspondingly). These two centers have the highest number of reviews in basic "commercial attractiveness" category (46,48 % and 46,5 %) with the lowest indicators in "aesthetic articulation of the place" (9,25 % и 5,52%). Meanwhile, "French Boulevard" has the lowest relative "commercial attractiveness" of the five examples, measuring at 38,02 %.

"Ave Plaza" has the least relative weight of "energization of activity" category in its reviews (16,89 %) due to the lack of possibilities for active entertainment, while still ranking positively because of availability of additional services and its connection to adjacent urban spaces.

IOP Conf. Series: Materials Science and Engineering **907** (2020) 012071 doi:10.1088/1757-899X/907/1/012071

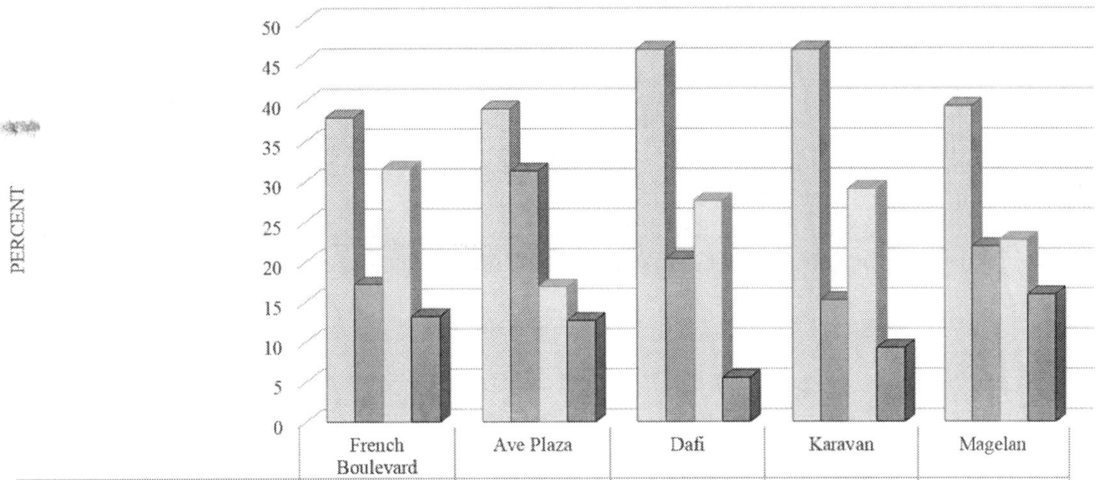


| | French Boulevard | Ave Plaza | Dafi | Karavan | Magelan |
|---|---|---|---|---|---|
| □ Commercial attractiveness | 38,02 | 39,11 | 46,5 | 46,48 | 39,45 |
| ■ Comfortability | 17,19 | 31,33 | 20,38 | 15,2 | 21,92 |
| □ Energization of activity | 31,6 | 16,89 | 27,6 | 29,07 | 22,74 |
| ■ Aesthetic articulation of the place | 13,19 | 12,67 | 5,52 | 9,25 | 15,89 |

**Figure 2.** Ratios of qualitative categories' mentions in reviews on five shopping centers of Kharkiv

Since the object of study limits the interpretation of results to a degree, the reviews on five foreign shopping centers are analyzed for comparison purpose (Figure 3). These reviews are generally positive (ranging from 89,21 to 91,68 %), which is expected for objects successful in their typological niche.

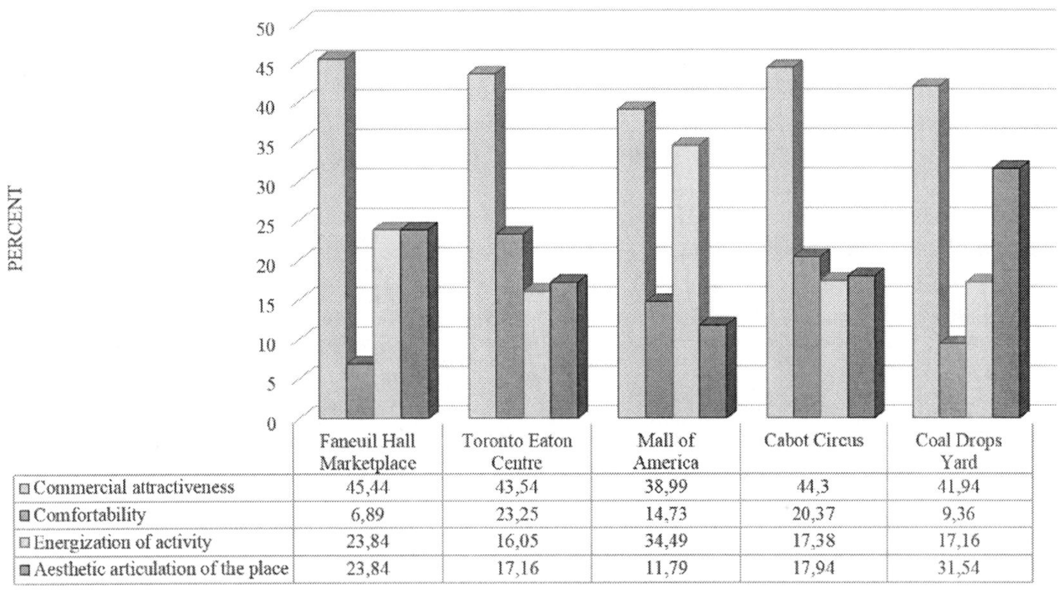


| | Faneuil Hall Marketplace | Toronto Eaton Centre | Mall of America | Cabot Circus | Coal Drops Yard |
|---|---|---|---|---|---|
| □ Commercial attractiveness | 45,44 | 43,54 | 38,99 | 44,3 | 41,94 |
| ■ Comfortability | 6,89 | 23,25 | 14,73 | 20,37 | 9,36 |
| □ Energization of activity | 23,84 | 16,05 | 34,49 | 17,38 | 17,16 |
| ■ Aesthetic articulation of the place | 23,84 | 17,16 | 11,79 | 17,94 | 31,54 |

**Figure 3.** Ratios of qualitative categories' mentions in reviews on five foreign shopping centers

The "comfortability" category received the least amount of positive reviews (76 % on average). The least positive reviews in category are on "Faneuil Hall Marketplace" (48,65 %, with low relative weight of a category measuring at 6,89 %), the most positive are on "Cabot Circus" (85,78 %, with significantly higher relative weight of category at 20,37 %). An "aesthetic articulation of the place" category has a relative weight of up to 31,54 % in comparison with other categories, which is significantly higher than numbers demonstrated in reviews on Kharkiv's centers. As a rule, reviews in this category are positive when present (94,71 % on average). However, a moderate but notable amount of negative reviews in the category (13,97 %) concerns "Mall of America", the prominent example of super-regional introverted shopping centers with sizable entertainment function. Aforementioned negative reviews generally criticize its standard reading on the place.

## 4. Discussion

Going by the data on typological characteristics and the situation of shopping centers in urban structure, we can deduce the existence of certain patterns of qualities' distribution specific to different shopping center types. The diagrams of qualities' distribution based on the data obtained from analysis show that distinct patterns are indeed present corresponding to shopping centers of specific types (Figure 4).

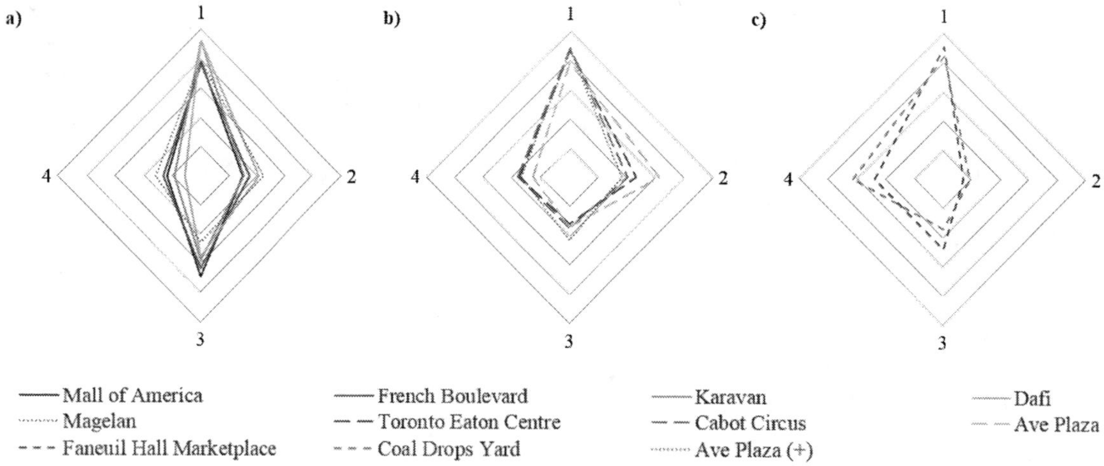


**Figure 4.** Three main patterns of visitors' preferences' distribution by qualitative categories (a – pattern for introverted type, b – pattern for integrated type, c – pattern for thematic type; 1 – "commercial attractiveness" category, 2 – "comfortability", 3 – "energization of activity", 4 – "aesthetic articulation of the place").

The similar distribution of qualities can be observed between "French Boulevard" (Kharkiv, Ukraine) and "Mall of America" (Bloomington, USA) in the process of comparing the patterns of visitors' preferences for shopping centers of introverted type (Figure 4a). This fact corresponds to the common logic of both centers' positioning as shopping centers with a large role of entertainment functions, designed for a prolonged but rarer visit. The patterns for "Dafi" and "Karavan" are similar enough as both these centers are of the same introverted type gravitating towards an out of city placement that B. Maitland defined as the "department store mall". However, both are characterized by somewhat lower role of "energization of activity" category, thus perceived as more focused on their primary commercial function.

It is also worth to note the comparably high level of positive reviews on the aforementioned three shopping centers of Kharkiv – "French Boulevard" with 89,15 %, "Dafi" with 90,02 % and "Karavan" with 92,29 %. All three are of introverted type. Combined with the above results it signifies the persisting demand for this shopping center type in Kharkiv. Both "Dafi" and "Karavan" are also situated on directly adjacent plots while having remarkably close characteristics in visitors' perception. Semantic analysis of visitors' reviews allows to suppose these centers' functioning as a singular system with mutually supporting qualities and offer. Both "Dafi" and "Karavan" have a sizably low amount of reviews mentioning "aesthetic articulation of the place" category. Meanwhile, the comparable success of "French Boulevard" with distinctly different distribution of qualities demonstrates the possibility of another approach to an introverted shopping center type's realization in the context of Ukraine.

"Magelan" presents a special case. It is also an introverted shopping center, going by its structure and out of town situation – but its pattern of qualities' distribution differs from those of the type. It has a relatively low amount of positive reviews in general. Visitors appreciate its "commercial attractiveness" in rather moderate numbers, while its "energization of activity" relative weight is not high due to its distant situation and lower accessibility in comparison with other centers. However, lower number of visitors in general has unexpectedly caused an increase in "aesthetic articulation of the place" reviews (15,89 % – highest relative amount among Kharkiv's centers), with visitors appreciating its calm and cozy atmosphere. Still, some reviewers remark on a negative effect of deserted spaces on their mood. This can serve as an evidence in favor of greater differentiation of qualities and types for future shopping centers in Kharkiv with the goal of orienting them towards different visitors' preferences.

While comparing the preferences' distribution patterns between examples of an integrated (urban) shopping center type (Figure 4b), the similarity between foreign examples is evident. Meanwhile, "Ave Plaza"'s pattern differs by greater attention to "comfortability" category, even taking into account that said category has relatively high values for all observed examples of the type. It is caused by an observable problem with paid character of center's sanitation services. Similar distortion of pattern due to negative reviews is observed in case of "Dafi" compared with other examples of introverted type. Going only by positive reviews, the corrected pattern for "Ave Plaza" (displayed on diagram as "Ave Plaza (+)") is closer to average pattern for the type. The level of visitors' satisfaction by shopping center's commercial offer is also relatively moderate.

Existing studies and literature also denote the existence of a third shopping center type by structural organization. It is an extroverted type of shopping center with an accentuated expressive and memorable environment. Examples of the type can be described by common definition of a "thematic shopping center", which also encompasses objects known as "festival malls" and "lifestyle centers". The prominent examples of the type analyzed by this study are "Faneuil Hall Marketplace" and "Coal Drops Yard". Similar to integrated urban shopping centers, they are incorporated into the city, but in a different way. First of all, they supplement urban environment with memorable, energetic, enlivened connective recreational spaces. Said spaces are usually open and adhere less by "comfortability" category, instead observably focusing on "aesthetic articulation of the place" (Figure 4c). Presently there are no examples of said shopping center type in Kharkiv. This is largely due to the fact that Kharkiv's shopping centers have not yet reached the level of development evident in western countries, wherein rising competition pushes owners into diversifying the existing shopping center types by concept and the way of use. Another possible cause is connected to seasonal limitations for open spaces' use imposed by Kharkiv's climate, when open parts of connective recreational space would see less traffic in colder months. However, "Coal Drops Yard" can be considered a successful shopping center despite facing similar limitations (90,81 % of positive reviews), which seems promising for the use of this shopping center type in the conditions of Kharkiv.

An important point in the context of results is a significantly lower attention to an "aesthetic articulation of the place" category demonstrated by Kharkiv's centers' visitors in comparison with foreign ones (Figure 5). On the one hand, Kharkiv does not have any example of thematic shopping centers whose perception is structured around this category (and only one example of an urban integrated shopping center). Nevertheless, Kharkiv's shopping centers have lower indicators in "aesthetic articulation of the place" even in comparison with foreign shopping centers of corresponding types. "Ave Plaza" has a relatively high level of visitors' attention to this category (12,67 %), while still inferior to its level for "Eaton Centre" (17,16 %) and "Cabot Circus" (17,94 %) of the similar integrated type. "Magelan" has a relatively large amount of mentions of the category (15,89 %), however only 2/3 of them are positive. "Karavan" and "Dafi" both has low level of category weight in its reviews, measuring at 9,25 and 5,52 % correspondingly. "French Boulevard" is a notable exception in this trend, having a rather high relative level of category mentions at 13,19 % while being itself evaluated positively by visitors (89,15 %). It is the only shopping center in Kharkiv that has a clear thematic concept defining both its logic of use and decorations, which is remarked positively upon by reviews. Subsequently, it can be stated that although an "aesthetic articulation of the place" plays a relatively small part in existing patterns of Kharkiv's visitors' preferences for the qualities of shopping centers, this development is not ubiquitous nor regionally ingrained, being probably caused by the present lack of prominent shopping center examples of the type that makes this category the focus of its presentation to visitors.

**Figure 5.** Comparison of visitors' preferences for different qualitative categories per shopping center (top row, left to right – "Mall of America", "Toronto Eaton Centre", "Cabot Circus", "Faneuil Hall Marketplace", "Coal Drops Yard", bottom row, left to right – "French Boulevard", "Karavan", "Dafi", "Magelan", "Ave Plaza").

## 5. Conclusions

An analysis of reviews on five shopping centers of Kharkiv shows three among them having a positive perception by visitors that is comparable to those of worlds' prominent examples. Said centers, "Dafi", "Karavan" and "French Boulevard", are all of the introverted shopping center type with a significant role of entertainment in its function. This indicates that the retail real estate market of Kharkiv has not yet reached a point of saturation with the type that would consequently lead to diversification of shopping centers' concepts and ways of use. Notable amount of negative reviews on "Ave Plaza" is caused by an observable problem in "comfortability" category, specifically the center lacking in convenience quality of its connective recreational space. "Magelan" has a wider spread of negative comments by categories, while demonstrating unexpected positive reviews in «aesthetic articulation of the place» category due to changes in the way and intensity of its usage as a recreational space. This result indicates potential possibilities for "Magelan" in case of its reformatting into another shopping center type with extended recreational function. In a broader sense it denotes the existence of hidden unrealized reserves, able to manifest spontaneously in the process of shopping center's usage depending on its localization in urban structure and other factors. Said unrealized reserves can become a resource for further customization and fine-tuning of shopping center's use, its restructuring or reconceptualization with the goal of increasing its attractiveness to visitors.

The distribution of reviews by categories per shopping center type confirms that presently there are four introverted shopping centers in Kharkiv, and only one of the integrated urban type. There is still no example of thematic type as of yet. Kharkiv's shopping centers that correspond well to their typological niche, as seen in comparison with prominent foreign examples, are characterized by generally high level of positive reviews. Two shopping centers with lower level of positive reviews have differing distributions of qualities compared with foreign examples of similar type.

Analysis diagnosed noticeably lower level of attention to "aesthetic articulation of the place" category in visitors' perception of the popular shopping centers of Kharkiv in comparison to perception of foreign examples. A notable exception is seen in "French Boulevard" which received a generally high amount of positive reviews (89,15 %) that signals a successful application of thematic concept in Kharkiv. Subsequently, the presence of a unifying theme as a narrative semantic resource assisting in identifying a unique image of a shopping center is considered underused in Kharkiv's situation.

An adoption of shopping center organization methods from western practice and its belated spread in Ukraine lead to the lag in new types' introduction. In this situation it is considered appropriate to take the progressive experience in global development of shopping centers into account with goal of evading known mistakes and issues. Monitoring the visitors' perception of existing shopping centers' environment is deemed relevant in this context. It will allow not only to identify their disadvantages, but also determine hidden potential resources and ways to optimize the spatial characteristics of their environment. Studies of this kind help to reveal and engage regional reserves in the process of development and diversification of shopping centers. Subsequently, the presence of a unifying theme as a narrative semantic resource assisting in identifying a unique image of a shopping center is considered underused in Kharkiv's situation.

**References**

[1] Google Maps Platform Terms of Service URL: https://cloud.google.com/maps-platform/terms
[2] Gosling D and Maitland B 1976 *Design and Planning of Retail Systems* (London: Architectural press)
[3] Maitland B 1985 *Shopping malls: Planning and design* (London: Construction Press)
[4] Raimers V and Clulow V 2009 Retail centres: It's time to make them convenient *Int. J. of Retail & Distribution Management* **37(7)** pp 541–562
[5] Stillerman J and Salcedo R 2012 Transposing the Urban to the Mall: Routes, Relationships and Resistance in Two Santiago, Chile, Shopping Centers *Journal of Contemporary Ethnography* **41(3)** pp 309–336
[6] Beyard M D, Corrigan M B, Kramer A, Pawlukiewicz M and Bach A 2006 *Ten principles for rethinking the mall* (Washington, D.C.: Urban Land Institute)
[7] Beddington N 1982 *Design for Shopping Centres* (London: Butterworth Scientific)
[8] Teller C 2008 Shopping Streets versus Shopping Malls – Determinants of Agglomeration Format Attractiveness from the Consumers' Point of View *International Review of Retail, Distribution and Consumer Research* **18(4)** pp 381-403
[9] Teller C and Reutterer T 2008 The Evolving Concept of Retail Attractiveness: what makes retail agglomerations attractive when customers shop at them? *Journal of Retailing and Consumer Services* **15(3)** pp 127-143
[10] Bagdare S and Jain R 2013 Measuring retail customer experience *International Journal of Retail & Distribution Management* **41(10)** pp 790-804
[11] Gomes R M and Paula F 2017 Shopping mall image: Systematic review of 40 years of research *International Review of Retail, Distribution and Consumer Research* **27(1)** pp 1–27
[12] Canter D 1977 *The Psychology of Place* (London: Architectural Press)

IOP Conf. Series: Materials Science and Engineering **907** (2020) 012072 doi:10.1088/1757-899X/907/1/012072

# Problems of Formation of Architectural and Spatial Composition Organization of the Environment of Donbas Mono City

**I V Honcharova[1], V P Myronenko[1] and O V Hubanov[2]**

[1]Department of Architectural Environment Design, Kharkiv National University of Civil Engineering and Architecture, Kharkiv, Ukraine,
[2]Department of Urban Planning, Donbass National Academy of Civil Engineering and Architecture, Kramatorsk, Ukraine

Innahoncharova37@gmail.com

**Abstract.** One of the city-forming factors for the development of the Donbas region is the creation of large full-cycle industrial complexes situated together as a group of several related industries. It allowed for the efficient use of urban lands and opened up wide range of opportunities for achievement of technological, compositional and spatial unity of industrial complexes. During industrial cities development, the residential districts was formed as large industrial complexes expanded. The historic centers of industrial cities are characterized by the grouping of cultural heritage sites around individual (usually the most powerful) industrial enterprises, most often coal mines, metallurgical and coke-chemical enterprises. The most advanced city planning ideas of the 20-50's were fully reflected in the planning structure and features of construction of industrial cities of Donbas. Many preserved and partially lost monuments of industrial architecture, as well as those that can claim such a status, are objects of protection of historical and cultural heritage, characteristic of the industrial type of cities. There are a number of specific problems inherent to cities of different categories. The major ones are those caused by significant discrepancy between the real state of the urban environment and its required state. Finding solutions to problems of corporate cities requires a multifaceted approach, as the prerequisites and characteristics of different cities can vary greatly due to historical and economic factors. Therefore, approaches to addressing the issue of corporate cities should be considered individually, considering global trends and regional peculiarities of individual regions.

## 1. Relevance

Donetsk region is the largest industrial district of Ukraine, which is very interesting in terms of history and practice of urban planning in Ukraine. However, it remains less explored than other regions and some aspects of its development are not reflected in the history of architecture and urban theory of Ukraine. Because of that, residents of other regions of Ukraine and the world at large perceive the value of the Donetsk region's cultural image as low.

In the second half of the nineteenth century, when the country's industrial development intensified, the first industrial boom in the East began. It was interrupted by the revolution of 1917, due to which the cities lost foreign investments and the active construction of the Donetsk region slowed down. The second industrial boom (1945-1970-ies) began after the end of World War II, when the damaged and destroyed factories were rebuilt accompanied by landscaping of the surrounding territories.

A combination of factors influenced the emergence and development of Donbas cities. It is established that the most significant factors were: 1) the presence of minerals; 2) the availability of cheap labor; 3) railroad construction. These factors contributed to the attraction of investments and the emergence of industry in the region. Over time, a large system of cities and towns, specialized by industry, was formed on the basis of it [1-3].

Most of these cities are characterized by the following factors: the presence of a low-power economic base; the narrowness of the industrial sphere; closure of enterprises; population outflow.

Today the city's historic centers are characterized by a grouping of cultural heritage sites around non-operating industrial enterprises. Objects and areas that used to carry out their traditional function (like housing and recreation) are undergoing significant changes. Because of this, the environment becomes depressed (as it carries a negative socio-psychological message for residents). Therefore, revitalization of the industrial buildings in post-industrial cities, preserving historic sites and changing the territories of industrial zones become an urgent issue.

## 2. Data analysis

The recognition of the industry as the main factor forming the city led to the emergence of the scientific term “industrial city” – a city of a special type with its own industrial specialization and corresponding development characteristics.

Corporate cities are widespread in all countries of the world that have begun industrialization. These cities were built around business enterprises, initially as working settlements ("industrial colonies" in Spain, "factory settlements" in Germany, "corporate cities" in the United States and Britain, "industrial mono city" in Eastern Europe).

Therefore, a mono city is an industrial center that is dominated by one enterprise (or industry). Other possible characteristics of a mono city include: isolation from large settlements; corporate paternalism; the dependence of a significant part of the population on the town-building enterprise (to varying degrees, but in all aspects – economic, cultural, emotional); economic interconnection of the enterprises of the city, formed after the town emergence [4, p. 8-10].

Typologically, mono city is divided by sectoral affiliation (mono-profile cities of light industry, mechanical engineering, etc.) [5], or by structural features of relations with the enterprise (mono profile satellite cities, cities with several city-forming enterprises, cities with single city-forming enterprise) [6].

The economic development of the Donetsk region is connected with the discovery of coal deposits in 1721 by Nikita Vepreisky, who represented civilian authority in the Bakhmut district, and Semen Chirkov, commander of the Bakhmut fortress guard battalion [7].

Active development of the region begins with the start of the reforms of Alexander II. Said reforms gave a powerful stimulus to the socio-economic development of both the empire as a whole and its individual regions. The prevalence of subsistence farming in the post-reform period created a market for capitalist production. In the second half of the 1860's of the XIX century, the rapid growth of the factory industry began. Peasant reform of 1861 contributed to the emergence of cheap labor, which contributed to the extremely high standard of its operation. Spatial forms become ambivalent, combining the features of advanced industrial cities and the development of peasant housing. However, the rapid pace of urbanization attracted foreign capital to the country, which bought the shares of Russian companies, financed domestic enterprises. From the end of the 1860's an intensive uplift of the Donbass coal industry began. Capitalist entrepreneurship has gained particular attention in connection with the construction of railways in the south of Ukraine. Because of this, the cities formed rather compactly, creating a large number of agglomerations (as typical for all of the highly urbanized areas of the world). In the early 1870's, the English capitalist J. Hughes built a large metallurgical plant in the Donetsk Basin. In the second half of the 1880s - early 1890s, new metallurgical plants were built in Ukraine, including the plant of the Donetsk Society in Druzhkivka [7]. The first industrial boom in the East began in the 2nd half of the 19th century and lasted until 1917.

In just a few years the steppe part of Ukraine has acquired industrial features. The natural landscape was being replaced by the industrial - formed by human production activities and subordinate to the needs of industry. Industry was perceived as a visual "highlight" of the region [4, p. 44]. The dominance of industrial objects as spatial dominants became evident.

The location of business enterprises and residences was entirely determined by the interests of entrepreneurs and other individuals seeking to capture the most convenient and profitable areas. The residential development was mostly typical, fulfilling only its main function. However, the external monotony fully corresponded to the advanced urban development trends of its time - a managed and controlled city.

In the early twentieth century, cities already had a more streamlined planning structure. The centers were built by professional architects and town planners, including those who were invited by the business owners from abroad (village Novo-Chaikine - architect Pushkaryov V I, village Zhovtneve - architect Trotenko V, village Radialne - architect Beketov O, social center Makiivka - architect E May, M Stam). For this reason, it is possible to trace the contrasting combination of the chaotic development of the periphery and the clearly defined blocks of the city center. This, in turn, formed a clear social structure of the settlement. The richer residents (management, officials) lived in the central orderly areas, while the poorer residents (workers and their families) lived in the periphery in the barracks or dugouts with mostly chaotic development. However, the entire disparate structure was held together by the factory as a center of gravity, and the urban space was oriented on its buildings (Figure 1, 2).

**Figure 1**. Fragment of Sloviansk city panorama. Postcard, 1902 (top) and current status, 2019 (bottom)

According to researcher Rafiyenko V, the main city-forming factor in the Donbas region was not national or cultural, but an economic one (with the enterprise as a center of community), while people as individuals were of no particular interest to anyone. The determining factor was their professional and business suitability [5, 8]. This is typical of most industrial cities in the world.

After the revolution of 1917, the active construction in Donetsk region slowed down. Cities lost foreign capital; many enterprises stopped working and were nationalized. The quality of the manufactured goods dropped, there was a brain drain and loss of technologies.

In the early 1930's, the concept of the "socialist city" began to be actively developed as a means of quickly solving a number of urgent problems. Among these problems were resettlement of the rapidly growing population of industrial areas, the provision of the production process with necessary resources, and the introduction of "new life" practices [4, p. 84-87]. The construction of new social sites began far from the existing settlements, and consequently the buildings were deprived of the urban context. In the late 1930s the territory of the Donbas became the region of the USSR with the largest area of landscapes, altered by human intervention (that will be called anthropogenic landscapes in the Soviet literature since 1970s) [4, p. 54].

**Figure 2.** Fragment of Kramatorsk city panorama. Photo card, 1920 (top) and current status, 2017 (bottom)

The second industrial boom began after the end of World War II, when the damaged and destroyed factories were rebuilt, landscaping was created, mining villages were reconstructed.

The industrial landscapes and the work of the workers themselves are ideological. Industrial landscapes are the object of aestheticization.

In the 1960s and 1970s, a special program for landscaping was launched. The ideology of landscaping was brought to the level of state instructions and included in the projects of construction of new cities. The cities were to be surrounded by 5 to 30 km of "green" zones.

Another manifestation of traditionalism and historical legitimation appeared in the post-war search for a «national» feature in architecture (the way Stalin saw it). It was in the individual house that the elements of national Ukrainian features of the region's architecture were searched for. Donbas village dwelling is closer to urban dwelling in its architectural and artistic image than the historically formed peasant house. This is accentuated by the presence of a veranda, increased height of the premises with increased window sizes, the use of factory roofing materials, etc. (However, many features of Ukrainian folk housing were preserved in its exterior) [4, p. 90].

The development of new urban areas was separated by significant distance from the historic center of the city. As a result, it became increasingly disordered and disjointed. Discussions about the future development of the Donbas started calling for following traditional features of the coal miner's settlements (proximity of housing to production, manor-building). Industrial mono city reflected their entire life cycle in district and urban spatial development (kindergarten, school, professional lyceum, factory, cultural house). This was supported by the growth of cities that already had a complete

IOP Conf. Series: Materials Science and Engineering **907** (2020) 012072 doi:10.1088/1757-899X/907/1/012072

compositional and spatial structure. Thus, each new district had a central industrial facility around which all infrastructure, social security facilities and residential quarters were built. These districts, in turn, were located outside the historic boundaries of cities with their central industrial site (often not in use yet). Consequently, industrial vacant lots formed in the centers of historic cities – as large territories littered with industrial debris.

The process of forming such industrial zones is traced to the master plans of cities. There it can be seen that residential development and social infrastructure of residential areas are formed around industrial zones (Figure 3, 4).

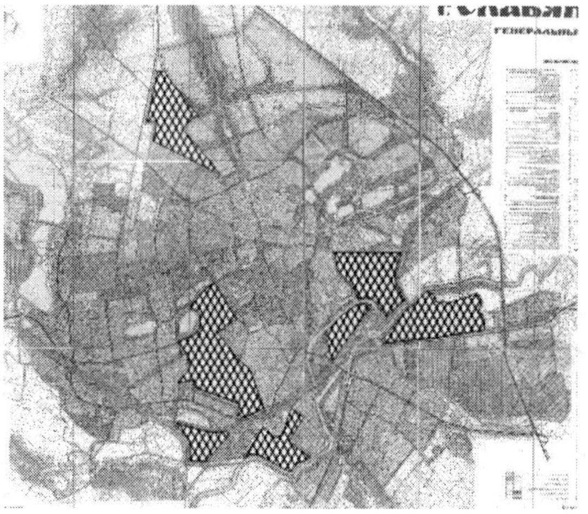

**Figure 3.** Master plan of the city of Sloviansk with the definition of large industrial zones

**Figure 4.** Master plan of the city of Kramatorsk with the definition of large industrial zones

In 2014, Donetsk was one of the most economically developed regions of Ukraine. Its main inter-branch complexes are: energy, metallurgical, mechanical engineering, chemical industry. The suburban type of agriculture is predominant. Well-developed transport and recreation complexes are present.

## 3. Research results

Donbas is characterized by the specifics of the formation of cities with a high level of urbanization, which have developed historically. Cities of Donbas are multifunctional and monofunctional industrial-administrative centers with developed functional-planning connections.

Identified problems of mono-cities can be structured into the following groups: urban (those that have arisen because of urban and architectural development of the urban environment), regional (caused by the formation of historical regional features), spatio-functional (as a result of volumetric spatial development), socio-economic. In addition, every problem identified has its impact on the architectural planning as well as compositional and spatial organization of the environment of industrial mono city. The problems and the level of their influence are clearly comprehensively presented in Table 1.

## 4. Conclusions

This study identifies and structures the problems of forming the environment of industrial mono city of Donbas. The classification of said problems is given, and the degree of their influence on the architectural-planning and spatial composition organization of the urban environment is determined. The state of the problem core of Donbas national mono-cities and their planning has been established.

The city is a complex system. The uniqueness of the city is not only in the elements of which it consists, but also in the nature of the interaction between these elements. The nature and mechanism of these relations in accordance with the principle of integrity is deemed interesting. According to it, the city can be considered as a holistic system. The location and relative position of all elements has its own rationale. Each element of this system has to be interchangeable and variable. At the same time, a complex system of interaction between urban environment facilities should not be linear. This eliminates the possibility of problems of single-industry towns when the exclusion of one element (factory) leads to the failure of the entire algorithm of the city's life.

**Table 1.** Problems of formation of environment of industrial mono city of Donbas

| № | Problems | Architectural planning | Compositional and spatial |
|---|---|---|---|
| Town planning | External monotony of rational planning | | |
| | The ideology of landscaping | | |
| | Orientation of urban space to the plant | | |
| | Distinct urban space | | |
| | Industry dominating as spatial dominants | | |
| | Social cities without urban planning context | | |
| | Loss of individuality of an image of an urban environment | | |
| | Use of a historical and regular grid of streets that do not correspond to the terrain | | |
| Regional | A large number of agglomerations | | |
| | Vacant industrial lots | | |
| | The combination of chaotic and ordered construction | | |
| | National features - mainly in the housing exterior/construction | | |
| | Prejudices about the need of a special way of life for workers and, consequently, avoiding new trends in architecture | | |
| | Submission to the ideological views of the past | | |
| Spatio-functional | Ambivalence of spatial forms | | |
| | Anthropogenicity of landscapes | | |
| | Absence of hierarchical tissue structure in superficiality and scale | | |
| | The presence of disturbed areas adjacent to the building or included in its structure | | |
| | Underdeveloped social infrastructure that corresponds to the time of its construction, but not to present conditions | | |
| Socio-economic | Clear social structure of settlement | | |
| | Reflection of the whole life cycle in the district and urban development | | |
| | Differences in the ordering of the central, middle and peripheral zones | | |
| | Poor quality of life and high social conflict | | |
| | Deterioration of the environmental situation | | |

## References

[1] Sholukh N V and Chernysh M A 2012 The cultural and historical environment in the industrial cities of Europe: the experience of conservation and regeneration *Scientific Bulletin of the Donbas National Academy of Civil Engineering and Architecture: Urban planning and spatial planning* **1** p 354-361

[2] Tabolina T V 2005 *Methods of architectural rehabilitation of the urban environment of Donbass, which contains disturbed areas* (Kharkiv: Kharkiv National University of Civil Engineering and Architecture)

[3] Shlehel K 2012 The return of European cities *Domestic notes* **3**

[4] Ilchenko M 2018 *Labor, exhaustion and success: industrial monocities of Donbass* (Lviv: Shmylovych enterprise) p 244

[5] Local Government Library 2004 *Ways to enhance the socio-economic development of single-industry cities of Russia* vol **54** (Moscow: Moscow Science Foundation) URL: http://demoscope.ru/weekly/2005/0193/biblio02.php

[6] Antipova E, Titov A 2016 The single-industry towns of Belarus: Differences in demographic and economic development *Journal of Settlements and Spatial Planning* **7** 125-136 doi: 10.19188/03JSSP022016.

[7] 2013 *Donetsk Region: Encyclopaedical Edition* (Kiev: Business Companion Publishing House) p 823

[8] Vermenych Y V 2015 Methodology. Historiography. Source studies. Donbass in the context of border theories: socio-humanitarian analysis *Ukrainian historical journal* vol **1** p 108-34

[9] Pavlov K V, Liashenko V I, Kotov E V and Mytrofanova I V 2016 Models of modernization of Donbass *Bulletin of Volgograd State University* vol **1** p 66-78

[10] Hubanov O V 2008 *Compositional and spatial principles of formation of ensembles of the 30-50s of the XX century. in the historical city centers of Donbass* (Kharkiv: Kharkiv National University of Civil Engineering and Architecture)

[11] Verheles A V 1991 *Features of urban development of the Donbass in the period from 1917 to 1941* (Moscow: Moscow State University Press)

IOP Conf. Series: Materials Science and Engineering **907** (2020) 012073 doi:10.1088/1757-899X/907/1/012073

# Principles of architectural and planning arrangements for school complexes on difficult terrain

**O Sleptsov, A Karban**

Department of Fundamentals of Architecture and Architectural Design, Kyiv National University of Construction and Architecture, Povitroflotsky Avenue 31, Kyiv, 03680 Ukraine

annakarban@ukr.net

**Abstract.** The article deals with issue of educational infrastructure in Ukraine. The authors analyzed recent research and publications with respect to development, design and construction of educational institutions. The article provides the rationale for the necessity to have school complexes on difficult terrain in Ukraine at this stage, given the needs of education, architecture and urban planning. The article provides the definition of school complexes on difficult terrain. It identifies specificities of building on difficult terrain and points out the advantages and disadvantages of shaping architectural setting of educational institutions given these conditions. The article lists and characterizes principles of shaping architectural and planning arrangements of these complexes. The distinction is made between individual principles directly linked to the object of research, school complexes on difficult terrain, and general ones applicable for all types of buildings. The study distinguishes artistic and composition principles that have an impact on forming school complexes on difficult terrain. The article also identifies approaches, means and techniques to apply identified principles.

## 1. Introduction

Environment plays crucial role in shaping personality of an individual. It is especially important for the children. We are responsible for the future of young generation and our state. It is important to provide decent conditions for the life, well-being and development of children. Unfortunately, at this stage most of the educational institutions in Ukraine do not meet modern requirements, schools need to be modernized, adjusted and sometimes completely overhauled. The country needs institutions that can become an alternative to most Ukrainian schools. Therefore, these institutions will reflect the values of ongoing educational reform in the country, modern architectural trends and construction technologies. They will promote development of new Ukrainian school and contemporary educational institutions architecture. All that prompted the need for a new type of educational institutions — school complexes on difficult terrain.

School complexes on difficult terrain are multi-profile multidisciplinary educational centers meant for temporary stay of people for the purpose of education, upbringing, spending leisure time, children physical development and sport, culture and entertainment, their nutrition, health care and exercising administrative and economic functions.

## 2. Analysis of recent research and publications

Issues and development of educational institutions buildings and their network in the system of agglomeration, development of architecture of educational institutions with unconditional teaching methods, the influence of information age in science and architecture as well as construction in challenging geological-engineering conditions have been subject of research of the following scientists: Oleg Sleptsov, Leonid Kovalskyi, Vadym Kutsevych, Helena Kovalska, Josif Karakis, Oksana Dyachok, Larysa Shuldan, etc.

Oleg Sleptsov conducted PhD research Designing *space planning structure of school buildings in challenging engineering and geological conditions (collapsible soil and undermining areas),*) [1]. A scientist and architect examine innovative approaches to organizing learning space and covers the issues of alignment of educational institutions in urbanized environment and compliance of educational buildings with the modern standards. He conducted serious theoretical research in architectural development of educational institutions in Ukraine and introduced it into practice. The book authored by Oleg Sleptsov «*Modern school architecture: gymnasium of the latest biotechnologies*» [2] can be considered a reference book on arranging educational institutions environment. Leonid Kovalskiy conducted theoretical research on architecture of educational institutions and had Doctoral thesis «*Problems of development of architecture of educational institutions buildings*» [3] and authored a publication «*Architecture of educational institutions buildings*». He has enjoyed cooperation with Oleg Sleptsov on designing, construction and operation of educational buildings in challenging engineering and geological conditions. Vadym Kutsevych made a significant contribution to theoretical issues of development that deal with legal basis for educational institutions design.

Research on general theoretical issues of development of modern education pedagogics, history of educational science, psychology and philosophy was conducted by Myhailo Levkivsky, Vasyl Madzigon, Valentyna Dorotuk, Lubov Tarabasova, Hans Brugelmann and Ken Robinson.

## 3. Objective of the article

The article is aimed at determining grounds for introducing a new type of educational institutions in Ukraine, identifying and defining principles and approaches to architecture and planning of school complexes on difficult terrain.

## 4. Results and discussions

Selection of a building site depends on several factors. In Ukraine, challenging terrains are virtually unused for mass building. Typically, there you can see private dwelling housed based on individual design or recreation zones with landscape design, otherwise, they mostly remain in their original condition. Difficult terrains have one big advantage: price of a plot of land. Normally it is considerably lower than that of the sites on the even land, therefore they are affordable for developers. Their disadvantage, though, is the complexity of construction works and the need in larger investments. Provided you pursue the right approaches to the development these areas, you can balance construction expenditures. For instance, if you choose hillside facing south, south east and east with sufficient insulation and visibility conditions, you can have an energy-efficient house and save on heating and electricity bills. Additional advantage that can be provided there are green terraces and easy drainage of surface waters. Sloping terrain has an effect on architecture, planning, structure of the buildings and their artistic composition. Thanks to latest construction and technological methods challenging terrains do not cause any major difficulties for architects and they can become an asset in case of right architectural and planning approaches employed.

The research proved that school complexes (centers) on challenging terrain can see several types of spatial organization, formation solutions and artistic composition image. These types are determined by specific features and are governed by common principles of development. The notion «principles» provides for a number of general requirements and rules impacting forming different projects.

You can distinguish nine individual principles of architectural and planning organization of school complexes on difficult terrain. They are characteristic of a specific object. You can also single out four general groups of requirements for all public buildings. Individual principles include «localization of a setting», «integration», «accessibility», «functional distribution», «adaptation», «ethical expressiveness», «environmental friendliness and energy efficiency», «progressiveness» і «visibility».

IOP Conf. Series: Materials Science and Engineering **907** (2020) 012073 doi:10.1088/1757-899X/907/1/012073

General requirements are based on the rules of functionality and technical feasibility, cost-effectiveness and artistic expressiveness. All these principles are closely interlinked and complement each other.

Principle of «localization of a setting» deals primarily with the choice of slope, its configuration and location with regard to cardinal direction as well as its location in relation to a populated (build-up) area. This principle serves the basis for categorizing/ distinguishing main and additional functions of the center/complex objects. Principle of «localization of a setting» is closely linked to the principles of «integration» and «accessibility».

«Integration» principle unites all the buildings of the complex and the landscape into one integral whole through the development of their interconnectivity and adjustment. These requirements are linked with the principle of «ethical expressiveness». They include visual interaction of nature and architecture.

«Accessibility» principle is crucial to find solution to transport and pedestrian connections horizontally and vertically on the slope, passages and driveways are arranged on the territory of the center and external accesses are provided for. In other words, the complex is getting connected with the existing transport infrastructure outside the construction object and inside the complex. Fire safety requirements are essential to the principle.

«Functional distribution» principle establishes the rules to determine functional zones on the project site and provides for their accessibility. That is the principle that provides the solution to the location and grouping of the school complex building depending on their functionality as set out by general layout and spatial planning structure– developing the center buildings separately. The principle governs the choice and location of points of entrance (from the top, in the middle or at the bottom of the hill), main and ancillary construction objects.

«Adaptation» principle is aimed at ensuring prospective development of the center thanks to internal and external objects transformation and tailoring buildings spatial and planning structure to the needs dictated by the time. The principle also provides for universal use of buildings internal space.

«Ethical expressiveness» is defined by architectural, composition and artistic solutions and artistic expressiveness. It is based on the combination of the function and form and philosophy of the architecture. Developing the exterior and reflecting its elements in the interior.

«Environmental friendliness and energy efficiency» secure sufficient greenery to meet various needs, including recreational. This principle requirements are part of site organization and contribute to the development of architectural and artistic character, create the best controlled environment, temperature and humidity control, radiological and wind regimes within the site and adjoining areas. The principle is based on effective energy use. It is achieved by installing solar cell panels and solar power generators and «passive houses».

«Progressiveness» means compliance with modern social conditions and needs. It is achieved by using cutting edge technologies, scientific, technical, architectural, artistic and engineering achievements.

«Visibility» will ensure realization of the intended use of the buildings. The principal provides for common stylistic solution of all school complexes /centers. This principle is closely intertwined with the fundamentals of «ethical expressiveness».

Since school complexes on challenged terrains are public buildings, they should meet general common principles of buildings design and development: functionality and technical feasibility, cost – effectiveness and artistic expressiveness.

«Functionality» principle means that buildings should be in full compliance with their intended use. Technical feasibility is to ensure building`s protection from the climatic and natural impact (snow load, wind, precipitation, temperature variations, etc.). The principle determines the choice of the most relevant and proper construction design for each public building type.

«Cost-effectiveness» provides for the choice and use of the most efficient way of buildings construction, with the least financial and human labor costs. To comply with the requirements construction costs should be estimated and recovered in the course of building upkeep and use.

Implementation of the abovementioned principles of school complexes construction on difficult terrain require engaging specific techniques and means. They interact with each other and are employed simultaneously by several principles. Key elements for «localization of a setting», «integration», «accessibility», «functional distribution» are the choice of specific spatial and construction volume solutions of the buildings design, their architectural and construction solutions. «Ethical expressiveness» is determined by the color scheme and construction materials used for the decor. «Visibility» principle is secured through the external display of understanding of the intended use of the building. Techniques employed in «adaptation» principle are spatial modification and transformation by using movable partitioning and transformer furniture. These techniques are partly applicable in the «progressiveness» principle. To implement «environmental friendliness and energy efficiency» principle we use solar and wind energy, environmentally friendly materials for construction and decoration. This is achieved through solar cell panels, solar power generators, «passive houses», vertical and horizontal landscaping, modern ventilating systems, air humidification and purification, creating the best controlled environment.

Except to the generally accepted rules and approaches that influence the establishment of educational institutions, the school complexes considered in this study have a number of individual ones. They are associated with the design area a complex terrain. Difficult terrain is the dominant environment, creates a silhouette of the building, affects the development of architectural and planning organization and artistic and compositional image.

When designing buildings on the terrain should be guided by two basic rules of interaction between architecture and landscape: «polarization» (separation) and «integration» (merger) [4] (Fig. 1). The first work due to the approaches of contrast - the opposition of architecture to the natural landscape, the accentuation of their differences, the second is a nuanced relationship - the fusion of terrain and buildings. The rules of «polarization» and «integration» have two types of formation: visual and construction. The visual is the maximum preservation of natural terrain, the approaches of contrast and nuance work only as visible. During this type of building rises above the natural landscape with the help of columns and supports, consoles are used. with the construction type, the territory undergoes significant changes, and the approaches of contrast and nuance are formed by the architecture itself. Buildings can go significantly deeper into the ground, to straight them you need to level or fill the ground cover, terracing the area.

The interaction of architecture and landscape according to the principles of «polarization» involves the use of materials that will bring dissonance between buildings and nature. This effect can be achieved through architectural and structural and finishing solutions. To do this, you should use metal, glass, bright colors, textures, open structures, tools that will clearly distinguish between natural and human activities.

In the case of organizing an artistic and compositional image according to the rules of «integration», the building should be disguised in a natural landscape. They must fit organically into the terrain with the help of terraces, ramps, stairs, retaining walls. One of the main methods of merging architecture and nature is the subordination of buildings to differences in terrain. In this case, the objects seem to repeat the slope, parallel to it. To achieve harmony using nuanced approaches, natural materials (stone, wood) and pastel, earth tones are used. Landscaping of vertical and horizontal surfaces of buildings plays an important role.

When searching for the most profitable solution, depending on the specific case, the principles of «integration» and «polarization» can be combined in one object. Successful examples of compliance with the rules of «polarization» and «integration»« in modern buildings located on difficult terrains are educational institutions: Lusitania Paz de Colombia School and Jean Moulin High School (Fig. 1. a, b). They reflect opposite directions to the organization of architectural objects on difficult terrains. Lusitania Paz de Colombia School - is a model of «polarization» with contrasting approaches, and Jean Moulin High School – merges with the environment through nuanced grounds.

Lusitania Paz de Colombia School was built in 2015 in Medellin, Colombia by architect Camilo Avellaneda. The project uses pure forms which are taken into account in order to divide the program into three buildings, thus creating open spaces linked to the surroundings. The articulation of the volumetric composition puts in evidence a double scale relation: a territorial one that establishes visual links along with the landscape; and a local one, where the spaces produced between the volumes, generate high impact relations in the urban and school life. The building is one of the best examples of modern eco-architecture, in addition, due to its progressive architectural and engineering solution, as well as the characteristic aesthetics is the hallmark of the city. The school building is located on a difficult terrain and its volume, which combines rectangular and cylindrical shapes, contrasts with the environment. In addition, the school has a light exterior facades, which further accentuates the building – distinguishes it among the greenery of the park, next to which it is located [5].

Jean Moulin High School was built in 2016 in Revin, France. This is a project of the architectural firm «Duncan Lewis Scape Architecture». The project of the entire school complex was based on a strong and shared concern for the rebuilding of the relationship between the built and the natural environment, long before it had become a trend among the new generation of architects. The building had to develop a relationship with the landscape, one that would help merge them together. The ambition of the architect was to mimic the mountain silhouette and to redress the building as far as possible in vegetation. The new school closely follows the topography of the land, it literally sticks to the bedrock which it occupies, from the crest of the plateau all the way to the residential area that separates it from the river. The building is located on a territory with steep hills and gentle slopes, surrounded by forest. The educational elements of the program are distributed across two long, low buildings (referred to as 'strips'). These buildings are arranged in terraces that provide all the classrooms wide panoramic views to the south west. Due to the stepped shape, the building repeats the shape of the slope, and green terraces «dissolve» it in its natural environment [6].

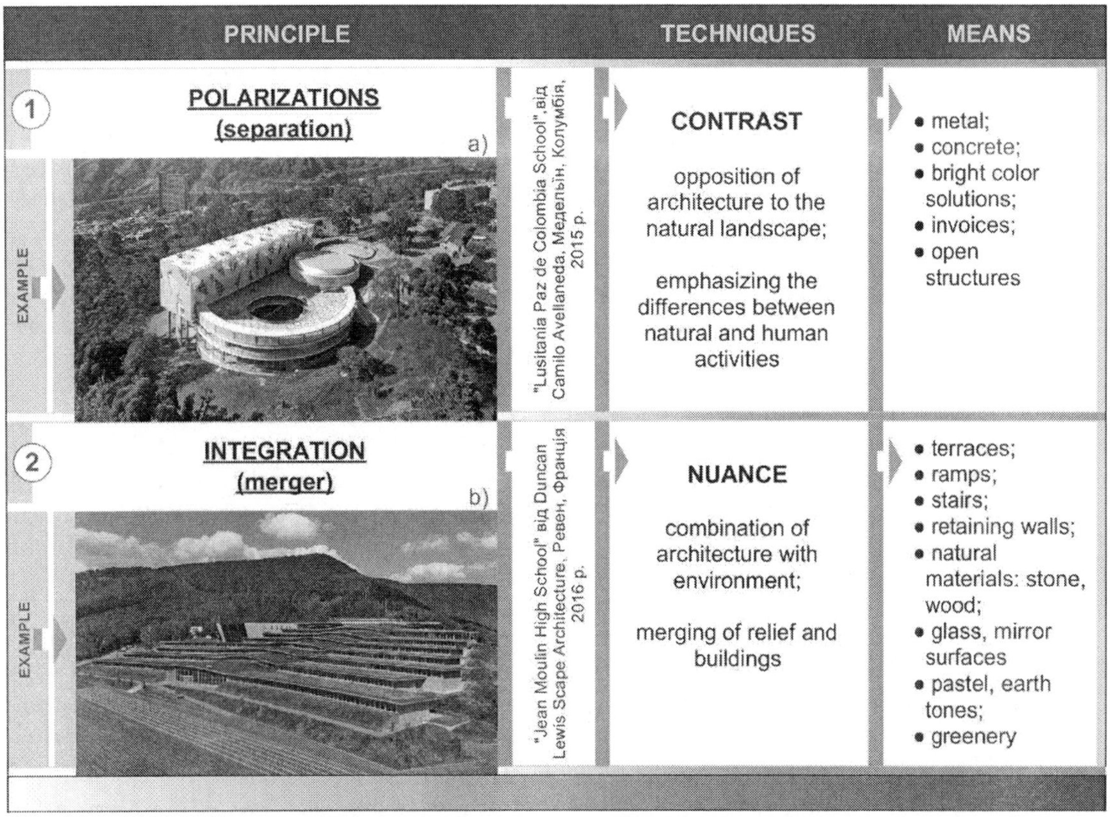


**Figure 1**. Interaction of architecture and landscape

An architect improves the setting to match the objectives and intended use of the building. The choices made have significant impact on arranging school complexes on challenging terrain. It can improve quality and efficacy of the learning process and knowledge and skills acquired.

IOP Conf. Series: Materials Science and Engineering **907** (2020) 012073 doi:10.1088/1757-899X/907/1/012073

The interaction of architecture and landscape directly forms the architectural, artistic and compositional principles and patterns. We can distinguish the principles «stylistic unity», «harmony with the setting», «universality of interior», «harmony of functions and forms», «content philosophy» and «including elements of exterior into interior». They affect the organization of premises and spaces of school complexes on difficult terrain.

«Stylistic unity» covers both exterior and interior decoration: finishing materials, colors, structural elements of a building used in all complexes to turn them into a showcase.

«Harmony with the setting» principle means making architectural object part of the environment based on contrast (contrasting architecture with the landscape with the focus on differences between the human –being and nature) or nuanced approaches (harmony of architecture with environment, blending of architecture with environment, unity of topography and buildings). Techniques engaged in accomplishing the goals are materials and colors as well as construction solutions: terraces, ramps- it all depends on the approach pursued. The principle is based on the proportion of architectural forms and setting, stylistic integration of forms and space, bioclimatology [7] (adaptation to the climate of the region, use of renewable sources of energy).

«Universality of interior» principle allows for the change, adjustment and space adaptation to respective needs. It enables open common space arrangement. To this end, it is necessary to use movable partitions and transformer furniture, use neutral colors for decoration. Open space zoning is accomplished by using different colors, lighting and difference in elevation of ceiling and floor.

Harmony of functions, form and functionality, unity of style in form and space are the techniques and methods of «harmony of functions and forms» principle. This is accomplished thanks to the construction materials, structural design, color scheme and decor.

Since school complexes on difficult terrain are developed for children, philosophy and artistic character play crucial role in finding best architectural solution when constructing educational establishment. That`s the core of «content philosophy» principle. Everything matters here: form, color, texture, theme, style and decor.

«Including elements of exterior into interior» is based on the interconnection of rooms and external space that blend, use of natural materials and beautification, glassed surfaces, «engaging the outside in».

Buildings architecture is of crucial importance for people`s perception and it shapes person`s attitude to the space. Scale relations, details, structural components, color, surface finish and, naturally, unity with the landscape around them are critical to developing harmonious impression.

Identified principle of architectural planning and artistic and structural organization are mandatory for developing school complexes on difficult terrain. They are applicable interchangeably and comprehensively, complementing each other.

## 5. Conclusions

Properly planned and well thought-through work of an architect ensures development of a comfortable environment and space for people tailored to their needs. In the course of working on the construction object it is necessary to take into account the setting of the construction object and its potential users. An architect shapes the respective character of buildings to be using relevant principles, techniques, location and planning, interior and exterior. Philosophy and design themes, choice of colors and decor play big role in creation of architectural objects. It is especially important for educational institutions, as children are more sensitive to the world around them.

One of the main tasks when designing school complexes is developing creative space enabling children personal development with the help of ergonomic, flexible, multifunctional interest-generating space. Ethical value of educational environment: lighting, colors, spatial relations, unity with nature are integral elements of the process.

Following the rules of architectural planning and artistic composition arrangements of school complexes on difficult terrain will ensure creation proper conditions to encourage learning process and assist children in their personal self–fulfillment and making informed decisions as to their future, thus, promoting smart integral communities.

In connection with the reform of education in Ukraine, the introduction of the program «New Ukrainian School» [8], there is a need for new architectural and construction solutions. Children need space and freedom of action both in school and indoors and outdoors. Freedom of action and space encourages development and does not restrain children in search of knowledge.

School complexes /centers on difficult terrain are technologically new educational institutions that have no analogues. They are designed not only to meet the functional and standardized needs and the generally accepted set of subjects provided by the reform of education in Ukraine, such educational complexes give children a choice and take into account their requests. School complexes on difficult terrain contribute to the formation and development of comprehensively developed individuals motivated by self-education and self-development, conscious, socially active citizens, patriots, able to ensure economic growth and cultural development of the country.

**References**

[1] Sleptsov O S 1986 *Designing space planning structure of school buildings in challenging engineering and geological conditions (collapsible soil and undermining areas)*, Doctoral Thesis, Kyiv National University of Construction and Architecture (Kyiv KievZNIIEP) p 230

[2] Sleptsov O S 2011 *Modern school architecture: gymnasium of the latest biotechnologies* (Kyiv, A+S) p 120

[3] Giyasov A 2018 Architectural-planning and constructive structures of residential buildings and buildings on complex terrain *IOP Conf. Series: Materials Science and Engineering Form-2018* **365** 022024 doi:10.1088/1757-899X/365/2/022024

[4] *Lusitania Paz de Colombia School* Retrieved from: https://www.archdaily.com/793564/lusitania-paz-de-colombia-school-camilo-avellaneda?ad_source=search&ad_medium=search_result_all

[5] *Jean Moulin High School* Retrieved from: https://www.archdaily.com/804930/jean-moulin-high-school-duncan-lewis-scape-architecture?ad_source=search&ad_medium=search_result_all

[6] Udomiaye E, Okon I U, Uzodimma O C, Patrick N 2018 Eco-friendly buildings: the architect's perspectives *International Journal of Civil Engineering, Construction and Estate Management* **6** (2) 14-26.

[7] *The new Ukrainian school* 2016 Edited by Gryshchenko M (Kiev) p 36 URL: http://mon.gov.ua/Новини %202016/08/21/2016-08-17- 3-.pdf

IOP Conf. Series: Materials Science and Engineering **907** (2020) 012074 doi:10.1088/1757-899X/907/1/012074

# Architectural and Urban Planning Features of Ecotopia

**A Fedak**

Department of Design and Architecture Fundamentals, Institute of Architecture, Lviv Polytechnic National University, Bandera str., 12, Lviv, 79013 Ukraine

ann.fedak67@gmail.com

**Abstract**. This article has been devoted to the analysis of the description of architecture in the works of utopian philosophers, and the identification of the main means by which they tried to ensure the environmental friendliness of their society and connection with nature. Utopian ideas and concepts arise as a result of dissatisfaction with the standard of living and contribute to the active development of society. The models of states described in utopian works serve as a source of ideas for possible scenarios for improving current systems. Ecotopies are a specific type of utopias which describe states with a clear and exclusively ecologically-oriented social order. This article presents an analysis of architectural and engineering means of communication with nature in the works by Plato, T. More, T. Campanella, Cyrano de Bergerac, D. Vairasse and V. Odoyevsky. The work of E. Callenbach, as the first example of ecotopia, is analyzed in detail, and the peculiarities of the architecture and urban structure of the cities of Ecotopia are highlighted.

## 1. Introduction

Considering the deterioration of the environment and climate change as a result of human activities, the principles of sustainable development are becoming highly important. This concept includes aspects of social policy, economics and ecology. The concept of sustainable development envisages ensuring quality living conditions for people and minimizing the negative impact on the environment to preserve and improve the environment in the future.

Creating a society that ensures social and economic equality of citizens is an integral attribute of utopian concepts. The term "utopia" comes from the title of Thomas More's work "Utopia" and means "a place that does not exist" [1]. As of today, there are a significant number of different types of utopian consciousness, the youngest of which is considered to be ecological [2]. For their part, works that contain descriptions of ideal states, where everyone lives in appropriate and equal conditions and, at the same time, pays considerable attention to ecology and the state of nature are called ecotopias. Ecotopia – is "an ecologically ideal region or form of society, generally viewed as imaginary" [3]. Such works include the works of E. Callenbach, M. Bookchin and others.

Architecture and urban planning in utopias serve as a basis for implementing ideas into reality. This explains the detailed description of the urban environment and modeling the lives of people in it by utopian philosophers.

It should be noted that among the specialized utopias within a particular field of human activity, architectural utopias are also distinguished. Architectural utopias are utopian concepts authored by architects and urban planners. The main task of these utopias is to develop a holistic image of the city and humanize the architectural space, following the social component of utopian ideas, and, as a consequence, ensure the possibility of their implementation. "Architectural utopias formed from socio-utopian ideals, images-goals realized by architecture" [4, p 4].

Partial implementation of the concept of sustainable development is possible due to the use of environmentally friendly materials, reducing the building area, providing buildings with environmentally friendly energy sources and more. In fact, qualitative transformations are feasible only with significant changes in people's lifestyles. According to the fact that one of the features of utopias is an attempt to predict the future or demonstrate an alternative version of the present, it is interesting to analyze ecotopias and describe the urban environment and architecture that they envisage.

A large number of scientific articles and books are devoted to the study of utopias. Most researchers analyze them primarily in terms of the socio-political sphere, although there are also works that consider the architectural and urban components.

The problems of coverage of architecture and urbanism in social-utopists' works have been outlined in works by H. Hradov (1968); S. Khan-Magomedov (2001); M. Chyutin (2006); T.Morrison, M.Rubin (2014); A. Romanova (2015); E. Eylers (2015) and others. For instance, Hradov provides schematic maps of the cities of T. More and T. Campanella, and describes the main aspects of the architecture of these ideal countries. Khan-Magomedov analyzed the images of architecture and urban structure in the works of Plato, T. More, T. Campanella, D. Vairasse and others.

From the architectural perspective, the most interesting part in the works of utopian philosophers is the description of the actual urban structure and the approach to the functional division of the city and buildings. Besides, it is worth mentioning the study by K. Rusiev (2015), who analyzes the images of architectural utopias in environmental science fiction films.

The analysis of the attitude to nature and ecology in utopian works is reduced mainly to the consideration of political laws and behavioral practices of the population. A political theorist Marius de Geus (1999) analyzed aspects of ecology in treatises written by T. More,H.-D. Thoreau, P.Kropotkin, W. Morris, E.Howard, B. F. Skinner, A. Huxley, E. Callenbach, and M.Bookchin [5].

A detailed analysis of the architecture of ecological settlements in utopias, both socio-political and architectural, is presented in the thesis by Pandit Silpirekha [6].

The aim of this article is to analyze the description of architecture in the works of utopian philosophers and to identify the main means by which they tried to ensure the environmental friendliness of their society and connection with nature.

## 2. Research Methodology

The main research method in this study was thematic analysis, which was used for analyzing qualitative data in treatises by Plato, T. More, T. Campanella, Cyrano de Bergerac, D. Vairasse and V.Odoyevsky. These novels were selected, as they are vivid and most common examples of utopian works. These treatises are referred to as "utopia of flight", they describe the structure of states located in remote areas from real territories. Aspects of architecture and urban planning are analyzed in more detail in the work of E. Callenbach – since this work is considered the first ecotopia.

In the analysis of these works, there were applied general scientific methods, namely empirical methods of operation: the thematic analysis of literature, documents; theoretical methods-operations: generalization, abstraction and comparison.

## 3. Results and discussions

### *3.1. Connection with nature in the treatises by Plato, T. More, T. Campanella, Cyrano de Bergerac, D. Vairasse and V. Odoyevsky*

The works of utopian philosophers emphasize the perfect social order, which extends to all possible levels of human life, including their interaction with nature. Thus, the utopian authors, even without paying much attention to it, still tried to demonstrate the mechanisms of combining architectural objects and citizens with ecology and nature. Even Plato (354 BC), describing the perfect state noted the need for equipment of primitive structures to collect rainwater. The proposal to collect precipitation for future use was quite common in the first works of utopian philosophers.

Thomas More in his "Utopia" (1516) described the island state of Utopia, the capital of which is the city of Amaurot. It is a square city with a grid plan with no private property. At the same time, describing the water supply system, the philosopher noted that the cities provide special containers for collecting rainwater. Such tanks allowed residents of those areas where it is not possible to supply water through the terrain through the general water supply network, to receive clean water for their own needs. Thomas More was succeeded by Tomazzo Campanella, who in his philosophical work "The City of the Sun" (1623) described an ideal state on an island in the Indian Ocean. The city of the Sun in plan has the form of a circle, and the street network is formed by a radial-circular scheme. The author points out that all residential buildings are located in a circle (the city itself consists of 7 rings) and have flat roofs. Similar to More, Campanella considers it appropriate to create a network of gutters to save rainwater, which can then be used [7, p 42].

Cyrano de Bergerac in his novel "Comical History of the States and Empires of the Moon " (1657), describing the imaginary cities on the Moon, mentions the "Ambulatory Towns". Such cities consist of separate buildings on wheels, which due to the use of wind force can move long distances. [8].

Severambian – a fictional perfect society on an island near Australia, described by Denis Vairasse in the utopia "History Of Sevarambes" (1675). The author emphasizes that all cities are planned according to a single principle, and consist of osmasia – a modular unit of the city, which are multifunctional square buildings. Sevarambes are very caring about nature and believe that the territory should be built only when the number of inhabitants of existing cities exceeds the norm. All cities have a grid plan and are designed as comfortable as possible for people. A large amount of greenery is provided, both on the patio and the streets. Continuing the theme of collecting rainwater, Vairasse proposes to install rainwater collectors not only on the streets but also on the roofs of buildings. This water enters the fountains, which are in each osmasies, in the bathroom and is used for watering plants. Taking care of the land, the sevarambes dug a large number of water canals to provide dry areas with water and invented a secret way to turn sand into fertile soil [9].

A more futuristic version of utopia was described by Vladimir Odoyevsky in his unfinished novel "The Year 4338: Petersburg Letters" (1835-1840). The described events take place in the year 4338, when high-speed underground trains and "flying aerostaties" became the main transport of cities. Given the significant development of technology, people have learned to have a strong influence on nature. The network of special pipes covers almost the entire surface of the earth, due to which warm air is delivered to residential buildings, gardens, and even to the airways. Scientists have learned to modify nature and the atmosphere through the use of technology (fans) and the mixing of chemical compounds. An artificial environment has been developed on the territory of the research center, which simulates the natural conditions of different parts of the globe, and allows to reproduce the flora and fauna of those regions. Due to the powerful technical development, the work does not pay much attention to the issue of ecology [10].

### *3.2. Features of architecture and urban planning in "Ecotopia" by E. Callenbach*

The first version of a truly ecological utopia is Ernest Callenbach's novel "Ecotopia: The Notebooks and Reports of William Weston" (1975), which describes the structure of the fictional country of Ecotopia. William Weston – is an American journalist who went to Ecotopia for 6 weeks to study their standard of living and behavioral habits, social and political system. The book itself consists of an alternation of small magazine columns that the journalist wrote for publication in a newspaper with pages from personal diary.

Callenbach emphasizes that "Ecotopia is not a "futurist" book that simply attempts to project trends. Nor does it portray a utopia—an imaginary country where everything is entirely perfect forever" [11, p 170]. Nevertheless, this work still has clear signs of utopianism, and is an example of an ecological model of an ideal social order, where people are equal to each other, regardless of sex or level of knowledge, and have a strong connection with nature. "Thus people can be just people, without our symbolic loading on sex roles" [11, p 33].

The country of Ectopia is located in the western United States and consists of Northern California, Oregon and Washington, which were separated from the common territory of the United States and gained their independence as a separate state (Figure 1).

**Figure 1.** Location of Ecotopia on the map of USA
(WA – Washington; OR- Oregon; NorCal - Northern California) [map by author]

At the time of the events described in the novel, Ecotopia has been independent for 19 years. The country is surrounded by a closed border and separated from the rest of America by the Sierra Nevada Mountain. The border is formed of a wooden fence with a large gate, near which is a small stone guard-house [11, p 5]. The author mentions such cities as San Francisco; Alviso; Redwood City; Healdsburg; Marshall city; Punta Gorda; Santa Cruz etc.

Reforms in the country began with the introduction of decentralization; normalization of the number of populations by reducing the birth rate; nationalization of agriculture; bans on the oil industry; and laws on the protection of forests and water [11, p 44]. After the declaration of independence, there was a period when the construction of only wooden buildings was allowed. Archibald Fir was the most famous architect of that period. "He wrote a remarkably influential study of earlier wood building, and helped lay down the specifications that plastics had to meet to be allowable as building materials" [11, p 97].

The main transport connecting different settlements is the train. The developed railway network covers almost the entire territory of the country. Trains have a futuristic design and each one "looks like a wingless airplane" [11, p 7]. In addition to trains and electric cars, they actively use water transport, for which harbors are arranged [11, p 25].

Unlike many utopian countries of its predecessors, the Ecotopia was "built" not on new lands, but on the territory already built up by cities and inhabited by inhabitants. As a result, there is a need to describe not only the mechanism of building new cities, but also the reorganization of the environment of existing cities following the idea of quality and appropriate living space.

Thus, the streets of San Francisco have changed from pedestrian streets to pedestrian, rich in trees, and recreation areas with fountains, sculptures, small gardens with benches and stalls. The carriageway of the main boulevard now consists of only two lanes, which are mostly used by electric taxis, delivery cars and minibuses. Public transport stops look like pavilions with a conical roof, where there are small shops with a press. The entrance to the vehicle has no steps, so the floor level of the stop corresponds to the floor level of the vehicle. There are no drivers, and all transport is electric and free for the population [11, p 11]. Given the proliferation of public transport, the car-free zone has expanded to almost all densely populated cities: private transport is prohibited for all residents. Public transport is frequent, although most residents have bicycles that are convenient to travel both in and between cities. The streets are clean and not littered with information garbage: not a significant number of street signs on the facades of buildings, and those that are, attached to the corner facades [11, p 14]. The streets are not lit at night, although this does not affect the crime rate.

In addition to the reconstruction of the street network, rivers - which years ago had sunk underground to form a sewer and water supply network, -were also raised. Ecotopians cleaned such reservoirs and raised them to the street level, arranging landscaped promenades, and forming attractive urban spaces. [11, p 12].

Urban space is actively used by all segments of the population, regardless of age. In addition, various fairs are held in the city squares, where you can buy farm products, etc. And on the stairs of the old town hall, for example, various performances can be held [11, p 88]. Most of the buildings of historic cities have become multifunctional: combined residential and public functions (shops, offices, catering establishments, etc.) [11, p 14]. Part of the housing stock was demolished to restore natural conditions: increase the area of forests [11, p 29].

Thus, it can be argued that urban spaces and the street network of existing cities are undergoing significant transformations and becoming people-oriented. In part, this resonates with ideas that were prevalent among architects and urban planners in the 1970s. The theme is "new urbanism" and reducing the use of cars by increasing the number and frequency of public transport, landscaping of urban spaces and the introduction of a mixed function of districts.

In Ecotopia, new settlements are emerging next to historic cities, which were immediately designed according to the ideology of sustainable development. The street network of such cities is formed not by a grid plan, but by narrow and winding streets. Public functions (shops, libraries, bakeries, grocery stores, workshops) - are mixed with residential buildings, which are built mostly of wood. “Though these structures are old-fashioned looking, they have pleasant small balconies, roof gardens, and verandas—often covered with plants, or even small trees. The apartments themselves are very large by our standards—with 10 or 15 rooms, to accommodate their communal living groups” [11, p 24]. The city has a very large number of greenery and mini parks, relatively high population density. Each new small town is located next to a train stop, which allows to form a network of connected cities. “Ideal number of citizens for an urban constellation is 40 - 50,000 – total number of people in satellite mini cities” [11, p 62].

This idea of small green rural settlements is reminiscent of Ebenezer Howard's idea of the Garden City. Due to decentralization and the formation of autonomous settlements with a large number of greeneries, which should be of a fixed size, residents would have to live in comfortable conditions and be in constant contact with nature.

The main building materials are wood and stone. In line with ecotopians desire to move to zero waste generation at the legislative level, many materials that cannot be plugged have been replaced by natural analogues and reduced use of paint, etc. [11, p 20].

Plastic produced in Ecotopia is plant-based and contains biological sources - is light and cheap, and recyclable. In this case, part of the plastic decomposes itself within a month. In addition to this type of plastic, more solid plastic is also made, which is not a subject to rotting or self-decomposition. It is used “for minibus bodies, “extruded houses,” coins, bottles, and mechanical objects of many kinds” [11, p 78].

“Extruded houses” are modular houses made from plastic which are easily transported due to their size. Each module has the form of a pipe, approximately 3 m high and 4 m wide, with a flat floor. Such modules form larger houses of different shapes, with different sets of windows and doors, thus very rarely repeated, even though they consist of manufactured solid elements. Such modular housing is much faster and easier to implement than wooden buildings, and more cost-effective. In addition, ecotopians sometimes build the main part of the building from wood or stone, and the required number of rooms is selected by such modules. Each module has special gutters and pipes, so such housing units can be equipped with everything you need. Instead of foundations, modules of such buildings are fastened by means of "large adjustable corkscrew devices which anchor each corner but leave the earth surface undisturbed" [11, p 125].

In addition, small factories even produce separate bathroom modules, equipped with environmentally friendly drainage of water and sewage. "A companion unit, a large plastic tank, is buried outside and connected by two flexible hoses. This, it turns out, is a septic tank, which not only digests sewage but produces methane gas in the process, which in turn operates the heater! The effluent that runs out the other end is not at all repulsive, but clear and excellent for watering gardens, so that ordinarily the garden is placed adjacent to the bathroom" [11, p 124].

These buildings are fully compliant and ensure all the needs of the residents, if necessary, it is possible to add or subtract rooms, carry the buildings on another site and so on. The massive spread of the production of such modules has affected the lack of the profession of an architect as such, because everyone can form a building for themselves, according to their own needs. So, "The community governments have design staffs for public buildings" [11, p 134].

This idea is reminiscent of an improved and more environmentally friendly version of the manufacture of prefabricated elements for residential buildings, which became popular in the 1920s. Standardized and unified construction of such buildings in the last century, though provided housing for a significant number of people, but created a monotonous urban space with itsproblems. The variant of production of standard modules described by E. Callenbach in "Ecotopia" recognizes individuality; after all, each inhabitant will have an opportunity to define where and how the module will stand. Thus, the author believes that all buildings will have a different look, although they consist of the same elements. An approximate view of "Extruded houses" is shown on Figure 2.

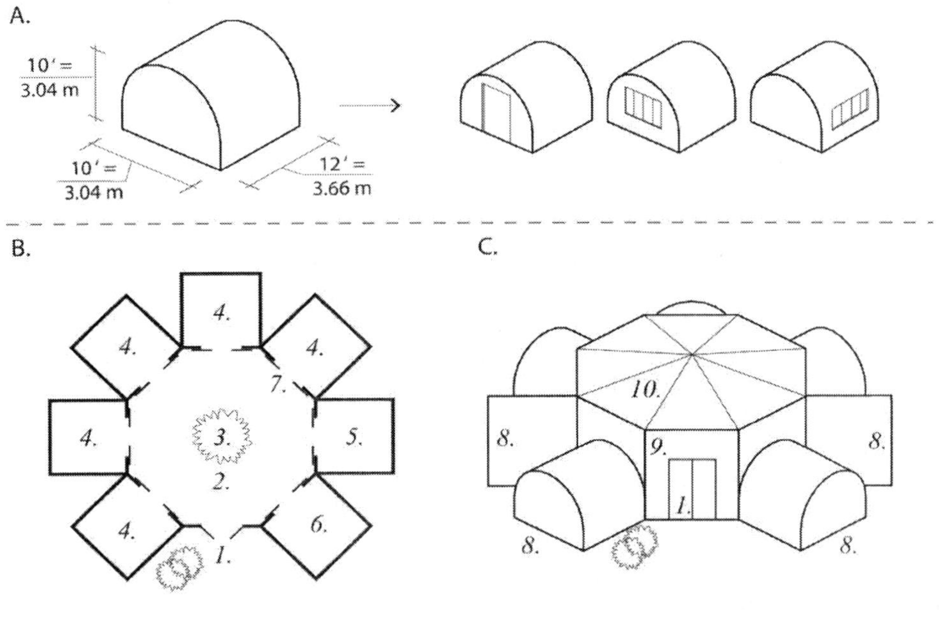


**Figure 2.** Image of "Extruded houses". A –typical module, and an example of possible variations in the location of windows and doors; B - Plan of a house built of standard tube modules; C - Axonometric image of "Extruded houses". 1. Entrance; 2. Kitchen-living room; 3. Tree; 4. Bedroom; 5. Bath room; 6. Cabinet; 7. Sliding doors; 8. Typical module; 9. Central part, made of stone; 10. Translucent dome [drawn by author]

The interiors of ecotopians residential buildings are very minimalist: there are no beds in residential buildings, residents sleep either directly on the floor or on a mattress; all wooden furniture, etc [11, p 81]. Lighting, although electric, but minimized, so if necessary, residents use candles [11, p 124]. Some houses have indoor green gardens. Spaces in public buildings also do not differ in rich decor: in the cabinets of ministers there are no tables for conferences or fashionable furniture, but only comfortable pillows on the floors and large windows. A similar principle is followed in the design of public transport. Each train consists of three cars, which do not have classic seats, but purely pillows on the floor, where you can comfortably sit or lie down, looking out the large windows.

Given that recycling is mandatory in the Ecotopia, all residents and businesses are required to sort their waste by appropriate categories. Yes, even in a hotel room or a train car there are three special pipes for different garbage [11, p 20].

Despite the first impression, ecotopians do not reject technical development, although meticulously select the technologies used in everyday life. One of these types of equipment is videophones, which are equipped with all houses and hotel rooms, and household appliances (refrigerators, dryers, etc.) are made as quiet as possible so as not to create noise pollution. In addition, the entire country, except for a few isolated rural areas, is covered by a network of cable that provides communication and television [11, p 38].

The energy here is exclusively ecological. The largest source of energy, at the time of the story, is a massive thermal-gradient power plant at Punta Gorda [11, p 102]. There were several gas and oil power plants in Ecotopia, but all of them were closed after the declaration of independence. Due to decentralization, smaller power stations are being built, which will supply energy exclusively where it is needed.

One of the acceptable sources of energy for them is the source of geothermal power and solar or wind energy. These sources of electricity are renewable and have less impact on nature than any other. For example, hydroelectric power plants, of which there is a significant number in the country, are not considered completely environmentally friendly, as they have a negative impact on water bodies and their inhabitants [11, p 103]. Ecotopians enjoy windmills and wind turbines on the roofs of buildings, which are common in both cities and remote areas. Solar panels are also attached to the roofs of buildings - silvered parabolic mirror about 30 feet in size.

Also, there is very common a massive photo-cells system. Solar energy is collected in special underground containers, from where it is supplied to residential premises for heating. Devices that provide this process are attached to the walls of buildings, which allow covering a larger area for the absorption of radiation [11, p 105].

Among the public buildings, the book mentions hotels [11, p 14], administrative institutions [11, p 24]. In many large public buildings, train stops are located underground, which allows you to easily get from one point to another in adverse weather conditions [11, p 24]. There are almost no workshops in Ecotopia, because things are made durable and reliable, and in case of minor issues, residents can fix them themselves [11, p 40]. The factories that operated before the country's secession have been modified since independence, although the buildings still perform their production functions [11, p 87]. However, the areas adjacent to the factories have been cleared and their natural appearance restored as much as possible.

Decentralization encourages citizens to build a large number of small hospitals and schools [11, p 62]. Schools occupy large areas, including streams and landscaping, parks. However, they do not have any permanent buildings, and all classes are held outdoors, or in small, temporary wooden houses for a maximum of 10 students [11, p 116]. A network of universities is widespread across the country, next to which are smaller research institutes [11, p 129]. There are also pharmacies in Ecotopia that sell prescription drugs only [11, p 96].

Due to the clear legislation and the "ideality" of society, there is no special need for prisons, although the author mentions that there are prisons, and prisoners are obliged to work there.

The main idea of all transformations of residential and public buildings and the urban structure was the desire of citizens to get as close as possible to nature, and minimize own negative impact on it. “But what matters most is the aspiration to live in balance with nature, "walk lightly on the land," treat the earth as a mother" [11, p 29].

The life in the communes described by Callenbach echoes the idea of "utopian communities" popular in the United States since the 18th century. Separating themselves from the world around them by certain borders, they sought to create their own state where all residents would feel comfortable living among like-minded people. Such communities were formed mostly on the religious or ideological community and failed due to the lack of a clear vector of economic development [12]. Thus, in contrast, Callenbach develops a clear model, detailing all levels of human life. The ideas of "Ectopia" have a significant impact on the perception of the ideas of ecological utopias, and on eco-cities and stable cities.

Although the Ecotopia society looks stable, it is constantly improving and changing: new laws are being adopted, technology is being improved, new nature reserves are being restored, and so on. This qualitatively distinguishes this work from the classical utopias, which exist in a timeless mode and do not involve any transformations.

Close to ecotopia is arcology - the development of urban concepts based on the relationship between architecture and ecology. The author of the arcology is Paolo Soleri (1969), who developed the project of the ecological city of Arcosanti. This project provides for a compact layout of the territory due to the construction of hyperstructures. According to the idea of arcology, cities should be built on the banks of rivers, lakes near forests, making it easier for people to access natural resources. The cities are a vertical city-commune, which would be located in a small area, with minimal interference in nature [13, 14].

Due to the deterioration of the environment, in modern architectural utopias, the issue of environmental stability is much more acute than it was a decade ago. As an example, the thesis of a Spanish architect Manuel Dominguez "Very Large Structure" is a city project located on a special platform that ensures its movement. At the same time, the city is fully self-sufficient in electricity and has waste processing stations, thus being a completely self-sufficient structure [15].

## Conclusions

Ecotopia is a model of a socio-political system, which takes into account the postulates of sustainable development, and a clear connection with nature. Utopias, as models of an ideal system, are characterized by attention to natural conditions and the established mechanism of functioning of all systems. With the development of science in general, the vision of utopians on how to properly interact with nature has changed.

The utopian works of Plato, T. More, T. Campanella and Denis Vairasse are focused on good natural conditions and soil fertility, which allows states to prosper economically. In cities, the number of which as well as the population is limited, special rainwater collectors are provided to collect precipitation for future use. The common thing to these works on the concept of sustainable development in general - is the control over population growth and development of the territory. The population must be equal to the capabilities of natural resources and the economy, only then will harmony be achieved in development.

In his utopia, Cyrano de Bergerac proposes to use the force of the wind to move his wandering cities. And Vladimir Odoyevsky, who is trying to predict the future – in 4338, considers it more appropriate to try to tame nature, and use technical development to reproduce and simulate natural conditions where it is needed.

A more detailed and comprehensive model of the state described by E. Callenbach in "Ecotopia" is a striking example of the introduction of environmental friendliness in design: a well-established public transport network, waste processing stations, renewable energy sources, and construction of exclusively environmentally friendly materials. The author describes in detail all spheres of life: from politics to agriculture. Ecotopia is a striking example of a well-established system that focuses on quality interaction between man and nature. Cities and buildings are designed according to the concept of sustainable development: green, pedestrian, and focused primarily on people, not transport. At the same time, the author emphasizes the importance of appropriate behavior of residents, because without "confessors" it is impossible to achieve the implementation of such a system.

Thus, ecological utopias involve the design of settlements taking into account the impact of architecture and urban planning on nature and ecology, which will be inhabited by conscious people willing to profess the idea of connection with nature.

**References**

[1] *English Oxford Living Dictionaries* [online] – Available at: https://en.oxforddictionaries.com/definition/utopia [Accessed 5 May 2020]
[2] Petrushenko O 2010 Utopia's modifications and the main kinds of utopian consciousness in contemporary world *Bulletin of Lviv Polytechnic National University. Series "Architecture"* № **661** (Lviv: Lviv Polytechnic Nat. Univ.) pp 80-84 (In Ukrainian)
[3] Ikonnikov A 2004 *Utopian thinking and architecture* (Moscow: Arkhitektura-S) p 400 (In Russian)
[4] de Geus M 1999 *Ecological Utopias: Envisioning the Sustainable Society* (Utrecht: International Books) p 319
[5] Silpirekha P 2017 *Architecture of ecotopia: A study of contemporary fictional and non-fictional narratives for development of self-sustaining communities* (India: Nagaland University)
[6] Linda S and Fedak A 2019 In search of the perfect form: the geometry of the city in the treatises by Plato, T. More, T. Campanella *Architectural studies* (Lviv: Lviv Polytechnic Nat. Univ) pp 35- 44
[7] Sirano de Berzherak 1971 The Other World: Comical History of the States and Empires of the Moon (In Russian) *Utopian novel of the 16th — 17th centuries* (Moscow: Fiction literature) pp 227-308
[8] Vairasse D 1971 History of Sevarambes (In Russian) *Utopian novel of the 16th — 17th centuries* (Moscow: Fiction literature) pp 309-448
[9] Odoyevskiy V 1926 *4338 god Petersburg letters* (Moscow: Ogonek) p 64 (In Russian)
[10] Callenbach E 2004 *Ecotopia* (California: Banyan Tree Books, Berkeley) p 172
[11] Diyeva T 2008 *The Phenomenon of Eco villages Worldwide and in Ukraine* (Kyiv: Scientific Notes NaUKMA) Vol **83** pp 77-82 (In Ukrainian)
[12] Grierson D 2016 Unfinished business at the urban laboratory: Paolo Soleri, Arcology, and Arcosanti *Open house international* Vol **41** No 4 pp 63-72
[13] Ruth A Rae 2016 Arcology, Arcosanti and the green urbanism vision *Open house international* Vol **41** No 4 pp 56-62
[14] Stott R 2013 *A Walking City for the 21st Century* - Available at: https://www.archdaily.com/443701/a-walking-city-for-the-21st-century/ [Accessed 10 May 2020]

IOP Conf. Series: Materials Science and Engineering **907** (2020) 012075 doi:10.1088/1757-899X/907/1/012075

# The Modern Trends of an Architectural Parterre Formation

**V P Dubinskiy [1] and O V Skorobohatko[2]**

[1]Department of architectural Design, Kharkiv National University of Civil Engineering and Architecture, Kharkiv, Ukraine
[2]Department of Architecture, O. M. Beketov National University of Urban Economy in Kharkiv, Kharkiv, Ukraine

vdubinskiy60@gmail.com

**Abstract.** Theoretical prerequisites for the appearance of term of "architectural parterre" with the aim of its further use have been considered in the article. The historical prerequisites for an architectural parterre formation in the urban environment are given and the main types of spatial elements of a city structure are enumerated. The peculiar features of human perception of the urban environment in an architectural parterre zone are described. The analysis of the conditions of a dwelling unit density, allowing an architectural parterre formation, has been conducted and the acceptable physical dimensions of an architectural parterre have been derived. A list of urban design elements of an architectural parterre structure has been revealed. Three aspects of increasing the quality of an architectural parterre, namely: ergonomics, ecology and esthetics have been derived on the basis of modern features in open spaces organization in the urban environment. Subsequently, modern trends of an architectural parterre formation have been revealed, namely: the engagement of the public into the formation of design tasks in the process of creating an architectural parterre spatial elements; removal of the carriageway from the streets or "de-mulling"; transforming spatial elements into "places" distinguished by their unique design solutions and architectural metaphor; creating a "hybrid zone" in front of a dwelling unit; providing multiple tasks of an architectural parterre spatial elements or "mixed use"; including of an architectural parterre spatial elements into the system of green corridors of an urban environment.

## 1. Introduction

Rapid expansion of urban territories and the population growth in the industrial era create specific conditions for urbanists when organizing the urban environment. The urban design appeared as a definite professional sphere aimed at the search of different designing measures for transforming certain districts in conditions of increasing requirements of a modern society to the urban environment. Now the urban design at all levels of its introduction tends to increase the quality of the urban environment in its entirety completely. The urban design transforms the existing spatial connections between buildings, streets, squares, parks and water ways. A dwelling unit density is considered to be one of the main criteria of the urban environment quality. It forms the image of the urban environment in separate urban units. The increase of a dwelling unit density often leads to the increase of the number of floors in buildings. Multistory building system forms environmental image, which cannot be perceived by a person entirely. Furthermore, the index of the number of inhabitants on a certain territory increases and it requires the creation of a hierarchic system of open spaces.

Organization of open spaces stimulate an increase of social activity in cities. For this purpose, special attention is paid to walking routes and landscape recreation spaces. The tendency of organizing green ways in cities aims at providing pedestrians with a whole network of landscape recreation objects. It is in the footway part of the urban environment that a person realizes a larger part of social processes. The most widespread way of providing safe environment is the removal of a traffic way from the streets. The main aim of creating a network of footway spaces in the urban environment is the integration of cultural and natural resources into the urban environment.

Because of the variety of social activities in a city it is relatively difficult to reveal exact indices of a dwelling unit density. Rather diverse way of living of contemporary people does not allow to calculate average indices of the impacts on the territory with acceptable accuracy. Urban design interferences become more local and aim at satisfying the needs of certain communities. The possibility of realizing social processes outside is one of the main criteria of urban spaces quality.

The relevance of the research:

The relevance of the research is determined by modern tendencies of footway spaces formation in large cities with high dwelling unit density and great number of floors in buildings. Therefore, the urban environment differs greatly from the natural surrounding medium. An architectural parterre may serve as transitional space between natural and anthropogenic urban territories. The formation of an adequate transitional space for a person will allow to reduce harmful influence of the urban environment on a person and to create comfortable conditions for realizing vital activities.

The aim of the research:

Which dwelling unit density index causes the formation of an architectural parterre and how are a dwelling unit density and physical dimensions of an architectural parterre connected?

Which elements of the urban design are included into an architectural parterre by reference to the peculiarities of an architectural parterre perception?

In which main aspects does the formation of an architectural parterre occur by reference to the main requirements to footway spaces formation in the urban environment?

## 2. Structure of the research

### *2.1 Methods of the research*

The research is based on the analysis of the theoretical and practical experience of footway spaces formation of the urban environment in conditions of a high dwelling unit density, in particular, an investigation of interrelation of a specific index of the population impact on the territory and its influence on the number of floors in buildings. The data obtained are compared on the basis of peculiarities of human perception of the urban environment. Subsequently, an average index of a dwelling unit density, appropriate for an architectural parterre formation, is revealed.

The description of the derived architectural parterre elements due to which an architectural parterre is formed, has been made on the basis of the analysis of the conditions of an architectural parterre formation, peculiarities of the perception and the main ways of its formation.

The specifics of the formation of open footway spaces in a modern urban environment and the prevailing tendencies in their organization have been analyzed. By reference to the tendencies in footway spaces organization and the elements of urban design, the generalization of the main aspects of increasing the quality of spatial elements of a modern city architectural parterre has been made.

Three modern trends of the formation of an architectural parterre of a modern city are derived on the basis of the data on space formation, perception peculiarities, elements of urban design in an architectural parterre, as well as aspects of increasing an architectural parterre quality.

### *2.2 Theoretical basis*

An architectural parterre as an environmental object appeared in the 1960s-1970s. The tendencies of the growth of cities due to an increase of the number of floors in buildings led to the creation of the system with optimal subjects and spaces from the point of view by their dimension. Le Corbusier wrote about the necessity of putting large skyscrapers *to the* background out of sight: “If skyscrapers raise their floors higher than 200 meters, the boulevards with the attached buildings with one, two or three floors with shops, restaurants, cafes … will be situated between these giant structures and the center of open spaces. A street will be reconstructed mainly with the help of elements, corresponding to a human scale" [1].

V. Glazychev characterizes "an urban parterre" as "the realm of urban design" in his book "Urban science". "We do not often have enough free time for perceiving a city entirely. More and more often a city is perceived by us like a room interior by a child who needs to raise his head to see the adults' faces or a plat-band. In other words, our attention is focused on the objects which are at eye level or a bit higher. It is the level, at which the architecture of buildings, if it is higher than two floors, is perceived only fragmentarily. It is the realm of urban design, which has been a natural constituent of architecture until recent times, but in the XX century it detached and started moving to the forefront" [2].

Forward thinking architects tried to erase traditional barriers between the interior space of a building and its environment, at least visually, psychologically (organic architecture by Frank Lloyd Wright, modernism by Mies van der Rohe and Alvar Aalto). Glazing of in-built floor-to-ceiling holes with minimal area of separating slips provided visual connection of spaces, while a slab of outer paving, included into the interior, became the extension of outer environment in the interior.

The quality of an urban landscape is the main factor of cities' quality perception. The physical character of a city is determined by the nature of its streets, squares and other open spaces from the point of view of their formation by means of including elements into the structure. The term "architectural parterre" is defined in the article as an environmental object in the structure of the urban environment with the systems of interior and exterior spaces, limited by the buildings facades at the level of two-three floors of buildings and including the elements of pedestrian streets and small open spaces, limited by a dwelling unit.

### *2.3 The formation of an architectural parterre*

The appearance of an architectural parterre is connected with the concept of increasing the number of floors and the density of cities. The increase in the number of floors has led to the formation of an architectural environment not proportioned with on a human scale. An architectural parterre spaces are different from the point of view of their functions and composition location in the urban environment.

The formation of an architectural parterre often occurs in the existing urban construction conditions. These conditions are: a dwelling unit density, dimensions of squares, the width of pedestrian parts of streets, etc. The width of pedestrian ways is determined on the basis of human scale, flows intensity and acceptable distances for different kinds of pedestrians' activities.

The main types of spatial elements of an architectural parterre in the urban environment are:

- Linear space. These are pedestrian streets, street elements outside the traffic area, footways inside a dwelling unit.
- A corner joint of two streets. It may have different compositional features: it may be an outer or inner corner, additional building, etc.
- Outdoor carol or a courtyard.
- Spaces protecting from external influence such as portico, gallery and others.
- Interior spaces, integrated into the outdoor space due to design solutions.

The peculiarities of an architectural environment perception

Depending on the dimensions of pedestrian spaces certain visual perception angles of the buildings facades and spatial elements of an architectural parterre occur. The perception of an architectural parterre occurs at two levels: optical and sensory. Different kinds of sensory receptors are used at each level of the urban environment perception.

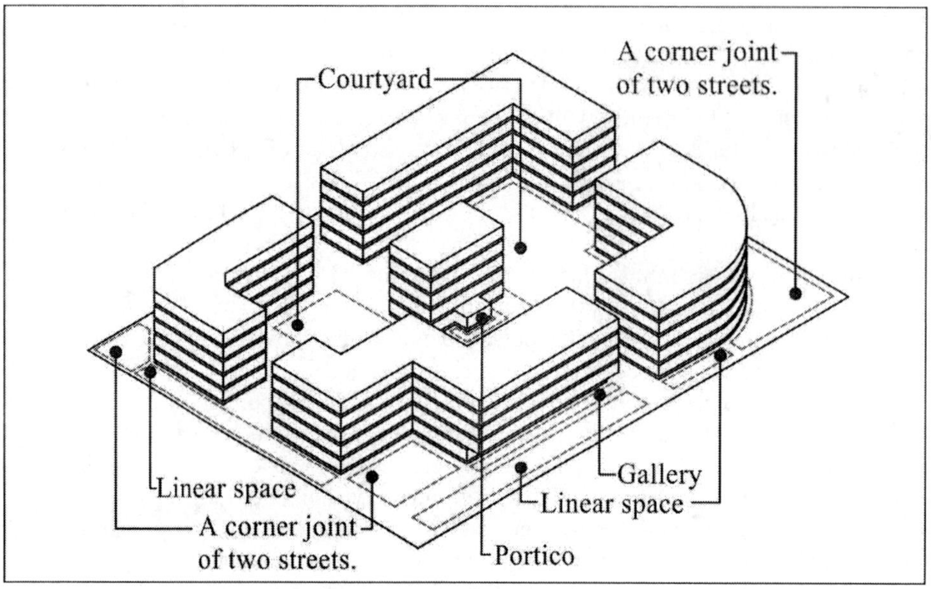


**Figure 1.** The main types of spatial elements of an architectural parterre.

At the sensory level of the environment perception peripheral vision and different kinds of analyzers are used. A person feels a touch, perceives olfactory and acoustic information. Such information may include: the texture and slope of paving, object elements when in touch, surrounding temperature, aerodynamic changes under the influence of the surrounding objects and others. When the peripheral vision is used, a floor line, located in 6-6,5 meters from a pedestrian, is usually perceived [3].

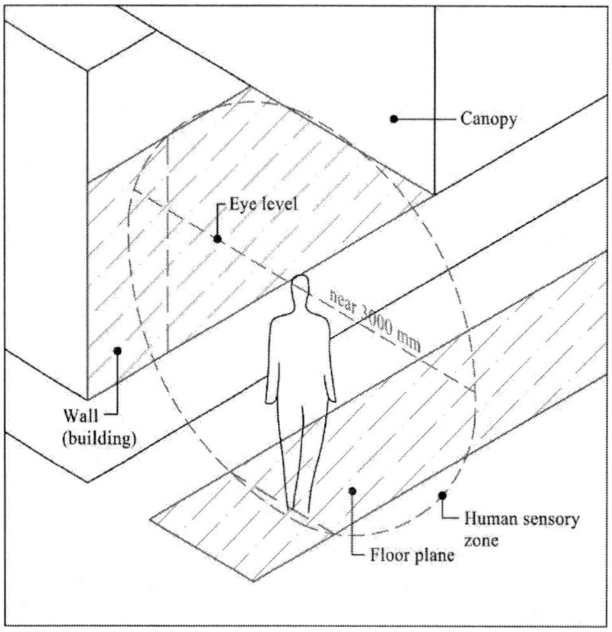


**Figure 2.** The sensory level of the environment perception.

At the optical level a zone of high visual activity is used. A person perceives the environment inside the angular range of 30° vertically and 60° horizontally. When a building is very high this angle allows to scan only the first 2-3 floors of a dwelling unit to the full extent, depending on the width of a pedestrian corridor and the height of floors.

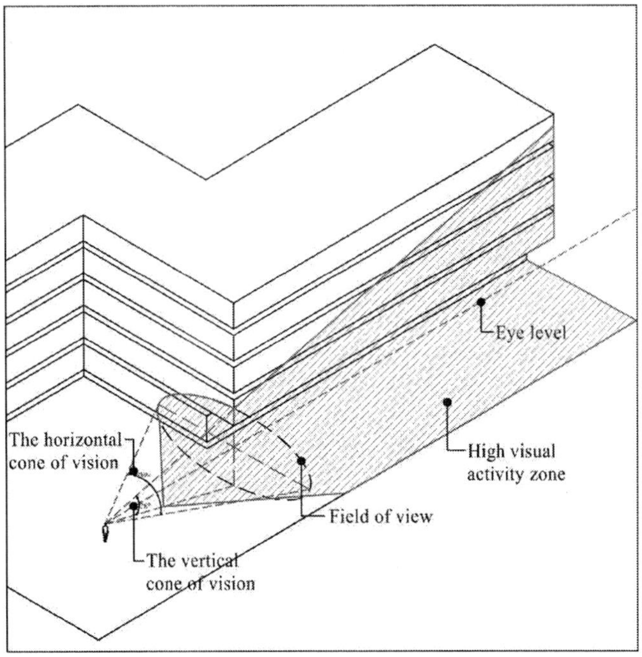


**Figure 3.** The optical level of the environment perception.

The perception of an architectural parterre may occur either vertically or horizontally, depending on the location of objects in reference to the visual conus of a person. The most comfortable conditions for human perception are those that can be perceived completely, at the same time having rather strong degree of space barriers. The perception of the urban environment occurs in the system of space barriers.

## 3. Results

### *3.1 Physical dimensions and dwelling unit density*

Physical dimensions of an architectural parterre are determined by the dimensions of its elements in the structure of an autonomous territorial formation in the urban environment. The spaces of an architectural parterre are public territories inside a dwelling unit of a city block with an accessible footway for everyone irrespective of the form of property.

The mechanisms of studying the demographic situation and the peculiarities of social activity of people in a certain territorial formation become more and more popular in modern methods of calculating a dwelling unit density of separate territories. These research aims at studying the most accurate pattern of the territory load. But for revealing the average indices of a dwelling unit density, when an architectural parterre is formed, the use of average data on the population proportion per territory is quite enough [4].

In conditions of a certain index of a dwelling unit density a certain number of floors is formed. In case of 100 people/ha the number of floors is 1-2 floors, 100-200 people/ha – 3-4 floors, 200-300 people/ha – 5 floors. When the index is about 500 people/ha the number of floors is 9-10, 700 and more – 15 floors and more. So, the formation of an architectural parterre can be revealed in conditions when the number of floors allows a person to have a high degree of space protection.

IOP Conf. Series: Materials Science and Engineering **907** (2020) 012075 doi:10.1088/1757-899X/907/1/012075

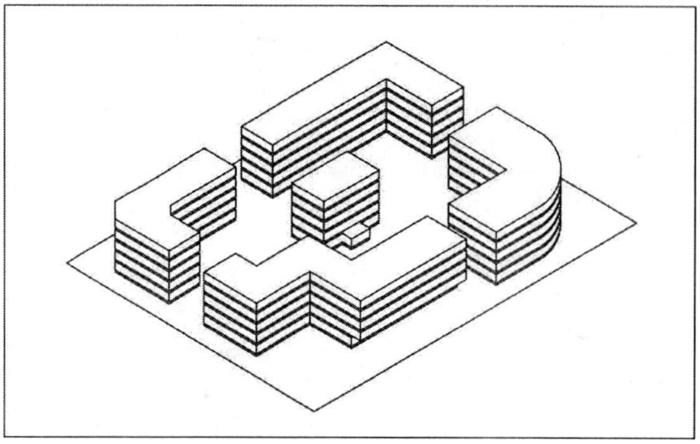

**Figure 4.** A dwelling unit, which density is 200-300 people/ha.

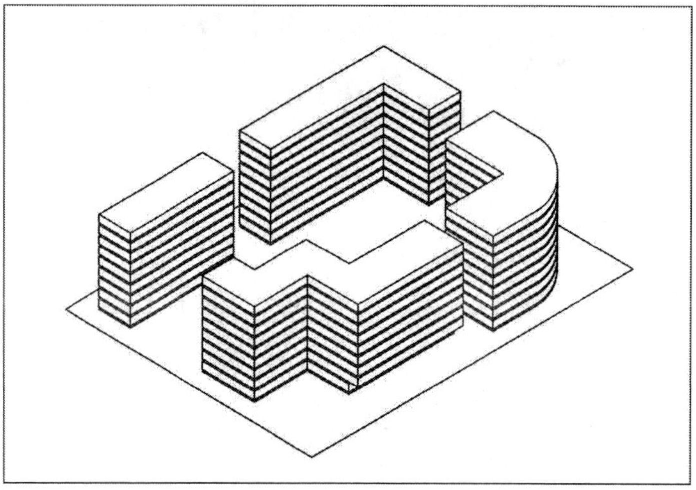

**Figure 5.** A dwelling unit, which density is 500 people/ha.

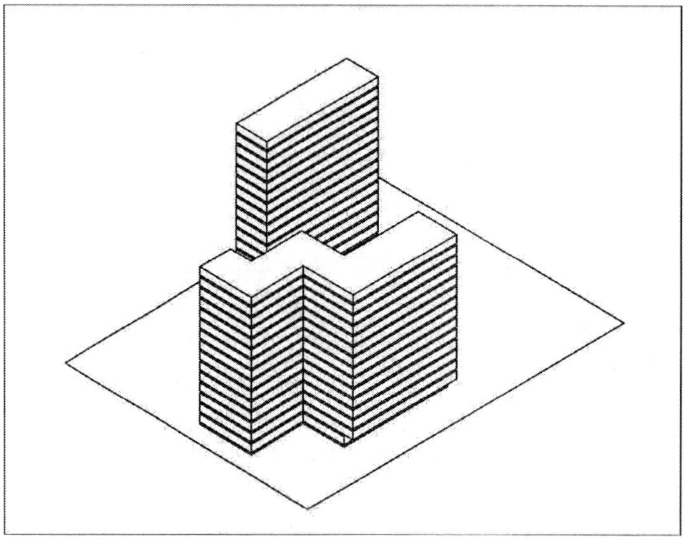

**Figure 6.** A dwelling unit, which density is 700 and more people/ha.

An architectural parterre starts its formation when the number of floors is five and more and when a dwelling unit density is 200-250 people/ha. In such cases the physical dimensions of an architectural parterre may vary on the average from 0,5 to 5 ha, depending on a dwelling unit density and the total area of open spaces, which can be referred to an architectural parterre.

### *3.2 Urban design elements in the formation of an architectural parterre*

Filling an architectural parterre with the elements of urban design depends on the types of human activities. Human activities in the urban environment often become diverse and unpredictable, cause some difficulties in any unification of the urban environment, and make it less urgent.

When an architectural parterre is formed the urban design occurs at the level of separate object and space formations in a city. The means of the formation are listed below:

Design solution of the facades at the level of the first two-three floors of a dwelling unit. In conditions of the diversity of human activities in a modern city the first floors of a dwelling unit with a high density are rather multifunctional. Marking of the facades at this level allows to limit the urban environment visually, which can be perceived entirely within a vertical angle of sight. In this case, attention is focused on the first levels of a dwelling unit and multistory buildings become less massive and heavy for visual perception. Design means are rather diverse in such situation. They may include a style contrast, structural marking, compositional accent etc.

Landscape design. The use of natural elements when forming an architectural parterre allows to reduce the harmful influence of anthropogenic environment on a person, as well as to create visualization of a natural environment familiar for a person. Natural elements are manifested in landscaping of an architectural parterre spaces, vertical gardening, container planting, water elements etc. Unlike the traditional ways of using a landscape design, it is implemented in an architectural parterre in conditions of prevalence of artificial territories over the natural ones. It leads to a great deficit of territories, applicable for the landscape gardening. The following means are used for solving this problem in an architectural parterre: vertical gardening, container planting, water elements, green barriers etc. Artificial landscape forms, requiring no additional care, are also widely used.

Small architectural forms. These elements form the most diverse content of the urban environment. They include shaded shelters, alcoves, fountains, public transport stops, kiosks, information stands, telephone booths etc. Considered as architectural monuments, works of landscape architecture and territorial improvements, small architectural forms can emphasize the urban environment. They are also used for decorating streets, main roads, parks, squares and yards and can fully reflect originality of natural environment, ethnic charm, uniqueness of a created object, as well as perform their function.

Street graphics. Street graphics include the works of fine arts, which are integrated into the urban environment. Such elements are the cheapest way of avoiding monotony and utility of the urban environment, wherein continuous end blank walls and commercial firms with simple facades become the main field for activities.

Urban sculpture. Sculpture compositions are often integrated into the structure of an architectural parterre alongside with small architectural forms and upgrading elements of territorial improvement. The materials and the stylistic component of sculptures are quite diverse and often depend on the area image in the formed town planning conditions. Sculptural forms create the image of an architectural parterre due to their thematic and semantic components and their scale, which is optimal for visual perception.

High technological objects. High technological objects are used in the structure of an architectural parterre for increasing comfortability, informativity and functional substance of the urban environment. Such objects include thermometers, clocks, TV screens, security cameras, cash machines, solar batteries etc. These elements become smaller in size, more dynamic, functional, wear-proof and informative. It is necessary to foresee the conditions for easy assembling and disassembling of these elements.

Street furniture. Street furniture includes benches, shop-windows attached to the facades, artificial barriers etc. The use of street furniture allows to provide the use of an architectural parterre for recreation and informal communication. The synthesis of object and urban design, emphasizing the image of the environment, is also provided. Ergonomics of these elements is considered to be the main criterion for creating of these elements.

Light and color design. Light and color design forms the image of the urban environment at night time, making it more artistic, safe and attractive from the commercial point of view. Lighting can emphasize the plastics of facades at the level of the first two-three floors of a dwelling unit, preserve the composition and image of the space elements of an architectural parterre. Façade lighting, street lighting and lighting of separate object forms as well as technical lighting are used for this purpose.

Means of visual communications. In the structure of an architectural parterre these means of the urban design are used for improving orientation and increasing the informativeness of the urban environment. These are advertisement, info panels, sign boards, type compositions. Rational use of means of visual communication make the common urban life processes.

*3.3 The main ways of increasing the quality of an architectural parterre*

According to modern standards in creating open spaces in the urban environment three main ways can be derived:

Universality of design. This approach to designing aims at providing full access and possibilities of the environment usage by the widest possible range of people [5].

An access to footways. Providing an access to footways aims at connecting the starting street elements, parking lots, recreational objects and buildings. Predetermined context of streets usually prescribes the strategies of designing, therefore, the urban design in such conditions stresses the correlation between space elements: interiors of buildings, recreation zones, parking lots, entrance groups etc [5].

Full level of access to public spaces. The system of providing an access to public spaces is the most applicable strategy of recreational territories organization. It aims at providing pedestrian transit due to multi-functionality and integral perception of recreational territories. A wide system of means of visual communications and spatial references is also used [5].

Among the main aspects of organizing a comfortable environment of an architectural parterre one can derive three main standards, namely ergonomics, esthetics and ecology.

Ergonomics. In an ergonomic aspect of the formation of an architectural parterre the possibility of creating a comfortable urban environment is studied. High indices of psychological and emotional comfort are achieved and the pedestrian space for performing human public activities is provided in this environment. Recreation spaces, rational system of visual communications and advertisement are created in an architectural parterre, a system of night lighting is organized and the objective high technological content along with street furniture are added.

Modern tendencies in the organization of an architectural parterre are becoming more focused on the creation of public spaces, with the goal of increasing the social activity in urban streets. An architectural parterre is a resource for the informal communication of people. A comfortable environment of an architectural parterre becomes interesting for the community and stimulates the social processes.

Esthetics. In the esthetic aspect, a complex of means of the urban design for creating a unique artistic solution is used when an architectural parterre is formed. The possibility of creating a semantic artistic solution, allowing to avoid unification and to diversify the urban environment, is also investigated. Esthetics is also applied in the images of an objective content under conditions of certain stylistics of an architectural parterre space.

Ecology. In the ecological aspect the application of a complex approach for protecting a person from outer external sources of pollution and preventing pollution from an architectural parterre as well as the use of alternative energy sources are researched. The use of safe building materials and lighting devices is also provided in this aspect.

In conditions of the urban environment, an architectural parterre becomes the space capable of combining cultural and natural urban space. In this case a rational system of footways, allowing a person to get from work or home to the system of natural recreational urban territories, is formed.

## 4. Conclusion

On the basis of the analysis, modern trends of an architectural parterre formation have been revealed, namely: the engagement of the public into the formation of design tasks in the process of creating an architectural parterre spatial elements; removal of the carriageway from the streets or "de-mulling"; transforming spatial elements into the "places" distinguished by their unique designing solutions and architectural metaphor; creating a "hybrid zone" in front of a dwelling unit; providing multiple tasks of an architectural parterre spatial elements or "mixed use"; including of an architectural parterre spatial elements into the system of green corridors of an urban environment.

The engagement of the public into the design of an architectural parterre space elements. The development dynamics, the existing quality and the demands in a modern city are in elaborations of an object level, architectural details and local design solutions on certain areas. An architectural parterre formation turns into cooperation, experience exchange and allocation of responsibilities in different processes. It is caused by the interaction of a public demand increase and scarcity of municipal budgets. Long-term plans and trends of a city development interact with the increase of a number of human and enterprises ideas. An architectural parterre formation is a relatively new sphere in urban planning. New techniques of an architectural parterre formation can be obtained while studying completed projects connected with the activities of different communities in an urban environment.

"De-mulling" or removal of the carriageway from the streets for widening a vehicle-free sector in a city. In conditions of automobilization, the problem of separating pedestrians from vehicles became urgent. People can't use a street in a way they had done it before. They got used to a new concept with pavements and crosswalks, while children were removed to the playgrounds. The formation of pedestrian precincts allows to use streets as public spaces, where people could gather and exchange information, their children could play and the streets themselves could turn into the extension of interior spaces.

"Place-making" or transformation of space into a place. Public spaces, maintaining human relations and economic interactions, are formed in an architectural parterre. These are constant processes occurring inside the space elements with functionality and environment improvement. All these processes occur due to people, their assembly areas and interpersonal relationships. In the process of "places" formation, all participants, including local residents, enterprises workers and local authorities can become co-designers and modifiers of an architectural parterre space object. This process aims at creating places with individual designing solutions and specific metaphora on the basis of an architectural parterre spatial elements.

The formation of a hybrid zone in front of a dwelling unit facades allows people to create personified spaces on the public area in front of their houses. In this way, they create a security zone between the private and public parts of a street. It can be achieved due to the agreements with the local authorities. With creation of a hybrid zone, people obtain the space where they can spend their free time and observe the urban life.

"Mixed use" or broad-spectrum use of an architectural parterre space elements. This trend forms the spaces for use by different groups of people, namely: age groups, professional communities, informal groups and others. Mixed use allows to provide the relevance of an architectural parterre space element in a long-term perspective.

"Green corridors" of modern cities form a recreation network of spaces from the spatial elements of a city by using landscape architecture. An architectural parterre is the first link in a chain of a "green belt" of an urban environment, uniting the cultural and natural urban resources. Due to this system a person has a pedestrian way to the natural recreation areas in an urban environment.

An architectural parterre formation provides the optimum vision response of a "green corridor" due to an accessible scale and conveniences of an urban design elements.

## References

[1] Corbusier L 1933 *City planning* (Moscow: Ogis-Isogis)
[2] Glazychev V 2008 *Urban studies* (Moscow: Europe)

[3] Harris C W and Dines N T 1998 *Time-saver standards for landscape architecture.* (United States: McGraw-Hill, Inc)
[4] Clarke P and Gheerawo R 2007 *Metricity: Exploring New Measures of Urban Density* (London: Royal College of Art, Helen Hamlyn Centre)
[5] Lang J 2005 *Urban design: A Typology of Procedures and Products* (Oxford: Elsevier)
[6] Fawcett A P 2007 *Architecture design notebook* (Oxford: Elsevier)
[7] Karssenberg H and Laven J 2016 *The City At Eye Level: Lessons For Street Plinths.* (Delft: Eburon Academic Publishers)
[8] Belyaeva E 1977 *The architectural and spatial environment of the city as an object of visual perception* (Moscow: Stroiizdat)
[9] Mikhailov S 2011 *Design of modern city: complex organization form-spatial environment.* (Moscow: All-Russian Research Institute of Technical Aesthetics)

# The formation of residential complexes in the structure of Ukrainian largest cities

**I V Ladigina, N G Dubina and E V Bizhko**

Department of Architectural Design, Kharkiv National University of Civil Engineering and Architecture, Sumska str, 40, Kharkiv, 61002, Ukraine

evgeniabiz@gmail.com

**Abstract.** Previous decades saw a new type of housing format emerge in Ukraine – namely, residential complexes (RC). Following ongoing integration with European Union, Ukraine becomes included in a global stage of urbanization process, therefore accepting all positive and negative consequences of this stage. From the point of structure and planning, it means the formation of largest global world cities with their peculiar organization, and the appearance of multifunctional complexes, which constitute a specific structure of «city within a city» and subsequently form megalopolis, an urban system of highest hierarchy, from its base.

## 1. Introduction

Previous decades saw a new type of housing appear in the real estate sector of Ukraine – namely, residential complexes. These complexes range from multifunctional to including a limited amount of services, from low-rise to high-rise, from utilizing territory of several hectares to being comprised of a few structures. Still, all the instances of this type are relatively isolated objects realized by non-government construction companies in accordance with «city within a city» trend, meaning they are symbolically connected with urban context in a certain way and are often chaotically emergent without connection to project documentation of higher hierarchical levels – such as city master plan and detailed plan of the territory. Residential complexes can be situated in any structural elements of a city, ranging from its core to its influence zone, but they are always connected to the city's transport infrastructure. Their emergence is usually dictated by economic expedience and availability of suitable territories for construction.

Taking aforementioned into account, it is deemed **relevant** by authors to identify the specifics of residential complexes' formation, as well as their pros and cons, in the larger city's structure as a realization of its self-organizing qualities under complex conditions of the modern stage of urbanization process.

**The novelty of work** is in studying urban complexes as modern structural elements of the city that form a new subsystem after being realized above certain numbers. Said subsystem is then capable of influencing the perspective formation of an entire urban system.

## 2. Few words about definitions

The concept of residential complex is recent in our country's practices and does not have a normatively fixed definition at this time.

Existing sources often define **residential complex (RC) as a single or, usually, several multi-apartment residential buildings united by common territory and common architectural style, thus forming a singular spatial and territorial unit**.

The residential complex is viewed as a singular complex of real estate that is able to satisfy potential buyers' needs in comfortable housing with developed household, cultural and social infrastructure, as well as car parking and landscaping of adjacent territories. The most often additions to these complexes are office and commercial functions.

The relevant sector of residential construction in Ukraine is developing rapidly in recent years. Substantial part of present residential complexes is situated in cities with population of million or above (Kyiv, Kharkiv, Odesa, Dnipro, Lviv). Most construction companies divide residential complexes into following classes: economy, comfort, business and elite. Meanwhile, according to State Building Codes (ДБН В.2.2-15:2019 «Будинки і споруди. Житлові будинки. Основні положення») the housing is divided into categories I and II.

The class of residential complex determines technology of its construction, assortment of building materials used, architectural specifics, services provided to tenants and the peculiarities of adjacent territory's landscaping. It also influences the value of housing area per square meter, which is especially important for investors.

**Economy-class residential complex** is put into operation with minimally allowed parameters denoted in state building codes. It is comprised of small-area apartments and usually situated in residential districts of the city. Nevertheless, it has a reasonably well developed internal and external infrastructure, apartments with scenic views and a well thought-out planning. Said complex can consist of several dozen buildings, formed in accordance with a «city within a city» structural concept.

**Comfort-class residential complex** is built with use of novel materials and propose a greater level of comfort than previous class, as its name imply. Its buildings are constructed with red bricks, porcelain blocks or gas-blocks.

**Business-class residential complex** is a new level of primary real estate that proposes an individual design project, additional amenities and a greater level of security. It can be built in the central part of the city with use of modern materials, or it can be situated far from downtown surrounded by green zone.

**Elite-class residential complex** marks the highest level in primary housing. It often includes such amenities as direct elevator access to apartment, panoramic terraces and the heightened level of security. It can be situated in the city center and have an improved level of décor. Other than that, said complex can provide enclosed adjacent territory, private kindergarten and unique small architectural forms in landscaping.

The modern residential complex may include, depending on its class, the following: residential areas, an administrative area with developer company's management, shops, cafeterias, pharmacies, educational and medical institutions, car parking (aboveground, underground, dedicated for guests), children playgrounds, recreation areas and sports areas.

The combination of different functions in residential complex leads to strict requirements to its structure and planning. An important concern is differentiating people circulation directed to the housing and to the public services, leading to the separation of corresponding vertical communications, lobbies, corridors and other communications servicing these distinct parts of the building.

New engineering communications are allotted to residential complexes therefore guaranteeing a stable functioning of sewerage, heating and plumbing systems. Modern residential complexes are often regarded as passive buildings.

Presently all residential complexes are designed with concierge post and video surveillance, some even having closed territory with barriers allowing access only to residents and their guests.

As noted, residential complexes are usually formed in the «city within a city» format.

«City within a city» means a complex of buildings that incorporates apartments and non-residential zones. In contradistinction to conventional residential buildings, this type of complex urban development takes into account the daily needs of prospective inhabitants. Therefore, the «city within a city» complex necessarily includes the following functions:

– *catering enterprises.* Cafeterias, restaurants, coffee houses;

– *sports facilities.* Street gyms, fitness centers, swimming pool, tennis tables, sports grounds, sports field;

– *recreation areas.* Small parks, cinemas, dance floors, galleries and other places for entertainment and leisure activities;

– *commercial enterprises.* Grocery stores, department stores, beauty salons, notary offices and other commercial establishments;

– *medical facilities.* Dental office, family doctor's office, first aid post, as well as pharmacies (necessary for most development types);

– *educational institutions.* Kindergartens, music, sports or general education schools, educational groups by interests.

Currently there are no specific regulatory requirements for «city within a city» development in Ukraine. As a rule, residential complex built in accordance with said concept occupies a large area and includes both residential and non-residential premises (Figure 1).

**Figure 1.** Residential complex «Comfort Town» in Kyiv, the largest and most complex urban development in the capital in accordance to «city within a city» concept

### 3. Results and Discussion

Residential complexes were probably derived from the building practice of Europe, which is rightfully considered pioneer in eco-city formation. Its new residential developments constructed in place of degraded territories since the middle of XX century were defined as the first sustainable settlements by Habitat II conference in Istanbul in 1996.

Presently, at the beginning of XXI century the trend of residential complexes adjacent territories' landscaping develops further in the countries of Western Europe. One of the most notable examples is the residential complex «CityLife Hadid Residences» in Milan, which makes an emphasis on ecological

IOP Conf. Series: Materials Science and Engineering **907** (2020) 012076 doi:10.1088/1757-899X/907/1/012076

aspect. RC is situated in the CityLife park zone, its territory is completely free of automobile movement while cars can reach its garages and parking spaces via an underground road system. Adjacent to residential complex are water landscape objects and a golf course (Figure 2).

**Figure 2.** Residential complex «CityLife Hadid Residences» in Milan, designed by Zaha Hadid

Ukrainian developers are gradually adopting European trends in residential development, combining it with their own traditions in residential complex construction. They often utilize bold and non-trivial approach to adjacent territories' landscaping. RC «Greenville Park» in Kyiv can serve as an example of said approach – a unique project that is constructed with goal of attaining a maximum level of life's comfort within (figure 3). Special attention is granted to the design of adjacent territory and recreation zones. The key distinction of this complex from others is a colossal enclosed park with an area of 8000 square meters. According to the project, it will include walking paths, cycling lanes, sports areas, playground etc.

Another important avenue of residential complex territorial development is its situation on the outskirts of natural zones. RC «Parkove Misto» can serve as an example of said approach as it is situated in Podilsky region of Kyiv on the territory of historic park «Kristerova Girka» (figure 4).

The massive enclosed green area of 17 hectares is occupied by landscaped inner spaces of the complex, lake cascades with unique flora, playgrounds and sports grounds. At the same time, the transport access to the inner territories of residential complex is limited.

Therefore, it can be stated that the formation of residential complexes in Ukrainian cities' structure is a peculiar form of response to external and internal challenges in society's development.

**Figure 3.** Business-class residential complex «Greenville Park», Kyiv

**Figure 4.** Comfort-class residential complex «Parkove Misto», Kyiv, 2015

By following the path of integration with European Union, Ukraine becomes involved in the global stage of urbanization process, accepting all its positive and negative consequences. Said consequences manifest themselves in the onset of a new technological post-industrial age, the advancement of technical and scientific progress, the prioritizing of tertiary and quaternary sectors of economy. Meanwhile, in the structural and planning respect it means the formation of largest global world cities with their peculiar organization, and the appearance of multifunctional complexes, which constitute a specific structure of «city within a city» and subsequently form megalopolis, an urban system of highest hierarchy, from its base.

Internal challenges are largely caused by the situation that has developed in Ukraine following the collapse of the Soviet Union. Nowadays Ukraine is facing a problem of restructuring Soviet cities whose functional peculiarities are stemming from decades-long industrial development priorities of economic complex. Other factors contributing to the problem are the mass typical residential development of the cities, the lag in cultural and social infrastructure development that is especially apparent in sub-centers of planning formations, as well as ecological challenges.

On the other hand, the development of local government, the utilization of technical and scientific advancements in the construction and engineering spheres, as well as the availability of global and European experience on overcoming post-industrial crisis, contribute to the optimistic outlook on the future development of Ukrainian cities. In this outlook, the formation of multifunctional complexes becomes an important avenue for renewal under current conditions.

In its turn, the formation of residential complexes is based upon a number of objective factors, which have developed in Ukraine to the present day.

Thus, **sociopolitical** factors are caused by situation when state delegates the provision of housing for its population to private and commercial construction companies.

**Socioeconomic** factors are based on the development of market relations, economic expediency of construction and the social stratification that continually deepens.

**Legislative and regulatory** factors indicate the constant lag and disassociation of regulatory framework from consumer needs, wherein the division of housing into categories I and II is maintained while in practice life dictates different requirements.

**Structural and planning** factors show the presence of free undeveloped territories on settlements' peripheries as well as in elements of planning districts' sub-centers that remain from disproportion in the development of Soviet period. They also indicate degrading territories in historic city centers and areas directly adjacent to it.

**Architectural** (spatial) factors denote the influence of global architecture's aesthetics realized in Ukraine with consideration of local conditions – namely, development of regulatory framework, construction and engineering systems as well as availability of investments.

**Engineering and ecological** factors mirror the development of scientific and technological progress that allows the use of modern technologies with the goal of realizing the sustainable development concept in design and construction. It is achieved by creation of closed cycle engineering systems, leading to passive buildings with zero impact on environment.

Thus it can be stated that the emergence and formation of residential complexes in the structure of Ukrainian largest cities is an objective process that corresponds to the global experience of postindustrial cities' structural transformation in a certain way.

Systematic approach to the formation of cities and their individual structural elements is legislated in Ukraine by General Settlement Scheme (Генеральна схема розселення) as an official state policy in urban planning. It allows the consideration of residential complexes as elements of city's self-organization process that emerge spontaneously in the bifurcation period and demonstrate the local, individual, pointed influence on its environment (the structure of an urban element). With its numbers increasing, said elements form the new urban subsystem that interposes itself on city's existing functional planning organization. It is characterized by both positive qualities and the manifestation of certain risk factors.

Among positive aspects is their exclusive architecture, comfortability, social security, environmental safety, connection to transport infrastructure and the presence of so-called «smart city» cells.

The risk factors of residential complexes include a somewhat chaotic character of their situation in city structure, the competition with existing architectural dominants in case of high-rise residential complexes and the social isolation that deepens the stratification of society.

## 4. Conclusions

Taking the aforementioned into account, it can be stated that the residential complexes' formation in largest cities' structure is an objective historical process that is observed under conditions of settlements reaching new level of post-industrial relations.

Despite the residential complexes' architectural, typological, engineering differences, they are characterized by several common traits: chaotic situation in city structure, social isolation and territorial insularity.

At the same time, it can be stated with confidence that residential complexes are manifestations of a causal way of post-industrial city's self-organizing process that is spontaneous in nature, wherein new organized urban structure emerges spontaneously as an effect of internal elements' cooperation.

Under such conditions, the need to transform the causal way of self-organization to an objective, purpose full one becomes paramount. It means transforming the new urban subsystem of residential complexes into a stable ordered structure of attractors under the influence of targeted external actions and factors.

## References

[1] Cherkes B Petryshyn H Konyk S 2019 Urbanizational Trends and Impulses in The Housing Consruction of Ukraine *IOP Conference Series: Materials Science and Engineering* Vol **471** Issue 9

[2] Fesenko D 2010 *The theory of architectural process: outline of a new paradigm* (Moscow: Magazine "Architectural vestnik") p 239

[3] Haken H 2003 *Secrets of nature. Synergetics: Doctrine of Interaction* (Moscow-Izhevsk: Institute of Computer Researches) p 320

[4] Kokarevich M N Lankin V G Voytsekhovskaya V Y 2019 Synergetics as heuristic background of modern architecture and urbanism *IOP Conference Series: Materials Science and Engineering* Vol **687** Issue 5

[5] Pronina T V 2020 Formation of the Appearance of Multistory Residential Complexes in Largest Cities *IOP Conf. Series: Materials Science and Engineering* Vol **753** Chapter 2

[6] Turkina V G 2008 The city as a self-organizing non-linear system *Izvestiya of Saratov University: The Series of Philosophy, Psychology, Pedagogy* Vol **8** Issue 2 pp 62-66

[7] West G 2017 *Scale: the universal laws of growth, innovation, sustainability, and the pace of life in organisms, cities, economies, and companies* (London, United Kingdom: Penguin Press) p 496

[8] Zhuykov S S 2011 Synergistic approach to studying the architectural process *Akademicheskij vestnik UralNIIproekt RAASN* No **3** (Yekaterinburg) pp 47-52

IOP Conf. Series: Materials Science and Engineering **907** (2020) 012077 doi:10.1088/1757-899X/907/1/012077

# Peculiarities of Formation of a Systems Approach in Domestic Urban Planning

**I V Ladigina, A O Rudenko**

Department of Architectural Design, Kharkiv National University of Civil Engineering and Architecture, Sumska str, 40, Kharkiv, 61002 Ukraine

*alinarudenko75@gmail.com*

**Abstract.** The formation of a systems approach to urban planning in Ukraine is considered, starting from the second half of the twentieth century. Territories with a high level of urban planning development were formed at the time as a result of the development of an industrial component of the production industrial complex, which became the basis for perspective formation of urban systems as real entities. The transition to post-industrial relations causes the increase in complexity of urban systems and their study as complex, open, non-linear, socio-territorial entities capable of self-organization as a real process that takes place in their structure.

## 1. Introduction

Starting from the second half of the nineteenth century, a concentration of production is observed in the largest cities of the most developed countries of Western Europe and the United States as a result of the formation and development of capitalist socio-economic relations. It is accompanied by rapid growth of the urban population and as a consequence by the chaotic growth of cities with an extensive territorial development, formation of urban agglomerations on their basis, that gradually merge among themselves, transforming into **«megapolises»**.

In the context of deepening the process of urbanization in the twentieth century, the necessity of finding new flexible methodological approaches that would provide opportunities for forecasting and managing the development of the largest cities, as well as centers and forms of settlement based on them, becomes paramount. This was facilitated by the emergence and dissemination in the 1960s and 1970s of a general scientific systems approach, an important methodological tool for the study of complex objects and interactions.

## 2. The urgency

The urgency of applying a systems approach in urban planning is obvious, since the basis of this discipline is the study of urban entities and their evolution as complex systems during the deployment of the process of urbanization in different states and regions.

The study of urban planning objects as a systems has a long tradition. V.I. Vernadsky's doctrine of the biosphere and noosphere; A.O. Grigoriev on the geographical envelope can serve as prominent examples; L.S. Berg on landscapes; M.M. Kolosovsky about territorial production complexes. From the systematic understanding of the evolution of urban networks emerged in 1930 - 1940, the founders of Soviet geography, O.A. Konstantinov (1934) and M.M. Baransky (1946), as well as a well-known specialist in urban planning V.G. Davidovich can be cited. The concept of urban systems of USSR and it`s regions in the 1960s and 1970s was developed by Yu. G. Saushkin; V.V. Pokshyshevsky; V. Sh. Jaoshvili; N.I. Blazhko; E.I. Pityurenko and other authors.

**Novelty** of the study in such conditions lies in the further development of the basic principles of the systems approach in their application to urban entities.

## 3. The main section

In Ukraine, the understanding of the city as a system was formed in the middle of the last century as a result of urbanization of territories. The post-war recovery period is generally coming to an end at that time. Industrialization is picking up the pace as the emphasis in the development of the socio-economic potential of state is placed on the development of the production complex, its industrial component.

The so-called intensively developed areas are manifested as real city-planning entities, which I. O. Fomin considers the starting point in the formation of urban systems in Ukraine settlement systems (SS according to the terminology of the 1980s), which are marked by two main features [1].

Firstly, intensively developed areas are characterized by the presence of negative qualities of the structural organization acquired in the process of formation. The irreversible process of increasing the degree of economic and urban planning usage of the territory led to the deterioration of the natural component state of the environment and impaired ecological balance. It could not determine the necessity of their planning reconstruction. Secondly, in intensively developed areas, systemic features of group forms of settlement appear earlier than in others, among which I. O. Fomin distinguishes the interrelation of parts, hierarchical structure and dynamism.

Urban planning activities, using architectural and planning tools (determined by I. O. Fomin), aimed at improving the quality of various components of the human environment. From these positions the environment is perceived as a series of interrelated objects of architectural and planning design, differing in size of territories, degree of economic development, level of urbanization, functional characteristics, quality of structural planning organization, etc. The most massive and traditional objects of urban planning are the cities and internal structural elements of the city – the main links of the human environment. The intensification of public production, the development and improvement of intercity communication facilities and the growth of diverse needs of the population stimulate the development of connections and interdependence of cities and other settlements. Cities are losing their autonomy and independence, increasingly turning into different by their meaning elements of larger and more complex urban structures. At the same time, the largest cities go beyond their formed administrative boundaries and unfold in space, drawing an increasing number of settlements into their sphere of influence. New large-scale forms of populated areas are emerging [1]. Agglomerations, urban areas, urban zones (in the terminology of the USSR), and megalopolises as urban (above-urban) systems are gradually formed. Back in the 1980s, Ukrainian scientists managed to study and describe the phenomenon of urban systems based on a systems approach, which began to be increasingly used in urban planning and manifested in the systematic interpretation of group forms of urban settlements.

The necessity for perspective improvement of existing forms of settlement seemed obvious due to the manifestation of previously acquired negative properties of their structural organization. It was reflected in long-term targeted programs - General and regional settlement schemes, respectively, on the territory of the Soviet Union and individual republics, that identified the priority of designing urban systems as the most relevant urban planning objects. Thus, at the end of the twentieth century, the concept of planned-regulated formation of settlement systems (SS) emerged.

In determining the ways and means of formation of the SS on the large territory of the Soviet Union, which was characterized by the great diversity of urban planning, economic, historical, national and other features of its individual parts, great attention was paid to local and regional conditions. Therefore, after the collapse of the USSR in the conditions of the formation of market relations in the newly created independent states, including Ukraine, the earlier scientific researches in the field of studying urban systems have become the foundation for the renewal of works on the improvement of settlement systems and the sustainable formation of cities. This was facilitated by the adoption of the Law of Ukraine "On the General Scheme of Planning of the Territory of Ukraine" (Kyiv, February 7, 2002, No. 3059-III), which defined the priorities and conceptual directions of planning and use of the territory of the country for the future. Furthermore, rethinking of understanding of the urbanization process and recognition of its objectivity and historicity by Ukrainian scientists in the end of 1990s assists in this renewal as well.

This approach made it possible to study the patterns and use specific positive experience of world urbanization in solving acute problems on the territory of Ukraine.

Moreover, the methodology of a systems approach that has expanded conceptual apparatus and has been in use for several decades, although it has no universal, generally accepted character and has no standardized interpretation because of the dynamic nature of processes in the field of human activity, basically allows it to be used for solving almost any task, including urban planning.

In 1999 U. L. Pivovarov defined the urban system as a spatial form of settlement of any taxonomic rank that develops around an urbanized nucleus [2]. For example, it is an autonomous city, an urban agglomeration, an urbanized area, an urbanized zone, a megalopolis, for example.

The study of the city as a system includes the use of a number of systems-representations (categories) [3], with the following being the main among them:

1. Structural representation associated with the selection of elements of the urban system and the connections between them. Thus, at a certain level of the city's development as a system, when the agglomeration formed on its basis reaches its "mature" development, a hierarchically organized planning structure of the city center appears from the historical core to the zone of influence, which is superimposed on the planning framework that provides links between individual structural elements.

The structure of the urban system is a set of interrelations between its elements that remains unchanged for a long time (at least during observation). As a rule, it is formed over a long historical period and has the most stable character in relation to its other elements.

The urban system as a whole exists because of the connections between its elements. It can be said that the connections make direct interaction between the elements (or subsystems) of the system, as well as with the elements and subsystems of the environment. Production, cultural and recreational connections are distinguished in the structure of the urban system labor.

Connections can be direct and reverse by their nature. For example, the unity of a city center with its surrounding territories may be characterized by so-called pendulum migration when labor resources commute daily to work in the city and back home to a suburban area within two hours' reach.

2. Functional representations of an urban system – an emphasizing of the set of functions (purposeful actions) of the system and its components aimed at achieving a specific goal. The existence of the urban system implies the vital activity of its functional planning organization – the complex formation of separate functional zones, these being production, rural and recreational, and their interaction, interpenetration and relationships.

The functional unity of settlements, or the integrity of the system, is determined on the basis of inter-urban connections, which are manifested in inter-urban socio-economic contacts, regular movements of the population for different purposes, as well as the transportation of raw materials, semi-finished products and final products required in the production process.

3. Macroscopic imagery – an understanding of the urban system as an indivisible whole, that interacts with the external environment. That is, the urban system is always considered as a holistic center of a larger hierarchical level in relation to its adjacent territories.

4. A microscopic representation, that is based on the consideration of the urban system as a set of interconnected elements and involves the discovery of the peculiarities of the environment of the structural elements of the city.

5. Hierarchical representation, based on the concept of city subsystems, which is obtained by decomposition of an urban system, that has systemic properties that should be distinguished from its element into smaller parts (in terms of the solving problem). The city system can be represented as sets of subsystems of different levels that make up the system hierarchy, which is closed at the bottom only by elements.

The hierarchical organization of the city as a system in the system of cities was considered in 1964 by the American economist geographer B. Berry.

A city is a complex system, as such it has elements, which in turn are systems acting as subsystems of its larger complex system. When its population exceeds 100 thousand people, the above-urban systems begin to form on its basis – namely, agglomeration, urbanized area, urbanized zone, megalopolis [4].

There are three main subsystems of the city at the most basic level. These are the population, the economic basis, the sphere of life. At the same time, the city is part of a system of higher hierarchical level.

6. A procedural representation that involves understanding of the urban system as a dynamic object, characterized by the sequence of its states over time. Organically applied to the analysis of the city genesis, its further development or degradation, which can be considered as a time-distributed system. In the study and further formation of the urban system, an urban education is considered within the existing administrative boundaries. In the context of agglomeration processes, wherein the industrialization of the city go beyond their traditional boundaries, there is a process of so-called extensive development of the settlement, and settlements of the suburban zone are gradually included in the structural elements of the agglomeration. The city begins to be considered as a singular entity with its suburban zone, and the new boundaries are determined on the basis of the unity of new structural formations, determined by means of connections by the method of taxonomic zoning. Gradually, a hierarchical structure is being formed including not only the urban agglomeration but also the territories external to it. Thus, a model of agglomeration and supra-agglomeration (suburban) systems emerges [5].

Depending on the impact on the environment and the nature of the interaction with other systems, the functions of the urban system may be arranged by their increasing rank:

- passive existence – can be observed in the case of non-prospective settlements, situated far enough from local centers of certain hierarchical levels in the absence of developed transport infrastructure;
- material for other systems – when it comes to the disappearance of non-prospective settlements (completion of "monocities" activity, resettlement after complete destruction in case of natural or man-made disasters, etc.);
- maintenance of higher rank systems – for example, supplying labor and other resources to a city center of the agglomeration from its area of influence;
- opposition to other systems (survival) – observed in the case of the formation of many center agglomerations, when individual centers compete to become the main one;
- absorption of other systems (expansion) – occurs in the process of formation of above-urban systems;
- transformation of other systems and environments (active role) – inherent in the largest cities that can be identified as global (world) cities.

The city is always considered an open system that allows the exchange of energy, matter and information (for example, energy supply, labor, innovation, etc.) with the environment.

In modern conditions of post-industrial society formation, the process of perspective emergence of urban establishment as systems is complicated. There is a stratification of the subject that affects the urban system for the researcher, developer of regulatory documentation, designer, manager, investor, etc. This raises the risk of losing the idea of the integrity of the urban system.

Urban planning development of cities as a system activity cannot be reduced to the development of project documentation only. Moreover, experience in the implementation of district planning projects and master plans for individual settlements shows that even in conditions of a planned economy, it was never possible to fully reach the project indicators, and violations in the master plans of the largest cities were counted in the dozens.

In the context of the global stage of the process of urbanization, the city is considered a complex synthetic object in which the historical, economic, social, political, environmental, biological, natural and scientific laws act coherently and thus determine the existence of common mechanisms of emergence and functioning of the city as a whole.

Attempts to describe the processes of emergence and development of the city, as well as the possibilities of their regulation, are often made with the help of a synergistic approach, which mostly responds to existing challenges. The city itself is considered by this approach as a complex open nonlinear, socio-territorial system capable of self-organization – a synergistic system.

According to V. G. Turkina, synergetics allows to describe many urban processes and phenomena from a single position, allows to study the internal laws of urban systems development, gives basis for certain forecasts related to urban dynamics [6]. The synergetic description, as a rule, shows that the whole arisen as a result of the action of synergetic mechanisms of dynamics, begins to have properties that were absent in its individual parts and elements. This is fully consistent with the idea of any city. The city is not a simple sum of its separate components, it represents absolutely new object possessing new qualities inherent in it as an integral organism.

Certainly, in the formation of synergistic ideas about the city, the phenomenon of self-organization comes to the fore. Self-organization of the system is the process of evolution of a complex system; the birth of a certain orderly structure with new qualities out of the physical chaos. It includes all processes of self-structuring, self-regulation and self-reproduction.

When it comes to the historical-genetic analysis of the development of society, particularly the formation of the settlement system and its individual elements, the knowledge of laws of self-organization allows us to interpret various social entities as social structures capable of self-organization.

Nonlinear description of the formation of urban systems at certain stages of their historical development involves the formation of different types of such structures, ranging from a single city to a network of cities, agglomerations, megalopolises, that are born, change and interact with each other according to nonlinear laws. Nonlinearity means that many solutions correspond to many ways of evolution of a nonlinear system; it indicates the variety, alternative, and irreversibility of possible ways of system evolution.

The uniqueness of each city is manifested in its genetic code, economic and geographical location, natural conditions, peculiarities of historical, socio-economic, structural and planning, compositional development, silhouette formation, and corresponds to non-linear ideas about cities that can exhibit many possible types of behavior in such conditions. That is, synergetics tries to identify real mechanisms and types of behavior that allow to consider the city as a special dissipative structure capable of self-organization, the emergence of which is associated with certain conditions of the environment that generates it. Dissipativeness (openness) of the city, in other words, means the ability of the system to exchange energy, matter, and (or) information with the outside world.

Cities as real systems that are connected to the environment by many economic, social, political, cultural and other ties, show their dissipation (openness). In the process of formation of above-urban systems, the city draws the suburban space into its limits and at the same time spreads the urban way of life into the adjacent territories.

Under the influence of external factors (political, economic, social, etc.), the city can change its mode of existence from prosperity to decline.

A condition that is steadily maintained by inflows of energy and substance from the outside; the property of an open inhomogeneous urban system, which actively interacts with the environment and is able to acquire dissipativity, is defined as a disequilibrium. However, its individual elements and subsystems may have characteristics that are quite different from the average.

For the urban system, disequilibrium is a rather characteristic condition. For example, the incomes of individual citizens or the housing supply of a single family are quite different from the average. The city is never isolated. The condition of the city, which is perceived as an "equilibrium" at deep examination is not such, but characterizes only a stable period of development of the nonequilibrium structure. According to synergistic consideration, disequilibrium is evaluated as a positive, creative factor of development. Achieving "equilibrium" for each city means the inability to create new socio-economic structures, is perceived as a process that leads to stagnation or even precedes the degeneration and dying [6].

Therefore, the conditions for the emergence and existence of a city system capable of self-organization are the fulfillment of the properties of nonlinearity and dissipation. In this case, self-organization is genetically related to the appearance of small fluctuations in the previous state – a random deviation of the system from equilibrium, when the changes in the external environment are not "noise" but the generator of new structures.

## 4. Research results

Features of self-organization of the urban system can be considered on the example of an industrial city. Its emergence is primarily based on the development of capitalist socio-economic relations as well as the scientific and technological progress as its consequence. The gradual transition from handicraft to machine production causes the location of industrial enterprises in the structure of cities, the formation of whole districts and zones on their basis, including storage, municipal and construction territories, as well as the development of transport and engineering infrastructure.

Territorially, the city is growing extensively, going beyond its historical boundaries and absorbing suburban territories and small settlements located therein. Subsequently, its clear structural and planning organization, including the historic core, city center, historical building area, new dispersed planning districts and suburban area, is revealed. All this indicates that the so-called "Big City" is formed with an adjacent area, which, due to the developed intensive connections, can be considered as a single formation. Parts of this formation are "mature" agglomeration, suburban zone, a zone of influence, a zone of joint interests of the center and periphery. The evolutionary stage is unfolding.

At this stage, the city already exhibits systemic qualities that can be described by specific determinations: cause-effect, functional, target, correlative ties and more.

Gradually, the negative effects of industrialization in the industrial city are brought to the fore. There is an uncontrolled development of suburban arrays of various functional purpose and, because of their overlap, the joining of suburban arrays between themselves and the central core of agglomeration, as well as the degradation or even disappearance of open green spaces, expansion of the city center to 100 - 200 square kilometers and more, the emergence of functional-landscape overlap, appearance of large centers of environmental pollution, "recreational urbanization". Fluctuations are increasing.

Even the most stable multifunctional cities show an increase in chaotic processes in their structure during the fall of the main indicators of industrial production in such conditions. This increase is characterized by the appearance of centers of degrading industrial areas.

Settlements formed during the industrial period as the "monocities" with a leading industrial function that provided jobs for up to 90% of the able-bodied population, are facing the problem of the continued existence of the settlement itself. With the increase of entropy (the growth of irreversible processes), losing its stability under the influence of fluctuations, the system enters a nonequilibrium state and approaches the bifurcation phase, which in a physical sense marks the point of branching of the paths of system's evolution.

1996 can be considered the point of bifurcation for the industrial city that determines its critical condition, when at the Habitat II conference held under the auspices of the UN in Istanbul, the exhaustiveness of the industrial society was declared and the concept of the sustainable development of civilization in the context of the post-industrial era formation was conceived. Of course, this "point" is rather conditional. It was preceded by a transition period for several decades, which saw appearance of socio-economic relations focused on the development of tertiary and quaternary sectors of the economy and the "new industrialization" based on the latest technologies. Some positive experience was gained in restructuring the degraded industrial territories in the structure of cities into the so-called "ecocites", which were recognized as the first sustainable urban settlements.

Therefore, the urban system undergoes two main stages in the process of self-organization - evolutional and bifurcational.

The study of the self-organization of the urban system testifies to its complete compliance with the methodological principles of the full-format synergetic modeling proposed by V. G. Budanov for application in the humanitarian sphere and interdisciplinary design. The theory proposes two structural principles of being - homeostatic and hierarchy – that characterize the phase of "order", the stable functioning of the system. Meanwhile, the principles of formation – nonlinearity, instability, openness, dynamic hierarchy, observability – characterize the phase of transformation, the renewal of the system, ways of destruction of the old order, the chaos of testing alternatives, and the birth of the new order [7].

Passing through the point of bifurcation, the urban system continues its development along one of the possible ways of evolution. Thus, the implementation of the concept of sustainable development in the conditions of the post-industrial technological era formation stimulates the restructuring of cities on the basis of humanization, providing the diverse needs of the population, preserving the natural environment. At the same time, improvements in the economic and political spheres of the country cause new human requirements for housing, social sphere, recreation, which in turn stimulates the development of design, construction, research, etc.

IOP Conf. Series: Materials Science and Engineering **907** (2020) 012077 doi:10.1088/1757-899X/907/1/012077

## 5. Conclusions

1 The systems approach in urban planning in Ukraine begins to be widely used in the second half of the twentieth century, when the emphasis on the development of socio-economic potential of the state were placed on the formation of an industrial component of the production complex, resulting in the formation of territories with high level of urban planning development, which became the basis for perspective formation of urban systems as real entities.

2 In the process of development of agglomeration processes the city begins to be considered as a system in the system of cities. That is, above-urban systems are formed on the basis of cities, and the city acquires a valid hierarchical organization.

3 In the conditions of increasing complexity of urban systems in the process of formation of post-industrial relations, first attempts to describe the processes of their emergence, development and perspective formation are made by means of a synergistic approach.

4 The city is considered a complex open nonlinear, self-organizing socio-territorial system - a synergistic system. The self-organization of the urban system is a real process that takes place in its structure and involves the evolutionary and bifurcational stages of formation.

## References

[1] Fomin I A 1986 *City in the system of populated places* (Kyiv: Budivelnik) p 111
[2] Pivovarov Yu L 1999 *Fundamentals of geo-urbanism. A manual textbook for student's higher institutions* (Moscow: VLADOS) p 232
[3] Arnold R D, Wade J P 2015 A Definition of Systems Thinking: A Systems Approach *Procedia Computer Science* Vol **44** pp 669-678
[4] Wirt L 1938 Urbanism as Way of Life *The American Journal of Sociology* Vol **44** No 1 pp 1–24
[5] Jencks N 1995 *Architecture of the Jumping Universe: A Polemic. How Complexity Science is Changing Architecture and Culture* (London, New York: Academic Press) p 192
[6] Turkina V G 2008 The city as a self-organizing non-linear system *Izvestiya of Saratov University: The Series of Philosophy, Psychology, Pedagogy* Vol **8** Iss 2 pp 62-66
[7] Zhuykov S S 2011 Synergistic approach to studying the architectural process *Akademicheskij vestnik UralNIIproekt* (Yekaterinburg: RAASN) Vol **3** pp 47-52
[8] Kokarevich M N, Lankin V G, Voytsekhovskaya V Y 2019 Synergetics as heuristic background of modern architecture and urbanism *IOP Conference Series: Materials Science and Engineering* Vol **687** Iss 5

IOP Conf. Series: Materials Science and Engineering **907** (2020) 012078 doi:10.1088/1757-899X/907/1/012078

# Problems of environmental safety of water disposal as a factor of sustainable urban development

**V. O. Iurchenko, E. S. Lebedeva, Yu. S. Levashova, A. V. Kovalenko**

Department of Life Activity Safety and Engineering Ecology, Kharkov National University of Civil Engineering and Architecture 40 Sumskaya Street, Kharkiv, Ukraine

elena.lebedeva0504@gmail.com

**Abstract.** This paper examines the environmental safety of water disposal systems – one of the life support systems for a population – as a condition for sustainable urban development. Gaseous emissions from such systems contain a number of highly hazardous substances. The concentration of formaldehyde – a hazard category 2 substance – in sewer network emissions (using direct measurement) and in the underroof space of sewer collectors (by a calculation) was experimentally investigated. The calculations were carried out based on the measurements of the concentration of formaldehyde in the emissions and the data collected during special laboratory experiments with polymer coatings, which accumulated organic gaseous compounds from being in the sewers. It has been proved that the concentration of formaldehyde in the underroof space and in sewer network emissions significantly exceeds the corresponding maximum permissible concentration. Cooling and installation of degassers in shafts were suggested as reasonable methods for protecting the city's atmospheric air against the emissions from sewer systems.

## 1. Introduction

The study of the problems of sustainable development of settlements, especially cities, nowadays is the key area of scientific substantiation of the main paradigm of the 21st century - sustainable economic, social and environmental development. Cities play a special role in achieving sustainable development, since the problems and opportunities of modern civilization are the most significant and noticeable there [1]. What is understood under the term "constant urban development" is a set of socially, economically and ecologically balanced changes in the socio-territorial system of the city, aimed at the fullest possible realization of all its potential's components and preventing the tendencies for the population's quality of life to deteriorate [2], which includes environmental safety as one of its elements. The stability of the city affects the ability to implement the principles of sustainable development of higher-level systems and depends on the sustainability of all the constituent elements included in the socio-economic system of the city.

The concept of sustainable development systematically connected the three main components of the development of society: economic, social and environmental. From an environmental point of view, sustainable development should ensure the integrity of biological and physical natural systems and their viability. Within the city, the viability of natural systems – first and foremost, people (the population of the city) – depends on the environmental safety of the living conditions. It is important, during the practical implementation of the principles of sustainable development of a city, not only to determine its resource provision (resources, their structure, cost, potential), but also to analyse the situation from the perspective of various types of activities, economic areas and their interaction [2, 3].

A big city changes all the environmental components - the atmosphere, the vegetation, the soil, the topography, the hydrographic network, the groundwaters and even the climate. This leads to great environmental problems caused by its high population concentration, industrial facilities, and urban infrastructure. The growth of the urban population, especially in recent decades, the rapid concentration and intensification of production and non-production activities have led to the fact that the environment in many cities around the world cannot satisfy all the biological and social requirements of modern man, including ensuring complete environmental safety.

In the context of the foregoing, the reliability and environmental safety of the functioning of city population's life support systems is of particular importance for the sustainable development of cities. One of the links of this system is wastewater disposal, which ensures the environmental safety of water usage and, in general, performs an extremely large-scale environmental function - protecting the natural environment from pollution by liquid anthropogenic wastes.

However, sanitation facilities are large-scale technical facilities. They exert an intense technogenic load on the environment and often create dangers for urban regions, mainly as a source of ingredient pollution of the urban atmosphere. Even when sewer networks operate in the trouble-free mode, their operation poses an environmental problem caused by the formation of toxic gaseous compounds (hydrogen sulphide, mercaptan, sulphur dioxide, carbon dioxide, methane, etc.) that pollute the atmosphere of urban regions through shafts and wells (Table 1) [4, 5].

**Table 1.** Contents of odour-generating substances in Australian sewers [4, 5]

| Substance | Concentration in Australian sewers, μg/m$^3$ |
|---|---|
| Benzene | 8.85 |
| Decane | 141.88 |
| Dimethyl disulphide | 8.72 |
| Dimethyl supplied | 65.4 |
| Diethyl sulphid | 1.12 |
| Diethyl disulphide | 0.15 |
| Ethyl mercaptan | 3.81 |
| Hydrogen sulphid | 1880 |
| Methyl mercaptan | 293 |
| Sulphur dioxide | 870–2600 |
| Toluene | 61.15 |
| Threechloromethane | 654.25 |
| Xylene | 183.7 |

A particular environmental hazard in terms of the excess of maximum permissible concentration (MPC) in gaseous emissions from sewer networks is hydrogen sulfide (substance hazard category 2) [6, 7]. In addition, it initiates the development of biogenic sulfuric acid corrosion on the arched part of concrete sewer pipelines, which drastically reduces the operational life of these objects [8]. In general, all sulfur compounds (including organic) found in emissions from sewer networks are environmentally hazardous. However, their concentration, except $H_2S$ and $SO_2$, is very low [4, 5].

However, a number of specialists found not only sulfur-containing compounds of a high hazard category in emissions from sewer networks. Unfortunately, among the carbon-containing gaseous compounds in the emissions from sewer networks there are also many potentially very hazardous substances (Table 1). When it comes to environmental safety of carbon-containing compounds in emissions from sewer networks, formaldehyde, which was discovered by a number of specialists, attracts great attention [4,9,10]. In terms of its effect on the human body, formaldehyde, like hydrogen sulfide, is a highly hazardous substance (substance hazard category 2) [11]. These substances have a pungent and unpleasant odour. Moreover, the MPC of all types of rationing for formaldehyde is significantly lower than the MPC for hydrogen sulfide. Formaldehyde is classified as a probable carcinogen for humans with a minimum single inhalation dose of 1.3 × 10-5 mg/m$^3$. According to foreign scientists, its concentration in emissions from sewer networks is 0.370 mg/m$^3$ [9].

The presence of this gas in emissions from sewer networks causes such concern because of the steady excess of its concentrations in the cities of Ukraine. [12, 13]. Moreover, such excesses are usually associated with the work of certain industrial enterprises, or vehicles. However, detailed studies in cities with a small number of industrial enterprises also found that the concentration of formaldehyde in atmospheric air was higher than the MPC. Upon conducting detailed studies and calculations, it was found that the contribution of vehicles to formaldehyde emissions was also very small (up to 5%) [13]. Potentially, air pollution is the most serious environmental problem for human health in the short and in the long run. It is more difficult to avoid exposure to air pollution than to water pollution [13].

The solution to the problem of protecting the urban environment from pollution by gaseous emissions from sewer networks is carried out in various areas of the project: standardization of industrial enterprises' discharges, operational solutions for the transportation of wastewater, the usage of degassing installations on sewer networks. However, the issue of effective removal of formaldehyde from sewer network emissions to ensure an appropriate level of environmental safety in the urban air hasn't been studied.

## 2. Objects and research methods

The object of this research is to assess the level of health hazard posed by the presence of gaseous compounds of the 2nd hazard class – $H_2S$ and formaldehyde – in sewer network emissions and some environmentally friendly means of solving this problem.

The studies were carried out in real conditions in May 2019 together with the employees of the operational service of PU "Kharkovvodokanal". The section of measurements on the sewer network of Kharkov is shown in figure 1.

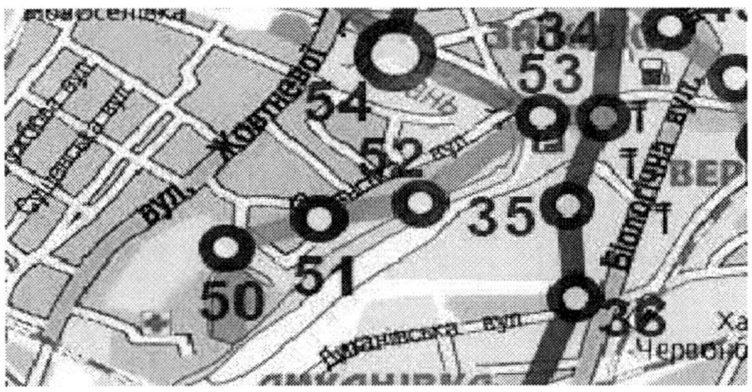


**Figure 1**. Section of the drainage channel (Stroyhydraulika) of Kharkov where the examined shaft is located (No. 52)

The measurements of the concentrations of $H_2S$, $SO_2$, CO, $CO_2$ and $CH_4$ were performed with universal portable gas analyser UG-2 and mine interferometer ShI-11 in the underroof space of the collector, where the device for sampling the air-gas mixture was lowered through the entire height of the shaft. The concentration of these substances is periodically measured by the analytical laboratory of PU "Kharkovvodokanal". The concentrations of formaldehyde and volatile organic compounds were determined at the outlet of the sewer shaft by using a digital formaldehyde detector, the WP6910 air quality analyser.

## 3. Results and discussion

The concentration of these substances in Kharkov sewer shafts was measured for the first time. The results of measuring the concentration of gaseous compounds in the underroof space of shaft No 52 are given in Table. 2.

As it is seen from the gathered data, the concentration of $H_2S$, $SO_2$, $CO_2$ and CO in the underroof space significantly exceeded $MPC_{working\ area}$. Thus, carrying out any kind of work in this shaft is unacceptable without personal protective equipment.

**Table 2.** Concentrations of polluting substances in the underroof space of the sewerage pipeline of the surveyed shaft

| Polluting substance | Concentrations of polluting substances in the underroof space, | $MPC_{working\ area}$, $mg/m^3$ |
|---|---|---|
| Hydrogen sulfide, $H_2S$ | 15* | 10 |
| Sulphur dioxide, $SO_2$ | 35* | 10 |
| Carbon oxide, CO | 0* | 20 |
| Carbon dioxide, $CO_2$ | 0,83* | - |
| Methane, $CH_4$ | 0,57** | 1500-1700 |

* - $mg/m^3$; ** - volume %

Using the data on the concentration of gaseous substances in the underroof space (table 2), we calculated the concentration of these contaminants in the emissions from the shaft (at the outlet of the shaft). For this calculation, an empirically established dependence (1, 2,3) of the concentration of $H_2S$ in sewer shafts and the height of the shaft was used.

$$N = 7{,}37\ H + 97{,}49, \quad (1)$$

where N - relative concentration of H2S at the output from the shaft with respect to the concentration in the underroof space, %; H – height of shaft, m (12).

$$N = \frac{C\ output \cdot 100}{C\ underroof\ space.}, \quad (2)$$

where C output - concentration of $H_2S$ at the output from the shaft, $mg/m^3$; C underroof space - concentrations of $H_2S$ in the underroof space, $mg/m^3$; 100 – conversion factor, %.

Hence:

$$C\ output = \frac{(7{,}37\ H + 97{,}49)C\ underroof\ space}{100} \quad (3)$$

This dependence could be used for an approximate calculation of concentrations of gases that have the same molecular mass as hydrogen sulphide, i.e., CO. Taking into account that the concentration of $SO_2$, CO and $CH_4$ is proportional to the dynamic of $H_2S$ in sewer networks [10], it can be used for a rough estimate of the concentration of these gases as well. Concentrations of polluting substances at the output from the surveyed shaft represented in Table 3.

**Table 3.** Concentrations of polluting substances at the output from the surveyed shaft

| Polluting substance | Concentrations of polluting substances in release from the shaft, $mg/m^3$ | $MPC_{daily\ average}$, $mg/m^3$ | Rate of MPC excess |
|---|---|---|---|
| Hydrogen sulphide, $H_2S$ | 1,5** | 0,008 | 188 |
| Sulphur dioxide, $SO_2$ | 3,5** | 0,05 | 70 |
| Carbon oxide, CO | 0** | 3 | |
| Carbon dioxide, $CO_2$ | 0,083** | - | |
| Methane, $CH_4$ | 458** | - | |
| Formaldehyde, $CH_2O$ | 1,99* | 0,003 | 663 |
| Volatile organic compounds | 4,88* | - | |

* - concentration values obtained using direct measurement at the shaft;
** - the concentration value calculated in accordance with the dependence determined by the authors

As it is seen from table 3, formaldehyde – a hazardous gas – poses the biggest threat as its concentration in the gaseous emissions from the observed sewer shaft is much higher than the MPC.

The concentrations of $H_2S$ and $SO_2$ also significantly exceed the MPC $_{\text{daily average}}$. This creates a special danger for citizens living in areas adjacent to sewer shafts, since all these gases belong to hazard class 2. Given that formaldehyde is close in its molecular weight to hydrogen sulfide, the concentration of formaldehyde at the outlet of the shaft (in the emission) can be used to calculate its approximate concentration in the underroof space of the sewer collector using the determined dependence (1). The concentration of formaldehyde in the underroof space (C underroof space) could be calculated using the equation:

$$C\ \text{underroof space} = \frac{C\ \text{output} \cdot 100}{(7{,}37H+97{,}49)\cdot} \quad (4)$$

where, 100 – conversion factor, %.

According to the calculation, this concentration is 19.9 mg/m³, while the MPC of formaldehyde in the atmospheric air of the working area is 0.5 mg/m³.

One can assume that such high concentrations of formaldehyde, which were found in the emissions from the observed shaft, are episodic and can rarely be observed. However, further studies performed in the shaft, where the concentration of hydrogen sulphide in the underroof space was approximately the same as this indicator in the underroof space of shaft No. 52, have proved that the presence of formaldehyde in the atmosphere of the underroof space of sewer collectors is not an accidental phenomenon. In the process of researching the effectiveness of various polymer concrete coatings in protecting concrete from corrosion processes, a special attention was paid to the intense saturation of these coatings with unpleasant odorous gases after being in the sewer networks for some time (3 months). The polymer coatings must have played the role of an adsorbent for certain types of gaseous compounds in the sewer networks.

During special laboratory experiments, quantitative indicators of formaldehyde emission from the coating surface were determined after being exposed in the sewer collector. For this, samples of concrete coated with a polyethylene coating (total coating area ~ 0.3 dm²) were placed in hermetically sealed desiccators (2 dm³) with a formaldehyde gas analyzer. The results of the measurements of the concentration of formaldehyde in the desiccator are shown in Figure 2.

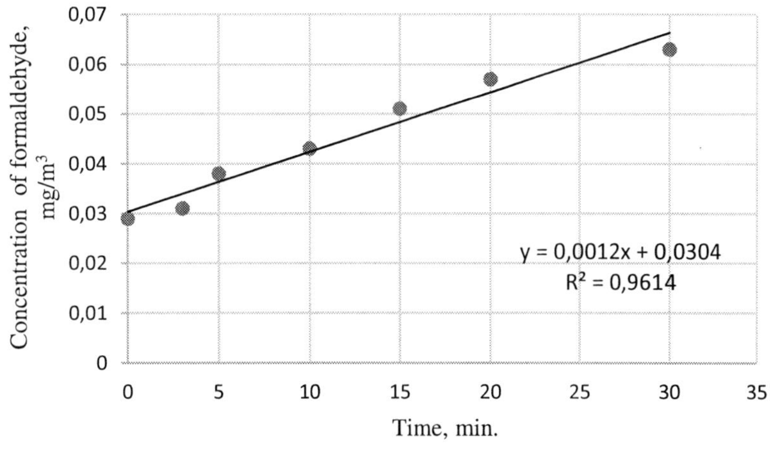


**Figure 2.** The accumulation of formaldehyde over samples of polymer coating after exposure in a sewer collector

As it is seen from the presented data (Figure 2), the concentration of formaldehyde in the desiccator steadily increased in a linear relationship with increasing exposure time. The specific emission of formaldehyde from the coating surface calculated using this data was 0.002-0.007 mg (m² h). Assuming that formaldehyde accumulated in this coating at a similar rate, then the concentration of formaldehyde in the underroof space could be calculated using the formula (5):

$$C_f = \frac{v \cdot T}{V}, \tag{5}$$

where $C_f$ - concentration of formaldehyde in the underroof space, mg/m$^3$; v - formaldehyde accumulation rate, 0,007 mg (m$^2$ hour); V – volume of the underroof space above 1 m2 of arch cover, m$^3$/m$^2$ (~ 1); T - exposure period, month (3) (2160 hours).

The concentration of formaldehyde calculated by the formula (3) is 15.12 mg/m$^3$. Since the experiment with a concrete sample coated with a protective coating was carried out some time after the actual exposure in the collector, formaldehyde emission from the coating may have been somewhat weakened. Therefore, it could be noted that the value of the concentration of formaldehyde in the underroof space calculated by the formula (5), corresponds to the value of the concentration of formaldehyde received using the formula (4).

The obtained data indicates a consistently high concentration of formaldehyde in the underroof space of the sewer networks. So, while carrying out repairs in sewer network collectors, employees of the maintenance and repair services of drainage networks are exposed to environmentally hazardous concentrations of formaldehyde and other hazardous gaseous substances, which negatively affects their performance and general physical condition. Measures must be taken to prevent maintenance workers from getting poisoned by sewer gases that can vary in composition.

Based on the established positive correlation between the concentration of hydrogen sulphide in sewer networks and the concentration of formaldehyde [10], it is safe to assume that the concentration of formaldehyde is very high in the emissions from the shafts with the concentration of hydrogen sulphide that is substantially higher than 1.5 mg/m$^3$ [6-9]. This should be taken into account when calculating the dispersion of gaseous substances in emissions from sewer shafts in the atmospheric air of cities. Such calculations were performed almost exclusively for H2S, as a hazardous component of such emissions [6]. It is worth noting that the $MPC_{daily\ average}$ for formaldehyde is 2.7 times lower than the $MPC_{daily\ average}$ for hydrogen sulphide.

Among the methods for protecting the urban atmospheric air against the emissions of $H_2S$ and formaldehyde (two gaseous compounds, which are hazard category 2 substances) from water disposal systems, the most promising are cooling and the installation of degasifiers on the shafts. These methods [6, 7] have shown their effectiveness in protecting atmospheric air in Kharkov against the H2S pollution and safety for the urban environment, including the videoecological aspect. The prospects of using wastewater cooling transported by networks to reduce formaldehyde emissions are quite good, since it is a product of microbiological metabolism and microbial destruction of wastewater pollution under anaerobic conditions. It is known that the rates of microbiological reactions follow the temperature dependence of Vant-Hoff: with an increase in temperature by 10 °C, the speed increases by a factor of 2–3. Therefore, just like in the case with $H_2S$, this effect halves the concentration of gaseous matter in the atmosphere of the underroof space, and, consequently, in the emissions from the shafts into the urban atmospheric air. The usage of a degasifier to remove formaldehyde from gaseous emissions from water disposal systems also has positive prospects, since the tests of biochemical treatment of emissions of a similar composition were able to demonstrate a high cleaning effect [14]. Moreover, the videoecological problems of such installations for cleaning gaseous emissions from sewer networks (degasifier) – visual discomfort when located among residential and historical buildings, parks, and recreational areas – have been successfully resolved as well [6]. Various design solutions can help to harmoniously blend in these technogenic elements (small architectural forms) with the urban environment.

In addition to this, not only do the suggested technical solutions (cooling of wastewater transported by sewage networks and degasifiers on sewer networks) effectively remove $H_2S$, but they also remove an even more environmentally hazardous gas – formaldehyde, which significantly increases their environmental and economic attractiveness.

## 4. Conclusions

One of the most important factors in ensuring sustainable development of cities is the security of the living environment of its inhabitants, which includes atmospheric air. One of the sources of urban air pollution by environmentally hazardous substances are water disposal facilities.

The study focused on the concentration of gaseous compounds of hazard class 2 – $H_2S$ and formaldehyde – in emissions from sewer shafts. It was shown that the concentration of formaldehyde in emissions from sewer shafts is many times greater than the $MPC_{daily\ average}$ for this compound.

Two technical solutions that have proven to be effective in the removal of $H_2S$ were deemed promising for the effective extraction of this compound from emissions from sewer shafts: cooling of wastewater transported by water disposal networks and degasifiers on sewer networks. Such technical solutions could significantly reduce the level of danger that such gaseous emissions from sewage networks pose for the urban atmosphere. They are environmentally friendly and safe in terms of their visual perception.

## 5. References

[1] Goptsiy D O 2015 *Scientific approaches to understanding the concept of sustainable development and directions of its use in urban development planning* URL: http://dspace.knau.kharkov.ua/jspui/bitstream/123456789/464

[2] Mozgovoi A A 2014 *Sustainable urban development: preconditions and contradictions* URL: http://ir.nmu.org.ua/bitstream/handle/123456789/148166/58-61.pdf?sequence=1/

[3] Kovalevskaya A V, Zelensky S V, Petrova R V 2017 Sustainable city development: the essence of the concept and prerequisites of implementation *Social economics* **53** (1) pp 69-75

[4] Pochwat K, Kida M, Ziembowicz S and Koszelnik P 2019 Odours in sewerage-a description of emissions and of technical abatement measures environments *Electronic Materials* **6** 89 doi:10.3390/environments6080089

[5] Shammay A, Sivret E C, Le-Minh N, Fernandez R L, Evanson I, Stuetz R M 2016 Review of odour abatement in sewer networks *Electronic Materials* **4** 3866–3881

[6] Iurchenko V, Lebedeva E, Brigada E 2016 Environmental Safety of the Sewage Disposal by the Sewerage Pipelines *Procedia Engineering* **134** pp 181-186

[7] Iurchenko V, Lebedeva E 2016 Reduction of environmental hazard of hydrogen sulfide emission from sewerage pipelines and visual pollution, formed by degassing units *Environmental problems* **1** (2) pp 155–158

[8] Iurchenko V, Brigada E 2014 Kinetic characteristics of microbial corrosion of concrete drainage networks *Water and ecology. Problems and solutions* **1** pp 51-61

[9] Hvitved-Jacobsen T, Vollertsen J and Nielsen A H 2005 Odor from sewer networks - processes and prediction *9th NORDIWA Conference on Collection and Treatment of Wastewater* (Stockholm, Sweden)

[10] Yan-li Zhu, Guo-di Zheng, Ding Gao and el. 2016 Odor composition analysis and odor indicator selection during sewage sludge composting *Journal of the Air & Waste Management Association* pp 930-940

[11] Dorogova V B, Taranenko N A, Rychagova O A 2010 Formaldehyde in the environment and its effect on the body (review) *Bulletin of the VSSC SB RAMS* **1**(71) pp 32–35

[12] Zerkalov D V 2007 *Ecological safety: management, monitoring, control* (CST, Dakor, Osnova) p 412

[13] Survey of Formaldehyde 2014 *A LOUS review project* The Danish Environmental Protection Agency p 73

[14] Kulikov N I, Annan A A, Kostik V V, Beldiy M G 1995 Installation for researching the processes of biochemical purification of industrial gas emissions *Chemistry and Water Technology* **17** (60) pp 621-624

IOP Conf. Series: Materials Science and Engineering **907** (2020) 012079 doi:10.1088/1757-899X/907/1/012079

# Creation of modern Orthodox churches in Ukraine - "as languages of Ukrainian autochthony" in the context of urban development"

**O S Sleptsov**[1], **Y U Dunaevskiy**[2]

[1]Department of Fundamentals of Architecture and Architectural Design, Kyiv National University of Construction and Architecture, Kyiv, 01001 Ukraine
[2]Department of Architecture of Buildings and Structures, Odessa State Academy of Civil Engineering and Architecture, Odessa, 65125, Ukraine

o.sleptsov@me.com

**Abstract.** The article "Creation of modern Orthodox churches in Ukraine - "as languages of Ukrainian autochthony" in the context of urban development" is devolved to analysis of the process of revival of Ukrainian church building in the early 90s of XX century and today. It is pointed current research problems and issues. Research papers, publications, abstracts, dissertations, books of recent years by various artists, architects, founders of Ukrainian indigenous architecture are studied. Three main periods of the national architecture of the Ukrainian state (the era of Kievan Rus, Ukrainian Baroque, and Ukrainian Art Nouveau) are described. It is noted that concepts, approaches and methods of designing modern architects of Ukraine. The paper presents architectural psychotypes of the customer of the project, the architect and the priest, and also the model of interaction and influence of sociocultural tendencies on architecture of temples. The paper is supplied with figures and tables. The research provides practical and methodological advice and conclusions on the modern design of temples in Ukraine.

## 1. Introduction

The development of Orthodox architecture in Ukraine at the end of the XX – beginning of the XXI century was influenced by seven decades of Soviet policy. The ban on the construction of churches, the closure of religious educational institutions, and the mass repression of the clergy resulted in the destruction of the system of implicit religious knowledge, which provoked an identity crisis in the Ukrainian Orthodox Church. The post-Soviet development of the Orthodox Church, characterized by a synthesis of ideas of Russian national revival and Orthodox identity, formed both the views of temple theorists, expressed in the desire to restore tradition and canon, and the dominant architectural aesthetics that developed in the Russian Empire in the second half of XX century [14].

With the collapse of the Soviet Union and the formation of independent states – the Russian Federation, the Republic of Belarus and Ukraine – there have been changes in relation to religious renewal. Democratic Ukraine was given the opportunity to fully realize freedom of religion and freedom of church organization. Article 35 of the new Constitution of Ukraine [1] states: "Everyone has the right to freedom of thought and religion. This right includes the freedom to profess any religion or not to practice any religion..." But freedom of religion in Ukraine does not exclude the most important thing

– the great ancient Orthodox tradition, which is the basis of the spiritual life of the Ukrainian people. Orthodoxy came to Kievan Rus in the IX century, and in 988 after the baptism of Prince Vladimir became the state religion of Kievan Rus.

A new landmark in the history of Orthodoxy in Ukraine was marked, first of all, by the scale of church building of all Orthodox denominations in the 1990s. The Cabinet of Ministers of Ukraine approved the program of restoration of outstanding monuments of history and culture of Ukraine by the resolution of April 23, 1999 № 700. To date, many church buildings have been erected, in the appearance of which the architects sought to express the architectural idea of the Orthodox Church in modern language, without going beyond the canonical requirements.

Modern church architecture of Ukraine is experiencing a period of revival and renewal. Seventy years of banning church building have led to an acute shortage of churches, loss of architectural skills and building culture. In the early 1990s, this situation led to the need to forcibly fill the shortage of religious buildings. Ukraine was experiencing a boom in sacred construction. The design of new, reconstruction and completion of existing temples have become an important feature of the architectural life of an independent country [2].

## 2. Analysis of recent research, publications and methods

### *2.1. Historiography of the problems of tradition and innovation in sacred autochthonous architecture*

In determining the sacred meaning of the identity of the Ukrainian church, the Holy Scriptures – the Bible, the works of the Holy Fathers of the Church, the works of religious scholars and art critics – were studied. The continuity of traditions in the temple architecture of Ukraine has interested such domestic researchers as: Linda S, Obidnyak M, Sleptsov O, Tretyak K, Chepelyk V, Lypynsky V, Trubetsky E, Florensky P, Gabrus T, Arabe V. In recent years, dissertations have been defended, which are related to this study: Ivashko Yu, 1997; Yatsiv M, 2002; Krivoruchko Yu, 2018; Goshovsky R, 2015; Pugachova M, 2015; Grigorovich L, 2007; Klimenko O, 2010. Potential interest in determining promising strategies for the evolution of church building, the author addresses the Belarusian researchers of modern Orthodox Church building: Ustinovich E., Balunenko I., Andreyuk A., Pospelovsky D. From Russian researchers addressed this issue – Mitrokhin N, Abbot Alexander ( Fedorov AN), Kuznetsov S, Buseva-Davidova I, Berta A, Kudryavtsev M. and Kudryavtseva T, Sukhanova A, Shchenkov A, Ukhnalev A. They cover important issues of Orthodox Church building. Positions such as: tradition and innovation in Orthodox architecture, national identity; design methods.

### *2.2. Research methods and objectives*

The relevance of the study lies in the need to identify promising ways to develop Orthodox Church building, based on the interpretation of centuries-old traditions of sacred architecture, mastering the worldview of modern society and conceptual ideas of modern architecture. The most acute problem is the search for expressive method of contemporary architecture that do not contradict the functional system of the Orthodox Church; which was formed due to the peculiarities of worship, established internal church traditions and the national specifics of Ukrainian cult architecture.

It is important to study, rethink and assess the prospects for the development of original architecture of modern Orthodox churches in Ukraine and the difficult processes that take place in architectural practice today and are related, in particular, to the need to harmonize the traditions of the Ukrainian Orthodox Church and modern design methods and construction technologies. After a break of almost seventy years, architects are again faced with the problem of the relationship of symbols, traditions, autochthony and innovation in such an area of architecture as temple construction.

## 3. Socio-economic and mental factors that affect the individual development of Ukrainian church building

Research in the field of psych type, national consciousness "archetype" of the Ukrainian mentality in-depth studies and covers the issues of identity of the Ukrainian nation.

There is no denying the well-known truth that national features are largely formed both by the natural environment and the type of folk mentality, both by objective factors and by the subjectivity of historical life and foreign policy conditions [4].

According to the psychological model of CG Jung [3], as present on table 1, there are four main groups of sociotypes in the hierarchy of socio-political relations, which are based on the tendency of individual peoples to logical or sensory, rational or intuitive, collective or individual self-affirmation.

**Table 1.** Socio-economic and mental factors that affect the individual development of Ukrainian church building

| Objective factors: | Subjective factors: |
|---|---|
| geographical and natural-climatic conditions of the environment; | socio-political system and historical living conditions (influences, contacts); |
| mentality; | features of material and spiritual culture; |
| natural-economic and resource bases of economy; | natural resource specifics of economic employment; |
| socio-ethnic structure of the people and the nation. | socio-ethnic trends of culture. |

But as an expression of the nation perceive the dominant type, the characteristics of which deduce the affiliation (of the whole people) to one of the following subtypes of self-realization:

1) Rational-logical, or patrimonial (male) – peoples of epic-historical type of culture, operational-evolutionary function.

2) Sensual-intuitive, or matrimonial (female) – peoples of the mythical-cyclical type of culture, conservative-traditional function.

Thus, according to the definition of belonging to the archetypes of spirituality, the majority of Ukrainians belong to the peoples with a pronounced ideal of the image-sensory model of worldview – poetic-intuitive, or matrimonial. According to the definition of belonging to the sociotype, Ukrainians in their ethnic integrity belong to ethical-intuitive introverts [6]. On the basis of the offered socionic analysis it is possible to allocate features of the Ukrainian national psychotype which become fundamental signs of national:

1) Ukrainian philosophical is based on the principles of "philosophy of the heart" – cordocentrism as the "wisdom" of the heart [7];

2) Ukrainian nature development is solved through urban planning implementation, not dominance in space;

3) Ukrainian spirituality (archetype) – on the pantheism (spiritualization) of the natural environment as a mental means of combining the real with the mystical [5];

4) Ukrainian aesthetics – on poeticization and synthesis of creative means [10];

5) Ukrainian folk life – on artistic "magic", ritual as a tradition [8].

The reasons for the predominant affiliation of Ukrainian church architecture to the "sensory-image" solution of the symbol [9] lie in the peculiarities of the fundamental foundations of the dominant (female) type of national mentality [10].

In folk architecture, these are the wooden temples that have preserved in their three-dimensional composition the figurative archetypes of the pre-Christian folk shrine [11] and which, under the influence of the state architecture of temples, have adapted to the centrally cruciform churches.

## 4. Historiography of the main architectural periods of the Ukrainian temple structure

On figure 1 are presented an important historical periods of Ukrainian church formation, namely three periods of independence and independent state-building periods, which caused three outbreaks of authentic architecture of the Ukrainian church and formed three figurative systems-archetypes of Ukrainian church architecture:

1) Ukrainian Byzantium – in Kievan Rus (P′yatnyts′ka Church in Chernihiv, St. Sophia Cathedral, 1037 - early 1040s). In these churches' Ukrainian features and differences from Byzantine churches are traced;
2) Ukrainian Baroque – with its two varieties: Hetman's Baroque and Cossack Baroque.
3) Ukrainian Art Nouveau – in the early twentieth century.

**Figure 1.** Historical periods of Ukrainian temple formation

A characteristic feature of each of these epochs is not only that these were the periods of the greatest flowering of Ukrainian architecture, but also that each of them was not completed historically, namely didn't move from one art system to another, but was suddenly stopped by public cataclysms [12].

After Vladimir's baptism, his return to Christianity with the help of Chersonese and Constantinople monks became widespread in Russia [13]. With this political act, Kievan Rus entered the family of European states as an Orthodox state. The Metropolitan, appointed by Constantinople, was in the

Cathedral of St. Sophia in Kiev. The cathedral was originally made of wood and burned repeatedly, until a stone temple was built in the 1037s and 1040s.

The cathedrals of the Vydubetsky (1070–1088) and Kyiv-Pechersk (1073–1079) monasteries were built. In Chernihivм – Borysohlibs′kyy Cathedral of the Kremlin (1120–1123), in 1030 the Preobrazhens′kyy Cathedral was founded.

Most of the temples of the Kievan Rus were based on Byzantine prototypes and archetypes – cross-domed one-, three-, five-nave with one, three, five altar apses from the east, four- or six-column one-, three-, five- , thirteen floors. A rich variety of specific solutions allows us to distinguish Kyiv, Galicia, Chernihiv and Pereyaslav architectural schools. Famous architects from the chronicles – Peter Milonig and Obadiah.

In 1238–1240, the Mongol-Tatars managed to conquer Kievan Rus. Kyiv was destroyed and lost its significance as a cultural and religious center. The last stronghold of Christianity in western Russia was the Galician-Volyn principality [13].

The turning point for Ukrainian history and culture was definitely the time that came shortly after the liberation war led by Bohdan Khmelnytsky in 1648–1657. Then, at the end of the 17th–18th centuries, folk traditions of wooden architecture began to be embodied in brick construction.

In the culture of Europe XVII – XVIII centuries were marked as the Baroque era. In architecture - Baroque was distinguished by the play of light and shadow, rich plastic decor, curved lines of facades, ingenuity and richness of forms. Established types of Ukrainian wooden churches get a brick look. Monumental buildings of cross-domed temples and tetraconch churches appear. An indispensable detail of Ukrainian baroque temples are the original pear-shaped ends of the baths, wavy pediments and portals with characteristic decor. According to folk tradition, brick buildings were painted white, sometimes blue.

The cities of Kyiv, Chernihiv, Pereyaslav, Poltava, Lubny, Kharkiv and many other cities and villages of the Hetmanate, Slobozhanshchyna receive new dominants - majestic baroque churches, monastery complexes, bell towers.

Wooden church architecture is also marked by a variety of architectural forms and details: "opasannya" (covered gallery around the church), attic, galleries, carved columns around the whole church or just the porch, network doorposts, beams, wrought iron doors, crosses. The development of church architecture brought after the three-tier five- and nine-tiers types of temples with five and nine domes. Five-tier (cruciform) churches with five tops were built in Galicia, Podillya, Hutsul, Kyiv, Chernihiv, Poltava, Slobozhanshchyna, which testifies to their all-Ukrainian character and uniqueness in the European context [9].

At the beginning of the XX century, with the growth of the national movement in all Ukrainian lands, the style of Ukrainian Art Nouveau spread.

"Ukrainian style", "Ukrainian architectural style", "Ukrainian folk style", "new Ukrainian style" and others. The concepts of "Ukrainian architecture" and "Ukrainian style" have much broader meanings, and Ukrainian architectural modernism as a separate style had its own internal trends. In this regard, VV Chepelyk notes: "In the history of Ukrainian architecture, no other stylistic phenomenon has been marked by such diversity, diversity and inconsistency in the name as Ukrainian Art Nouveau". Beginning with the first publications on this topic in the 1900s and ending with the works of historians of recent years, the search for the optimal name, which is able to fully reveal the nature and features of the phenomenon under study.

Several examples violate the homogeneity of uncharacteristic stylistic forms for Ukrainian church buildings built in the late nineteenth and early twentieth centuries in Ukraine. These are St. George's Church-Mausoleum on the Cossack graves near country Berestechko in Volyn (1910 - 1914, architect V. Maksimov, paintings by I. Izhakevich), Church of St. John the Baptist, country Pryluky, Volyn (early XX century, architect S. Tymoshenko). These temples are attempts to revive national Ukrainian motifs in Art Nouveau architecture.

Another bright and the only cult building in the world, erected in the national traditions of Ukrainian Art Nouveau was built in 1911-14 in honour of the meeting of representatives of Poltava province with

IOP Conf. Series: Materials Science and Engineering **907** (2020) 012079 doi:10.1088/1757-899X/907/1/012079

Emperor Nicholas II during the celebration of the 200th anniversary of the Battle of Poltava in 1909. The chapel is a single-chamber building, square in plan (4.1 m x 4.1 m). The walls of the chapel are lined with ceramic tiles with numerous ornamental stripes, inserts. The gables are filled with plot panels. The eastern pediment is decorated with lush Ukrainian floral ornaments.

At the present transitional stage, the original basis of creativity can only be a skilful innovative use of the phenomenon of folk traditions of Ukrainian church architecture. But the society of the Catholic world took a new approach to the intellectual and aesthetic "challenge" of technological civilization.

## 5. Architectural innovative solutions of modern Catholic and Protestant churches

On figure 2 are presented the changes that sacred architecture underwent in the first half of the twentieth century were introduced by the Catholic Church at the Second Vatican Council, opened by Pope John XXIII on October 2, 1962 in Rome. The purpose of the Council was to comprehend the place of the Church in the modern world, to renew the cult and religious life of Christians "figure 2". The need to involve the faithful in worship is postulated in the "Constitution on the Holy Liturgy": "All parishioners must take full conscious and active participation in the liturgy" [14].

**Figure 2.** Innovations in modern Catholic and Protestant sacred architecture.

The Vatican Council supported the rejection of the tendency to use "decoration". The purpose of the architect is primarily to create a building where it is possible to worship in accordance with the requirements of the cult. The majesty and monumentality of ancient religious buildings are replaced by the simplicity of architectural and artistic design, moderate use of decor and limiting the number of sacred images that can distract the parishioner, prevent their concentration on worship [14].

For example, the altar was raised, one of the traditional elements of the church, the rejection of which was recommended at the Second Vatican Council. [15]. The expressive language of the international style is based on the rejection of "architectural excesses" dictated by the specifics of regional culture, climate and relief, for the sake of minimalism of clear geometric shapes, straight lines, flat surfaces made of concrete, metal and glass.

Innovations in modern Catholic and Protestant sacred architecture are understood as excessively secular and fundamentally unsuitable for the needs of Orthodox worship in modern church building because of the desire to emphasize the originality of the Orthodox Church, to emphasize its differences from Catholic and Protestant models.

The methods of designing current architects and architectural bureaus in the context of modern temple building in Ukraine are relevant for further analysis of the research issues.

For the creation of today's architecture of the Ukrainian church is important not only the scientific basis for studying the history of formation and development of church architecture, but also a sense of this sacredness. After all, the architecture of the Ukrainian church should not be a repetition of the best examples, but a search, creativity, based not only on knowledge but also on feeling. Feelings depend on their formation, creative potential, energy, experience and knowledge, as well as the author's life hobbies in literature, culture, music.

## 6. The method of architectural style of modern Orthodox churches of Ukraine

On figure 3 are presented the main methods of designing religious buildings are related to the stylization of the historical prototype of temples of different types of interpretation of the author. There are several types of stylized architectural design approaches, namely:

1) I Architectural styling method – accurate, or close copying, primarily of the architectural elements of the temples, rather than the image as a whole.

IOP Conf. Series: Materials Science and Engineering **907** (2020) 012079 doi:10.1088/1757-899X/907/1/012079

2) II Architectural styling method (female) – greatly increases some elements, reducing others, and even somewhat distorts the overall shape to achieve a holistic impression.

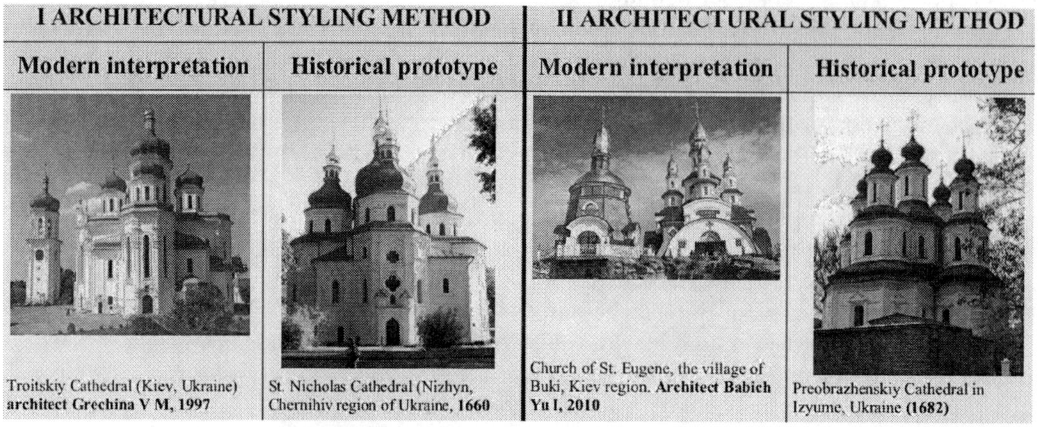


**Figure 3.** Architectural styling method.

Widespread notion of the existence of a strict canon of temple, supported by a popular stereotype about the regulation of all aspects of Orthodox worship, including iconography and sacred architecture. Among practicing architects, the canon is often understood not as a model with outstanding artistic qualities, but as a system of rigid norms, subject to Christian dogma, which regulates all aspects of the external and internal structure of the church. But based on the research of many scientists and architects, including I Busev-Davidova, the existence of a rigid canon is denied: "the canon in the literal sense as a set of well-established rules in temple building has never been and does not exist" [15]. This point of view is presented in regulations that are applied on a voluntary basis in order to comply with the state system of standards.

The method used by architects is the concept of the tradition of Orthodox Church building, as features of the architectural and spatial organization of the cult object are largely determined by its function, carried out by worship. But for the convenience of a successful project combination of the relationship of dogma and established symbolic sets of n researcher of cult architecture M. Kessler introduced the term "canonical tradition" of Orthodox Church building, ie universal compositional-symbolic structure of the Orthodox Church and the system of proportional relations of its elements [16].

Table 2 presents the main directions of designing methods of modern Orthodox churches in Ukraine. Modern Orthodox architects, despite the wave of using the same Russian architectural forms in their projects since the beginning of the restoration of church construction in our country, are increasingly turning to the Ukrainian national style.

**Table 2.** The main directions of design methods of modern Orthodox churches in Ukraine.

| |
|---|
| Imitation approach to the use of unique traditions of Ukrainian Baroque and Neo-Baroque |
| Imitation approach to the use of traditional Ukrainian folk and wooden church architecture |
| Imitation approach to the use of modern architectural forms and details aimed at the latest Protestant and Catholic churches |
| Imitation approach to other stylistic and typical features of cult architecture (neo-Russian, neo- or pseudo-Byzantine, postmodern, traditional style, eclectic) |

1) The first direction is characterized by the use of central and symmetrical architectural and figurative composition, the dominance of its central volume, the use of curved gables, which is characteristic of Baroque architecture.

2) The second direction is characterized by the use of central and symmetrical architectural composition, alternating walls and roof fractures, tiers of tops and their configuration, the slope of the outer walls, the use of modified window frames using national carvings, and brick ornaments in the plane of the wall.

3) The third direction is characterized by the use of cult construction experience of foreign architects of Ukrainian origin, who worked in the period of anti-clericalism, and – the experience of the modern spirit of church building in the Catholic and Protestant world, particularly in Europe.

4) The fourth direction is characterized by the use of historicism, and those areas that existed before 1917, when the development of temple formation was interrupted.

On figure 4 are presented the modern architects of Ukraine, namely: – Sleptsov O, Lositsky I, Rybenchuk M, Sivenky R, Zhezherin V, Chepelyk V.

**Formation of the architecture of modern Ukrainian churches in the country**

| ARCHITECT | Imitation approach to the use of unique traditions of Ukrainian Baroque and Neo-Baroque | Imitation approach to the use of traditional Ukrainian folk and wooden church architecture | Imitation approach to the use of modern architectural forms and details aimed at the latest Protestant and Catholic churches | Imitation approach (neo-Russian, pseudo-Byzantine, postmodern, eclectic and other derivatives) |
|---|---|---|---|---|
| O.S. Sleptsov |  Church of the Holy Apostles Peter and Paul, Kyiv, 2009 |  Temple and ethnographic museum complex of the Holy Great Martyr Demetrius of Thessalonica "Ukrainian Village". with. Buzova, Kyiv-Sviatoshynskyi district, Kyiv region 2009 | |  Church of the Holy Prophet of God Elijah Project: * 1998-2008, implementation: since 1999 Slavutych, Kyiv region. |
| YU. Losyts'kyy |  Church of the Holy Apostles Peter and Paul, Kyiv, 2009 |  Church of St. equal to the Apostolic great. Princess Olga and the chapel-bell tower | |  Church of St. Andrew in Vyshhorod |
| V. Zhezherin | | |  Preobrazhenskiy Cathedral (Kiev, Teremki) | |
| M. Ribenchuk |  Church of the Voznesinaya on Levandivka. | |  Church of St. Luigi Orione in Lviv. |  Church of Saints Vladimir and Olga in Birki |

**Figure 4.** Formation of the architecture of modern Ukrainian churches in the country

Numerous religious buildings built either by immigrants from Ukraine or by foreigners of Ukrainian origin after 1917 abroad. The greatest flowering of national church architecture abroad came in the second half of the XX century in the United States, Canada and Australia. Ukrainian architects in the diaspora sought to preserve their identity in foreign lands, to express or idealize their native architectural

IOP Conf. Series: Materials Science and Engineering **907** (2020) 012079 doi:10.1088/1757-899X/907/1/012079

national traditions in the sacred. Among the influential luminaries-architects are: Zhuk R, Kodak Yu, Osadtsa A, Mazurkevich C, Nimtsiv MD and many other artists. Table 3 present several basic methods of designing sacred architecture.

**Table 3.** Methods of architects in the Ukrainian Diasporas of the world

| |
|---|
| Mixing (eclectic) of traditional compositional characteristics of the Byzantine style, folk wooden architecture and architecture of the Ukrainian Baroque period with the use of modern building structures and materials; |
| application in the spatial and architectural-planning structure of certain elements of historic temples; |
| processing of historical styles (Princely era, Ukrainian baroque, Ukrainian modernism). |

On Figure 5 are presented the main sacred objects of architects in the Ukrainian Diasporas of the world. After analysing the church architecture of the Ukrainian Diasporas, it turns out that most of the architects are original in terms of original three-dimensional solutions, although they show the use of traditional national methods of religious buildings. This is due to the fact that the sacred architecture of the diaspora developed in the context of world sacred culture, despite 70 years of anti-clericalism. In contrast to the Ukrainian practice of temple building, which is being revived, in the architecture of Ukrainian temples of the diaspora, their three-dimensional solution corresponds to modern forms of cultural identity of the time in the world, but sometimes contradicts the national mentality.

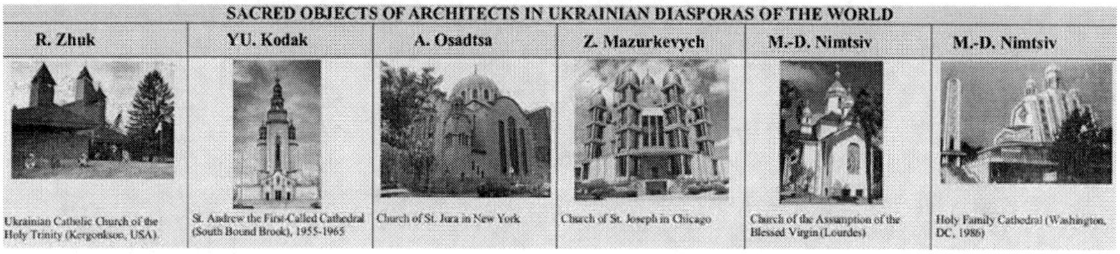


**Figure 5.** Sacred objects of architects in the Ukrainian Diasporas of the world

One of the main problems of modern design, particularly in Ukraine, is that modern church architecture is a young specialty, and very often inexperienced not only customers but also the architects themselves. In the best case, the fate of the architectural and figurative existence of the temple is decided by the abbot, but, as a rule, the customer also tries to dictate his will. It is good if the developer or future abbot has some idea of church architecture and art. And if he has no such ideas and they are based only on his taste, which was also formed in Soviet times, when churches were not built, then, of course, the situation is more complicated.

On figure 6 are presents the dependence of the architecture of a modern church on the canon, the customer or consumer of the architecture of the church, government and secular culture.

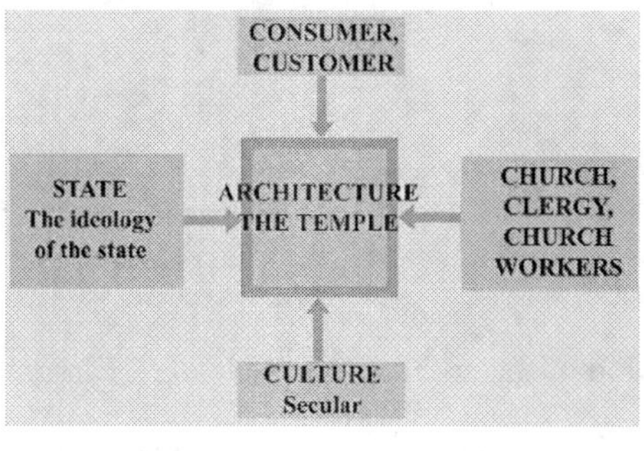


**Figure 6.** The process of forming the sphere of dependence of the temple architecture.

## 7. Recommendations for the practical use of research results

The results obtained during the study can be used in research papers on the problems of modern Ukrainian religious architecture. The actual methods of combining traditional and innovative elements in the modern form of the church, fitting the architectural object into the global and local context identified in the study can be introduced into the design practice of Ukrainian church building. Actual methods of designing religious buildings that take into account the traditions of sacred architecture, sociotypes and mentality can be used in historical and theoretical refresher courses for practicing architects.

## 8. Conclusion

For modern trends in church architecture is characterized and important not so much compliance with traditional historical styles, as any stylization, as a sense and understanding of symbolism and sacredness, as key categories of architectural image of modern Ukrainian churches. Life needs an adequate modern architectural expression for sacred objects, especially in new areas of cities, where the traditional image of new temples is at odds with the language of modern space. Taking into account all key aspects of the historical development of Ukrainian church building and modern circumstances, it is possible to develop a model of an autochthonous church for Ukrainian society, so practical advice for architects will be not to copy other people's projects, but to think and create Ukrainian church architecture.

## 9. References

[1] The Verkhovna Rada of Ukraine 1996 *Constitution of Ukraine Art* **35** (Kyiv: Prosvita) p 14-15
[2] Halyshych R 2002 *Ukrainian Church Architecture and Monumental and Decorative Art Abroad* (Lviv: Spolom) p 6
[3] Sharp D 1996 *Personality Types (Jungian Model)* p 69
[4] Obidnyak M M 2003 Architecture as an expression of logical and sensory worldview *Bulletin of Lviv Polytechnic National University* **486** p 60
[5] Kulchytskyi O 1992 *Worldview of the Ukrainian* (Kyiv: Ukrainian Thought) p 57
[6] Produn M 1994 *Nation of Golden Collars* (Ternopil) p 125-140
[7] Chyzhevsky D 1983 *Essays on the history of philosophy in Ukraine (*Munich) p 136-156
[8] Voropay O 1991 *Customs of our people* (Kyiv) p 48-50
[9] Kryvoruchko Yu I 2001 *Sacral architecture* (Kyiv: Encyclopedia of modern Ukraine) p 723-730
[10] Onatsky Is 1992 *Ukrainian emotionality* (Kyiv: Ukrainian thought) p 36
[11] Taras I 2003 Experience of studying the source of Ukrainian wooden sacred architecture *Ethnographic notebooks* № **1-2** p 279
[12] Tscherkes B, Linda S 2009 Tausende Kirchen für die Ukraine *Bauwelt* **43** p 32–35
[13] Kudryavtsev M, Kudryavtseva T 1994 Russian Orthodox Church. Symbolic language of architectural forms *To the light* **17** p 60
[14] Balunenko I I 2015 *Architecture of modern Orthodox churches of Belarus: traditions and pestilent context* (Minsk)
[15] Kieckhefer R 2004 *Theology in stone: church architecture from Byzantium to Berkeley* (New York: Oxford Univ. Press) p 372
[16] Kesler M Yu 2006 The principle of the canonical tradition in Orthodox urban planning *Econ. Vestn.* **11** pp 17-21

**Acknowledgement**
The authors would like to thank to Professor, Doctor of Technical Science from Kharkiv National University of Civil Engineering and Architecture, the co-chair of the organizing committee V P Sopov for providing additional materials and constructive advice.

IOP Conf. Series: Materials Science and Engineering **907** (2020) 012080 doi:10.1088/1757-899X/907/1/012080

# The conceptual framework of sustainable development and the role of environmental and technogenic safety in achieving it

**O Hunchenko, O Voloshkina, I Korduba, M Kravchenko and P Stefanovych**

Department of Occupational Safety and Environment, Kyiv National University of Construction and Architecture, 31 Povitroflotsky Avenue, 03037, Kyiv, Ukraine

gunchenko.oksana@gmail.com

**Abstract.** The article deals with the issues related to the environmental aspects of the development of economic production performance and their impact on the sustainability of production development. Reaching a certain level of technogenic safety and the ability to influence it through a number of organizational and technical measures are the factors for improving environmental safety. The formation of the optimal direction and value of the «vector of sustainable development» (VSD) which includes the «security vector» (SV) that currently may be limited by the economic development and stability indicators, the requirements of environmental legislation and the quality of the ecological fund, the labour legislation, the level of the technological effectiveness and the environmental compatibility of production, the quality of labour potential, the health and well-being standards, the social aspects of development and protectability, the level of understanding of personal safety and personal responsibility for the safety of others were justified. The directions of risk formation and technological and environmental safety reduction for the construction industry, among which the main ones are those that are being formed at the level of the global world economic and political processes, the state, the construction industry itself, the local communities, the external organizations and contractors and a specific construction organization were outlined.

## Introduction

Throughout the ages, humanity has been constantly facing the global problems such as exhaustive resources, climate change, food and safe drinking water insecurity, overpopulation, military conflicts, as well as social problems related to unemployment, poverty, gender inequality, low levels of remuneration, the non-compliance of working conditions to the declared safety level and the social requirements for the riskogenics in everyday life. In addition to these stable risk formation factors, it should be taken into account that the modern world is changing rapidly both in everyday life and in technological processes. There are new materials, technologies, principles of human interaction with the natural environment, modern forms of employment and progressive methods of protecting humanity from negative factors of different origin.

Human being is constantly interacting with the natural environment and trying to make the living conditions more secure and protected from natural threats. However, these activities are too controversial since human beings exploit natural resources (such as forest plantations, soils, water resources, clean air, flora, fauna, mineral resources, etc.) to maintain their personal life and health, to achieve decent living conditions and economic growth in different industries in amounts that exceed the ability of natural resources to recover or the possibility of maintaining them in sufficient quantities to ensure humanity a comfortable life in the distant future. The threatening dynamics of human interaction with the environment reveal two major problems in this direction [1]. Firstly, there is a strong belief that all natural resources can be divided into exhaustible and inexhaustible, but it is possible to assert that owing to the pace of human impact on the environment and the irresponsible natural resource management, the inexhaustibility of resources is only a matter of time and new dangerous technologies. Secondly, a retrospective analysis of the human achievements demonstrates the geometric growth of the amounts of consumable natural resources in the way of increasing the technological effectiveness of production that can constitute a danger taking into consideration the world population growth.

**Presentation of basic material of the research.** Nowadays four fundamental laws of ecology outlined by Barry Commoner in the last century are undeniable [2]. The basic idea of these laws is as follows:
- everything is connected to everything else;
- everything must go somewhere;
- nature knows best;
- there is no such thing as a free lunch.

These postulates accurately emphasize that any kind of anthropogenic or anthropological impact on the environment will inevitably lead to certain transformations, and these changes will not cause global disaster only if appropriate and correct measures are taken. Due to the fact that in view of the capitalization of all components and using the principles of wealth accumulation by any means necessary, even if it means jeopardizing peace and security of future generations, a technologically advanced society has actually ensured the rapid growth in production and has completely disregarded the safety and health of their descendants.

The dangerous feature of imprudent environmental behavior is the ultimate impact branching that is not limited to the biological characteristics of the species, the territorial boundaries or temporal limitations. The environmental problems of certain limited territories, regions or countries impede economic growth in one way or another and require from all professionals in natural resource management to carry out their activities based on the opportunities of the natural environment to recover resources expended by them, to take measures to regulate the overwhelming impact, to limit the volume of unsustainable productions and to restrict significantly the development pressure on ecosystems. Another direction in reducing human pressure on the natural environment is to increase the quality and the rate of its recovery, to improve the existing natural conditions, to plan the activities that rely on the overuse of natural resources considering the nature of the impact of such activities in the distant future, to use the alternative resources that have not been used before with the increased benefits of all natural sources.

Therefore, the global human-nature interaction strategy should be harmonized and balanced in order to create the safe and comfortable conditions for human existence and constructive activity while maintaining the current state of the environment to the greatest possible extent and undertaking the purposeful activities necessary for its improvement. The level of harmonization and interpenetrating influence is definitely limited by a certain contour of regulation which includes the main goals of social development, the criteria for determining the achievement of a certain level of their realization and the presence of mechanisms for their achievement. Overcoming the contradictions between human and nature, which is still often considered only as a resource base, should be based on the most advanced technologies, the most effective strategies, the best available production practices and the professional labour potential. The global goal for modern society is to create the safest environment in which the high-quality objects, goods, services, infrastructure elements etc. are created with minimal impact on the environment in order to preserve the progressive vector of social development and create the preconditions for appearance, comfortable existence and progressive development of future generations based on the principles of humanism and full responsibility.

The sustainable development strategy (SDS) which is defined as an organizational principle that meets the needs of the present without compromising the ability of future generations to meet their own needs is fully consistent with these criteria.

Sustainable development combines three basic components such as environment, economy and society and this concept was proposed by the economist Rene Passet in 1979 and was further developed until it became the main strategy of social and economic development at the beginning of the XXI century. This was preceded by the adoption of a number of interstate and national documents [3-10]:

- the Declaration of the United Nations Conference on the Human Environment, or Stockholm Declaration (1972);
- the Rio Declaration on Environment and Development, or Rio Declaration (1992);
- «Agenda-21» (1992);
- the Pan-European Biological and Landscape Diversity Strategy (1995);
- the Guiding Principles for Sustainable Spatial Development of the European Continent (CEMAT) (2000);
- the Lisbon Strategy, also known as the Lisbon Agenda or Lisbon Process (2000);
- United Nations General Assembly, «The Future We Want» Resolution 66/288 (2012);
- Europe 2020 (2010 p.);
- the 2030 Agenda for Sustainable Development (2017).

The gradual development of this concept has led to the extension and receiving of the new influential emphases such as the assessment of the impact of political transformations, cultural heritage and administrative decisions on the sustainability of states, public associations, campaigns, and so on. However, what is certain is that humanity is striving for the maximum level of safety (social, political, environmental, technogenic) which can be provided only with the proper attention to all these areas rather than focusing only on one of the strategies.

The level of human development and sustainability was increasing together with the expanding of the worldview and the introduction of additional elements such as «nature», «technology», «information» in the «human-human» system and as a result it has moved from a relatively low «human-nature» level of relations to higher level, i.e. to the sustainable development.

Nowadays the sustainable development principles are crucial in the issues of security of the country and they encourage the formation of national development strategies in the light of these aspects. The Sustainable Development Strategy «Ukraine 2020» has been in place since 2016 and it requires the implementation of a number of reforms and programs, including:

- workplace relations reform;
- welfare reform;
- environmental conservation program;
- energy efficiency program;
- a healthy lifestyle and longevity program [11].

Extremely ambitious levels of country development strategy indicators are outlined in this strategy and they are designed to double GDP per person (up to $ 16,000), to increase the average life expectancy by three years according to the World Bank, to reduce GDP energy intensity to 0,2 toe (tonnes of oil equivalent) per $ 1,000, and so on. However, recently updated legislative acts such as The Draft Law of Ukraine on the Strategy of Implementation of the Balanced Development Model of Ukraine by 2030 and the Draft Law of Ukraine on the Sustainable Development Strategy for Ukraine by 2030 have already been developed and are under consideration in the Verkhovna Rada of Ukraine. There are both optimistic thorough views on the subject [12] and skeptical slogans that claim that Ukraine lacks the sustainable development concept and strategy [13].

Such a range of statements is resulted from the complexity of the problem of combining the three main vectors of influence on the sustainability and the need to create a conscious responsibility for personal and financial security, personal development among the average citizens; and for economic growth, decent working conditions, preconditions for gender equality and economic justice among the entrepreneurs, leaders of the public union, party or country by shaping their activities and making administrative decisions on principles of environmental consciousness and public virtues, by promoting the preservation of biodiversity and recovering the expended natural resources on the basis of the need to preserve environmental and social conditions for future generations at present or even higher level.

Jay Forrester, Dennis Meadows, Mihajlo D Mesarovic, Eduard Pestel, William Nordhaus, Jan Tinbergen and others are the foreign scientists, innovators who drew public attention to the inevitability of the end of the human race in the condition of further development of society in the industrial paradigm without taking into account technogenic and anthropogenic impact on the environment, and who created the preconditions for the formation of the concept of sustainability and its further development [14]. Further development of the sustainability concept has been reflected in the global documents of the UN General Assembly which were mentioned above.

In her scientific work, Z A Mynhaleva suggested to present the entire range of existing approaches to sustainable development in the form of three-dimensional space with axes of coordinates as «human», «economy», «nature» [14] (Figure 1). The author claims, «The main approaches and theories of sustainable development that are grouped on the basis of qualitative and quantitative assessment as the «min» and «max» values of the main classification indices, that is, the index of the dominant (leading) element in the relation (which is the main index), the subordination index (which depends on something) and the index prevailing in the set of recommendations for policy formulation in the field of industrial and socio-economic development, nature protection with regard to sustainable development of society. The dashed lines in Figure 1 show the developmental trajectories of conceptual approaches to sustainable development in view of the importance of all sides of such triad as «human-economy-nature»: the straight arrow that points to «max» for all three elements reflects the trajectory of the sustainable development concept with equal inclusion of all elements in the relation system».

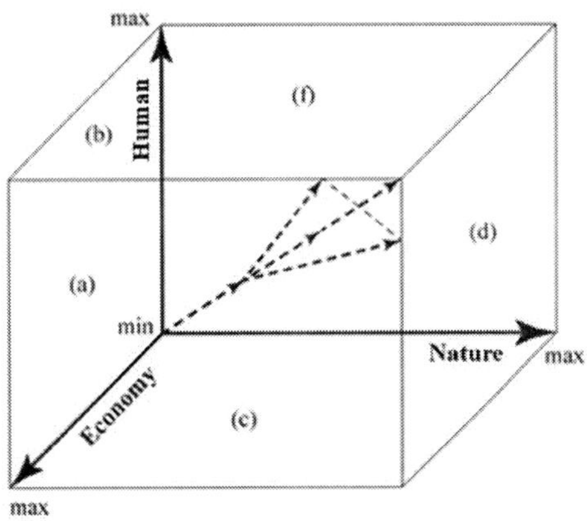


**Figure 1.** A variety of approaches to the combination of such elements as «human-economy-nature»: a) - the theory of full subordination of nature to the human and production interests; b - the theory of equal well-being for all; c - the theory and approaches of resource conservation, their rational management, environmental safety of production; d - the theory of the full convergence with nature, the rejection of industrial production; e – the theory of economic growth based on the environmental principles; f - the theory of implementation of all environmental standards, healthy lifestyle.

It is definitely a global, and perhaps even slightly utopian desire to achieve the maximum level of economic development in the most secure, healthy, equitable social environment and ecologically stable and balanced natural environment, but there is no other way, there are only prospects for expanding our focus and influence with regard to the new risks that arise with increasing economic performance and humanity's demands for comfort and safety.

Thus, it can be assumed that the optimal direction and value of the «vector of sustainable development» (VSD), which includes the «security vector» (SV) currently may be limited by the economic development and stability indicators, the requirements of environmental legislation and the quality of the ecological fund, the labour legislation, the level of the technological effectiveness and environmental compatibility of production, the quality of labour potential, the health and well-being standards, the social aspects of development and protectability, the level of understanding of personal safety and personal responsibility for the safety of others.

From a mathematical point of view, the length of VSD $|\overrightarrow{a}|$ (Figure 2) is determined by the components $a_x$, $a_y$, $a_z$ and is calculated as follows:

$$\vec{a} = (a_x; a_y; a_z)\text{, i.e. } |\overrightarrow{a}| = \sqrt{a_x{}^2 + a_y{}^2 + a_z{}^2}, \quad (1)$$

and the length of the SV $|\overrightarrow{a_1}|$, is calculated according to the formula:

$$\overrightarrow{a_1} = (a_{x1}; a_{y1}; a_{z1})\text{, i.e. } |\overrightarrow{a_1}| = \sqrt{a_{x1}{}^2 + a_{y1}{}^2 + a_{z1}{}^2}, \quad (2)$$

where $a_x$ – the level of environmental sustainability; $a_y$ – the level of economic growth; $a_z$ – the level of social sustainability; $a_{x1}$ – the level of environmental safety; $a_{y1}$ – the level of economic sustainability; $a_{z1}$ – the level of human security in the production process and in everyday life.

In general, all these indices should be summarized considering the end result i.e. the improvement of health and the increase in the life expectancy in conditions that ensure the basic needs of humanity such as the availability of safe drinking water, clean air, food, safe accommodation and working environment, self-realization.

Undoubtedly, the environmental compatibility of relations in the system of «human-technology-nature» is the foundation for sustainable development that is focused on risk management in the national health system, which itself requires the financing of environmentally oriented services both from the state and from participants of production activities and social sphere. The environmental component of sustainability requires society to develop measures to meet human needs while maintaining the planetary life-support systems (sustainable use of natural resources and water resources, use of renewable energy, biomass and biodiversity conservation, avoidance of hazardous industrial and household emissions).

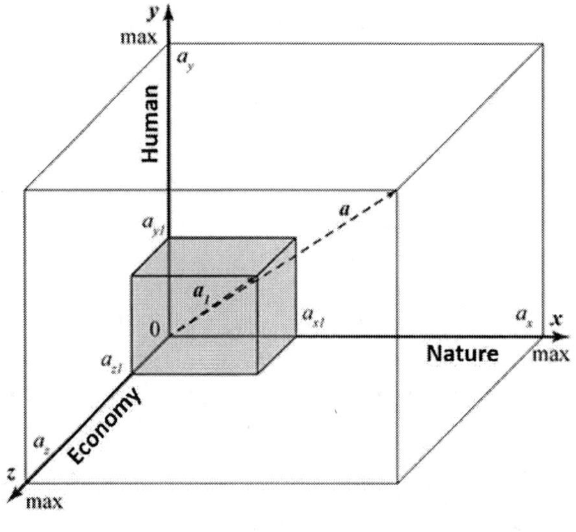


**Figure 2.** The definition of the «vector of sustainable development» and «security vector».

The precarious situation arises when natural capital is consumed faster than can be restored, and human capital suffers significant losses in the form of deterioration of living conditions, health, safety, and self-realization. Sustainable development is based on such human activity during which natural resources are used only in accordance with the natural rate of their recovery. It is theoretically possible to design the future of mankind where a strategy of destructive and negligent attitude towards the environment in which it is impossible to create the conditions for human existence and the birth of future generations is implemented. That is to say, environmental degradation causes the destructive processes that provoke an increase in mortality among the population and make it impossible for humans to have an impact on the degraded environment when passing a critical point.

It should also be noted that the environmental impact is closely related to technologies, modes of their use and the scale of their mass production. Therefore, the most destructive factor of global environmental degradation is definitely a narrow-minded human who concentrates only on economic component of sustainable development and strives only for financial stability without taking into account significant external impact factors, so this human will always face the risks and will need significant financial investments and the ability to respond promptly and to direct efforts in the right direction in order to avoid, localize these risks or to reduce their negative and dangerous impact that generally has a synergistic effect.

The efficiency and sustainability of individual sectors of the economy strongly depend on the risks that are developed at different levels of impact. For the Ukrainian construction industry, the main levels of medium, significant and high risks formation are:

- interstate level;
- state level;
- level of Ukrainian construction industry;
- local community level;
- level of a specific construction organization;
- level of the external organizations and contractors.

The modern strategies and ways to influence the risk have two main directions:

- reduction of probability (impossibility) of danger occurrence;
- reducing the consequences of a dangerous event that has occurred.

It will be relevant to implement certain proactive strategies for increasing the sustainability of construction industry of Ukraine since it has such extensive levels of risk formation. The main elements of these strategies are presented in Table 1.

Strategies for risk management in Ukrainian construction industry summarized in Table 1 allow to embrace the levels of medium, significant and high risk. Preventive actions at these levels range from developing strategies for the construction industry, a local community, and a specific enterprise to simple and definite solutions at the lower administrative levels. One of the current strategies for technogenic and environmental risk management in the construction industry is the introduction of modern and safe green technologies, the so-called «green construction». Reducing of risks with the introduction of green construction technologies contributes to:

- reduction of dangerous technogenic and environmental load from corporate activities and the financial portfolio for its localization;
- reduction of the budget for overcoming the effects of negative residual environmental impact from construction industry;
- reduction of waste (including hazardous one) generated at the stage of production, operation and disposal of construction materials, equipment, tools and facilities;
- improvement of working conditions;
- enhancement of occupational safety;
- reduction of long-term risks connected with the occurrence of occupational diseases among construction workers;
- increase of the social importance of the industry;
- increase of positive.

Strategies for risk management in Ukrainian construction industry summarized in Table 1 allow to embrace the levels of medium, significant and high risk. Preventive actions at these levels range from developing strategies for the construction industry, a local community, and a specific enterprise to simple and definite solutions at the lower administrative levels.

IOP Conf. Series: Materials Science and Engineering **907** (2020) 012080 doi:10.1088/1757-899X/907/1/012080

**Table 1.** Strategies for Risk Management in Construction

| Level of risk formation | Strategies for risk management |
|---|---|
| Interstate level | - Analysis of current global economic and political tendencies to the changes in the basic economic indicators of developed countries and economic sectors.<br>- Investigation of impact of the changes in the national economic activity on the stability and profitability of the construction industry.<br>- Analysis of existing strategies for risk and crisis management used in different economic sectors by successful enterprises of Ukraine and other countries.<br>- Prediction about the level and nature of the impact of transformations in the world economy on the construction industry based on the national peculiarities and global policy aspects of influence.<br>- Development of the strategies to protect the construction industry from the impact of global economic and political factors, taking into account the best global practices and national characteristics. |
| State level | - Development and implementation of legislation to support the risk-oriented production development strategy.<br>- Implementation of effective mechanisms to control the technogenic and environmental safety of production.<br>- Implementation of state programs for the development and support of economic sectors that are the driving force of economic growth.<br>- Equal conditions in the market and the inevitability of punishment of offenders.<br>- Encouragement of long-term cheap investments in the industry. |
| Local community level | - Preparation and implementation of a community development plan, taking into account the environmental, technogenic and social components.<br>- Support of the construction industry as the basis to ensure sustainable community development.<br>- Creation of preconditions for development of labour potential of the region.<br>- Increase in the level of social standards and safety for community. |
| Construction industry level | - Preparation and implementation of a strategic plan for sustainable development of the construction industry.<br>- Formation of request and financing of scientific research results on the current areas of industrial development.<br>- Creation of the development concept of labour potential and production capacities for the construction industry.<br>- Promotion of advanced technologies and their integration into the global professional community.<br>- Support of small businesses and expansion of regional manufacturing complex.<br>- Constructive dialogue with all players of the construction services market. |
| Level of the external organizations and contractors | - Employment of strategy of trusted partnerships.<br>- Long-term cooperation agreements only with companies that promote the values of safety, environmental compatibility and corporate social responsibility.<br>- Secure and fast logistics.<br>- Regular dialogue and cooperation with contracting organizations.<br>- Constant analysis of alternatives, advantages and disadvantages for existing contractors. |
| Level of a specific construction organization | - Risk-oriented approach is the basis of construction company organization.<br>- Engagement of trusted and reliable contracting organizations.<br>- Availability of own technological developments.<br>- Introduction of modern and safe green technologies.<br>- Accurate staff selection, their constant training and development.<br>- Raising of social standards in the enterprise.<br>- Security leadership. |

One of the current strategies for technogenic and environmental risk management in the construction industry is the introduction of modern and safe green technologies, the so-called «green construction». Reducing of risks with the introduction of green construction technologies contributes to:

- reduction of dangerous technogenic and environmental load from corporate activities and the financial portfolio for its localization;
- reduction of the budget for overcoming the effects of negative residual environmental impact from construction industry;
- reduction of waste (including hazardous one) generated at the stage of production, operation and disposal of construction materials, equipment, tools and facilities;
- improvement of working conditions;
- enhancement of occupational safety;
- reduction of long-term risks connected with the occurrence of occupational diseases among construction workers;
- increase of the social importance of the industry;
- increase of positive public attitude to the industry.

This approach provides more opportunities for sustainability of the construction industry due to reduction of production risks for workers by using environmentally-friendly materials and their processing and production techniques, reducing the costs on recycling of environmentally hazardous packaging materials and waste generated at the stage of production, employment and further recycling of materials, increasing the competitive position of industry and the incomes of a specific construction organization as a result of the higher demand for environmental compatibility of buildings and social facilities and the constant growth in such demand from society, changing attitudes to the labour potential of the state and caring attitudes towards the health of the population of the state.

**Conclusions.** Thus, it would be efficient to create a specific behavioral strategy both for individuals in everyday life and for businesses in all sectors of the economy in order to reduce the negative impact of the above-listed range of risks on the environment and to preserve the natural potential of the planet through the introduction of environmental technologies and the restriction on the direct use of natural resources with low growth rate. If the aim of the society in the XX century was to form various forms of activity and consumption, then the aim of the XXI century is to optimize these forms through an exclusively ecological worldview.

The impact on the natural environment is the the most significant within the production environment since it has the dangerous consequences for the health condition and human reproduction both in terms of production itself and as a secondary factor outside the enterprise (toxic emissions, pollution of drinking water sources, hydrological phenomena, changes in biodiversity, etc.). Direct and indirect (secondary) environmental production factors create a synergistic effect and require equal attention and a comprehensive solution to this problem. Human also needs as much attention as the natural environment, being an element in the «human-technology-nature» system and having certain limits in stability and adequacy of actions in stressful and dangerous situations. Human often causes hazardous or emergency situations even with high equipment safety and controlled minimal environmental impact, and therefore leads to the large-scale environmental disasters with significant financial losses, human casualties or land use restrictions for indeterminate or limited period. These strategies for risk management in the construction industry of Ukraine are created on the basis of the main directions of promoting the sustainable development of a state and they allow to increase the stability of the construction industry of Ukraine and to improve its economic indicators through a number of precautionary measures. The introduction of green construction techniques is a success strategy for the construction industry that will provide an organic transition from a state of danger to a state of security, and from indifference to interest and responsibility.

## References

[1] Strukova M N 2016 Maintenance of the balance in the system of «human-nature» relations as a necessary condition for the sustainable development of society *Proc. of the International scientific-practical conference "Medical, social and psychological aspects of industrial agglomeration safety"* Available at: https://elibrary.ru/item.asp?id=25855172 (in Russian)

[2] Kokitko A A 2018 Environmental safety in the context of sustainable development *Proc. of the 5th International Congress "Environmental Protection. Balanced Nature Management"* (Lviv, Lviv Polytechnic) pp 186–187 (in Ukrainian)
[3] The Declaration of the United Nations Conference on the Human Environment. Available at: https://zakon.rada.gov.ua/laws/show/995_454. (in Russian)
[4] *Agenda 21* Available at: http://www.r-komitet.ru/s_i_d/agenda21/index.htm. (in Russian)
[5] *United Nations General Assembly "The Future We Want" Resolution 66/288* Available at: https://rio20.un.org/sites/rio20.un.org/files/a-conf.216-l-1_russian.pdf.pdf (in Russian)
[6] *The Pan-European Biological and Landscape Diversity Strategy* Available at: https://zakon.rada.gov.ua/laws/show/994_711 (in Ukrainian)
[7] *The Guiding Principles for Sustainable Spatial Development of the European Continent* Available at: https://rm.coe.int/168070018e (in Ukrainian)
[8] *Lisbon European council 23 and 24 march 2000. Presidency conclusions* Available at: http://www.europarl.europa.eu/summits/lis1_en.htm
[9] *Europe 2020: A strategy for smart, sustainable and inclusive growth* Available at: https://eulaw.ru/content/novaya-evropejskaya-strategiya-evropa-2020/ (in Russian)
[10] *Transforming our world: the 2030 Agenda for Sustainable Development* Available at: https://sustainabledevelopment.un.org/post2015/transformingourworld
[11] *Sustainable Development Strategy "Ukraine 2020"* Available at: https://zakon.rada.gov.ua/laws/show/5/2015#n10 (in Ukrainian)
[12] Rudenko L G 2012 Ukrainian national report on the status of implementation of the provisions of «Agenda for the XXI Century» in Ukraine for a ten-year period (Kyiv) p 345 (in Ukrainian)
[13] *The task of geography in implementing the paradigm of sustainable development and goals in Ukraine in 2030* Available at: https://don.kyivcity.gov.ua/files/2018/3/30/1.pdf. (in Ukrainian)
[14] Mingaleva Zh A *The ideas of sustainable development: an «apple of discord» or a platform for unification?* Available at: https://www.econ.msu.ru/sys/raw.php?o=44760&p=attachment. (in Russian)

IOP Conf. Series: Materials Science and Engineering **907** (2020) 012081 doi:10.1088/1757-899X/907/1/012081

# Integrated management systems as a basis for sustainable development

**Okcana Hunchenko**

Department of Occupational Safety and Environment, Kyiv National University of Construction and Architecture, 31 Povitroflotsky Avenue, Kyiv, 03037 Ukraine

*hunchenko.om@knuba.edu.ua*

**Abstract.** The issue of the urgency of the development and implementation of an integrated type of systems into management systems that combine mechanisms of safety management and hygiene of production processes and environmental safety were considered. Considering the purposes of sustainable development that form the priority directions of development of society, economy, industry and nature management, five main blocks that are responsible for the development of such areas as "SOCIETY", "HEALTH", "FINANCE", "LABOR PROTECTION", "ECOLOGY" are identified. All of these directions are cemented and have an extensive spectrum for cooperation and development, thanks for their combination, based on the security aspect. Therefore, among the sustainable development purposes there is one more parameter that influences the achievement of the set purposes, the factor "SAFETY" that integrates all the previously defined directions. An effective security strategy today is a risk-oriented approach based on risk assessment and selected strategies for influencing them. The using of integrated systems of enterprise management that combine the impact on the safety and hygiene of production processes and environmental safety of production, especially in industries with remote environmental impacts and significant health consequences for employees engaged in these industries allows to consider the issues of assessing existing and predicting the emergence of new risks more weighed, more reasonably and with less amount of resources.

The concept of sustainable development is aimed at the harmonious development of the three components - economic growth, environmental well-being and social equality. Today 17 sustainable development purposes (SDP) and 168 indicators of their achievement have been substantiated [1].

Amomg 17 SPD 4 – directly related with the state of health of nation and specifically a person – SPD 2, 3, 6 and 11; 6 – connected with the safety of production processes and professional activities – SPD 3, 5, 8, 9, 11 and 12; 7 – show interest in environmental improving – SPD 6, 7, 11-15 and 14 – focused on social inequality overcoming– SPD 1-12, 16, 17 (Figure 1).

A graphical interpretation of the SPD by the directions of influence of «HEALTH», «LABOR PROTECTION», «ENVIRONMENTAL SAFETY» and «SOCIETY», is shown in Figure 1 and presentation of these data is in the form of a table (Table 1), taking into account the «FINANCES» component and identify purposes that integrate three, four or all five directions and create an impact on economic stability to maximize their value by separating them as a component of "SAFETY" in the issue of achievement of sustainable development.

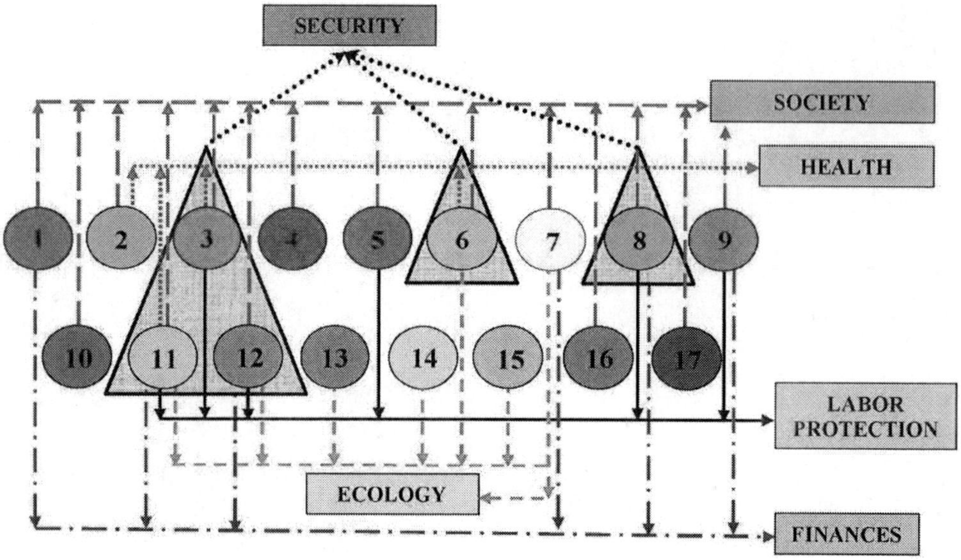


**Figure 1.** Orientation of sustainable Development Goals: "HEALTH", "LABOR PROTECTION", "ECOLOGY", "SOCIETY", «FINANCES» and "SECURITY"

**Table 1.** Purposes of sustainable development and their impact

| № SPD | SPD | H | LP | E | S | F | St |
|---|---|---|---|---|---|---|---|
| **1.** | Poverty overcoming. | | | | + | + | |
| **2.** | Hunger overcoming, achieving food safety, improving nutrition, promotion of sustainable development of agriculture. | + | | | + | | |
| **3.** | Good health, a healthy lifestyle, well-being for all. | + | + | | + | | + |
| **4.** | Comprehensive, quality education during the whole life. | | | | + | | |
| **5.** | Gender equality, women's empowerment. | | + | | + | | |
| **6.** | Availability and rational using of water resources, proper sanitation. | + | | + | + | | + |
| **7.** | Inexpensive, reliable, sustainable, clean, modern energy sources. | | | + | + | + | |
| **8.** | Decent work, full and productive employment, economic growth. | + | + | | + | + | + |
| **9.** | Industrialization, innovations, infrastructure development. | | + | | + | + | |
| **10.** | Reducing inequalities within and between countries. | | | | + | | |
| **11.** | Openness, safety, sustainability and environmental sustainability of cities and communities. | + | + | + | + | + | + |
| **12.** | Rational responsible consumption and production. | | + | + | + | + | + |
| **13.** | Overcoming the effects of climate change and struggling with factors that provoke it. | | | + | | | |
| **14.** | Preservation of marine resources. | | | + | | | |
| **15.** | Protection and recovery of land ecosystems. | | | + | | | |
| **16.** | Peace, fairness, justice. | | | | + | | |
| **17.** | Global partnership for SD. | | | | + | + | |

*Note.* Table 1 uses the following abbreviations: H - "HEALTH", LP - "LABOR PROTECTION", E - "ECOLOGY", S - "SOCIETY", F - "FINANCE", St - "SAFETY".

IOP Conf. Series: Materials Science and Engineering **907** (2020) 012081 doi:10.1088/1757-899X/907/1/012081

The issue of professional selection, medical accompainment (periodic medical checkups and medical conclusion before employment), implementation of health-saving methods at work and in life [2], using of effective physiologically sound modes of work and rest, eco-friendly materials and environmentally friendly production processes is connected with the improvement of safety of the person and society in general. It will be much easier to achieve the desired effect by using integrated management systems that combine directions of labour protection and ecological safety into a single system for managing ecological and technological safety of production.

The concept of ecological and technological safety of production is based on the integrated perception of ecological and technogenic components of human and environmental hazards within the framework of performing socially actualized production activities at high-risk objects.

In conditions of economic crisis and orientation of all developed countries on achieving of purposes of sustainable development let us consider which role in this process is played by labor protection and environmental safety. Among the priorities of the state is the operation of a development strategy, the creation of a legislative base, control and fiscal bodies, government programs to support businesses and certain sectors of the economy, the development of progressive sectors of the economy and financial institutions. In developed countries, the development of state-level sustainable development strategies began in the 1990s, Sweden, the United Kingdom, Switzerland and Ireland became the flagships of this process, and have been reviewed since then [3]. In EU countries, the principle of sustainable development is enshrined in the founding Amsterdam treaty of the EU in 1997 and recognized as key to all politics of EU countries. Therefore, any sectoral or regional development strategies in these countries are subordinated to the achievement of the global sustainable development purposes.

In 2015 Ukraine on the way of the euro integration process also adopted a declarative document Sustainable strategy «Ukraine 2020»[4], the main purpose of which was the introduction of European standards in life and achievement leading positions in the world. Achieving this purpose should be based on the implementation of four main vectors:

the vector of development - the ensuring sustainable development of the state, carrying out structural reforms and, as a consequence, raising living standards;

the security vector - the providing guarantees for the security of the state, business and citizens, protection of investments and private property, where particular attention should be paid to the safety of life and human health;

the vector of responsibility – providing guarantees that every citizen, regardless of race, color, political, religious and other beliefs, gender, ethnic and social background, property status, place of residence, linguistic or other characteristics, will have access to a high-quality education, health care system public and private sector services and other services;

the vector of pride – the ensuring of mutual respect and tolerance in society, pride for own state, its history, culture, science, sport.

This document had a scientific basis for the formation and was widely discussed at the stage of adoption and further implementation [5-8] in the scientific community and at the public level, which created the preconditions for the development of the project of the law of Ukraine about the Strategy of sustainable development of Ukraine until 2030 [9], adoption which will enable the implementation of sustainable development principles in all spheres of human life and integrate the programs of development of economic sectors and regions.

In the social sphere, the creation of preconditions for the formation and development of labor potential is one of the priority tasks. Akulov M.G. in his work [10] provides the following definition of labor potential - it is the total social capacity for work, the potential capacity of society, its resources of labor. Not only able-bodied people employed in the industry but also those who are just educated and those who have already completed their professional activities and retired are included into labor potential. The main factors that affect the quality of labor potential are shown in Figure 2.

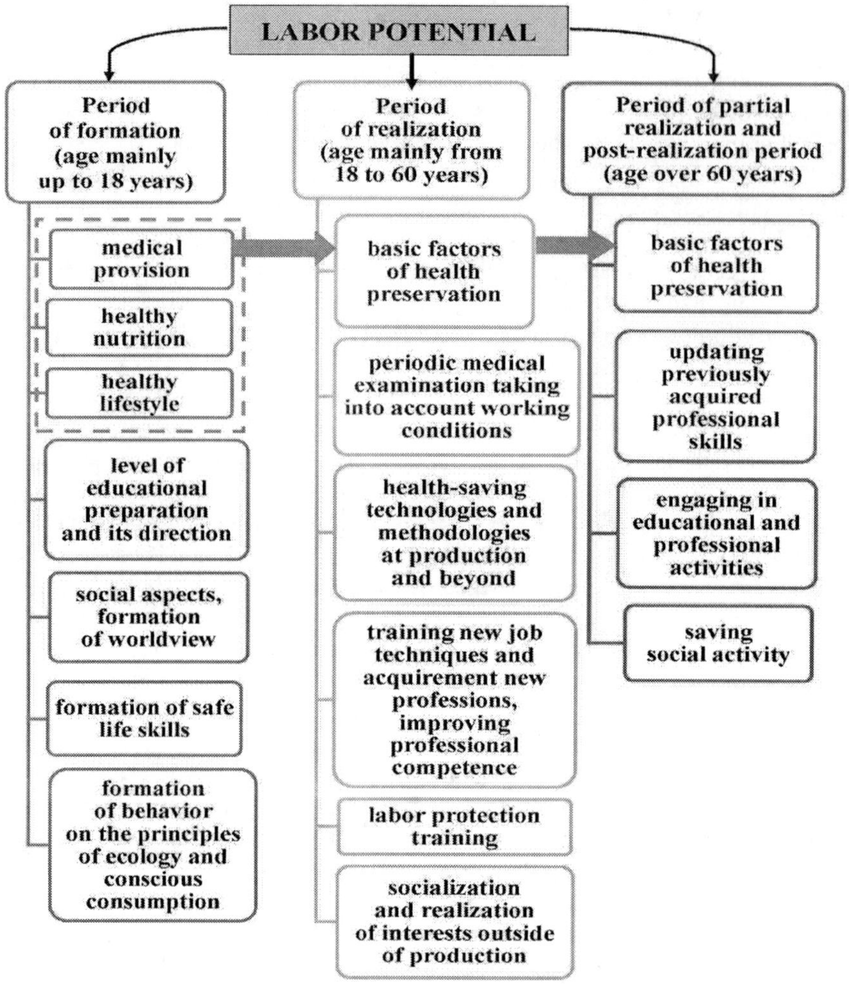


**Figure 2.** Factors of formation and realization of labor potential.

Among the main factors that depend on the state the attention should be paid to the creation of powerful medical, social, educational and safety components, and for the person the question of personality formation and comprehensive realization in society is always relevant.

Employers also need to be involved in shaping their work potential and utilizing it, modernizing production and requesting professionals with specific levels of education, skills, competencies, health, and more. Effective and rational use of the labor potential of the state is not possible without the creation of safe working conditions and measures for restoration of working capacity, as well as the preservation of biological diversity and cleanliness of ecological space.

If in the context of production, safety and social issues of labor relations are not taken into account or updated from time to time, then the results of such neglect are accompanied by social and economic consequences in the form of occupational diseases, injuries, deaths of employees or other people, disruption of the stability of the enterprise. Each of the areas of formation, realization, and post-realization of labor potential requires financing and constant updating, modernization and control, but without this activity, it is impossible to achieve stability and sustainable development, therefore the issues of occupational safety and environmental safety are important in the field of health protection. nations are integral levers of achieving sustainable development (Figure 3).

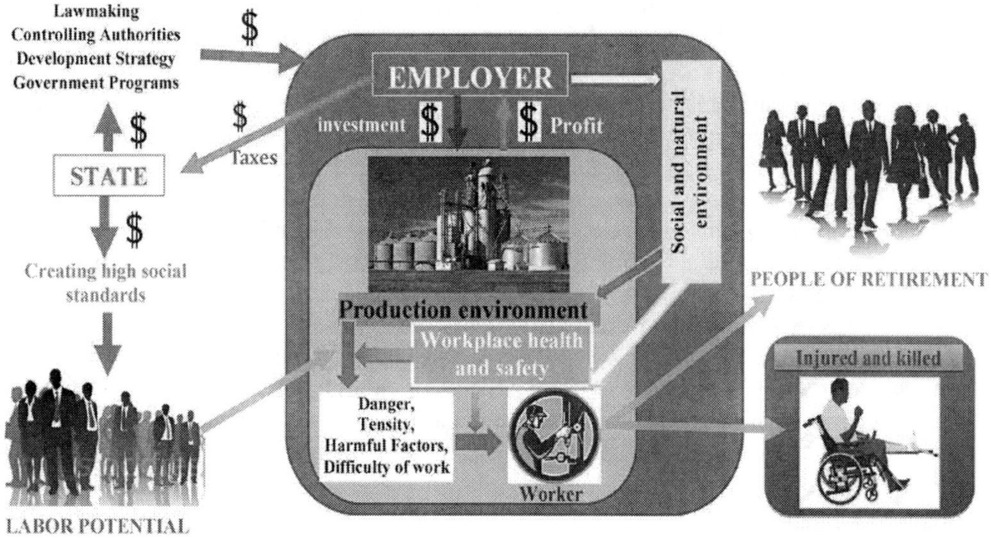


**Figure 3.** The role of labor protection and environmental safety in the concept of sustainable development

Consideration should also be given to the behavior of the personality in the public space and aspects of the propensity for danger or deviant behavior.

A number of bad habits of a person, which represent the danger of a social nature can be attributed to deviant behavior [11].

Social hazards can be classified according to certain characteristics. By nature, the following groups of hazards can be identified:

a) dangers associated with mental impact on a person (blackmail, fraud, theft, mobbing, etc.);

b) dangers of physical violence (robbery, banditry, terror);

c) dangers associated with the using of substances that destroy the human body (drug addiction, alcoholism, smoking);

d) disease risks (AIDS, sexually transmitted diseases, hepatitis, tuberculosis, viral diseases, etc.);

e) suicide risk.

By the scale of events, social hazards can be divided into: local; regional; global. By gender and age, the social hazards that characterize children, youth, women, men, and elderly people are distinguished. By their organization, social hazards can be accidental and intentional. The basis of the definition of social dangers that are caused by a low spiritual level is based on the values and components of society and human. There are two value components, the relationship between which characterizes the state of social life.

The first value component is the value of the culture of society. The second value component is the value orientation of the individual. The connection between these two extreme components of culture is the most important cementing and stimulating beginning of all social life. Conversely, breaking this link defines a deep spiritual crisis. On the background of changing landmarks, society suffers from the social dangers that have caused the change and loss of common values and orientations of a large population. Certain social hazards form "risk groups" in the human environment, which affect the state of society by increasing the number of criminal offenses, pulling in all the ranks new members of a healthy part of society, influencing the health of the people around them, worsening the gene pool of the nation. The above-mentioned negative phenomena in society create a negative circle, the causes of which, in most cases, are connected.

Escalation of circumstances also causes poor conditions of working, living and mode of life. The whole complex of circumstances increases the degree of social tension. Social tension deepens the development and spread of bad habits. Bad habits disorganize a person, make his will weak, reduce working capacity, make health worse and make life expectancy shorter. The earlier they are formed, the more detrimental they are and the more difficult it is to get rid of them. These habits bring a lot of troubles and suffering, causing great economic, social and moral harm to society. As social problems, they also affect the financial sphere of the state. In such circumstances, the forms and sizes of social deviations (crime, suicide, alcoholism, drug addiction, etc.) increase sharply. These and many other forms of perverted behavior in conditions of social control system can be threatening to society.

Many people, facing material problems, can not overcome them, and this is not due to the weakness of their character, the lack of willpower, and the lack of the necessary conditions to help overcome these problems. It causes disinterest, indifferent attitude to the results of work, deterioration of the material and psychological state of the staff, causes indiscriminate behavior, which leads to serious consequences on production and does not contribute to the awareness of environmental behavior.

Risks are always formed as a result of human activity or inactivity, and the highest levels of their manifestation are characteristic of the combination of the homosphere (space in which human is located) and the noxosphere (the space in which dangers are formed) and the greater the interpenetration of the two components, the higher the probability of danger and the degree of influence. To reduce the risk of adverse events that have a negative impact on humans, the ideal situation would be when the two components do not intersect or where security measures are used, adequate to the level of the threat, and they are proactive.

The technogenic, environmental, social impacts are so closely intertwined and must always be considered in the light of the economic context that it is time to combine multi-vector efforts to address problems in a particular direction to achieve stability. The main factor and driving force for achieving this purpose is the creation of integrated management systems. Economic challenges, demographic situation, the disappearance of sustainable markets and the appearance of new forms of employment are shaping new principles of doing business: «the production of quality products/services with minimal environmental impact and maximum protection for the consumer and the employee who makes it according to needs and interest of present and future».

## Conclusions

Employers are interested in using management strategies that allow them to generate sustainable profits and to increase them. Today, markets are oversaturated with goods and the struggle for the consumer is not in the plane of increasing the quantity of goods and services, but on the plane of improving their quality, accessibility and safety for the consumer, society or the environment. New tendencies in the development of the labor market determine the using of new strategies for the selection and training of staff and professional growth, taking into account economic and social transformations and adjustments in different areas of influence.

The using of integrated management systems allows us to solve problems of formation and development of risks more fundamentally and economically more expedient, with rational using of the resource both human and material. It also gives the opportunity to analyze the existing and future risks and strategies for their impact, taking into account scientific and technical developments and best practices that were previously used individually and did not always lead to a synergistic effect. Important is the fact that such systems allow to make effective decisions in the context of limited response time and insufficient funding for environmental, occupational and man-made safety measures.

## References

[1] *Sustainable Development Purposes: Ukraine. National report* 2017 URL: http://un.org.ua/images/SDGs_NationalReportUA_Web_1.pdf

[2] Prysiazhniuk S, Oleniev D *et al* 2019 Formation of heath preserving competence of students of higher educational institutions of information technologies specialties *International Journal of*

IOP Conf. Series: Materials Science and Engineering **907** (2020) 012081 doi:10.1088/1757-899X/907/1/012081

*Applied Exercise Physiology* vol **8** (3.1) p 274-283 URL: http://ijaep.com/Journal/vol.8.3.1.pdf
[3] Rudenko D G, Marushevsky G B, Tymochko T V 2016 Ukraine's sustainable development strategy: development and the road to recognition *Ecological bulletin* **6** p 2-4
[4] *Sustainable development strategy "Ukraine 2020"* URL: https://zakon.rada.gov.ua/laws/show/5/2015#n10
[5] *Scientific principles of development of the strategy of sustainable development of Ukraine 2012* (Odessa: IPMEER of NAS of Ukraine) p 714
[6] *State of implementation of position of «daily routine for the XXI century» in Ukraine (2002-2012)* 2014 edited by L G Rudenko (Kiev: Academic periodicals) p 359
[7] Grechko T K, Lisovsky S A, Romanyk S A, Rudenko L G 2015 *Public administration in providing of sustainable (balanced) development* (Kherson: Grin D.S.) p 264
[8] Dobrovolsky V V, Bezsonov E M 2016 Global sustainable development purposes until 2030 as a basis for local government and business *Ecological bulletin* **6** p 5-9
[9] *The project law of Ukraine about the Strategy of sustainable development of Ukraine until 2030* URL: http://search.ligazakon.ua/l_doc2.nsf/link1/JH6YF00A.html
[10] Akulov M G *et al* 2012 *The economy of labor and social labor relations* (Kiev: Center for Educational Literature) p 328
[11] Hunchenko O M, Valchenko O I 2014 Social hazards – a factor of state instability *Digest of the scientific and technical conferences of students, postgraduates, lecturers and scientists* p 67-68

IOP Conf. Series: Materials Science and Engineering **907** (2020) 012082 doi:10.1088/1757-899X/907/1/012082

# Formation of atrium spaces in the urban environment

**P A Solobay and O V Vakhnichenko**

Department of Architectural Design, Kharkiv National University of Civil Engineering and Architecture, Sumska str, 40, Kharkiv, 61002 Ukraine

avakhnichenko@gmail.com

**Abstract.** This article addresses characteristics of interactions between an atrium space and the urban environment. Renewal and transformation of urban areas over time become urgent problems which solution affects further development and preservation of the uniqueness of environment. The urban environment serves as a basis for a new spatial structure – an atrium space – formed as a centre of social activity. It contributes to the improvement of the level of comfort of the existing urban built-up area by including additional elements of the environment, new functional centres and communication linkages in the urban interior. The formation of an atrium space involves the development of a design solution based on the proposed principles of composition and planning, including "Concentration", "Combination", "Integration" and the "City Hall" principles. In the high-density urban environment, it is advisable to apply the "New-in-the-Old" and "New Inside" principles. The use of atrium spaces in the urban environment opens up new opportunities for existing building system and its components without compromising buildings' uniqueness.

## 1. Introduction

Over time, there is a growing need for local or global transformations of the urban space that will solve the problems of harmonization of the environment. The problem of the renewal of a functional and aesthetic components of historical city centres becomes especially critical.

The development and application of the principles of composition and planning will contribute to the high-quality transformation of the spatial urban environment. In the current state of urban development, the integration of an atrium space as a new component is a way to form additional communication linkages and aesthetic features. Its integration in the existing urban environment stimulates the development of new functional areas, with are linked with each other through certain processes.

The existing urban environment forms the basis for an atrium space created as an active public area promoting the development of multidisciplinary components in the structure of urban built-up areas. Spaces inside building blocks, courtyards, major and minor streets can be renewed this way, to provide new development incentives and land use options. An atrium space combines branched parts of the urban area and forms a special object-spatial environment. Due to its connection with the surrounding environment it combines both interior and exterior properties.

## 2. Identification of the problem

Modern architects face two main challenges related to the urban environment. First is the preservation of an existing urban built-up area with its authenticity, originality and established traditions. Second is the introduction of new functions and three-dimensional components into the existing built-up area.

To highlight a certain fragment of the existing architectural environment, the following parameters can be defined as typological units: boundaries of the area planning, compositional characteristics, habitual life of residents and sustainable social processes, and the identification of a place for public [1, p. 112].

Due to interventions affecting only interior or intermediate spaces, functional renewal of the urban environment does not disturb a usual appearance of historically formed streets and at the same time contributes to the expansion of human activities. That is, the development goes "from inside out", forming new spatial elements [2, p. 294]. Very often those fragments of the urban environment that are located inside building blocks, are left behind. However, after all, those possess a great potential for further development of the environment, expanding its functional and compositional boundaries. In addition, a high occupancy of cities should be compensated by a sufficient number and quality of buffer zones, especially in the central parts of the cities forming centres of business, administrative and economic activities of the city. It is expedient to place functional combinations in such zones to meet various needs of the public, and the inclusion of an atrium space can connect such zones with each other. An atrium space in urban environment promotes communication between people, and therefore should be made as comfortable as possible. It forms some kind of a community centre, which concentrates a set of functions. The level of comfort is primarily defined by a correct selection of the key criteria, which include: - a space created for public communication; - a developed set of functions; - an artistic expressiveness and meaningfulness; - rational planning and zoning of the area where the atrium space is located; - taking into account the characteristics serving for a psychological comfort; - working with the local community to identify people's preferences, priorities and needs; - solving local and city-wide problems of the improvement and development of the urban environment.

Observing changes of the urban environment, at a certain stage one may notice, how mobile and flexible it is in relation to the modern requirements of the society. A comprehensive approach to the problem of renewal of the urban environment leads to a high-quality result, which is preceded by the following stages of work: - a comprehensive research investigation; - a development of concepts based on the key conclusions of the previous investigation; - design proposals for a three-dimensional solution of an atrium space; - public discussions on the project; - financial investments to contribute to the full implementation of the project.

Therefore, it should be noted that the process of including new components in the urban environment should be delicate and well balanced. In particular, the formation of an atrium space provides a possibility of combining the urban space with the surrounding built-up area.

The balance between open spaces and the existing built-up area can be achieved with a combination of different styles separated by decades or centuries of construction. The new building should be considered as a part of the historical centre of the city, which is organically included in it without using stylistic patterns of the past [3, p. 186].

Urban public spaces must have their own identity and to combine in different ways. An identity can be defined: dynamic meaning of a site, landscape and city landscape elements, sites with a specific functional filling etc [4, p. 106]. Cites today – are monotone patterns of static elements which are missing flexibility and elasticity, visual and fiscal features which are concordant with functions and technologies. Our cities must change according to their social and economic status, get rid of "temporality" flaws in their worst manifestations [5].

Atrium spaces' applying in cities with big population and acute shortage of public spaces are especially important. In this case an atrium space can combine functions of a public zone as part of urban space and serve as a transport hub [6]. The urban environment is assumed by an observer to include an area that fills in the space between buildings and structures. Some types of such areas are assumed as independent spatial units (squares, parks, public gardens, etc.), others as an integral part of

IOP Conf. Series: Materials Science and Engineering **907** (2020) 012082 doi:10.1088/1757-899X/907/1/012082

the urban space (roads and sidewalks). The formation of an atrium space is aimed at balancing the overall appearance of the environment, i.e. at combining individual spatial components into a single compositional framework. Thus, it transforms certain areas of the urban environment, changing it at different levels, including a complete or partial restructuring. In this context, an atrium space allows to achieve a balance between the functional and artistic-compositional aspects of individual buildings, the natural component and the various spatial urban elements. An atrium space is an integral unit which creates a type of the environment for the development of various forms of public life which provide new ways of spatial organization of urban centres and result from organization of an atrium.

## 3. Result

At this stage, a critical task is to develop fundamental solutions aimed at integration of an atrium and urban spaces. Therefore, in this context, an atrium is considered as a key connecting element of the environment. It is also an object that plays a role of a buffering spatial component of the urban environment, capable of uniting parts into a single whole forming a unique object-spatial environment.

Forming atrium spaces in the urban environment requires application of a number of principles of composition and planning, which form the background for architectural and urban planning decisions, those include concentration, combination, integration, and the "City Hall" principles. The use of one or another principle in the investigation of an object of the urban environment leads to formation of some kind of an initial template or scenario for further work. The development of principles for integration of different urban objects should contribute to the effectiveness of project activities and lead to the unity of separate objects. An atrium space is a new unit to the existing urban built-up area, so it should be noted that it not only combines the ongoing processes in the formed environment, but also renews functional indicators of such environment. [7].

*The Concentration Principle* provides for the enrichment of the object-spatial environment within a certain local zone of the urban built-up area. It is characterized by the compaction of three-dimensional and planning structures of the space, bringing in order and achieving functional saturation of narrow interior spaces of urban blocks and completeness of the three-dimensional-spatial solution (Figure 1).

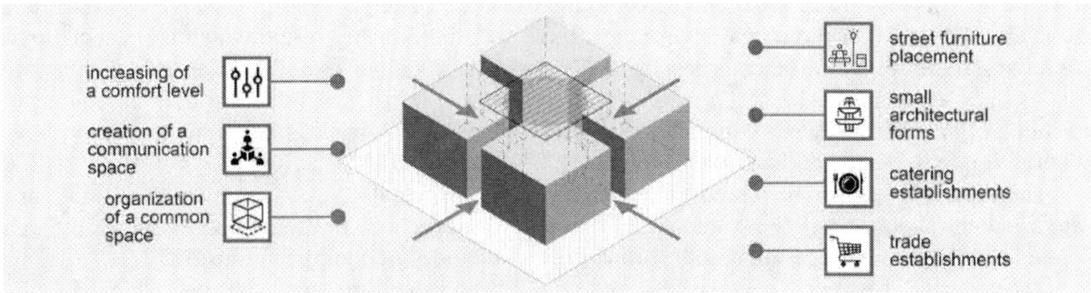


**Figure 1.** The Concentration Principle

This principle provides for:

- concentration of functions (introduction of new objects and combination of the existing ones through their organization within an integral spatial planning solution);
- organization of a closed-type atrium space (a space limited by existing buildings, most commonly – a small courtyard, the space of which is formed by several buildings);
- increasing the level of social activity (using the area as a common public space, stimulating social communications);
- improvement of an atrium space as a part of the urban interior (placement of street furniture and installations for a smooth transition from the outdoor environment to the interior, aesthetic appeal).

*The Combination Principle* is used to combine two or more local spaces. It involves a sequential transition from one object to another and is used in cases where it is necessary to combine discrete areas of the urban environment into a single system (Figure 2). This principle provides for:

- organization of a system of spaces that are linked with each other (a combination of three-dimensional and planning solutions of one object with another, a common composition of ideas);

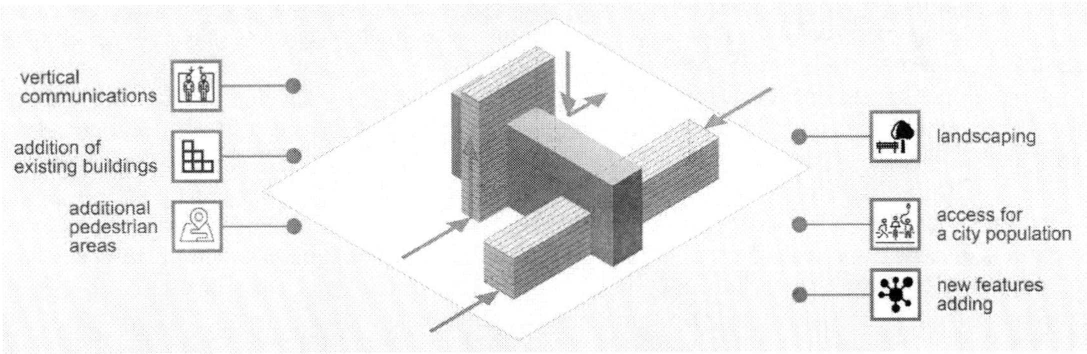


**Figure 2.** The Combination Principle

- a multilevel organization of communication zones (use of horizontal and vertical communications for a seamless transition);
- consolidation of branched sections of the urban space (complex solution of an atrium space for a certain area, usually within one block);
- structural reorganization of space (arrangement of additional volume of space, complete or partial reconstruction of objects of the built-up area due to the inclusion of an atrium space);
- functional saturation (a set of functional points that support social activity in a particular environment);
- site improvement (street furniture, installations, landscaping, all of which are located at different levels and complement the environment);
- organization of additional pedestrian connections (the multilevel organization of an atrium space allows the distribution of pedestrian zones and inclusion of additional communication nodes in the space).

*The Integration Principle* establishes a connection between different types of the space. It provides for adaptation of new spatial structures to the existing urban environment and allows renewal of the urban environment without violating its integrity, characterizes actions aimed at the recovery of the environment and filling it with new qualities (Figure 3).

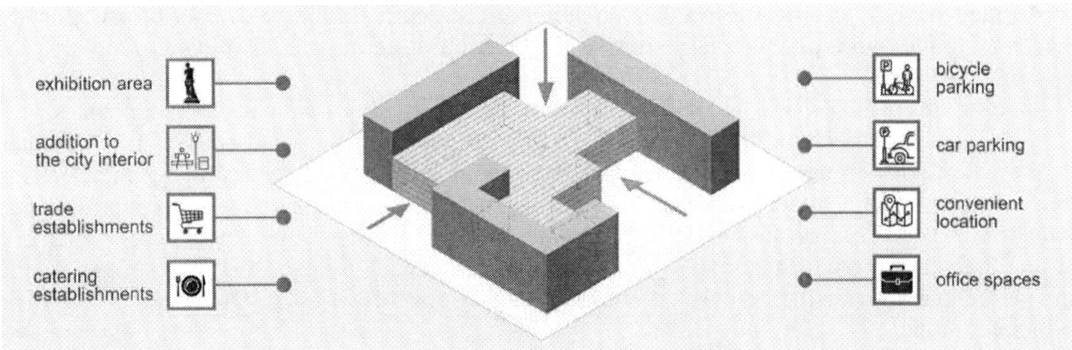


**Figure 3.** The Integration Principle

This principle provides:

- formation of an atrium space inside building blocks (use of an intra-block space as a part of the urban interior);
- providing an active public space (inclusion of various social functions, organization of a multi-purpose place, mobility of the space);
- expansion of spatial features of the existing built-up area (integration of an atrium space makes it possible to expand the useful area and structural volumes of public buildings and zones);
- organization of a recreational area (with an atrium space green recreational zones accessible to all residents can be organized within the built-up area);
- a multipurpose use of an atrium space (as a public centre with a diverse set of functional areas, such as a trading zone, food court, exhibition space, recreational area, etc. It is also appropriate to arrange office spaces there).

*The «City Hall» Principle* provides for the placement of socially active space within the city structure. The term "City Hall" refers to the space, in which a functional node of a citywide importance needs to be organized. This principle should be applied in cases where there is a need not only to combine the surrounding built-up area with a new function, but also to organize interconnections of pedestrian and transport routes (Figure 4).

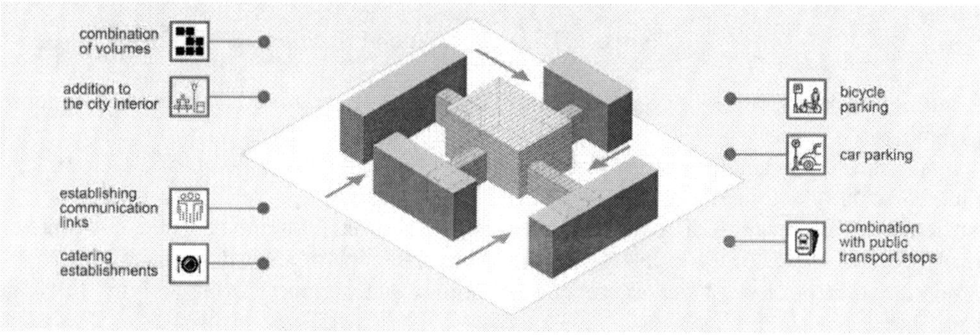


**Figure 4.** The «City Hall» Principle

This principle provides for:

- an efficient use of yard spaces within blocks (rational space planning, improvement of aesthetic features and the comfort in general);
- a place for organizing public events (an atrium space filled with mobile elements, which can be adapted to different tasks);
- organization of a functional node (object filling, which is subject to a clear functional scenario, possible combination of an atrium space with public transport);
- arrangement of a compositional core (creation of a structure forming a compositional basis of the surrounding built-up area and contributing to the creation of an integral, harmonious and artistically expressive solution);
- a recreational area within a block (landscaping, creation of play areas and recreation zones).

«The need for an organic combination of the interior with the nature and protrusion of the interior to the external environment gives rise to the development of a method destroying the boundaries between the external and internal space» [8, p. 286]. Therefore, the construction of a composite model, which is saturated during the development of the project, helps to achieve a balanced order of the entire three-dimensional structure [9]. When an atrium is created in the historic centre of the city and the dense built-up area, it is also advisable to apply the «New-in-the-Old» and «New Inside» principles [10].

*The «New-in-the-Old» Principle* is aimed at expanding the functionality of the existing built-up area and renewal of its quality characteristics. This principle is based on placing the internal space within the established built-up area of the city (Figure 5). The principle provides for:

- a creation of an integral composition between the old and the new architecture (following compositional rules during formation of atrium spaces in the existing built-up area);

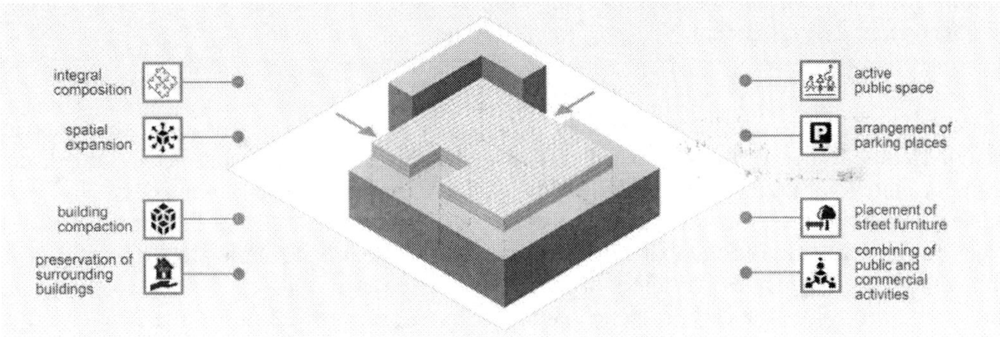


**Figure 5.** The «New-in-the-Old» Principle

- an introduction of spatial structures (adding new elements to the existing built-up area, in particular, spatial expansion of historic city centres);
- a functional expansion (introduction of new functions that stimulate intensification of activities in obsolete urban spaces);
- a compaction of the spatial planning structure (obtaining additional areas for the surrounding built-up area or individual buildings);
- improving the quality of the environment (the environment is complemented by new elements of urban design, creating recreational areas with outdoor furniture, architectural installations and art objects).

*The «New Inside» Principle* contributes to creation of a renewed environment without violating the integrity of the existing one. The application of this principle to the creation of an atrium space allows to expand spatial and functional capabilities of a dense urban built-up area (Figure 6). The principle provides for:

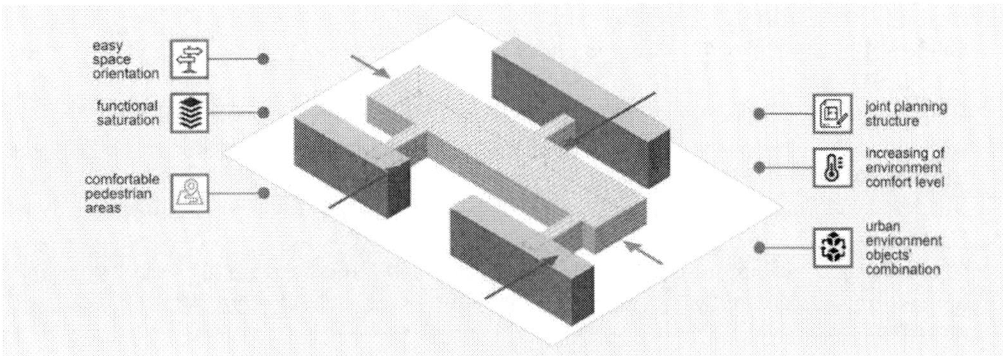


**Figure 6.** The «New Inside» Principle

- obtaining new qualitative characteristics of the space (improving the comfort of the urban environment in a dense built-up area, setting up and arranging space within urban blocks);
- establishing a single planning structure for the existing and new objects (thorough a detailed study and analysis of the existing built-up area to obtain a high-quality design solution for a specific area of the city);
- expanding spatial capabilities of the urban built-up area (inclusion of an atrium space as a new spatial component possessing a certain set of characteristics);

- integration or branching of functions (formation of an atrium space provides a possibility of functional distribution within a dense urban built-up area).

Each principle should be appropriate to the specific situation, i.e. it should ensure proper functioning of the environment, support socio-cultural processes, provide compositional integrity and promote economic development.

## 4. Conclusion

Preserving unique historical features of city centres is a task of a top priority in the renewal of the urban environment that includes searching for appropriate architectural solutions and the discovery of new opportunities for urban areas and objects without compromising their uniqueness.

Using an atrium space for the renewal of the urban environment allows the creation of additional zones of public attraction in the existing urban structure. The urban interior formed with the creation of an atrium space, becomes an object of public use. The design process involves solving of urban, architectural, artistic and social problems, since the existing urban built-up area is renewed with the inclusion of new elements.

It is always necessary to analyse different options and apply those principles that are the most relevant and rational for a particular situation. An architectural model of the urban environment exists, first of all, in a static form which makes it possible to include both active and passive components. The modern urban environment has to provide a high level of comfort and the inclusion of an atrium space there should in addition create centres for the development of a public life.

Urban environment transformation using atrium space forms an updated look of the building. In relation to existing urban environment, atrium space is a new architectural unit that integrates into a street space, that is a linear type of space, and a courtyard is an accentuated area between architectural objects. In other words, what we have as a starting point for urban development reorganization of itself. An atrium space design in existing urban environment involves extensive analysis and careful study of a specific situation. After all, thanks to a comprehensive approach, it is possible to conduct quality research and effective atrium space adaptation into urban environment.

## References

[1] Timokhin V O, Shebek N M, Malik T V ect. 2010 *Basics of architectural environment design* (Kyiv: Osnova) p 400
[2] Gutnov A E 1985 *World of Architecture: The Language of Architecture* (Moscow: Molodaya Gvardiya) p 351
[3] Walter Gropius 1971 *Scope of total architecture* (Moscow: Iskusstvo) p 287
[4] Carmona M, Tiesdell S 2007 *Urban Design Reader* (Architectural Press is an imprint of Elsevier) p 375
[5] Maki F 1964 *Investigation in collective form* (St. Louis, School of Architecture, Washington University) p 87
[6] Charlie Q L, Xue Luming Ma and Ka Chuen Hui 2012 Indoor 'Public' Space: A study of atria in mass transit railway (MTR) complexes of Hong Kong *URBAN DESIGN International* **17** (2) 87–105
[7] Vakhnichenko O V 2019 *Compositional and planning principles of formation of atrium spaces in the conditions of urban environment revitalization* (Kharkiv: Kharkiv National University of Civil Engineering and Architecture) p 195
[8] Novikova E B 1991 *Public buildings interior: Artistic concerns* (Moscow: Strojizdat) p 368
[9] Solobay P A 2001 *Structurally-functional and composition modeling of educational complexes (on University example)* (Kharkiv: Kharkiv State Technical University of Construction and Architecture) p 18
[10] Solobay P A 2008 Fundamentals and principles of reconstruction of the architectural environment of universities *Visnik Kharkivskoyi derzhavnoyi akademiyi dizajnu i mistecztv. Mistecztvoznavstvo. Arkhitektura* **1** p 114-121

IOP Conf. Series: Materials Science and Engineering **907** (2020) 012083 doi:10.1088/1757-899X/907/1/012083

# Modern trends at natural and wastewater treatment plants reconstruction

**S Martynov, O Kvartenko, V Kovalchuk, A Orlova**

Department of Water Supply, Sewerage and Drilling, National University of Water and Environmental Engineering, Rivne, Ukraine

s.y.martynov@nuwm.edu.ua

**Abstract.** This article discusses ways to intensify the natural water treatment plants operation using foamed polystyrene as a contact and filtering medium for bioreactors and filters. The reagent schemes with polystyrene foam filters are offered for the surface water treatment. The improved physicochemical and biological methods for slightly acidic and neutral groundwater preparation are recommended to use. The use of ferrobacteria metabolism products for the ammonium nitrogen oxidation in a slightly acidic medium was investigated. The x-ray spectral analysis of bio-minerals matrix structures was conducted. The reasons of unsatisfactory technical condition of sewage treatment plants in Ukraine were analyzed. The ways of increasing their work efficiency using sedimentation aerotanks of large hydraulic height and using the simultaneous nitrification-denitrification were offered.

## I. Introduction

The provision of quality drinking water for the population and ecosystem preservation needs modern innovative solutions in the field of working plants reconstruction and building new water preparation and wastewater treatment plants [1].

The groundwater chemical composition is formed under the many natural and anthropogenic factors influence: temperature, $O_2$ and $CO_2$, pH and Eh, changes in routes, water levels and water consumption, microbial life and man-made load. Therefore, groundwater refers to complex multicomponent systems, characterized by different pH – Eh values, hydrocarbonate alkalinity, salt content, chromaticity, contain heavy metal ions, easily oxidizable organic compounds, humic acids, dissolved gases, and compounds containing nitrogen $NH_4^+$, $NO_2^-$, $NO_3^-$), phenols, surfactants, phosphates, microflora.

The technogenic component of contaminations is based on the infiltration water inflow from the earth surface water and surface water reservoirs. It is known [2] that 80% of household wastewater drains are treated at mechanical and biological treatment plants which were built in 1960-1980. The technological equipment of such plants is outdated and worn-out and does not allow water to be treated to normative parameters. The wastewater treatment technologies require the innovative changes [1, 3-6]. This problem is especially acute in cities with population less than 200 thousand people, where the number of unsatisfactory treated wastewater is on average from 44 to 63% [2, 7]. It should also be noted that 43% of urban settlements and 97% of rural settlements do not have centralized wastewater systems.

The discharges of insufficiently treated urban and industrial wastewater, sludge storages, areas of sediment accumulation from household wastewater treatment plants, surface discharge wastewater from cities and industrial plants, which treatment requires the modern technologies introduction [3, 4] bring the particular danger for the environment and for water sources. In particular, every year up to 40 million tonnes of sediment are accumulated at areas, which decomposition products gradually infiltrate the underground aquifers.

Considering that 24.0% of the main groundwater aquifers in Ukraine are classified as conditionally protected and another 36.0% is unprotected, the role of water treatment plants in the water supply systems of settlements is greatly increased and requires the modern integrated methods use [8, 9].

Therefore, a complex problem solution of sanitary and environmental rehabilitation in settlements through the reconstruction of working and building of new natural and wastewater treatment plants, using modern methods, resource saving and energy saving technologies and equipment, using modern mathematical modelling of processes are the urgent tasks of today [10-12].

## 2. Materials and Methods

We had conducted a long-term study of water preparation and wastewater treatment processes by experimental and theoretical methods. The plants for technological processes of water treatment research on a macro scale were used in the laboratory of the department of water supply, sewage and drilling (Figure 1) to get mode data and on a micro scale to obtain the model parameters. A considerable number of experimental data was accumulated during the inspection of the working treatment plants, in the commissioning process and long-term operation of the treatment plants which were reconstructed according our recommendations.

The titration, potentiometric and photocolorimetric methods in the certified laboratory of the department were used to determine the contamination concentrations in water. Moreover, the water physicochemical parameters were determined by independent certified laboratories. The electron microscopy of bio-minerals matrix structures samples was performed by X-ray spectroscopy analysis with a Quanta 200 scanning electron microscope (USA), equipped by a XDA spectrometer for microanalysis (EDAX). Water consumption was determined by ultrasonic flowmeters or by volumetric method.

The studies were carried out at the pilot plant (Figure 2) which was included the 2000 $dm^3$ inlet water tank (No 1), the 500 $dm^3$ reaction flask (No 2), the Kamovsky pump (No 6) for air pumping from the reaction flask and pumping the solution under block diagrams, flasks with removable paper filter "blue ribbon" and the ceramic filter (No 4), the MM-5 magnetic stirrer (No 3), the filtrate collection tank (No 5) and rubber hose systems that connected the separate compartments into the one technological scheme. The dissolved oxygen concentration was 0.5-0.8 $mg/dm^3$.

The experiment planning and mathematical statistics methods using the MS Excel statistical package and the MathCAD appendix were used during the preparation and processing of experimental studies.

## 3. Results and Discussions

Our long-term experience has allowed us to develop a number of resource-efficient technological schemes for water treatment, both in new building and reconstruction the existing water supply and sewage systems. We recommend the polystyrene foam filters use, which are firstly proposed at the Department of Water Supply, Sewage and Drilling (NUWEE, Rivne) during water filtration at granular filters application. The polystyrene foam filter is a container with foamed polystyrene which is in flooded state with the help of the retarded net. Foam polystyrene medium is made directly at the water treatment plant by foaming the polystyrene product or produced at plants of thermal insulation boards in special foaming devices. The polystyrene foam medium can be single-layer or double-layer.

The floating foam polystyrene filters with ascending filtration are increasingly introduced into the practice of water treatment [13]. In a constructive way the floating foam polystyrene filters are easier than filters with heavy medium (sand, zeolite, etc.). The treated water is collected in the filtration space during the filtration and during washing such water moves downstream and washes the medium. It is enough to close the latch of outlet water supply and open the washed latch to the necessary value. As a rule, the distribution system is not arranged over the polystyrene foam medium and its functions are performed by the retaining net. The distribution system should be at the bottom of the filter: tubular with round holes, hole or pipe with holes and additional sloping walls, only one hole with the outlet into the pipe at small filter sizes.

a) b)

**Figure 1.** Laboratory plant for water iron removal at polystyrene foam filter with the increasing bed of suspended sediment a) filtering column; b) water preparation system with high iron concentration.

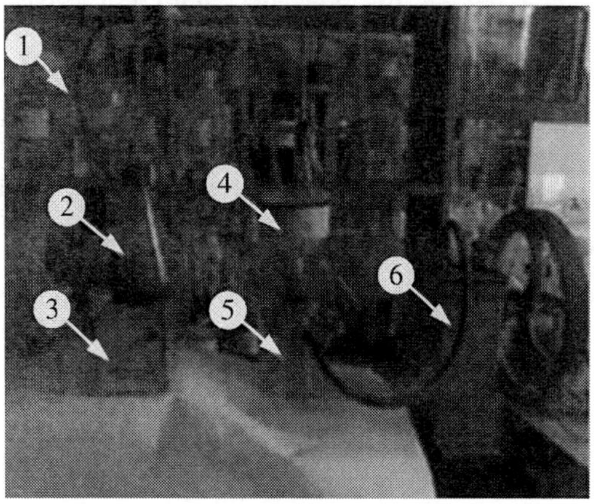


**Figure 2.** Scheme of the pilot plant for study the kinetics of the iron compounds oxidation in suspended sediment bed from bio-minerals matrices, iron bacteria and iron hydroxide.

The polystyrene foam filters can be used in two-stage reagent schemes for water clarification and discoloration for its final treatment after settling tanks or clarifiers with suspended sediment, or in one-stage schemes as contact polystyrene foam filters, at water iron removal schemes, for the surface water pre-treatment, for the wastewater tertiary treatment [14].

First, the source water is carefully treated with sodium hypochlorite for the organic substances oxidation in reagent schemes of water clarification and discoloration and then the coagulant solution is introduced into the water to provide the hydrophobic particles coagulation.

The sorption medium bed should be filled up to remove odors and flavors into the underfilter space of the polystyrene foam filter, forming a polystyrene foam-carbon filter so. Sorption capacity of coal AG-3 allows using it without regeneration up to one year. High uniformity of coal AG-3 provides favorable hydraulic modes of filters operation, especially during washing.

The aeration and subsequent filtration is recommended to use for water iron removal with sufficient alkalinity and high pH. The ordinary polystyrene foam filter is used in the outlet water (Figure 3) at the small iron ions concentration, and at high iron ions concentrations, it is used with a growing suspended sediment bed and polystyrene foam medium [14].

**Figure 3.** The plant for physicochemical groundwater iron removal with polystyrene foam filters.

**Figure 4.** The plant for biochemical groundwater treatment.

The reconstruction of water iron removal plants with low alkaline reserve and low pH is another perspective direction of work intensification by the biochemical treatment method in complex with physicochemical methods (Figure 4) [15]. The biochemical oxidation method application is made possible by the widespread distribution of ferro-bacteria in groundwater in different regions of the world [16-18].

So, the treating process was completed within 35 minutes at slightly acidic groundwater treating under laboratory conditions with cations $Fe^{2+}$ 3.0-3.6 mg/dm$^3$ concentrations and sediment volume of 1 cm$^3$ (Figure 5, curve 1). The value pH of 6.2 and hydrocarbonate alkalinity to 2.5 mmol/dm$^3$ increasing was occurred with the soda ash solution of 10 mg/dm$^3$ addition that resulted to the biochemical processes intensification by an average of 20% and the completion of the process within 25 minutes (Figure 5, curve 2). The kinetics of $Fe^{2+}$ cations removal from the solution coincides with the kinetics of the process at adding the soda ash solution of 10 mg/dm$^3$ to the system at the hydrocarbon alkalinity value increasing in the system to 3.7-3.8 mmol/dm$^3$ (Figure 5, curve 3).

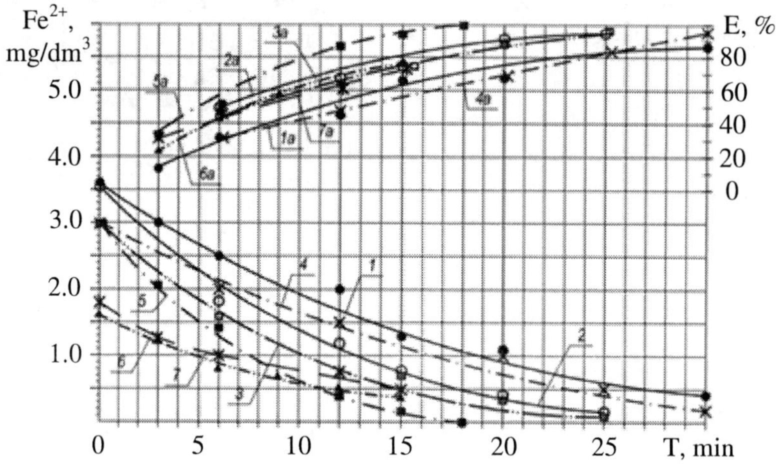


**Figure 5.** Kinetics and efficiency of weakly acidic and neutral groundwater treatment with the cations concentration of $Fe^{2+}$ 1.6-3.6 mg/dm$^3$ in the matrix structures bed of ferro-bacteria depending on the contact time and inlet water quality parameters:
1– pH 5.5; The alkalinity 0.8 mmol/dm$^3$; $Fe^{2+}$ 3.6 mg/dm$^3$; 2 – pH 6.2; The alkalinity 2.5 mmol/dm$^3$; $Fe^{2+}$ 3.6 mg/dm$^3$; $Na_2CO_3$ 10 mg/dm$^3$; 3 – pH 6.8; The alkalinity 3.74 mmol/dm$^3$; $Fe^{2+}$ 3.0 mg/dm$^3$; 4 – pH 6.0; The alkalinity 1.0 mmol/dm$^3$; $Fe^{2+}$ 3.0 mg/dm$^3$; 5– pH 7.4; The alkalinity 3.4 mmol/dm$^3$; $Fe^{2+}$ 3.0 mg/dm$^3$; $Na_2CO_3$ 10 mg/dm$^3$; 6 – pH 7.1; The alkalinity 3,4 mmol/dm$^3$; $Fe^{2+}$ 1.6 mg/dm$^3$; 7 – pH 7.1; The alkalinity 3.4 mmol/dm$^3$; $Fe^{2+}$ 1.8 mg/dm$^3$.

The analysis of the formed matrix structures of bio-minerals was achieved due to the X-rays diffraction, light and scanning electron microscopy (SEM). Consideration of the general spectral analysis data (Figure 6) shows that a large number of O, N, C, Al, Si, and Fe atoms are present in the sediment. The carbon atoms and nitrogens presence testified about the biological component of the bio-minerals matrix structures.

The $Al^{3+}$ та $Fe^{3+}$ ions presence has aroused an interest in terms of the metals adsorption process with variable valence. Metals were electrostatically bound to both the anionic surface of the cell wall and the organic polymers that were exuded by the cells on the outside and accumulated on their developed surface. The white impregnations were observed on the catalytic film surface considering the electronic images, which are surface-adsorbed matrix structures of Al compound, which accents the structure activity to the metals adsorption from the water environment [19].

The ammonium nitrogen removal in bioreactors can be represented by several mechanisms, depending on the water quality parameters and the microorganisms' qualitative composition in groundwater and in contact medium of the bioreactor. The process of $NH_4^+$ ions sorption on the matrix structures surface of bio-minerals and on the bacteria cells can be considered in neutral and near neutral groundwater with the hemolitoautotrophic bacteria of the genus Gallionella presence, due to the presence of functional groups: $PO_4^{-3}$, $COO^-$, $OH^-$ on their surface, which is confirmed by our X-ray spectral analysis of matrix structures.

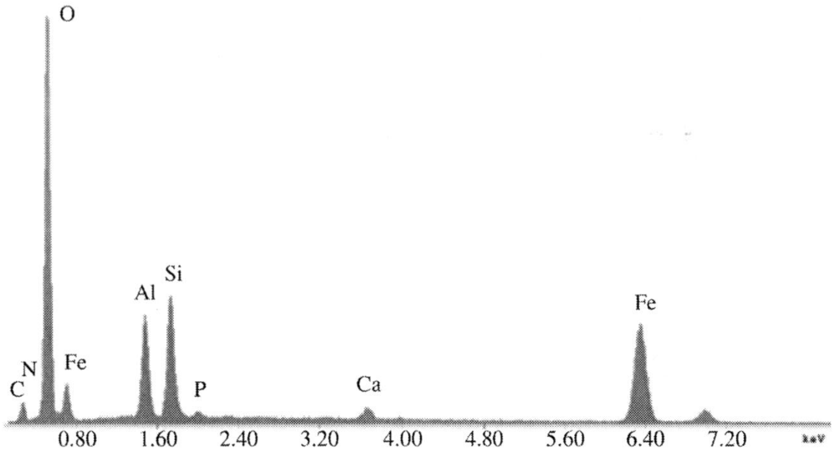


**Figure 6.** X-ray spectral analysis of the granule surface of filtering medium at the water treatment plant, which works by combined biological and physicochemical methods.

The following mechanism for the ammoniacal nitrogen with hydroxyl radicals $^{\bullet}OH$ interaction which are formed in the system according to the Fenton reaction is possible at underground slightly acidic waters treatment, containing genus Lertothrix bacteria in contact medium of the bioreactor

$$Fe^{2+} + H_2O_2 \rightarrow Fe^{3+} + OH^{\cdot} + OH^{-}. \quad (1)$$

With the reaction of dissolved ammonium nitrogen presence in underground ferruginous waters, it is also possible to oxidize it with hydroxyl radicals formed as a result of the bacteria activity of the genus Leththrix, by the mechanism [20]:

$$[NH_4^+ + {}^{\bullet}OH] \rightarrow [{}^{\bullet}NH_2 + H_2O_2] \rightarrow [{}^{\bullet}NHOH] \rightarrow [NO_2^-] \rightarrow [NO_3^-] \quad (2)$$

The dissolved organic compounds (RH) oxidation, in particular lightly oxidized organic compounds, also was recorded during our study at natural groundwater due to technological schemes: bioreactor-filter; bioreactor – liming – coagulation – filtration. The their oxidation mechanism in a weakly acidic environment with the $Fe^{2+}$ ions, ferrobacteria, peroxide hydrogen, hydroxyl radicals presence can be described by the following chemical reactions

$$^{\bullet}OH + RH \rightarrow H_2O + R^{\bullet}, \quad (3)$$

$$2OH^{\bullet} \rightarrow H_2O_2, \quad (4)$$

$$R + H_2O_2 \rightarrow ROH + OH^{\bullet}, \quad (5)$$

$$R + O_2 \rightarrow ROO^{\bullet}, \quad (6)$$

$$ROO^{\bullet} + RH \rightarrow ROOH + R^{\bullet}. \quad (7)$$

The unsatisfactory technical condition of wastewater treatment plants creates a significant technogenic danger for Ukraine groundwater aquifers, as was noted above. Mainly this is the result of the long duration of their exploitation and imperfect technologies, with the use of materials and equipment, which quality does not correspond to the wastewater aggressive nature. The reconstruction of such wastewater treatment plants usually involves the outdated equipment replacement to newer, which responds to modern requirements, the implementation of planned or major repairs of individual facilities.

In particular, increasing the efficiency of large waste removal from wastewater is carried out by installing mainly filtered step grilles that are widely produced by industry. Their small transparency which increases the retarded waste amount and the easy waste removal from wastewater with the subsequent pressing possibility are the peculiarity of such grilles.

The use of the aerated grit chambers is an effective method of the increasing sand retention by 1.5 times. In addition, it provides a low content of organic impurities in the retarded sediment, its ash content is 5-10%. Aerated grit chambers can be used to retain fats, petroleum products and as a biocoagulator at the excess active sludge adding. In addition, the wastewater aeration is also provided.

The most effective method to improve the primary sedimentation tanks work is to add reagents into the wastewater. However, such technology use requires the additional mixers construction, flocculation chambers and reagent facilities. Conversion of primary sedimentation tanks to biocoagulant mode with the excess activated sludge addition or biofilm allows the additional removal of organic pollutants from wastewater due to biosorption. There is a positive experience of combining settling processes in one settling tank with subsequent separation of the suspended solids together with excess activated sludge by flotation (flotation biocoagulation). Conversion of primary sedimentation tanks into flotation biocoagulators allows to reduce the primary wastewater treatment duration up to 30-40 min, and to reduce $BOD_5$ of wastewater entering into the aeration tank by 30-40%. At the same time, there is also a decrease in concentrations of detergents, fats, petroleum products of heavy metal ions and other contaminants. The moisture content of the removed flotation sludge is 94-96%. In recent years, thin-layer blocks of polymeric materials have been effectively used to intensify the initial sedimentation at the treatment plants reconstruction. Thin-layer blocks can significantly increase the wastewater treatment plants productivity and at the same time to increase the suspended solids retention efficiency.

One of the most effective methods of increasing the aeration tanks oxidizing power is to increase the mass of sludge involved in wastewater treatment process. The activated sludge mass increasing can be achieved by three methods. The first method is carried out in aeration tanks with fixed or moving-bed to increase the concentration of biomass (MBBR, “vija”, “cleaning glass brush”). The second method provides sludge mixes separation at special filter partitions or membranes (MBR). The third method involves the separation of concentrated sludge mixture by thin-layer sedimentation tanks, pressure flotation cells or applying secondary settling tanks of large hydraulic height.

The growth of the sludge mass requires an increase in oxygen supply in the aeration tanks for the endogenous respirations needs. In the aeration tank with a high concentration of activated sludge the aerators with high oxidizing ability, mechanical, pneumatic-mechanical and jet aerators can be used. However, mechanical and pneumatic-mechanical aerators have significant disadvantages; they are the insufficient degree of mixing volume aeration zone; the limited capacity of the working area of one aerator; the special equipment to service the aerators; the low reliability because of possible damage to the engine or gearbox; the possible water pollution for grease lubrication. The main disadvantage of jet aerators is small penetration depth in sludge mixture flow of the working liquid captured air bubbles. However, the rational organization of flow in jet aeration tank can successfully provide mixing sludge, as evidenced by their positive experience with long-term use in many wastewater treatment plants.

At the wastewater treatment plants reconstruction is important to increase the productivity of wastewater sludge stabilization and dewatering facilities. For example, increasing of aerobic digesters productivity can be achieved by their work transferring into autothermophilic mode. At the aerobic digestion of sludge with less than 95% humidity the amount of heat released in the decay of volatile part organic matter is sufficient for self-heating of the sludge to a temperature of 55-65°C. As a result, the duration of the process is reduced in several times, and therefore the productivity of existing facilities increases. Due to the nitrifying bacteria death, the demand for oxygen is reduced and the sludge is pasteurized. The heat of the heated sludge can be utilized.

The transfer of sludge fermentation process into digesters from mesophilic to thermophilic mode allows to increase their productivity by about two times. It is advisable to supply the resulting gas to cogeneration units to produce heat and electricity.

IOP Conf. Series: Materials Science and Engineering **907** (2020) 012083 doi:10.1088/1757-899X/907/1/012083

The most effective devices for mechanical sludge dewatering in our time are filter presses and dehydrators.

The implementation of the above measures allows increasing the efficiency of wastewater treatment, but cannot guarantee the maintenance of modern requirements for the quality of treatment in principle, mainly for the nitrogen and phosphorus compounds content [2]. In practice, two technologies are commonly used, MLE (modified Ludzak-Etinger process) and A2O (anaerobic-anoxic-oxic process), which have been proven for many years and are reliable in exploitation. The first technology also involves the phosphates reagent removal and the second involves the treatment with the biological removal of nitrogen and phosphorus compounds. The facilities are calculated according to the ATV-131E standards.

The practical implementation of these technologies in the reconstruction of existing wastewater treatment plants can be carried out in two ways (Figure 7).The first way involves the arrangement of anaerobic, aerobic and anoxic zones on the basis of individual existing elements of the wastewater treatment plant, primary or secondary sedimentation tanks, aeration tanks, etc. In the anaerobic and anoxic zones, the mixing of the sludge mixture should be carried out using electromechanical stirrers.

The second way is advisable to use at high treated wastewater consumptions and large sized aeration tanks. In this case, it is advisable to arrange the aerobic and anoxic zones in the aeration tanks by constructing a vertical partition that will divide the volume of the aeration tank into two parts by the oxidation ditches type. In this case, it is advisable to circulate the sludge mixture using electromechanical mixers and aeration by one or more sections arranging with aerators at the bottom.

The treatment facilities of the meat processing plant are an example of modern technologies use at the wastewater treatment plants reconstruction (Figure 8). The aeration tank of high hydraulic height with surface jet aeration was used for biological wastewater treatment in this case. The effective ammonium nitrogen removal by simultaneous nitrification-denitrification is occurred due to the formation of the aerobic zone in the upper part of the aeration tanks and anoxide zone in the lower part [2].

**Figure 7.** Reconstruction of wastewater treatment plant with productivity 7000 $m^3$/day.

**Figure 8.** Reconstruction of meat processing wastewater treatment plant.

## 4. Conclusions

It is possible to conclude, that the presence of such factors as the multicomponent nature of natural and wastewater, with the impurities ability to change their phase-dispersed state under the influence of physical and chemical factors, allows the wide range of techniques and methods use for the water treatment plants reconstruction based on the analytical review results and our own theoretical and experimental studies results. The most suitable methods, processes and apparatus choice should be conducted using the following current trends in the field of natural water and wastewater treatment:

- it is recommended to use foamed polystyrene of different fractions as contact medium of bioreactors and filtering medium of clarifying filters;
- the foam polystyrene filters may be used in the natural surface water tertiary treatment in two-stage reagent schemes after settling tanks or clarifiers with suspended sediment;
- the aeration method with the next filtration on polystyrene foam filters is recommended to use to purify groundwater from low concentrations of iron ions, sufficient alkalinity and pH. It is advisable

to use a combined technology with a growing suspended sediment bed and polystyrene foam medium at the high concentration of iron ions;
- to use a complex of biological and physicochemical methods for purification for the treatment of aggressive weakly acidic and near neutral groundwater with low alkaline reserve;
- to conduct the correction of the pH-Eh values of the medium, the value of hydrocarbonate alkalinity, the concentration of inorganic carbon by adding a soda ash solution to activate biological processes of iron removal in bioreactors, which allows to improve the treatment efficiency up to 30%
- the ferrobacterial metabolism products ($H_2O_2$, $O^{\bullet}$, $OH^{\bullet}$) use for ammonium nitrogen oxidation and easily oxidizing organic compounds in a weakly acidic medium on the matrix structures surface in the pore space of contact bioreactors medium;
- using the ability of bio-minerals matrix structures to adsorb metal ion with variable valence;
- the conversion of primary sedimentation tanks into flotation biocoagulators mode allows to reduce $BOD_5$ of wastewater with simultaneous decreasing concentrations of detergents, fats, petroleum products of heavy metal ions and other contaminants.
- it is recommended to use MBBR and MBR type bioreactors to increase the aeration tanks oxidizing power by the activated sludge mass increasing;
- it is possible to use technologies MLE and $A_2O$ during the reconstruction of existing wastewater treatment plants, which allow for simultaneous removal of nitrogen and phosphorus compounds.

**References**

[1] Moshynsky V and Riabova O 2013 *NATO Science for Peace and Security Series C: Environmental Security* **128** 125-135
[2] Kovalchuk V 2002 *Wastewater treatment: textbook* (Rivne: Rivne Printing House)
[3] Epoyan S, Karahiaur A, Volkov V and Babenko S 2018 *Eastern-European Journal of Enterprise Technology* **1/10** (91) 62-69
[4] Epoyan S M *et al* 2010 *Water supply and purification of natural waters: textbook* (Kharkiv: Factor)
[5] Moshynsky V 2001 *Water Engineering and Management* **148** (2)
[6] Oleynik A and Airapetian T 2017 *Eastern-European Journal of Enterprise Technologies* **4/10** (88) 4-11
[7] Oleynik A, Kalugin Yu and Airapetian T 2017 *Eastern-European Journal of Enterprise Technologies* **6/10** (90) 17-23
[8] Klymenko N A, Kozyatnyk I P and Savchyna L A 2010 *Water Research* **44** (18) 5316-5322
[9] Svietlieishaya E M, Mitchenko T E, Astrelin I M 2014 Journal of Water Chemistry and Technology **36** (1) 47-56
[10] Polyakov V, Kravchuk A, Kochetov G and Kravchuk O 2019 *EUREKA: Physics and Engineering* **1** 2019 28-42
[11] Polyakov V L 2012 *Journal of Water Chemistry and Technology* **34** (2) 65-78
[12] Martynov S, Kunytskyi S and Orlova A 2017 *Eastern-European Journal of Enterprise Technologies* **5/10**(89) 19-26
[13] Martynov S *et al* 2017 *Journal of Water and Land Development* **39** 93-99
[14] Orlov V, Martynov S and Kunitskiy S 2016 *Journal of Water and Land Development* **31** 119-122
[15] Kvartenko O, Sabliy L, Kovalchuk N and Lysytsya A 2018 *Journal of Water and Land Development* **39** 77-82
[16] Mouchet P 1992 *Journal of the American Water Works Association* **84** 158-167
[17] Kunoh T, Kunoh H and Takada J 2015 *Journal of Microbial & Biochemical Technology* **7** 419-426
[18] Miot J *et al* 2009 *Applied and Environmental Microbiology*. **75** 5586-5591
[19] Kvartenko A *et al* 2017 *Eastern-European Journal of Enterprise Technologies* **4/10** (88) p 37-43
[20] Huang Li *et al* 2008 *Environmental science & technology* **42** (21) 8070-8075

# Effect of technological parameters on clarification low-concentration suspension by dissolved air flotation

**S Epoyan, O Syrovatsky, O Haiduchok and A Titov**

Department of Water Supply, Sewage and Hydraulics, Kharkiv National University of Civil Engineering and Architecture, Kharkiv, Ukraine

vkg.knuca@ukr.net

**Abstract.** In this work, the technological parameters that can significantly effect on a quality of clarification low-concentration suspension by dissolved air flotation were considered. These parameters are: particle size, air bubble size, water rise velocity and air concentration in water. The size of solid particles influences on the probability of their collision with bubbles. The water rises velocity change time of contact between solid particles and bubbles in the working area, which also effects on the probability. Increasing the air concentration in a tank improves the quality of water clarification, because more bubbles are formed there. These factors will be taken into account in further studies to find the optimal dimensions of flotation tank for better cleaning low-concentration suspension.

## 1. Introduction

The high level of development of the city is determined by the possibility of supplying water with high drinking quality. Unfortunately, for Ukraine there is a big problem to supply water with such quality to customers. Due to the uneven distribution of surface water sources in the country it is difficult to supply water to the consumer in the required quantity [1]. Reservoirs or storage ponds, which collect water, were built in order to achieve the necessary volume of water for citizens and industry [2]. The disadvantage of these facilities is the deterioration of hydrochemical parameters of surface water. In many cases, the quality of source water is inappropriate [3]. This is due to discharge of insufficiently treated wastewater or surface runoff, which contains contaminants, into surface water.

In our country, a traditional technological scheme of clarification a low-concentration suspension is a two-stage scheme (settling tank-filter), or a scheme with contact illuminators [4]. Advantages of contact illuminators are that water with reagents (coagulants) almost instantly come into contact with the loading of the illuminator and these facilities are easy to use. The disadvantages are low rate of filtration and the deterioration of the formation flakes in the cold season, which leads to a decrease in the efficiency of water clarification. Therefore, it is necessary to intensify the methods of clarification low-concentration suspension, or to introduce new methods and devices that could effectively clarify water to the quality [5, 6]. One of such method is dissolved air flotation. This technology allows due to the different physicochemical properties of the phases to isolate suspended solids from the low-concentration suspension. The physical and chemical properties of air bubbles help to remove suspended solids, algae, metal hydroxide deposits and natural organic matter from the water [7-9]. Using facilities with dissolved air flotation process will reduce the area of treatment plants [10]. Unfortunately, the influence of technological parameters on the efficiency of the removal suspended particles by size $10^{-6}$- $10^{-3}$ m remains poorly understood.

The aim of this work is to study the technological parameters which can effect on dissolved air flotation process for clarifying low-concentration suspension and to research the possibility of their control.

## 2. Material and Methods

We have improved the mathematical model to research influence of the technological parameters on flotation process. It is based on the following equations and dependencies [11]:

1. Transfer concentration of suspended solids by the water stream equation:

$$\frac{\partial C}{\partial t} + W\frac{\partial C}{\partial z} = D_C\frac{\partial^2 C}{\partial z^2} - bKC, \tag{1}$$

The equation of mass transfer that determine the distribution of concentration of suspended solids on the height of the working part in flotation tank:

$$\frac{\partial C(z,t)}{\partial t} + W\frac{\partial C(z,t)}{\partial z} = D_C\frac{\partial^2 C(z,t)}{\partial z^2} - J(z,t), \tag{2}$$

where t is the time, s; z is the vertical coordinate, m; W is the water rise velocity, m/s; C(z, t) is the suspended solid concentration in water, kg/m$^3$; $D_C$ is the diffusion coefficient of the concentration of suspended solids, m$^2$/s; J(z,t) is the flow of suspended solid concentration from the liquid phase into the gaseous phase, kg/(m$^3$·s).

According to the ideas about the kinetics of the flotation process, the flow J is proportional to the concentration to bubble (air) concentration and the suspended solid concentration in the water:

$$J(z, t) = b(z, t)KC(z, t), \tag{3}$$

where b is the volumetric concentration of bubbles in water; K is the flotation process constant [11].

2. The concentration transfer equation is:

$$\frac{\partial b(z,t)}{\partial t} + \frac{\partial\left((W + u_{bub}(z,t))b(z,t)\right)}{\partial z} = D_b\frac{\partial^2 b(z,t)}{\partial z^2}, \tag{4}$$

where $u_{bub}$(z, t) is the bubble rise velocity in stationary water, m/s. The initial bubble velocity can be found by the Stokes equation; $D_b$ is the diffusion coefficient of bubble concentration, m$^2$/s.

The change in the water rise as a result of compression of the flow by bubbles is taken into account:

$$W(z,t) = \frac{Q}{F}\left(1 - \frac{\pi}{4}\left(\frac{6b}{\pi}\right)^{\frac{2}{3}}\right)^{-1}. \tag{5}$$

3. Suspended particle concentration on bubbles is determined by the mass transfer equation:

$$\frac{\partial S(z,t)}{\partial t} + \frac{\partial\left((W + u_{bub}(z,t))S(z,t)\right)}{\partial z} = KC(z,t) \tag{6}$$

Using this improved mathematical model for the clarification low-concentration suspension by dissolved air flotation, we have studied the effect of the main parameters on the facility.

As the main technological parameters, which influenced on the process, we accepted:

- size of suspended solid (r);
- size of bubble ($d_{bub}$);
- average of water rise velocity (W);
- air concentration in source water ($b_0$).

We decided to take the quality of cleaning as a criterion of efficiency. It is calculated as the ratio of the suspended solid concentration in purified ($C_1$) and source water ($C_0$).

Figure1 and Figure 2 are presented examples of calculations the quality of cleaning for average sizes of solid particles and bubbles. These parameters we varied in ranges: r = 1-5 μm; $d_{bub}$ = 10-30 μm. The calculations are performed with the following fixed parameters: average water rise velocity W = 5 mm/s, the volumetric concentration of bubbles in source water $b_0$ = 0,1 $m^3/m^3$. The working height of flotation tank is 3 m.

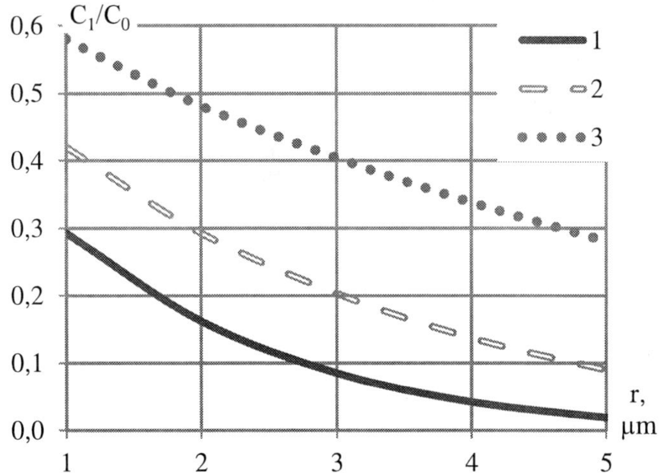


1 – $d_{bub}$ = 15 μm; 2 – $d_{bub}$ = 20 μm; 3 – $d_{bub}$ = 30 μm

**Figure 1.** Dependence of the quality of clarifying on the bubble size with a particle size from 1 to 5 μm

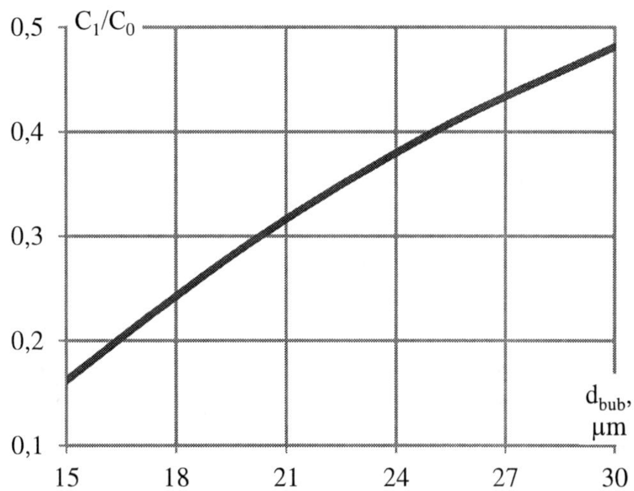


**Figure 2.** Dependence of the quality of clarifying on the bubble size

The data, which presented in Figure 1, show that with decreasing particle size of the contaminant, the quality of cleaning deteriorates sharply. As the size of the suspended solid decreases, the probability of encountering the bubble and the solid phase fraction decreases, as well as its capture by the bubble. The radius of the suspended solid depends on the capture coefficient and, accordingly, the flotation process constant.

When diameter of the bubble decreases, then the cleaning process improves. A larger number of bubbles is formed from one unit of the volume of dissolved air. The probability of their encounter with the suspended solid particles increases and the constant of the flotation process increases.

The Figure 3 presents an example of calculating the dependence of the quality of cleaning on the average water rise velocity in the working area and the average diameter of the bubbles. If the cross-sectional area of the floater does not change in height, the flow rate will be the same. The water rise velocity we varied in the range of 2-7 mm/s.

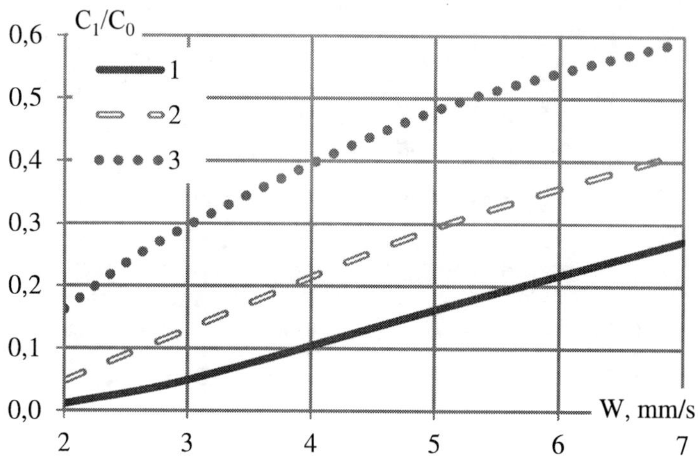


1 – $d_{bub}$ = 15 μm; 2 – $d_{bub}$ = 20 μm; 3 – $d_{bub}$ = 30 μm

**Figure 3.** Dependence of the quality of cleaning on water rise velocity

When the rise velocity decreases, then the quality of cleaning increases too. The time, when bubbles are in the working zone, increases, and the probability of attachment bubbles with particles increases. The dependence between the quality of cleaning and the rise velocity, for bubbles with a diameter of 20 μm in the range of the parameter W, is almost linear (line 2, Figure 3). For larger bubbles with low water rise velocity the cleaning quality improves (line 3, Figure 3), and for smaller bubbles it slows down (line 1, Figure 3).

The Figure 4 shows an example of calculating the dependence of the quality of cleaning on air concentration in source water and average bubble size ($d_{bub}$). The value of air concentration varied in the range of 0.03-0.12.

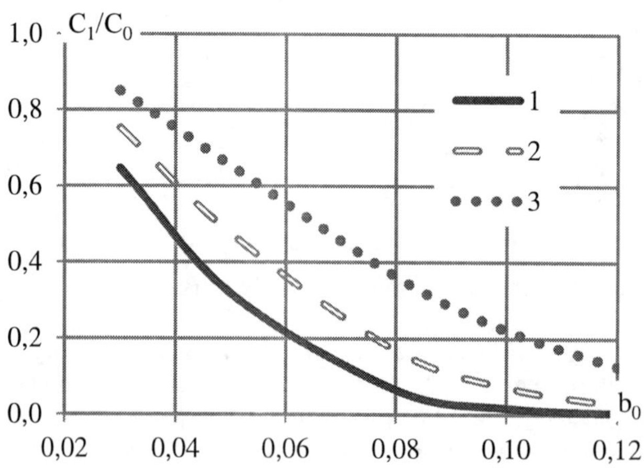


1 – $d_{bub}$ = 15 μm; 2 – $d_{bub}$ = 20 μm; 3 – $d_{bub}$ = 30 μm

**Figure 4.** Dependence of the quality of cleaning on air concentration in source water and bubble size

IOP Conf. Series: Materials Science and Engineering **907** (2020) 012084 doi:10.1088/1757-899X/907/1/012084

These data show that if air concentration increase in source water, then the cleaning quality improves dramatically. In this case, a larger number of bubbles are formed. The process of mass transfer of suspended solids from the liquid phase to the gaseous phase goes more intensively.

The slowdown in the growth of cleaning quality at large values of $b_0$, especially for small bubbles, can be explained that the air concentration and the number of bubbles increase, but the hydrodynamics of the flow change slightly.

The Figure 5 is presented the dependence between water rise velocity in tank, where flow is squeezed by bubbles, ($W_{sq}$) and water rise velocity without bubbles in tank ($W_0$).

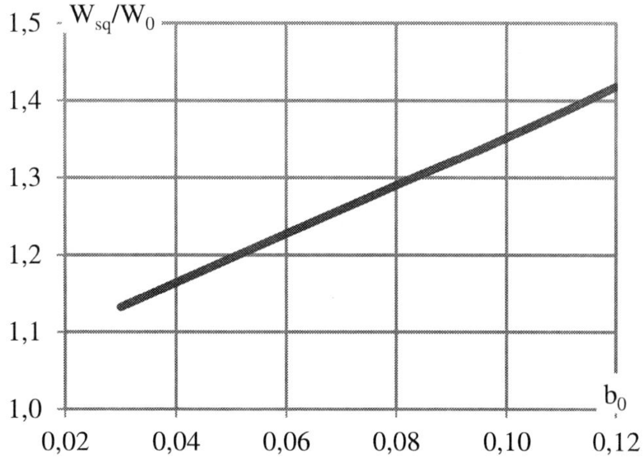


**Figure 5.** Dependence between water rise velocity with bubbles ($W_{sq}$) and without bubbles in tank ($W_0$)

In this case, the compression of the flow with bubbles increases, the water rise velocity increases and the probability of contact with the solid particles decreases.

## 3. Conclusion

In this work, the technological parameters of clarification low-concentration suspension by dissolved air flotation were considered. We have analysed dependences and decided that technological parameters have a significant impact on the flotation process. In the process of designing new dissolved air facilities, it is possible to control and change: water rise velocity and air concentration in the source water. During operation, it is necessary to change the water rise velocity and air concentration too. Reducing the rise velocity requires more flotation tanks. This is increase area of water treatment plant. To increase the air concentration possible when water is saturated with more gas under greater pressure, but this requires higher energy consumption.

Our next step will be to research the impact of design parameters on the flotation process and develop a design of flotation tank, which would provide positive impact of these parameters on the quality of cleaning without increasing capital and operating costs.

## Reference

[1] Havrylenko O 2008 *Ecogeography of Ukraine* (Kyiv: Znannia) p 646
[2] Iatsyk A 2007 *Water resources: use, protection, restoration, management* (Kyiv: Geneza) p 360
[3] Vasenko O 2016 The assessment of the current ecological condition of the Dnieper basin by hydrochemical parameters *The problems of environmental protection and ecological safety* (Kharkiv: UkrNDIEP) **38** 43-60
[4] Zhurba M and Govorova Z 2010 *Water supply: Water quality improvement* (Moscow: ASV) Vol **2** p 544

[5] Epoyan S, Karahiaur A, Volkov V and Babenko S 2018 Research into the influence of vertical drainage elements on the operational efficiency of rapid filters *Eastern-European Journal of Enterprise Technology* **1/10**(91) 62-69
[6] Martynov S, Kunytskyi S and Orlova A 2017 Simulation study of surface water purifying through a polystyrene foam filter *Eastern-European Journal of Enterprise Technologies* **5/10**(89) 19-26
[7] Zabel T 1985 The advantages of dissolved air flotation for water treatment *American Water Works Association* **77**(5) 42-46
[8] Crossley I and Valade M 2006 A review of the technological developments of dissolved air flotation *Journal of Water Supply: Research and Technology* **55**(8) 479-491
[9] Wang Y, Zhuo S, Yang Y and Li N 2012 Application of chitosan in removing algae by dissolved air flotation *Advanced Materials Research* **347-353** 1911-1916
[10] Haarhoff J 2008 Dissolved air flotation: Progress and prospects for drinking water treatment *Journal of Water Supply: Research and Technology- AQUA* **57**(8) 555-567.
[11] Haiduchok O, Syrovatsky O, Karahiaur A and Kostenko S 2020 Mathematical model for clarifying low-concentration suspension by dissolved air flotation *Proceedings of the 2nd International Conference on Building Innovations* **73**

IOP Conf. Series: Materials Science and Engineering **907** (2020) 012085 doi:10.1088/1757-899X/907/1/012085

# Acoustic loading in modern city as negative factor of sustainable development

**A I Samokhvalova, V O Iurchenko, N G Onyshchenko and N O Kosenko**

Department of Life Safety and Environmental Engineering, Kharkiv National University of Civil Engineering and Architecture, Sumska street, 40, Kharkiv, 61002 Ukraine,

samohvalova_anya@mail.ua

**Abstract.** At the present time, acoustic contamination is one of the largest ecological problems, which people face with in a big city. It is impossible to eliminate road traffic noise, thus regulation and limitation of environmental noise pollution are important and obligatory actions. Noise can negatively affect human health, vegetal and animal world. Its harm depends on intensity, frequency and duration. Under the current conditions noise control is technically complicated, complex and economically wildly expensive. The study of noise level was performed in the territories, boarded on main traffic arteries of Kharkiv without rail transport and with rail transport. The studies showed that the roads with rail transport are 7-14 dB louder than the roads without it. The average noise level on highways of the city ranged 80-94 dB, besides on some of the highways the average noise level did not reach the statutory level at 100 m from traffic roads. Such noise load level in the urban environment requires implementation of solutions, directed at its reduction. The impact of the noise barrier and green planting on noise emission is studied.

## 1. Introduction

In recent year conception of steady development gets considerable distribution in the entire countries the world as one of leading global problems humanity that is examined in many aspects and foreshortening. Introduction of strategy sustainable development conditioned by the necessity of stabilizing the technogenic loading on a natural environment, and also decision of complex ecological problems in the conditions of socio-economic increase. Priority direction of sustainable development is optimization vital functions of humanity in the conditions of safe natural environment and harmonious relations both into society and between separate associations.

It is changed by a man, degraded an artificial urban environments have a complex effect on a health of population as a result of air pollution, deficit of sunbeam, water, and also stress factors, that is predefined by the tense rhythm of life, congestion of population, by insufficiency of green plantations and others like that. Some problems have already reached a global nature and need to be addressed urgently to create a "safe" environment for humans and their generations.

Now there is a tendency to expansion the areas of acoustic discomfort on built-up territories. Imperfection of normatively-legal base, absence of economic instruments adjusting of possible sound-levels, is the cause of increasing acoustic pollution of cities. As the result of development of industry, transport and infrastructure in modern person's life, new kinds of noise appear. With each passing year the noise level in big cities is constantly growing due to expansion in the number of vehicles that move constantly in the entire territory of the city.

Over the past 30 years in all big cities the noise level has increased by 12-15 dB, and subjective loudness has increased by 3-4 times. Furthermore, the noise level is close to the red lines – 80 dB in the modern urban areas with heavy vehicle traffic. Approximately 30% of urban population suffers from noise. This interferes normal sleep, rest, reduces working efficiency, affecting central nervous system, the noise causes changes in heart function, fatigue of a bode as a whole, increases blood pressure, sometimes leads to amblyacousia [1].

By its impact on a human body the noise is more harmful than chemical pollution. The noise slowed down efficiency by 15-20%, significantly raised excess incidence. Excessive street noise causes headaches in 80% of people, memory impairment and neurosis in 52%. The noise causes ageing in 30% of city residents and reduces duration of life by 8-12 years [2, 3]. For this reason, responsible and active fight with environmental noise is an integral part of the problem of human society cooperation.

Major noise sources in the city are [3-8]: means of transport (automobile, rail and air transport); industrial enterprises; railway branches and highways; garages, parking areas, gas stations and service stations; places of mass gathering (stadiums, trade fairs, shopping centers, markets, children's playgrounds, etc.); transformer plants; noise of industrial enterprises; at home additional noise from household appliances, radio and television equipment, minor home repair works; inhabitants themselves.

The noise regime of the city is related with its planning, which determines the organization of the traffic route and the possibility of zoning the territory with the appropriate placement of individual objects on it. This creates a certain level of noise on the streets, in recreation areas and houses for various purposes.

Different sources of technology-related noise make an important contribution in sound environment of the city, but still the main source in the city is ground automobile and rail transport. It causes 80-90% of the street noise level. When the scales of motor-car motion grow, the zones of acoustic discomfort increase considerably. On many streets of large cities, noise from public transport exceeds the permissible level by 25-35 dBA [9].

Cars make the largest contribution to the overall sound power of transport noise. They are sources of low-frequency noise, which has a high permeability. Increasing the amount of freight and public transport by 13% in traffic flows increases the noise by 1 dB [10]. The noise level at intersections is 3-6 dB higher than in areas with constant traffic [11].

Therefore, the problem of combating traffic noise has an environmental, social and economic significance. The intensity of road traffic noise depends on many factors such as: the condition of roads, road widths and the distance of residential buildings from the axis of the roadway.

All noise sources in size (relatively to geometrical parameters of the source) are divided into 3 types [4, 10, 11]:

- dotted (transformer plant, ventilation unit, automobile, plant, children's playground, etc.);
- linear (highway transport flow with traffic intensity 5000-6000 vehicles per hour, train that moves, etc.);
- spatial (industrial areas, logistic centers, bus stations, automobile enterprises, transport junctions, etc.).

Depending on distribution of sound energy over time, noise occurs [14]:

- constant – noise, the sound level of which changes over time by no more than 5 dBA, measured by a noise meter),

- intermittent – the sound level changes over time by more than 5 dBA.

In turn, intermittent noise occurs [14]:

- intermittent – intermittent noise, the sound level of which periodically drops sharply to the level of background noise, the duration of the intervals during which the level remains constant, is 1 s or more,

- impulsive – intermittent noise, which consists of one or more sound pulses, the duration of each of them is less than 1 s;

- hesitating – noise, the sound level of which changes continuously over time.

Depending on the geographical distribution of the city, the noise sources are divided into: noise sources of residential and away from residential territories [4]. The following noise sources are found on the residential territory of the city: traffic flows, rail transport; some industrial and communal and storage enterprises; railway branches and motorways; parking areas, garages, gas stations and service stations; dance, concert venues; sports, public utility sites; transformer plants; children's playgrounds; trading floors. Furthermore, there are noise sources inside buildings.

## 2. Unresolved issues

By research of the impact of traffic noise on the environment is devoted to the works of the following domestic and foreign scientists [3]: E P Samoilyuk, G L Osipov, O O Kruze, E Y Yudin, B G Prutkova, P I Pospelova, M I Ivanova, MacKenzie Leo Davis, P M Nelson, M Kapusta and others.

Each person has a different reaction to the noise: some people perceive it "calmly", while it produces strong annoying effect for others. If a person lives long enough in a place with high noise levels, it can cause not only psychological but also physiological changes that are a direct threat to human life. Prolonged exposure to noise causes increased morbidity. Intensive noise reduces a person's attention, increases the number of errors when performing any work.

As under the conditions of strong urban noise constant tension of an acoustic analyzer arises, its permanent action may not only negatively influence on sense of hearing, but also cause such harmful results as: ringing in the ears, headache, quinism, sickliness, irritation, excessive fatigue, rise of blood pressure, even hearing loss.

Furthermore, excessive noise may cause nervous exhaustion, psychic despondency, vegetative neurosis, stomach ulcer, disorders of endocrine, central nervous and cardiovascular systems [2]. In the result so called "noise sickness" appears. It occurs in uprising of headache, sickliness, irritation, which are often followed by temporary loss of hearing.

Most of inhabitants of big cities, who constantly receive noise loads, suffer from noise sickness. Such noise interferes people to work and rest, interferes normal sleep, reduces labor productivity.

The noise affects the human body, not only directly, but also indirectly. Its influence depends on the level and nature, duration of action, as well as on individual characteristics of a person: temperament, state of health, age [3, 9, 12, 13, 15–20].

The greatest hazard of noise is that it is able to be accumulated and each time strongly to affect the nervous system, suppressing it. Some people lose their hearing even after a short noise exposure relatively to increased intensity.

Noises that are particularly unfavorable for the human body are sharp, unstable, unexpected and such that they repeat irregularly. Numerous studies have confirmed the fact that noise refers to general physiological stimuli, which under certain circumstances can affect most organs and systems of the human body.

The action of noise different levels is characterized as follows [6, 9]:

1. Noise up to 50 dBA (usually does not have a harmful effect on a person in the course of his work).
2. Noise of 50-60 dBA (may have a psychological effect, which is manifested in the deterioration of mental activity, weakening of attention, speed of reaction, difficulty working with arrays of information, etc.).
3. Noise 65-90 dBA (possible physiological effects: heart rate increases, blood pressure increases, blood vessels narrow, that worsens the supply of organs blood).
4. Noise of 90 dBA and above (can lead to functional disorders in organs and systems of the human body: deteriorating activity of the stomach and intestines, decreased auditory sensitivity, tinnitus, nausea, headache).
5. Noise of 120 dBA and above (there is a mechanical impact on the hearing organ, which is manifested in the violation of connections between individual parts of the inner ear; even a rupture of the eardrum is possible, there is an impact not only on the hearing organs but also on the whole body, sound waves, penetrating through the skin, cause mechanical vibrations of body tissues, resulting in the destruction of nerve cells, rupture of small vessels, etc.).

In Ukraine, the standard permissible noise level in the environment is 55/45 dBA (day/night) for the residential area, 40 dBA for the recreation area, 65/55 dBA (day/night) for the first line of buildings from the railway track [3, 8].

The objective of the experimental researches was to determine the level of acoustic load in urban areas (for example, Kharkiv) and the protective effectiveness from it by means of the noise barrier.

## 3. Main part

In the work, the studies were performed to determine the noise levels in Kharkiv in the afternoon in the territories, boarded on highways without rail transport – Sumska street, Pushkinska street, Myronosytska street, Alchevskykh street, Dynamivska street, and to highways with rail transport – Poltavskyi Shlyakh street, Klochkivska street, Moscovskiy avenue. To perform measurements, the noise level meters VShV-003 and FLUS MT-901A were applied. The difference between the values of the noise load performed by various measuring devices was 1-3 dB. At each section, measurements were performed 5 times, the tables show the average results. The streets were chosen taking into account the traffic intensity, type of surface and the distance of the carriageway to the residential construction, taking into account the presence of trees near the highway and the location of the tram tracks.

The results of the performed measurements are presented in table (refer with: Table 1, Table 2).

**Table 1.** The research results of acoustic load in the nosiest streets of Kharkiv without rail transport

| Place of research | Road material | Traffic intensity, [car/day] | Noise level at a distance from carriageways, [dB] | | | |
|---|---|---|---|---|---|---|
| | | | Distance, [m] | | | |
| | | | 1 | 10 | 20 | 50 |
| Sumska street | paving stone blocks | 31064 | 94/94 | 86/85 | 78/78 | 70/71 |
| Sumska street | asphalt | 31041 | 85/85 | 80/80 | 76/75 | 64/63 |
| Pushkinska street | asphalt | 23040 | 84/83 | 79/78 | 72/70 | 61/60 |
| Myronosytska street | asphalt | 7856 | 80/82 | 74/75 | 68/69 | 50/50 |
| Alchevskykh street | asphalt | 8624 | 82/83 | 74/75 | 64/65 | 55/54 |

* in the numerator measurements were performed by means of the sound level meter VShV - 003, in the denominator – by means of the sound level meter FLUS MT-901A

As it is seen from the results of measuring of the acoustic load on streets with heavy traffic density without rail transport (refer with: Table 1), the noise level at 1 m from the carriageways (asphalt coating) depends on the traffic density and ranges from 80 to 94 dB (at the sanitary standards 55 dB at daytime). As it was determined, the noise level measurements in Sumska street, stone block pavement causes increase of the noise level about 10%. Even at 20 m from the roadway, the standard values of noise rates in the examined streets are not achieved. But at 50 m in the streets with less traffic density (Myronosytska and Alchevskykh) the noise is within the standard values. The highest noise level is formed in Sumska street at the sections with carriageways from stone block pavement that has hollow spots.

In the course of the work, the noisiest streets with the rail transport were examined, specifically: Poltavskyi Shlyakh street, Klochkivska street, Moscovskiy avenue (refer with: Table 2). The results showed that the acoustic load level is over the limits even at 50 m from the roadway, and even much less next to the carriageways.

The researches have shown (refer with: Table 1, Table 2) that roads with rail transport are 7-14 dBA louder than roads without it. As it is seen from the results of measuring the acoustic load in places with heavy traffic density, including with rail transport, the noise level at 1 m from the carriageways is 90- 94 dB. Moving from the carriageways, the noise is certainly damped (and the acoustic load is reduced), but still it does not reach an environmentally-friendly safe level even at 50 m from the carriageways.

**Table 2.** The research results of acoustic load in the nosiest streets of Kharkiv with rail transport

| Place of research | Road material | Noise level at a distance from carriageways, [dB] | | | |
|---|---|---|---|---|---|
| | | Distance, [m] | | | |
| | | 1 | 10 | 20 | 50 |
| Poltavskyi Shlyakh street | asphalt | 92 / 94 | 80 / 83 | 75/ 76 | 61/ 63 |
| Klochkivska street | asphalt | 94 / 93 | 80/ 80 | 70/ 70 | 57/ 60 |
| Moscovskiy avenue | asphalt | 94 / 94 | 85/ 85 | 78 / 78 | 70/ 69 |

* in the numerator measurements were performed by means of the sound level meter VShV-003, in the denominator – by means of the sound level meter FLUS MT-901A

Reduction of the acoustic load in these areas of the city needs modernization of the tram depot. Modernization includes the following works: installation of a depreciated floor, false bags with sound absorbers closing the wheels, replacement of tram carts, rubber tram wheels, replacement of double-glazed windows, replacement of door cylinders with silent ones, modernization of electronics.

The determined noise level rates in the territory of Kharkiv are in line with the data of other researchers: average sound levels on highways of big cities – 73-83 dB (maximum – 90-95 dB), in residential buildings along the main roads – 62-77 dB [3].

Generally, the determined excess of permissible noise levels in the streets of Kharkiv is caused by high traffic density in the streets cobbled with pebbles, existence of tram tracks and the movement of tram cars in the general traffic flow. According to the research results, it can be argued that the streets where residential buildings are located in the immediate vicinity of the automobile road with the coating type - paving stone blocks have the highest noise levels, which, as known, interfere the normal life of the residents of these streets and negatively affect their functional status. The determined rates indicate the necessity to improve the environmental safety of residents of these streets through the implementation of special solutions to reduce acoustic load.

It is known that green planting, shaped in the form of special soundproofing strips, can reduce the noise level by 8-10 dBA. Trees, which are planted close to each other and surrounded by dense bushes, significantly reduce the level of man-made noise and improve the urban environment. And particularly the green planting in Kharkiv is the most common method of protection from the noise level.

In the course of the work the noise-proofing properties of green planting along the carriageways of the city and the noise screen in Dynamivska street (refer with: Table 3) were examined. As indicated by the data presented, the green planting in Kharkiv in the noisiest central streets of the city do not have any noise protection properties because the trees on the sidewalks are planted in one or two rows at a 4-6 m from each other, while the tree crowns cannot create enough protective mass dense even at an age when all the tree properties are fully presented (the average height of the trees is 7-8 m, and the crown is at a height of at least 3.5 m); for street planting commonly standard trees are used, that allows freely spread of the sound to residential buildings. But the usage of a noise shield allows reducing noise in the examined area by 10 dB. Therefore, the installation of such screens on large and noisy highways will significantly reduce the acoustic load. According to the foreign experience, the noise screens protect surrounding houses from noise, as well as crowded places (public transport stops, parks). The installation of such constructions is economically feasible in densely populated areas where road location at a distance from residential and office buildings is not possible. Thus, even when passing in the immediate vicinity of a busy highway – it is possible to create a quiet, residential area.

Nowadays in Ukraine there is a significant number of normative-legal acts of ecological, sanitary-hygienic, transport, administrative and other legislation, which regulate issues related to protection of the population from dangerous noise influences. Also, to solve the problem of noise reduction, it is important to predict the noise background of cities.

**Table 3.** The research results of acoustic load in the nosiest streets of Kharkiv with the green planting and noise screen

| Place of research | Kind of protection | Noise level, [dB] | |
|---|---|---|---|
| | | before protective belt | after protective belt |
| Sumska street | Green planting (1-2 rows) | 85 | 84 |
| Pushkinska street | Green planting (1-2 rows) | 84 | 84 |
| Myronosytska street | Green planting (1-2 rows) | 80 | 80 |
| Klochkivska street | Green planting (1-2 rows) | 94 | 93 |
| Dynamivska street | Noise screen | 80 | 70 |

Protection methods against external urban noise usually have a complex character. In order to reduce the acoustic load level in the city, it is necessary to use several noise protections measures together [3, 5, 9]:

- technological (development of more advanced vehicles and industrial plants relating to acoustic);
- constructional and acoustic (increasing of sound-proofing of building envelopes, shielded and architectural and planning solutions for buildings taking into account noise influences, arranging of noise-proof windows in buildings, removing of residential space from the ground floors of buildings and arranging of uninhabited facilities there (commercial, assembly space, offices, food outlets, etc.)
- city planning (development of roadside clear zones, functional noiseproofing zoning of urban areas, usage of green planting, acoustic shielding (natural and artificial relief elements, roadside barrier, retaining and special protective shields-walls)
- administrative and organizational (rational organization of traffic flows, differentiation of transport types as per special lanes (as per speed), their separation at various levels, prohibition of sound car signals, time regulatory actions of loud sources or prohibition of their work, restriction of the movement of trucks and motorcycles in certain city areas and as per time, removing of noisy enterprises outside the residential districts).

An effective preventive measure in the noise control is application of vehicles with minimal noise level. An important action in the noise control is proper planning of residential places, construction of belt highways that unload the city center, and reduction of the number of road intersections in the territory of a residential area. Furthermore, construction projects should provide maximum protection of residential districts from noise by means of development of natural and artificial screens, green planting (trees planted close to each other, surrounded by dense bushes, significantly reduce the industrial noise level and improve the urban environment), transfer of engineering structures under the ground.

## 4. Conclusion

1. The noise control in the city is one of the most pressing challenge of human habitat protection. It is aimed at reduction of noise level and involves a wide range of various general and special activities.
2. The noise rating in the daytime in 8 streets of Kharkiv, which differed by availability/ absence of rail transport, traffic density, road material, availability/absence of green planting, was carried out.
3. It was established that the average noise level in the territories boarded on the city highways without rail transport at 1 m from the roadway ranges from 80 to 94 dB, and on the highways with rail transport – from 90 to 94 dB. The standard values of the noise level (55 dB) were reached only at 50 m from the carriageway and only in streets without rail transport, and with low traffic density (Myronosytska and Alchevskykh).
4. It was found out that usage of a noise screen reduces the noise level in Dynamivska street by about 13%. The green planting along the carriageway of 4 city streets in the studied sections practically did not reduce the noise. This was stipulated by the large distances between the trees, the high height of the crowns and lack of a dense protective mass.

## 5. References

[1] Jariwala H J, Syed H S, Pandya M J, Gajera Y M 2017 Noise Pollution & Human Health: A Review *Proceeding of Conference "Noise and Air Pollution: Challenges and Opportunities"* 1- 4

[2] *The health effects of environmental noise* 2018 (Department of Health, Canberra) p 96

[3] Samokhvalova A I, Onyshchenko N G, Ponomarov K S and Nesterenko O V 2018 Investigation of the impact of noise sources in an urban environment *SHEI "Dnieper. state Academy of Arts and Architecture"* **105** pp 88–94

[4] Zakharchenko M V, Orlov M V, Golubev A K and et al. 1996 *Safety of Life in Everyday Conditions of Production, Everyday Life and in Emergency Situations*: *Educ. Manual* (Kyiv: IZMY) p 196

[5] Gupta S, Ghatak Ch 2011 Environmental noise assessment and its effect on human health in an urban area *International Journal of Environmental Sciences* **1** (7) p 1954-1964

[6] Kucheryavy V P 2002 *Urban ecology*: Textbook (Lviv: Svit) p 439

[7] Timchenko O I 1993 Noise as a Physical Factor That Is Able Potentially to Influence on Chromosome Integrity *Proceeding of Scientific Conference of Ukrainian Hygienic Centre* p 163– 164 [in Russian]

[8] DBN B.1.1-31: 2013 Protection of territories, buildings and structures from noise. Kyiv 2014 [in Ukrainian]

[9] Samokhvalova A I and Onyshchenko N G 2019 Research of the problem of acoustic pollution of cities *Proceeding of the XVII International scientific-practical Internet-conference "Problems and perspectives of development modern science in the countries of Europe and Asia" Collection of scientific works* (Pereyaslav-Khmelnytsky) pp 19–22

[10] Danilko V K 2013 *Statistics of road transport ecology* (Zhytomyr) p 172

[11] Vetoshkin A G 2012 *Theoretical foundations of environmental protection: Textbook* (Moscow: Abris) p 397

[12] Zhelibo E P, Zaverukha N M and Zarzarny V V 2001 *Life Safety* (Kyiv: Caravel) p 320

[13] Dzhigirey V S and Zhititsky V T 2001 *Life Safety, fourth ed.* (Lviv: Poster) p 256

[14] Denisov V N and Rogalev V A 2005 *Problems of greening road transport* (Kyiv: Prosit) p 311

[15] *The Protection of the Environment Operations (Noise Control) Regulation* 2017 (Sydney: Environment Protection Authority) p 58

[16] Vnukova N V 2011 Assessment of acoustic pollution of the roadside territory *Eastern European Journal of Advanced Technology* **52** pp 42–47

[17] GOST 23337-78 (ST SEB 2600-80) Noise. Noise measurement methods in residential areas and in residential and public buildings [in Russian]

[18] Gutarevich Y F, Zerkalov D V, Govorun A G, Korpach A O and Merzhievskaya L P 2002 *Ecology of road transport: Textbook* (Kyiv: Aristei) p 162

[19] Babisch W 2002 The noise/stress concept, risk assessment and research needs *Noise & Health* **4**(16) p 1–11

[20] Voloshin I A 2013 Noise protection on highways. Problems and directions of their solution *Modern technologies of construction and operation of highways* pp 76–82

# AUTHOR INDEX


Abyzov, V A ... 275
Amelina, N O ... 285
Anopko, D V ... 247
Arutyunov, V ... 218
Aznaurian, I O ... 304
Bazeliuk, N L ... 275
Belichenko, O A ... 411
Berdnyk, O Yu ... 285
Bielka, V V ... 200, 210
Bikov, R ... 454
Bizhko, E V ... 569
Blahovestova, O O ... 62
Blinova, M Y ... 10
Bondarenko, D ... 218, 359
Bondarenko, O P ... 434
Bondarenko, Yu V ... 200
Borodai, A S ... 487
Borodai, D S ... 487
Borodai, S P ... 487
Borodai, Ya O ... 487
Borysenko, A S ... 527
Bulhakova, T V ... 173
Buriak, O ... 102
Butnik, S V ... 210
Butskaya, L N ... 341
Chayka, D O ... 328
Cherkasova, E T ... 520
Chub, A N ... 494
Dansheva, S O ... 62
Danylov, S M ... 135
Danylov, Serhii ... 141
Dedenova, O ... 359
Demina, O ... 359
Dolinina, O E ... 20
Dubina, N G ... 569
Dubinskiy, V P ... 128, 559
Dunaevskiy, Y U ... 590
Dvorkin, L ... 191
Dzhalalov, M N ... 210
Dziubenko, V ... 480
Emeljanova, I A ... 328
Epoyan, S ... 379, 631
Fedak, A ... 550
Filippova, Daria ... 151
Fomenko, O O ... 135
Fomenko, Oksana ... 141
Fursovych, M ... 191
Gabitov, A I ... 354, 374
Gaisin, A M ... 374
Galinsky, O M ... 427
Galkin, A ... 366
Gasan, Yu G ... 313
Gayevyi, Y ... 102
Gelevera, A G ... 261
Gella, O. ... 122
Glyva, V A ... 334
Golovchenko, A ... 102
Gorbyk, O O ... 1
Gots, V I ... 256, 261, 386
Haiduchok, O ... 379, 631
Holubchak, K T ... 113
Honchar, O A ... 247, 275
Honcharova, I V ... 536
Hovorukha, I V ... 210
Hryhorenko, O ... 454
Hubanov, O V ... 536
Hunchenko, Okcana ... 609
Hunchenko, O ... 600
Hurkalenko, V A ... 341
Iakymenko, M V ... 200
Ilchenko, S A ... 10
Isupova, M I ... 38
Iurchenko, V O ... 637
Iurchenko, V.O ... 583
Ivanova, N V ... 86
Izbash, A M ... 135
Izbash, Andrii ... 141
Izbash, M Yu ... 467
Kabus, O V ... 292
Kachemtseva, L ... 122
Karban, A ... 543
Kazimahomedov, I ... 403
Khaidukov, V ... 159
Khoroian, N ... 122
Klapchenko, V I ... 304, 313
Kochevykh, M O ... 247, 275
Kondrachenko, E V ... 449
Kondrachenko, V I ... 449
Kondrashchenko, V I ... 393
Konstantynovskyi, O P ... 418
Korduba, I ... 600
Korkh, O I ... 467
Korol, E I ... 473
Korshunova, N N ... 20
Kosenko, N O ... 637
Kostyuk, T ... 218

Kovalchuk, A V ........ 418
Kovalchuk, V ........ 623
Kovalenko, A. V. ........ 583
Kozlovskyi, O V ........ 505
Krasnianskyi, G Iu ........ 304, 313
Kravchenko, M ........ 600
Kreizer, I I ........ 494
Krivenko, P V ........ 418
Krot, O P ........ 321
Krot, O Y ........ 321
Kubko, A Y ........ 1
Kudryashova, I ........ 28
Kushnierova, L O ........ 247
Kuzmych, V ........ 95
Kuznetsova, I O ........ 313
Kvartenko, O ........ 623
Kysil, Olga ........ 167
Kysil, S S ........ 173
Laboda, M A ........ 443
Ladigina, I V ........ 569, 576
Lapovska, S D ........ 304, 313
Lastivka, O V ........ 285, 386
Latorets, K V ........ 341
Lebedeva, E S ........ 328
Lebedeva, E.S. ........ 583
Levashova, Yu.S. ........ 583
Levchenko, L O ........ 334
Lushnikova, N ........ 191
Maimeskul, O V ........ 181
Makarenko, O V ........ 341
Malovanyy, M S ........ 500, 514
Mamon, A S ........ 181
Martynenko, A ........ 28
Martynov, S ........ 623
Martyshova, L S ........ 45
Maystrenko, A A ........ 285
Mazur, T M ........ 473
Mezhenna, Nataliia ........ 151
Mikhalchenko, Sergii ........ 167
Mileikovskyi, V ........ 480
Mironenko, V ........ 70
Mokryy, V I ........ 514
Molodid, O S ........ 427
Monfort, I B ........ 443
Morozova, E B ........ 20
Myronenko, O V ........ 55
Myronenko, V P ........ 55, 86, 536
Nagurskiy, O A ........ 500
Nedoseko, I V ........ 268
Nesen, A A ........ 128
Nikiforova, S. ........ 122
Nikolaev, A P ........ 449
Nosovsky, Yu L ........ 443
Oksak, S ........ 366
OKulaienko, O ........ 292
Oliynyk, E O ........ 460
Onyshchenko, N G ........ 637
Orlova, A ........ 623
Palchyk, P P ........ 256
Palchyk, S P ........ 256
Panova, O V ........ 334
Pechertsev, O O ........ 62
Pershina, L O ........ 341
Petropavlovsky, O N ........ 261
Petropavlovskyi, O M ........ 418
Petrovska, Yu ........ 95
Petrushka, I M ........ 514
Petrushka, K.I. ........ 514
Pipa, V V ........ 443
Plakhotnikov, K ........ 359
Plokhuta, R O ........ 427
Plugin, Andrii ........ 239
Plugin, A ........ 218
Pluhin, O ........ 218
Pohrebennyk, V D ........ 514
Popov, Y ........ 454
Posatskyi, B S ........ 473
Protsenko, E M ........ 347
Protsenko, E ........ 70
Pushkarova, K K ........ 275
Pushkarova, Kateryna ........ 239
Pyrig, Y ........ 366
Radomska, M M ........ 334
Rakhimov, R Z ........ 268
Remizova, O ........ 28
Riazanov, A A ........ 268
Riazanov, A N ........ 268
Rogozina, N V ........ 261
Rolnik, L Z ........ 354
Rudenko, A O ........ 576
Rudenko, I I ........ 418
Runova, R F ........ 443
Rusanova, M V ........ 181
Ryabushko, A V ........ 292
Ryazanova, V A ........ 354, 374
Safronova, O O ........ 173
Saienko, L ........ 403
Saienko, N ........ 454
Salem, S ........ 403
Salov, A S ........ 354, 374
Samokhvalova, A I ........ 637
Semyroz, N H ........ 173
Sharikina, N V ........ 427
Shemet, R M ........ 200

Shevtsova, G V ........ 1
Shishkin, Alexander ........ 298
Shishkina, Alexandra ........ 298
Shyshko, N S ........ 393
Skorobohatko, O V ........ 559
Sleptcov, Oleg ........ 141
Sleptsov, O S ........ 113, 173, 590
Sleptsov, O. ........ 159, 543
Smeshko, V V ........ 261
Smolenska, S A ........ 527
Solobay, P A ........ 616
Soloviy, K M ........ 500
Sopov, D ........ 70
Sopov, V P ........ 292, 393, 467
Spirande, K V ........ 200
Starkova, O ........ 359
Stefanovych, P ........ 600
Subota, D Yu ........ 328
Sukhanevych, Maryna ........ 239
Sukhorukov, G ........ 379
Synelnikov, S D ........ 500
Syrovatsky, O ........ 631
Telyma, S V ........ 460
Timofeev, A A ........ 374
Timofeev, V A ........ 354
Titov, A ........ 631
Tkachenko, O ........ 480
Tkachenko, T ........ 480
Tolmachov, D S ........ 411
Tolmachov, S M ........ 411
Tomin, O O ........ 386
Tomlins, R ........ 113
Tovbych, Valery ........ 167
Tsapko, A Yu ........ 434
Tsapko, Yu V ........ 434
Tsymbalova, T A ........ 55
Tykhenko, O M ........ 334
Tymchuk, I S ........ 500
Tymoshenko, S A ........ 256, 386
Usherov-Marshak, O V ........ 292
Vakhnichenko, O V ........ 616
Viatkin, V A ........ 210
Vihdorovich, O ........ 102
Vildman, I M ........ 460
Vinnichenko, V I ........ 268, 321
Vnukova, N V ........ 505
Volkov, V ........ 379
Voloshkina, O S ........ 460
Voloshkina, O ........ 600
Volosyuk, M A ........ 347
Yefremov, Sergii ........ 231
Yesipov, A O ........ 86
Yihe, Zhang ........ 393
Younis, B ........ 403
Zadorozhnyi, A A ........ 226
Zavorina, A A ........ 520
Zhelnovach, G M ........ 505
Zhitkovsky, V ........ 191
Zolkina, Ye ........ 454
Zymina, Svitlana ........ 78